Edited by
Gabriel Cristóbal, Peter Schelkens,
and Hugo Thienpont

Optical and Digital Image Processing

Related Titles

Ackermann, G. K., Eichler, J.

Holography

A Practical Approach

2007
ISBN: 978-3-527-40663-0

Ersoy, O. K.

Diffraction, Fourier Optics and Imaging

2006
ISBN: 978-0-471-23816-4

Acharya, T., Ray, A. K.

Image Processing

Principles and Applications

2005
E-Book
ISBN: 978-0-471-74578-5

Singer, W., Totzeck, M., Gross, H.

Handbook of Optical Systems

Volume 2: Physical Image Formation

2005
ISBN: 978-3-527-40378-3

*Edited by Gabriel Cristóbal, Peter Schelkens,
and Hugo Thienpont*

Optical and Digital Image Processing

Fundamentals and Applications

WILEY-VCH Verlag GmbH & Co. KGaA

The Editors

Dr. Gabriel Cristóbal
Instituto de Optica (CSIS)
Imaging and Vision Dept.
Serrano 121
28006 Madrid
Spain

Prof. Dr. Peter Schelkens
Vrije Universiteit Brussel
Department of Electronics and
Informatics (ETRO)
Pleinlaan 2
1050 Brussels
Belgium

Prof. Hugo Thienpont
Vrije Universiteit Brussel
Brussels Photonics Team (B-PHOT)
Pleinlaan 2
1050 Brussels
Belgium

■ All books published by Wiley-VCH are carefully produced. Nevertheless, authors, editors, and publisher do not warrant the information contained in these books, including this book, to be free of errors. Readers are advised to keep in mind that statements, data, illustrations, procedural details or other items may inadvertently be inaccurate.

Library of Congress Card No.: applied for

British Library Cataloguing-in-Publication Data
A catalogue record for this book is available from the British Library.

Bibliographic information published by the Deutsche Nationalbibliothek
The Deutsche Nationalbibliothek lists this publication in the Deutsche Nationalbibliografie; detailed bibliographic data are available on the Internet at <http://dnb.d-nb.de>.

© 2011 WILEY-VCH Verlag GmbH & Co. KGaA, Boschstraße. 12, 69469 Weinheim

All rights reserved (including those of translation into other languages). No part of this book may be reproduced in any form – by photoprinting, microfilm, or any other means – nor transmitted or translated into a machine language without written permission from the publishers. Registered names, trademarks, etc. used in this book, even when not specifically marked as such, are not to be considered unprotected by law.

The copyright for the Matlab and other software codes printed in this book remains with the author.

Composition Laserwords Private Ltd., Chennai, India
Printing and Binding betz-druck GmbH, Darmstadt
Cover Design Adam Design, Weinheim

Printed in Singapore
Printed on acid-free paper

ISBN: 978-3-527-40956-3

Foreword

Optical and Digital Image Processing

There is a tendency these days for scientists and engineers to be highly specialized. It is therefore a pleasure to see a book covering a truly broad set of topics. Granted that all the topics relate in one way or another to the field of optics, broadly interpreted; however, within that broad category, this book certainly covers a breadth of subjects.

The first element of breadth lies in the joint coverage of both optical signal processing and digital signal processing. In fact, many modern signal processing systems depend on both optics and digital technologies. Images are usually the entity to be processed, and most often these images are formed by optical systems. The methods for processing such images are numerous and diverse, depending in large part upon the application.

At one time, optical analog processing held sway as the fastest method for performing linear operations on 2D signals, but the relentless progress in digital processing, a consequence of Moore's law, has displaced optical processing in many applications. However, the most interesting and complex optical systems often entail some optical preprocessing followed by digital manipulation. Good examples are found in the field of adaptive optics, in which optical methods for wavefront sensing are followed by digital methods for determining appropriate changes for an adaptive mirror.

The subject matter covered in this book ranges over many topics, which can be broadly classified as follows: (i) fundamentals of both optics and digital signal processing; (ii) optical imaging, including microscopy and holography; (iii) image processing including compression, deconvolution, encryption, and pattern recognition; (iv) signal representation, including time-frequency, spline, and wavelet representations; and (v) miscellaneous applications, including medical imaging and displays. The authors are drawn internationally, thus allowing a window into the research interests of scientists and engineers in many countries.

As mentioned above, it is refreshing to see such breadth under one cover. This book should provide interesting and informative reading to those wishing to see the broad picture of image processing and its applications through an international lens.

Joseph W. Goodman

Contents

Preface *XXIX*
List of Contributors *XXXIX*
Color Plates *LI*

1 Fundamentals of Optics *1*
Ting-Chung Poon and Jung-Ping Liu
1.1 Introduction *1*
1.2 The Electromagnetic Spectrum *1*
1.3 Geometrical Optics *3*
1.3.1 Ray Transfer Matrix *3*
1.3.2 Two-Lens Imaging System *6*
1.3.3 Aberrations *8*
1.4 Maxwell's Equations and the Wave Equation *9*
1.5 Wave Optics and Diffraction *11*
1.6 Fourier Optics and Applications *14*
1.6.1 Ideal Thin Lens as Optical Fourier Transformer *14*
1.6.2 Imaging and Optical Image Processing *17*
1.6.3 Optical Correlator *19*
1.7 The Human Visual System *21*
1.8 Conclusion *23*
References *23*

2 Fundamentals of Photonics *25*
Erik Stijns and Hugo Thienpont
2.1 Introduction *25*
2.2 Interference and Diffraction *25*
2.2.1 Interference *25*
2.2.2 Diffraction *26*
2.2.2.1 Diffraction at a One-Dimensional Slit *26*
2.2.2.2 Diffraction at a Circular Aperture *27*
2.2.3 Resolution *28*
2.2.3.1 Angular Resolution *28*
2.2.3.2 Spatial Resolution *29*

2.2.4	Coherence	29
2.2.4.1	Temporal or Longitudinal Coherence	29
2.2.4.2	Transverse or Spatial Coherence	30
2.3	Terms and Units: The Measurement of Light	30
2.3.1	Introduction: Radiometry versus Photometry	30
2.3.2	Radiometric Terms and Units	30
2.3.2.1	Radiant Energy	30
2.3.2.2	Radiant Flux	31
2.3.2.3	Radiant Flux Density	31
2.3.2.4	Radiant Intensity	31
2.3.2.5	Radiance	32
2.3.2.6	Radiant Exposure	32
2.3.3	Photometric Terms	33
2.3.3.1	Spectral Terms	33
2.3.3.2	Spectral Sensitivity of the Eye	33
2.3.3.3	Luminous Terms	33
2.3.4	Photometric Units	34
2.3.4.1	Other Visual Terms and Units	34
2.4	Color	35
2.4.1	Introduction	35
2.4.2	The Spectrum of Light	36
2.4.3	Tristimulus Theory	36
2.4.3.1	The Tristimulus	36
2.4.3.2	The 1931 CIE Standard	38
2.4.3.3	CIE 1976 UCS Diagram	39
2.4.4	Theory of the Opponent Colors	40
2.4.4.1	Describing the Visual Observations	40
2.4.4.2	Saturation or Chroma	41
2.4.4.3	Hue	41
2.4.4.4	The CIELAB Diagram	42
2.5	Basic Laser Physics	43
2.5.1	Introduction	43
2.5.2	Normal or Spontaneous Emission of Light	43
2.5.3	Absorption	44
2.5.4	Stimulated Emission of Light	44
2.5.5	Amplification	45
2.5.6	Basic Setup	45
2.6	Basic Properties of Laser Light	46
2.6.1	Laser Light Has One Direction	47
2.6.2	Laser Light Is Monochromatic	47
2.6.3	Laser Light Is Coherent	47
2.6.4	Laser Light Is Intense	47
2.7	Conclusions	48
	References	48

3	**Basics of Information Theory** *49*	
	Michal Dobes	
3.1	Introduction *49*	
3.2	Probability *49*	
3.2.1	Several Events *50*	
3.2.2	Conditional Probabilities: Independent and Dependent Events *51*	
3.2.3	Random Variable *52*	
3.2.4	Distribution Function *52*	
3.2.5	Discrete Distribution *53*	
3.2.6	Continuous Distribution *53*	
3.2.7	Expected Value *54*	
3.3	Entropy and Mutual Information *54*	
3.3.1	Historical Notes *55*	
3.3.2	Entropy *55*	
3.3.2.1	Some Properties of Entropy *55*	
3.3.3	Joint Entropy *56*	
3.3.4	Mutual Information *60*	
3.3.5	Kullback–Leibler Divergence *62*	
3.3.6	Other Types of Entropies *62*	
3.4	Information Channel *62*	
3.4.1	Discrete Channel *63*	
3.4.2	Channel Capacity *63*	
3.4.3	Symmetric Channel *64*	
3.4.4	Binary Symmetric Channel *65*	
3.4.5	Gaussian Channel *65*	
3.5	Conclusion *66*	
	Appendix 3.A: Application of Mutual Information *67*	
	References *68*	
4	**Fundamentals of Image Processing** *71*	
	Vaclav Hlavac	
4.1	Introduction *71*	
4.2	Digital Image Representation *73*	
4.2.1	Topological and Metric Properties of Images *74*	
4.2.2	Brightness Histogram *77*	
4.3	Image Filtering Paradigm *78*	
4.4	Frequency Domain *80*	
4.4.1	1D Fourier Transform *80*	
4.4.2	2D Fourier Transform *85*	
4.5	Filtering in the Image Domain *90*	
4.6	Conclusions *96*	
	References *96*	

5	**Joint Spatial/Spatial-Frequency Representations** *97*
	Gabriel Cristóbal, Salvador Gabarda, and Leon Cohen
5.1	Introduction *97*
5.2	Fundamentals of Joint Representations *98*
5.2.1	Notation *99*
5.2.2	The Wigner Distribution *100*
5.2.2.1	Marginals *101*
5.2.2.2	Inversion *101*
5.2.2.3	Translation Invariance *101*
5.2.2.4	Product of Images *102*
5.2.2.5	Overlap of Two Images *102*
5.2.2.6	Real Images *102*
5.2.2.7	Cross Wigner Distribution *103*
5.3	Other Distributions *103*
5.3.1	The Spectrogram *104*
5.3.2	The Analytic Image *104*
5.4	The Pseudo-Wigner–Ville Distribution (PWVD) *105*
5.4.1	1D-Smoothed PWVD *105*
5.4.2	1D Directional PWVD *105*
5.4.3	2D-Smoothed PWVD Definition and Implementation *108*
5.5	2D Log-Gabor Filtering Schemes for Image Processing *110*
5.6	Texture Segmentation *112*
5.7	Hybrid Optical–Digital Implementation *114*
5.8	Conclusions *116*
	Acknowledgments *116*
	References *116*

6	**Splines in Biomedical Image Processing** *119*
	Slavica Jonic and Carlos Oscar Sanchez Sorzano
6.1	Introduction *119*
6.2	Main Theoretical Results about Splines *120*
6.2.1	Splines as Interpolants and Basis Functions *120*
6.2.1.1	Tensor Product Splines *120*
6.2.1.2	Polyharmonic Splines *127*
6.2.2	Splines for Multiscale Analysis *129*
6.3	Splines in Biomedical Image and Volume Registration *131*
6.4	Conclusions *132*
	References *133*

7	**Wavelets** *135*
	Ann Dooms and Ingrid Daubechies
7.1	Introduction *135*
7.2	Chasing Sherlock Holmes: How to Scrutinize an Image *139*
7.2.1	Classical Fourier Analysis *140*
7.2.2	Forces of Nature *141*

7.3	A Natural Evolution: The Continuous Wavelet Transform	142
7.4	Theory into Practice: The Discrete Wavelet Transform	143
7.5	Mallat and Meyer Digging Deeper: Multiresolution Analysis	144
7.5.1	Examples	146
7.6	Going to Higher Dimensions: Directional Transforms	148
7.6.1	Separable Transforms	148
7.6.2	Dual-Tree Complex Wavelet Transform	149
7.6.3	Shearlets	151
7.7	Conclusion	152
	References	153
8	**Scale-Space Representations for Gray-Scale and Color Images**	155
	Iris U. Vanhamel, Ioannis Pratikakis, and Hichem Sahli	
8.1	Introduction	155
8.2	Background	156
8.2.1	Definitions	156
8.2.2	Axioms and Properties	157
8.2.2.1	Fundamental Axioms	158
8.2.2.2	Properties	160
8.2.2.3	Morphological Properties	161
8.2.3	PDE Equation-Based Formulation	162
8.2.3.1	Classification	162
8.2.3.2	Well-Posedness	163
8.2.3.3	Solving the PDE	163
8.2.4	Variational Formulation	165
8.3	Representation	165
8.3.1	Gaussian Scale Space	165
8.3.2	Variable Conductance Diffusion	167
8.3.3	Geometry-Driven Scale Space	171
8.3.4	Mathematical Morphology Scale Space	173
8.4	Conclusions	176
	References	176
9	**Spatial Light Modulators (SLMs)**	179
	Philip M. Birch, Rupert Young, and Chris Chatwin	
9.1	Introduction	179
9.2	Types of SLM	180
9.2.1	Liquid Crystal	180
9.2.1.1	Nematic Liquid Crystal SLMs	181
9.2.1.2	Frame Rates	183
9.2.1.3	Temperature Effects	184
9.2.1.4	Twisted Nematic LC-SLM	184
9.2.1.5	Modulation Methods	186
9.2.1.6	Ferroelectric	187
9.2.1.7	Addressing Methods	189

9.2.2	Multiple Quantum-Well SLMs	*191*
9.2.3	Mirror Devices	*192*
9.2.3.1	Amplitude Modulators	*192*
9.2.3.2	Phase Modulators	*193*
9.3	Fully Complex Modulation Methods	*194*
9.4	Applications	*196*
9.5	Conclusions	*197*
	References	*198*
10	**Holographic Visualization of 3D Data**	*201*
	Pierre-Alexandre Blanche	
10.1	Introduction	*201*
10.2	Reproducing the Amplitude and the Phase	*203*
10.3	Different Types of Holograms	*207*
10.3.1	Transmission versus Reflection	*207*
10.3.2	Denisyuk Hologram	*209*
10.3.3	Color Reproduction	*210*
10.3.3.1	Reflection	*211*
10.3.3.2	Transmission	*212*
10.3.4	Phase versus Amplitude: The Diffraction Efficiency	*213*
10.3.5	Surface Relief Holograms	*214*
10.3.6	Thin versus Thick Hologram	*215*
10.4	Holographic Approximations	*215*
10.4.1	Rainbow Hologram	*216*
10.4.2	Holographic Stereogram	*217*
10.5	Dynamic Holography	*220*
10.5.1	Holographic Cinematography	*220*
10.5.2	Real-Time Integral Holography	*222*
10.5.3	Holographic Video	*223*
10.6	Conclusion	*224*
	Acknowledgment	*225*
	References	*225*
	Further Reading	*226*
11	**Holographic Data Storage Technology**	*227*
	Kevin Curtis, Lisa Dhar, and Pierre-Alexandre Blanche	
11.1	Introduction	*227*
11.2	Holographic Data Storage Overview	*228*
11.2.1	Drive Architecture	*231*
11.2.2	Consumer Drive Implementation	*233*
11.3	Tolerances and Basic Servo	*234*
11.4	Data Channel Overview	*236*
11.5	Materials for Holography	*237*
11.5.1	Silver Halide Photographic Emulsion	*238*
11.5.2	Photopolymers	*239*

11.5.3	Dichromated Gelatin	*240*
11.5.4	Miscellaneous Recording Media	*241*
11.5.4.1	Photothermoplastics	*241*
11.5.4.2	Photochromics	*242*
11.5.4.3	Photorefractive	*242*
11.6	Material for Data Storage	*243*
11.7	Media for Holographic Data Storage	*246*
11.8	Conclusions	*246*
	References	*247*

12	**Phase-Space Rotators and their Applications in Optics**	*251*
	José A. Rodrigo, Tatiana Alieva, and Martin J. Bastiaans	
12.1	Introduction	*251*
12.2	Signal Representation in Phase Space: The Wigner Distribution	*252*
12.2.1	Description of Partially Coherent Light	*252*
12.2.2	Wigner Distribution	*253*
12.3	Matrix Formalism for the Description of Phase-Space Rotations	*255*
12.4	Basic Phase-Space Rotators for Two-Dimensional Signals	*257*
12.5	Optical System Design for Phase-Space Rotators and their Experimental Implementations	*260*
12.6	Applications of Phase-Space Rotators in Optics	*264*
12.7	Conclusions	*269*
	Acknowledgments	*269*
	References	*269*

13	**Microscopic Imaging**	*273*
	Gloria Bueno, Oscar Déniz, Roberto González-Morales, Juan Vidal, and Jesús Salido	
13.1	Introduction	*273*
13.2	Image Formation: Basic Concepts	*274*
13.2.1	Types of Image	*274*
13.2.2	Image Formation in the Optical Microscope	*274*
13.2.3	Light	*275*
13.2.4	Resolution	*275*
13.3	Components of a Microscopic Imaging System	*276*
13.4	Types of Microscopy	*277*
13.4.1	Bright-Field Microscopy	*278*
13.4.2	Phase Contrast Microscopy	*279*
13.4.3	Dark Contrast Microscopy	*280*
13.4.4	Differential Interference Contrast (DIC) Microscopy	*281*
13.4.5	Fluorescence Microscopy	*282*
13.4.6	Confocal Microscopy	*283*
13.4.7	Electron Microscopy	*283*
13.5	Digital Image Processing in Microscopy	*284*
13.5.1	Image Preprocessing	*284*

13.5.2	Image Enhancement	286
13.5.3	Segmentation	287
13.5.4	Classification	289
13.6	Conclusions	292
	Acknowledgments	292
	References	292

14 Adaptive Optics in Microscopy 295
Martin J. Booth

14.1	Introduction	295
14.2	Aberrations in Microscopy	296
14.2.1	Definition of Aberrations	296
14.2.2	Representation of Aberrations	297
14.2.3	Effects of Aberrations in Microscopes	298
14.2.4	Sources of Aberrations in Microscopes	300
14.2.5	Effects of the Numerical Aperture	300
14.3	Principles of Adaptive Optics	301
14.3.1	Methods for Aberration Measurement	303
14.3.2	Aberration Correction Devices	305
14.4	Aberration Correction in High-Resolution Optical Microscopy	307
14.4.1	Microscope Configurations	307
14.4.2	Point-Scanning Microscopes	308
14.4.3	Widefield Microscopes	309
14.5	Aberration Measurement and Wavefront Sensing	312
14.5.1	Direct Wavefront Sensing in Microscopy	312
14.5.2	Indirect Wavefront Sensing	314
14.6	Control Strategies for Adaptive Microscopy	317
14.6.1	Choice of Signal for Wavefront Sensing	318
14.6.2	Aberration Dynamics	318
14.6.3	Field-Dependent Aberrations	319
14.7	Conclusion	320
	Acknowledgments	321
	References	321

15 Aperture Synthesis and Astronomical Image Formation 323
Anna Scaife

15.1	Introduction	323
15.2	Image Formation from Optical Telescopes	324
15.3	Single-Aperture Radio Telescopes	326
15.4	Aperture Synthesis	327
15.4.1	Principles of Earth Rotation Aperture Synthesis	327
15.4.2	Receiving System Response	330
15.5	Image Formation	333
15.5.1	Derivation of Intensity from Visibility	333
15.5.1.1	Full-Sky Imaging	335

15.5.2	Deconvolution Techniques 337
15.5.2.1	The CLEAN Algorithm 338
15.5.3	Maximum Entropy Deconvolution (The Bayesian Radio Astronomer) 339
15.5.4	Compressed Sensing 342
15.6	Conclusions 343
	References 343

16	**Display and Projection** 345
	Tom Kimpe, Patrick Candry, and Peter Janssens
16.1	Introduction 345
16.2	Direct View Displays 345
16.2.1	Working Principle 345
16.2.2	Transmissive Displays 346
16.2.3	Emissive Displays 348
16.2.3.1	CRT Display 348
16.2.3.2	Plasma Display 349
16.2.3.3	LED Display 350
16.2.3.4	OLED Display 350
16.2.4	Reflective Displays 351
16.2.4.1	Reflective LCD 352
16.2.4.2	Electronic Paper (e-Paper) 352
16.3	Projection Displays 353
16.3.1	Basic Concepts and Key Components 353
16.3.2	Projector Architectures 356
16.3.2.1	Three-Panel Transmissive 357
16.3.2.2	Three-Panel Reflective 357
16.3.2.3	One-Panel Reflective DLP with UHP Lamp 358
16.3.2.4	One-Panel Reflective LCoS with High-Brightness LEDs 359
16.3.2.5	Three-Panel Grating Light Valve Projector 359
16.4	Applications 362
16.4.1	Medical Imaging Displays 362
16.4.1.1	Medical LCD Displays 362
16.4.1.2	Calibration and Quality Assurance of Medical Display Systems 362
16.4.2	Other Applications 364
16.5	Conclusion 366
	References 366

17	**3D Displays** 369
	Janusz Konrad
17.1	Introduction 369
17.2	Planar Stereoscopic Displays 370
17.2.1	Stereoscopic Displays with Glasses 371
17.2.1.1	Spectral Filtering 371
17.2.1.2	Light Polarization 373

17.2.1.3	Light Shuttering	*375*
17.2.2	Autostereoscopic Displays (without glasses)	*375*
17.3	Planar Multiview Displays	*378*
17.3.1	Active Multiview 3D Displays	*378*
17.3.2	Passive Multiview 3D Displays	*379*
17.4	Signal Processing for 3D Displays	*381*
17.4.1	Enhancement of 3D Anaglyph Visualization	*381*
17.4.2	Ghosting Suppression in Polarized and Shuttered 3D Displays	*382*
17.4.3	Anti-Alias Filtering for Multiview Autostereoscopic Displays	*384*
17.4.4	Luminance/Color Balancing for Stereo Pairs	*387*
17.4.5	Intermediate View Interpolation	*389*
17.5	Conclusions	*393*
	Acknowledgments	*394*
	References	*394*

18 Linking Analog and Digital Image Processing *397*
Leonid P. Yaroslavsky

18.1	Introduction	*397*
18.2	How Should One Build Discrete Representation of Images and Transforms?	*398*
18.2.1	Signal Discretization	*399*
18.2.2	Imaging Transforms in the Mirror of Digital Computers	*401*
18.2.3	Characterization of Discrete Transforms in Terms of Equivalent Analog Transforms	*402*
18.2.3.1	Point Spread Function and Frequency Response of a Continuous Filter Equivalent to a Given Digital Filter	*403*
18.2.3.2	Point Spread Function of the Discrete Fourier Analysis	*406*
18.3	Building Continuous Image Models	*408*
18.3.1	Discrete Sinc Interpolation: The Gold Standard for Image Resampling	*408*
18.3.1.1	Signal Recovery from Sparse or Nonuniformly Sampled Data	*409*
18.3.2	Image Numerical Differentiation and Integration	*411*
18.4	Digital-to-Analog Conversion in Digital Holography. Case Study: Reconstruction of Kinoform	*414*
18.5	Conclusion	*417*
	References	*418*

19 Visual Perception and Quality Assessment *419*
Anush K. Moorthy, Zhou Wang, and Alan C. Bovik

19.1	Introduction	*419*
19.2	The Human Visual System	*420*
19.3	Human-Visual-System-Based Models	*422*
19.3.1	Visual Difference Predictor (VDP)	*423*
19.3.2	Visual Discrimination Model (VDM)	*423*
19.3.3	Teo and Heeger Model	*423*

19.3.4	Visual Signal-to-Noise Ratio (VSNR)	424
19.3.5	Digital Video Quality Metric (DVQ)	424
19.3.6	Moving Picture Quality Metric (MPQM)	424
19.3.7	Scalable Wavelet-Based Distortion Metric for VQA	425
19.4	Feature-Based Models	425
19.4.1	A Distortion Measure Based on Human Visual Sensitivity	425
19.4.2	Singular Value Decomposition and Quality	425
19.4.3	Curvature-Based Image Quality Assessment	426
19.4.4	Perceptual Video Quality Metric (PVQM)	426
19.4.5	Video Quality Metric (VQM)	426
19.4.6	Temporal Variations of Spatial-Distortion-Based VQA	427
19.4.7	Temporal Trajectory Aware Quality Measure	427
19.5	Structural and Information-Theoretic Models	427
19.5.1	Single-Scale Structural Similarity Index (SS-SSIM)	428
19.5.2	Multiscale Structural Similarity Index (MS-SSIM)	428
19.5.3	SSIM Variants	429
19.5.4	Visual Information Fidelity (VIF)	429
19.5.5	Structural Similarity for VQA	430
19.5.6	Video VIF	430
19.6	Motion-Modeling-Based Algorithms	430
19.6.1	Speed-Weighted Structural Similarity Index (SW-SSIM)	431
19.6.2	Motion-Based Video Integrity Evaluation (MOVIE)	431
19.7	Performance Evaluation and Validation	432
19.8	Conclusion	435
	References	435

20 Digital Image and Video Compression 441
Joeri Barbarien, Adrian Munteanu, and Peter Schelkens

20.1	Introduction	441
20.2	Typical Architecture	441
20.3	Data Prediction and Transformation	442
20.3.1	Removing Data Redundancy	442
20.3.2	Spatial Prediction	443
20.3.3	Spatial Transforms	444
20.3.4	Color/Spectral and Multiple-Component Prediction and Transforms	445
20.3.5	Temporal Redundancy Removal by Motion Estimation	445
20.3.5.1	Motion-Compensated Prediction	445
20.3.5.2	Improvements over the Basic Approach	447
20.4	Quantization	449
20.4.1	Principle	449
20.4.2	Lloyd–Max Quantizers	450
20.4.3	Embedded Quantization	451
20.5	Entropy Coding	452

20.5.1	Huffman Coding	*452*
20.5.2	Arithmetic Coding	*453*
20.6	Image and Volumetric Coding	*455*
20.6.1	Generic Image Coding	*455*
20.6.2	JPEG	*456*
20.6.3	JPEG 2000	*456*
20.7	Video Coding	*457*
20.7.1	H.261	*458*
20.7.2	H.264/AVC	*459*
20.8	Conclusions	*460*
	Acknowledgments	*460*
	References	*460*

21 Optical Compression Scheme to Simultaneously Multiplex and Encode Images *463*
Ayman Alfalou, Ali Mansour, Marwa Elbouz, and Christian Brosseau

21.1	Introduction	*463*
21.2	Optical Image Compression Methods: Background	*464*
21.3	Compression and Multiplexing: Information Fusion by Segmentation in the Spectral Plane	*466*
21.4	Optical Compression of Color Images by Using JPEG and JPEG2000 Standards	*470*
21.4.1	Optical JPEG Implementation Results	*472*
21.4.2	Optical and Digital JPEG Comparison	*473*
21.4.3	Optical JPEG2000 Implementation	*474*
21.5	New Simultaneous Compression and Encryption Approach Based on a Biometric Key and DCT	*474*
21.6	Conclusions	*480*
	References	*481*

22 Compressive Optical Imaging: Architectures and Algorithms *485*
Roummel F. Marcia, Rebecca M. Willett, and Zachary T. Harmany

22.1	Introduction	*485*
22.1.1	Organization of the Chapter	*486*
22.2	Compressive Sensing	*486*
22.3	Architectures for Compressive Image Acquisition	*488*
22.3.1	Coded Apertures	*490*
22.3.2	Compressive-Coded Apertures	*492*
22.4	Algorithms for Restoring Compressively Sensed Images	*494*
22.4.1	Current Algorithms for Solving the CS Minimization Problem	*494*
22.4.2	Algorithms for Nonnegativity Constrained $\ell_2 - \ell_1$ CS Minimization	*496*
22.4.3	Model-Based Sparsity	*497*
22.4.4	Compensating for Nonnegative Sensing Matrices	*498*

22.5	Experimental Results	499
22.6	Noise and Quantization	502
22.7	Conclusions	502
	Acknowledgments	503
	References	503

23 Compressed Sensing: "When Sparsity Meets Sampling" 507
Laurent Jacques and Pierre Vandergheynst

23.1	Introduction	507
23.1.1	Conventions	508
23.2	In Praise of Sparsity	508
23.3	Sensing and Compressing in a Single Stage	510
23.3.1	Limits of the Shannon–Nyquist Sampling	510
23.3.2	New Sensing Model	511
23.4	Reconstructing from Compressed Information: A Bet on Sparsity	512
23.5	Sensing Strategies Market	515
23.5.1	Random sub-Gaussian Matrices	516
23.5.2	Random Fourier Ensemble	516
23.5.3	Random Basis Ensemble	517
23.5.4	Random Convolution	517
23.5.5	Other Sensing Strategies	518
23.6	Reconstruction Relatives	518
23.6.1	Be Sparse in Gradient	518
23.6.2	Add or Change Priors	519
23.6.3	Outside Convexity	520
23.6.4	Be Greedy	520
23.7	Some Compressive Imaging Applications	521
23.7.1	Compressive Imagers	521
23.7.2	Compressive Radio Interferometry	523
23.8	Conclusion and the "Science 2.0" Effect	524
23.8.1	Information Sources	524
23.8.2	Reproducible Research	525
	Acknowledgments	526
	References	526
	Further Reading	527

24 Blind Deconvolution Imaging 529
Filip Šroubek and Michal Šorel

24.1	Introduction	529
24.2	Image Deconvolution	530
24.3	Single-Channel Deconvolution	534
24.4	Multichannel Deconvolution	539
24.5	Space-Variant Extension	542
24.6	Conclusions	546

Acknowledgments 547
References 547

25 Optics and Deconvolution: Wavefront Sensing 549
Justo Arines and Salvador Bará
25.1   Introduction 549
25.2   Deconvolution from Wavefront Sensing (DWFS) 550
25.3   Past and Present 551
25.4   The Restoration Process 552
25.4.1 Estimating the Wave Aberration with Gradient-Based Wavefront Sensors 553
25.4.2 PSF and OTF: Estimation Process and Bias 558
25.4.3 Significance of the Restoration Filter 559
25.4.4 Resolution of the Restored Image: Effective Cutoff Frequency 560
25.4.5 Implementation of the Deconvolution from Wavefront Sensing Technique 562
25.5   Examples of Application 563
25.5.1 Astronomical Imaging 563
25.5.2 Eye Fundus Imaging 564
25.6   Conclusions 567
Acknowledgments 568
References 568
Further Reading 569

26 Image Restoration and Applications in Biomedical Processing 571
Filip Rooms, Bart Goossens, Aleksandra Pižurica, and Wilfried Philips
26.1   Introduction 571
26.2   Classical Restoration Techniques 574
26.2.1 Inverse Filter and Wiener Filter 574
26.2.1.1 Inverse Filter 574
26.2.1.2 Wiener Filter 575
26.2.2 Bayesian Restoration 577
26.2.2.1 Gaussian Noise Model 578
26.2.3 Poisson Noise Model 580
26.2.3.1 Richardson–Lucy Restoration 580
26.2.3.2 Classical Regularization of Richardson–Lucy 581
26.3   SPERRIL: Estimation and Restoration of Confocal Images 583
26.3.1 Origin and Related Methods 583
26.3.2 Outline of the Algorithm 584
26.3.2.1 Noise Reduction 584
26.3.2.2 Deblurring Step 585
26.3.2.3 SPERRIL as RL with a Prior? 585
26.3.3 Experimental Results 586
26.3.3.1 Colocalization Analysis: What and Why? 586
26.3.3.2 Experimental Setup 586

26.4	Conclusions 589
	Acknowledgment 589
	References 590

27	**Optical and Geometrical Super-Resolution** 593
	Javier Garcia Monreal
27.1	Introduction 593
27.2	Fundamental Limits to Resolution Improvement 594
27.3	Diffractive Optical Super-Resolution 595
27.3.1	Optical System Limitations and Super-Resolution Strategy 595
27.3.2	Nonholographic Approaches 597
27.3.2.1	Time Multiplexing 597
27.3.2.2	Angular Multiplexing 600
27.3.2.3	Multiplexing in Other Degrees of Freedom 601
27.3.3	Holographic Approaches 602
27.3.3.1	Holographic Wavefront Coding 602
27.3.3.2	Multiplexed Holograms 603
27.3.3.3	Digital Holography 604
27.3.3.4	Axial Super-Resolution 607
27.4	Geometrical Super-Resolution 608
	References 611

28	**Super-Resolution Image Reconstruction considering Inaccurate Subpixel Motion Information** 613
	Jongseong Choi and Moon Gi Kang
28.1	Introduction 613
28.2	Fundamentals of Super-Resolution Image Reconstruction 614
28.2.1	Basic Concept of Super-Resolution 614
28.2.2	Observation Model 616
28.2.3	Super-Resolution as an Inverse Problem 617
28.2.3.1	Constrained Least Squares Approach 617
28.2.3.2	Bayesian Approach 618
28.2.4	The Frequency Domain Interpretation 621
28.3	Super-Resolution Image Reconstruction considering Inaccurate Subpixel Motion Estimation 623
28.3.1	Analysis of the Misregistration Error 623
28.3.2	Multichannel-Regularized Super-Resolution Image Reconstruction Algorithm 624
28.3.3	Experimental Results 628
28.4	Development and Applications of Super-Resolution Image Reconstruction 631
28.4.1	Super-Resolution for Color Imaging Systems 632
28.4.2	Simultaneous Enhancement of Spatial Resolution and Dynamic Range 634
28.4.3	Super-Resolution for Video Systems 636

28.5	Conclusions 640
	Acknowledgments 640
	References 641

29	**Image Analysis: Intermediate-Level Vision** 643
	Jan Cornelis, Aneta Markova, and Rudi Deklerck
29.1	Introduction 643
29.2	Pixel- and Region-Based Segmentation 645
29.2.1	Supervised Approaches 646
29.2.1.1	Classification Based on MAP (Maximizing the A posteriori Probability) 646
29.2.2	Unsupervised Approaches 647
29.2.2.1	*K*-means Clustering 648
29.2.2.2	Expectation-Maximization (EM) 649
29.2.3	Improving the Connectivity of the Classification Results 650
29.2.3.1	Seeded Region Growing 651
29.2.3.2	Mathematical Morphology 651
29.3	Edge-Based Segmentation 652
29.3.1	Optimal Edge Detection and Scale-Space Approach 654
29.4	Deformable Models 654
29.4.1	Mathematical Formulation (Continuous Case) 655
29.4.2	Applications of Active Contours 657
29.4.3	The Behavior of Snakes 658
29.5	Model-Based Segmentation 661
29.5.1	Statistical Labeling 662
29.5.2	Bayesian Decision Theory 662
29.5.3	Graphs and Markov Random Fields Defined on a Graph 662
29.5.4	Cliques 663
29.5.5	Models for the Priors 663
29.5.6	Labeling in a Bayesian Framework based on Markov Random Field Modeling 663
29.6	Conclusions 664
	References 664

30	**Hybrid Digital–Optical Correlator for ATR** 667
	Tien-Hsin Chao and Thomas Lu
30.1	Introduction 667
30.1.1	Gray-Scale Optical Correlator System's Space–Bandwidth Product Matching 669
30.1.2	Input SLM Selection 671
30.2	Miniaturized Gray-Scale Optical Correlator 673
30.2.1	512×512 GOC System Architecture 673
30.2.2	Graphic User Interface of the GOC System 674
30.2.3	Gray-Scale Optical Correlator Testing 675
30.2.4	Summary 676

30.3	Optimization of OT-MACH Filter	*677*
30.3.1	Optimization Approach	*677*
30.4	Second Stage: Neural Network for Target Verification	*681*
30.4.1	Feature Extraction Methods	*682*
30.4.1.1	Horizontal and Vertical Binning	*682*
30.4.1.2	Principal Component Analysis	*684*
30.4.2	Neural Network Identification	*686*
30.4.2.1	Neural Network Algorithm	*686*
30.5	Experimental Demonstration of ATR Process	*687*
30.5.1	Vehicle Identification	*687*
30.5.2	Sonar Mine Identification	*689*
30.6	Conclusions	*690*
	Acknowledgments	*692*
	References	*692*
31	**Theory and Application of Multispectral Fluorescence Tomography**	*695*
	Rosy Favicchio, Giannis Zacharakis, Anikitos Garofalakis, and Jorge Ripoll	
31.1	Introduction	*695*
31.2	Fluorescence Molecular Tomography (FMT)	*696*
31.2.1	FMT Principle	*696*
31.2.2	Theoretical Background	*697*
31.2.2.1	Optical Parameters	*698*
31.2.2.2	The Diffusion Equation	*699*
31.2.2.3	Solutions of the Diffusion Equation for Infinite Homogeneous Media	*699*
31.2.2.4	The Excitation Source Term	*700*
31.2.2.5	The Fluorescence Source Term	*700*
31.2.2.6	The Born Approximation for the Excitation Term	*702*
31.2.2.7	Boundary Conditions	*702*
31.2.2.8	Inverse Problem	*703*
31.2.2.9	The Normalized Born Approximation	*703*
31.2.3	Experimental Setup	*705*
31.3	Spectral Tomography	*706*
31.3.1	Spectral Deconvolution	*707*
31.4	Multitarget Detection and Separation	*709*
31.4.1	Multicolor Phantom	*709*
31.4.1.1	*In vitro* Fluorophore Unmixing	*709*
31.4.1.2	Methodology	*709*
31.4.2	*In vivo* Study	*711*
31.4.2.1	*In vivo* Fluorophore Unmixing	*711*
31.4.2.2	Methodology	*711*
31.5	Conclusions	*712*
	References	*713*

32	**Biomedical Imaging Based on Vibrational Spectroscopy** 717
	Christoph Krafft, Benjamin Dietzek, and Jürgen Popp
32.1	Introduction 717
32.2	Vibrational Spectroscopy and Imaging 718
32.2.1	Infrared Spectroscopy 718
32.2.2	Raman Spectroscopy 720
32.2.3	Coherent Anti-Stokes–Raman Scattering Microscopy 721
32.3	Analysis of Vibrational Spectroscopic Images 723
32.3.1	Preprocessing 723
32.3.1.1	Quality Test 723
32.3.1.2	Denoising 724
32.3.1.3	Background and Baseline Correction 724
32.3.1.4	Normalization 724
32.3.1.5	Image Compression 725
32.3.2	Exploratory Image Analysis 725
32.3.2.1	Classical Image Representations 725
32.3.2.2	Principal Component Analysis 726
32.3.3	Image Segmentation: Cluster Analysis 728
32.3.4	Supervised Image Segmentation: Linear Discriminant Analysis 729
32.4	Challenges for Image Analysis in CARS Microscopy 730
32.4.1	Particle Identification in Nonlinear Microscopy 731
32.4.2	Numerical Reduction or Suppression of Nonresonant Background 732
32.4.3	Outlook – Merging CARS Imaging with Chemometrics 734
32.5	Biomedical Applications of Vibrational Spectroscopic Imaging: Tissue Diagnostics 734
32.6	Conclusions 736
	Acknowledgments 736
	References 736

33	**Optical Data Encryption** 739
	Maria Sagrario Millán García-Varela and Elisabet Pérez-Cabré
33.1	Introduction 739
33.2	Optical Techniques in Encryption Algorithms 740
33.2.1	Random Phase Mask (RPM) and Phase Encoding 740
33.2.2	Double-Random Phase Encryption (DRPE) 741
33.2.3	Resistance of DRPE against Attacks 746
33.2.4	Encryption Algorithms Based on Real (Phase-Only and Amplitude-Only) Functions 748
33.2.5	Holographic Memory 749
33.2.6	Wavelength Multiplexing and Color Image Encryption 750
33.2.7	Fresnel Domain 751
33.2.8	Fractional Fourier Transforms 753
33.3	Applications to Security Systems 755
33.3.1	Optical Techniques and DRPE in Digital Cryptography 755

33.3.2	Multifactor Identification and Verification of Biometrics	756
33.3.3	ID Tags for Remote Verification	759
33.4	Conclusions	765
	Acknowledgments	765
	References	765

34 Quantum Encryption 769
Bing Qi, Li Qian, and Hoi-Kwong Lo

34.1	Introduction	769
34.2	The Principle of Quantum Cryptography	770
34.2.1	Quantum No-Cloning Theorem	770
34.2.2	The BB84 Quantum Key Distribution Protocol	771
34.2.3	Entanglement-Based Quantum Key Distribution Protocol	774
34.2.4	Continuous Variable Quantum Key Distribution Protocol	776
34.3	State-of-the-Art Quantum Key Distribution Technologies	777
34.3.1	Sources for Quantum Key Distribution	777
34.3.1.1	Single-Photon Source	777
34.3.1.2	EPR Photon Pair	778
34.3.1.3	Attenuated Laser Source	778
34.3.2	Quantum State Detection	779
34.3.2.1	Single-Photon Detector	779
34.3.2.2	Optical Homodyne Detector	780
34.3.3	Quantum Random Number Generator	781
34.3.4	Quantum Key Distribution Demonstrations	781
34.3.4.1	QKD Experiments through Telecom Fiber	781
34.3.4.2	QKD Experiments through Free Space	782
34.4	Security of Practical Quantum Key Distribution Systems	783
34.4.1	Quantum Hacking and Countermeasures	783
34.4.2	Self-Testing Quantum Key Distribution	784
34.5	Conclusions	785
	Acknowledgments	786
	References	786

35 Phase-Space Tomography of Optical Beams 789
Tatiana Alieva, Alejandro Cámara, José A. Rodrigo, and María L. Calvo

35.1	Introduction	789
35.2	Fundamentals of Phase-Space Tomography	790
35.3	Phase-Space Tomography of Beams Separable in Cartesian Coordinates	793
35.4	Radon Transform	794
35.5	Example: Tomographic Reconstruction of the WD of Gaussian Beams	796
35.6	Experimental Setup for the Measurements of the WD Projections	798
35.7	Reconstruction of WD: Numerical and Experimental Results	800
35.8	Practical Work for Postgraduate Students	802

35.9	Conclusions *807*	
	Acknowledgments *807*	
	References *807*	

36	**Human Face Recognition and Image Statistics using Matlab** *809*	
	Matthias S. Keil	
36.1	Introduction *809*	
36.2	Neural Information-Processing and Image Statistics *811*	
36.2.1	Possible Reasons of the k^{-2} Frequency Scaling *811*	
36.2.1.1	Occlusion *813*	
36.2.1.2	Scale Invariance *813*	
36.2.1.3	Contrast Edges *813*	
36.2.2	The First Strategy: Spatial Decorrelation *814*	
36.2.3	The Second Strategy: Response Equalization ("Whitening") *816*	
36.3	Face Image Statistics and Face Processing *818*	
36.3.1	Face Image Dataset *818*	
36.3.2	Dimension of Spatial Frequency *819*	
36.4	Amplitude Spectra *820*	
36.4.1	Window Artifacts and Corrected Spectra *821*	
36.4.2	Whitened Spectra of Face Images *821*	
36.4.2.1	Slope Whitening *822*	
36.4.2.2	Variance Whitening *822*	
36.4.2.3	Whitening by Diffusion *824*	
36.4.2.4	Summary and Conclusions *826*	
36.5	Making Artificial Face Recognition "More Human" *826*	
36.6	Student Assignments *827*	
	References *828*	

37	**Image Processing for Spacecraft Optical Navigation** *833*	
	Michael A. Paluszek and Pradeep Bhatta	
37.1	Introduction *833*	
37.2	Geometric Basis for Optical Navigation *835*	
37.2.1	Example: Analytical Solution of *r* *837*	
37.3	Optical Navigation Sensors and Models *837*	
37.3.1	Optics *837*	
37.3.2	Imaging System Resolution *838*	
37.3.3	Basic Radiometry *839*	
37.3.4	Imaging *840*	
37.3.4.1	Noise and Performance Factors *841*	
37.3.4.2	Data Reduction *844*	
37.3.4.3	Error Modeling *845*	
37.4	Dynamical Models *845*	
37.5	Processing the Camera Data *847*	
37.6	Kalman Filtering *847*	
37.6.1	Introduction to Kalman Filtering *847*	

37.6.2	The Unscented Kalman Filter	848
37.7	Example Deep Space Mission	850
37.7.1	Sensor Model	851
37.7.2	Simulation Results	854
37.8	Student Assignment	855
37.9	Conclusion	856
	References	857

38 ImageJ for Medical Microscopy Image Processing: An Introduction to Macro Development for Batch Processing *859*

Tony Collins

38.1	Introduction	859
38.2	Installation	859
38.2.1	Add-Ons	860
38.3	Plugin Collections	861
38.4	Opening Images	861
38.5	Developing a Macro	862
38.5.1	Getting Started – Manually Planning the Analysis	862
38.5.2	Automating the Analysis: Recording Menu Commands	863
38.5.3	Measuring Intensity	864
38.5.4	Basic Macro programming	865
38.5.5	Batch Processing	866
38.5.5.1	Making a "Function"	866
38.5.5.2	Processing a List of Files with a "for" Loop	868
38.5.5.3	Filtering Files with an "if" Statement	869
38.5.6	Adding a Dialog Box	869
38.5.6.1	Creating a Dialog	869
38.5.6.2	Assigning Values to Variables from the Dialog	870
38.6	Further Practical Exercises	872
38.7	Important Websites	872
	Appendix 38.A: Analyzing a Single Image	872
	Appendix 38.B: Including Intensity Measurements	873
	Appendix 38.C: Making a Function	873
	Appendix 38.D: Batch Processing a Folder	873
	Appendix 38.E: Adding a Dialog and Batch Processing a Folder	874
	Appendix 38.F: Batch Processing Subfolders	875
	References	877

Index *879*

Editors

Gabriel Cristóbal received his degree in electrical engineering from Universidad Politécnica de Madrid (Spain) in 1979. Thereafter, he obtained a PhD degree in telecommunication engineering at the same University in 1986. He held several research positions there between 1982 and 1989. During 1989–1991 he was a visiting scholar at the International Computer Science Institute (ICSI) and, from 1989 to 1992, an assistant research engineer at the Electronic Research Lab (UC Berkeley). During the period 1995 to 2001, he headed the Department of Imaging and Vision at the Instituto de Optica Spanish Council for Scientific Research (CSIC). He is currently a research scientist at the same institute. His current research interests are joint representations, vision modeling, multidimensional signal processing, and image quality assessment. He has been responsible for several national and EU research and development projects. He has published more than 125 papers in international journals, monographs, and conference proceedings. He has been a senior member of the IEEE Signal Processing Society since 1996, a member of the Optical Society of America (OSA), EURASIP Spanish liaison officer during the period 2009–2010 and a member of the ISO/IEC JTC1/SC29/WG1 (JPEG2000) and WG11 (MPEG) committees.

Correspondence address:
Instituto de Optica (CSIC)
Serrano 121, 28006 Madrid, Spain
Tel: +34-91-561-6800 x942319; FAX: +34-91-564-5557
Email: gabriel@optica.csic.es; Alternative email: gabriel.cristobal@gmail.com

Peter Schelkens received his degree in electronic engineering in VLSI-design from the Industriële Hogeschool Antwerpen-Mechelen (IHAM), Campus Mechelen. Thereafter, he obtained an electrical engineering degree (MSc) in applied physics, a biomedical engineering degree (medical physics), and, finally, a PhD degree in applied sciences from the Vrije Universiteit Brussel (VUB). Peter Schelkens currently holds a professorship at the Department of Electronics and Informatics (ETRO) at the Vrije Universiteit Brussel (VUB) and, in addition, a postdoctoral fellowship with the Fund for Scientific Research – Flanders (FWO), Belgium. Peter Schelkens is a member of the scientific staff of the Interdisciplinary Institute for Broadband Technology (www.IBBT.be), Belgium. Additionally, since

1995, he has also been affiliated to the Interuniversity Microelectronics Institute (www.IMEC.be), Belgium, as scientific collaborator. He became a member of the board of councilors of the same institute recently. Peter Schelkens coordinates a research team in the field of multimedia coding, communication, and security and especially enjoys cross-disciplinary research. He has published over 200 papers in journals and conference proceedings, and standardization contributions, and he holds several patents. He is also coeditor of the book, "The JPEG 2000 Suite," published in 2009 by Wiley. His team is participating in the ISO/IEC JTC1/SC29/WG1 (JPEG), WG11 (MPEG), and ITU-T standardization activities. Peter Schelkens is the Belgian head of delegation for the ISO/IEC JPEG standardization committee, editor/chair of part 10 of JPEG 2000: "Extensions for Three-Dimensional Data" and PR Chair of the JPEG committee. He is a member of the IEEE, SPIE, and ACM, and is currently the Belgian EURASIP Liaison Officer.

Correspondence address:
Vrije Universiteit Brussel – Interdisciplinary Institute for Broadband Technology
Department of Electronics and Informatics (ETRO)
Pleinlaan 2, 1050 Brussels, Belgium
Tel: +32 2 6291681; FAX: +32 2 6291700
Email: peter.schelkens©vub.ac.be

Hugo Thienpont is a full professor at the Faculty of Engineering of the Vrije Universiteit Brussel (VUB). He chairs the Applied Physics and Photonics Department and is director of its photonics research group B-PHOT, which he built over the years and which today counts about 50 scientists, engineers, and administrative and technical staff. He graduated as an electrotechnical engineer with majors in applied physics in 1984, and received his PhD in applied sciences in 1990, both at the VUB. Over the years, Hugo and his team have made research efforts in various fundamental and applied research topics, most of them in the domain of microphotonics and micro-optics. With the results of this work, he has authored around 200 SCI-stated journal papers and around 400 publications in international conference proceedings. He has edited more than 15 conference proceedings and authored 7 chapters in books. He was invited to or was keynote speaker at more than 50 international conferences and jointly holds 20 patents.

His research work has been internationally recognized by way of several awards. In 1999, he received the International Commission for Optics Prize ICO'99 and the Ernst Abbe medal from Carl Zeiss for "his noteworthy contributions in the field of photonics and parallel micro-optics." In 2003, he was awarded the title of "IEEE LEOS Distinguished Lecturer" for serving as international lecturer from 2001–2003 on the theme "Optical Interconnects to Silicon Chips." In 2005, he received the SPIE President's Award 2005 for meritorious services to the Society and for his leadership in photonics in Europe. In 2006, he was nominated SPIE Fellow for his research contributions to the field of micro-optics and microphotonics. In 2007, he received the award "Prof. R. Van Geen" for his scientific achievements during his research career at VUB and is nominated as EOS Fellow. In October 2007, he received the International Micro-Optics Award MOC'07 from the Japanese Optical Society. In 2008, he obtained the prestigious status of Methusalem top scientist from the Flemish government for his research track record in photonics.

Hugo Thienpont is also appreciated by his peers for his service to the photonics community. Indeed, Hugo has been member of many technical and scientific program committees of photonics-related conferences organized by international societies such as SPIE, IEEE, OSA, EOS, and ICO. One of his major achievements is the conception and initiation of SPIE's flagship symposium in Europe, "Photonics Europe." Hugo has been general chair of this pan-European conference, which was held in Strasbourg for the years from 2004 until 2008 and in Brussels since 2010, drawing more than 2000 attendees. He has served as associate editor of *Optical Engineering* and *Opto-Electronics Review* and was guest editor of several special issues on Optics in Computing and on Optical Interconnects for applied optics and the *IEEE Journal of Selected Topics in Quantum Electronics*. Since 2008 he has been a member of the editorial board of the online journal *SPIE Reviews*. He currently serves on the board of directors of SPIE and is a member of the Board of Stakeholders of the Technology Platform Photonics21, a high-level advisory board for optics and photonics in EC FP 7.

Correspondence address:
Vrije Universiteit Brussel
Department of Applied Physics and Photonic – Brussels Photonics Team (B-PHOT)
Pleinlaan, 2, 1050 Brussels, Belgium
Tel: +32 2 7916852; FAX: +32 2 6293450
Email: hthienpo©b-phot.org

Preface

> *Good composition is like a suspension bridge – each line adds strength and takes none away.*
> Robert Henri
>
> *It should be possible to explain the laws of physics to a barmaid.*
> Albert Einstein

In recent years, Moore's law has fostered the steady growth of the field of digital image processing, though computational complexity remains a significant problem for most of the digital image processing applications. In parallel, the research domain of optical image processing has also matured, potentially bypassing the problems digital approaches are facing and bringing in new applications. This book covers the fundamentals of optical and image processing techniques by integrating contributions from both research communities to enable resolving of bottlenecks that applications encounter nowadays, and to give rise to new applications. Therefore, this book, "Optical and Digital Image Processing – Fundamentals and Applications," has a broad scope, since, besides focusing on joint research, it additionally aims at disseminating the knowledge existing in both domains. A precedent of the current book can be traced back to the mid-1980s when one of the coeditors (G. Cristobal) organized a series of annual courses on the topic of "Optical and Digital Image Processing" (see Figure 1).

In 2008, a joint conference on optical and digital image processing was organized for the first time as part of the SPIE Photonics Europe meeting in Strasbourg. Later on in 2010, a second conference was organized at SPIE Photonics Europe in Brussels on the same topic.

Image processing using optical or digital approaches are mature fields covered by many monographs. However, in the literature, a gap exists in terms of monographs that cover both domains. The current, interdisciplinary book is intended to be a valuable source reference to bridge the gap between the optical and digital worlds and to enable better communication between them. The artwork on the cover of this book serves as a good illustration of this idea. In addition to traditional aspects of optics and digital image processing, this book includes the state-of-the-art methods and techniques currently used by researchers as well as the most significant applications. It is necessary to emphasize that a book that covers both optical and

Figure 1 The book "Tratamiento Óptico and Digital de Imágenes" (in Spanish), edited by G. Cristóbal and M.A. Muriel and published in 1984 by the Publications department of the ETS Ing. Telecomunicacion (Polytechnic Univ. of Madrid), ISBN: 84-7402-156-1, can be considered as a predecessor to this book.

digital domains including the fundamentals, the current state of the art, and a selected range of applications has not been published so far.

The book has been structured in five different parts:

- Chapters 1–4 introduce the basic concepts in optics, photonics, information theory, and digital image processing.
- Chapters 5–17 include the basic methodologies and techniques that serve as the foundation upon which the remainder of the book has been built.
- Chapter 18 serves as a bridge between the analog and digital image processing approaches.
- Chapters 19–34 include a selection of the most representative applications following an optical and/or digital approach. In most of the cases, the editors' intention has been to illustrate how the different applications areas have been solved by optical, digital, or even hybrid optical–digital approaches.
- Chapters 35–38 describe four hands-on projects allowing the reader to experiment with some of the described techniques.

This book can be used as a textbook in a two-semester course on the topic, primarily in computer science and electrical engineering departments but also in physics and medicine. This book will be a valuable reference for physicists, engineers, computer scientists, and technologists, in general, due to its integration in a single monograph of information that is usually spread across many sources.

For the reader's convenience, there is an accompanying website with supplementary material at www.wiley-vch.de. It contains selected MATLAB codes, testing images, and errata.

Gabriel Cristóbal, Peter Schelkens, and Hugo Thienpont
The Editors

Acknowledgments

We would like to express our appreciation for the quality of chapters delivered by the authors and for their efforts to keep the chapter length within the given limits. This project could not have been achieved without the valuable contributions made by a significant number of experts in the field from both the academia and industry. We are grateful to them for their willingness to contribute to this groundbreaking resource. Special thanks to Prof. J. Goodman for agreeing to write the foreword for this book.

We would like to extend thanks to all the Wiley VCH members and in particular to Nina Stadthaus, who helped us in managing the project, and to Val Molière for her enthusiastic support.

February 2011 *Gabriel Cristóbal, Peter Schelkens, and Hugo Thienpont*
The Editors

List of Contributors

Tatiana Alieva
Universidad Complutense
de Madrid
Facultad de Ciencias Fìsicas
Avda. Complutense s/n
28040 Madrid
Spain

Ayman Alfalou
ISEN Brest
Institut Supérieure de
l'Electronique et du Numérique
20 rue Cuirassé Bretagne
C.S. 42807
29228 Brest Cedex 2
France

Justo Arines
Universidade de
Santiago de Compostela
Departamento de Física Aplicada
(área de Óptica)
Escola Universitaria de Óptica e
Optometría (Campus Vida)
15782 Santiago de Compostela
(A Coruña)
Spain

and

Universidad de Zaragoza
Facultad de Ciencias
Depto. De Física Aplicada
Pedro Cerbuna 12
50009 Zaragoza
Spain

Salvador Bará
Universidade de
Santiago de Compostela
Departamento de Física
Aplicada (área de Óptica)
Escola Universitaria de Óptica e
Optometría (Campus Vida)
15782 Santiago de Compostela
(A Coruña)
Spain

Joeri Barbarien
Vrije Universiteit Brussel
Interdisciplinary Institute for
Broadband Technology (IBBT)
Department of Electronics and
Informatics (ETRO)
Pleinlaan 2
1050 Brussels
Belgium

List of Contributors

Martin J. Bastiaans
Eindhoven University of Technology
Department of Electrical Engineering
P. O. Box 513
5600 MB, Eindhoven
The Netherlands

Pradeep Bhatta
Princeton Satellite Systems
33 Witherspoon Street
Princeton, NJ 08542-3207
USA

Philip M. Birch
University of Sussex
Engineering and Design
Sussex House
Brighton BN1 9RH
UK

Pierre-Alexandre Blanche
University of Arizona
College of Optical Sciences
1630 E. University Blvd.
Tucson, AZ 85721
USA

Martin J. Booth
University of Oxford
Department of Engineering Science
Parks Road
Oxford OX1 3PJ
UK

Alan C. Bovik
University of Texas at Austin
Department of Electrical and Computer Engineering
Austin, TX 978712
USA

Christian Brosseau
Laboratoire en Sciences et Technologies de l'Information de la Communication et de la Connaissance
6 avenue Victor Le Gorgeu
C.S. 93837
29238 Brest Cedex 3
France

Gloria Bueno
Universidad de Castilla-La Mancha
E.T.S.Ingenieros Industriales
Avda. Camilo Jose Cela s/n
13071 Ciudad Real
Spain

Alejandro Cámara
Universidad Complutense de Madrid
Facultad de Ciencias Físicas
Avda. Complutense s/n
28040 Madrid
Spain

María L. Calvo
Universidad Complutense de Madrid
Facultad de Ciencias Físicas
Avda. Complutense s/n
28040 Madrid
Spain

Patrick Candry
Barco NV
Pres. Kennedypark 35
8500 Kortrijk
Belgium

Tien-Hsin Chao
California Institute of Technology
Jet Propulsion Laboratory
Mail Stop 303-300
4800 Oak Grove Drive
Pasadena, CA 91109-8099
USA

Chris Chatwin
University of Sussex
Engineering and Design
Sussex House
Brighton BN1 9RH
UK

Jongseong Choi
Yonsei University
Department of Electrical and
Electronic Engineering
134 Sinchon-dong
Seodaemun-gu
Seoul 120-749
South Korea

Leon Cohen
Hunter College and
Graduate Center
695 Park Avenue
New York, NY 10021
USA

Tony Collins
McMaster University
Department Biochemistry and
Biomedical Sciences HSC 4H21A
McMaster Biophotonics Facility
1200 Main St. W. Hamilton
ON L8N 3Z5
Canada

Jan Cornelis
Vrije Universiteit Brussel (VUB)
Department of Electronics and
Informatics (ETRO)
Pleinlaan 2
1050 Elsene
Belgium

and

Interdisciplinary Institute for
Broadband Technology (IBBT)
Zuiderpoort Office Park
Gaston Crommenlaan 8 (box 102)
9050 Ghent-Ledeberg
Belgium

Gabriel Cristóbal
Imaging and Vision Department
Instituto de Óptica (CSIC)
Serrano 121
28006 Madrid
Spain

Kevin Curtis
InPhase Technologies Inc.
2000 Pike Road
Longmont, CO 80501-6764
USA

Ingrid Daubechies
Duke University
Box 90320
Durham, NC 27708-0320
USA

Rudi Deklerck
Vrije Universiteit Brussel (VUB)
Department of Electronics and
Informatics (ETRO)
Pleinlaan 2
1050 Elsene
Belgium

and

Interdisciplinary Institute for
Broadband Technology (IBBT)
Zuiderpoort Office Park
Gaston Crommenlaan 8 (box 102)
9050 Ghent-Ledeberg
Belgium

Oscar Déniz
Universidad de Castilla-La
Mancha
E.T.S.Ingenieros Industriales
Avda. Camilo Jose Cela s/n
13071 Ciudad Real
Spain

Lisa Dhar
InPhase Technologies Inc.
2000 Pike Road
Longmont, CO 80501-6764
USA

Benjamin Dietzek
Friedrich Schiller-University Jena
Institute of Physical Chemistry
Lessingstr. 10
07743 Jena
Germany

Michal Dobes
Palacky University
Computer Science Department
Tr. Svobody 26
771 46 Olomouc
Czech Republic

Ann Dooms
Vrije Universiteit Brussel (VUB),
Department of Electronics and
Informatics (ETRO)
Pleinlaan 9
1050 Brussels
Belgium

Marwa Elbouz
ISEN Brest
Institut Supérieure de
l'Electronique et du Numérique
20 rue Cuirassé Bretagne
C.S. 42807
29228 Brest Cedex 2
France

Rosy Favicchio
Foundation for Research and
Technology–Hellas
Institute of Electronic
Structure and Laser
Vassilika Vouton
71110 Heraklion, Crete
Greece

Salvador Gabarda
Imaging and Vision Department
Instituto de Óptica (CSIC)
Serrano 121
28006 Madrid
Spain

Anikitos Garofalakis
Foundation for Research and
Technology–Hellas
Institute of Electronic
Structure and Laser
Vassilika Vouton
71110 Heraklion, Crete
Greece

and

Institut d'Imagerie Biomédicale (I2BM)
Service Hospitalier Frédéric Joliot
Laboratoire d'Imagerie Moléculaire Expérimentale - INSERM U803
4 place du général Leclerc
CEA, DSV
91401 Orsay Cedex
France

Roberto González-Morales
Universidad de Castilla-La Mancha
E.T.S.Ingenieros Industriales
Avda. Camilo Jose Cela s/n
13071 Ciudad Real
Spain

Bart Goossens
Ghent University
Department of Telecommunications and Information Processing (TELIN-IPI-IBBT)
St-Pietersnieuwstraat 41
9000 Gent
Belgium

Zachary T. Harmany
Duke University
Department of Electrical and Computer Engineering
Box 90291
3463 FCIEMAS
Durham, NC 27708
USA

Vaclav Hlavac
Czech Technical University
Faculty of Electrical Engineering
Department of Cybernetics
Karlovo namesti 13
121 35 Prague 2
Czech Republic

Laurent Jacques
Université catholique de Louvain (UCL)
Information and Communication Technologies
Electronics and Applied Mathematics (ICTEAM)
Place du Levant
1348 Louvain-la-Neuve
Belgium

and

Signal Processing Laboratory (LTS2)
Swiss Federal Institute of Technology (EPFL)
1015 Lausanne
Switzerland

Peter Janssens
Barco NV
Pres. Kennedypark 35
8500 Kortrijk
Belgium

Slavica Jonic
Centre National de la récherche scientifique (CNRS)
IMPMC-UMR 7590
Campus Jussieu
Tour 22/23–5e étage
Case courier 115
4 place Jussieu
75252 Paris Cedex 05
France

Moon Gi Kang
Yonsei University
Department of Electrical and Electronic Engineering
134 Sinchon-dong
Seodaemun-gu
Seoul 120-749
South Korea

Matthias S. Keil
University of Barcelona
Basic Psychology Department
Campus Mundet Passeig de la
Vall d'Hebron 171
08035 Barcelona
Spain

and

University of Barcelona
Institute for Brain Cognition
and Behavior (IR3C)
Campus Mundet Passeig de la
Vall d'Hebron 171
08035 Barcelona
Spain

Tom Kimpe
Barco NV
Pres. Kennedypark 35
8500 Kortrijk
Belgium

Janusz Konrad
Boston University
Department of Electrical and
Computer Engineering
8 Saint Mary's Street
Boston, MA 02215
USA

Christoph Krafft
Institute of Photonic
Technology Jena
Albert-Einstein-Straße 9
07745 Jena
Germany

Jung-Ping Liu
Feng Chia University
Department of Photonics
100 Wenhwa Rd
Taichung 40724
Taiwan

Hoi-Kwong Lo
University of Toronto
Center for Quantum Information
and Quantum Control
Toronto, Ontario
Canada

and

University of Toronto
Department of Electrical and
Computer Engineering
10 King's College Road
Toronto
Ontario M5S 3G4
Canada

and

University of Toronto
Department of Physics
Toronto
Ontario M5S 1A7
Canada

and

University of California
Kavli Institute for
Theoretical Physics
Kohn Hall
Santa Barbara, CA 93106
USA

Thomas Lu
California Institute of Technology
Jet Propulsion Laboratory
Mail Stop 303-300
4800 Oak Grove Drive
Pasadena, CA 91109-8099
USA

List of Contributors

Ali Mansour
Curtin University of Technology
Department of Electrical and
Computer Engineering
GPO Box U1987
Perth, WA 6845
Australia

Roummel F. Marcia
University of California, Merced
School of Natural Sciences
5200 N. Lake Road
Merced, CA 95343
USA

Aneta Markova
Vrije Universiteit Brussel (VUB)
Department of Electronics and
Informatics (ETRO)
Pleinlaan 2
1050 Elsene
Belgium

and

Interdisciplinary Institute for
Broadband Technology (IBBT)
Zuiderpoort Office Park
Gaston Crommenlaan 8 (box 102)
9050 Ghent-Ledeberg
Belgium

Maria Sagrario Millán García-Varela
Universitat Politécnica de
Catalunya
Departament d'Óptica i
Optometria
Violinista Vellsolá 37
08222 Terrassa
Spain

Anush K. Moorthy
University of
Texas at Austin
Department of Electrical and
Computer Engineering
Austin, TX 78712
USA

Javier Garcia Monreal
Universitat de Valencia
Departamento de Optica
c/Dr. Moliner 50
46100 Burjassot
Spain

Adrian Munteanu
Vrije Universiteit Brussel
Interdisciplinary Institute for
Broadband Technology (IBBT)
Department of Electronics and
Informatics (ETRO)
Pleinlaan 2
1050 Brussels
Belgium

Michael A. Paluszek
Princeton Satellite Systems
33 Witherspoon Street
Princeton, NJ 08542-3207
USA

Elisabet Pérez-Cabré
Universitat Politécnica de
Catalunya
Departament d'Óptica i
Optometria
Violinista Vellsolá 37
08222 Terrassa
Spain

Wilfried Philips
Ghent University
Department of
Telecommunications and
Information Processing
(TELIN-IPI-IBBT)
St-Pietersnieuwstraat 41
9000 Gent
Belgium

Aleksandra Pižurica
Ghent University
Department of
Telecommunications and
Information Processing
(TELIN-IPI-IBBT)
St-Pietersnieuwstraat 41
9000 Gent
Belgium

Ting-Chung Poon
Virginia Tech
Bradley Department of Electrical
and Computer Engineering
Blacksburg, VA 24061
USA

Jürgen Popp
Institute of Photonic
Technology Jena
Albert-Einstein-Straße 9
07745 Jena
Germany

and

Friedrich Schiller-University Jena
Institute of Physical Chemistry
Lessingstr. 10
07743 Jena
Germany

Ioannis Pratikakis
National Center for Scientific
Research 'Demokritos'
Institute of Informatics and
Telecommunications
Computational Intelligence
Laboratory
P.O. BOX 60228
153 10 Agia Paraskevi
Athens
Greece

and

Democritus University of Thrace
Department of Electrical and
Computer Engineering
12 Vas. Sofias Str.
67100 Xanthi
Greece

Bing Qi
University of Toronto
Center for Quantum Information
and Quantum Control
Toronto, Ontario
Canada

and

University of Toronto
Department of Electrical and
Computer Engineering
10 King's College Road
Toronto
Ontario M5S 3G4
Canada

Li Qian
University of Toronto
Center for Quantum Information
and Quantum Control
Toronto, Ontario
Canada

and

University of Toronto
Department of Electrical and
Computer Engineering
10 King's College Road
Toronto
Ontario M5S 3G4
Canada

Jorge Ripoll
Foundation for Research and
Technology–Hellas
Institute of Electronic Structure
and Laser
Vassilika Vouton
71110 Heraklion, Crete
Greece

José A. Rodrigo
Instituto de Óptica (CSIC)
Imaging and Vision Department
Serrano 121
28006 Madrid
Spain

Filip Rooms
Ghent University
Department of
Telecommunications and
Information Processing
(TELIN-IPI-IBBT)
St-Pietersnieuwstraat 41
9000 Gent
Belgium

Hichem Sahli
Vrije Universiteit Brussel
Department of Electronics and
Informatics (ETRO)
Pleinlaan 2
1050 Brussels
Belgium

and

Interuniversitair Micro-
Elektronica Centrum (IMEC)
Kapeldreef 75
3001 Leuven
Belgium

Jesús Salido
Universidad de Castilla-La
Mancha
E.T.S.Ingenieros Industriales
Avda. Camilo José Cela s/n
13071 Ciudad Real
Spain

Carlos Oscar Sánchez Sorzano
National Center of Biotechnology
(CSIC)
Biocomputing Unit
c/Darwin 3
28049 Cantoblanco
Madrid
Spain

Anna Scaife
University of Cambridge
Cavendish Laboratory
JJ Thomson Avenue
Cambridge CB3 0HE
UK

Peter Schelkens
Vrije Universiteit Brussel
Interdisciplinary Institute for
Broadband Technology (IBBT)
Department of Electronics and
Informatics (ETRO)
Pleinlaan 2
1050 Brussels
Belgium

Michal Šorel
Academy of Sciences of the
Czech Republic
Institute of Information
Theory and Automation
Department of Image Processing
Pod Vodárenskou věží 4
182 08 Prague 8
Czech Republic

Filip Šroubek
Academy of Sciences of the
Czech Republic
Institute of Information
Theory and Automation
Department of Image Processing
Pod Vodárenskou věží 4
182 08 Prague 8
Czech Republic

Erik Stijns
Vrije Universiteit Brussel
Brussels Photonics Team
(B-PHOT)
Pleinlaan 2
1050 Brussels
Belgium

Hugo Thienpont
Vrije Universiteit Brussel
Brussels Photonics Team
(B-PHOT)
Pleinlaan 2
1050 Brussels
Belgium

Pierre Vandergheynst
Signal Processing Laboratory
(LTS2)
Swiss Federal Institute of
Technology (EPFL)
Station 11
1015 Lausanne
Switzerland

Iris U. Vanhamel
Vrije Universiteit Brussel
Department of Electronics and
Informatics (ETRO)
Pleinlaan 2
1050 Brussels
Belgium

Juan Vidal
Universidad de
Castilla-La Mancha
E.T.S.Ingenieros Industriales
Avda. Camilo José Cela s/n
13071 Ciudad Real
Spain

Zhou Wang
University of Waterloo
Department of Electrical and
Computer Engineering
200 University Avenue West
Ontario N2L 3G1
Canada

Rebecca M. Willett
Duke University
Department of Electrical and
Computer Engineering
Box 90291
3463 FCIEMAS
Durham, NC 27708
USA

Leonid P. Yaroslavsky
Tel Aviv University
School of Electrical Engineering
Department of Physical
Electronics
Ramat Aviv
Tel Aviv 69978
Israel

Rupert Young
University of Sussex
Engineering and Design
Sussex House
Brighton BN1 9RH
UK

Giannis Zacharakis
Foundation for Research and
Technology–Hellas
Institute of Electronic Structure
and Laser
Vassilika Vouton
71110 Heraklion, Crete
Greece

Color Plates

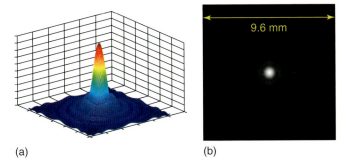

(a) (b)

Figure 1.11 (a) 3D plot of a diffraction pattern at $z = 80$ cm (Fraunhofer diffraction), $|\psi_p(x, y; z = 80 \text{ cm})|$. (b) Gray-scale plot of $|\psi_p(x, y; z = 80 \text{ cm})|$. (This figure also appears on page 15.)

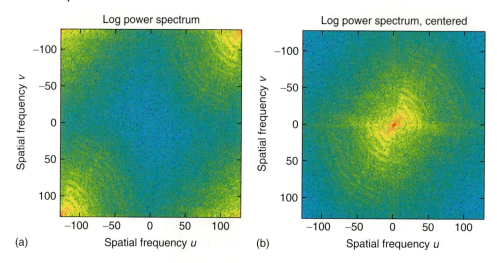

Figure 4.19 Power spectrum displayed as an intensity image. Lighter tones mean higher values. (a) Noncentered. (b) Centered. (This figure also appears on page 89.)

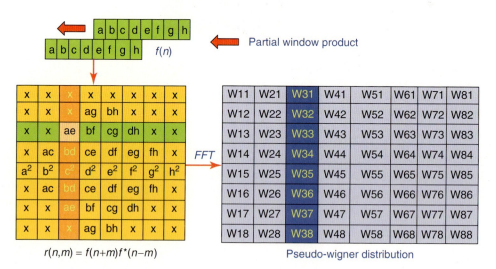

Figure 5.1 Procedure to compute the 1D directional PWVD corresponding to one pixel. (Reproduced with permission of the Society of Photo-Optical Instrumentation Engineers © 2009 SPIE.) (This figure also appears on page 106.)

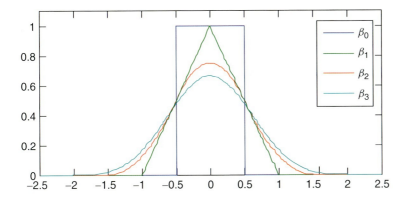

Figure 6.1 B-Splines Plot of the first four B-splines (degrees 0, 1, 2, and 3). (This figure also appears on page 123.)

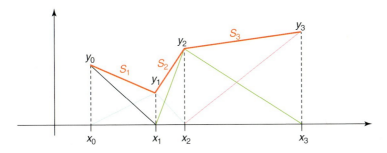

Figure 6.2 Spline interpolation with irregularly spaced samples. Example of B-spline (degree 1) interpolation of irregularly spaced samples. (This figure also appears on page 126.)

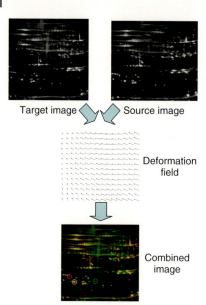

Figure 6.4 Elastic registration of 2D gels Elastic registration of two 2D protein gels (source and target) using the method proposed in [33]. The deformation field represents the continuous deformation required to convert the source into the target. The combined image shows the target image in the red channel and the warped source in the green channel. Note that there are red spots (proteins expressed in the target image and not expressed in the source image), green spots (just the opposite), and yellow spots (proteins equally expressed in both images). (This figure also appears on page 133.)

(a) (b) (c) (d)

Figure 8.3 Example of scale-space representations for "Lena" color image using (a) Gaussian scale space with $t =$ 1.11, 8.22, 22.35, 60.76, 165.15 (top to bottom). (b) Edge-affected variable conductance diffusion with $t =$ 1.11, 60.76, 448.92, 3317.12, 24510.40 (top to bottom). (c) Mean curvature motion with $t =$ 1.66, 18.30, 60.76, 201.71, 669.72 (top to bottom). (d) Levelings (Eq. 8.42) with $t =$ 1.66, 18.30, 60.76, 201.71, 669.72 (top to bottom). (This figure also appears on page 171.)

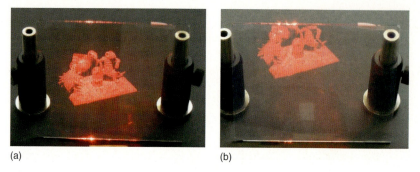

(a)   (b)

Figure 10.5 Picture of a hologram taken from different angles. The hologram is read out with a red laser light at 633 nm. (Hologram courtesy of Pierre Saint Hilaire.) (This figure also appears on page 206.)

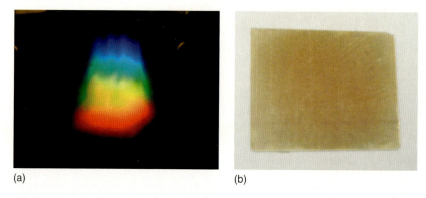

(a)   (b)

Figure 10.6 (a) Picture of the same hologram from Figure 10.5 replayed with white light. (b) The same hologram on the top of a diffusive background. (Hologram courtesy of Pierre Saint Hilaire.) (This figure also appears on page 207.)

Color Plates | LVII

 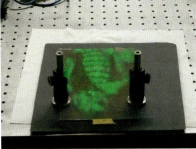

Figure 10.7 Picture of the same hologram as Figure 10.5 positioned in a different orientation; see the text for details. (Hologram courtesy of Pierre Saint Hilaire.) (This figure also appears on page 208.)

(a)   (b)

Figure 10.10 Picture of a Denisyuk hologram replayed with (a) a halogen point source and (b) fluorescent tubes. (Hologram courtesy of Arkady Bablumian.) (This figure also appears on page 210.)

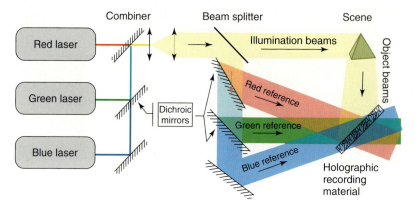

Figure 10.13 Recording a color transmission hologram. (This figure also appears on page 212.)

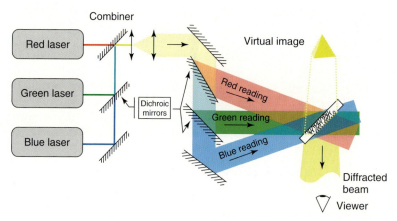

Figure 10.14 Reading a color transmission hologram. (This figure also appears on page 213.)

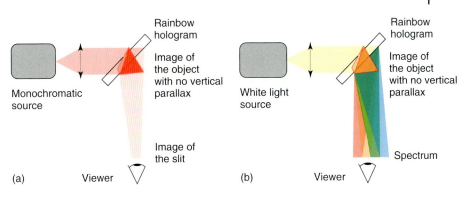

Figure 10.17 Reconstructing the image from a rainbow hologram (a) with a monochromatic source and (b) with a white light source. (This figure also appears on page 217.)

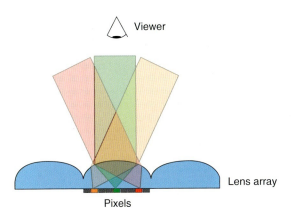

Figure 10.18 Principle of integral photography. When the viewer moves left or right, he/she views different pixels due to the lens array. (This figure also appears on page 218.)

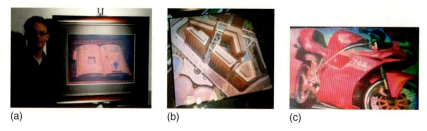

(a) (b) (c)

Figure 10.20 Pictures from holographic stereograms. (a) "Tractatus Holographis" from Jaques Desbiens, printed by RabbitHoles (*www.rabbitholes.com*). (b) Computer-generated building by Zebra Imaging (*www.zebraimaging.com*). (c) Ducati 748 from Geola (*www.geola.lt*). (This figure also appears on page 220.)

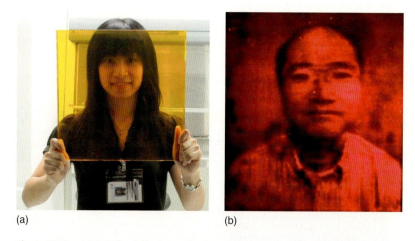

(a) (b)

Figure 10.22 (a) A 12″ × 12″ photorefractive screen. (b) Example of 3D telepresence using a photorefractive polymer screen. (This figure also appears on page 222.)

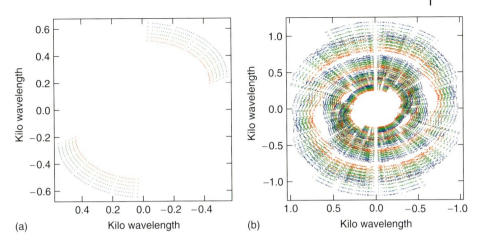

Figure 15.4 (a) The uv coverage of a single baseline in the AMI array. The six frequency channels are shown with the lowest frequency channel in red and the highest frequency channel in dark blue. (b) The uv coverage of the whole AMI array. (This figure also appears on page 332.)

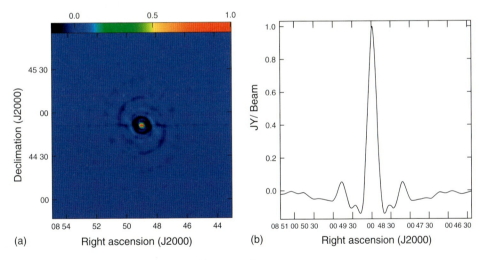

Figure 15.5 (a) A map of the synthesized beam resulting from the uv coverage shown in Figure 15.4(b). (b) A horizontal profile through the synthesized beam. (This figure also appears on page 333.)

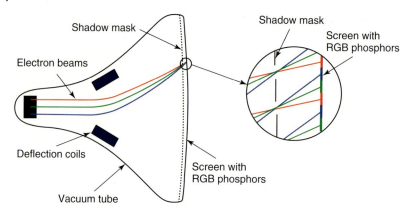

Figure 16.2 Working principle of a CRT display. (This figure also appears on page 348.)

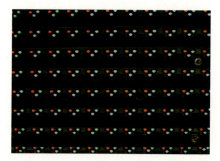

Figure 16.4 LED wall: each pixel consists of a red, green, and blue LED. (This figure also appears on page 350.)

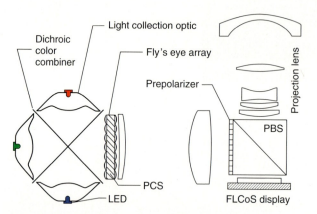

Figure 16.14 Ferroelectric LCoS (FLCoS) architecture [16]. (This figure also appears on page 360.)

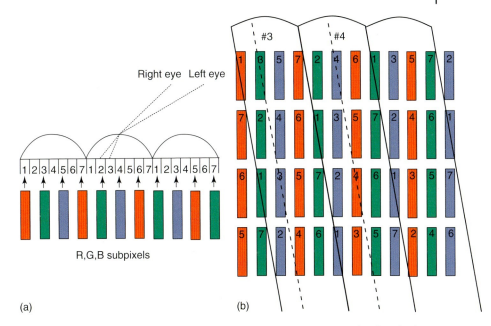

Figure 17.4 Illustration of the principle of operation of lenticular seven-view display: (a) single-row cross section with marked locations where views 1–7 are perceived; and (b) impact of lenticule slant on the spatial distribution of same-view pixels (a dashed line passes through subpixels that are visible from one location in front of the screen; 1–7 indicate the views from which R, G, or B value is rendered). (This figure also appears on page 380.)

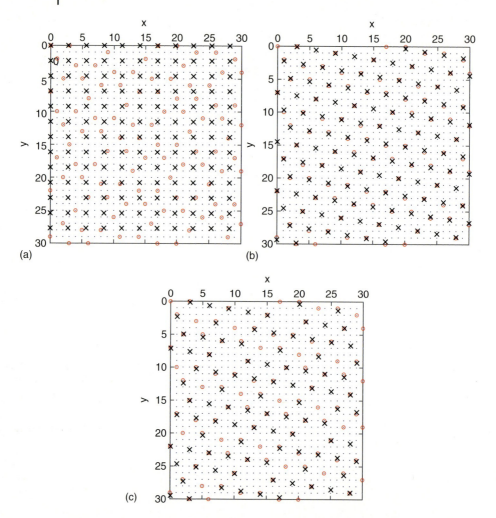

Figure 17.6 Approximation of single-view subpixels using (a) orthogonal lattice; (b) nonorthogonal lattice; and (c) union of 21 cosets. Dots denote the orthonormal RGB screen raster, circles (○) denote red subpixels of this raster that are activated when rendering one view ($\mathcal{V}$), while crosses (×) show locations from model Ψ. (From Konrad and Halle [1]. Reproduced with permission of the Institute of Electrical and Electronics Engineers © 2007 IEEE.) (This figure also appears on page 385.)

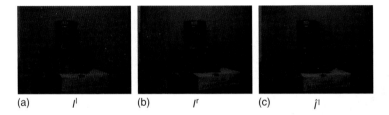

(a) I^l (b) I^r (c) $\hat{I}^l$

Figure 17.8 Example of color balancing in a stereo pair captured by identical cameras but with different exposure parameters: (a) original left view I^l; (b) original right view I^r; and (c) left view after linear transformation from Eq. (17.4). (This figure also appears on page 388.)

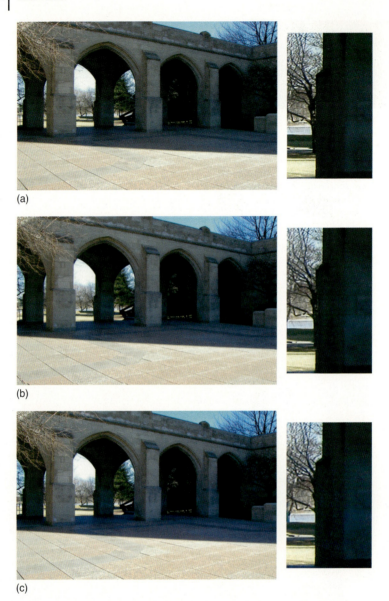

Figure 17.10 Example of intermediate view interpolation for images captured by a stereo camera with 3-inch baseline using disparity estimation based on an optical flow model [28] and simple linear interpolation from Eq. (17.7): (a) original left view; (b) reconstructed intermediate view; and (c) original right view. On the right are magnified windows from the middle of the image, highlighting the accuracy of interpolation (e.g., partially occluded tree). (This figure also appears on page 392.)

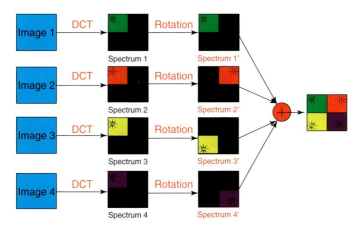

Figure 21.7 Synoptic diagram of the proposed compression and encryption approach: (a) Input image; (b) DCT(s) of the input images; (c) DCT(s) of the input images with rotation; (d) compressed and encrypted spectra. (This figure also appears on page 475.)

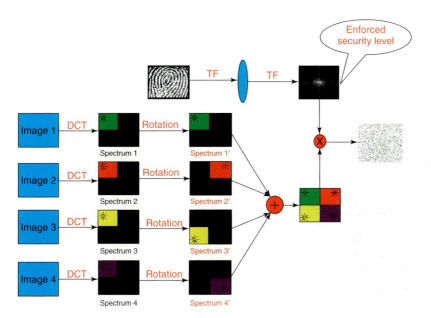

Figure 21.8 Synoptic diagram of the enhanced compression and encryption system using two security levels: (a) Input image; (b) DCT(s) of the input images; (c) DCT(s) of the input images with rotation; (d) compressed and encrypted spectra using one security level; (e) compressed and encrypted spectra using two security levels. (This figure also appears on page 476.)

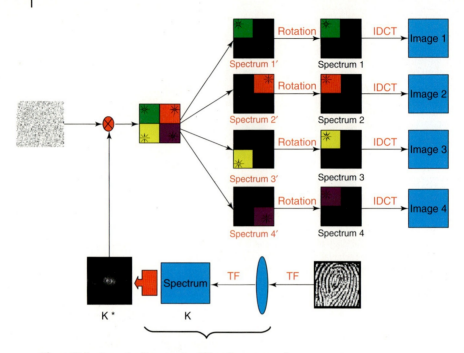

Figure 21.9 Synoptic diagram describing the system of decompression and decryption using two security levels: (a) Compressed and encrypted spectra using two security levels; (b) decompressed and encrypted spectra using one security level; (c) DCT(s) of the input images with rotation; (d) output image. (This figure also appears on page 477.)

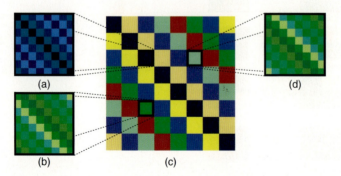

Figure 22.3 The matrix $\mathcal{F}^{-1}\mathbf{C_H}\mathcal{F}$ in (c) is block-circulant with circulant blocks. Enforcing symmetry, that is, requiring the diagonal blocks (a) to be symmetric and opposing nondiagonal blocks (e.g., those denoted by (b) and (d)) to be transposes of each other, produces a transfer function $\hat{\mathbf{H}}^{CCA}$ that is symmetric about its center, which translates to a real-valued point-spread function, and, consequently, a physically realizable-coded aperture pattern $\mathbf{H}^{CCA}$. (This figure also appears on page 493.)

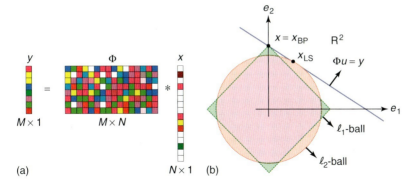

Figure 23.1 (a) Visual interpretation of the "compressed" sensing of a signal x that is sparse in the canonical basis ($\Psi = I$). (b) Explanation of the recovery of 1-sparse signals in $\mathbb{R}^2$ with BP compared to a least square (LS) solution. (This figure also appears on page 513.)

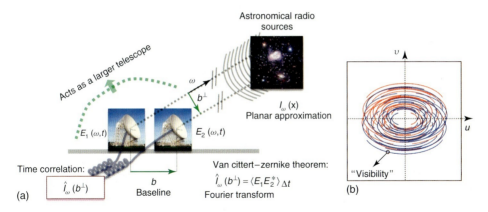

Figure 23.3 Radio Interferometry by aperture synthesis. (Left) General principles. (Right) Visibility maps in the Fourier plane. (This figure also appears on page 524.)

Figure 24.4 Single-channel blind deconvolution of real data: (a) input blurred and noisy image, (b) image and PSF reconstruction by Fergus's method, (c) image and PSF reconstruction by Shan's method, and (d) TV-based nonblind deconvolution using the PSF estimated by Fergus's method. (This figure also appears on page 538.)

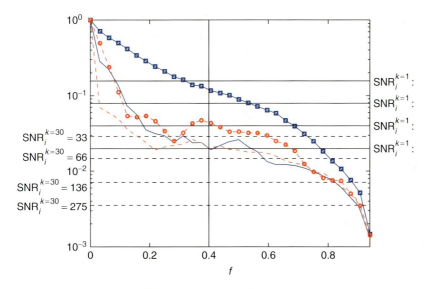

Figure 25.5 Example of application of the criterion for the determination of the system-effective cutoff frequency. See Ref. [14] for more details. (This figure also appears on page 561.)

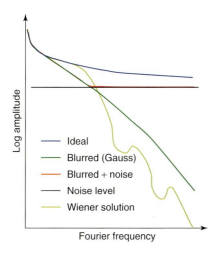

Figure 26.3 Spectra of ideal image, degraded image, and Wiener restored image. (This figure also appears on page 576.)

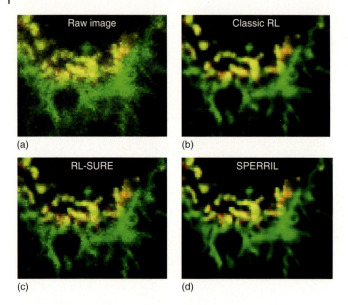

Figure 26.5 Detailed views of the different restoration results for a cell from the test set. (a) The raw image; (b) the results after classical RL; (c) the result after RL-SURE; and (d) the result of SPERRIL. Note that (b) and (c) still contain noise, while (d) maintains similar sharpness and the noise is better suppressed. (This figure also appears on page 588.)

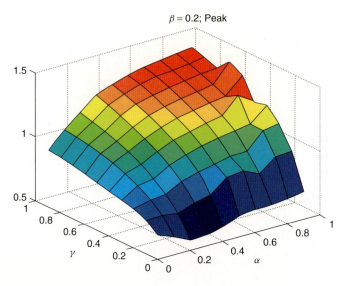

Figure 30.9 3D plot of peak value versus α versus γ with constant β. (This figure also appears on page 679.)

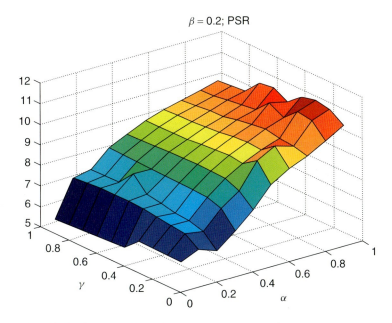

Figure 30.10 3D plot of PSR value versus α versus γ with constant β. (This figure also appears on page 680.)

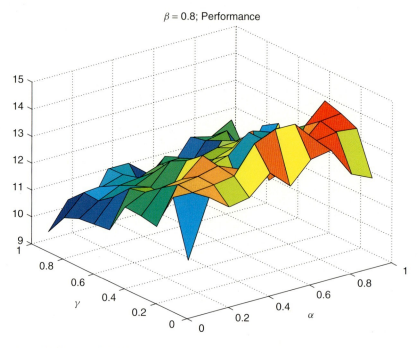

Figure 30.11 Sample PSR output from permutation procedure for filter optimization. Filter performance as a function of parameters α and γ, for $\beta = 0.8$. (This figure also appears on page 680.)

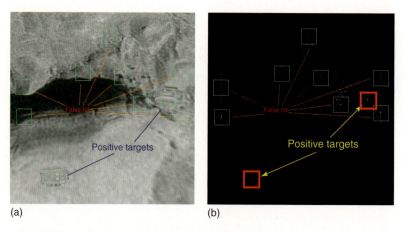

Figure 30.19 (a) An input image with objects embedded and (b) the correlation plane shows the detection of the object and the false alarm of the clutters. (This figure also appears on page 689.)

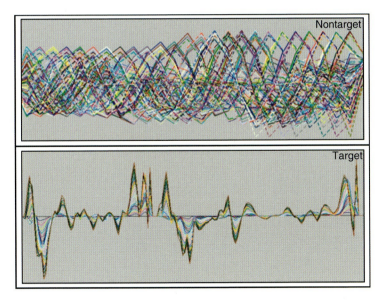

Figure 30.20 Examples of the target and nontarget gradient features for the neural network. (This figure also appears on page 690.)

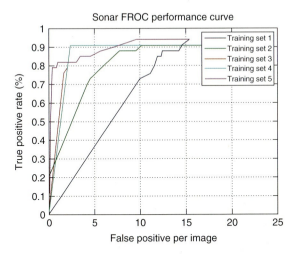

Figure 30.21 FROC curve of sonar targets. (This figure also appears on page 691.)

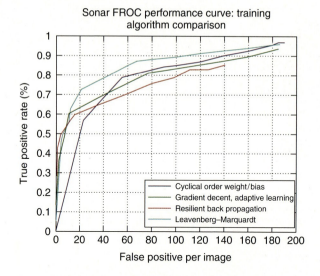

Figure 30.22 Comparing the performance of different NN learning algorithms. (This figure also appears on page 691.)

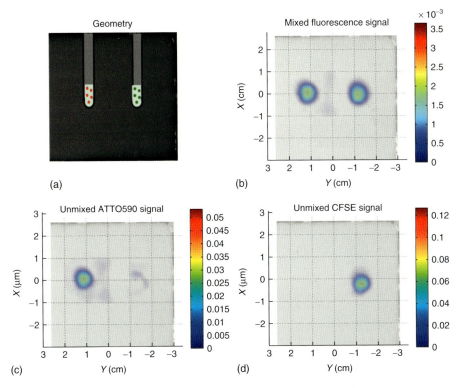

Figure 31.5 *In vitro* unmixing of CFSE and ATTO590 in a phantom model. (a) Geometry image of the phantom and position of the capillary sample tubes. Equal concentrations of the two fluorophores were imaged in two spectral channels (green and red). (b) The mixed signal in the red channel detects two tubes of similar intensities. (c, d) After spectral unmixing, the red and the green components were separated: the unmixed signals for CFSE and ATTO590 could then be individually rendered. The CFSE unmixed signal was approximately twofold stronger than the ATTO590, as would be expected from their spectral profiles. (This figure also appears on page 710.)

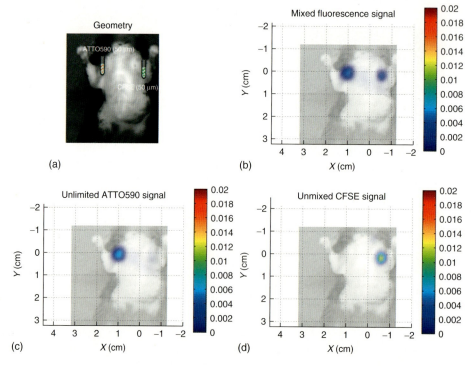

Figure 31.6 *In vivo* unmixing of CFSE and ATTO590 capillary tubes placed subcutaneously in a C57bl mouse and imaged with the FMT. (a) Geometry image of the mouse showing the position in which the capillary sample tubes were inserted. Equal concentrations (50 µM) of the two fluorophores were imaged in two spectral regions (green and red). (b) The mixed signal in the red channel correctly detects two tubes and the CFSE tube emission appears weaker than the ATTO590. (c,d) After spectral unmixing, the red and the green components were separated: the unmixed signals for both dyes were then rendered. The CFSE unmixed signal was approximately twofold stronger than the ATTO590, consistent with what was shown for the *In vitro* data. (This figure also appears on page 712.)

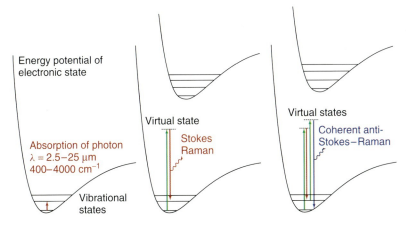

Figure 32.1 Energy transitions of IR, Raman, and CARS spectroscopy. IR spectroscopy is a one-photon process in which vibrations are probed by absorption of a photon. Spontaneous Raman spectroscopy is a two-photon process in which the first photon excites the molecule to a virtual state and upon emission of a second photon the molecule returns to an excited vibrational state. CARS microscopy is a four-photon process, involving two pump photons and one probe photon before emission of an anti-Stokes photon. (This figure also appears on page 719.)

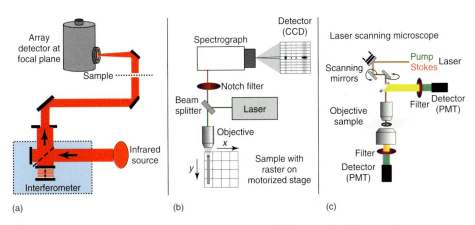

Figure 32.2 Principle of Fourier-transform infrared imaging using focal plane array detectors (a). Point-mapping and line-mapping modes in Raman imaging (b). In point-mapping, the laser is focused to a spot (blue) and spectra are registered sequentially in x and y dimensions, whereas in line-mapping mode the laser generates a line (gray) and the spectral information is registered sequentially in x direction and parallel in y direction using both dimensions of the CCD detector. Diagram of a CARS microscope operating in epi and forward detection (c). The short-pulse laser system was omitted for simplicity. (This figure also appears on page 720.)

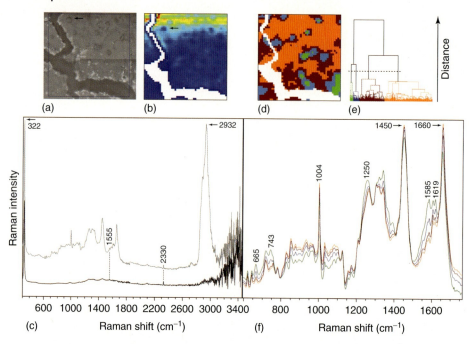

Figure 32.3 Photomicrograph of a liver tissue section on calcium fluoride substrate, with a box indicating the sampling area (a). Intensity plot of the band near 2932 cm^{-1} of the 49 × 49 Raman image, with a step size of 2.5 μm (b). Unmanipulated Raman spectrum (gray trace) and background spectrum representing the crack (black trace) (c). Color-coded class membership of K-means clustering (d) and dendrogram of hierarchical clustering (e) from the liver Raman image. Four-cluster-averaged Raman spectra are overlaid for comparison (f). (This figure also appears on page 727.)

Color Plates | LXXXI

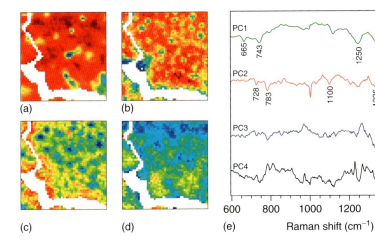

(a) (b) (c) (d) (e)

Figure 32.4 Illustration of PCA and the relation between scores t, loadings p, average $\bar{x}$, and data matrix X of N samples with k variables (a). Three-dimensional space with a straight line fitted to the points as an example of a one-component PC model. The PC score of an object t_i is its orthogonal projection on the PC line (b). Score plots of PC1 (c), PC2 (d), PC3 (e), PC4 (f), and loadings (g) of a PCA from the liver Raman image in Figure 32.3. Labeled negative values of PC1 are correlated with spectral contributions of heme, negative values of PC2 with nucleic acids. (This figure also appears on page 728.)

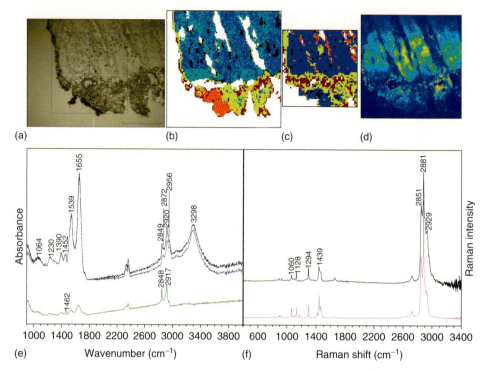

Figure 32.6 Photomicrograph (a), FTIR image (b), Raman image (c), and CARS image (d) of an unstained colon tissue section on calcium fluoride slide showing the transition from muscle (top) to mucosa (bottom). Droplets are resolved as black clusters in Raman and FTIR images and as red spots in the CARS image. IR difference spectrum in (e) (green = black minus blue cluster) and Raman spectrum in (f) show high fatty acid content in droplets. Raman spectrum of unsaturated fatty acids is included as reference (magenta trace). (This figure also appears on page 735.)

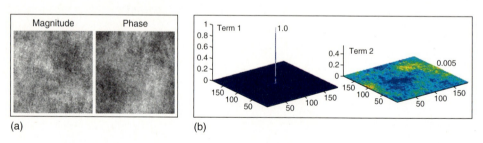

Figure 33.11 (a) Magnitude and phase distributions of the encrypted image $\psi(x)$. Its noiselike appearance does not reveal the content of any primary image. (b) Output intensity distributions. (This figure also appears on page 757.)

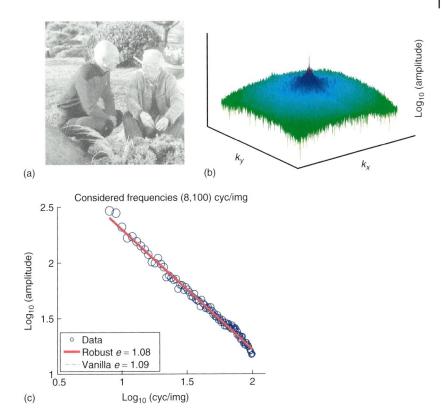

Figure 36.2 Amplitude spectrum of a natural image. (a) A natural or photographic image. (b) In a double-logarithmic representation, the power spectrum of many natural images falls approximately with the square of spatial frequency, which has been interpreted as a signature of scale invariance (Eq. (36.1)). The power spectrum is equal to the squared amplitude spectrum shown here. (c) Hence, the slope of a line fitted to the orientation-averaged amplitude spectrum falls with the inverse of spatial frequency k^e with $e \approx -1$. See Figure 36.6 for further details. (This figure also appears on page 812.)

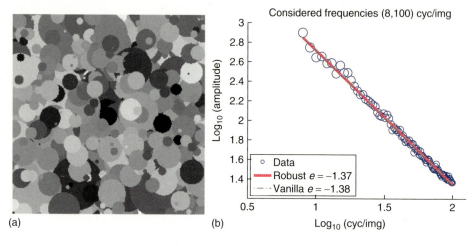

Figure 36.3 Occluding disks. The k^{-1} fall off of spectral amplitudes in the context of scale invariance and occlusion, respectively. (a) A "naturally looking" artificial image composed of mutually occluding disks of many sizes and intensity values. (b) The lines fitted to the orientation-averaged log–log amplitude spectrum. The fitting procedure is explained in Section 36.4.1.1. (This figure also appears on page 814.)

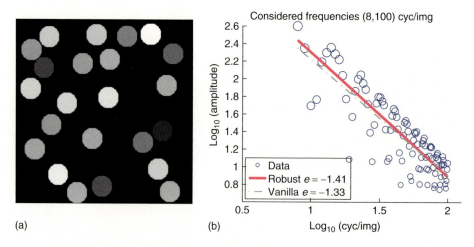

Figure 36.4 Single disks. The k^{-1} fall off of spectral amplitudes in the context of contrast edges. (a) An artificial image composed of nonoccluding, equally sized disks with various intensity values. The important point to realize here is that this image thus does not have the typical properties of natural images, that is, scale invariance, occlusion, perspective, and so on. (b) Despite being completely different, the lines fitted to the orientation-averaged amplitude spectrum have slope values that are consistent with natural image statistics (cf previous figure). (This figure also appears on page 815.)

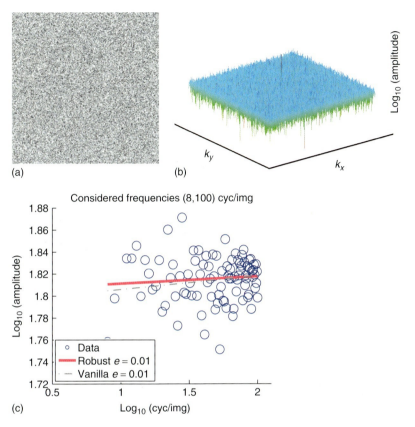

Figure 36.6 White noise: (a) in a white noise image, each 2D spatial frequency (k_x, k_y) occurs with the same probability. (b) 2D amplitude spectrum: this is equivalent to a flat amplitude spectrum. (c) 1D amplitude spectrum: the slope of a line fitted to orientation-averaged spectral amplitudes measures the flatness of the spectrum. With white noise, the *spectralslope* is approximately zero (legend symbol $e \equiv \alpha$ denotes slopes). Details on the fitting procedure can be found in Section 36.4.1.1. Notice that although white noise contains all frequencies equally, our perception of the image is nevertheless dominated by its high frequency content, indicating that the visual system attenuates all lower spatial frequencies. (This figure also appears on page 817.)

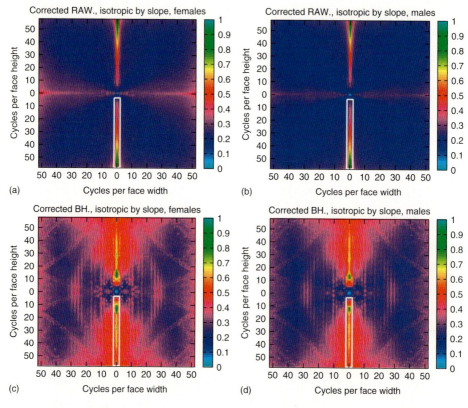

Figure 36.9 Slope-whitened mean amplitude spectra. (a,b) Slope-whitened, corrected mean raw spectra do not reveal amplitude enhancement as a consequence of whitening – (a = females, b = males). (c,d) However, when the mean-corrected BH spectra are whitened, clear maxima appear at horizontal feature orientations (marked by a white box), at ≈ 10–15 cycles per face height. Thus, the maxima were induced by internal face features. (This figure also appears on page 820.)

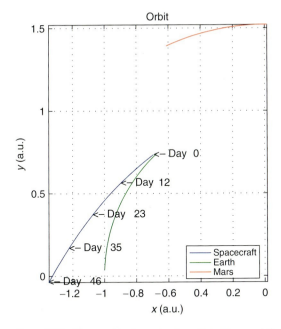

Figure 37.5 Planet orbit about the Sun. Only Mars and the Earth are used in the simulation. (This figure also appears on page 855.)

1
Fundamentals of Optics

Ting-Chung Poon and Jung-Ping Liu

1.1
Introduction

In this chapter, we discuss some of the fundamentals of optics. We first cover the electromagnetic (EM) spectrum, which shows that the visible spectrum occupies just a very narrow portion of the entire EM spectrum. We then discuss geometrical optics and wave optics. In geometrical optics, we formulate the propagation of light rays in terms of matrices, whereas in wave optics we formulate wave propagation using the Fresnel diffraction formula, which is the solution to Maxwell's equations. Fourier optics and its applications in optical image processing are then discussed. One of the important optical image processing applications in optical correlation is subsequently explained. Finally, we discuss the human visual system and end this chapter with a brief conclusion.

1.2
The Electromagnetic Spectrum

James Clerk Maxwell (1831–1879) used his famous Maxwell's equations to show that EM waves exhibit properties similar to those of visible light. After continuous studies of optical and EM waves in the last century, we have finally understood that visible light is only a narrow band in the EM spectrum. Figure 1.1 shows the complete EM spectrum. It consists of regions of radio waves, microwaves, infrared (IR), visible light, ultraviolet (UV), X-rays, and γ-rays.

Visible light is the EM radiation that can be observed by human eyes directly. Because of the various receptors in human eyes (see Section 1.7), people can identify EM radiation of different wavelengths by their colors. Visible light can be roughly separated into violet (390–455 nm), blue (455–492 nm), green (492–577 nm), yellow (577–597 nm), orange (597–622 nm), and red (622–780 nm). Beyond the red end of the visible band lies the IR region. It contains the near infrared (NIR, 0.78–3 µm), the middle infrared (MIR, 3.0–6.0 µm), the far infrared (FIR, 6.0–15 µm), and the extreme infrared (XIR, 15–1000 µm). There are also other nomenclatures

---

Optical and Digital Image Processing: Fundamentals and Applications, First Edition. Edited by Gabriel Cristóbal, Peter Schelkens, and Hugo Thienpont.
© 2011 Wiley-VCH Verlag GmbH & Co. KGaA. Published 2011 by Wiley-VCH Verlag GmbH & Co. KGaA.

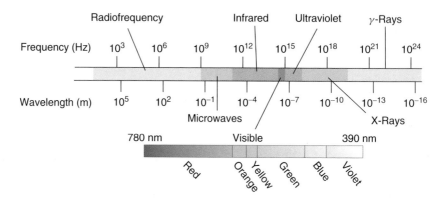

Figure 1.1 The electromagnetic spectrum.

and division schemes, depending on the purposes of the applications and the users [1]. The IR imaging technique is applied in astronomical measurements, thermographs, heat targets, and noninvasive medical imaging of subcutaneous tissues. The band merging into the violet end of the visible light is the UV region. Its spectrum range is roughly between 1.0 and 390 nm. UV radiation is used as the light source for lithography to achieve narrower linewidth. Sometimes, the visible band, together with parts of the UV and IR regions, is classified as the *optical region*, in which we use similar ways to produce, to modulate, and to measure the optical radiation.

At the long-wavelength end of the EM spectrum are microwaves and radio waves. Their ranges are from 3×10^{11} to 10^9 Hz, and from 10^9 Hz to very low frequency, respectively. The radiations in this region exhibit more wave properties, and are usually applied to remote sensing and communications. Although antennae are used to produce and to receive microwaves and radio waves, we can still perform imaging in this region, especially for astronomical measurements [2]. It is worth noting that the spectrum range between FIR and microwaves, namely, from 0.1 to 10 THz, is called the *terahertz band (THz band)*. The THz band covers the millimeter wave (MMW, 0.03–0.3 THz), the submillimeter wave (SMMW, 0.3–3 THz), and the FIR regions. Terahertz radiation is highly absorbed by water and reflected by metals. Thus terahertz imaging has unique applications in nondestructive testing, biomedical imaging, and remote sensing.

At the short-wavelength end of the EM spectrum are the X-rays and γ-rays. X-rays, ranging from 1.0 nm to 6.0 pm, can pass through the soft tissues easily and are thus widely used in applications in medical imaging. The X-ray lasers and their related imaging techniques have been studied for decades and are still under development [3]. Since the wavelength of X-rays is shorter than UV rays, the success of X-ray imaging techniques will lead to a greater improvement of image resolution. γ-Rays have the shortest wavelength and thus they possess very high photon energy ($>10^4$ eV) and behave more like particles. γ-Rays have been applied in medical imaging because of their excellent depth of penetration into soft tissues and their high resolution. For example, Technetium-99m or ^{99m}Tc is the most commonly used γ-ray-emitting radionuclide in radiopharmacology, and it produces γ-radiation with principal photon energy of 140 keV. Another nuclear medicine technique called *positron emission tomography (PET)* is also an imaging technique of 511 keV γ-radiation resulting from the annihilation of positron–electron pairs of a radionuclide.

1.3
Geometrical Optics

In *geometrical optics*, light is treated as particles called *photons* and the trajectories that these photons follow are termed *rays*. Hence geometrical optics is also known as *ray optics*. Ray optics is based on *Fermat's principle*, which states that the path that a light ray follows is an extremum in comparison to nearby paths. Usually the extremum is a minimum. As a simple example shown in Figure 1.2, the shortest distance, that is, the minimum distance, between two points A and B is along a straight line (solid line) in a *homogeneous medium* (medium with a constant refractive index). Hence the light ray takes the solid line as a path instead of taking the nearby dotted line. By the same token, according to Fermat's principle, we can derive the laws of reflection and refraction [4]. The law of refraction, which describes a light ray entering from one medium, characterized by refractive index n_i, into another medium of refractive index n_t, is given by

$$n_i \sin \theta_i = n_t \sin \theta_t \tag{1.1}$$

where θ_i and θ_t are the angles of incidence and transmission (or refraction), respectively, as shown in Figure 1.3. The angles are measured from the normal NN' to the interface MM', which separates the two media.

1.3.1
Ray Transfer Matrix

To describe ray propagation through optical systems comprising, for instance, a succession of lenses all centered on the same axis called the *optical axis*, we can use matrix formalism if we consider *paraxial rays*. Figure 1.4 depicts the system coordinates and parameters under consideration. We take the optical axis along

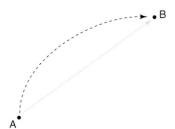

Figure 1.2 A light ray takes the shortest distance, a straight line (solid line), between two points in a homogeneous medium.

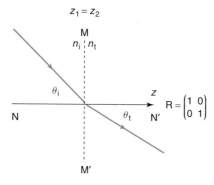

Figure 1.3 Law of refraction and its transfer matrix **R**.

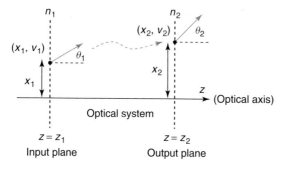

Figure 1.4 Input and output planes in an optical system.

the z-axis, which is the general direction in which the rays travel. Paraxial rays lie only in the x–z plane and are close to the z-axis. In other words, paraxial rays are rays with angles of incidence, reflection, and refraction at an interface, satisfying the small-angle approximation in that $\tan\theta \approx \sin\theta \approx \theta$ and $\cos\theta \approx 1$, where the angle θ is measured in radians. *Paraxial optics* deals with paraxial rays. Hence in paraxial optics, the law of refraction simplifies to

$$n_i\theta_i = n_t\theta_t \tag{1.2}$$

A ray at a given x–z plane may be specified by its *height x* form the optical axis and by its angle θ or *slope* that it makes with the z-axis. The convention for the angle is anticlockwise positive measured from the z-axis. The height x of a point on a ray is taken positive if the point lies above the z-axis and negative if it is below the z-axis. The quantities (x, θ), therefore, represent the coordinates of the ray for a given z-plane. It is customary to replace the corresponding angle θ by $v = n\theta$, where n is the refractive index at the z-constant plane. Therefore, as shown in Figure 1.4, the ray at $z = z_1$ passes through the input plane with input ray coordinates $(x_1, n_1\theta_1)$ or (x_1, v_1). If the output ray coordinates at $z = z_2$, the output plane, are $(x_2, n_2\theta_2)$ or (x_2, v_2), we can relate the input coordinates to the output coordinates by a 2 × 2 matrix as follows:

$$\begin{pmatrix} x_2 \\ v_2 \end{pmatrix} = \begin{pmatrix} A & B \\ C & D \end{pmatrix} \begin{pmatrix} x_1 \\ v_1 \end{pmatrix} \quad (1.3)$$

The above *ABCD* matrix is called the *ray transfer matrix*, which can be made up of many matrices to account for the effects of the ray passing through various optical elements such as lenses. Equation (1.3) is equivalently given by

$$x_2 = Ax_1 + Bv_1 \quad (1.4a)$$
$$v_2 = Cx_1 + Dv_1 \quad (1.4b)$$

Hence, the law of refraction given by Eq. (1.2) can be written as

$$\begin{pmatrix} x_2 \\ v_2 \end{pmatrix} = \begin{pmatrix} 1 & 0 \\ 0 & 1 \end{pmatrix} \begin{pmatrix} x_1 \\ v_1 \end{pmatrix} = \mathbf{R} \begin{pmatrix} x_1 \\ v_1 \end{pmatrix} \quad (1.5)$$

where $v_2 = n_2\theta_t = n_t\theta_t$ and $v_1 = n_1\theta_i = n_i\theta_i$. Therefore, the ABCD matrix for the law of refraction is $\mathbf{R} = \begin{pmatrix} 1 & 0 \\ 0 & 1 \end{pmatrix}$ and Figure 1.3 summarizes the matrix formalism for the law of refraction. From Eq. (1.5), we see that $x_2 = x_1$ as the heights of the input and output rays are the same. Also, $v_2 = v_1$, which gives the law of refraction, $n_t\theta_t = n_i\theta_i$. We also notice that the input and output planes in this case are the same plane at $z = z_1 = z_2$.

Figure 1.5 shows two more useful ray transfer matrices: the *translation matrix* **T** and the *thin-lens matrix* **L**. The translation matrix describes the ray undergoing a translation of distance d in a homogeneous medium characterized by n and the

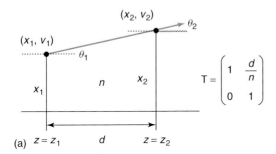

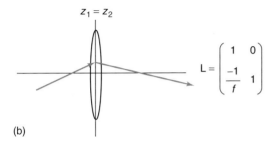

Figure 1.5 (a) Translation matrix and (b) lens matrix.

matrix equation is given as follows:

$$\begin{pmatrix} x_2 \\ v_2 \end{pmatrix} = \begin{pmatrix} 1 & \frac{d}{n} \\ 0 & 1 \end{pmatrix} \begin{pmatrix} x_1 \\ v_1 \end{pmatrix} = \mathbf{T} \begin{pmatrix} x_1 \\ v_1 \end{pmatrix} \quad (1.6a)$$

where $v_2 = n\theta_2$, $v_1 = n\theta_1$, and

$$\mathbf{T} = \begin{pmatrix} 1 & \frac{d}{n} \\ 0 & 1 \end{pmatrix} \quad (1.6b)$$

Note that when the ray is undergoing translation as shown in Figure 1.5a in a homogeneous medium, $\theta_1 = \theta_2$ and therefore $v_1 = v_2$ as given by Eq. (1.6a). From Eq. (1.6a), we also see that $x_2 = x_1 + dv_1/n = x_1 + d\theta_1$. Hence x_2 is $d\theta_1$ higher than x_1, consistent with the situation shown in Figure 1.5a. When a thin converging lens of focal length f is involved, the matrix equation is

$$\begin{pmatrix} x_2 \\ v_2 \end{pmatrix} = \begin{pmatrix} 1 & 0 \\ \frac{-1}{f} & 1 \end{pmatrix} \begin{pmatrix} x_1 \\ v_1 \end{pmatrix} = \mathbf{L} \begin{pmatrix} x_1 \\ v_1 \end{pmatrix} \quad (1.7a)$$

where **L** is the *thin-lens matrix* given by

$$\mathbf{L} = \begin{pmatrix} 1 & 0 \\ \frac{-1}{f} & 1 \end{pmatrix} \quad (1.7b)$$

By definition, a lens is thin when its thickness is assumed to be zero and hence the input plane and the output plane have become the same plane or $z_1 = z_2$ as shown in Figure 1.5b. We also have $x_1 = x_2$ as the heights of the input and output rays are the same for the thin lens. Regarding the slope, from Eq. (1.7a), we have $v_2 = -\frac{1}{f}x_1 + v_1$. For $v_1 = 0$, that is, the input ray is parallel to the optical axis, $v_2 = -\frac{1}{f}x_1 = -\frac{1}{f}x_2$. For positive x_1, $v_2 < 0$, since $f > 0$ for a converging lens. For negative x_1, $v_2 > 0$. Hence all input rays parallel to the optical axis converge behind the lens to a point called the *back focal point*, that is, at a distance f away from the lens. Note that for a thin lens, the front focal point is also a distance f away from but in front of the lens. For the input ray coordinates given by $(x_1, x_1/f)$ and according to Eq. (1.7), the output ray coordinates are $(x_2 = x_1, 0)$, which implies that all output rays will be parallel to the optical axis as $v_2 = 0$. This is the case that all rays passing through the front focal point of a lens will give rise to parallel output rays. The output plane that contains the back focal point is called the *back focal plane* and, similarly, the plane that contains the front focal point is the *front focal plane*.

1.3.2
Two-Lens Imaging System

Figure 1.6 shows a more complicated optical system as an example. The system is a two-lens imaging system with lenses L_1 and L_2 of focal lengths f_1 and f_2, respectively, and the two lenses are separated by a distance $f_1 + f_2$. The object is placed in the front focal plane of lens L_1. We can relate the input ray coordinates (x_1, v_1) and the output ray coordinates (x_2, v_2) using the ray transfer matrices.

1.3 Geometrical Optics

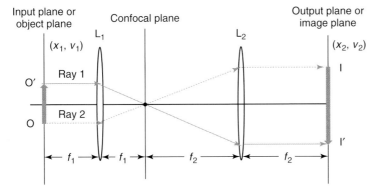

Figure 1.6 Two-lens imaging system.

Therefore, from Eq. (1.6), we first relate the input coordinates (x_1, v_1) on the input plane to the ray coordinates (x'_1, v'_1), located on a plane immediately before the lens L_1 by

$$\begin{pmatrix} x'_1 \\ v'_1 \end{pmatrix} = \begin{pmatrix} 1 & f_1 \\ 0 & 1 \end{pmatrix} \begin{pmatrix} x_1 \\ v_1 \end{pmatrix} \quad (1.8)$$

where we have assumed that the two-lens system is immersed in air, that is, $n = 1$. The distance of ray transfer is $d = f_1$. Now we relate (x'_1, v'_1) to (x''_1, v''_1), the ray coordinates located on a plane immediately after the lens L_1 by

$$\begin{pmatrix} x''_1 \\ v''_1 \end{pmatrix} = \begin{pmatrix} 1 & 0 \\ \frac{-1}{f_1} & 1 \end{pmatrix} \begin{pmatrix} x'_1 \\ v'_1 \end{pmatrix} \quad (1.9)$$

where we have used the lens matrix in Eq. (1.7) with $f = f_1$. Now by substituting Eq. (1.9) into Eq. (1.8), we can relate the input ray coordinates (x_1, v_1) to the ray coordinates (x''_1, v''_1) just after the lens L_1:

$$\begin{pmatrix} x''_1 \\ v''_1 \end{pmatrix} = \begin{pmatrix} 1 & 0 \\ \frac{-1}{f_1} & 1 \end{pmatrix} \begin{pmatrix} 1 & f_1 \\ 0 & 1 \end{pmatrix} \begin{pmatrix} x_1 \\ v_1 \end{pmatrix} \quad (1.10)$$

Note that the overall system matrix so far is expressed in terms of the product of two matrices written in order from right to left as the ray transverses from left to right on the optical axis. By the same token, we can show that the input ray coordinates (x_1, v_1) and the final output ray coordinates (x_2, v_2) on the image plane are connected using five ray transfer matrices (three translation matrices and two-lens matrices):

$$\begin{pmatrix} x_2 \\ v_2 \end{pmatrix} = S \begin{pmatrix} x_1 \\ v_1 \end{pmatrix} \quad (1.11a)$$

where **S** is *system matrix* of the overall system and is given by

$$S = \begin{pmatrix} 1 & f_2 \\ 0 & 1 \end{pmatrix} \begin{pmatrix} 1 & 0 \\ \frac{-1}{f_2} & 1 \end{pmatrix} \begin{pmatrix} 1 & f_1+f_2 \\ 0 & 1 \end{pmatrix} \begin{pmatrix} 1 & 0 \\ \frac{-1}{f_1} & 1 \end{pmatrix} \begin{pmatrix} 1 & f_1 \\ 0 & 1 \end{pmatrix} \quad (1.11b)$$

1 Fundamentals of Optics

The system matrix $\mathbf{S}$ can be simplified to

$$\mathbf{S} = \begin{pmatrix} -\frac{f_2}{f_1} & 0 \\ 0 & -\frac{f_1}{f_2} \end{pmatrix} \tag{1.12}$$

and Eq. (1.11a) becomes

$$\begin{pmatrix} x_2 \\ v_2 \end{pmatrix} = \begin{pmatrix} -\frac{f_2}{f_1} & 0 \\ 0 & -\frac{f_1}{f_2} \end{pmatrix} \begin{pmatrix} x_1 \\ v_1 \end{pmatrix} \tag{1.13}$$

From the above equation, we can find, for example, $x_2 = -\frac{f_2}{f_1} x_1$; and the lateral magnification M of the imaging system is found to be

$$M = \frac{x_2}{x_1} = -\frac{f_2}{f_1} \tag{1.14}$$

Since both the focal lengths of the converging lenses are positive, $M < 0$ signifies that the image formed is real and inverted as shown in Figure 1.6. In addition, if $f_2 > f_1$, as is the case illustrated in Figure 1.6, we have a magnified image. In the figure, we show a ray diagram, illustrating an image formation of the system. A ray (Ray 1 in the Figure 1.6) coming from O' parallel to the optical axis hits the image point I' and a ray (Ray 2) from the object point O parallel to the optical axis hits the image point I.

1.3.3
Aberrations

In the preceding section, we have introduced the basic concepts of an imaging system in which a point object will produce a point image using paraxial rays. In real optical systems, nonparaxial rays are also involved in image formation and the actual image departs from the ideal point image. In fact, these rays form a blur spot instead of a single point image. This departure from ideal image formation is known as *aberrations*.

Aberrations are usually due to the deviation from the paraxial approximation, but this is not always so. We also have defocus aberration and chromatic aberration because these aberrations also result in a blurred image spot. However, the former is due to the mismatch between the observation plane and the image plane, while the latter is due to the various refractive indices of the same material at different wavelengths.

To systematically analyze the aberrations, we can expand the sine of the angle between the ray and the optical axis, that is, $\sin\theta \approx \theta - \theta/3! + \theta/5! - \cdots$. The first expansion term is also what we use in the paraxial approximation (Eq. (1.2)), which leads to an ideal image point due to an ideal object point. Aberrations that come about from neglecting the second term in the expansion are called *third-order aberrations*, while the fifth-order aberrations arise from neglecting the third term in the expansion. Usually, in aberration analysis we only need to consider the third-order aberrations. The effect of fifth-order and higher order aberrations are

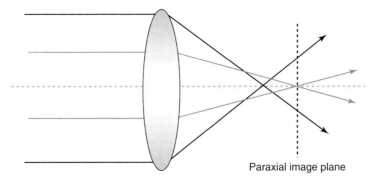

Figure 1.7 Spherical aberration on a single lens.

complicated and minor, and these are only considered in high-precision optical systems. Spherical aberration is one of the five third-order aberrations, and it is easily found in a low-cost single lens. We can consider the imaging of a general spherical lens, in which a point object is set at infinity as shown in Figure 1.7. The light rays near the center of the lens are converged to the paraxial image plane (back focal plane). The light rays passing through the periphery of the lens also converge, but the image point deviates from the paraxial image point. As a result, the image of a point in the image plane is not an infinitesimal point but a blurred spot. Spherical aberration is a function of the lens profile, the lens thickness, the lens diameter, and the object distance (or equivalently, the image distance). The easiest way to reduce spherical aberration is to shorten the effective diameter of the lens. However, the light intensity is reduced accordingly. Alternatively, spherical aberration can be minimized by using an aspheric surface (departing from spherical) on a single lens or by using several lenses.

Besides spherical aberration, there are also coma, astigmatism, field curvature, and distortion that comprise the third-order aberrations. The readers may consult Chapters 13 and 25 of this book or other books for more details [2, 5–7]. Finally, it may be noted that even if the optical system is free from all the third- and high-order aberrations, we still cannot accomplish an infinitesimal image point because of diffraction effects (see Section 1.5).

1.4
Maxwell's Equations and the Wave Equation

In geometrical optics, we treat light as particles. In *wave optics*, we treat light as waves. Wave optics accounts for wave effects such as interference and diffraction. Maxwell's equations form the starting point for wave optics:

$$\nabla \cdot \mathbf{D} = \rho_v \tag{1.15}$$

$$\nabla \cdot \mathbf{B} = 0 \tag{1.16}$$

$$\nabla \times \mathbf{E} = -\frac{\partial \mathbf{B}}{\partial t} \tag{1.17}$$

$$\nabla \times \mathbf{H} = \mathbf{J} = \mathbf{J_C} + \frac{\partial \mathbf{D}}{\partial t} \qquad (1.18)$$

where, in these equations, we have four vector quantities called *electromagnetic fields*: the electric field strength **E** (V m^{-1}), the electric flux density **D** (C m^{-2}), the magnetic field strength **H** (A m^{-1}), and the magnetic flux density **B** (Wb m^{-2}). The vector quantity $\mathbf{J_C}$ and the scalar quantity ρ_v are the current density (A m^{-1}) and the electric charge density (C m^{-3}), respectively, and they are the sources responsible for generating the EM fields. In addition to Maxwell's equations, we have the constitutive relations:

$$\mathbf{D} = \varepsilon \mathbf{E} \qquad (1.19)$$

and

$$\mathbf{B} = \mu \mathbf{H} \qquad (1.20)$$

where ε and μ are the permittivity (F m^{-1}) and permeability (H m^{-1}) of the medium, respectively. In this chapter, we take ε and μ to be scalar constants, which is the case for a linear, homogeneous, and isotropic medium such as in vacuum. Using Eqs (1.15–1.20), we can derive a wave equation in **E** or **B** [8]. For example, the wave equation in **E** is

$$\nabla^2 \mathbf{E} - \mu\varepsilon \frac{\partial^2 \mathbf{E}}{\partial t^2} = \mu \frac{\partial \mathbf{J_C}}{\partial t} + \frac{1}{\varepsilon}\nabla \rho_v \qquad (1.21)$$

where $\nabla^2 = \frac{\partial^2}{\partial x^2} + \frac{\partial^2}{\partial y^2} + \frac{\partial^2}{\partial z^2}$ is the Laplacian operator in Cartesian coordinates. For a source-free medium, that is, $\mathbf{J_C} = 0$ and $\rho_v = 0$, Eq. (1.21) reduces to the homogeneous wave equation:

$$\nabla^2 \mathbf{E} - \mu\varepsilon \frac{\partial^2 \mathbf{E}}{\partial t^2} = 0 \qquad (1.22)$$

Note that $v = 1/\sqrt{\mu\varepsilon}$ is the velocity of the wave in the medium. Equation (1.22) is equivalent to three scalar equations, one for every component of **E**. Let

$$\mathbf{E} = E_x \mathbf{a}_x + E_y \mathbf{a}_y + E_z \mathbf{a}_z \qquad (1.23)$$

where $\mathbf{a}_x$, $\mathbf{a}_y$, and $\mathbf{a}_z$ are the unit vectors in the x, y, and z directions, respectively. Equation (1.22) then becomes

$$\left(\frac{\partial^2}{\partial x^2} + \frac{\partial^2}{\partial y^2} + \frac{\partial^2}{\partial z^2} \right)(E_x \mathbf{a}_x + E_y \mathbf{a}_y + E_z \mathbf{a}_z) = \mu\varepsilon \frac{\partial^2}{\partial t^2}(E_x \mathbf{a}_x + E_y \mathbf{a}_y + E_z \mathbf{a}_z)$$

$$(1.24)$$

Comparing the $\mathbf{a}_x$-component on both sides of the above equation gives us

$$\frac{\partial^2 E_x}{\partial x^2} + \frac{\partial^2 E_x}{\partial y^2} + \frac{\partial^2 E_x}{\partial z^2} = \mu\varepsilon \frac{\partial^2 E_x}{\partial t^2}$$

Similarly, we have the same type of equation shown above for the E_y and E_z components by comparing other components in Eq. (1.24). Hence we can write a compact equation for the three components as follows:

$$\frac{\partial^2 \psi}{\partial x^2} + \frac{\partial^2 \psi}{\partial y^2} + \frac{\partial^2 \psi}{\partial z^2} = \mu\varepsilon\frac{\partial^2 \psi}{\partial t^2} \tag{1.25a}$$

or

$$\nabla^2 \psi = \mu\varepsilon\frac{\partial^2 \psi}{\partial t^2} \tag{1.25b}$$

where ψ may represent a component, E_x, E_y, or E_z, of the electric field **E**. Equation (1.25) is called the *scalar wave equation*.

1.5
Wave Optics and Diffraction

In the study of wave optics, we start from the scalar wave equation. Let us look at some of the simplest solutions. One of the simplest solutions is the plane wave solution:

$$\psi(x, y, z, t) = A\exp[i(\omega_0 t - \mathbf{k}_0 \cdot \mathbf{R})] \tag{1.26}$$

where ω_0 is the oscillating angular frequency (rad s^{-1}), $\mathbf{k}_0 = k_{0x}\mathbf{a}_x + k_{0y}\mathbf{a}_y + k_{0z}\mathbf{a}_z$ is the propagation vector, and $\mathbf{R} = x\mathbf{a}_x + y\mathbf{a}_y + z\mathbf{a}_z$ is the position vector. The magnitude of $\mathbf{k}_0$ is called *wavenumber* and is $|\mathbf{k}_0| = k_0 = \sqrt{k_{0x}^2 + k_{0y}^2 + k_{0z}^2} = \omega_0/v$ with v being the velocity of the wave in the medium given by $v = 1/\sqrt{\mu\varepsilon}$. If the medium is free space, $v = c$ (the speed of light in vacuum) and $|\mathbf{k}_0|$ becomes the wavenumber in free space. Equation (1.26) is a plane wave of amplitude A, traveling along the $\mathbf{k}_0$ direction. If a plane wave is propagating along the positive z-direction, Eq. (1.26) becomes

$$\psi(z, t) = A\exp[i(\omega_0 t - k_0 z)] \tag{1.27}$$

Equation (1.27) is a complex representation of a plane wave and since the EM fields are real functions of space and time, we can represent the plane wave in real terms by taking the real part of ψ to get

$$\text{Re}\{\psi(z, t)\} = A\cos(\omega_0 t - k_0 z) \tag{1.28}$$

For a plane wave incident on an aperture or a diffracting screen, that is, an opaque screen with some openings allowing light to pass through, the field distribution exiting the aperture or the diffracted field is a little more complicated to find. To tackle the diffraction problem, we need to find the solution of the scalar wave equation under some initial condition. Let us assume the aperture is represented by a transparency function, $t(x, y)$, located on the plane $z = 0$ as shown in Figure 1.8. A plane wave of amplitude A is incident on it. Hence at $z = 0$, according to Eq. (1.27), the plane wave immediately in front of the aperture is given by $A\exp(i\omega_0 t)$. The field immediately after the aperture is

$$\psi(x, y, z = 0, t) = At(x, y)\exp(i\omega_0 t) = \psi_p(x, y; z = 0)\exp(i\omega_0 t)$$
$$= \psi_{p0}(x, y)\exp(i\omega_0 t) \tag{1.29}$$

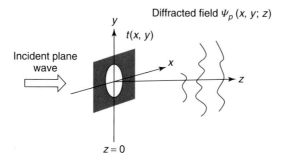

Figure 1.8 Diffraction geometry: $t(x, y)$ is the diffracting screen.

where we have assumed that the aperture is infinitively thin. $\psi_p(x, y; z = 0)$ or $\psi_{p0}(x, y)$ is called the *initial condition*. In the present case, the initial condition of the complex wave is given by $\psi_{p0}(x, y) = A \times t(x, y)$, the amplitude of the incident plane wave multiplied by the transparency function of the aperture. To find the field distribution at a distance z away from the aperture, we model the solution in the form of

$$\psi(x, y, z, t) = \psi_p(x, y; z) \exp(i\omega_0 t) \tag{1.30}$$

where $\psi_p(x, y; z)$ is the unknown to be found. In optics, $\psi_p(x, y; z)$ is called a *complex amplitude*. In engineering, it is known as a *phasor*. To find $\psi_p(x, y; z)$, we substitute Eq. (1.30) into Eq. (1.25). With the given initial condition $\psi_{p0}(x, y)$, we find [8]

$$\psi_p(x, y; z) = \exp(-ik_0 z) \frac{ik_0}{2\pi z} \int_{-\infty}^{\infty} \int_{-\infty}^{\infty} \psi_{p0}(x', y')$$
$$\times \exp\left\{\frac{-ik_0}{2z}[(x - x')^2 + (y - y')^2]\right\} dx' dy' \tag{1.31}$$

Equation (1.31) is called the *Fresnel diffraction formula* and describes the Fresnel diffraction of a "beam" during propagation and having an initial complex amplitude given by $\psi_{p0}(x, y)$. The Fresnel diffraction formula has been derived under the following conditions: (i) z must be many wavelengths away from the aperture and (ii) it should be valid under the paraxial approximation, that is, $z^2 \gg x^2 + y^2$. The Fresnel diffraction formula can be written in a compact form if we make use of the convolution integral

$$g(x, y) = g_1(x, y) * g_2(x, y) = \int_{-\infty}^{\infty} \int_{-\infty}^{\infty} g_1(x', y') g_2(x - x', y - y') dx' dy' \tag{1.32}$$

where $*$ denotes convolution of two functions $g_1(x, y)$ and $g_2(x, y)$. We also define a function

$$h(x, y; z) = \exp(-ik_0 z) \frac{ik_0}{2\pi z} \exp\left[\frac{-ik_0}{2z}(x^2 + y^2)\right] \tag{1.33}$$

With Eqs (1.32) and (1.33), the Fresnel diffraction formula can be written simply as

$$\psi_p(x, y; z) = \psi_{p0}(x, y) * h(x, y; z) \tag{1.34}$$

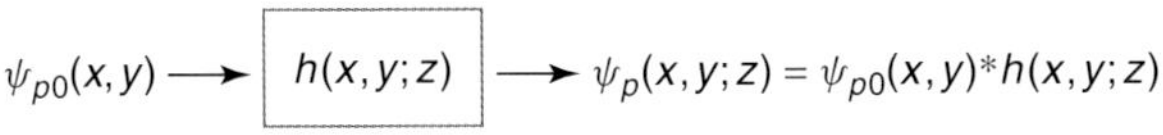

1.5 Wave Optics and Diffraction

$$\psi_{p0}(x,y) \longrightarrow \boxed{h(x,y;z)} \longrightarrow \psi_p(x,y;z) = \psi_{p0}(x,y) * h(x,y;z)$$

Figure 1.9 Block diagram summarizing Fresnel diffraction.

Hence Fresnel diffraction can be considered as a linear and space-invariant system [8] with input given by $\psi_{p0}(x,y)$ and with the system's impulse response given by $h(x,y;z)$. In optics, $h(x,y;z)$ is called *free-space spatial impulse response*. Figure 1.9 is a block diagram, which summarizes the Fresnel diffraction of a beam with initial profile $\psi_{p0}(x,y)$.

If we have a situation such that the calculation of the diffraction pattern is at distances far away from the aperture, Eq. (1.31) can be simplified. To see how this can be done, let us complete the square in the exponential function and then rewrite Eq. (1.31) to become

$$\psi_p(x,y;z) = \exp(-ik_0 z)\frac{ik_0}{2\pi z}\exp\left[\frac{-ik_0}{2z}(x^2+y^2)\right]\int_{-\infty}^{\infty}\int_{-\infty}^{\infty}\psi_{p0}(x',y')$$

$$\times \exp\left\{\frac{-ik_0}{2z}[(x')^2+(y')^2]\right\}\exp\left[\frac{ik_0}{z}(xx'+yy')\right]dx'dy' \quad (1.35)$$

In the above integral, ψ_{p0} is considered as the "source," and therefore the coordinates x' and y' can be called the *source plane*. In order to find the field distribution ψ_p on the observation plane, or the x–y plane, we need to have the source multiplied by the two exponential functions as shown inside the integrand of Eq. (1.35), and then to integrate over the source coordinates. The result of the integration is then multiplied by the factor $\exp(-ik_0z)\frac{ik_0}{2\pi z}\exp\left[\frac{-ik_0}{2z}(x^2+y^2)\right]$ to arrive at the final result on the observation plane given by Eq. (1.35). Note that the integral in Eq. (1.35) can be simplified if the approximation below is true:

$$\frac{k_0}{2}[(x')^2+(y')^2]_{max} = \frac{\pi}{\lambda_0}[(x')^2+(y')^2]_{max} \ll z \quad (1.36)$$

The term $\pi[(x')^2+(y')^2]_{max}$ is like the maximum area of the source and if this area divided by the wavelength is much less than the distance z under consideration, the term $\exp\left\{\frac{-ik_0}{2z}[(x')^2+(y')^2]\right\}$ inside the integrand can be considered to be unity, and hence Eq. (1.35) becomes

$$\psi_p(x,y;z) = \exp(-ik_0z)\frac{ik_0}{2\pi z}\exp\left[\frac{-ik_0}{2z}(x^2+y^2)\right]\int_{-\infty}^{\infty}\int_{-\infty}^{\infty}\psi_{p0}(x',y')$$

$$\times \exp\left[\frac{ik_0}{z}(xx'+yy')\right]dx'dy' \quad (1.37)$$

Equation (1.37) is the *Fraunhofer diffraction formula* and is the limiting case of Fresnel diffraction. Equation (1.36) is therefore called the *Fraunhofer approximation* or the *far field approximation* as diffraction is observed beyond the distance of near-field (Fresnel) diffraction.

1.6
Fourier Optics and Applications

Fourier optics is a term used by many authors to describe the use of Fourier transform to analyze problems in wave optics [8]. We first introduce the definition of Fourier transform and then we formulate wave optics using the transform. The two-dimensional spatial Fourier transform of a signal $f(x, y)$ is given by

$$\mathcal{F}\{f(x,y)\} = F(k_x, k_y) = \int_{-\infty}^{\infty}\int_{-\infty}^{\infty} f(x,y)\exp(ik_x x + ik_y y)]dxdy \quad (1.38)$$

and its inverse Fourier transform is

$$\mathcal{F}^{-1}\{F(k_x, k_y)\} = f(x,y) = \frac{1}{4\pi^2}\int_{-\infty}^{\infty}\int_{-\infty}^{\infty} F(k_x, k_y)\exp(-ik_x x - ik_y y)]dk_x dk_y \quad (1.39)$$

where the transform variables are spatial variables, x, y (in m), and spatial radian frequencies, k_x, k_y (in rad m^{-1}). We can now rewrite the *Fresnel diffraction formula* (see Eq. (1.35)) in terms of Fourier transform:

$$\psi_p(x, y; z) = \exp(-ik_0 z)\frac{ik_0}{2\pi z}\exp\left[\frac{-ik_0}{2z}(x^2 + y^2)\right]$$
$$\times \mathcal{F}\left\{\exp\left[\frac{-ik_0}{2z}(x^2 + y^2)\right]\psi_{p0}(x, y)\right\}_{k_x=\frac{k_0 x}{z}, k_y=\frac{k_0 y}{z}} \quad (1.40)$$

Similarly, the Fraunhofer diffraction formula given by Eq. (1.37) can be written as

$$\psi_p(x, y; z) = \exp(-ik_0 z)\frac{ik_0}{2\pi z}\exp\left[\frac{-ik_0}{2z}(x^2 + y^2)\right]\mathcal{F}\{\psi_{p0}(x,y)\}_{k_x=\frac{k_0 x}{z}, k_y=\frac{k_0 y}{z}} \quad (1.41)$$

Figure 1.10 shows the simulation of Fresnel diffraction of a circular aperture function circ(r/r_0), that is, $\psi_{p0}(x,y) = $ circ(r/r_0), where $r = \sqrt{x^2 + y^2}$ and circ(r/r_0) denotes a value 1 within a circle of radius r_0 and 0 otherwise. The used wavelength for simulations is 0.6 μm. Since $\psi_p(x, y; z)$ is a complex function, we plot its absolute value in the figures. Physically, the situation corresponds to the incidence of a plane wave with unit amplitude on an opaque screen with a circular opening of radius r_0 as $\psi_{p0}(x,y) = 1 \times t(x,y)$ with $t(x,y) = $ circ(r/r_0). We would then observe the intensity pattern, which is proportional to $|\psi_p(x, y; z)|^2$, at distance z away from the aperture. In Figure 1.11, we show Fraunhofer diffraction. We have chosen the distance of 80 cm so that the Fraunhofer approximation from Eq. (1.36) is satisfied.

1.6.1
Ideal Thin Lens as Optical Fourier Transformer

An ideal thin lens is a phase object, which means that it will only affect the phase of the incident light. For an ideal converging lens having a focal length f, the phase function of the lens is given by

$$t_f(x, y) = \exp\left[\frac{ik_0}{2f}(x^2 + y^2)\right] \quad (1.42)$$

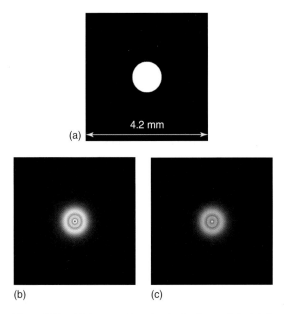

Figure 1.10 (a) Aperture function in the form of circ(r/r_0), $r_0 = 1$mm. (b) Diffraction at $z = 7$ cm (Fresnel diffraction), $|\psi_p(x, y; z = 7 \text{ cm})|$. (c) Diffraction at $z = 8$ cm (Fresnel diffraction), $|\psi_p(x, y; z = 8 \text{ cm})|$.

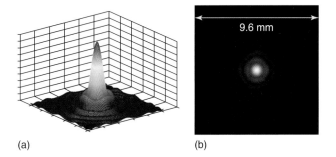

Figure 1.11 (a) 3D plot of a diffraction pattern at $z = 80$ cm (Fraunhofer diffraction), $|\psi_p(x, y; z = 80 \text{ cm})|$. (b) Gray-scale plot of $|\psi_p(x, y; z = 80 \text{ cm})|$. (Please find a color version of this figure on the color plates.)

where we have assumed that the lens is of infinite extent. For a uniform plane wave incident upon the lens, we can calculate the field distribution in the back focal plane of the lens using Eq. (1.34). To employ Eq. (1.34), we need to find the initial condition of the complex wave, which is given by $\psi_{p0}(x, y) = A \times t(x, y)$, the amplitude of the incident plane wave multiplied by the transparency function of the aperture. In the present case, $A = 1$ and the transparency function of the aperture is now given by the lens function $t_f(x, y)$, that is, $t(x, y) = t_f(x, y)$. Hence

$\psi_{p0}(x, y) = A \times t(x, y) = 1 \times t_f(x, y) = t_f(x, y)$. The field distribution at a distance f away from the lens, according to Eq. (1.34) with $z = f$ in $h(x, y; z)$, is then given by

$$\psi_p(x, y; f) = t_f(x, y) * h(x, y; f)$$
$$= \exp\left[\frac{ik_0}{2f}(x^2 + y^2)\right] * \exp(-ik_0 f)\frac{ik_0}{2\pi f}\exp\left[\frac{-ik_0}{2f}(x^2 + y^2)\right] \quad (1.43)$$

The above equation can be shown on evaluation to be proportional to a delta function, $\delta(x, y)$ [8], which is consistent with the geometrical optics that already states that all input rays parallel to the optical axis converge behind the lens to a point called the *back focal point*. The discussion thus far in a sense justifies the functional form of the phase function of the lens given by Eq. (1.42). We now look at a more complicated situation shown in Figure 1.12a, where a transparency $t(x, y)$ illuminated by a plane wave of unit amplitude is located in the front focal plane. We want to find the field distribution in the back focal plane. In Figure 1.12b, we describe the process of finding the field using a block diagram. Assuming that $t(x, y)$ is illuminated by a plane wave of unity, the field immediately after $t(x, y)$ is given by $1 \times t(x, y)$. The resulting field then undergoes Fresnel diffraction at a distance f, and according to Figure 1.9, $1 \times t(x, y)$ is the input to the block $h(x, y; f)$ as shown in Figure 1.12a. The diffracted field, $t(x, y) * h(x, y; f)$, is now immediately in front of the lens with a phase function given by $t_f(x, y)$, and hence the field after the lens is $[t(x, y) * h(x, y; f)] \times t_f(x, y)$. Finally, the field at the back focal plane is found using Fresnel diffraction one more time for a distance f, which is shown in Figure 1.12b. The resulting field is

$$\psi_p(x, y; f) = \{[t(x, y) * h(x, y; f)]t_f(x, y)\} * h(x, y; f) \quad (1.44)$$

The above equation can be worked out to become [8], apart from some constant,

$$\psi_p(x, y; f) = \mathcal{F}\{t(x, y)\}\bigg|_{k_x = \frac{k_0 x}{f}, k_y = \frac{k_0 y}{f}} = T\left(\frac{k_0 x}{f}, \frac{k_0 y}{f}\right) \quad (1.45)$$

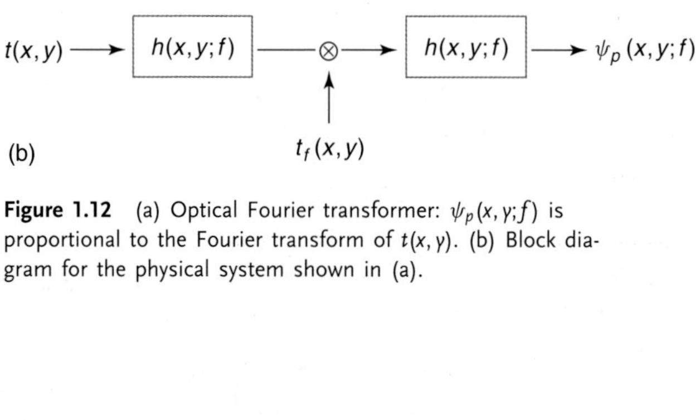

Figure 1.12 (a) Optical Fourier transformer: $\psi_p(x, y; f)$ is proportional to the Fourier transform of $t(x, y)$. (b) Block diagram for the physical system shown in (a).

where $T\left(\frac{k_0 x}{f}, \frac{k_0 y}{f}\right)$ is the Fourier transform or the *spectrum* of $t(x,y)$. We see that we have the exact Fourier transform of the "input," $t(x,y)$, on the back focal plane of the lens. Hence an ideal thin lens is an optical Fourier transformer.

1.6.2
Imaging and Optical Image Processing

Figure 1.12a is the backbone of an optical image processing system. Figure 1.13 shows a standard image processing system with Figure 1.12a as the front end of the system. Figure 1.13 is the same optical system we have studied before in geometrical optics (Figure 1.6) except for the additional transparency function, $p(x,y)$, on the confocal plane. $p(x,y)$ is called the *pupil function* of the optical system. Now we analyze the system using the wave optics approach. On the input plane, we have an image in the form of a transparency, $t(x,y)$, which is assumed to be illuminated by a plane wave. Hence, according to Eq. (1.45), we have its spectrum on the back focal plane of lens L_1, $T\left(\frac{k_0 x}{f_1}, \frac{k_0 y}{f_1}\right)$, where T is the Fourier transform of $t(x,y)$. Hence the confocal plane of the optical system is often called the *Fourier plane*. The spectrum of the input image is now modified by the pupil function, and the field immediately after the pupil function is $T\left(\frac{k_0 x}{f_1}, \frac{k_0 y}{f_1}\right) p(x,y)$. According Eq. (1.45) again, this field will be Fourier transformed to give the field on the image plane as

$$\psi_{pi} = \mathcal{F}\left\{T\left(\frac{k_0 x}{f_1}, \frac{k_0 y}{f_1}\right) p(x,y)\right\}_{k_x = \frac{k_0 x}{f_2}, k_y = \frac{k_0 y}{f_2}} \tag{1.46}$$

which can be evaluated, in terms of convolution, to give

$$\psi_{pi}(x,y) = t\left(\frac{x}{M}, \frac{y}{M}\right) * \mathcal{F}\{p(x,y)\}_{k_x = \frac{k_0 x}{f_2}, k_y = \frac{k_0 y}{f_2}}$$

$$= t\left(\frac{x}{M}, \frac{y}{M}\right) * P\left(\frac{k_0 x}{f_2}, \frac{k_0 y}{f_2}\right)$$

$$= t\left(\frac{x}{M}, \frac{y}{M}\right) * h_C(x,y) \tag{1.47}$$

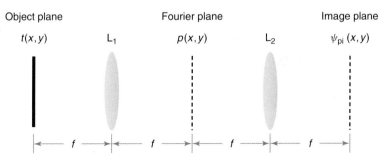

Figure 1.13 Optical image processing system.

where $M = -\frac{f_2}{f_1}$ as is the case for the two-lens imaging system in Figure 1.6 and P is the Fourier transform of p. From Eq. (1.47), we can recognize that

$$h_C(x,y) = \mathcal{F}\{p(x,y)\}\Big|_{k_x=\frac{k_0 x}{f_2}, k_y=\frac{k_0 y}{f_2}} = P\left(\frac{k_0 x}{f_2}, \frac{k_0 y}{f_2}\right) \tag{1.48}$$

is the point spread function in the context of a linear and spatial invariant system. In optics, it is called the *coherent point spread function*. Hence, the expression given by Eq. (1.47) can be interpreted as the result of a linear and spatial invariant system in that the scaled image of $t(x, y)$ is processed by an impulse response given by Eq. (1.48). The impulse response, and therefore the image processing capabilities, can be varied by simply changing the pupil function, $p(x, y)$. For example, if we take $p(x, y) = 1$, which means we do not modify the spectrum of the input image, then, $h_C(x, y)$, according to Eq. (1.48), becomes a delta function and the output image from Eq. (1.47) is $\psi_{pi}(x, y) \propto t\left(\frac{x}{M}, \frac{y}{M}\right) * \delta(x, y) = t\left(\frac{x}{M}, \frac{y}{M}\right)$. The result is an image scaled by M, consistent with the result obtained from geometrical optics.

If we now take $p(x, y) = \text{circ}(r/r_0)$, then, from the interpretation of Eq. (1.46), we see that for this kind of chosen pupil, filtering is of low-pass characteristic as the opening of the circle on the pupil plane only allows physically the low spatial frequencies to go though. Figure 1.14 shows examples of low-pass filtering. In Figure 1.14a,b, we show the original of the image and its spectrum, respectively. In Figure 1.14c,e, we show the filtered images, and their low-pass filtered spectra are shown in Figure 1.14d,f, respectively, where the low-pass filtered spectra are obtained by multiplying the original spectrum by $\text{circ}(r/r_0)$. Note that the radius r_0 in Figure 1.14d is larger than that in Figure 1.14f. In Figure 1.15, we show high-pass filtering examples where we take $p(x, y) = 1 - \text{circ}(r/r_0)$.

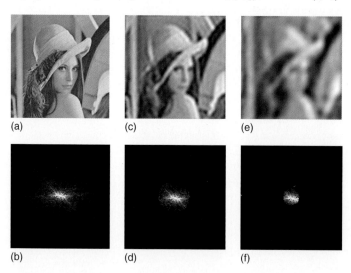

Figure 1.14 Low-pass filtering examples: (a) original image, (b) spectrum of (a), (c), and (e) low-pass images, (d) and (f) spectrum of (c) and (e), respectively.

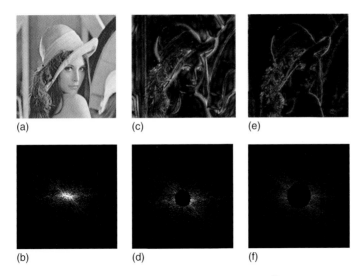

(a) (c) (e)
(b) (d) (f)

Figure 1.15 Same as in Figure 1.14, but with high-pass filtering.

1.6.3
Optical Correlator

The correlation of two images is one of the most important mathematical operations in pattern recognition [9]. However, as the images become more and more complex and large in size, the calculation becomes time consuming for a digital computer. Optical processing provides an alternative to digital processing as it offers greater speed. For images $g_1(x, y)$ and $g_2(x, y)$, the correlation between them is given by

$$g_1(x,y) \otimes g_2(x,y) = \int_{-\infty}^{\infty} \int_{-\infty}^{\infty} g_1^*(x',y')g_2(x+x',y+y')\mathrm{d}x'\mathrm{d}y' \tag{1.49}$$

where $g_1^*(x, y)$ is the complex conjugate of $g_1(x, y)$ and $\otimes$ denotes the operation of correlation. The role of *optical correlators* is to implement Eq. (1.49) optically. In this section, we describe an optical correlator called the *joint-transform correlator*. Figure 1.16 shows a standard optical correlator. $g_1(x, y)$ and $g_2(x, y)$ are the two images to be correlated, and they are in the form of transparencies, which are illuminated by plane waves. They are separated by a distance of $2x_0$ in the front focal plane of Fourier transform lens L_1 as shown in Figure 1.16. The so-called joint-transform power spectrum, $\mathrm{JTPS}(k_x, k_y)$, on the back focal plane of lens L_1 is given by

$$\mathrm{JTPS}(k_x, k_y) = |\mathcal{F}\{g_1(x+x_0, y)\} + F\{g_2(x-x_0, y)\}|^2 \tag{1.50}$$

The expression in Eq. (1.50) is essentially the intensity pattern on the focal plane. Expanding Eq. (1.50), we have

$$\begin{aligned}\mathrm{JTPS}(k_x, k_y) = &|\hat{g}_1(k_x, k_y)|^2 + |\hat{g}_2(k_x, k_y)|^2 + \hat{g}_1^*(k_x, k_y)\hat{g}_2(k_x, k_y)\exp(\mathrm{i}2k_0x_0)\\ &+ \hat{g}_1(k_x, k_y)\hat{g}_2^*(k_x, k_y)\exp(-\mathrm{i}2k_0x_0)\end{aligned} \tag{1.51}$$

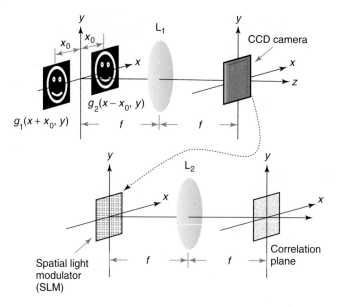

Figure 1.16 Optical correlator.

where $\hat{g}_1(k_x, k_y)$ and $\hat{g}_2(k_x, k_y)$ are the Fourier transform of g_1 and g_2, respectively with $k_x = k_0 x/f$ and $k_y = k_0 y/f$, where k_0 is the wavenumber of the plane wave. The joint-transform power spectrum (JTPS) is now detected by a CCD camera, whose output is fed to a 2D *spatial light modulator*. A 2D spatial light modulator is a device with which one can imprint a 2D pattern on a laser beam by passing the laser beam through it (or by reflecting the laser beam off it) [10]. A liquid crystal TV (LCTV) (upon suitable modification) is a good example of a spatial light modulator. In fact, we can think of a spatial light modulator as a real-time transparency because one can update 2D images on the spatial light modulator in real time without developing films into transparencies. The readers may refer to Chapter 9 of this book for more details of spatial light modulators. Once the JTPS is loaded to the spatial light modulator; we can then put the spatial light modulator in the front focal plane of Fourier transform lens L_2 as shown in Figure 1.16. The back focal plane of lens L_2 is the correlation plane as it shows all the correlation outputs. To see that, we take the Fourier transform of JTPS.

$$\mathcal{F}\left\{\text{JTPS}\left(\frac{k_0 x}{f}, \frac{k_0 y}{f}\right)\right\}\bigg|_{k_x = \frac{k_0 x}{f}, k_y = \frac{k_0 y}{f}} = C_{11}(-x, -y) + C_{22}(-x, -y)$$
$$+ C_{12}(-x - 2x_0, -y) + C_{21}(-x + 2x_0, -y) \quad (1.52)$$

where

$$C_{mn}(x, y) = g_m \otimes g_n \quad (1.53)$$

with $m = 1$ or 2, and $n = 1$ or 2. $C_{mn}(x, y)$ is the autocorrelation when $m = n$, and is the cross-correlation when $m \neq n$. Therefore, if the two images g_m and g_n

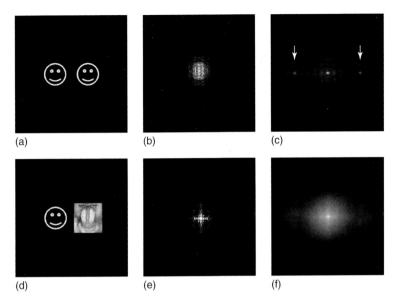

Figure 1.17 Simulation results for optical correlator shown in Figure 1.16. (a) Input patterns, (b) JTPS of the two input patterns, (c) output on the correlation plane, (d–f) are the same as in (a–c) but with the input patterns not identical.

are the same, besides a strong peak at the origin of the correlation plane due to the first two terms in Eq. (1.52), we have two strong peaks centered at $x = \pm 2x_0$. Figure 1.17a–c shows the input patterns, the JTPS (on the CCD camera plane), and the output on the correlation plane, respectively. Note that since the two input patterns are identical, we have two bright correlation spots (denoted by white arrows in the Figure 1.17c) on both sides of the center spot. Figure 1.17d–f is similar to Figure 1.17a–c, respectively but the two input patterns are different. Note that there are no discernible bright spots on both sides of the output in the center of the correlation plane in the case of mismatched patterns in the input plane.

1.7
The Human Visual System

Figure 1.18 is a schematic diagram of a human eye. The cornea and the crystalline lens together serve as a converging lens in the imaging system. To control the entering light flux, there is an iris between the cornea and the lens. The diameter of the iris can vary from about 2 to 8 mm. The crystalline lens is connected with several ciliary muscles in order to alter the shape of the crystalline lens so that the effective focal length of the lens can be adjusted. The eye is relaxed when it is focused on an infinitely distant object. When a person looks at an object, his brain will automatically control the ciliary muscles to focus the image on the

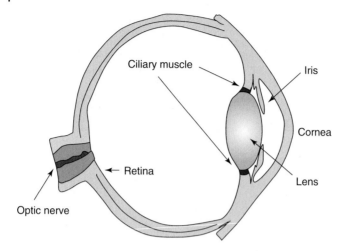

Figure 1.18 Cross section of the human eye.

retina clearly. The retina is a film that consists of numerous light-sensitive cells. Its function is to transform the illuminating light to an electric signal – like the CCD sensor in a digital camera. The electric signal is delivered to the brain via the optic nerve, and is finally read and analyzed by the brain.

The human eye can measure not only light strength but also colors because the retina contains two classes of receptors, the rods and the cones. Rods are very sensitive to weak light, but cannot identify colors. In contrast to the rods, the cones are insensitive to the weak light but can identify colors. This is because there are three different types of cones, and they are respectively sensitive to red light, green light, and blue light. The sensitivity spectrum of a normal human eye is roughly between 390 and 780 nm. Owing to the various responses of the receptors, it is interesting that the most sensitive wavelength of human eye is 555 nm in daytime, but is 507 nm at night. Some people's retina is short of one or more types of cones, and thus they cannot identify colors correctly. This is called *"color blindness."* Although most people own complete cones and rods, their color experiences are usually different from one another. This is one of the major challenges in color engineering [11, 12].

There are several kinds of abnormal eyes: nearsightedness, farsightedness, and astigmatism. Nearsightedness and farsightedness are both due to the defocus aberration of the eye. In the former case, the image of an object that is far away is formed in front of the retina, and in the latter case the image is formed in the back of the retina. Astigmatism of the eye is different from astigmatism aberration that we mentioned in Section 1.3.3. For an eye with astigmatism, its cornea is asymmetric. Thus for a distant point object, the image formed on the retina will be an asymmetric and elliptic spot.

The above-mentioned abnormal eyes result in blurred images, which can be corrected by glasses [2]. Note that the iris serves as an aperture stop of the lens so that aberrations can be adjusted. People with nearsightedness or farsightedness see clearer in daytime (smaller iris) than at night (larger iris). They are also used to squinting for clearer vision.

1.8 Conclusion

This chapter has provided some of the fundamentals and applications in optical image processing. When diffraction is of no concern in the imaging systems, we can use geometrical optics to explain the imaging process using matrix formalism. However, the matrix formalism is only applicable for paraxial rays. To consider the effect of diffraction, we need to use wave optics. Wave optics can be formulated in terms of the Fresnel diffraction formula, which can be reformulated using Fourier transforms. Hence image formation and image processing can be described in terms of Fourier transforms, which is the basis of what is known as *Fourier optics*. However, the use of Fourier optics is only applicable under the paraxial approximation. Nevertheless, Fourier optics is widely used and one of its most important applications is in the area of optical correlation. We have discussed an optical correlation architecture called *joint-transform correlator*, which has provided an important example on how one can use Fourier optics in the treatment of optical image processing.

References

1. Hudson, R.D. (1969) *Infrared System Engineering*, John Wiley & Sons, Inc., New York.
2. Hecht, E. (2002) *Optics*, Addison-Wesley, San Francisco.
3. Hecht, J. (2008) The history of the X-ray laser. *Opt. Photonics News*, **19** (4), 26–33.
4. Poon, T.-C. and Kim, T. (2006) *Engineering Optics with MATLAB*, World Scientific, Singapore.
5. Smith, W.J. (2008) *Modern Optical Engineering*, SPIE Press, New York.
6. Mahajan, V.N. (1998) *Optical Imaging and Aberrations*, SPIE Press, Bellingham.
7. Born, M. and Wolf, E. (1999) *Principles of Optics*, Cambridge University Press, Cambridge.
8. Poon, T.-C. (2007) *Optical Scanning Holography with MATLAB*, Springer, NewYork.
9. Yu, F.T.S. and Jutamulia, S. (1998) *Optical Pattern Recognition*, Cambridge University Press, Cambridge.
10. Poon, T.-C. et al. (1998) Spatial light modulators-research, development, and applications: introduction to the feature issue. *Appl. Opt.*, **37**, 7471.
11. Wyszecki, G. and Stiles, W.S. (1982) *Color Science*, John Wiley & Sons, Inc., New York.
12. Shovell, S.K. (2003) *The Science of Color*, Elsevier, Amsterdam.

2
Fundamentals of Photonics

Erik Stijns and Hugo Thienpont

2.1
Introduction

This is a short introductory chapter about photonics. It is not our aim to provide an overview of the broad field covered by photonics. Such a goal would be completely unrealistic in a short chapter such as this one. It would also not be useful to the readers of this book, since they are mainly interested in image processing and not in photonics as such. Nevertheless, we are of the opinion that an introductory chapter explaining some basic concepts of photonics could be helpful.

The concepts of *interference* and *diffraction*, for example, are widely used and applied in different chapters of this book. Most readers are surely familiar with it. However, newcomers with a limited background in physics may need some explanation on the basic physics concepts. Section 2.2 of this chapter is intended for them.

We also realized that a number of researchers in the field are not familiar with the *terms* and *units* used in the measurement of light. Therefore, we added Section 2.3.

Colors and their measurements are so important in image processing and are used throughout this book that we also found it necessary to give a short introduction on this topic.

And finally, we added a section on *lasers*, because lasers are still the basic instruments for photonics.

2.2
Interference and Diffraction

2.2.1
Interference

Around 1800, Thomas Young performed an experiment that proved the wave character of light (Figure 2.1). He illuminated a slide in which two holes had been punched. Those two holes act as point sources of light. According to Huygens'

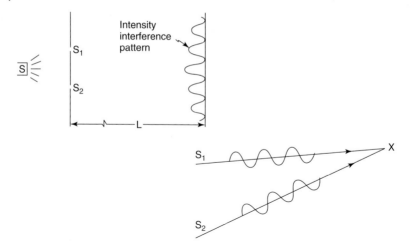

Figure 2.1 Interference between two waves.

principle, each of those points can be considered as the source of a spherical wave. In other words, each point emits light in all directions. Consequently, every point X on a distant screen receives two light waves, one from each hole. The electric field at each individual point on the screen is the algebraic sum of the two incident electric fields. When both waves have the same frequency, and when they are in phase, the two waves will amplify each other. One speaks then of *constructive interference* and the point X is a bright point [2].

On the other hand, when the two waves have the same frequency but are in opposition, the total field is always zero, and the point P is a dark point. This is called *destructive interference*.

Moreover, one can verify by calculation that the overall light distribution on the screen is a sinusoidal one.

This same phenomenon of interference can also explain the light pattern one observes when a grating is illuminated by a plane wave. A grating does not have two holes (as in Young's experiment) but has a very large number of slits: typically a few hundred per millimeter. Each point on the distant screen receives thousands of waves. At those points at which all waves are in phase, a bright spot appears. Such points are much brighter than in Young's experiment because many more waves are added together. In all other points on the screen all those waves average to zero. The conclusion is that the pattern generated by a grating at a distant screen consists of a few bright spots on a (average) dark background.

2.2.2
Diffraction

2.2.2.1 Diffraction at a One-Dimensional Slit

Assume that a single slit of width d is illuminated by a collimated beam (Figure 2.2). Each point of this slit can now be considered as a point source that (according to

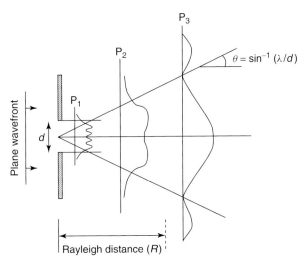

Figure 2.2 Diffraction at a 1-dimensional slit.

Huygens' principle) sends light rays in all directions. On a distant screen, all these light rays will interfere. The calculation of the light pattern on the screen is now very complicated, because there is a *continuum* of light sources. The final light field in each point is consequently given by an *integral* over the slit, the so-called diffraction integral. These calculations show that the incident light beam will spread out when propagating behind the slit. This spreading is called *diffraction* of light, and the light distribution at the screen is called the *diffraction pattern* of the slit. When the screen is at a very large distance from the slit, larger than the Rayleigh distance, the pattern is called the *far-field or the Fraunhofer diffraction pattern*. At a smaller distance one speaks of *near-field or Fresnel diffraction pattern*. Near-field patterns are characterized by fast oscillations called *ringing*.

2.2.2.2 Diffraction at a Circular Aperture

Consider now a circular aperture of diameter D in an opaque screen, which is illuminated by a collimated beam of wavelength λ (Figure 2.3). The diffraction pattern on a distant screen can be calculated along the lines given above. It consists of a central disc of radius $r = 1.22\, \lambda\, d/D$, surrounded by fainter rings. This pattern is called the *Airy pattern*.

Assume now that this aperture is illuminated by a *convergent* beam, which converges toward a point at a distance f behind the aperture (Figure 2.4). The calculations show that the diffraction pattern is once again an Airy pattern of radius r given by the same formula. There is, however, an essential difference between the two setups: indeed, here f should *not* to be very large! An equivalent setup is realized when a single lens is illuminated by a collimated beam. The Airy pattern in the focal plane has a radius $r = 1.22\, \lambda f/D$ [3].

Those formulas allow for the calculation of the diffraction pattern in different optical systems or instruments.

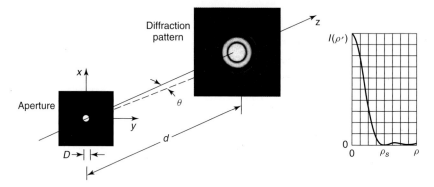

Figure 2.3 Diffraction at a circular aperture.

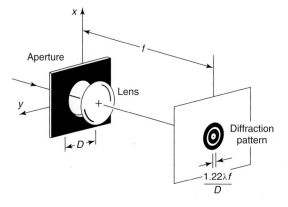

Figure 2.4 Diffraction by a convergent wave at a circular aperture.

2.2.3
Resolution

It is not possible to eliminate diffraction in optical systems or instruments; consequently, each image point is always surrounded by a diffraction disc. When two image points are so close to each other so that their diffraction disks overlap, the two points are "not resolved." Hence diffraction always limits the resolution of an optical system.

2.2.3.1 Angular Resolution

Assume that the circular aperture of Figure 2.3 is illuminated by *two* parallel beams, which form an angle $\Delta\Phi$; let us say, for example, that you look at two stars with an angular separation of $\Delta\Phi$ (Figure 2.5). Both objects give an Airy diffraction pattern. Following the criterion of Rayleigh, we say that the two images are "just resolved" when the maximum of one of the Airy disks coincides with the first zero of the other one. This happens when $\Delta\Phi = 1.22\,\lambda/D$. Consequently, we say that the *angular resolution* of an optical system [6] is

$$\Delta\Phi_{\text{resol}} = 1.22\,\lambda/D \tag{2.1}$$

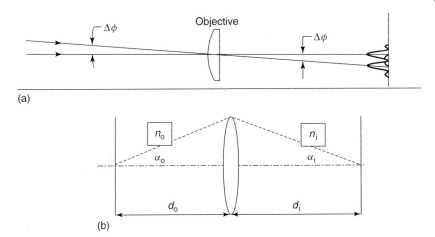

Figure 2.5 Angular (a) and spatial (b) resolution.

2.2.3.2 Spatial Resolution

Consider now an optical system that images an object (in a transverse object plane at a distance d_o) into a plane image at a distance d_i. By applying the previous formulas, one can calculate that the spatial resolution in the image plane is given by

$$x_{i,resol} = 0.61\, \lambda/NA_{image} \tag{2.2}$$

in which NA_{image} is the numerical aperture in the image space, that is,

$$NA_{image} = n_i \sin \alpha_i \tag{2.3}$$

The resolution in the object plane is

$$x_{o,resol} = 0.61\, \lambda/NA_{obj} \tag{2.4}$$

with

$$NA_{obj} = n_o \sin \alpha_o \tag{2.5}$$

2.2.4 Coherence

2.2.4.1 Temporal or Longitudinal Coherence

The basic idea of interference (Figure 2.1) is only valid when both waves behave regularly, that is, when they are *coherent*. Real light beams, however, consist of small light bursts; hence they are only coherent over a limited time interval, the so-called *coherence time* τ_{coh}. During that time interval, they run a distance called the *coherence length* L_{coh} which is, of course, given by $L_{coh} = c\tau_{coh}$. Although these coherence values are statistical averages, a simple picture can be used. It is as if the light beam consists of a succession of coherent wavelets of length L_{coh}. The light wave remains coherent over a distance L_{coh}. Then the overall phase suddenly

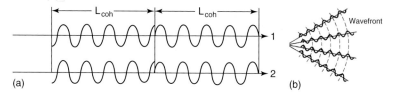

Figure 2.6 Longitudinal (a) and transverse (b) coherence.

jumps, destroying the coherence at that moment, after which we once again have a coherent wavelet of length L_{coh} (Figure 2.6) [4].

Typical values for the coherence length are about 1 µm for sunlight, about 1 mm for a spectral lamp, and typically 20 cm for a gas laser.

2.2.4.2 Transverse or Spatial Coherence

A point source emits spherical waves. According to Huygens' model, each point of this spherical wavefront is the origin of a "secondary" wavelet. Because all points on the wavefront have the same phase, those secondary wavelets are in phase with each other; in other words, they are coherent (Figure 2.6). One says that such a light wave is *transverse* (or *spatially*) *coherent*. This is true – to a good approximation – with light beams from a laser. Wavefronts that are emitted by other sources, for instance, a light bulb, are transverse *incoherent* [5].

2.3
Terms and Units: The Measurement of Light

2.3.1
Introduction: Radiometry versus Photometry

This section is about the measurement of light. In general two types of units are used.

In *radiometry* all quantities are measured in the appropriate SI units: energy in joules, power in watts, and so on.

For visual applications, on the other hand, other types of units are needed. Indeed, the (human) eye has only a limited spectral sensitivity: UV or IR photons cannot be seen. The measurement system that takes this into account is called *photometry*.

2.3.2
Radiometric Terms and Units

2.3.2.1 Radiant Energy

Radiant energy (Q_e) is the total *energy*, measured in joules, emitted by a light source, or falling on an irradiated object; the latter could possibly be a detector.

Note that all radiometric terms bear a subscript e (from "energy")

2.3.2.2 Radiant Flux
Radiant flux (Φ_e) is the total *power* (in watts) emitted by the light source or falling on an object. It is given by

$$\Phi_e = \frac{\partial Q_e}{\partial t} \tag{2.6}$$

2.3.2.3 Radiant Flux Density
Radiant flux density is the radiant flux divided by the area S. It is measured in watts per square meter. When this is measured at the light source, it is called the *excitance* and written as M_e; when it is measured at an illuminated surface (which could be the surface of a detector), it is called *irradiance* E_e (Figure 2.7).

$$M_e \quad \text{or} \quad E_e = \frac{\partial \Phi_e}{\partial S} \tag{2.7}$$

2.3.2.4 Radiant Intensity
The radiant flux density is not a good parameter to describe a point source. For point sources it is better to consider the power within a *cone* of light (Figure 2.8); this is called the *radiant intensity*. It is denoted by the symbol I_e, and measured in watts per steradian.

$$I_e = \frac{\partial \Phi_e}{\partial \Omega} \tag{2.8}$$

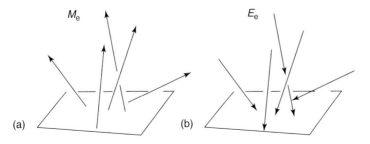

Figure 2.7 Radiant flux density at a source (a) or at an irradiated surface (b).

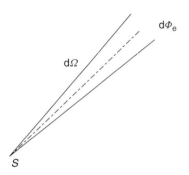

Figure 2.8 Intensity of a point source.

2.3.2.5 Radiance

Consider now an extended source, of area dA, that emits light in all directions. How does one calculate the power entering an optical system, as shown in Figure 2.9a?

That power is proportional to the solid angle $d\Omega$ at which the optical system is seen from the center of the emitting surface.

Because this surface is not a point source, each small part of dA will contribute its own cone (Figure 2.9b). Consequently, the flux is also proportional to the surface area dA. When the surface is not perpendicular to the direction of viewing, then the optical system only sees the *projected* area $dA \cos\theta$. The proportionality constant is called the *radiance* L_e. This leads to the definition of radiance:

$$L_e = \frac{\partial^2 \Phi_e}{\partial \Omega \, \partial A \cos\theta} \tag{2.9}$$

Radiance is measured in watts per steradian per square meter.

2.3.2.6 Radiant Exposure

One knows that some detectors do not measure the incident *power* per square meter, but rather the total incident *energy* per square meter (in joules per square meter) which comes down to the integrated power; this is, for example, the case with an optical film used in cameras. Therefore, one defines the *radiant exposure* H_e as

$$H_e = \frac{\partial Q_e}{\partial A} \tag{2.10}$$

which is equivalent to

$$H_e = \int E_e dt \tag{2.11}$$

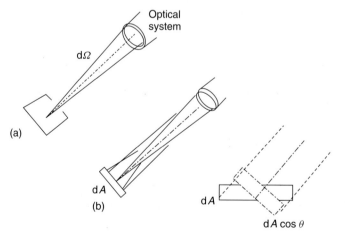

Figure 2.9 Radiance: the flux is proportional to the solid angle (a) and to the projected area (b).

2.3.3
Photometric Terms

2.3.3.1 Spectral Terms

For each of the above quantities one can define a corresponding *spectral* quantity. Consider, for example, the flux. In Figure 2.10, we measure the flux through a filter, which is assumed to be a perfect bandpass filter: the transmission equals 1 in a rectangular window of width $d\lambda$ around central wavelength λ. The detector now measures only a *small* flux $d\Phi_e$, which is (for small values of $d\lambda$) proportional to $d\lambda$. The proportionality constant is called the *spectral radiant flux* $\Phi_{e,\lambda}$. Hence

$$\Phi_{e,\lambda} = \frac{\partial \Phi_e}{\partial \lambda} \qquad (2.12)$$

In principle, this should be expressed in watts per meter (W m^{-1}), but mostly one uses milliwatts per nanometer (mW nm^{-1}).

If one takes another filter, at another central wavelength, then one will find another value for the spectral flux. In other words, the spectral flux depends on the wavelength. This function is called the *spectral flux density* or simply the *spectrum* of the light source.

Of course, the total flux is the integral of the spectral flux.

What we have done for flux can be repeated for each of the other radiometric terms that we have introduced above – one can then speak of a spectral radiance, spectral intensity, and so on.

2.3.3.2 Spectral Sensitivity of the Eye

The eye has only a very limited spectral range: we can only observe colors of which the wavelength varies between about 400 and 700 nm, with a maximum sensitivity at 555 nm. By measuring the relative spectral sensitivity of the eye of a large number of people, one has established a "standard sensitivity curve" of the human eye. This is written $y_{(\lambda)}$ and is shown in Figure 2.11.

2.3.3.3 Luminous Terms

The spectral radiant flux of a typical lamp extends far beyond the visible spectrum, and emits much more infrared than visible light. In order to find the range in which we can see with our eyes, we have to multiply this spectrum by the relative sensitivity curve of the eye. We call this the *visual spectral flux*, and we write it with

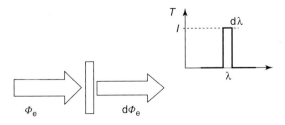

Figure 2.10 Definition of spectral radiant flux.

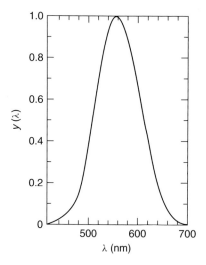

Figure 2.11 Sensitivity curve of a "standard" human eye.

a subscript v. Hence

$$\Phi_{v,\lambda}(\lambda) = y(\lambda)\Phi_{e,\lambda}(\lambda) \tag{2.13}$$

The total visual flux can then be defined by integrating Eq. (2.13). However, in order to be able to define appropriate units, this is multiplied by a constant K. The total visual flux is then

$$\Phi_v = K \int \Phi_{v,\lambda}(\lambda) d\lambda = K \int y(\lambda)\Phi_{e,\lambda}(\lambda) d\lambda \tag{2.14}$$

This is called the *luminous flux*. The constant K depends, of course, on the choice of the units and on the calibration.

2.3.4
Photometric Units

To emphasize their difference from radiometric quantities, and also for historical reasons, photometric units have different names than the radiometric ones.

The basic unit is the unit of luminous or visual flux, which is called the *lumen*. An optical beam has a luminous flux of 1 lm if, by definition, it has a total power of 1.4705 mW($= 1/680$ W) at a wavelength of 555 nm. In other words, the constant K in Eq. (2.14) equals 680 lm W^{-1} at $\lambda = 555$ nm.

2.3.4.1 Other Visual Terms and Units

The scheme we have developed for flux can also be applied to all other radiometric terms. So each radiometric quantity corresponds to a *visual* or *photometric* quantity, which is represented by the same symbol, but with a subscript v instead of e. These are summarized in Table 2.1.

Table 2.1 Summary of radiometric and photometric terms and units.

Physical meaning	Radiometry		Photometry	
Energy	Radiant energy Q_e	J	Luminous energy Q_v	talbot = lm · s
Power	Radiant flux Φ_e	W	Luminous flux Φ_v	lumen (lm)
Power per area	Radiant flux density	W m^{-2}	Luminous flux density	lux (lx) = lm m^{-2}
	Radiant excitance M_e		Luminous excitance M_v	
	Irradiance E_e		Illuminance E_v	
Power per solid angle	Radiant intensity I_e	W sr^{-1}	Luminous intensity I_v	candela (cd) = lm/sr
Power per solid angle and per area	Radiance L_e	W sr^{-1}m^{-2}	Luminance L_v	nit = cd m^{-2} = lx sr^{-1}
Energy per area	Radiant exposure H_e	J m^{-2}	Light exposure H_v	lx · s

Starting from the basic definition of lumen, one can go "down" to talbot, or go "up" to lux, candela, nit, or to lux.second.

2.4 Color

2.4.1 Introduction

The world around us is a colored world. All objects we observe have a color, even if it is only dull gray. We are used to describing these colors with everyday words. But is it also possible to give a *quantitative* value to them? Can we also calculate "how much" two colors differ from each other? It is clear that a positive answer to these questions is important for different branches of science and technology, and not only for the coating industry. That is what we are going to discuss in this section.

In order to observe an object, we have to illuminate it by a light source. It is clear that, one way or other, the properties of the light source will also influence the final appearance of the object. However, we are not going to analyze this in detail. We simply look at the final appearance of the object, without questioning whether those properties are inherent to the object itself, or whether they are coming from the light source.

2.4.2
The Spectrum of Light

Light is an electromagnetic phenomenon. In its simplest form a light wave can be described as a harmonic wave:

$$E(z, t) = A_0 \cos (kz - \omega t) \tag{2.15}$$

which is illustrated in Figure 2.12.

The distance between two crests is called the *wavelength* λ; this wavelength determines the color of the light wave. Visual colors have wavelengths between (about) 400 and 700 nm.

Real light sources do not emit such simple waves as in Figure 2.12; their space–time variations are much more complicated. However, we know that we can write every complicated function as a superposition of simple harmonics. The power density per nanometer of the component at wavelength λ is written as $P_\lambda(\lambda)$. This function $P_\lambda(\lambda)$ is called the *spectrum* of the light source (Figure 2.13).

When the spectrum is given, then the complete chromatic content of the light is known. In Chapters 31 and 32 the reader can find out how these spectra are measured and used.

2.4.3
Tristimulus Theory

2.4.3.1 The Tristimulus

Although a spectrum gives complete information about the chromatic content of a light beam, it is not always the most convenient way to describe it. The main reason is that our eyes do not work the way a spectrometer does, but in a completely different manner. So for visual observations, one needs a completely different way to describe colors [7].

The eye works as a kind of camera, which projects the image of an object onto a light sensitive layer at the back of it, which is called the *retina*.

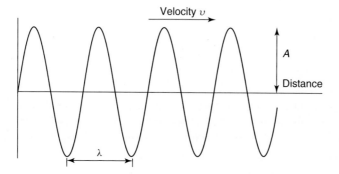

Figure 2.12 A simple harmonic wave.

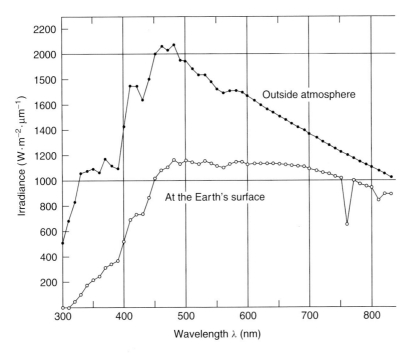

Figure 2.13 Spectrum of sunlight.

It turns out that there are two different types of cells. In the center of the retina, in the so-called *fovea*, one finds the *cones*; the rest of the retinal area is filled with *rods*.

The rods are very sensitive to light: when a (small) amount of light is absorbed by a rod, it sends an electrical pulse to the brain. The wavelength of the incident light does not matter; in other words: rods are *in*sensitive to color. Were one to use only the rods to look at the world around, it would be dull gray.

Cones, on the other hand, are sensitive to color. Some cones are mainly sensitive to red light; we could call them *R-cones*. Others are sensitive to green-yellow (*G-cones*) or to blue light (*B-cones*). If, for instance, blue light is incident on the eye, only the B-cones will send an electric signal to the brain; G- and R-cones will do nothing. The magnitude of this signal depends on the wavelength of the incident light. This relative sensitivity is described by a function we denote by $\bar{z}(\lambda)$, and which we call the *color-matching function* for blue light (Figure 2.14).

Assume now that we shine a light beam with spectral distribution $R(\lambda)$ on this B-cone. The amount of light at wavelength λ in the beam is given by the value of R at the wavelength λ, which is written as $R(\lambda)$. Consequently, the signal going to the brain is proportional to the product $R(\lambda)\overline{z(\lambda)}$. For finding the total signal that this cone sends to the brain, we have to add all contributions at the different wavelengths λ, and hence the signal is proportional to $\int R(\lambda)\overline{z(\lambda)}\,d\lambda$. This is called

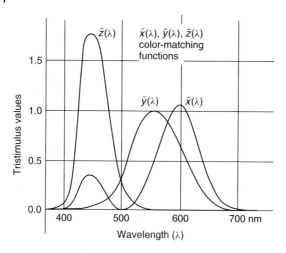

Figure 2.14 The three color-matching functions.

a *stimulus*; here it is the Z-stimulus. This stimulus is consequently defined by

$$Z = \int K\bar{z}(\lambda)\, R(\lambda)\, d\lambda \tag{2.16}$$

The proportionality constant k depends, among other things, on the power of the incident light.

In a similar way, the relative sensitivity of the G- (respectively, R-) cones are described by color-matching functions $\bar{y}(\lambda)$ and $\bar{x}(\lambda)$, see Figure 2.14. These functions have been established by averaging the spectral sensitivity of the eyes of a very large number of people.

The green and the red stimuli are then

$$Y = \int k\bar{y}(\lambda)\, R(\lambda)\, d\lambda \tag{2.17}$$

$$X = \int k\bar{x}(\lambda)\, R(\lambda)\, d\lambda \tag{2.18}$$

2.4.3.2 The 1931 CIE Standard

If we are only interested in the *relative* color *distribution* (which is mostly the case), then we can normalize the expressions (2.16–2.18) and at the same time get rid of the constant k.

So we define

$$\begin{aligned} x &= \frac{X}{X+Y+Z} \\ y &= \frac{Y}{X+Y+Z} \\ z &= \frac{Z}{X+Y+Z} \end{aligned} \tag{2.19}$$

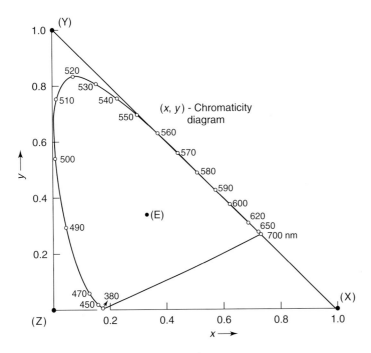

Figure 2.15 The 1931 CIE chromaticity diagram.

We see that $x + y + z = 1$; hence we need to know only *two* variables (e.g., x and y) to know the complete color distribution. Of course we then have no information about the total power. Because each color is now represented by only two variables, we can represent this in a two-dimensional diagram. This was introduced in 1931 by CIE, the "Commission Internationale de l'Eclairage," and is since then known as the *1931 CIE chromaticity diagram* (see Figure 2.15).

At the bottom left, we have $x \approx 0$ and $y \approx 0$; hence (because $x + y + z = 1$) $z \approx 1$ and this gives blue light. At the top, $y \approx 1$ and, consequently, we have green light, and at the bottom right, we find red light. The circumference of this diagram gives the "pure" colors, whereas inside we have mixtures. The very center ($z = y = 0.33$) represents white light, because here the three colors are evenly mixed.

2.4.3.3 CIE 1976 UCS Diagram

The 1931 CIE chromaticity diagram was introduced in 1931 and has been used from then on to describe and compare colors. Soon it became clear that this diagram has a major inconvenience – it is not "uniform." This means that the difference one observes visually between two different colors is not proportional to the Euclidean distance between the corresponding points in the diagram. And this is a major inconvenience if the diagram is used to compare different colors. This problem was solved in 1976 by introducing a new diagram, which is calculated from the 1931 diagram by a linear transformation of coordinates (Figure 2.16).

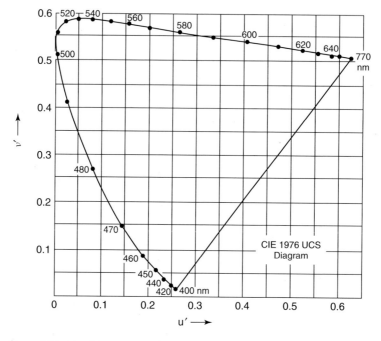

Figure 2.16 The CIE 1976 UCS diagram.

Instead of using the x and y coordinates, one introduces coordinates u' and v' as

$$u' = \frac{4x}{-2x + 12y + 3}$$
$$v' = \frac{9y}{-2x + 12y + 3} \tag{2.20}$$

This diagram is known as the *CIE 1976 UCS diagram* (UCS stands for *uniform chromaticity space*). This is now a real uniform diagram. Indeed, the visual difference between two colors is now proportional to the Euclidean distance measured between the corresponding points.

2.4.4
Theory of the Opponent Colors

2.4.4.1 Describing the Visual Observations

When we look at a colored object, the cones in our eyes produce three stimuli, which are sent to our brain. The brain does not measure these signals directly, with a kind of voltmeter. No, our brain processes them in a very complicated way. It was only at the end of the nineteenth century that people started to understand more or less what was happening. Ewald Hering (1835–1918) made an important contribution with his "opponent colors" theory.

He had observed that, when describing colors, some combinations of colors are often used, whereas others are never used. For example, we do combine green with blue; different kinds of blue-green colors are common. We also combine green with yellow, and we know different hues of green-yellow. But we never combine green with red. We never say that an object has a green-red color. Hering concluded that green and red are colors which – at least in our brains – "exclude" each other. In a similar way, yellow and blue exclude each other – we never say that an object has a yellow-blue color.

So in order to translate those observations into quantitative parameters, one defines two new variables, called the *chromatic coordinates* a^* and b^*. Parameter a^* is more or less "red minus green." The maximum value of a^* corresponds with pure red, while the minimum value gives pure green. In between, we go smoothly from pure red to pure green. In the middle of the interval we have no color ("achromatic"); the object is gray.

Variable b^* is more or less defined as "yellow minus blue." The maximum value of b^* gives a pure yellow color, the minimum value gives pure blue. In the middle we find gray.

After numerous experiments, the following definitions were agreed upon:

$$a^* = 500 \left(\sqrt[3]{\frac{X}{X_n}} - \sqrt[3]{\frac{Y}{Y_n}} \right) \tag{2.21}$$

$$b^* = 200 \left(\sqrt[3]{\frac{Y}{Y_n}} - \sqrt[3]{\frac{Z}{Z_n}} \right) \tag{2.22}$$

X, Y, and Z are the three stimuli as defined previously; X_n, Y_n, Z_n are the tristimulus values of a perfect diffuser.

2.4.4.2 Saturation or Chroma

Pure colors are situated at the rim of the chromatic diagram (they were also situated at the rim in the previous color diagrams). These pure colors are called *fully saturated* colors. The more one moves to the center, the more the "pure" colors are mixed with other ones, and the less saturated the result.

Hence one defines the saturation or the *chroma* (as it is usually called) as the distance from the center to the representative point. The formal definition is

$$C_{ab}^* = \sqrt{a^{*2} + b^{*2}} \tag{2.23}$$

2.4.4.3 Hue

The color varies along the rim of the disk. In other words: the color is defined by the *angular* position of the point. In scientific literature one does not speak of color; instead one uses the word "hue."

The formal quantitative definition is then

$$h_{ab} = arctg\left(\frac{b^*}{a^*}\right) \tag{2.24}$$

2.4.4.4 The CIELAB Diagram

In order to give a complete description of the light beam, it would also be useful to have a parameter that describes its *power*. Hence, one introduced a third parameter, which is then called the *lightness parameter L**. After numerous experiments, the following definition was agreed upon:

$$L^* = 116 \sqrt[3]{\frac{Y}{Y_n}} - 16 \tag{2.25}$$

Consequently, one now has a three-dimensional space, describing the power of the light beam and at the same time its chromatic content. This 3D representation is called the *L*a*b* color space*, also referred to as the *CIELAB color space* (Figure 2.17).

The color difference between any two colors in the *L*a*b** color space is proportional to the Euclidean distance between the corresponding points. It is denoted by ΔE^* and given by

$$\Delta E^* = \sqrt{(\Delta L^*)^2 + (\Delta a^*)^2 + (\Delta b^*)^2} \tag{2.26}$$

The numerical values in the definition of a^*, b^*, and L^* are chosen such that the minimum discernible color difference corresponds to $\Delta E^* = 1$.

The *L*a*b** color space is one of the most popular color spaces used for measuring color. It is widely used in different fields of application and in a number of color-measuring instruments.

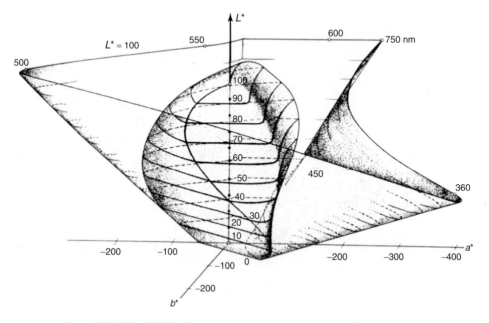

Figure 2.17 The CIELAB color space.

2.5
Basic Laser Physics

2.5.1
Introduction

The acronym LASER stands for *light amplification by stimulated emission of radiation*. Hence a laser is an *amplifier* of *light*: a light beam entering a laser will be amplified. The word *laser* also explains how this amplification is realized: by *stimulated emission*. The same process of stimulated emission is also responsible for the special properties of laser light. Hence it is worthwhile to study it in more detail [1].

We begin with a few words about the interaction between light and matter in general, before giving more details on the phenomenon of stimulated emission. In principle, one should use quantum mechanics to describe matter (atoms) and light; but then things soon become very complicated. Fortunately, there also exist much simpler models, which are sophisticated enough to reveal the properties we need.

One of these models was introduced in 1916 by Albert Einstein. In order to describe thermodynamic equilibrium between matter and light in a blackbody radiator, he considered atoms as systems with discrete energy levels. In this section, we limit the description to a two-level system with energy values $E1$ and $E2$, see Figure 2.18. The interaction with light is then described as absorption, spontaneous emission, or the (newly introduced) effect of stimulated emission.

2.5.2
Normal or Spontaneous Emission of Light

When an atom is excited into an energy level above its ground state, it soon returns to the lower level; the excess energy is emitted, often as a light wave or a photon. Its frequency f is given by Planck's equation $hf = E2 - E1$; here h is Planck's constant. Light generated by spontaneous emission has the following properties:

1) It is emitted in all directions. Indeed, there is no reason that a specific direction should be favored.
2) It is emitted in a broad spectral line. There is indeed no line narrowing, and the emitted light shows all line broadening effects (natural, Doppler, and/or homogeneous)

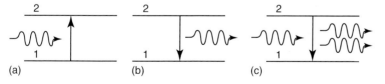

Figure 2.18 (a) Absorption; (b) spontaneous emission; and (c) stimulated emission in a two-level atom.

3) There is no coherence. There is no relationship between the phases of the different light waves; one says that the light is incoherent.

All light sources, other than lasers, emit their light incoherently, for example,

- thermal light sources (light bulb)
- nuclear (the sun)
- chemical (a fire)
- electrical (TL-tube (tube luminescent), LED).

2.5.3
Absorption

When a light wave (or a photon) meets an atom in its ground state, the light can be absorbed, on condition that Planck's equation $E2 - E1 = hf$ is fulfilled. As can be expected, the rate of absorption is proportional to the incident light power and to the number of atoms $N1$ in level 1. The proportionality constant is called the *cross section* σ. As a result, a light beam transmitted through matter is absorbed and its power will exponentially decrease. This is the well-known Lambert--Beer's law for (linear) absorption.

2.5.4
Stimulated Emission of Light

When a light wave is incident on an atom that is already excited, the light cannot be absorbed anymore. But when the "resonance" condition is fulfilled (i.e., when $E2 - E1 = hf$), the atom is set to oscillate, and quickly returns to its ground state, with emission of a new photon or a wave. The new light wave has the

- same frequency
- same direction
- same phase

as the incoming wave.

Hence the new wave adds coherently to the incident wave. In other words, the incident light wave is *amplified*.

The rate of stimulated emission is proportional to

- the incident light power
- the population $N2$ of level 2
- the cross section σ for stimulated emission.

It is worth mentioning that the same symbol σ is used for the cross section of stimulated emission as we did for absorption. Indeed, in 1916, Albert Einstein had already proved that both cross sections are equal.

2.5.5
Amplification

Up to now we have analyzed the three phenomena of absorption, spontaneous emission, and stimulated emission separately, but in a real system, they occur together. Which of the three phenomena will ultimately be dominant depends on the relative values of $N1$ versus $N2$. In thermal equilibrium, there are almost no atoms in level 2, and only absorption takes place. Emission is possible when $N2 > 0$, and hence the atomic system should be *excited* or *pumped* in order to emit light. Then spontaneous and stimulated photons will be emitted, but because, usually, $N2 \ll N1$, the stimulated photons are almost immediately reabsorbed. Hence the system, as a whole, emits only *spontaneous* photons.

If we want to realize amplification, stimulated emission should be the dominant phenomenon. This can only occur when $N2 > N1$. This situation is completely "abnormal," as it is just the opposite of what it is in equilibrium; hence it is called *population inversion*. It is only then that amplification can be realized; of course, it will help if, moreover, the cross section σ and the incident light power also have large values. This had already been predicted by Albert Einstein in 1916, but it was only in 1960 that this was realized experimentally for the first time. Indeed, it turned out to be extremely difficult to produce population inversion.

The process by which population inversion is created is called the *pumping process*.

What will happen when a light beam is incident on a slice of matter in which population inversion is realized? First of all, we can see that – in this scenario – we can neglect the *spontaneous* emission; indeed, spontaneous photons are emitted in all directions and are completely incoherent. Hence they only contribute *noise* to the passing light beam. Stimulated emission, on the other hand, amplifies the incident beam – we thus obtain laser action!

2.5.6
Basic Setup

An atomic system in which a population inversion is realized can amplify an incoming light beam, and hence can act as a laser. However, it turns out that the amplification is rather small – in a typical He–Ne laser, it is only a few percent; so the laser beam will be rather weak. People soon found a solution: when the amplifier is enclosed between two mirrors, the light is reflected back and forth, so that it passes tens or hundreds of times through the amplifier. These mirrors form the laser *resonator* or laser *cavity;* they ultimately shape the (transverse) distribution of the laser beam. By adding these mirrors, which act as a feedback system, we have transformed the light *amplifier* into a light *oscillator* (Figure 2.19), but one continues to speak of a LASER, and not of a LOSER (light oscillation by stimulated emission of radiation).

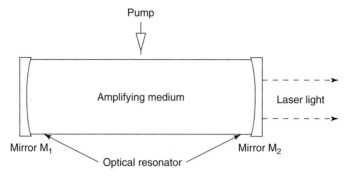

Figure 2.19 The basic elements of a laser: amplifier, pump, and resonator.

Of course, a part of the laser light should be coupled out of the cavity, in order to obtain a useful laser beam. Therefore, one of the mirrors (the *front* mirror) is made partially transparent.

It should be remembered that the state of population in the amplifier is not a "normal" situation, and certainly not one of equilibrium. Hence, one should constantly pump new energy into the amplifier, in order to maintain the population inversion. This finally gives the basic building blocks of each laser: an amplifying medium, a pump system, and a resonator.

2.6
Basic Properties of Laser Light

In Section 2.5.4, we have summarized the main properties of stimulated emission: the new wave has the

- same frequency
- same direction
- same phase

as the incoming wave.

So we would expect that the laser light would be

- monochromatic
- collimated
- coherent.

Although laser light has indeed a better monochromaticity, directionality, and coherence than any other light source, it is not as perfect as one would expect (or hope for). Let us take a closer look at these properties.

2.6.1
Laser Light Has One Direction

In principle, the stimulated photons have the same direction as the stimulating ones; hence, one would expect to find a perfect parallel beam of light out of the laser. However, because of the wave character of light, the beam undergoes diffraction, which implies that it diverges (slightly).

It is possible to estimate the order of magnitude of the divergence angle. A perfect parallel beam (i.e., a beam consisting of perfect plane waves) passing through an aperture of diameter D undergoes a divergence θ of about λ/D (see Figure 2.2), which gives a divergence of about 1 mrad.

2.6.2
Laser Light Is Monochromatic

Stimulated emission can only occur when the frequency f of the incoming beam obeys Planck's law $\Delta E = hf$. This same equation also gives the frequency of the emitted photon. Hence, one would expect laser light to be perfectly monochromatic. However, the laser is not perfectly monochromatic, although it is true that no other light source has better monochromaticity than a laser.

In lasers with a large linewidth (e.g., in Fabry-Pérot type semiconductor lasers), the linewidth is typically a few nanometers, whereas in lasers with small linewidths (as in gas lasers), it goes down to a few picometers.

2.6.3
Laser Light Is Coherent

As the stimulated photon is in phase with the stimulating one, one would expect the laser light to be perfectly coherent; this is, however, not realistic. As already discussed in Section 2.2.4 of this chapter, coherence lengths of gas lasers are typically 10 cm, whereas for semiconductor lasers they are orders of magnitude smaller.

2.6.4
Laser Light Is Intense

Most lasers do not deliver that much power, but their power is concentrated in a very narrow beam; hence the "concentration" of light is enormous. This is described by the concept of *intensity*. Intensity is, by definition, the power divided by the solid angle (see Section 2.3.2). But what can be said about the numerical values? A light bulb with a nominal electrical power of 100 W gives about 10 W optical power in a solid angle of 4π sr, hence its intensity is about 1 W sr^{-1}. On the other hand, a laser beam of about 1 mW (typical for a low-power He–Ne laser) radiates in a solid angle of about 10^{-6} sr; hence it has an intensity of about 10^{+3} W sr^{-1}. Therefore, although the optical output power of the laser is 10 000 times smaller than that of the light bulb, its intensity is about 1000 times stronger. For this reason, lasers beams are far more dangerous than other light beams.

2.7
Conclusions

We have introduced some basic concepts of photonics that are systematically used in the subsequent chapters of this book. This included basic concepts such as interference and diffraction, the *terms* and *units* to measure light, *colors* and their measurements, and finally *lasers*, because of their importance as basic instruments for photonics.

References

1. Milonni, P.W. and Eberly, J.H. (1988) *Lasers*, John Wiley & Sons, Inc.
2. Smith, F.G. and King, T.A. (2000) *Optics and Photonics*, John Wiley & Sons, Inc.
3. Saleh, B.E.A. and Teich, M.C. (1991) *Fundamentals of Photonics*, John Wiley & Sons, Inc.
4. Gasvik, K.J. (1987) *Optical Metrology*, John Wiley & Sons, Inc.
5. Shulman, A.R. (1970) *Optical Data Processing*, John Wiley & Sons, Inc.
6. Klein, M.V. and Furtak, T.E. (1986) *Optics*, 2nd edn, John Wiley & Sons, Inc.
7. Wyszecki, G. and Stiles, W.S. (2000) *Color Science*, John Wiley & Sons, Inc.

3
Basics of Information Theory
Michal Dobes

3.1
Introduction

Information theory has been successfully applied in different areas of signal and image processing. Sampling and the theory of signal transfer are such very well known areas. Other areas where information theory is utilized are image registration and image recognition.

This chapter provides a short overview of information theory and probability. The chapter covers basic topics such as probability, entropy, mutual information, their relationship, and applications. Signal transfer through the information channel is also explained as a part of information theory. Examples of the application of information theory are given in Appendix 3.A.

Section 3.2 describes the basics of probability. Section 3.3 describes entropy, mutual information, and their relationship. Section 3.4 describes transfer of the signal through the information channel. Examples of the application of mutual information for image registration and iris recognition are given in Appendix 3.A.

3.2
Probability

In the theory of probability, we study random events. Probability space is a triplet $(S, \mathbf{F}, p)$, where $\mathbf{F}$ is a class (field) of subsets of the sample space S and p is a measure. A class $\mathbf{F}$ is a field of subsets of sample space S, which has the following properties:

$$A \in \mathbf{F} \Rightarrow S \backslash A \in \mathbf{F} \wedge \{A_i\}_{i=1}^{\infty} \subset \mathbf{F} \Rightarrow \bigcup_{i=1}^{\infty} A_i \in \mathbf{F} \tag{3.1}$$

$\mathbf{F}$ is called σ-algebra of events or a *class of random events* A. The complement of A is $\overline{A} = S \backslash A$ (also $\overline{A} \in \mathbf{F}$). An event A is a set of outcomes from a sample space S. A single-element subset of the sample space is called an *elementary event*. The elementary event is denoted as a. An empty set (subset) is called an

Optical and Digital Image Processing: Fundamentals and Applications, First Edition. Edited by Gabriel Cristóbal, Peter Schelkens, and Hugo Thienpont.
© 2011 Wiley-VCH Verlag GmbH & Co. KGaA. Published 2011 by Wiley-VCH Verlag GmbH & Co. KGaA.

impossible event. Evidently, an empty set $\emptyset \in \mathbf{F}$ and $S \in \mathbf{F}$. A probability measure $p: \mathbf{F} \to \langle 0, 1 \rangle$ is defined on $\mathbf{F}$ such that $p(\emptyset) = 0$, $p(S) = 1$, $p(A) \in \langle 0, 1 \rangle$, and $p\left(\bigcup_{i=1}^{\infty} A_i\right) = \sum_{i=1}^{\infty} p(A_i)$ if $A_1, A_2, \ldots$ are disjunctive events.

For example, 32 cards represent a 32-element space; the event "heart ace" includes the one-element subset, but the event "ace" includes a four-element subset.

A probability p assigned to an event A, denoted as $p(A)$, is a chance that the event A occurs. The probability p is a real number between 0 and 1. An impossible event is the event that never occurs. The probability of an impossible event is $p(A) = 0$. A certain event is the event that always occurs. The probability of a certain event is $p(A) = 1$. The event $\overline{A}$ (not A) is called a *complement* (or opposite) to the event A.

If the event A has m possible outcomes, then the event $\overline{A}$ has $n - m$ possible outcomes from totally n possible outcomes. It is obvious that $p(\overline{A}) = 1 - p(A)$.

3.2.1
Several Events

Two events are called *mutually exclusive* or *disjunctive* if the occurrence of one event excludes the occurrence of the other event (they cannot happen together). Mutual exclusivity of more than two events is understood analogically – if one event happens, any other event cannot happen. The events A_i, A_j are disjunctive if $A_i \cap A_j = \emptyset$ for $i \neq j$.

When tossing a coin, one toss can yield either a head or a tail, but not both simultaneously. As it was already mentioned, the probability of the occurrence of two mutually exclusive events $p(A \cup B)$ is the sum of probabilities of the individual events [1, 2]:

$$p(A \cup B) = p(A) + p(B) \qquad (3.2)$$

Analogically, the probability of the union of an arbitrary number of mutually exclusive events is the sum of probabilities of individual events:

$$p\left(\bigcup_{i}^{\infty} A_i\right) = \sum_{i=1}^{\infty} p(A_i) = p(A_1 \cup A_2 \cup \cdots) = p(A_1) + p(A_2) + \cdots \qquad (3.3)$$

Example

What is the probability that one roll of a dice gives either 1 or 6? When rolling a dice, one roll can yield 1, 2, 3, 4, 5, or 6. The events are mutually exclusive. The probability that the outcome of one roll is either 1 or 6 is the sum of their individual probabilities:

$$p(A_1 \cup A_6) = p(1) + p(6) = \frac{1}{6} + \frac{1}{6} = \frac{1}{3}$$

The relation (Eq. (3.3)) is called a *sum of probabilities rule*. The notation $(A_1 + A_2 + \cdots)$ is sometimes used for the union $(A_1 \cup A_2 \cup \cdots)$ of disjunctive events.

Suppose that the events A and B are not mutually exclusive, then

$$p(A \cup B) = p(A) + p(B) - p(A \cap B) \tag{3.4}$$

Simply said, outcomes that occurred in A and also in B cannot be counted twice and the probability $p(A \cap B)$ must be subtracted from the sum of probabilities.

3.2.2
Conditional Probabilities: Independent and Dependent Events

Let consider several examples and think about the observation space. A box contains five white balls and four black balls. Consider several cases.

First example: Suppose that we draw a ball from a box and after each draw we return the ball back to the box. Clearly, the next draw does not depend on the result of the previous draw. Given the order, what is the probability that we draw a white ball in the first draw and a black ball in the second draw? The events are independent and $p = \frac{5}{9} \cdot \frac{4}{9} = \frac{20}{81}$.

Second example: Let us draw two balls simultaneously (two balls together in one draw): What is the probability that one ball is white and the other is black? The number of ways of drawing k unordered outcomes from n possibilities (i.e., combinations) is $C(n, k) = \frac{n!}{k!(n-k)!}$. The number of possible couples of black balls is the number of combinations $C(4, 2) = \binom{4}{2} = \frac{4!}{2!(4-2)!} = 6$, namely, {1, 2}, {1, 3}, {1, 4}, {2, 3}, {2, 4}, and {3, 4}. The number of white couples is $C(5, 2) = \binom{5}{2} = \frac{5!}{2!(5-2)!} = 10$, namely, {1, 2}, {1, 3}, {1, 4}, {1, 5}, {2, 3}, {2, 4}, {2, 5}, {3, 4}, {3, 5}, and {4, 5}.

Since all possible number of combinations of two balls from nine is $C(9, 2) = \binom{9}{2} = 36$, the number of possible mixed couples is $C(9, 2) - C(4, 2) - C(5, 2) = 36 - 6 - 10 = 20$. Alternatively, we see that each of four black balls can be picked up with any of five white balls, which is $4 \times 5 = 20$ possibilities of drawing two different balls. The probability of drawing one black ball and one white ball is $p = \frac{20}{36}$.

In the third example, consider dependent events: one draws one ball first and its color is known, and then the second ball is drawn. What is the probability that the second ball is black if the first ball was white? The probability that the second ball is black is influenced by the outcome of the first event (A). Probability of drawing a white ball (event A) in the first draw is $p(A) = \frac{5}{9}$. Then, four white and four black balls remain in the box and the probability of drawing a black ball in the second draw (event B) is $p_A(B) = \frac{4}{8} = \frac{1}{2}$. The probability $p_A(B)$ is called a *conditional probability* and is defined as

$$p_A(B) = \frac{p(A \cap B)}{p(A)} \tag{3.5}$$

The event B is conditioned by the occurrence of the event A. The probability of the event $A \cap B$ is $p(A \cap B) = p(A)p_A(B) = \frac{5}{9} \cdot \frac{1}{2} = \frac{5}{18}$.

It follows from the definition that

1) $0 \leq p_A(B) \leq 1$ and $p_A(B) = 1$ if $A \subset B$ or B is a certain event; $p_A(B) = 0$ if A and B are mutually exclusive or B is impossible.
2) If the events $B_1, B_2, l \ldots, B_k$ are mutually exclusive, then $p_A(B_1 \cup B_2 \cup \cdots \cup B_k) = p_A(B_1) + p_A(B_2) + \cdots + p_A(B_k)$.
3) $p_A(\bar{B}) = 1 - p_A(B)$.
4) $p(A \cap B) = p(A)p_A(B) = p(B)p_B(A)$ and $\frac{p(A)}{p_B(A)} = \frac{p(B)}{p_A(B)}$.

It is obvious that sum of events $A \cap B$ and $\bar{A} \cap B$ is equal to the event B. A and $\bar{A}$ are mutually exclusive, $p(A \cap B) = p(A)p_A(B)$, $p(\bar{A} \cap B) = p(\bar{A})p_{\bar{A}}(B)$, and, therefore, $p(B) = p(A \cap B + \bar{A} \cap B) = p(A \cap B) + p(\bar{A} \cap B)$.

Analogically, when the experiment has k mutually exclusive results $A_1, A_2, \ldots, A_k$ and $\bigcup_{i=1}^{k} A_i = S$, that is, $\sum_{i=1}^{k} p(A_i) = 1$, then

$$p(B) = p(A_1)p_{A_1}(B) + p(A_2)p_{A_2}(B) + \cdots + p(A_k)p_{A_k}(B) \qquad (3.6)$$

One useful theorem that is sometimes applied when solving image recognition problems is Bayes' theorem (or Bayes' law). The theorem discusses how posterior probabilities can be expressed in terms of prior probabilities.

Let $A_1, A_2, \ldots, A_k \in \mathbf{F}$, $\{A_i\}_{i=1}^{k}$ are mutually exclusive, and $\bigcup_{i=1}^{k} A_i = S$, that is, $\sum_{i=1}^{k} p(A_i) = 1$. Then

$$p_B(A_i) = \frac{p_{A_i}(B)p(A_i)}{\sum_{j}^{k} p_{A_i}(B)p(A_j)} \qquad (3.7)$$

Usually, the values of $p(A_i)$ and $p_{A_i}(B)$ $i = 1, \ldots, k$ are known in practice, and Bayes' theorem enables one to determine posterior probability of A_i conditioned by B: $p_B(A_i)$.

3.2.3
Random Variable

Let us consider a sample space S and another measurable space (called the *state space*). A real function X that assigns values to the elements a from a sample space S to the state space is called a *random variable*:

$$X : S \rightarrow \Re^1 \ \{\forall x \in \Re\}\{a \in S : X(a) \leq x\} \in \mathbf{F} \qquad (3.8)$$

Simply said, a random variable maps the sample space to real numbers.

3.2.4
Distribution Function

The distribution of the probability of a random variable X can be described by the so-called (cumulative) distribution function:

$$F(x) = p(X \leq x) \qquad (3.9)$$

The distribution function F specifies the probability that the random variable X takes the value less than or equal to x. The distribution function is a nondecreasing, right-continuous function with countable number of discontinuities, $F(-\infty) = 0$ and $F(\infty) = 1$. For every real $x_1 < x_2$, the probability that a random variable takes values from some interval can be expressed through its distribution function:

$$p(x_1 < X \le x_2) = F(x_2) - F(x_1) = p(X \le x_2) - p(X \le x_1) \tag{3.10}$$

A random variable can take values from some countable set of values $\{x_1, x_2, \ldots\}$ or it can take arbitrary value from a given interval of real values. Accordingly, we speak about a discrete random variable or a continuous random variable.

3.2.5
Discrete Distribution

A discrete random variable has a discrete distribution if a probability $p(X = x) \ge 0$ is assigned to every x_i from a countable set $\{x_1, x_2, \ldots\}$ and $\sum_{x_i} p(X = x) = 1$. For a discrete random variable, the distribution function is

$$F(x) = \sum_{x_i \le x} p(X = x_i), \quad -\infty < x < \infty \tag{3.11}$$

A function $p(X = x)$ is called a *probability mass function* and it is used to characterize the distribution at each point x (i.e., x_i, $i = 1, 2, \ldots$).

3.2.6
Continuous Distribution

A random variable has a continuous distribution if there exists a real nonnegative function $f(x)$, so that for every real x the distribution function $F(x)$ can be expressed as

$$F(x) = \int_{-\infty}^{x} f(t)dt, \quad -\infty < x < \infty \tag{3.12}$$

The function $f(x)$ is called a *probability density function (PDF)*. It describes the density of probability at each point in the sample space and at each point where the derivative of the distribution function exists [1]:

$$f(x) = \frac{dF(x)}{dx} \tag{3.13}$$

Obviously, it follows that for every real $x_1 < x_2$,

$$p(x_1 < X \le x_2) = F(x_2) - F(x_1) = \int_{x_1}^{x_2} f(x)dx \tag{3.14}$$

also $\int_{-\infty}^{\infty} f(x)dx = 1$.

3.2.7
Expected Value

A probability distribution of a random variable is fully described by its distribution function $F(x)$. Alternatively, it can be described by a mass probability function $p(X = x)$ in the case of a discrete random variable or by a PDF in the case of a continuous random variable. Random variables are fully determined by such information. In some cases, not all information is required but a simple description would be convenient – some summary expressed as one number (or more numbers) called a *characteristic of random variable*. The typical example is a description using moments.

One of the most important characteristics is the mean, or the first moment of a random variable X. It is called *expected value* or *expectation*. The expected value of a discrete random variable X with probabilities $p(X = x_i)$ for $x_1, x_2, \ldots$ and satisfying $\sum_{x_i} p(X = x) = 1$ is the probability-weighted sum of possible outcomes:

$$E(X) = \sum_i x_i p(x_i) \tag{3.15}$$

An expectation is an average of values when performing independent repetitions of an experiment infinitely many times. It is important to stress that it is not the typical outcome but the mean of random variable values. For example, a roll of a dice can yield a value between 1 and 6, and the expectation is $E(X) = \sum_i p(x_i) x_i = \frac{1}{6}(1 + 2 + \cdots + 6) = 3.5$

In the case of a continuous random variable, the expected value E is an integral. A random variable X with a PDF $f(x)$ has the expected value

$$E(X) = \int_{-\infty}^{\infty} x f(x) dx \tag{3.16}$$

if $\int_{-\infty}^{\infty} x f(x) dx$ exists. In some cases, such as the Cauchy or Lorentzian distribution, the integral does not exist [1]. In fact, the Cauchy distribution has no mean, variance, and moments defined.

3.3
Entropy and Mutual Information

The term *entropy* was introduced in physics in connection with the theory of heat (thermodynamics) in the nineteenth century. Later, Shannon [3] elaborated the information theory where the concept of entropy that was applied on random events, respectively, random variables. Entropy is considered to be a measure of "disorder." The greater the disorder, the greater is the entropy and the less predictable are the events.

3.3.1
Historical Notes

Shannon's considerations are expressed in the following lines. Let us consider an event A that can yield K possible results $a_1, a_2, \ldots, a_K$ with a probability $p(a_i)$ of each individual result a_i. Correspondingly, let us assume a random variable X, yielding K possible values $x_1, x_2, \ldots, x_K$. The possible results are also called *outcomes* or *states*. Shannon decided to quantify the amount of uncertainty by some function $f(K)$. Having K possible results of a random variable X, it is obvious that the amount of uncertainty should grow with K. For $K = 1$, the event is certain, and there is no uncertainty and, therefore, the function should be $f = 0$. The value of f was supposed to be higher for greater k. Consider two independent variables X and Y, where X has K possible results and Y has L possible results. The compound experiment (X, Y) has KL possible results. Shannon suggested that the total uncertainty should be expressed as a sum of uncertainties of individual independent events, that is, $f(KL) = f(K) + f(L)$. This can be satisfied by applying a logarithm $(\log(KL) = \log K + \log L)$. Also, $\log K$ is zero for $K = 1$ and it grows with K. Let the uncertainty of the event be $\log(K)$, and each individual result has a probability $p(x_i) = \frac{1}{K}$; then, each individual result should have the uncertainty $\frac{1}{K} \log(K) = -\frac{1}{K} \log(\frac{1}{K})$. These considerations of Shannon led to the definition of information entropy as a measure of an average uncertainty [2, 3].

3.3.2
Entropy

Entropy is defined as

$$H(X) = -\sum_{i=1}^{K} p(x_i) \log_2 p(x_i) \tag{3.17}$$

It is obvious that the entropy can be expressed in terms of the expected value E of $\log \frac{1}{p(X)}$:

$$H(X) = E \log \left(\frac{1}{p(X)}\right) = \sum_{1}^{K} p(x_i) \log \left(\frac{1}{p(x_i)}\right) \tag{3.18}$$

We stress that the entropy depends on the probability distribution of a random variable X and not on the specific values of X. That is, it is not the values but their probabilities that are important.

3.3.2.1 Some Properties of Entropy

Entropy is nonnegative. When the logarithm is base 2, the result of entropy is in bits, which is the abbreviation for binary units. When the logarithm is base e, the result of entropy is in nats (natural units). The base of the logarithm can be changed without a loss of generality. Let c, d be different logarithm bases; then, because $\log_d p = \log_d c \log_c p$, it follows that

$$H_d(X) = \log_d cH_c(X) \tag{3.19}$$

To facilitate reading of more complex formulas later in the text, log is used instead of $\log_2$ when necessary.

Let $X = 0$ with probability p and $X = 1$ with probability $1 - p$. Then,

$$H(X) = H(p) = -p \log_2 p - (1-p) \log_2(1-p) \tag{3.20}$$

The properties of the function $H(p)$ are illustrated in Figure 3.1. When $p = 0$, there is no uncertainty, $H(p) = -p \log(p) \underset{p \to 0}{=} 0$, and the entropy is 0. When $p = 1$, there is also no uncertainty, and $H(p) = -1 \log(1) = 0$.

It can be seen that if the probability value is close to 0 or 1, the uncertainty of an experiment as well as the entropy are low. The maximal entropy $H(p) = 1$ is for $p = 1/2$; the uncertainty is maximal. It can be proved that if the random variable has K possible states (values) with probabilities $p(x_i) = \frac{1}{K}, i = 1, 2, \ldots, K$, the entropy is maximal:

$$H(p) = \sum_{i=1}^{K} -\frac{1}{K} \log\left(\frac{1}{K}\right) = \log(K) \tag{3.21}$$

3.3.3
Joint Entropy

Consider a joint (compound) experiment where two random variables (X, Y) take place.

To express the joint entropy, we distinguish two cases:

- X and Y are mutually independent, and the result of Y does not depend on the result of X and

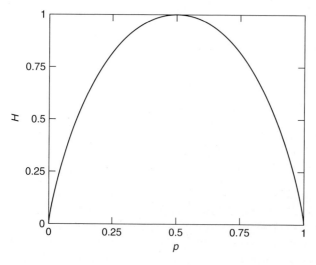

Figure 3.1 Illustration of $H(p) = -p \log_2 p - (1-p) \log_2(1-p)$.

- Y is conditioned by X, and the result of Y is influenced by the result of X (or, similarly, that X is conditioned by Y).

The joint experiment (X, Y) has KL possible results $x_1 y_1, x_2 y_1, \ldots, x_K y_L$, and the joint entropy of such a compound experiment is

$$
\begin{aligned}
H(X, Y) &= \sum_{i,j} -p(x_i y_j) \log p(x_i y_j) \\
&= -p(x_1 y_1) \log p(x_1 y_1) - p(x_2 y_1) \log p(x_2 y_1) -, \cdots, \\
&\quad -p(x_1 y_2) \log p(x_1 y_2) - p(x_2 y_2) \log p(x_2 y_2) -, \cdots, \\
&\quad -p(x_K y_L) \log p(x_K y_L)
\end{aligned}
\tag{3.22}
$$

First, consider the case where X and Y are mutually independent. In the case that X and Y are independent, it holds that $p(x_i, y_j) = p(x_i)p(y_j)$ as discussed in the previous section. By substituting $p(x_i)p(y_j)$ for $p(x_i, y_j)$ into Eq. (3.22) and making simple mathematical arrangements, we obtain

$$
\begin{aligned}
H(X, Y) = &-\sum_{i=1}^{K} p(x_i)[p(y_1) + p(y_2) + \cdots + p(y_L)] \log p(x_i) \\
&-[p(x_1) + p(x_2) + \cdots + p(x_K)] \sum_{j=1}^{L} p(y_j) \log p(y_j)
\end{aligned}
\tag{3.23}
$$

Considering that $[p(y_1) + p(y_2) + \cdots + p(y_L)] = 1$, as well as $[p(x_1) + p(x_2) + \cdots + p(x_k)] = 1$, we conclude that

$$
\begin{aligned}
H(X, Y) &= -\sum_{i=1}^{K} p(x_1) \log p(x_1) - \sum_{j=1}^{L} p(y_j) \log p(y_j) \\
&= H(X) + H(Y)
\end{aligned}
\tag{3.24}
$$

The joint entropy of the experiment composed of two independent random variables is the sum of entropies of individual experiments: $H(X, Y) = H(X) + H(Y)$.

Now, suppose that the result of Y depends on the result of X (X and Y are not independent). As discussed in the previous section, it holds that $p(x_i y_j) = p(x_i) p_{x_i}(y_j)$, where $p_{x_i}(y_j)$ is the conditional probability and $i = 1, 2, \ldots, K, j = 1, 2, \ldots, L$, that is, probability of y_j conditioned by x_i. Then, the joint entropy is

$$
\begin{aligned}
H(X, Y) &= \sum_{i=1}^{K} \sum_{j=1}^{L} -p(x_i) p_{x_i}(y_j) \log p(x_i) p_{x_i}(y_j) \\
&= -p(x_1) p_{x_1}(y_1)[\log p(x_1) + \log p_{x_1}(y_1)] \\
&\quad -p(x_1) p_{x_1}(y_2)[\log p(x_1) + \log p_{x_1}(y_2)] - \cdots
\end{aligned}
$$

$$-p(x_1)p_{x_1}(y_L)[\log p(x_1) + \log p_{x_1}(y_L)] - \cdots$$

$$\vdots$$

$$-p(x_2)p_{x_2}(y_1)[\log p(x_2) + \log p_{x_2}(y_1)] - \cdots$$

$$\vdots$$

$$-p(x_K)p_{x_K}(y_L)[\log p(x_K) + \log p_{x_K}(y_L)] \quad (3.25)$$

We rearrange Eq. (3.25) as

$$H(X, Y) = \sum_{i=1}^{K}\{-p(x_i)p_{x_i}(y_1)[\log p(x_i) + \log p_{x_i}(y_1)]$$
$$-p(x_i)p_{x_i}(y_2)[\log p(x_i) + \log p_{x_i}(y_2)] - \cdots -$$
$$-p(x_i)p_{x_i}(y_L)[\log p(x_i) + \log p_{x_i}(y_L)]\}$$

$$= \sum_{i=1}^{K} -p(x_i)[p_{x_i}(y_1) + p_{x_i}(y_2) + \cdots + p_{x_i}(y_L)]\log p(x_i)$$

$$+ \sum_{i=1}^{K} p(x_i)[-p_{x_i}(y_1)\log p_{x_i}(y_1)$$
$$-p_{x_i}(y_2)\log p_{x_i}(y_2) - \cdots - p_{x_i}(y_L)\log p_{x_i}(y_L)] \quad (3.26)$$

The expression within the square brackets in the last sum in Eq. (3.26) represents the conditional entropy $H_{x_i}(Y)$ conditioned by individual outcomes x_i:

$$H_{x_i}(Y) = -p_{x_i}(y_1)\log p_{x_i}(y_1) - p_{x_i}(y_2)\log p_{x_i}(y_2) - \cdots - p_{x_i}(y_L)\log p_{x_i}(y_L)$$

$$= \sum_{j}^{L} -p_{x_i}(y_j)\log p_{x_i}(y_j) \quad (3.27)$$

Considering that $[p_{x_i}(y_1) + p_{x_i}(y_2) + \cdots + p_{x_i}(y_L)] = 1$, it follows that

$$H(X, Y) = -\sum_{i=1}^{K} p(x_i)\log p(x_i) + \sum_{i=1}^{K} p(x_i)H_{x_i}(Y) = H(X) + H_X(Y) \quad (3.28)$$

The term $H_X(Y)$ is called *conditional entropy*, and it is defined as

$$H_X(Y) = -\sum_{i=1}^{K} p(x_i) \sum_{j}^{L} p_{x_i}(y_j) \log p_{x_i}(y_j) \quad (3.29)$$

It can be easily found that

$$H_X(Y) = -\sum_{i=1}^{K} \sum_{j=1}^{L} p(x_i, y_j) \log p_{x_i}(y_j) \quad (3.30)$$

The conditional entropy satisfies $0 \leq H_X(Y) \leq H(Y)$. In the case that X and Y are independent $H_X(Y) = H(Y)$, and the expression $H(X, Y) = H(X) + H_X(Y)$ reduces to $H(X, Y) = H(X) + H(Y)$. Generally, it holds that $H_X(Y) \neq H_Y(X)$ as it can be seen from the example below.

Example

Let us consider a couple of random variables (X, Y) with the following joint probability distribution p(x, y) (Table 3.1):

The marginal distribution p(X) of the variable X corresponds to the summation of probabilities through individual columns, and it is given by $p(X) = (\frac{5}{8}, \frac{3}{8})$. The marginal distribution p(Y) of the variable Y corresponds to the summation of probabilities through individual rows, and it is given by $p(Y) = (\frac{3}{4}, \frac{1}{4})$. Conditional probabilities corresponding to the above joint distribution are $p_{x_i}(y_j) = \frac{p(x_i, y_j)}{p(x_i)}$:

$$p_{x_1}(y_1) = \frac{\frac{1}{2}}{\frac{5}{8}} = \frac{4}{5}, \quad p_{x_1}(y_2) = \frac{\frac{1}{8}}{\frac{5}{8}} = \frac{1}{5}, \quad p_{x_2}(y_1) = \frac{\frac{1}{4}}{\frac{3}{8}} = \frac{2}{3}, \quad p_{x_2}(y_2) = \frac{\frac{1}{8}}{\frac{3}{8}} = \frac{1}{3}$$

Similarly $p_{y_j}(x_i) = \frac{p(x_i, y_j)}{p(y_j)}$:

$$p_{y_1}(x_1) = \frac{2}{3}, \quad p_{y_1}(x_2) = \frac{1}{3}, \quad p_{y_2}(x_1) = \frac{1}{2}, \quad p_{y_2}(x_2) = \frac{1}{2}$$

$$H(X) = \sum_{i=1}^{2} -p(x_i) \log p(x_i) = -\frac{5}{8} \log \frac{5}{8} - \frac{3}{8} \log \frac{3}{8} = 0.9544 \text{ bit}$$

$$H(Y) = \sum_{j=1}^{2} -p(y_j) \log p(y_j) = -\frac{3}{4} \log \frac{3}{4} - \frac{1}{4} \log \frac{1}{4} = 0.8113 \text{ bit}$$

$$H_Y(X) = -\sum_{j=1}^{2} p(y_j) H_{Y=y_j}(X) = -\sum_{j=1}^{2} p(y_j) \sum_{i=1}^{2} p_{y_j}(x_i) \log p_{y_j}(x_i)$$

$$= -\frac{3}{4} \left(\frac{2}{3} \log \frac{2}{3} + \frac{1}{3} \log \frac{1}{3} \right) - \frac{1}{4} \left(\frac{1}{2} \log \frac{1}{2} + \frac{1}{2} \log \frac{1}{2} \right) = 0.9387 \text{ bit}$$

$$H_X(Y) = -\sum_{i=1}^{2} p(x_i) H_{X=x_i}(Y) = -\sum_{i=1}^{2} p(x_i) \sum_{j=1}^{2} p_{x_i}(y_j) \log p_{x_i}(y_j)$$

$$= -\frac{5}{8} \left(\frac{4}{5} \log \frac{4}{5} + \frac{1}{5} \log \frac{1}{5} \right) - \frac{3}{8} \left(\frac{2}{3} \log \frac{2}{3} + \frac{1}{3} \log \frac{1}{3} \right) = 0.7956 \text{ bit}$$

Table 3.1 Joint probability distribution p(x, y).

p(x, y)	x_1	x_2
y_1	$\frac{1}{2}$	$\frac{1}{4}$
y_2	$\frac{1}{8}$	$\frac{1}{8}$

Obviously, $H_X(Y) \neq H_Y(X)$ as one can see: $H_Y(X) = 0.9387$ bit, $H_X(Y) = 0.7956$ bit.

Finally, we show that $H(X, Y) = H(X) + H_X(Y) = H(Y) + H_Y(X)$.

$$H(X, Y) = \sum_{i=1}^{2} \sum_{j=1}^{2} -p(x_i y_j) \log p(x_i y_j)$$

$$= -\frac{1}{2} \log \frac{1}{2} - \frac{1}{4} \log \frac{1}{4} - \frac{1}{8} \log \frac{1}{8} - \frac{1}{8} \log \frac{1}{8} = \frac{7}{4} = 1.75 \text{ bit}$$

$$H(X, Y) = H(X) + H_X(Y) = 0.9544 + 0.7956 = 1.75 \text{ bit}$$

$$H(X, Y) = H(Y) + H_Y(X) = 0.8113 + 0.9387 = 1.75 \text{ bit}$$

Note, that this example is not a proof, but a plain illustration.

3.3.4
Mutual Information

The entropy describes the amount of information about a random variable itself. In the previous section, we proved that if two experiments or random variables (X, Y) are mutually related, the joint entropy is $H(X, Y) = H(X) + H_X(Y) = H(Y) + H_Y(X)$, where $H_X(Y)$ is the amount of uncertainty of Y when the result of X is known. The more Y depends on the result of X, the smaller is the amount of uncertainty of Y. To resolve how the result of the experiment X lowers the uncertainty of Y, we compute the difference $I(X, Y) = H(Y) - H_X(Y)$. Such a difference is called *mutual information*. Mutual information can be expressed as a difference of entropy itself and its conditional entropy or as a difference of the sum of entropies and their joint entropies:

$$I(X, Y) = H(Y) - H_X(Y)$$
$$I(X, Y) = H(X) - H_Y(X)$$
$$I(X, Y) = H(X) + H(Y) - H(X, Y) \qquad (3.31)$$

The above three expressions are equivalent as we show in the following text. The last expression is frequently used. Often, the joint distribution of a couple of random variables is known, and it is easier to compute the joint entropy $H(X, Y)$ than conditional entropies $H_Y(X)$, or, $H_X(Y)$. To show the equivalence of the above three expressions, we rewrite the last expression as

$$I(X, Y) = H(X) + H(Y) - H(X, Y)$$

$$= \sum_i -p(x_i) \log p(x_i) + \sum_j -p(y_j) \log p(y_j) - \sum_i \sum_j -p(x_i, y_j) \log p(x_i, y_j)$$

$$= \sum_i \sum_j -p(x_i, y_j) \log p(x_i)$$

$$+ \sum_i \sum_j -p(x_i, y_j) \log p(y_j) - \sum_i \sum_j -p(x_i, y_j) \log p(x_i, y_j)$$

$$= \sum_i \sum_j p(x_i, y_j) \log \frac{p(x_i, y_j)}{p(x_i) p(y_j)} \quad (3.32)$$

In Eq. (3.32), we used the fact that $p(x_i) = \sum_j p(x_i, y_j)$, and $p(y_j) = \sum_i p(x_i, y_j)$. Finally, because $p_{y_j}(x_i) = \frac{p(x_i, y_j)}{p(y_j)}$ and $p_{x_i}(y_j) = \frac{p(x_i, y_j)}{p(x_i)}$, we see that

$$I(X, Y) = \sum_i \sum_j p(x_i, y_j) \log \frac{p(x_i, y_j)}{p(x_i) p(y_j)}$$

$$= \sum_i \sum_j -p(x_i, y_j) \log p(x_i) - \sum_i \sum_j -p(x_i, y_j) \log p_{y_j}(x_i)$$

$$= \sum_i -p(x_i) \log p(x_i) - \sum_i \sum_j -p(x_i, y_j) \log p_{y_j}(x_i)$$

$$I(X, Y) = H(Y) - H_X(Y) \quad (3.33)$$

Analogically, we can prove that $I(X, Y) = H(X) - H_Y(X)$. Note that the mutual information is a symmetric quantity $I(X, Y) = I(Y, X)$. The relationship between the mutual information, entropy, and conditional entropy is summarized in Figure 3.2 using Venn diagrams.

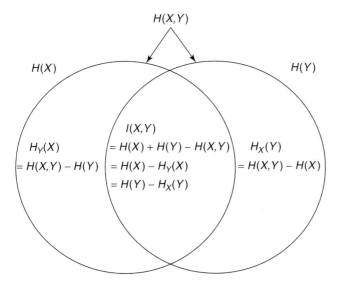

Figure 3.2 Venn diagrams of the relationship between mutual information, entropy, and conditional entropy of two random variables.

3.3.5
Kullback–Leibler Divergence

The first row in Eq. (3.33) corresponds to the definition of mutual information $I(X, Y)$ in terms of Kullback–Leibler distance (KLD). KLD is the distance (or a divergence) between two probability mass functions defined on the same set $\{x_1, x_2, \ldots, x_k\}$. Let us consider two probability mass functions $p_1(x)$ and $p_2(x)$. KLD is defined as

$$D(p_1\|p_2) = \sum_x p_1(x) \log \frac{p_1(x)}{p_2(x)} \quad (3.34)$$

The term divergence is often used instead of the term distance because KLD is not symmetric, generally $D(p_1\|p_2) \neq D(p_2\|p_1)$. KLD is also called *relative entropy* [2]. Mutual information is a special case of KLD defined as the distance between the joint distribution $p(x, y)$ and the product of distributions $p(x)p(y)$:

$$I(X, Y) = \sum_i \sum_j p(x_i, y_j) \log \frac{p(x_i, y_j)}{p(x_i)p(y_j)} \quad (3.35)$$

3.3.6
Other Types of Entropies

Besides the Shannon entropy, other types of diversity measures have been introduced. We present the Rényi entropy as an example of a generalized entropy defined as

$$H^\alpha(X) = \frac{1}{1-\alpha} \log \sum_i^n p_i^\alpha \quad (3.36)$$

where $\alpha \geq 0$ is the order of the Rényi entropy and p_i are the probabilities of $x_1, x_2, \ldots, x_n$. Higher values of α cause the most probable events to be preferred and lower values cause that the events are more equally likely. When $\alpha \to 1$, Rényi entropy converges to Shannon entropy [4, 5].

3.4
Information Channel

An important question in communication is how fast a specific amount of data can be transferred, and in the sequel text we mention the basic types of communication channels. A message containing symbols (values) $x_i \in X, i = 1, 2, \ldots, K$ from the allowable alphabet is coded by a coder. Each symbol x_i of a message that is to be passed through a channel has some probability $p(x_i)$. After the message is passed through an information channel, the received message is decoded as symbols $y_j \in Y, j = 1, 2, \ldots, L$ (Figure 3.3).

If the channel is not influenced by noise or errors, there is one-to-one mapping between the input and output values. In the case of noise, some symbols are not

3.4 Information Channel

Figure 3.3 Information channel.

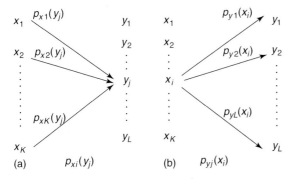

Figure 3.4 Situation in a discrete noisy channel and the conditional probabilities.

transferred correctly. Clearly, $p(x_i)$ and $p(y_j)$ are not sufficient to describe a transfer of information through the channel. A conditional probability $p_{x_i}(y_j)$ (respectively, $p_{y_j}(x_i)$) describes the channel transfer. The conditional probability $p_{x_i}(y_j)$ describes the probability that y_j was received when x_i was sent. Analogically, $p_{y_j}(x_i)$ describes the probability that x_i was received when y_j was sent. The situation is illustrated in Figure 3.4.

3.4.1
Discrete Channel

Discrete channel is characterized by a triplet $\{X, Y, [p_{x_i}(y_j)]\}$, where $x_i \in X$ is a finite set of symbols from the input alphabet, $y_j \in Y$ is a finite set of output symbols from the output alphabet, and $[p_{x_i}(y_j)]$ is a matrix describing conditional probabilities that characterize the properties of the channel transfer.

3.4.2
Channel Capacity

Usually, we are interested in what is the maximum flow of data that can be passed through the channel. The amount of information that passes through the channel is

$$I(X, Y) = H(X) - H_Y(X) = H(Y) - H_X(Y) \tag{3.37}$$

Conditional entropies $H_X(Y)$ and $H_Y(X)$, respectively, give the amount of lost information. A channel rate is defined as

$$C_R = vI(X, Y) \tag{3.38}$$

where $v = 1/t$ is the average speed of transmitting one symbol, and t is the average time of transferring one symbol through the channel. The maximal speed is usually given by the physical properties of the channel and cannot be increased above a certain limit. The other term $I(X, Y)$ can be optimized by a suitable coding. Therefore, the maximum channel capacity is

$$C = v \max_{p(X)} \big[I(X, Y) \big] \tag{3.39}$$

The speed is often omitted when expressing the channel capacity, and Eq. (3.39) is simplified to

$$C = \max_{p(X)} \big[I(X, Y) \big] \tag{3.40}$$

In the case of the noiseless channel that has no information loss, the channel capacity $C = \max_{p(X)} H(X)$. Having the alphabet x_i, $i = 1, 2, \ldots, K$, we already know that the entropy is maximal when all symbols have the same probability $p(x_i) = \frac{1}{K}$, $i = 1, 2, \ldots, K$, and, consequently, the maximal channel capacity can never be greater than $C = \log K$.

If the channel is affected by noise, then the optimal $p(X)$ should be found, and it depends on the matrix of conditional probabilities $[p_{x_i}(y_j)]$. In practice, some types of channels have properties that make the computation more feasible, such as a symmetric channel.

3.4.3
Symmetric Channel

A channel is symmetric with respect to its input if the rows of the matrix $[p_{x_i}(y_j)]$ are the permutation of probabilities $p_1, p_2, \ldots, p_K$. It means that it affects each input symbol by the same manner. If the assumption holds also for the columns of the matrix $[p_{x_i}(y_j)]$, the channel is symmetric with respect to the output. The channel that is symmetric with respect to the input and output is called a *symmetric channel*, for example,

$$[p_{x_i}(y_j)] = \begin{bmatrix} p_1 & p_3 & p_2 \\ p_3 & p_2 & p_1 \\ p_2 & p_1 & p_3 \end{bmatrix} \tag{3.41}$$

The probabilities $p_1, p_2, \ldots, p_K$ in the matrix are denoted as p_j, $j = 1, 2, \ldots, K$, and $K = L$. To find the capacity of a symmetric channel, we must determine the entropy $H(Y)$ and the conditional entropy $H_X(Y)$ in the expression

$$C = \max_{p(Y)} (H(Y) - H_X(Y)) \tag{3.42}$$

The entropy $H(Y)$ is maximal if all symbols y_j, $j = 1, \ldots, L$, appear with the same probability, and $H(Y)$ cannot be greater than

$$H_{\max}(Y) = \log(L) \tag{3.43}$$

The rest of the channel capacity is influenced by the conditional entropy $H_X(Y)$ that can be expressed as

$$H_X(Y) = -\sum_{i=1}^{K} p(x_i) \sum_{j=1}^{L} p_{x_i}(y_j) \log p_{x_i}(y_j) \quad (3.44)$$

The elements $p_{x_i}(y_j)$ in rows and columns of the matrix are permutations of probabilities $(p_1, p_2, \ldots, p_L)$, denoted as p_j $j = 1, \ldots, L$, and $K = L$. Therefore, maximum of distinguishable output symbols is $K = L$. The sum of probabilities $\sum_{i=1}^{K} p(x_i) = 1$, and the conditional entropy simplifies to

$$H_X(Y) = -\sum_{j=1}^{L} p_j \log p_j \quad (3.45)$$

The capacity of the symmetric channel is, therefore,

$$C_{\max} = \log(L) + \sum_{j=1}^{L} p_j \log p_j \quad (3.46)$$

3.4.4
Binary Symmetric Channel

In practice, a binary symmetric channel is often used as a convenient model. A binary symmetric channel has the alphabet of two symbols: 0 and 1. The probability that either the symbol 0 or 1 was transferred correctly is p. The probability of incorrect transfer is $1 - p$, that is, 0 was sent and 1 was received or 1 was sent and 0 was received. The matrix of conditional probabilities is

$$[p_{x_i}(y_j)] = \begin{bmatrix} p & 1-p \\ 1-p & p \end{bmatrix} \quad (3.47)$$

Assuming that log is base 2, the capacity of the symmetric binary channel is

$$C_{\max} = \log(2) + p \log p + (1-p) \log(1-p)$$
$$= 1 + p \log p + (1-p) \log(1-p) \quad (3.48)$$

The capacity is maximal when p is 1 or 0. If $p = \frac{1}{2}$, then the symbols are transferred totally at random and the channel capacity is zero (Figure 3.5).

3.4.5
Gaussian Channel

It is known that the signal composed of many signals with different distributions is close to normal (Gaussian) distribution. This is a consequence of the central limit theorem [6]. Signals that can be modeled by a Gaussian distribution play an important role in practice. A signal with an additive Gaussian noise is, therefore, a suitable model:

$$Y_i = X_i + Z_i \quad (3.49)$$

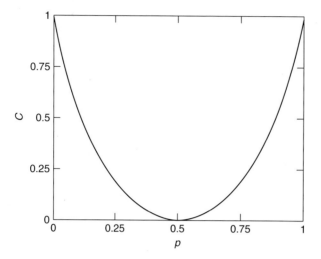

Figure 3.5 The capacity C of a binary symmetric channel depending on p.

X_i is the input, Y_i is an output, and Z_i is the noise. It is assumed that the noise Z is stationary and has a zero mean and a variance $\sigma^2 = N$. (N represents the noise power.) In addition, it is assumed that the signal and noise are statistically independent. Finally, the input power is constraint; having an input code word $(x_1, x_2, \ldots, x_n)$, the average power of an input signal is

$$S \leq \frac{1}{n} \sum_{i=1}^{n} x_i^2 \tag{3.50}$$

The capacity of the Gaussian channel can be computed as $C = \max_{p(x): EX^2 \leq S} I(X, Y)$. Without going into further details, it can be derived [2] that the maximal capacity of a power constrained channel is

$$C = \frac{1}{2} \log\left(1 + \frac{S}{N}\right) \text{ bits} \tag{3.51}$$

The ratio $\frac{S}{N}$ is called *signal-to-noise ratio* and denoted as SNR.

3.5
Conclusion

This chapter has provided an overview of the basics of probability and information theory. The usefulness of the application of information theory has been demonstrated with respect to the application of mutual information for image registration and recognition [10, 11]. Mutual information is able to perceive deep relationships between two signals or images. For example, this enables the registration of images from different sources [7–9], using mutual information as a similarity or quality measure.

The information theory is a rather broad theory that covers different topics related to computer science. For further readings and a deeper understanding of the information theory, we recommend several books [1, 2].

Appendix 3.A: Application of Mutual Information

Image Registration

Mutual information is applied in many image-processing applications. Image registration and recognition are among the most interesting applications. The advantage of mutual information (e.g., over linear correlation) is that it enables to register (align) images from different sources such as magnetic resonance, computer tomography, and positron emission tomography [7–9]. Mutual information was successfully applied for iris recognition [10] and iris registration [11].

Unfortunately, the computation of mutual information is not very straightforward. During the registration, the images are subject to geometric transformations. The transformations result in the coordinate values ceasing to be integers. Subsequent interpolation of intensity values result in the intensity values being real numbers. In a discrete case, the probability that two specific outcomes will occur in an event (such as the intensity $I = i$ will occur in the first image in a position (x, y) and the intensity $J = j$ will occur in the second image in a position (x, y)) is characterized by the joint probability mass function $p(I = i, J = j)$, which is practically represented by a two-dimensional joint histogram. The construction of the joint histogram, which is an important part of the whole computation chain, must be resolved. The nearest neighbor interpolation is generally insufficient. Bilinear interpolation gives one interpolated intensity value. The joint histogram is incremented by one at the position corresponding to this rounded value. More precise updating of the joint histogram called *partial volume interpolation* was suggested by Maes and Collignon [8, 12]. The principle of the partial volume interpolation consists in updating four values in the joint histogram (corresponding to four neighbors in the two-dimensional case) by the computed four weights.

For a thorough description of image registration using mutual information, see Refs [7]–[9]. For a survey of image registration methods, see Ref. [13].

Iris Recognition

A different method of computation of mutual information that is suitable for a smaller amount of data was applied in Refs [10], [11] during the iris registration and recognition. The amount of available data was very less to apply the traditional method of computation of mutual information. Therefore, the joint histogram was divided nonequidistantly and the computation of mutual information was done as if it were in continuous case in accordance with Refs [14]–[16]. Such a solution enables to operate with a lower amount of data, which was just the case with iris images

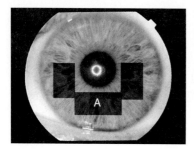 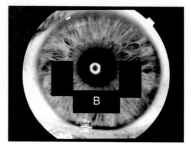

Figure 3.A.1 Iris recognition images A and B and the part of the image (dark) from which the mutual information was computed.

where only a small part of the image was involved in recognition (Figure 3.A.1). Iris images were then recognized based on the specific threshold value of mutual information. The success of the method was over 99%. Iris localization has been described in Ref. [16]. For a survey of current methods in iris recognition, see Ref. [17].

References

1. Papoulis, A. and Pillai, S.U. (2002) *Probability, Random Variables and Stochastic Processes*, 4th edn, McGraw-Hill.
2. Cover, T.M. and Thomas, J.A. (1991) *Elements of Information Theory*, John Wiley & Sons, Ltd., Chichester.
3. Shannon, C.E. and Weaver, W. (1963) *The Mathematical Theory of Communication*, The University of Illinois Press, Urbana, Chicago, London.
4. Karmeshu, J. (2003) *Entropy Measures, Maximum Entropy Principle and Emerging Applications*, Springer-Verlag, New York.
5. Rényi, A. (1996) *Proceedings of the Fourth Berkeley Symposium on Mathematical Statistics and Probability*, vol. 1, University of California Press, Berkeley, CA, pp. 547–561.
6. Feller, W. (1968) *An Introduction to Probability Theory and its Applications*, vol. 1, 3rd edn, John Wiley & Sons, Inc., New York.
7. Viola, P. and Wells, W. (1997) Alignment by maximisation of mutual information. *Int. J. Comput. Vis.*, **24** (2), 137–154.
8. Maes, F. *et al.* (1997) Multimodality image registration by maximization of mutual information. *IEEE Trans. Med. Imaging*, **16** (2), 187–198.
9. Pluim, J.P.W., Maintz, J.B.A., and Viergever, M.A. (2000) Interpolation artefacts in mutual information-based image registration. *Comput. Vis. Image Underst. (Elsevier)*, **77** (9), 211–232.
10. Dobes, M. Jr. *et al.* (2004) Human eye iris recognition using the mutual information. *Optik (Elsevier)*, **115** (9), 399–405.
11. Dobes, M. (2008) Eye registration using the fast mutual information algorithm, Proceedings of the EURASIP, Biosignal.
12. Collignon, A. (1998) Multimodality medical image registration by maximization of mutual information, Ph.D. thesis, Catholic University of Leuven, Leuven, Belgium.
13. Zitova, B. and Flusser, J. (2003) Image registration methods. A survey. *Image Vis. Comput.*, **21**, 977–1000.
14. Darbellay, G.A. (1999) An estimator for the mutual information based on a

criterion for independence. *J. Comput. Stat. Data Anal.*, **32**, 1–17.
15. Darbellay, G.A. and Vajda, I. (1999) Estimation of the information by an adaptive partitioning of the observation space. *IEEE Trans. Inf. Theory*, **45** (4), 1315–1321.
16. Dobes, M. *et al.* (2006) Human eye localization using the modified Hough transform. *Optik (Elsevier)*, **117** (10), 468–473.
17. Daugman, J. (2007) New methods in iris recognition. *IEEE Trans. Syst. Man Cybern.*, **37**, 1167–1175.

4
Fundamentals of Image Processing
Vaclav Hlavac

4.1
Introduction

In this chapter, the fundamentals of digital image processing are briefly summarized to support the chapters that follow.

The concept of a digital image and its properties are first explained. This is followed by an introduction to the Fourier transform, which is representative of a much larger class of linear integral transformations used in digital image processing. Finally, basic linear and nonlinear filtering operators are introduced.

A two-dimensional image is the result of the projection of a three-dimensional scene on the camera sensor or the retina of the eye. The pinhole camera model (also camera obscura) describes the geometry of the projection as illustrated in Figure 4.1.

The image plane has been reflected with respect to the XY plane to avoid negative coordinates. The values x, y, and z are coordinates of the point **X** in a 3D scene. The distance f from the pinhole to the image plane is commonly called the *focal length*. The projected point **u** has coordinates $(-u, -v)$ in the 2D image plane. (u, v) are the coordinates in the reflected image plane, which can easily be derived from similar triangles in Figure 4.1.

$$u = \frac{xf}{z}, \quad v = \frac{yf}{z} \tag{4.1}$$

Hence, the formation of the image on the camera sensor can be described by a simplified physical model. The observed intensity on a camera photo sensor is caused by reflecting part of the energy provided by the illuminant toward the camera photosensitive chip. Consider a simplified configuration with the single-point light source at infinity and exclude the case in which the illuminant shines directly on the photosensitive sensor. The observed intensity in a photosensitive chip pixel is caused by the reflection of the illuminant light by a small surface patch of a 3D scene. The observed light intensity in a sensor pixel depends on (i) the direction from the scene surface patch toward the illuminant, (ii) the direction from the surface patch toward the observer (pixel on the camera photosensitive chip), (iii) the

---

Optical and Digital Image Processing: Fundamentals and Applications, First Edition. Edited by Gabriel Cristóbal, Peter Schelkens, and Hugo Thienpont.
© 2011 Wiley-VCH Verlag GmbH & Co. KGaA. Published 2011 by Wiley-VCH Verlag GmbH & Co. KGaA.

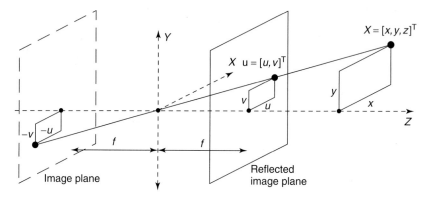

Figure 4.1 The pinhole model.

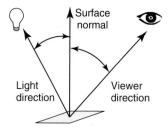

Figure 4.2 Observed image intensity depends on mutual angles between three directions: the direction from the small surface patch to the viewer, the light source, and surface normal.

surface reflection properties, and (iv) the inclination of the surface patch described by the normal vector to the surface (see Figure 4.2). This informally introduced *physical process of image formation* is often and successfully used in *computer graphics* to generate realistic images of scenes for which full 3D models have been created.

Computer vision aims at solving the *inverse task* to the above-described image formation process. This inverse task is very hard and ill posed, because only a single value, the image intensity, is observed (measured by the camera) and at least four parameters should be determined even in the above-mentioned simplified configuration. In most practical cases, computer vision is unable to provide inversion of the image formation process. Instead, the pattern recognition methodology is applied, in which *regions of interest* in the 2D image are sought, outlined, and classified, that is, interpreted with respect to some model aiming at understanding *semantics* of the 3D scene from its 2D projection. This means that the regions of interest in the image are assigned labels corresponding to the objects of interest in the observed scene. This *image interpretation* process depends on the specific application, its semantics, and the a priori knowledge available about it. Image segmentation methods serve this purpose. References [1] and [2] are two digital image processing textbooks of the many that are available.

4.2 Digital Image Representation

An *image* can be modeled by a continuous *image function* of two variables $f(x, y)$ where (x, y) are coordinates in the image plane. If a time sequence of images is taken, then the image function $f(x, y, t)$ depends on three variables, where t represents the time variable. For reasons of simplicity, this chapter restricts itself to gray-level images.

In *discretization*, the image x, y coordinates are sampled in the spatial domain in a chosen grid. Two questions have to be dealt with when sampling in the 2D plane. The first question asks what should be the *distance between the samples*. Naturally, the closer the samples the better is the approximation of the original continuous image function. Simultaneously, dense sampling needs a lot of memory to store the image. However, the well-known known Shannon's sampling theorem provides a simple answer to the question of how dense the grid has to be: the sampling frequency has to be at least twice as high as the highest frequency of interest in the sampled image function or, roughly speaking, the distance between samples ($\approx 1/$sampling frequency) has to be at least half of the size of the smallest detail of interest in the image.

The second question related to discretization in the 2D plane is the *spatial arrangement of samples* (see Figure 4.3). Two sampling grids are typically used: the *square grid* (rectangular grid in a little more general case) and the *hexagonal grid*. Even though the hexagonal grid has an obvious mathematical priority to the square raster, the square grid is used in most cases. The advantage of the hexagonal grid is its isotropy, that is, a pixel has the same Euclidean distance to all its neighbors.

Quantization describes the other needed process to create a digital image. In digital imaging, the luminance values are typically quantized utilizing a limited set of values. Most often, regular quantization, that is, division of the intensity range to equal sized bins, is applied. For photographs, this ranges between 256 (8-bit) and 1024 (10-bit) values for high dynamic range imagery. The reason for this limited number of discretization levels is the limited local contrast sensitivity of the human eye. It is important though to remark that as image resolution increases, the amount of discretization levels will also need to increase in order to accommodate local fitting of human perception. Therefore, accommodating high dynamic range representations has received significant attention lately.

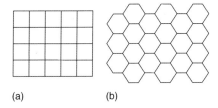

(a)   (b)

Figure 4.3 Commonly used sampling grid configurations: (a) square and (b) hexagonal.

The rectangular (square) grid has one obvious advantage – to represent a related image as a matrix. The row, column coordinates correspond to x, y spatial coordinates in the image and the value of the matrix element gives the value of the image function, that is, brightness. Each element of a matrix representing the image is called a *pixel*, which is the acronym from the term *picture element*.

4.2.1
Topological and Metric Properties of Images

Some metric and topological properties of the digital image are not very intuitive. The distance between two points with coordinates (i, j) and (h, k) may be defined in several different ways. We distinguish, for example, the Euclidean distance D_E, the "city block" distance D_4, and the "chessboard" distance D_8.

The well-known *Euclidean distance* D_E is defined as

$$D_E((i,j),(h,k)) = \sqrt{(i-h)^2 + (j-k)^2} \tag{4.2}$$

The distance between two points can also be expressed as the minimum number of elementary steps in the digital grid needed to move from the starting point to the end point. If only horizontal and vertical moves are allowed, the distance D_4 is obtained. The distance D_4 is also called *city block* distance:

$$D_4((i,j),(h,k)) = |i-h| + |j-k| \tag{4.3}$$

If moves in diagonal directions are allowed in the discretization grid, we obtain the distance D_8, often called the *chessboard* or *Chebyshev distance*:

$$D_8((i,j),(h,k)) = \max\{|i-h|, |j-k|\} \tag{4.4}$$

Pixel *adjacency* is an important topological concept in digital images. Any two pixels (p, q) are called *4-neighbors* if they have a distance $D_4(p, q) = 1$. Analogously, *8-neighbors* are two pixels with $D_8(p, q) = 1$. Both 4-neighbors and 8-neighbors are illustrated in Figure 4.4.

Pixel adjacency leads to another important concept: the *region*, which is a set consisting of several adjacent pixels (a connected set in set theory). More intuitively, a *path* or curve from pixel P to pixel Q can be defined as a sequence of points $A_1, A_2, \ldots, A_n$, where $A_1 = P$, $A_n = Q$, and A_{i+1} is a neighbor of A_i, $i = 1, \ldots, n-1$. A region is a set of pixels in which there is a path between any pair of its pixels, all of whose pixels also belong to the set.

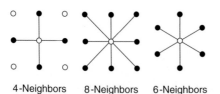

4-Neighbors   8-Neighbors   6-Neighbors

Figure 4.4 Illustration of 4-neighborhood, 8-neighborhood, 6-neighborhood (on a hexagonal grid).

If there is a path between any two pixels in the set of pixels, then these pixels are called *contiguous*. We can also say that a region is a set of pixels in which each pair of pixels is contiguous. The relation "to be contiguous" is reflexive, symmetric, and transitive and thus induces a decomposition of the set (the image, in our case) into equivalence classes (regions). Figure 4.5 illustrates a binary image decomposed by the relation "contiguous" into three regions. The interior of the regions is filled with different patterns to ease visualization.

Assume that R_i are disjoint regions in the image, which are induced by the relation "to be contiguous". Assume also that, to avoid special cases, the regions do not touch the image bounds. Let R be the union of all regions R_i; R^C be the set complement of R with respect to the image. The subset of R^C that is contiguous with the image bounds is called the *background*, and the remainder of the complement R^C is called *holes*. A region is called *simple contiguous* if there are no holes in it. Equivalently, the complement of a simply contiguous region is contiguous.

The concept of "region" explores only the property "to be contiguous". In practical situations, secondary properties are often attached to regions that rely on image data interpretation. Such regions to which image features and semantical concepts are attached are called *objects*. The process that identifies such objects in an image is called *segmentation* and is covered in Chapter 29.

Regions can be divided into two classes: *convex regions* and *nonconvex regions*. This property is used to describe shapes qualitatively. If any two points within a region are connected by a straight line segment, and the whole line lies within the region, then this region is convex. Regions that are not convex are called *nonconvex*. Convexity and nonconvexity are illustrated in Figure 4.6.

Convexity leads to another natural concept: *convex hull*, which is the smallest convex region containing a nonconvex region, Figure 4.7. Imagine a thin rubber band pulled around the object; the shape of the rubber band provides the convex hull of the object.

An object can be represented by a collection of its topological components. The set inside the convex hull that does not belong to an object is called the *deficit of*

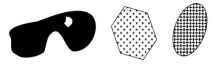

Figure 4.5 Decomposition of the image into regions as a result of the "to be contiguous" relation.

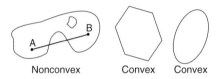

Figure 4.6 Convex and nonconvex regions.

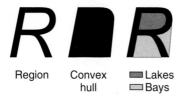

Figure 4.7 Region described qualitatively using topological components with respect to a convex hull; illustration of lakes and bays.

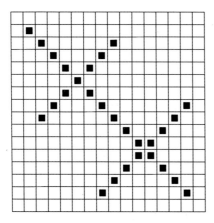

Figure 4.8 4-Neighborhood and 8-neighborhood lead to paradoxes. It can happen that obviously crossing lines do not have a common pixel.

convexity. This can be split into two subsets: *lakes* (dark gray) are fully surrounded by the object; and *bays* (light gray) are contiguous with the border of the convex hull of the object.

Neighborhood and contiguity definitions on the square grid create interesting paradoxes. Figure 4.8 shows two digital line segments with 45° slope. If 4-connectivity is used, the lines are not contiguous at each of their points. An even worse conflict with intuitive understanding of line properties is also illustrated; two perpendicular lines intersect in one case (upper left intersection) and do not intersect in the other (lower right), as they do not have any common point (i.e., their set intersection is empty).

The concept of a region introduces another important concept in image analysis–*border* (also boundary). The border of a region R is the set of pixels within the region that have one or more neighbors outside R; that is, it corresponds to a set of pixels at the bound of the region. This definition of border is sometimes referred to as the *inner border* to distinguish it from the *outer border*, that is, the border of the background (i.e., its complement) of the region. Inner and outer borders are illustrated in Figure 4.9.

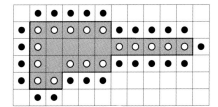

Figure 4.9 Borders, 4-neighborhood case. Inner borders of a region are shown as white circles and the outer borders are shown as black circles.

An *edge* is a local property of a pixel and its immediate neighborhood. The edge is the gradient of the image function, that is, a vector given, for example, by a magnitude and direction [3].

Note that a difference exists between a border and an edge. The border is a global concept related to a region, whereas an edge expresses local properties of an image function. Borders and edges are related as well. One possibility of finding boundaries is chaining the significant edges (points with high gradient of the image function).

4.2.2
Brightness Histogram

The image *brightness histogram* is the most prominent example of a global image characteristic. Besides being intuitive, the histogram is often used in digital photography to judge image illumination conditions. The histogram opens the door to probabilistic methods for characterizing images. Suppose that the brightness value in each pixel is a random variable independent of values in other pixels. It then makes sense to ask the following question: what is the probability that a certain pixel will have the brightness value z? Here the classical approach is considered, in which the probability is estimated as a ratio between the number of occurrences of a particular phenomenon and the number of all possible occurrences. The histogram of an image with L gray levels is represented by a one-dimensional array with L elements. The probability density function $p_1(z; x, y)$ indicates the probability that pixel (x, y) has brightness z. Dependence on the position of the pixel is not of interest here. Entries of the histogram are simply obtained as the number of pixels in the image of value z, $0 \leq z \leq L$. The histogram is an estimate of the probability density function $p_1(z)$. The histogram is illustrated in Figure 4.10. The gray-scale image on the left in Figure 4.10 is cropped from the image of Prague Astronomic Clock, which was finished in the year 1410 and has since been serving on the facade of the Prague Old Town Hall. The low resolution, 256 × 256, of the image is chosen deliberately to better visualize phenomena explicated in this section.

At this point, the main concepts needed to discuss digital images have been introduced and the following sections discuss how images can be processed.

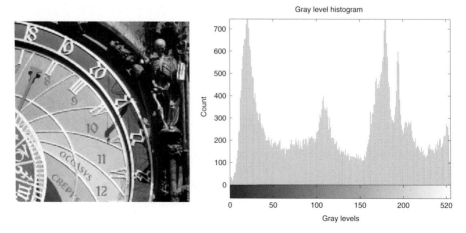

Figure 4.10 The intensity image and its histogram.

4.3
Image Filtering Paradigm

The filtering paradigm has a long history in mathematics and technology. Several disciplines benefited from the introduction of "dynamics," that is, modeling of transitional phenomena using differential equations. The active elements in such systems can be generically addressed as *filters*.

In image processing, the image is represented as a two-dimensional function. A filter modifies the input image and produces the output image using a transition process. Both linear and nonlinear filters are used for image processing. The image filtering paradigm allows expression of the image discretization process, noise suppression, finding of informative places in the image based on detection of sharp changes in image intensity (edges), and so on.

In this context, it is important to introduce the concept of linearity since it allows to express the filters in the time domain using convolution. The notion of *linearity*, which relates to vector (linear) spaces where commonly matrix algebra is used, is first recalled. Linearity also concerns more general elements of vector spaces, for instance, functions. The *linear combination* permits the expression of a new element of a vector space as a sum of known elements multiplied by coefficients (scalars, usually real numbers). A general linear combination of two vectors **x** and **y** can be written as $a\mathbf{x} + b\mathbf{y}$, where a and b are scalars.

Consider a mapping $\mathcal{L}$ between two linear spaces. The mapping is called *additive* if $\mathcal{L}(\mathbf{x} + \mathbf{y}) = \mathcal{L}\mathbf{x} + \mathcal{L}\mathbf{y}$ and *homogeneous* if $\mathcal{L}(a\mathbf{x}) = a\mathcal{L}\mathbf{x}$ for any scalar a. The mapping $\mathcal{L}$ is *linear* if it is additive and homogeneous. Equivalently, a linear mapping satisfies $\mathcal{L}(a\mathbf{x} + b\mathbf{y}) = a\mathcal{L}\mathbf{x} + b\mathcal{L}\mathbf{y}$ for all vectors **x**, **y** and scalars a, b, that is, it preserves linear combinations.

To introduce the concept of *convolution*, let us assume 1D time-dependent signals for reasons of simplicity. The convolution is an integral that expresses the amount of overlap of one function $f(t)$ as it is shifted over another function $h(t)$. A 1D convolution $f * h$ of functions f, h over a finite range $[0, t]$ is given by

$$(f * h)(t) \equiv \int_{-\infty}^{\infty} f(t) h(t - \tau) \, d\tau = \int_{-\infty}^{\infty} f(t - \tau) h(t) \, d\tau \tag{4.5}$$

The convolution integral has bounds $-\infty, \infty$. Here, we can restrict ourselves to the interval $[0, t]$, because we assume zero values for negative coordinates.

$$(f * h)(t) \equiv \int_{0}^{t} f(t) h(t - \tau) \, d\tau \tag{4.6}$$

Let f, g, and h be functions and a be a scalar constant. Convolution satisfies the following properties:

$$f * h = h * f \tag{4.7}$$
$$f * (g * h) = (f * g) * h \tag{4.8}$$
$$f * (g + h) = (f * g) + (f * h) \tag{4.9}$$
$$a(f * g) = (af) * g = f * (ag) \tag{4.10}$$

The derivative of a convolution is given by

$$\frac{d}{dx}(f * h) = \frac{df}{dx} * h = f * \frac{dh}{dx} \tag{4.11}$$

The convolution can be generalized to higher dimensional functions. The convolution of 2D functions f and h is denoted by $f * h$, and is defined by the integral

$$(f * h)(x, y) = \int_{-\infty}^{\infty} \int_{-\infty}^{\infty} f(a, b) h(x - a, y - b) \, da \, db \tag{4.12}$$

$$= \int_{-\infty}^{\infty} \int_{-\infty}^{\infty} f(x - a, y - b) h(a, b) \, da \, db \tag{4.13}$$

$$= (h * f)(x, y)$$

In digital image analysis, the *discrete convolution* is expressed using sums instead of integrals. A digital image has a limited domain on the image plane. However, the limited domain does not prevent the use of convolutions as their results outside the image domain are zero. The convolution expresses a linear filtering process using the filter h; linear filtering is often used in local image preprocessing and image restoration.

Linear operations calculate the resulting value in the output image pixel $g(i, j)$ as a linear combination of image intensities in a local neighborhood $\mathcal{O}$ of the pixel $f(i, j)$ in the input image. The contribution of the pixels in the neighborhood $\mathcal{O}$ is weighted by coefficients h:

$$f(i,j) = \sum_{(m,n) \in \mathcal{O}} h(i - m, j - n) g(m, n) \tag{4.14}$$

Equation (4.14) is equivalent to discrete convolution with the kernel h, which is called a *convolution mask*. Rectangular neighborhoods $\mathcal{O}$ are often used with an odd number of pixels in rows and columns, enabling specification of the central pixel of the neighborhood.

4.4
Frequency Domain

It has already been stated that commonly used representation for 1D signals is the *time domain*, and for image functions (2D signals), it is the *spatial domain* (pixels). There is another popular alternative to these representations, the *frequency domain* (frequency spectra), which transfers signals, images, and filters into another representation. This is one-to-one mapping in theory. Mathematics provides several linear integral transformations for this purpose. We constrain ourselves to Fourier transform because it is fundamental, relatively simple, intuitive, and represents the core idea of other transforms too.

In the frequency domain representation, the image is expressed as a linear combination of basis functions of a linear integral transform. For instance, the Fourier transform uses sines and cosines as basis functions. If linear operations are involved in the spatial domain (an important example of such a linear operation is the already discussed convolution), then there is a one-to-one mapping between the spatial and frequency representations of the image. Advanced signal/image processing goes beyond linear operations, and these nonlinear image processing techniques are mainly used in the spatial domain.

Image filtering is the usual application of a linear integral transform in image processing. Filtering can be performed in either the spatial or the frequency domains, as illustrated in Figure 4.11. In the frequency domain, filtering can be seen as boosting or attenuating specific frequencies.

4.4.1
1D Fourier Transform

We begin by reviewing the simpler 1D Fourier transform, and discuss 2D Fourier transform in the next section. The 1D Fourier transform F transforms a function $f(t)$ (e.g., dependent on time) into a frequency domain representation, $\mathcal{F}\{f(t)\} = F(\xi)$, where $\xi [\text{Hz} = \text{s}^{-1}]$ is a frequency and $2\pi\xi \, [\text{s}^{-1}]$ is an angular frequency. The complex function F is called the *(complex) frequency spectrum* in which it is easy to visualize relative proportions of different frequencies. For instance, the sine

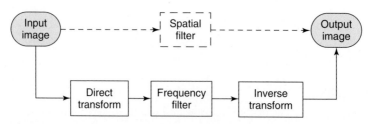

Figure 4.11 Two ways of equivalently processing images using linear operations, in the spatial and in the frequency domains.

wave has a simple spectrum consisting of a single spike for positive frequencies, indicating that only a single frequency is present.

Let i be the usual imaginary unit. The continuous Fourier transform $\mathcal{F}$ is given by

$$\mathcal{F}\{f(t)\} = F(\xi) = \int_{-\infty}^{\infty} f(t)\, e^{-2\pi i \xi t}\, dt \tag{4.15}$$

The inverse Fourier transform $\mathcal{F}^{-1}$ is then

$$\mathcal{F}^{-1}\{F(\xi)\} = f(t) = \int_{-\infty}^{\infty} F(\xi)\, e^{2\pi i \xi t}\, d\xi \tag{4.16}$$

The conditions for the existence of the Fourier spectrum of a function f are as follows

1) $\int_{-\infty}^{\infty} |f(t)|\, dt < \infty$, that is, f has to decrease faster than an exponential curve.
2) f can only have a finite number of discontinuities in any finite interval.

The Fourier transform always exists for digital signals (including images) as they are bounded and have a finite number of discontinuities. It will be seen later that if we use the Fourier transform for images, we have to adopt the assumption that images are periodic. The fact that images are not typically periodic presents problems that are discussed later.

To understand what Eq. (4.15) means, it is useful to express the inverse Fourier transform as a Riemannian sum:

$$f(t) \doteq (\cdots + F(\xi_0)\, e^{2\pi i \xi_0 t} + F(\xi_1)\, e^{2\pi i \xi_1 t} + \cdots)\, \Delta\xi \tag{4.17}$$

where $\Delta\xi = \xi_{k+1} - \xi_k$ for all k. The inverse formula shows that any 1D function can be decomposed as a weighted sum (integral) of many different complex exponentials. These exponentials can be separated into sines and cosines (also called *harmonic functions*) because $e^{i\omega} = \cos\omega + i\sin\omega$. The decomposition of $f(t)$ into sines and cosines starts with some basic frequency ξ_0. Other sines and cosines have frequencies obtained by multiplying ξ_0 by increasing natural numbers. The coefficients $F(\xi_k)$ are complex numbers in general and give both the magnitude and phase of the elementary waves.

The Fourier transform exhibits predictable symmetries. Recall the notion of even, odd, and conjugate symmetric functions, illustrated in Table 4.1.

Note that any 1D function $f(t)$ shape can always be decomposed into its even and odd parts $f_e(t)$ and $f_o(t)$

$$f_e(t) = \frac{f(t) + f(-t)}{2}, \quad f_o(t) = \frac{f(t) - f(-t)}{2} \tag{4.18}$$

The ability to form a function from its even and odd parts is illustrated in Figure 4.12.

The symmetries of the Fourier transform and its values are summarized in Table 4.2 (without proof; for details, see Ref. [4]).

Table 4.3 summarizes some elementary properties of the transform, easily obtained by manipulation with the definitions of Eq. (4.15).

Table 4.1 Concepts of even, odd, and conjugate symmetric functions (denoted by a superscript *).

Even	$f(t) = f(-t)$	
Odd	$f(t) = -f(-t)$	
Conjugate symmetric	$f(\xi) = f^*(-\xi)$	$f(\ 5) = 4 + 7i$ $f(-5) = 4 - 7i$

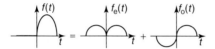

Figure 4.12 Any 1D function can be decomposed into its even and odd parts.

Table 4.2 Fourier transform symmetries if $f(t)$ is a real function.

Real $f(t)$	Values of $F(\xi)$	Symmetry of $F(\xi)$
General	Complex	Conjugate symmetric
Even	Only real	Even
Odd	Only imaginary	Odd

Table 4.3 Properties of the Fourier transform.

Property	$f(t)$	$F(\xi)$
Linearity	$af_1(t) + bf_2(t)$	$a\ F_1(\xi) + b\ F_2(\xi)$
Duality	$F(t)$	$f(-\xi)$
Convolution	$(f * g)(t)$	$F(\xi)\ G(\xi)$
Product	$f(t)\ g(t)$	$(F * G)(\xi)$
Time shift	$f(t - t_0)$	$e^{-2\pi i \xi t_0} F(\xi)$
Frequency shift	$e^{2\pi i \xi_0 t} f(t)$	$F(\xi - \xi_0)$
Differentiation	$\frac{df(t)}{dt}$	$2\pi i \xi\ F(\xi)$
Multiplication by t	$t\ f(t)$	$\frac{i}{2\pi} \frac{dF(\xi)}{d\xi}$
Time scaling	$f(a\ t)$	$\frac{1}{|a|} F(\xi/a)$

Some other properties are related to areas under the function f or its Fourier representation F. The DC offset (DC from direct current) is $F(0)$ and is given by the area under the function f:

$$F(0) = \int_{-\infty}^{\infty} f(t)\, dt \qquad (4.19)$$

and a symmetric property holds for the inverse formula. The value of $f(0)$ is the area under the frequency spectrum $F(\xi)$,

$$f(0) = \int_{-\infty}^{\infty} F(\xi)\, d\xi \qquad (4.20)$$

Parseval's theorem equates the area under the squared magnitude of the frequency spectrum and squared function $f(t)$. It can be interpreted as saying that the signal "energy" in the time domain is equal to the "energy" in the frequency domain. Parseval's theorem is often used in physics and engineering as another form of a general physical law of energy conservation. The theorem states (for a real function f, which is our case for images, the absolute value can be omitted)

$$\int_{-\infty}^{\infty} |f(t)|^2\, dt = \int_{-\infty}^{\infty} |F(\xi)|^2\, d\xi \qquad (4.21)$$

Figures 4.13, 4.14, and 4.15 show some properties of Fourier transforms of simple signals.

Let Re c denote the real part of a complex number c and let Im c be its imaginary part. The formulas describing four function spectrum definitions are as follows:

$$\begin{aligned}
\text{Complex spectrum} \quad & F(\xi) = \text{Re}\left(F(\xi)\right) + i\,\text{Im}\left(F(\xi)\right) \\
\text{Amplitude spectrum} \quad & |F(\xi)| = \sqrt{\text{Re}\left(F^2(\xi)\right) + \text{Im}\left(F^2(\xi)\right)} \\
\text{Phase spectrum} \quad & \phi(\xi) = \arctan\left(\text{Im}\left(F(\xi)\right)/\text{Re}\left(F(\xi)\right)\right) \quad \text{if defined} \\
\text{Power spectrum} \quad & P(\xi) = |F(\xi)|^2 = \text{Re}\left(F^2(\xi)\right) + \text{Im}\left(F^2(\xi)\right)
\end{aligned} \qquad (4.22)$$

It can be seen from Figures 4.14 and 4.15 that time signals of short duration or quick changes have wide frequency spectra and vice versa. This is a manifestation

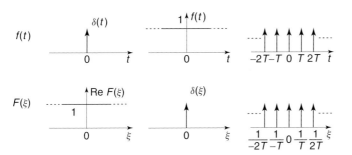

Figure 4.13 1D Fourier transform of the Dirac pulse, constant-value Dirac pulse, and infinite sequence of Dirac pulses.

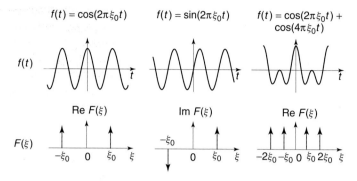

Figure 4.14 1D Fourier transform of the sine, cosine, and mixture of two cosines with different frequencies.

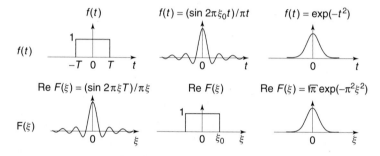

Figure 4.15 1D Fourier transform of the idealized rectangular pulse of length $2T$ in the time domain gives the spectrum $(2 \cos 2\pi \xi T)/\xi$. Symmetrically, the idealized rectangular spectrum corresponds to an input signal of the form $(2 \cos 2\pi \xi_0 t)/t$. The right column shows that a Gaussian pulse has the same form as its Fourier spectrum.

of the *uncertainty principle*, which states that it is impossible to have a signal that is arbitrarily narrow in both time and frequency domains (see Chapter 7 for more details).

If we need to process a *nonstationary signal*, one option is to divide it into smaller pieces (often called *windows*) and assume that outside these windows the signal is periodic. This approach is called the *short time Fourier transformation* (STFT) and was first introduced by Gabor in 1946. The STFT has been used in many disciplines, such as speech recognition. Unfortunately, mere cutting of the signal by nonoverlapping rectangular windows is not good as it introduces discontinuities that induce wide bandwidth in the frequency domain. This is why the signal at the bounds of the local window is smoothly damped to zero by, for example, a Gaussian or Hamming window (see Chapter 5).

The Fourier spectrum expresses global properties of the signal (as information of the speed of signal changes), but it does not reveal the time instant in which such a change appears. On the other hand, the time domain precisely represents as to what happens at certain instants but does not represent global properties of the signal. There are two ways to step toward having a little of both global frequency properties and localization. The first is the STFT, and the second is the use of different basis functions in the linear integral transformation which are less regular than sines and cosines. The wavelet transformation is one example (see Chapter 7).

Computers deal with discrete signals: the discrete signal $f(n)$, $n = 0 \ldots N-1$, is obtained by equally spaced samples from a continuous function f. The discrete Fourier transform (DFT) is defined as

$$F(k) = \frac{1}{N} \sum_{n=0}^{N-1} f(n) \exp\left(-2\pi i \frac{nk}{N}\right) \tag{4.23}$$

and its inverse is defined as

$$f(n) = \sum_{k=0}^{N-1} F(k) \exp\left(2\pi i \frac{nk}{N}\right) \tag{4.24}$$

The spectrum $F(k)$ is periodic with period N.

The DFT, if computed from its definition for the samples discretized into N samples, see Eqs (4.23) and (4.24), has time complexity $\mathcal{O}(N^2)$. The result can be calculated much faster if the fast Fourier transformation (FFT) algorithm is used. This algorithm was proposed in the early 1940s, and depends on the number of samples used to represent a signal being a power of two. The basic trick is that a DFT of length N can be expressed as a sum of two DFTs of length $N/2$, consisting of odd or even samples. This scheme permits the calculation of intermediate results in a clever way. The time complexity of the FFT is $\mathcal{O}(N \log N)$ (all signal processing textbooks provide details).

4.4.2
2D Fourier Transform

The 1D Fourier transform can be easily generalized to 2D [4]. An image f is a function of two coordinates (x, y) in a plane. The *2D Fourier transform* also uses harmonic functions for spectral decomposition. The 2D Fourier transform for the continuous image f is defined by the following integral:

$$F(u, v) = \int_{-\infty}^{\infty} \int_{-\infty}^{\infty} f(x, y) e^{-2\pi i(xu+yv)} \, dx \, dy \tag{4.25}$$

and its inverse transform is defined by

$$f(x, y) = \int_{-\infty}^{\infty} \int_{-\infty}^{\infty} F(u, v) e^{2\pi i(xu+yv)} \, du \, dv \tag{4.26}$$

Parameters (u, v) are called *spatial frequencies*. The function f on the left-hand side of Eq. (4.26) can be interpreted analogously to the 1D case (see Eq. (4.17)), that is, as a linear combination of basis functions, simple periodic patterns $e^{2\pi i(xu+yv)}$. The real and imaginary components of the pattern are cosine and sine functions. The complex spectrum $F(u, v)$ is a weight function that represents the influence of the elementary patterns.

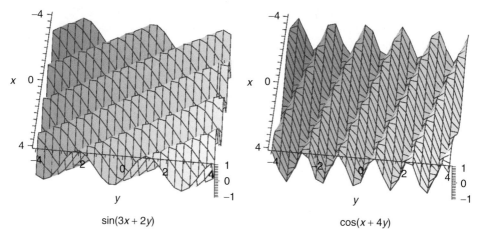

Figure 4.16 Illustration of Fourier transform basis functions.

Let us illustrate the 2D Fourier basis functions pictorially. Two examples are shown in Figure 4.16. The shape of the basis functions invokes the analogy with corrugated iron, of course, with varying frequency and changing direction.

More complicated 2D functions can be approximated after linearly combining basis functions. This is the essence of the 2D Fourier transformation from the practitioner's point of view. The effect of linear combination is illustrated in Figure 4.17.

Equation (4.25) can be abbreviated to

$$\mathcal{F}\{f(x,y)\} = F(u,v)$$

From the image processing point of view, the following properties (corresponding to the 1D case) are easily derived:

– Linearity:

$$\mathcal{F}\{a f_1(x,y) + b f_2(x,y)\} = a\, F_1(u,v) + b\, F_2(u,v) \qquad (4.27)$$

– Shift of the origin in the image domain:

$$\mathcal{F}\{f(x-a, y-b)\} = F(u,v)\, e^{-2\pi i(au+bv)} \qquad (4.28)$$

– Shift of the origin in the frequency domain:

$$\mathcal{F}\{f(x,y)\, e^{2\pi i(u_0 x + v_0 y)}\} = F(u - u_0, v - v_0) \qquad (4.29)$$

– If $f(x,y)$ is real valued then

$$F(-u, -v) = F^*(u,v) \qquad (4.30)$$

The image function is always real valued and we can thus use the results of its Fourier transform in the first quadrant, that is, $u \geq 0$, $v \geq 0$, without loss of generality. If, in addition, the image function has the property $f(x,y) = f(-x,-y)$, then the result of the Fourier transform $F(u,v)$ is a real function.

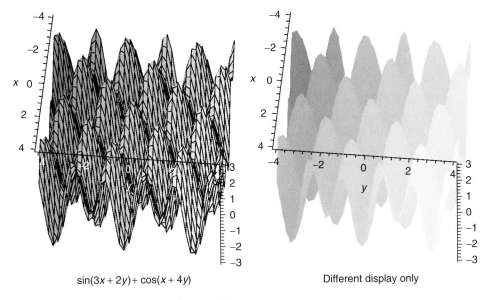

Figure 4.17 A function of any complexity can be represented as a linear combination of the Fourier transform basis functions.

- Duality of the convolution: Convolution, Eq. (4.12), and its Fourier transform are related by

$$\mathcal{F}\{(f * h)(x, y)\} = F(u, v)\, H(u, v)$$
$$\mathcal{F}\{f(x, y)\, h(x, y)\} = (F * H)(u, v) \qquad (4.31)$$

This is the *convolution theorem*.

The 2D Fourier transform can be used for discrete images too: integration is changed to summation in the respective equations. The discrete 2D Fourier transform (spectrum) is defined as

$$F(u, v) = \frac{1}{MN} \sum_{m=0}^{M-1} \sum_{n=0}^{N-1} f(m, n)\, \exp\left[-2\pi i \left(\frac{mu}{M} + \frac{nv}{N}\right)\right]$$
$$u = 0, 1, \ldots, M-1; \quad v = 0, 1, \ldots, N-1 \qquad (4.32)$$

and the inverse Fourier transform is given by

$$f(m, n) = \sum_{u=0}^{M-1} \sum_{v=0}^{N-1} F(u, v)\, \exp\left[2\pi i \left(\frac{mu}{M} + \frac{nv}{N}\right)\right]$$
$$m = 0, 1, \ldots, M-1; \quad n = 0, 1, \ldots, N-1 \qquad (4.33)$$

Considering implementation of the DFT, note that Eq. (4.32) can be modified to

$$F(u,v) = \frac{1}{M} \sum_{m=0}^{M-1} \left[\frac{1}{N} \sum_{n=0}^{N-1} \exp\left(\frac{-2\pi i n v}{N}\right) f(m,n) \right] \exp\left(\frac{-2\pi i m u}{M}\right)$$
$$u = 0, 1, \ldots, M-1; \quad v = 0, 1, \ldots, N-1 \tag{4.34}$$

The term in square brackets corresponds to the one-dimensional Fourier transform of the mth line and can be computed using the standard FFT procedures (assuming that N is a power of two). Each line is substituted with its Fourier transform, and the one-dimensional DFT of each column is computed.

Periodicity is an important property of DFT. A periodic transform F and a periodic function f are defined as follows:

$$\begin{array}{ll} F(u,-v) = F(u, N-v), & f(-m,n) = f(M-m, n) \\ F(-u,v) = F(M-u, v), & f(m,-n) = f(m, N-n) \end{array} \tag{4.35}$$

and

$$F(aM+u, bN+v) = F(u,v), \quad f(aM+m, bN+n) = f(m,n) \tag{4.36}$$

where a and b are integers.

The outcome of the 2D Fourier transform is a complex-valued 2D spectrum. Consider the input gray-level image (before the 2D Fourier transform was applied) with intensity values in the range, say, [0, ..., 255]. The 2D spectrum has the same spatial resolution. However, the values in both real and imaginary parts of the spectrum usually span a bigger range, perhaps millions. The existence of real and imaginary components and the range spanning several orders of magnitude makes the spectrum difficult to visualize and also to represent precisely in memory because too many bits are needed for it. For easier visualization, the range of values is usually decreased by applying a monotonic function, for example, $\sqrt{|F(u,v)|}$ or $\log|F(u,v)|$.

It is also useful to visualize a centered spectrum, that is, with the origin of the coordinate system (0, 0) in the middle of the spectrum. This is because centering has the effect of placing the low frequency information in the center and the high frequencies near the corners–consider the definition given in Eq. (4.32).

Assume that the original spectrum is divided into four quadrants (see Figure 4.18a). The small gray-filled squares in the corners represent positions of low frequencies. Owing to the symmetries of the spectrum, the quadrant positions can be swapped diagonally and the locations of low frequencies appear in the middle of the image (see Figure 4.18b).

The spectrum is demonstrated in Figure 4.20. The original spectrum is of size 256 × 256 in 256 gray levels. Figure 4.19 illustrates the spectrum. The image Figure 4.19a demonstrates the noncentered power spectrum and the image

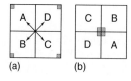

Figure 4.18 Centering of the 2D Fourier spectrum places the low frequencies around the origin of the coordinates. (a) Original spectrum. (b) Centered spectrum with the low frequencies in the middle.

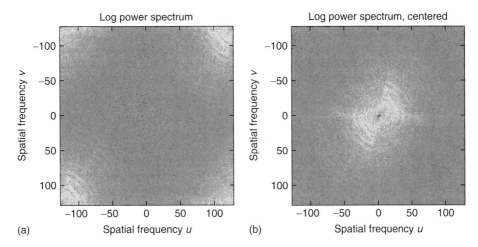

Figure 4.19 Power spectrum displayed as an intensity image. Lighter tones mean higher values. (a) Noncentered. (b) Centered. (Please find a color version of this figure on the color plates.)

Figure 4.20 Input image in the spatial domain is assumed periodic. Notice induced discontinuities on the borders of the image which manifests badly in the Fourier spectrum.

Figure 4.19b the centered power spectrum. The latter option is used more often. The range of the spectrum values has to be decreased to allow an observer to perceive the spectrum better; $\log P(u, v)$ is used here. For illustration of the particular range of this power spectrum, the pair of (minimum, maximum) for the $P(u, v)$ is $(2.4 \times 10^{-1}, \; 8.3 \times 10^6)$ and the (minimum, maximum) for $\log P(u, v)$ is $(-0.62, \; 6.9)$.

A distinct light cross can be seen in the centered power spectrum Figure 4.21b. This cross is caused by discontinuities on the limits of the image while assuming periodicity. These abrupt changes are easily visible in Figure 4.20.

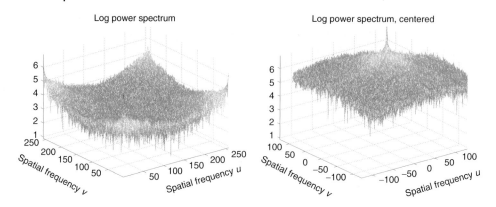

Figure 4.21 Power spectrum displayed as height in a 3D mesh; lighter tones mean higher values. (a) Noncentered and (b) centered.

The use of the Fourier transform in image analysis is pervasive. It can assist in noise filtering, in the detection of edges by locating high frequencies (sharp changes) in the image function; it also has applications in restoring images from corruption, fast matching using the convolution theorem, boundary characterization, image compression, and several other areas.

4.5
Filtering in the Image Domain

A basic hint is first given as to how filtering can be used in the image domain. Let us look at related issues from two viewpoints: (a) from the application perspective where usually two groups of methods are outlined: noise filtering and gradient operators (edge detection); (b) from the linearity/nonlinearity viewpoint. For the latter, we mainly deal with linear methods and mention the nonlinear robust noise filtering (median filtering) as an example.

We discuss filtering methods using a small neighborhood of a current pixel in an input image to produce a new brightness value in the output image. In addition, for simplicity, we consider only shift-invariant filters that perform the same operation at all positions.

Gradient operators suppress low frequencies in the frequency domain. Noise is often high frequency in nature; unfortunately, a gradient operator increases the noise level simultaneously. Smoothing and gradient operators have conflicting aims. The more sophisticated filtering algorithms permit one to perform smoothing and edge enhancement simultaneously, at least to a certain degree.

Linear operations calculate the resulting value in the output using convolution (see Eq. (4.14)). The filter is determined by coefficients in the convolution mask. Rectangular neighborhoods $\mathcal{O}$ are often used with an odd number of pixels in rows and columns, enabling specification of the central pixel of the neighborhood.

Local preprocessing methods typically use very little a priori knowledge about the image contents. Smoothing filters benefit if some general knowledge about image degradation is available, for example, statistical parameters of the noise. The choice of the local filter, the size and shape of the neighborhood $\mathcal{O}$, depend strongly on the size of objects in the processed image. If objects are rather large, an image can be enhanced by smoothing of small degradations. However, the filtering operations often do not take objects in the image into consideration, because they would need to interpret the image. This is, clearly, a chicken-and-egg problem.

Noise filtering (suppression, image smoothing) aims at suppressing image noise by exploring the statistical redundancy in the data. Let us make the usual assumption that the image is degraded by the random additive noise independent of the image. Noise is also assumed to have a zero mean and some standard deviation σ. If n samples from this random population are available, then it is natural to estimate the signal value as the mean value of available random samples. The positive aspect is that the standard deviation of the filtered result decreases by dividing by the factor $\sqrt{n}$. This fact comes from the rather general central limit theorem in statistics. The negative aspect is that the mean value as the estimate of the value with suppressed noise is rather sensitive to outlier data. However, this can to some degree be circumvented by using robust statistics. If several instances of the same static scene can be captured, then the assumptions made earlier are often fulfilled. In this case, the average value of the same pixel in several tens, hundreds, or thousands of images that are stacked one on the other is taken.

What happens if we are in the common situation in which only a single noise-corrupted image is at hand? Theoretically, nothing can be done. However, a pragmatic way is very often considered here. We assume that the pixel neighboring the pixel currently processed by the local filtering operator has the same intensity value as the current pixel. The mean or its more robust variant is taken as the filter result. Obviously, such an approach works for pixels in regions with homogeneous or slowly varying intensity and fails at the border between two regions with different intensities. In the latter case, the unwanted blurring effect occurs.

Let us illustrate the linear local filtering by a very small neighborhood of 3×3 size. The filter is given by the following convolution mask coefficients (see Eq. (4.14)).

$$h_1 = \frac{1}{9}\begin{bmatrix} 1 & 1 & 1 \\ 1 & 1 & 1 \\ 1 & 1 & 1 \end{bmatrix}, \quad h_2 = \frac{1}{10}\begin{bmatrix} 1 & 1 & 1 \\ 1 & 2 & 1 \\ 1 & 1 & 1 \end{bmatrix}, \quad h_3 = \frac{1}{16}\begin{bmatrix} 1 & 2 & 1 \\ 2 & 4 & 2 \\ 1 & 2 & 1 \end{bmatrix} \quad (4.37)$$

The significance of the pixel in the center of the convolution mask (h_2 case) or its 4-neighbors (h_3 case) is sometimes increased, as it better approximates the properties of noise with a Gaussian probability distribution.

An example illustrates the effect of this noise suppression. Images with low resolution (256×256) with 256 gray levels were chosen deliberately to show the discrete nature of the process. Figure 4.22a shows an original image cropped from the photo of the Prague Astronomic Clock; Figure 4.22b shows the same image with superimposed additive noise with Gaussian distribution; Figure 4.22c shows the

Figure 4.22 Noise and averaging filters. (a) Original image, (b) superimposed noise (random Gaussian noise characterized by the zero mean and the standard deviation 0.01), (c) 3 × 3 averaging, and (d) 7 × 7 averaging.

result of averaging with a 3 × 3 convolution mask (Eq. (4.37), left matrix). Noise is significantly reduced and the image is slightly blurred. Averaging with a larger mask (7 × 7) is demonstrated in Figure 4.22d, where the blurring is much more serious.

There are two commonly used smoothing filters whose coefficients gradually decrease to have the values close to zero at the limit of the window. This is the best way to minimize spurious oscillations in the frequency spectrum, cf. the uncertainty principle (Chapter 7). These filters are the Gaussian and the Butterworth filters [4].

Robust filtering uses two ideas. The first idea is finding the appropriate neighborhood of the current pixel in which image function value is homogeneous. This selection process is, of course, nonlinear. Having the selected neighborhood set, the second idea is used. The most probable value of the image function is calculated in a more clever way than averaging (which is sensitive to outliers). People's common sense uses robust estimates too, for example, in sports in which the "artistic impact" plays a role as in figure skating or ski jumping. Suppose that five referees decide. In ski jump, the best and the worst marks are not taken into account and the average from remaining three gives the result. Statistics has a simpler way of doing the job, called median filtering. In probability theory, the median divides the higher half of a probability distribution from the lower half. For a random variable x, the median M is the value for which the probability of the outcome $x < M$ is 0.5. The median of a finite list of real numbers can be found

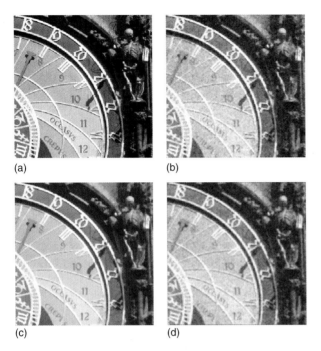

Figure 4.23 Noise and averaging filters. (a) Original image, (b) superimposed noise (random Gaussian noise characterized by the zero mean and the standard deviation 0.01), (c) 3 × 3 median filter of the original image (a), and (d) 7 × 7 median filter of the noisy image (b).

by ordering the values and selecting the middle one. Lists are often constructed to be odd in length to secure uniqueness. In image processing, median filtering is a nonlinear smoothing method that reduces the blurring of edges.

The effect of median filtering in the 3 × 3 neighborhood is demonstrated in Figure 4.23. The input images are the same as in Figure 4.22a and b. The result of median filtration of the original image is given in Figure 4.22c. The effect of the median on noisy image is given in Figure 4.22d.

Edge detection is a generic name for methods detecting and quantifying abruptness of the intensity change in the image. If the 2D image function $f(x, y)$ is viewed as an elevation map of a natural landscape then edge detection methods measure the steepness of a slope and its orientation in each point. The gradient of the image function, partial derivative according to x and y, $f_x(x, y)$ and $f_y(x, y)$, provides this information.

The approximations of a gradient are needed in the discretized images. Gradient operators fall under three categories:

1) Operators approximating derivatives of the image function using differences. Some of them are rotationally invariant (e.g., Laplacian) and thus are computed from one convolution mask only. Others, which approximate first derivatives,

use several masks. The orientation is estimated on the basis of the best matching of several simple patterns.
2) Operators based on the zero crossings of the image function second derivative (e.g., Marr–Hildreth [5] or Canny edge detectors [6]).
3) Operators that attempt to match an image function to a parametric model of edges.

Let us illustrate the methods from the first category only. Individual gradient operators that examine small local neighborhoods are convolutions and can be expressed by convolution masks. Operators that are able to detect edge direction are represented by a collection of masks, each corresponding to a certain direction.

The Prewitt operator (similarly Kirsch, Robinson, or Sobel), and some other operators, approximates the first derivative. The gradient is estimated in eight (for a 3 × 3 convolution mask) possible directions, and the convolution result with the greatest magnitude indicates the gradient direction. Larger masks are possible. Only the first three 3 × 3 masks for each operator are presented to save space; the other five masks can be created by simple rotation. The Prewitt operator is given by

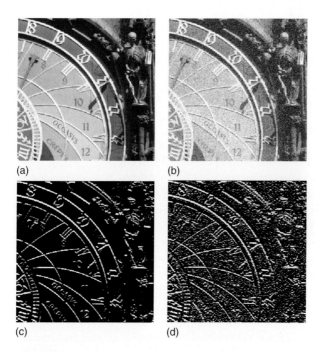

Figure 4.24 Prewitt edge detector. (a) Original image, (b) superimposed noise (random Gaussian noise characterized by the zero mean and the standard deviation 0.01), (c) Prewitt vertical edges calculated on the image (a), and (d) Prewitt vertical edges calculated on the noisy image (b).

$$h_1 = \begin{bmatrix} 1 & 1 & 1 \\ 0 & 0 & 0 \\ -1 & -1 & -1 \end{bmatrix}, \quad h_2 = \begin{bmatrix} 0 & 1 & 1 \\ -1 & 0 & 1 \\ -1 & -1 & 0 \end{bmatrix}, \quad h_3 = \begin{bmatrix} -1 & 0 & 1 \\ -1 & 0 & 1 \\ -1 & 0 & 1 \end{bmatrix}, \quad \ldots$$
(4.38)

The direction of the gradient is given by the mask giving maximal response. This is also the case for all the following operators approximating the first derivative.

The performance of the Prewitt operator is demonstrated in Figure 4.24.

The Sobel operator is another often used tool for edge detection.

$$h_1 = \begin{bmatrix} 1 & 2 & 1 \\ 0 & 0 & 0 \\ -1 & -2 & -1 \end{bmatrix}, \quad h_2 = \begin{bmatrix} 0 & 1 & 2 \\ -1 & 0 & 1 \\ -2 & -1 & 0 \end{bmatrix}, \quad h_3 = \begin{bmatrix} -1 & 0 & 1 \\ -2 & 0 & 2 \\ -1 & 0 & 1 \end{bmatrix}, \quad \ldots$$
(4.39)

The Sobel operator is often used as a simple detector of horizontality and verticality of edges, in which case only masks h_1 and h_3 are used. If the h_1 response is y and the h_3 response x, we might then derive edge strength (magnitude) as

$$\sqrt{x^2 + y^2} \quad \text{or} \quad |x| + |y| \qquad (4.40)$$

and direction as arctan(y/x). The outcome of the Sobel operator is demonstrated in Figure 4.25.

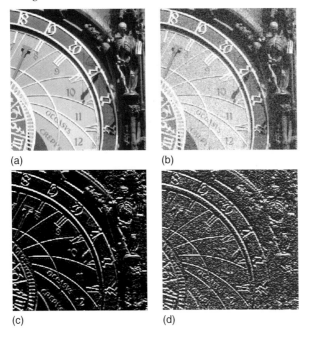

Figure 4.25 Sobel edge detector. (a) Original image, (b) superimposed noise (random Gaussian noise characterized by the zero mean and the standard deviation 0.01), (c) Sobel vertical edges calculated on the image (a), and (d) Sobel vertical edges calculated on the noisy image (b).

4.6
Conclusions

This chapter briefly introduced a number of fundamentals of digital image processing, starting with the digital image representation, topological and metric properties of images, and global image characteristics such as the brightness histogram. Thereafter, the concept of convolution in support of image filtering was introduced, facilitating the introduction of a 2D Fourier transform and its properties. Finally, this chapter covered spatial filtering allowing for the introduction of basic denoising and edge detection operators.

References

1. Gonzalez, R.C., Woods, R.E., and Eddins, S. (2004) *Digital Image Processing*, 2nd edn, Prentice Hall, New York.
2. Burger, W. and Burge, M.J. (2008) *Digital Image Processing: An Algorithmic Approach Using Java*, Springer Verlag, Berlin, Germany.
3. Šonka, M., Hlaváč, V., and Boyle, R.D. (2007) *Image Processing, Analysis and Machine Vision*, 3rd edn, Thomson Learning, Toronto, Canada.
4. Bracewell, R.N. (2004) *Fourier Analysis and Imaging*, 1st edn, Springer-Verlag.
5. Marr, D. (1982) *Vision – A Computational Investigation into the Human Representation and Processing of Visual Information*, Freeman, San Francisco.
6. Canny, J.F. (1986) A computational approach to edge detection. *IEEE Trans. Pattern Anal. Mach. Intell.*, **8**, 679–698.

5
Joint Spatial/Spatial-Frequency Representations
Gabriel Cristóbal, Salvador Gabarda, and Leon Cohen

5.1
Introduction

Over the last 20 years, joint spatial/spatial-frequency representations have received attention in the fields of image processing, vision, and pattern recognition. They could be more useful than the general techniques of spatial filtering of images, and therefore the possibilities of information filtering would be improved. Some researchers have considered the Wigner–Ville distribution (WVD) as a the basic distribution from which the rest of distributions can be derived [1–3]. The Wigner distribution was introduced by Wigner [4] as a phase space representation in quantum mechanics, and it gives a simultaneous representation of a signal in space and spatial-frequency variables. Subsequently, Ville derived the same distribution in the area of signal processing that Wigner had proposed several years earlier [5]. The WVD can be considered as a particular case of the occurrence of a complex spectrogram in which the sliding window function is the function itself. It belongs to a large class of bilinear distributions known as the *general class* in which each member can be obtained by choosing different kernels of a generalized bilinear distribution [6]. The WVD has a set of desirable properties as was first formulated by Claasen and Mecklenbrauker [7]. However, the presence of cross terms has limited its use in some practical applications, because cross terms can have a peak value as high as twice that of the autocomponents [8]. The WVD of a 2D image is a 4D function that involves Fourier transformation for every point of the original image. For many years, the computational burden associated with this fact limited the use of the WVD only to 1D signal processing applications. Although the WVD was initially proposed for continuous variables, various definitions exist for discrete variable functions. In the case of image applications, two problems arise in practice in the computation of the discrete WVD. First, the aliasing problem can be reduced by smoothing the original image using low-pass filtering. One additional problem that appears in practice is the spectral dispersion or leakage due to the window size, specially relevant in the case of small window sizes. To reduce leakage, it is necessary to introduce spatial apodization or truncation filters in order to minimize the spurious

Optical and Digital Image Processing: Fundamentals and Applications, First Edition. Edited by Gabriel Cristóbal, Peter Schelkens, and Hugo Thienpont.
© 2011 Wiley-VCH Verlag GmbH & Co. KGaA. Published 2011 by Wiley-VCH Verlag GmbH & Co. KGaA.

side lobes of the sinc function as a consequence of the windowing effect (remembering that the Fourier transform of a square window is a sinc function). In numerical implementations, a discrete WVD definition called *pseudo-WVD* is required.

The rest of the chapter is organized as follows. In Section 5.2, we first discuss the continuous WVD and its main properties. Next, in Section 5.3, two alternative numerical implementations are described as most suitable for image processing applications: (i) the 1D directional pseudo-Wigner–Ville distribution (PWVD) and (ii) the 2D-smoothed PWVD. Texture segmentation results are illustrated in Section 5.4 through principal component analysis (PCA) of the PWVD. In Section 5.5, a hybrid optical–digital system to compute the PWVD is described. Finally, conclusions are drawn in Section 5.6.

5.2
Fundamentals of Joint Representations

The fundamental idea of applying joint representations of position and spatial frequency (wave number) to images is that in an image, the spatial-frequency content maybe change from place to place, and formulating the relevant mathematical and physical ideas leads to both insight and practical methods to understand and manipulate images. For an image as a whole, the spatial-frequency content is given by the absolute square of the Fourier transform. However, this gives no indication as to how frequencies change from point to point. What is needed is a joint spatial distribution where one can see the spectral content at a particular point. The aim of this section is to briefly review some of the fundamental concepts in formulating this notion.

To crystalize the issue, consider a two-dimensional image that we symbolize as $f(x, y)$. We will work with normalized images. That is, we assume that

$$\int_{-\infty}^{\infty}\int_{-\infty}^{\infty} |f(x,y)|^2 dx dy = 1 \tag{5.1}$$

Note that we have allowed for the possibility of complex images, the notion and need for defining a complex image from a real one are discussed later. One can think of $|f(x, y)|^2$ as the density or energy in an image at position x, y. If we take the Fourier transform

$$F(k_1, k_2) = \frac{1}{2\pi} \int_{-\infty}^{\infty}\int_{-\infty}^{\infty} f(x,y) e^{-i(k_1 x + k_2 y)} dx dy \tag{5.2}$$

with its inverse given by

$$f(x,y) = \frac{1}{2\pi} \int_{-\infty}^{\infty}\int_{-\infty}^{\infty} F(k_1, k_2) e^{i(k_1 x + k_2 y)} dk_1 dk_2 \tag{5.3}$$

then $|F(k_1, k_2)|^2$ is an indication of the spatial frequencies that exist in the image $f(x, y)$.

The basic objective is to devise a joint density of the four variables x, y, k_1, k_2 in space and spatial frequency. One can set up the issue as follows. If we have an image $f(x, y)$ and its Fourier transform $F(k_1, k_2)$, then the instantaneous power is

$$|f(x,y)|^2 = \text{density (or intensity) per unit area at position } x, y \qquad (5.4)$$

and the density in frequency, called the *energy density spectrum*, is

$$|F(k_1, k_2)|^2 = \text{density (or intensity) per unit frequency area}$$
$$\text{at frequency } k_1, k_2 \qquad (5.5)$$

What one seeks is a joint density, $P(x, y, k_1, k_2)$, so that

$$P(x, y, k_1, k_2) = \text{the density (or intensity) at position } x, y \text{ and frequency } k_1, k_2 \qquad (5.6)$$

Ideally the joint density should satisfy

$$\iint P(x, y, k_1, k_2)\, dx dy = |F(k_1, k_2)|^2 \qquad (5.7)$$

$$\iint P(x, y, k_1, k_2)\, dk_1 dk_2 = |f(x, y)|^2 \qquad (5.8)$$

which can be called the *spatial and spatial-frequency marginal conditions*. If we choose a P that satisfies the marginals, then it also follows that

$$\iint P(x, y, k_1, k_2)\, dk_1 dk_2\, dx dy = 1 \qquad (5.9)$$

The above conditions do not force a particular joint density and many have been studied, which we discuss subsequently. Perhaps the best known joint distribution is the Wigner distribution. We derive and discuss its properties, and that will give us a framework for discussing others.

5.2.1
Notation

In the above discussion, we have written the equations in explicit coordinate form but it is generally simpler to use vector notation. In particular, we use the following notation. Boldface symbols denote vectors:

$$\mathbf{k} = (k_1, k_2) \qquad (5.10)$$
$$\mathbf{x} = (x, y) \qquad (5.11)$$

Dot products are denoted in the usual manner as

$$\mathbf{k} \cdot \mathbf{x} = k_1 x + k_2 y \qquad (5.12)$$

Using vector notation, we can write the above equations in a more compact manner. For the joint density, we write

$$P(\mathbf{x}, \mathbf{k}) = P(x, y, k_1, k_2) \qquad (5.13)$$

then, for example, for the marginal conditions may be written as

$$\int P(\mathbf{x}, \mathbf{k}) d\mathbf{x} = |F(\mathbf{k})|^2 \qquad (5.14)$$

$$\int P(\mathbf{x}, \mathbf{k}) d\mathbf{k} = |f(\mathbf{x})|^2 \qquad (5.15)$$

We also write Fourier transform pairs as

$$F(\mathbf{k}) = \frac{1}{2\pi} \int f(\mathbf{x}) e^{-i\mathbf{x}\cdot\mathbf{k}} d\mathbf{x} \tag{5.16}$$

$$f(\mathbf{x}) = \frac{1}{2\pi} \int F(\mathbf{k}) e^{i\mathbf{x}\cdot\mathbf{k}} d\mathbf{k} \tag{5.17}$$

In the above equations, we have used a single integral for the sake of neatness; the number of integrals is indicated by the number of differentials. Also, integrals range from $-\infty$ to ∞ unless indicated otherwise.

5.2.2
The Wigner Distribution

The Wigner distribution is the prototype distribution and, besides the spectrogram that is discussed later, is the most used distribution. It is defined by

$$W(\mathbf{x}, \mathbf{k}) = \left(\frac{1}{2\pi}\right)^2 \iint f^*(\mathbf{x} - \tau/2) f(\mathbf{x} + \tau/2) e^{-i\tau\cdot\mathbf{k}} d\tau \tag{5.18}$$

It can also be expressed in terms of the spectrum

$$W(\mathbf{x}, \mathbf{k}) = \left(\frac{1}{2\pi}\right)^2 \iint F^*(\mathbf{k} + \theta/2) F(\mathbf{k} - \theta/2) e^{-i\theta\cdot\mathbf{x}} d\theta \tag{5.19}$$

It is of some interest to express it explicitly,

$$W(x, y, k_1, k_2) = \left(\frac{1}{2\pi}\right)^2 \iint f^*(x - \tau_x/2, y - \tau_y/2) f(x + \tau_x/2, y + \tau_y/2)$$

$$e^{-ik_1\tau_x - k_2\tau_y} d\tau_x d\tau_y \tag{5.20}$$

Note that the Wigner distribution is the Fourier transform of $f^*(\mathbf{x} - \tau/2) f(\mathbf{x} + \tau/2)$. Now, in standard Fourier analysis, the energy density spectrum is the Fourier transform of the autocorrelation function, that is,

$$|F(\mathbf{k})|^2 = \left(\frac{1}{2\pi}\right)^2 \int R(\tau) e^{-i\tau\cdot\mathbf{k}} d\tau \tag{5.21}$$

where $R(\tau)$ is the deterministic autocorrelation function given by

$$R(\tau) = \int f^*(\mathbf{x}) f(\mathbf{x} + \tau) d\mathbf{x} = \int f^*(\mathbf{x} - \tau/2) f(\mathbf{x} + \tau/2) d\mathbf{x} \tag{5.22}$$

One tries to generalize this by writing

$$W(\mathbf{x}, \mathbf{k}) = \left(\frac{1}{2\pi}\right)^2 \int R_x(\tau) e^{-i\tau\cdot\mathbf{k}} d\tau \tag{5.23}$$

The Wigner distribution (and others) can be thought of as generalizing Eq. (5.21) with the autocorrelation function being replaced by a local autocorrelation function defined by

$$R_t(\tau) = f^*(\mathbf{x} - \tau/2) f(\mathbf{x} + \tau/2) \tag{5.24}$$

We now discuss some of the basic properties of the Wigner distribution.

5.2.2.1 Marginals
The Wigner distribution satisfies the marginals Eqs (5.14) and (5.15). In particular,

$$\int\int W(\mathbf{x},\mathbf{k})d\mathbf{k} = \left(\frac{1}{2\pi}\right)^2 \int\int\int\int f^*(\mathbf{x}-\tau/2)f(\mathbf{x}+\tau/2)\,e^{-i\tau\cdot\mathbf{k}}\,d\tau\,d\mathbf{k} \quad (5.25)$$

$$= \int\int f^*(\mathbf{x}-\tau/2)f(\mathbf{x}+\tau/2)\,\delta(\tau)\,d\tau \quad (5.26)$$

$$= |f(\mathbf{x})|^2 \quad (5.27)$$

and this holds similarly for the other marginal, Eq. (5.15). Since the Wigner distribution satisfies the marginals, it follows that the energy of the image is preserved:

$$\int W(\mathbf{x},\mathbf{k})\,d\mathbf{x}\,d\mathbf{k} = 1 \quad (5.28)$$

5.2.2.2 Inversion
A fundamental property of the Wigner distribution is that it is uniquely related to the image. In other words, if we have the Wigner distribution of an image, the image is uniquely recoverable up to a constant phase factor. To see this, notice that the Wigner distribution, Eq. (5.18), is the Fourier transform of $f^*(\mathbf{x}-\tau/2)\,f(\mathbf{x}+\tau/2)$ with respect to τ; hence, taking its inverse we have

$$f^*(\mathbf{x}-\tau/2)f(\mathbf{x}+\tau/2) = \int\int W(\mathbf{x},\mathbf{k})e^{i\tau\cdot\mathbf{k}}\,d\mathbf{k} \quad (5.29)$$

Letting $\mathbf{x} = \tau/2$, one obtains

$$f(\mathbf{x}) = \frac{1}{f^*(0)}\int\int W(\tau/2,\mathbf{k})e^{i\tau\cdot\mathbf{k}}\,d\mathbf{k} \quad (5.30)$$

which shows how $f(\mathbf{x})$ can be recovered from $W(\mathbf{x},\mathbf{k})$.

5.2.2.3 Translation Invariance
Suppose we consider a new image, $g(\mathbf{x})$, defined by

$$g(\mathbf{x}) = f(\mathbf{x}+\mathbf{x}_0) \quad (5.31)$$

The Wigner distribution of g is

$$W_g(\mathbf{x},\mathbf{k}) = \left(\frac{1}{2\pi}\right)^2 \int\int g^*(\mathbf{x}-\tau/2)f(\mathbf{x}+\tau/2)\,e^{-i\tau\cdot\mathbf{k}}\,d\tau \quad (5.32)$$

$$\left(\frac{1}{2\pi}\right)^2 \int\int f^*(\mathbf{x}+\mathbf{x}_0-\tau/2)f(\mathbf{x}+\mathbf{x}_0+\tau/2)\,e^{-i\tau\cdot\mathbf{k}}\,d\tau \quad (5.33)$$

Therefore, we have that

$$W_g(\mathbf{x},\mathbf{k}) = W_f(\mathbf{x}+\mathbf{x}_0,\mathbf{k}) \quad (5.34)$$

which shows that if one translates an image, the Wigner distribution is translated by the same amount. Similarly, if one translates the spectrum

$$G(k) = F(k + k_0) \tag{5.35}$$

then the Wigner distribution is similarly translated

$$W_g(x, k) = W_f(x, k + k_0) \tag{5.36}$$

5.2.2.4 Product of Images
Suppose we have an image that is the product of two images

$$h(x) = f(x)g(x) \tag{5.37}$$

then one can show that the corresponding Wigner distributions are related by way of

$$W_h(x, k) = \int W_f(x, k') W_g(x, k - k') dk' \tag{5.38}$$

Similarly, if one has a product of two spectra

$$H(k) = F(k)G(k) \tag{5.39}$$

then the Wigner distribution of $h(x)$ is given by

$$W_h(x, k) = \int W_f(x', k) W_g(x' - x, k) dx' \tag{5.40}$$

5.2.2.5 Overlap of Two Images
In comparing images, it is often desirable to calculate the overlap function between two images. In particular, the overlap of two images equals the overlap of their respective distributions.

$$\int f^*(x)g(x) dx = \iint W_f(x, k) W_g(x, k) dx dk \tag{5.41}$$

5.2.2.6 Real Images
If an image is real, the spectrum satisfies

$$F(k) = F(-k) \tag{5.42}$$

and the Wigner distribution satisfies

$$W(x, k) = W(x, -k) \tag{5.43}$$

Also, if the spectra is real, then

$$W(x, k) = W(-x, k) \tag{5.44}$$

5.2.2.7 Cross Wigner Distribution

Suppose we have two images $f(\mathbf{x})$ and $g(\mathbf{x})$. It is useful to define the cross Wigner distribution:

$$W_{fg}(\mathbf{x}, \mathbf{k}) = \left(\frac{1}{2\pi}\right)^2 \iint f^*(\mathbf{x} - \tau/2)\, g(\mathbf{x} + \tau/2)\, e^{-i\tau \cdot \mathbf{k}}\, d\tau \tag{5.45}$$

Now, consider an image composed of the sum of two images:

$$h(\mathbf{x}) = f(\mathbf{x}) + g(\mathbf{x}) \tag{5.46}$$

Substituting into Eq. (5.18), we have

$$W_{fg}(\mathbf{x}, \mathbf{k}) = W_f(\mathbf{x}, \mathbf{k}) + W_g(\mathbf{x}, \mathbf{k}) + W_{fg}(\mathbf{x}, \mathbf{k}) + W_{gf}(\mathbf{x}, \mathbf{k}) \tag{5.47}$$

$$= W_f(\mathbf{x}, \mathbf{k}) + W_g(\mathbf{x}, \mathbf{k}) + 2\,\mathrm{Re}\, W_{fg}(\mathbf{x}, \mathbf{k}) \tag{5.48}$$

Thus, the Wigner distribution of the sum of two images is not the sum of the respective Wigner distributions but there is an additional term $2\,\mathrm{Re}\, W_{fg}(\mathbf{x},\mathbf{k})$. Terms of this type are called *cross terms*.

5.3
Other Distributions

There have been many other distributions that have been studied and many of them have advantages over the Wigner distribution. All bilinear distributions can be obtained from the general class

$$C(\mathbf{x}, \mathbf{k}) = \left(\frac{1}{4\pi^2}\right)^2 \iiint f^*(\mathbf{u} - \tau/2)\, f(\mathbf{u} + \tau/2)\, \phi(\theta, \tau)$$

$$\times\, e^{-i\theta \cdot \mathbf{x} - i\tau \cdot \mathbf{k} + i\theta \cdot \mathbf{u}}\, d\mathbf{u}\, d\tau\, d\theta \tag{5.49}$$

where $\phi(\theta, \tau)$ is a two-dimensional function called the *kernel*, which characterizes the distribution. In terms of the spectrum, the general class is

$$C(\mathbf{x}, \mathbf{k}) = \left(\frac{1}{4\pi^2}\right)^2 \iiint F^*(\mathbf{u} + \theta/2)\, F(\mathbf{u} - \theta/2)\, \phi(\theta, \tau)$$

$$\times\, e^{-i\theta \cdot \mathbf{x} - i\tau \cdot \mathbf{k} + i\tau \cdot \mathbf{u}}\, d\mathbf{u}\, d\tau\, d\theta \tag{5.50}$$

The advantage of this formulation is that the properties of the distributions can be studied by way of the kernel $\phi(\theta, \tau)$. For example, to insure that the marginals are correct, one chooses kernels as indicated:

$$\iint C(\mathbf{x}, \mathbf{k})\, d\mathbf{x} = |F(\mathbf{k})|^2 \qquad \text{if } \phi(0, \tau) = 1 \tag{5.51}$$

$$\iint C(\mathbf{x}, \mathbf{k})\, d\mathbf{k} = |f(\mathbf{x})|^2 \qquad \text{if } \phi(\theta, 0) = 1 \tag{5.52}$$

5.3.1
The Spectrogram

Historically, the most important joint distribution is the spectrogram developed in the period of the Second World War. The idea behind the spectrogram is that if we are interested in the properties of the image at position **x**, one emphasizes the image at that position and suppresses the image at other positions. This is achieved by multiplying the image by a window function, $h(\mathbf{x})$, centered at **x**, to produce a modified image, $f(\tau)\, h(\tau - \mathbf{x})$ where τ is the running variable. Since the modified image emphasizes the image around the position **x**, the Fourier transform will reflect the distribution of spatial frequencies around that position. Taking its Fourier transform, commonly called the *short space Fourier transform*, we have

$$F_x(\mathbf{k}) = \frac{1}{2\pi} \int e^{-i\tau \cdot \mathbf{k}} f(\tau)\, h(\tau - \mathbf{x})\, d\tau \tag{5.53}$$

The magnitude square of $F_x(\mathbf{k})$ gives the distribution

$$P_{SP}(\mathbf{x}, \mathbf{k}) = \left| \frac{1}{2\pi} \int e^{-i\tau \cdot \mathbf{k}} f(\tau)\, h(\tau - \mathbf{x})\, d\tau \right|^2 \tag{5.54}$$

which is called the *spectrogram*. The spectrogram does not satisfy the marginals; however, for proper choice of window it does so approximately. It mixes properties of the image and window, but by appropriate choice of the window different aspects of the image can be revealed. We emphasize that there is total symmetry between window and image and that if the window is not chosen carefully, the result is that we use the image to study the window!

5.3.2
The Analytic Image

In the case of one-dimensional time signals, the issue is that given that instantaneous frequency is the derivative of the phase of a complex signal, how do we then form a complex signal from a real signal. Gabor gave such a procedure, and in particular his procedure is the following. For the real signal, $s(x)$, take the Fourier transform of the signal $S(k)$

$$S(k) = \frac{1}{\sqrt{2\pi}} \int_{-\infty}^{\infty} s(x)\, e^{-ikx}\, dx \tag{5.55}$$

and set the negative part of the spectrum equal to zero. Then, the complex signal, $z(x)$, whose spectrum is composed of the positive frequencies of $S(k)$ only, is given by

$$z(x) = 2\, \frac{1}{\sqrt{2\pi}} \int_{0}^{\infty} S(k)\, e^{ikx}\, dx \tag{5.56}$$

The factor 2 is inserted so that the real part of the analytic signal is $s(x)$. This is called the *analytic signal*. A similar formalism can be applied to spatial signals instead of temporal signals. However, for two-dimensional spatial images, we run into difficulty as we explain in Section 5.4.3.

5.4
The Pseudo-Wigner–Ville Distribution (PWVD)

5.4.1
1D-Smoothed PWVD

In numerical implementations, one can consider the so-called smoothed PWVD, by using two smoothing windows: a spatial-averaging window and a spatial-frequency averaging window. The use of two windows for tracking nonstationarity in 1D signals by the PWVD was proposed in Ref. [9] and later extended to 2D signals in Ref. [3]. Frequency smoothing is tantamount to spatial windowing and vice versa. In the current implementation, a spatial windowing operation together with a spatial smoothing (averaging) operation is performed. Although the term pseudo-Wigner distribution is used when smoothing is done only in the frequency domain following Refs [3, 9], we use the same term even in the case of smoothing in both domains. In the following sections, we refer to $g(\cdot)$ as the *spatial-averaging window* and $h(\cdot)$ as the *spatial window*. This double smoothing operation produces an improvement in the cross-term reduction. Decrease in the spatial window size produces an increment in the spatial resolution and, therefore, a reduction, in the presence of cross terms (since auto-terms corresponding to pixels outside the reduced spatial windows are eliminated), at the cost of broadening the auto-terms. Increasing the spatial-averaging window size has the effect of attenuating the cross terms at the cost of decreasing the spatial resolution and increasing computational cost [10]. The PWVD of a 1D signal $f(n)$ can be defined by

$$W_f(n, k) = \sum_{m=-N+1}^{N-1} [h(m)]^2 e^{-i2mk} \sum_{l=-M+1}^{M-1} g(l) f[n+1+m] f^*[n+1-m] \quad (5.57)$$

where n and k are the discrete spatial and discrete frequency variables, respectively, and M and N the spatial and spatial-frequency window size, respectively. The discrete PWVD of a sampled 1D signal $f(n)$ is periodic in the frequency variable with period π, that is, $W_f(n, k) = W_f(n, k + \pi)$. However, the signal's Fourier spectrum periodicity is 2π. This can result in aliasing, unless analytic signals are considered, or in signal oversampling by a factor of 2. In practice, the PWVD of the analytic signal of f is used to overcome the principal discretization problem: aliasing and cross terms [11].

5.4.2
1D Directional PWVD

Spatial-frequency information of an image can be extracted by associating the gray-level spatial data with one of the well-known space–frequency distributions [2]. In such a case, any specific pixel n of the image can be associated with a vector containing its 1D PWVD, calculated in a local neighborhood, by means of a small window of length N. The use of a windowed 1D transform for a 2D

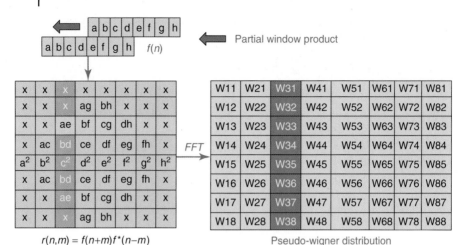

Figure 5.1 Procedure to compute the 1D directional PWVD corresponding to one pixel. (Reproduced with permission of the Society of Photo-Optical Instrumentation Engineers © 2009 SPIE.) (Please find a color version of this figure on the color plates.)

signal can be justified by considering three main aspects of the problem. First, by using a 1D PWVD, data can be arranged in any desired direction over a 2D image; second, calculation time is greatly diminished compared to a 2D version of the PWVD; and third, the 1D PWVD is an invertible function, allowing, in this way, for information to be totally preserved. A discrete approximation of the Wigner distribution proposed by Claasen and Mecklenbrauker [7], similar to Brenner's expression [12], has been used here:

$$W_f(n,k) = 2 \sum_{m=-\frac{N}{2}}^{\frac{N}{2}-1} f[n+m] f^*[n-m] e^{-2i\left(\frac{2\pi m}{N}\right)k} \quad (5.58)$$

In Eq. (5.58), n and k represent the spatial and frequency discrete variables, respectively, and m is a shifting parameter, which is also discrete. Here, f is a 1D sequence of data from the image, containing the gray values of N pixels, aligned in the desired direction. Equation (5.58) can be interpreted as the discrete Fourier transform (DFT) of the product $r(n,m) = f[n+m] f^*[n-m]$ (Figure 5.1). Here, f^* indicates the complex conjugate of f. This equation is limited to a spatial interval (the PWVD's window), allowing information to be extracted locally. By scanning the image with a 1D window of N pixels, that is, by sliding the window to all possible positions over the image, the full pixel-wise PWVD of the image is produced. The window can be tilted in any direction to obtain a directional distribution. Figure 5.2 shows a visualization of the PWVD for the same point of an image but for two different orientations (0 and 90°, respectively). The window size to produce the PWVD was 8 pixels. However, any other arbitrary orientation of the window is

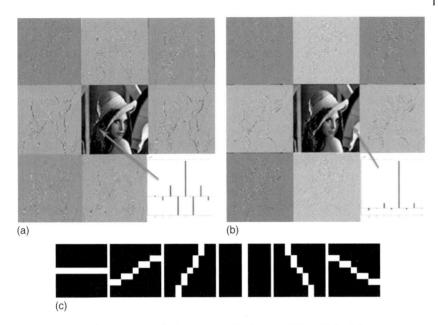

Figure 5.2 Pseudo-Wigner distribution corresponding to a point of the feathers region of Lena. (a) PWVD using a window oriented to 0°. (b) PWVD using a window oriented to 90°. Each pixel in the image has a vector $w = [w_{-N/2} \cdots w_{N/2-1}]$ representing its PWVD. The different blocks arrange elements with the same label. (c) Each image represents the different pixel arrangements for calculating the 1D PWVD along six different orientations. (Reproduced with permission of the Society of Photo-Optical Instrumentation Engineers © 2009 SPIE.)

possible, which allows one to provide a selective directional analysis of the images. It is worth noting that the discretizing process implies a loss of some properties of the continuous WVD. One important property preserved within the definition given by Eq. (5.58) is the inversion property, which is a very desirable feature for the recovery of the original signal, and which allows local filtering operations on the images under consideration. According to the inversion property [7], the even samples can be recovered from Eq. (5.58) from the following expression [13]:

$$f(2n)f^*(0) = \sum_{m=-\frac{N}{2}}^{\frac{N}{2}-1} W(n,m) e^{-2i\left(\frac{2\pi m}{N}\right)n} \qquad (5.59)$$

and the odd samples can be recovered from

$$f(2n-1)f^*(1) = \sum_{m=-\frac{N}{2}}^{\frac{N}{2}-1} W(n,m) e^{-2i\left(\frac{2\pi m}{N}\right)n} \qquad (5.60)$$

To get the original values, we have to perform an inverse DFT for recovering of the function $r(n, m)$, which gathers all the information included in the original discrete signal. Equation (5.59) is obtained by writing $n = k$ in $r(n, m)$, which implies that the values $f(2n)f^*(0)$ are on the main diagonal of the matrix. Similarly, Eq. (5.60) results from taking $m = n - 1$ in the product function $r(n, m)$; thus the odd sample values are located above the main diagonal. To recover the exact values of the samples, we have to divide the diagonal values by $f^*(0)$ and $f^*(1)$, respectively [13]. Although the signs of the samples are undetermined because of the product sign rule, they can always be considered positive, because we are dealing with digital images of real positive gray-value levels. From Eqs (5.59) and (5.60), it can be shown that n varies in the interval $[-N/4, N/4 - 1]$, because of the factor 2, which affects the left-hand side of both equations. The 2D directional PWVD can be considered as a generalization of the 1D case, but to reduce the computational cost to a minimum, the 1D PWVD is used as defined before, that is, by taking 1D directional windows of $N = 8$ pixels.

5.4.3
2D-Smoothed PWVD Definition and Implementation

The 1D-smoothed PWVD defined in Section 5.4.1 can be extended for 2D signals as follows

$$PWVD(n_1, n_2, k_1, k_2) = \sum_{m=-N_2+1}^{N_2-1} \sum_{l=-N_1+1}^{N_1-1} h_{N_1 N_2}(m, l) \sum_{r=-M_2+1}^{M_2-1}$$

$$\times \sum_{s=-M_1+1}^{M_1-1} g_{M_1 M_2}(r, s) f(n_1 + r + m, n_2 + s + l)$$

$$\times f^*(n_1 + r - m, n_2 + s - l) e^{-j2(mk_1 + lk_2)} \quad (5.61)$$

where $f(\cdots)$ now represents a 2D discrete input image. The PWVD can be considered as a particular occurrence of a complex spectrogram, where the function $f(\cdots)$ is chosen as the shifting window function. As mentioned in the case of the 1D-smoothed PWVD (Section 5.4.1), in practice the PWVD of the analytic signal of f (instead of the PWVD of f) is used to overcome the principal discretization problem: aliasing and cross terms. Aliasing (or frequency foldover) in discrete signals is caused by an inadequate sampling rate (i.e., a rate below the Nyquist limit, which is twice the highest frequency of the signal). Two methods are used to overcome this problem: (i) oversampling (or interpolating) by a factor of 2, and (ii) frequency-filtering the regions that cause aliasing (low-pass filtering). The latter can substantially suppress artifacts, but at the cost of significantly reducing the spatial-frequency support. For dimensions greater than one, the definition of an analytic image for 2D signals is not unique [14, 15]. The extension of the analytic signal concept to multidimensional signals has been done by Hahn [16]. Several 2D analytic signals have been proposed elsewhere with the aim of reducing aliasing artifacts but retaining the main PWVD properties. However, most of the methods fail to produce a substantial reduction in the aliasing terms and cross terms. In Ref. [10], a new analytic image (Figure 5.3a) has been proposed, which can be considered as a combination of two previous definitions (Figure 5.3b,c) [14, 15]. The region

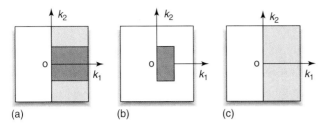

Figure 5.3 Three different analytic image definitions.
(a) Analytic image defined in Ref. [10]; (b) "1/4 domain method" defined in Ref. [14]; (c) analytic image defined in Ref. [15]. (Reproduced from [10] with permission of Institute of Electrical and Electronics Engineers © 1998 IEEE.)

of support of the new analytic image (Figure 5.3a) is the same as that proposed in Ref. [15] (Figure 5.3c) but it separates the region considered into two parts (depicted in Figure 5.3a by two different shadings). Although this method requires the computation of two PWVDs corresponding to the two regions, it still gives a substantial computational cost reduction when compared with interpolation-based methods. These can be easily implemented with a zero-padding operation, and therefore with higher computational cost, by computing an PWVD with four times the number of samples. Here, we consider a 2D analytic signal that may avoid most of the shortcomings of the previous definitions. This method is based in zeroing out the negative side of the spectrum as in Figure 5.3c. To do this, we split the positive region into two subregions [see Figure 5.3a] and take a PWVD from each subregion. The potential drawbacks of the analytic signal defined here are the introduction of additional sharp filter boundaries and boundary effects that might lead to ringing, besides the creation of a frequency variant PWVD. Ringing has been reduced by smoothing all the sharp transitions with a raised-cosine (Hanning) function. With regard to the cross-terms issue, it must be noted that the removal of the negative frequencies of the analytic signal also eliminates the cross terms between positive and negative frequencies. In addition to that, cross terms have oscillatory terms of relatively high frequencies that can often be reduced by PWVD low-pass filtering but at the expense of auto-term broadening. The method described here outperforms the traditional ones in that it removes cross terms between higher and lower spatial-frequency regions but without auto-term broadening.

The performance of the PWVD based on the described analytic image has been evaluated for some test images. Figure 5.4c shows a multicomponent test image composed of the sum of two cosine "zone plates" Figure 5.4a,b plus a small direct current (DC) term. Figure 5.5a (insets E–H) shows the results of the PWVD computation corresponding to the new analytic image described here, in comparison with two other alternative analytic signals shown in Figure 5.5(insets J–L and N–P, respectively). The four points of interest are shown in Figure 5.5a by the corresponding line intersections. For the first point corresponding to a low- to mid-range frequency, the three methods provide similar results but for the three

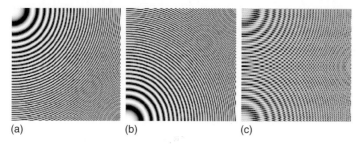

Figure 5.4 (a) Circular chirp "zone plate." Frequency variation is in the upward direction; (b) circular chirp "zone plate" $\cos(r^2)$. Frequency variation is in the downward direction; (c) image sum of (a) plus (b) plus a small DC term (10% of maximum). This is an example of a multicomponent image used here for testing. (Reproduced from Ref. [10] with permission of Institute of Electrical and Electronics Engineers © 1998 IEEE.)

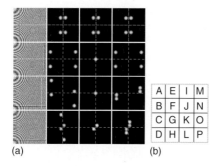

Figure 5.5 PWVD computation through the new analytic image [see (b) for legend]. (a) PWVD computation (spatial raised-cosine window size = 15 × 15; spatial-averaging window size = 5 × 5) of a composite of two $\cos(r^2)$ image "zone plates" (image size = 128 × 128) at several points of interest (indicated by the line intersection). The second column (E–H insets) corresponds to the analytic signal described here. Note either the presence of aliasing or the absence of spectrum' terms in the other two analytic signals (third (I–L insets) and fourth column (M–P insets) respectively. (b) Legend of (a). (Reproduced from [10] with permission of Institute of Electrical and Electronics Engineers © 1998 IEEE.)

other points corresponding to a higher frequency region, the two other analytic signal methods either fail to represent such frequency range or introduce aliasing.

5.5
2D Log-Gabor Filtering Schemes for Image Processing

The PWVD can be considered as a particular case of the complex spectrogram (i.e., a windowed Fourier transform), where the shifting window is the signal itself (complex conjugated). Both the spectrogram and the Wigner–Ville distribution belong to the general class. When the window function is a Gaussian, we obtain a "Gaussian" complex spectrogram as the inner product of the input signal and a localized Gabor function. The decomposition of a signal into its projections, on a set of displaced and modulated versions of a kernel function, appears in quantum optics and other areas of physics. The elements of the Gabor basis are the coherent states associated with the Weyl–Heisenberg group that sample the phase space of the representation. During the last 15 years, oriented bandpass multiresolution transforms have arisen with increased importance thanks to the development of steerable pyramids [17], Gabor multiresolutions [18], complex-valued wavelets [19, 20], curvelets [21], and contourlets [22], to name a few. Gabor functions are Gaussian multiplied by complex exponentials. Thus, in the Fourier domain, it is a Gaussian that is shifted from the origin. Gabor functions have a number of interesting mathematical properties: first, they have a smooth and infinitely differentiable shape. Second, their modulus is monomodal, that is, they have no side lobes, either in space or in the Fourier domain. Third, they are optimally joint-localized in space and frequency [23]. For such good properties, they have been proposed as ideal functions for signal and image processing. Classical Gabor filters give rise to significant difficulties when implemented in multiresolution. Filters overlap more importantly in the low frequencies than in the higher ones, yielding a nonuniform coverage of the Fourier domain. Moreover, Gabor filters do not have zero mean; they are then affected by DC components. The log-Gabor filters lack DC components and can yield a fairly uniform coverage of the frequency domain in an octave-scale multiresolution scheme [24].

In the spatial domain, complex log-Gabor filters are Gaussian modulated by a complex exponential. They have then both a real (cosine) and an imaginary (sine) part. The set of log-Gabor filters are normally set up along $n_s = 6$ scales and $n_t = 6$ to 8 orientations [25]. The choice of the number of orientations is motivated by the cortical simple cells' orientation sensitivity, which has been evaluated as 20–40°, which would imply the requirement of around 5–9 orientations to cover the 180° of the plane. The filters are defined in the log-polar coordinates of the Fourier domain as Gaussians that are shifted from the origin:

$$G_{pk}(\rho, \theta, p, k) = \exp\left(-\frac{1}{2}\left(\frac{\rho-\rho_k}{\sigma_\rho}\right)^2\right) \exp\left(-\frac{1}{2}\left(\frac{\theta-\theta_{(pk)}}{\sigma_\theta}\right)^2\right)$$

with
$$\begin{cases} \rho_k = \log_2(N) - k \\ \theta_{pk} = \begin{cases} \frac{\pi}{n_t} p & \text{if } k \text{ is odd} \\ \frac{\pi}{n_t}(p+\frac{1}{2}) & \text{if } k \text{ is even} \end{cases} \\ (\sigma_\rho, \sigma_\theta) = 0.996(\sqrt{\frac{2}{3}}, \frac{1}{\sqrt{2}}\frac{\pi}{p}) \end{cases}$$
(5.62)

where (ρ, θ) are the log-polar coordinates (octave scales), k indexes the scale, p is the orientation and N is the size of a $N \times N$ square image. The pair (ρ_k, θ_{pk}) corresponds to the frequency center of the filters and $(\sigma_\rho, \sigma_\theta)$ corresponds to the angular and radial bandwidths, respectively. One of the main advantages of the current scheme is the construction of the low-pass and high-pass filters [25]. This provides a complete scheme approximating a flat frequency response, which is obviously beneficial for applications such as texture synthesis, image restoration, image fusion, and image compression, among others, where the inverse transform is required. A Matlab toolbox that implements such log-Gabor filtering scheme is

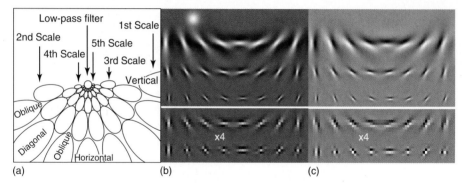

Figure 5.6 (a) Schematic contours of the log-Gabor filters in the Fourier domain with five scales and eight orientations (only the contours at 78% of the filter maximum are drawn); (b) the real part of the corresponding filters is drawn in the spatial domain. The two first scales are drawn at the bottom, magnified by a factor of 4 for a better visualization. The different scales are arranged in lines and the orientations are arranged in columns. The low-pass filter is drawn in the upper left part; (c) the corresponding imaginary parts of the filters are shown in the same arrangement. Note that the low-pass filter does not have an imaginary part. (Reproduced with permission of Springer Science © 2006 Springer.)

available from the authors [26]. The filters defined by Eq. (5.62) cover only one side of the Fourier domain (see Figure 5.6a). It is worth considering them in Fourier as the sum of a symmetric component and an antisymmetric component. Thus, these components sum up their amplitude on one side of the Fourier domain and cancel out themselves on the other side. This explains why in the spatial domain, the filters have both a real part (with cosine shape due to the symmetric component) and an imaginary part (with sine shape due to the antisymmetric component). A single log-Gabor filter defined in Fourier by Eq. (5.62) yields both a real and an imaginary part in the spatial domain. Both parts can be seen for the five-scale, eight-orientation scheme in Figure 5.6b and c, respectively.

5.6
Texture Segmentation

In this section, we illustrate the advantages of using a high-resolution joint representation given by the PWVD with an effective adaptive PCA through the use of feedforward neural networks. The task considered is texture segmentation for monitoring and detection of inhomogeneities. Texture is the term employed to characterize the surface of an object, and it is one of the important features used in object identification and recognition of an image. The term texture could be defined as the arrangement or spatial distribution of intensity variations in an image [27]. The two major characteristics of a texture are its coarseness and directionality. Since the spatial-frequency domain representation contains explicit information about the spatial distribution of an image, one expects to obtain useful textural features in the frequency domain. Many researchers have used a reduced set of filters or combinations of filters, for example, Gaussian or Gabor filters, for segmenting textures. Super and Bovik used a set of 72 filters for estimating the moments of the frequency distribution at each point [28]. However, there exist some applications, for example, shape from texture, that require a more dense sampling as provided by the PWVD or the spectrogram [29, 30]. This is because in that application the surface orientation of textures' surfaces causes slight shifts in frequency. Detection of defective textures and cracks are other challenging applications in which a denser set of filters could be more appropriate.

On the other hand, the PWVD of a 2D image increases the amount of redundant information; so, there is a need for data compression. Neural networks-based methods can be combined with spectral-based techniques as the PWVD for decorrelating such redundancy. A precedent of the current approach can be found in Ref. [31] in which an optical–digital approach for image classification by taking a Karhunen–Loève transform (KLT) expansion of Fourier spectra was proposed. This method can be considered as an extension of the modular learning strategy to 2D signals described in Ref. [32]. Many other examples can be found in the literature [33, 34]. The main idea is to consider a local transform, that is, the PWVD, as an N-dimensional feature detector followed by a neural-based PCA as a fast and adaptive spectral decorrelator. Neural networks-based methods provide an input–output mapping that can be considered as a nonparametric estimation of the data, that is, the knowledge of the underlying probability distribution is not required. In this way, we produce a substantial reduction in the volume of data due to the fact that after the PCA extraction we still have a local (but optimally reduced) representation.

The computational procedure consists in the following steps. First, the PWVD is computed by picking at random $N = 500$ points (i.e., 3% of the whole image size). This value provides enough accuracy for the subsequent PCA stage extraction. A small spatial window size ($h = 15 \times 15$ pixels) and a spatial-averaging window size (5×5 pixels) were considered. Points close to the image limits are excluded in order to avoid bordering effects. Prior to the PWVD computation, it is necessary to remove the global DC component of the texture in order to obtain a better discrimination. One of the distinct advantages of the PWVD over other joint representations is that it is a real-valued function, not necessarily positive. In this application, we have considered the absolute value of the PWVD in the subsequent stages of processing. The second step is to use the PWVD spectra previously generated for training a cascade recursive least squares (CRLS)-based neural network [35]. The CRLS network performs a data reduction process called *deflation*, that is, the network carries out the extraction process indirectly from the available errors. The CRLS method accomplishes the PCA extraction process in a serial way, that is, the training of a specific neuron is only triggered once the previous one has converged. In this way, the network automatically provides the number of eigenvectors for a given error. After the CRLS network convergence, the PWVD of each point is passed through the CRLS network that performs the data compression. The compressed data (i.e., the KLT coefficients) are used in the last step for classification. For the

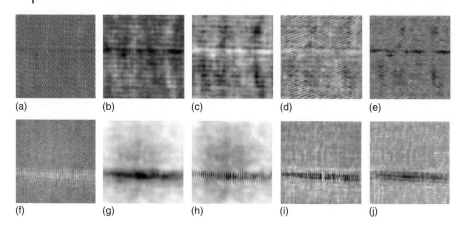

Figure 5.7 (a) Fabric texture with a *mispick* defective region along the middle row; (b–d) First three eigenvalue images after PWVD computation; (e) fused result of the first six eigenvalue images; (f) another *mispick* failure; (g–i) first three eigenvalue images after PWVD computation; (j) fused result of the first six eigenvalue images. Note that the detection of the defective regions can simply be obtained by binarization of the eigenvalue images. The fusion procedure, which is based on the 1D directional PWVD (Section 5.4.2), is described in Ref. [38].

classification process, we used a *K*-means clustering algorithm [36] or a simple thresholding of the KLT coefficients.

The described method has been tested for detecting the presence of defective areas in fabric textures (Figure 5.7) [37]. This is a very challenging problem for which other linear filtering methods do not give similar precision for such defect detection.

Besides the texture segmentation described here, we have successfully applied the 1D directional PWVD in other scenarios such as multifocus fusion [38], image denoising [39], noise modeling [40], image quality assessment [41], isotrigon texture modeling [42], and image restoration [43].

5.7
Hybrid Optical–Digital Implementation

Combining electronic technology with optical systems as a means of applying the fast processing ability and parallelism of optics to a wider range of problems was proposed by Huang and Kasnitz and subsequently developed by Davis and Mitchie [44]. The WVD computation can be done by using optical processors and storing the information in the computer for further analysis and processing. A hybrid optical–digital processor was proposed by Cristóbal *et al.* [45] on the basis of the optical setup developed by Bamler and Glunder [46]. Such a hybrid processor was subsequently improved by Gonzalo *et al.* [47]. Figure 5.8a presents the schematic diagram of the hybrid processor where the WVD computation is generated through the optical processor and stored in the computer for further analysis. Figure 5.8a

5.7 Hybrid Optical–Digital Implementation

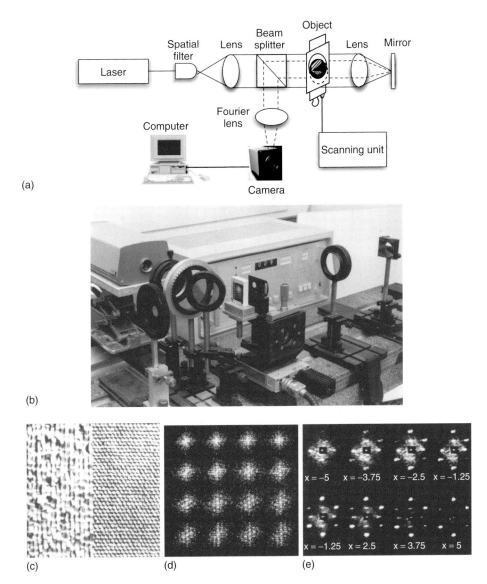

Figure 5.8 (a) Schematic diagram of the hybrid optical–digital processor; (b) experimental arrangement of the hybrid processor; (c) raffia–cotton canvas texture pair; (d) PWVD corresponding to eight points near the border of the two pairs; (e) spectra obtained with the hybrid processor corresponding to the same points as in (b). Note that the DC region has been masked to avoid overexposure. (Reproduced with permission of the Optical Society of America © 1989 OSA.)

and b shows the experimental arrangement of the optical system, which is connected through a charge-coupled device (CCD) camera to a digital image processor. A laser beam is spatially filtered to reduce noise, and then collimated. The beam goes through a beam splitter and illuminates the object transparency. The introduction of a lens and a plane mirror allows one to obtain the same object transmittance after reflection from the mirror, but rotated 180° with respect to the original one. A new transmission by the transparency yields the product image, which after reflection from the beam splitter, is Fourier transformed by a lens. In this way, the WVD for the center of the object transparency is obtained. Figure 5.8c–e presents the results obtained with the hybrid processor in the example of raffia–cotton texture transitions. The spectra are very similar to those generated via digital computation, except for the noise introduced by the camera acquisition process.

5.8
Conclusions

This chapter described the fundamentals of spatial/spatial-frequency representations paying special attention to the WVD and its properties. Two different digital implementations have been discussed: the 1D directional PWVD and the 2D-smoothed PWVD. The use of a small 1D window for the 1D directional PWVD analysis greatly decreases the computational cost versus a full 2D-smoothed PWVD analysis, while providing a pixel-wise spectral information of the data. Another joint representation based on a log-Gabor wavelet transform has been sketched here, which allows exact reconstruction and strengthening of the excellent mathematical properties of the Gabor filters. We show how spatial/spatial-frequency methods can be successfully applied to the analysis of image texture contents, in particular for applications such as monitoring and detection of inhomogeneities. Finally, a hybrid optical–digital processing system that allows one to compute the WVD optically has been described and further processing can be performed through the digital system.

Acknowledgments

The authors would like to thank A. Kumar for providing the fabric texture examples that are shown in this chapter. This work has been supported by the following grants: TEC2007-67025/TCM, TEC2007-30709E, and the Office of Naval Research.

References

1. Bartelt, K., Brenner, H.O., and Lohmann, A. (1980) The Wigner distribution and its optical production. *Opt. Comm.*, **32**, 32–38.
2. Jacobson, L. and Wechsler, H. (1988) Joint spatial/spatial-frequency representation. *Signal Proc.*, **14**, 37–68.
3. Reed, T. and Wechsler, H. (1990) Segmentation of textured images and Gestalt organization using spatial/spatial-frequency

representations. *IEEE Trans. Pattern Anal. Mach. Intell.*, **12**, 1–12.
4. Wigner, E. (1932) On the quantum correction for thermodynamic equilibrium. *Phys. Rev.*, **40**, 749–759.
5. Ville, J. (1948) Theorie et applications de la notion de signal analitique. *Cables Transm.*, **2A**, 61–74.
6. Cohen, L. (1966) Generalized phase-espace distribution functions. *J. Math. Phys.*, **7**, 781–786.
7. Claasen, T. and Mecklenbrauker, W. (1980) The Wigner distribution. a tool for time-frequency signal analysis. Part I. Continuous-time signals. *Philips J. Res.*, **35**, 217–250.
8. Jones, D. and Parks, T. (1992) A resolution comparison of several time-frequency representations. *IEEE Trans. Signal Proc.*, **40**, 413–420.
9. Martin, W. and Flandrin, P. (1985) Detection of changes of signal structure by using the Wigner-Ville spectrum. *Signal Proc.*, **8**, 215–233.
10. Hormigo, J. and Cristobal, G. (1998) High resolution spectral analysis of images using the Pseudo-Wigner distribution. *IEEE Trans. Signal Process.*, **46**, 1757–1763.
11. Zhu, Y. and Goutte, R. (1995) Analysis and comparison of space/spatial-frequency and multiscale methods for texture segmentation. *Opt. Eng.*, **34**, 269–282.
12. Brenner, K.H. (1983) A discrete version of the Wigner distribution function. Poster presented at the Proceedings of EURASIP, Signal Processing II: Theories and Applications, pp. 307–309.
13. Gonzalo, C. *et al.* (1989) Spatial-variant filtering through the Wigner distribution function. *Appl. Opt.*, **28**, 730–736.
14. Suzuki, H. and Kobayashi, F. (1992) A method of two-dimensional spectral analysis using the Wigner Distribution. *Electron. Commun. Jpn.*, **75**, 1006–1013.
15. Zhu, Y., Goutte, R., and Amiel, M. (1993) On the use of a two-dimensional Wigner-Ville distribution for texture segmentation. *Signal Proc.*, **30**, 205–220.
16. Hahn, S. (1992) Multidimensional complex signals with single-orthant spectra. *Proc. IEEE*, **80**, 1287–1300.
17. Simoncelli, W., Freeman, E.P., and Heeger, D.J. (1992) Shiftable multiscale transforms. *IEEE Trans. Inf. Theory*, **38**, 587–607.
18. Gross, M. and Koch, R. (1995) Visualization of multidimensional shape and texture features in laser range data using complex-valued Gabor wavelets. *IEEE Trans. Visual. Comput. Graph.*, **1**, 44–59.
19. Kingsbury, N. (2001) Complex wavelets for shift invariant analysis and filtering of signals. *J. Appl. Comput. Harmon. Anal.*, **10**, 234–253.
20. Portilla, J. and Simoncelli, E. (2000) A parametric texture model based on joint statistics of complex wavelet coefficients. *Int. J. Comput. Vis.*, **40**, 49–70.
21. Starck, J., Candes, E., and Donoho, D. (2002) The curvelet transform for image denoising. *IEEE Trans. Image Proc.*, **11**, 670–684.
22. Do, M. and Vetterli, M. (2005) The contourlet transform: an efficient directional multiresolution image representation. *IEEE Trans. Image Proc.*, **14**, 2091–2106.
23. Gabor, D. (1946) Theory of communication. *J. Inst. Electr. Eng.*, **93**, 429–457.
24. Field, D. (1987) Relation between the statistics of natural images and the response properties of cortical cells. *J. Opt. Soc. Am. A*, **4**, 2379–2394.
25. Fischer, S. *et al.* (2007) Self-invertible 2D Log-Gabor wavelets. *Int. J. Comput. Vis.*, **75**, 231–246.
26. Fischer, S., Redondo, R., and Cristóbal, G. (2009) How to construct log-Gabor filters? *Digit. CSIC*, http://hdl.handle.net/10261/14763.
27. Haralick, R. (1979) Statistical and structural approaches to texture. *Proc. IEEE*, **67**, 786–804.
28. Super, B. and Bovik, A. (1995) Shape from texture using local spectral moments. *IEEE Trans. Pattern Anal. Mach. Intell.*, **17**, 333–343.
29. Krumm, J. (1993) *Space frequency shape inference and segmentation of 3D surfaces*, Ph.D. thesis, Carnegie Mellon University, Robotics Institute.
30. Malik, J. and Rosenholtz, R. (1994) *Computing local surface orientation and shape from texture for curved surfaces.*

Tech. Rep. UCB/CSD 93/775, Computer Science Division (EECS), University of California, Berkeley.
31. Gorecki, C. (1991) Surface classification by an optoelectronic implementation of the Karhunen-Loève expansion. *Appl. Optics*, **30**, 4548–4553.
32. Haykin, S. and Battacharya, T. (1997) Modular learning strategy for signal detection in a nonstationary environment. *IEEE Trans. Signal Proc.*, **45**, 1619–1637.
33. Abeyskera, S. and Boashash, B. (1991) Methods of signal classification using the images produced by the Wigner-Ville distribution. *Pattern Recognit. Lett.*, **12**, 717–729.
34. Haykin, S. (1996) Neural networks expand SP's horizons. *Signal Proc. Mag.*, **13**, 24–49.
35. Cristóbal, G. and Hormigo, J. (1999) Texture segmentation through eigen-analysis of the Pseudo-Wigner distribution. *Patt. Rec. Lett.*, **20**, 337–345.
36. Theodoridis, S. and Koutroumbas, K. (2003) *Pattern Recognition*, 2nd edn, Elsevier.
37. Kumar, A. and Pang, G. (2002) Learning texture discrimination masks. *IEEE Trans. Syst. Man Cybern.*, **32**, 553–570.
38. Gabarda, S. and Cristóbal, G. (2005) Multifocus image fusion through Pseudo-Wigner distribution. *Opt. Eng.*, **44**, 047001.
39. Gabarda, S. et al. (2009) Image denoising and quality assessment through the Renyi entropy. *Proc. SPIE*, **7444**, 744419.
40. Gabarda, S. and Cristóbal, G. (2007) Generalized Rényi entropy: a new noise measure. *Fluctuations Noise Lett.*, **7**, L391–L396.
41. Gabarda, S. and Cristóbal, G. (2007) Image quality assessment through anisotropy. *J. Opt. Soc. Am. A*, **24**, B42–B51.
42. Gabarda, S. and Cristóbal, G. (2008) Discrimination of isotrigon textures using the Rényi entropy of Allan variances. *J. Opt. Soc. Am. A*, **25**, 2309–2319.
43. Gabarda, S. and Cristóbal, G. (2006) An evolutionary blind image deconvolution algorithm through the Pseudo-Wigner distribution. *J. Vis. Commun. Image Represent.*, **17**, 1040–1052.
44. Davis, L. and Mitchie, A. (1981) Edge detection in textures, in *Image Modelling*, Academic Press, pp. 95–109.
45. Cristóbal, G., Bescos, J., and Santamaria, J. (1989) Image analysis through the Wigner distribution function. *Appl. Opt.*, **28**, 262–271.
46. Bamler, R. and Glunder, H. (1983) The Wigner distribution function of two dimensional signals. Coherent optical generation and display. *Opt. Acta*, **30**, 1789–1803.
47. Gonzalo, C. et al. (1990) Optical-digital implementation of the Wigner distribution function: use in space variant filtering of real images. *Appl. Opt.*, **29**, 2569–2575.

6
Splines in Biomedical Image Processing
Slavica Jonic and Carlos Oscar Sanchez Sorzano

6.1
Introduction

The most common definition of splines is that they are piecewise polynomials with pieces smoothly connected together. To obtain a continuous representation of a discrete signal in one or more dimensions, one commonly fits it with a spline. The fit may be exact (interpolation splines) or approximate (least-squares or smoothing splines) [1]. By increasing the spline degree, one progressively switches from the simplest continuous representations (piecewise constant and piecewise linear) to the traditional continuous representation characterized by a bandlimited signal model (for the degree equal to the infinity). The traditional Shannon's sampling theory recommends the use of an ideal low-pass filter (antialiasing filter) when the input signal is not bandlimited. In the spline sampling theory, the Shannon's antialiasing filter is replaced with another filter specified by the spline representation of the signal [2].

The most frequently used splines are B-splines because of the computational efficiency provided by their short support. It has been shown that B-spline basis functions have the minimal support for a given order of approximation [3]. Cubic B-splines offer a good trade-off between the computational cost and the interpolation quality [1]. Also, B-splines are the preferred basis function because of their simple analytical form that facilitates manipulations [1]. Other interesting properties are that they are maximally continuously differentiable and that their derivatives can be computed recursively [1]. Thanks to the separability property of B-splines, the operations on multidimensional data can be performed by a successive processing of one-dimensional (1D) data along each dimension [1]. Besides, they have multiresolution properties that make them very useful for constructing wavelet bases and for multiscale processing [1, 4]. Because of all these properties, many image processing applications take advantage of B-splines.

In the first part of this chapter, we present the main theoretical results about splines which are of use in biomedical imaging applications (Section 6.2). First, we show how interpolation is related to sampling and the posterior reconstruction of the original signal from samples. When talking about spline interpolants, we can

Optical and Digital Image Processing: Fundamentals and Applications, First Edition. Edited by Gabriel Cristóbal, Peter Schelkens, and Hugo Thienpont.
© 2011 Wiley-VCH Verlag GmbH & Co. KGaA. Published 2011 by Wiley-VCH Verlag GmbH & Co. KGaA.

distinguish between tensor product splines and polyharmonic splines. The former is based on the tensor product of two 1D spline functions, while the latter is based on the use of the so-called radial basis functions. Finally, we review the multiscale properties of splines that allow one to address problems in a coarse-to-fine fashion. These multiscale approaches are not only faster but also usually more robust to noise and yield smoother functionals when optimization problems are involved. In the second part of the chapter (Section 6.3), we illustrate the applications of splines in biomedical image processing by showing their use in rigid-body and elastic image and volume registration.

6.2
Main Theoretical Results about Splines

In this section, we present the main results of the spline theory that are used in biomedical imaging applications.

6.2.1
Splines as Interpolants and Basis Functions

6.2.1.1 Tensor Product Splines

The presentation of the theoretical properties of the splines is done in 1D space. However, the extension to two-dimensional (2D) space is readily performed using the tensor product. For instance, if $\varphi_{1D}(x)$ is a 1D spline, the 2D tensor product spline is defined as $\varphi_{2D}(x, y) = \varphi_{1D}(x)\varphi_{1D}(y)$.

The Interpolation Context Given a discrete set of measurements (x_i, y_i), interpolating is the art of "filling in the gaps" with a continuous function $y = f(x)$ such that we meet the constraints imposed by our measurements, that is, $y_i = f(x_i)$. In biomedical imaging, our measurements are typically image values, $z_i = I(x_i, y_i)$, with z_i being the gray value of our image, and (x_i, y_i) being its location in space. For color images, we could decompose the color into three components (red, green, and blue; hue, saturation, and value; etc.); each one would impose a constraint of the same kind as the one for gray values. Interpolating is important to know the value of our image between known pixel values. This is useful for rotations, translations, unwarpings, demosaicking, downsampling, upsampling, and so on.

Given N measurements, we can estimate a polynomial of degree $N - 1$ (it has N coefficients and, therefore, N degrees of freedom) passing through all these measurements. In fact, this polynomial is unique and it is given by Newton's general interpolation formula:

$$f(x) = a_0 + a_1(x - x_0) + a_2(x - x_0)(x - x_1) + \cdots \\ + a_{N-1}(x - x_0)(x - x_1)\cdots(x - x_{N-2}) \qquad (6.1)$$

where a_n represents the divided differences defined as

$$a_0 = y_0$$
$$a_n = \sum_{i=0}^{n} \frac{y_i}{\prod_{\substack{j=0 \\ j \neq i}}^{n} (x_i - x_j)} \qquad (6.2)$$

The use of polynomials for interpolation is justified because they are simple to manipulate, differentiate, or integrate; it is also justified by a theorem of Weierstrass that states that any continuous function on a closed interval can be approximated uniformly to any degree of accuracy by a polynomial of a sufficient degree [5].

Regularly Spaced Interpolation: The Generalized Sampling Theorem However, polynomials are not the only functions interpolating the dataset. If the x_i points are regularly distributed ($x_i = iT$, for some integer i and a sampling rate T), then Whittaker showed that the series

$$C(x) = \sum_{i=-\infty}^{\infty} y_i \operatorname{sinc}\left(\frac{x - x_i}{T}\right) \qquad (6.3)$$

also interpolates the input measurements [6] $\left(\operatorname{sinc} \text{ is defined as } \operatorname{sinc}(x) = \frac{\sin(\pi x)}{\pi x}\right)$. $C(x)$ is called the *cardinal function*. Shannon [7] realized that this representation was unique for any function whose maximum frequency is smaller than $\frac{1}{2T}$ Hz. This is the famous Shannon sampling theorem which is valid for any function that is bandlimited.

The sampling theorem can be extended to a larger space of functions: the Hilbert space of L_2 functions, that is, all functions that are square integrable in the Lebesgue sense ($\|f\|^2 = \langle f, f \rangle < \infty$, where the inner product between two real functions is defined as $\langle f, g \rangle = \int_{-\infty}^{\infty} f(x)g(x)dx$). The set of bandlimited functions is a subset of L_2. The sampling theorem is generally formulated in this space as

$$C(x) = \sum_{i=-\infty}^{\infty} c_i \varphi_i(x), \qquad (6.4)$$

where c_i are some coefficients that have to be computed from the y_i input data, and $\varphi_i(x)$ is a shifted version of a basis function $\varphi(x)$ ($\varphi_i(x) = \varphi(x - x_i)$). In the particular case of bandlimited functions, the basis function used is $\varphi(x) = \operatorname{sinc}(\frac{x}{T})$ and the set of all bandlimited functions is spanned by the family of functions $\varphi_i(x)$. In other words, any bandlimited function can be expressed by a linear combination of the infinite set of $\varphi_i(x)$ functions as the sampling theorem shows. From now on, we drop the argument of functions as long as there is no confusion.

Let us consider any function φ in L_2, and the subspace V generated by its translations with step T:

$$V = \left\{ f(x) = \sum_{i=-\infty}^{\infty} c_i \varphi_i(x) : c_i \in l_2 \right\} \qquad (6.5)$$

where l_2 is the space of all square-summable sequences. It can be easily proved that if a set of functions φ_i in a Hilbert space is orthonormal (i.e., $\langle \varphi_i, \varphi_j \rangle = \delta_{i-j}$), then the projection of any function f in L_2 onto the subspace V is

$$\tilde{f} = P_V f = \arg\min_{g \in V} \|f - g\| = \sum_{i=-\infty}^{\infty} \langle f, \varphi_i \rangle \varphi_i \qquad (6.6)$$

Note that

$$\langle f, \varphi_i \rangle = \int_{-\infty}^{\infty} f(x)\varphi(x - x_i)dx = f(x) * \varphi(-x)\big|_{x=x_i} \qquad (6.7)$$

that is, the inner product $\langle f, \varphi_i \rangle$ can be easily computed by sampling the linear convolution of $f(x)$ and $\varphi(-x)$ at $x = x_i$. In fact, the antialiasing filter used to effectively limit the frequency of a signal before sampling corresponds exactly to the computation of this inner product. Note that if f is already bandlimited, then the convolution of $f(x)$ with $\varphi(-x)$ (whose frequency response is just a rectangle with a maximum frequency of $\frac{1}{2T}$) is $f(x)$ and, thus, $c_i = y_i$.

The problem of using $\varphi(x) = \text{sinc}\left(\frac{x}{T}\right)$ is that the sinc decays very slowly and the convolutions needed for the computation of the inner product are impractical. Thus, one might set off in search of new functions φ with more computationally appealing properties. First of all, one could relax the orthonormality condition and may ask only that the set $\{\varphi_i\}$ defines a Riesz basis, that is,

$$A \|c_i\|_{l_2}^2 \le \left\|\sum_{i=-\infty}^{\infty} c_i \varphi_i(x)\right\|_{L_2}^2 \le B \|c_i\|_{l_2}^2 \quad \forall c_i \in l_2 \qquad (6.8)$$

for some constants A and B depending on φ. Orthonormality is a special case when $A = B = 1$. The left-hand side condition assures that the basis functions are linearly independent, while the right-hand side condition assures that the norm of f is bounded and, therefore, the subspace V is a valid subspace of L_2. An important requirement for the basis is that it is able to represent any function with any desired level of accuracy by simply diminishing T. It can be shown that this condition can be reformulated as the partition of unity [1]:

$$\sum_{i=-\infty}^{\infty} \varphi_i(x) = 1 \quad \forall x \qquad (6.9)$$

i.e. the sum of all the basis functions (shifted versions of φ) is a constant function.

The function $\varphi(x) = \text{sinc}\left(\frac{x}{T}\right)$ meets all these requirements and is, by far, the most widely known function due to the sampling theorem, besides the fact that it is infinitely supported, making convolutions more complicated. However, other functions also meet these requirements. The shortest function meeting these is the box function

$$\varphi(x) = \beta_0 \left(\frac{x}{T}\right) = \begin{cases} 1 & |x| < \frac{T}{2} \\ 0 & |x| > \frac{T}{2} \end{cases} \qquad (6.10)$$

which is actually the cardinal B-spline of degree 0. The cardinal B-spline of degree n is simply obtained by convolving $\beta_0(x)$ with itself n times. For instance, $\beta_1\left(\frac{x}{T}\right)$ is the triangular function defined between $-T$ and T, and $\beta_2\left(\frac{x}{T}\right)$ is a parabolic

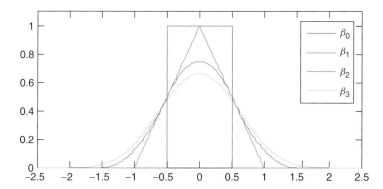

Figure 6.1 B-Splines Plot of the first four B-splines (degrees 0, 1, 2, and 3). (Please find a color version of this figure on the color plates.)

function defined between $-\frac{3}{2}T$ and $\frac{3}{2}T$. In general, $\beta_n\left(\frac{x}{T}\right)$ is an evenly symmetric, piecewise polynomial of degree n defined for $|x| < \frac{n+1}{2}T$. The following equations show the cardinal B-splines of degree 1, 2, and 3 for $T = 1$ (see also Figure 6.1):

$$\beta_1(x) = \begin{cases} 1 - |x| & |x| \leq 1 \\ 0 & |x| > 1 \end{cases} \tag{6.11}$$

$$\beta_2(x) = \begin{cases} \frac{3}{4} - |x|^2 & |x| \leq \frac{1}{2} \\ \frac{1}{2}\left(|x| - \frac{3}{2}\right)^2 & \frac{1}{2} < |x| \leq \frac{3}{2} \\ 0 & |x| > \frac{3}{2} \end{cases} \tag{6.12}$$

$$\beta_3(x) = \begin{cases} \frac{2}{3} + \frac{1}{2}|x|^2(|x| - 2) & |x| \leq 1 \\ \frac{1}{6}(2 - |x|)^3 & 1 < |x| \leq 2 \\ 0 & |x| > 2 \end{cases} \tag{6.13}$$

In general, the cardinal B-spline of degree n can be expressed as

$$\beta_n(x) = \frac{1}{n!} \sum_{k=0}^{n+1} (-1)^k \binom{n+1}{k} \left(x - \left(k - \frac{n+1}{2}\right)\right)_+^n \tag{6.14}$$

where $(x)_+^n$ is the one-sided n-th power of x, $(x)_+^n = \begin{cases} x^n & x \geq 0 \\ 0 & x < 0 \end{cases}$.

Cardinal B-splines are piecewise polynomials, because they can be represented by polynomials of degree n on $|x|$ that are different for each interval $i\,T \leq |x| < (i+1)\,T$. All these functions meet the aforementioned requirements and therefore can be used as basis functions for the representation of signals. Moreover, they are well localized and compactly supported in real space (consequently, infinitely supported in Fourier space), making computations in real space affordable.

The representation of signals using cardinal B-splines is intimately related to interpolation: the use of β_0 is equivalent to nearest neighbor interpolation and the use of β_1 is equivalent to linear interpolation (bilinear interpolation when the 2D tensor product B-spline is constructed). β_3 has proved to be a good compromise between computational complexity, compact support in real space, and approximation error.

At this point, we have already shown that the sampling theorem is a particular case of an interpolation problem and also shown how it can be generalized for L_2 functions instead of working exclusively with bandlimited functions. The question is how to compute the representation coefficients c_i in Eq. (6.4). We have seen a possibility for orthonormal basis in Eq. (6.7). However, cardinal B-splines are not orthonormal in general (only cardinal B-splines of degree 0 are orthogonal). In the general case, the calculation of the c_i coefficients is similar to the orthonormal case. The only difference is that, instead of using the basis φ_i itself, we have to use the dual basis $\tilde{\varphi}_i$:

$$\tilde{f} = P_V f = \sum_{i=-\infty}^{\infty} \langle f, \tilde{\varphi}_i \rangle \varphi_i \tag{6.15}$$

The dual basis is uniquely defined by the biorthogonality condition $\langle \tilde{\varphi}_i, \varphi_j \rangle = \delta_{i-j}$ and it also inherits the shift invariant property of the original basis function, $\tilde{\varphi}_i(x) = \tilde{\varphi}(x - x_i)$. However, we still do not have a clear way of computing the c_i's because we do not have a close-form formula for the dual basis. It can be proved [2] that the Fourier transform of $\tilde{\varphi}$ is given by

$$\widehat{\tilde{\varphi}}(j\Omega) = \frac{\widehat{\varphi}(j\Omega)}{\sum_{k=-\infty}^{\infty} |\widehat{\varphi}(j(\Omega + 2\pi k))|^2} \tag{6.16}$$

where $\widehat{f}(j\Omega)$ represents the continuous Fourier transform of the function $f(x)$. Fortunately, Unser and coworkers [8, 9] derived a very efficient way of computing these coefficients using standard digital filters. This is actually the way of producing these coefficients. These filters are derived in 1D. However, they are easily extended to n D. In the case of images, these filters are run individually on each row of the image, producing a new image of coefficients over the horizontal axis, x. Then, they are run on each column of the new image, finally producing the coefficients of the 2D tensor product B-spline. Once the coefficients are produced, images are treated as if they were continuous functions, although they are stored as a discrete set of cardinal B-spline coefficients.

Now, we may wonder if we could design a basis function based on B-splines such that $c_i = y_i$. This would be an interpolating spline and we would be back to a situation similar to the interpolation scheme presented in the sampling theorem, Eq. (6.3). The following function is such an interpolating spline:

$$\varphi_{\text{int}}(x) = \sum_{i=-\infty}^{\infty} q_{\text{int}}[i]\varphi_i(x) \tag{6.17}$$

where $q_{\text{int}}[i]$ is the l_2 sequence defined as the inverse Z transform of $Q_{\text{int}}(z) = \frac{1}{\sum_{k=-\infty}^{\infty} \varphi(kT)z^{-k}}$, and $\varphi(x) = \beta_n\left(\frac{x}{T}\right)$.

Approximation Error: How Far are We from the Truth? An important concept related to the generalized sampling theorem explained above is how well it reproduces any function f. Let us call f_T the approximation with a given sampling rate T. This problem has been studied by the approximation theory, proving that the following three statements are equivalent:

1) Let f be a sufficiently smooth function (f belongs to the Sobolev space W_2^L, i.e., its L first derivatives belong to L_2). Then, as T approaches 0, there exists a constant C independent of f such that $\|f - f_T\| \leq CT^L \|f^{(L)}\|$.
2) The first L moments of φ are constants, that is, $\sum_{i=-\infty}^{\infty} (x - x_i)^m \varphi_i(x) = \mu_m$ for $m = 0, 1, \ldots, L-1$.
3) The first L monomials can be exactly represented, that is, for each monomial x^m ($m = 0, 1, \ldots, L-1$), there exist constants $c_i \in l_2$ such that $x^m = \sum_{i=-\infty}^{\infty} c_i \varphi_i(x)$.

An important consequence of this result is that the approximation error of different basis functions depends mostly on their design, that is, given two polynomials of the same degree, one of them may approach smooth functions more quickly than the other. L is called the *order of approximation* and entirely depends on the moments of φ or the number of monomials that can be represented. In particular, B-splines of degree n have an order of approximation of $L = n + 1$. Another important consequence is that in order to converge to function f as $T \to 0$, the basis function φ must have $L \geq 1$, or in other words, that it fulfills the partition of unity. It is well known [10] that windowed sincs (which are commonly used as a solution to the infinite support of the sinc function) do not meet this condition. One may also try to design the φ family such that they have order of approximation L with a minimum support. This is how the MOMS (maximal-order interpolation of minimal support) set of functions is designed. It turns out that these functions are linear combinations of the cardinal B-spline of degree L and its derivatives [3].

The reader interested in the generalized sampling theorem and this approach to interpolation may gain information from Refs [1, 2, 10–12].

Back to Irregular Interpolation Problems So far, we have already introduced cardinal B-splines, dual splines, and interpolation splines. In fact, splines are a broad family of functions of which we have only seen those used with a regular spacing. In general, a function $S_n(x)$ is a spline of degree n if (i) it is defined by piecewise polynomials of degree at most n; (ii) it is of class C^{n-1}, that is, it has $n-1$ continuous derivatives even at the points joining the different polynomial pieces [13].

An alternative approach to splines is based on the idea of curve interpolation and knots instead of the idea of sampling. This other approach allows for irregularly spaced samples in a much more direct way. Let us assume that we are interpolating a real function in the interval $[a, b]$ with a piecewise polynomial. We subdivide this interval into N adjacent pieces such that each piece is defined in the interval $[x_i, x_{i+1}]$. The subdivision is such that $x_0 = a$, $x_N = b$, $x_i < x_{i+1}$, and $[a, b] = \bigcup_{i=0}^{N-1} [x_i, x_{i+1}]$. The input samples (x_i, y_i) are called *knots* and they are fixed points through which the interpolated polynomial must pass. Note that knots need not to be equally spaced in the interval [a, b]. We have to be specially careful in selecting the interpolating polynomials such that not only the spline is continuous but also all its derivatives up to the order of $n-1$ are continuous, even at the knots where the function on the left-hand side of the knot is defined by a certain polynomial, and on the right-hand side it is defined by a different polynomial (see Figure 6.2).

With $N+1$ knots, the spline is split into N intervals. A spline of degree n has $n+1$ coefficients in each interval; therefore, the spline has $(n+1)N$ degrees of freedom. Let us call $S_i(x)$ the polynomial of degree n in each interval ($i = 1, 2, ...N$). This spline must satisfy the following:

1) Interpolation of the knot values: $2N$ degrees of freedom

$$S_1(x_0) = y_0$$
$$S_i(x_i) = y_i = S_{i+1}(x_i) \quad i = 1, 2, ..., N-1 \quad (6.18)$$
$$S_N(x_N) = y_N$$

2) Continuity of the $n-1$ derivatives: $(n-1)(N-1)$ degrees of freedom

$$\begin{aligned} S_i^{(1)}(x_i) &= S_{i+1}^{(1)}(x_i) & i &= 1, 2, ..., N-1 \\ S_i^{(2)}(x_i) &= S_{i+1}^{(2)}(x_i) & i &= 1, 2, ..., N-1 \\ &\vdots \\ S_i^{(n-1)}(x_i) &= S_{i+1}^{(n-1)}(x_i) & i &= 1, 2, ..., N-1 \end{aligned} \quad (6.19)$$

However, there are still $n-1$ unfixed degrees of freedom. Depending on the way these degrees of freedom are defined, different kinds of splines are defined. For instance, a natural spline of degree $n = 3$ is one of the most common cases, in which the second derivative at the extremes is set to 0, $S_i^{(2)}(x_0) = S_i^{(2)}(x_N) = 0$. There are efficient algorithms for the solution of the resulting equation system [14].

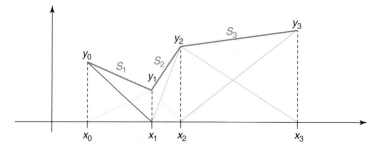

Figure 6.2 Spline interpolation with irregularly spaced samples. Example of B-spline (degree 1) interpolation of irregularly spaced samples. (Please find a color version of this figure on the color plates.)

The expression of the polynomial of degree n in any of the intervals $[x_i,\ x_{i+1})$ can be recursively constructed as

$$S_{i,0}(x) = \begin{cases} 1 & x_i < x \le x_{i+1} \\ 0 & \text{otherwise} \end{cases}$$
$$S_{i,l}(x) = \frac{x-x_i}{x_{i+l}-x_i} S_{i,l-1}(x) + \frac{x_{i+l+1}-x}{x_{i+l+1}-x_i} S_{i+1,l-1}(x)$$
(6.20)

where $S_{i,l}$ is the B-spline of degree l in the interval $[x_i,\ x_{i+1})$. It is common to repeat the extremes of the knots as many times as the finally desired spline degree. For example, the actual list of knots for degree 3 would be $\{x_0, x_0, x_0, x_0, x_1, x_2, ..., x_{N-1}, x_N, x_N, x_N, x_N\}$. Because of the repeated knots we admit by convention in the previous formula that $\frac{0}{0} = 0$.

The polynomial in each one of the subintervals is of the general form

$$S_i(x) = \sum_{l=0}^{n} a_l x^l \qquad (6.21)$$

In this expansion, we have used the fact that the set $\{1,\ x,\ x^2,\ ...,\ x^n\}$ is a basis of the polynomials of degree n. However, this is not an orthonormal basis, which causes numerical instabilities when solving for the a_l coefficients. Alternatively, we could have used any other basis of the polynomials of degree n, $\{P_0(x),\ P_1(x),\ P_2(x),\ ...,\ P_n(x)\}$:

$$S_i(x) = \sum_{l=0}^{n} a_l P_l(x) \qquad (6.22)$$

Employing different polynomial basis gives rise to different kinds of splines: Bernstein polynomials are used in the Bézier splines; Hermite polynomials are used in the Hermite splines; basic splines are used in the B-splines; and so on.

6.2.1.2 Polyharmonic Splines

The extension to several dimensions can be done by the tensor product of 1D splines as seen in the previous section, or by the specific design of the so-called polyharmonic splines, among which the most famous is the thin-plate spline.

Let us assume that we have a set of input multivariate data points $(\mathbf{x}_i,\ y_i)$ we would like to interpolate ($\mathbf{x}_i \in \mathbb{R}^d$, $y_i \in \mathbb{R}$). The goal would be to find a hypersurface $y = f(\mathbf{x})$ such that the surface contains the input data points.

We look for our interpolant in the Beppo-Levi space $BL^{(2)}(\mathbb{R}^d)$, that is, the space of all L_2 functions from $\mathbb{R}^d$ to $\mathbb{R}$, such that its second derivative is also in L_2. This space is large enough to contain a suitable interpolator. In this space, we can define the rotationally invariant seminorm:

$$\|f\|^2 = \int_{\mathbb{R}^d} \sum_{\eta=1}^{d} \sum_{\xi=1}^{d} \frac{\partial^2 f}{\partial x_\eta \partial x_\xi} d\mathbf{x} \qquad (6.23)$$

Duchon [15] showed that the interpolant minimizing the just introduced seminorm in $BL^{(2)}(\mathbb{R}^d)$ is of the form

$$f(\mathbf{x}) = p_m(\mathbf{x}) + \sum_{i=1}^{N} c_i \phi_{d,k}(\|\mathbf{x} - \mathbf{x}_i\|) \tag{6.24}$$

where $\|\mathbf{x}\|$ denotes the standard Euclidean norm in $\mathbb{R}^d$, N is the number of measurements in the input data, c_i is a set of coefficients that need to be determined, k is any integer such that $2k > d$ (in fact, the approximation order is $k - \frac{d}{2}$ [16]; therefore, on the one hand, it is interesting to choose high values of k although, on the other hand, these result in numerical instabilities in the determination of the spline coefficients as is seen later; a trade-off between these two goals must be achieved), $\phi_{d,k}$ is a radial basis function, and $p_m(\mathbf{x})$ is a polynomial in $\mathbf{x}$ of degree at most m, which is given by $m = k - \lceil \frac{d}{2} \rceil$. The radial basis function $\phi_{d,k}$ is

$$\phi_{d,k}(r) = \begin{cases} r^{2k-d} \log(r) & \text{for even } d \\ r^{2k-d} & \text{for odd } d \end{cases} \tag{6.25}$$

The function $\phi_{2,2} = r^2 \log(r)$ is the so-called thin-plate spline in $\mathbb{R}^2$ and the minimization of the seminorm in $\mathbb{R}^2$ can be understood as the minimization of the bending energy of a thin sheet of metal that interpolates the input data points. In $\mathbb{R}^3$, $\phi_{3,2}$ is called the *biharmonic spline* and $\phi_{3,3}$ is called the *triharmonic spline*.

The interpolation equations $f(\mathbf{x}_i) = y_i$ do not fully determine the function f (note that we have N input data points, but N coefficients c_i for the radial basis functions and $\sum_{p=0}^{m} \binom{d}{p}$ coefficients for the polynomial). In fact, the polynomial of degree m comes from the fact that it is in the kernel of the seminorm and, therefore, the addition of any polynomial of degree m is "invisible" to the seminorm minimization (for instance, for the thin-plate spline case, $m = 1$, and all polynomials of degree 1 have null second derivatives). In this way, we have to impose extra conditions which generally are

$$\sum_{i=1}^{N} c_i q(\mathbf{x}_i) = 0 \tag{6.26}$$

for all polynomials $q(\mathbf{x})$ of degree at most m (in the case of the thin-plate spline, we would have to use $q(\mathbf{x}) = 1$, $q(\mathbf{x}) = x_1$, and $q(\mathbf{x}) = x_2$). Let us assume that the set of coefficients of the polynomial p are written in vector form as $\mathbf{p}$. Then, the polyharmonic interpolation can be solved by the following equation system

$$\begin{pmatrix} \Phi & P \\ P^t & 0 \end{pmatrix} \begin{pmatrix} \mathbf{c} \\ \mathbf{p} \end{pmatrix} = \begin{pmatrix} \mathbf{y} \\ \mathbf{0} \end{pmatrix} \tag{6.27}$$

where $\mathbf{c}$ and $\mathbf{y}$ are column vectors with the c_i coefficients and the y_i measurements, respectively. Φ is the $N \times N$ system matrix corresponding to the measurements, i.e., $\Phi_{ij} = \phi_{d,k}(\|\mathbf{x}_i - \mathbf{x}_j\|)$ and P is a matrix related to some basis of polynomials up to degree m. Let $\{p_1, p_2, ..., p_l\}$ be such a basis; then $P_{ij} = p_j(\mathbf{x}_i)$. For example, for the thin-plate spline case, we could define $p_1(\mathbf{x}) = 1$, $p_2(\mathbf{x}) = x_1$, and $p_3(\mathbf{x}) = x_2$, but any other basis of polynomials of degree at most 1 would do. Note that the size of matrix P is $N \times l$ (in the thin-plate spline, $l = 3$).

This equation system is usually ill-conditioned due to the nonlocal nature of the $\phi_{d,k}(r)$ functions (they are not locally supported, instead they grow to infinity with growing r). Moreover, the complexity of solving the equation system depends on the number of sample points, N. Finally, the evaluation of the polyharmonic spline involves as many operations as the input data points (although fast algorithms have been developed for tackling this latter problem [17]). For solving the problem of the ill-conditioning of the equation system, a localization of the polyharmonic spline can be done. This is a process in which the noncompactly supported $\phi_{d,k}(r)$ function is substituted as a weighted sum of compactly supported functions (for instance, B-splines). For further information on this technique the reader is referred to Refs [18, 19].

In the case of noisy data, we can relax the interpolation condition by replacing it by a least-squares approximation condition:

$$f^* = \arg \min_{f \in BL^{(2)}(\mathbb{R}^d)} \lambda \|f\|^2 + \frac{1}{N} \sum_{i=1}^{N} (y_i - f(\mathbf{x}_i))^2 \tag{6.28}$$

It can be proved [20] that the coefficients **c** and **p** are the solutions of the following linear equation system:

$$\begin{pmatrix} \Phi - 8N\lambda\pi I & P \\ P^t & 0 \end{pmatrix} \begin{pmatrix} \mathbf{c} \\ \mathbf{p} \end{pmatrix} = \begin{pmatrix} \mathbf{y} \\ \mathbf{0} \end{pmatrix} \tag{6.29}$$

One may wonder why these radial basis functions are called splines; at least, they do not seem to fit our previous definition of piecewise polynomial functions. The solution is a slight modification of our concept of spline (particularly, the requirement of being piecewise). Let us consider the triharmonic ($\phi_{3,3} = r^3$) functions. It is clear that it is a cubic polynomial in r, and its second derivative is continuous everywhere. Therefore, it is a cubic spline.

The reader interested in this topic is referred to Refs [16, 21, 22].

6.2.2
Splines for Multiscale Analysis

The multiscale capabilities of splines come in two different flavors: spline pyramids and spline wavelets. Each one of these approaches exploits a different feature of splines that makes them suitable for multiresolution analysis. Because of the space limitations imposed for the chapter, we only describe the multiresolution spline pyramids here. The reader interested in spline wavelets is referred to Refs [1, 23].

Spline Pyramids Let us assume that we know the representation of a certain function $f(x)$ with B-splines of odd degree n and a sampling rate T:

$$f(x) = \sum_{i=-\infty}^{\infty} c_i \beta_n \left(\frac{x}{T} - i\right) = \left(\sum_{i=-\infty}^{\infty} c_i \delta(x - i)\right) * \beta_n \left(\frac{x}{T}\right)$$

$$= c_{\scriptscriptstyle\amalg}(x) * \beta_n \left(\frac{x}{T}\right) \tag{6.30}$$

One may wonder how a finer representation of f would be. For this, we consider the relationship between an odd degree B-spline and its contraction by M to a final sampling rate $\frac{T}{M}$. It can be shown [1] that

$$\beta_n\left(\frac{x}{T}\right) = \sum_{k=-\infty}^{\infty} h_k \beta_n\left(\frac{x}{\frac{T}{M}} - k\right) = h_{\text{U}}(x) * \beta_n\left(\frac{x}{\frac{T}{M}}\right) \quad (6.31)$$

that is, we can decompose a wide B-spline of degree n as the weighted sum of thinner B-splines of the same degree. For any M and n, the Z transform of the weight sequence h_k is

$$H(z) = z^{\frac{(M-1)(n+1)}{2}} \frac{1}{M^n} \left(\sum_{m=0}^{M-1} z^{-m}\right)^{n+1} \quad (6.32)$$

The case $M = 2$ plays an important role in the design of wavelets, and the corresponding property is called the 2-scale relationship. In case of B-splines of degree 1, the two scale relationship is simply

$$\beta_1\left(\frac{x}{T}\right) = \frac{1}{2}\beta_1\left(\frac{x}{\frac{T}{2}} + 1\right) + \beta_1\left(\frac{x}{\frac{T}{2}}\right) + \frac{1}{2}\beta_1\left(\frac{x}{\frac{T}{2}} - 1\right) \quad (6.33)$$

and, in general, for splines of odd degree n and $M = 2$, we have

$$h_k = \begin{cases} 2^{-n}\binom{n+1}{k+\frac{n+1}{2}} & |k| \leq \frac{n+1}{2} \\ 0 & |k| > \frac{n+1}{2} \end{cases} \quad (6.34)$$

Substituting the expression of the wide B-splines by the weighted sum of fine B-splines, we obtain

$$f(x) = \left((\uparrow_M \{c_{\text{U}}(x)\} * h)_{\text{U}}(x)\right) * \beta_n\left(\frac{x}{\frac{T}{M}}\right) \quad (6.35)$$

In other words, to obtain a finer representation of a signal, we simply have to upsample its B-spline coefficients and convolve them with a finite weight sequence depending on the scaling factor M and the spline degree. Note that the function represented with splines at the finer resolution is exactly the same as the original one. No interpolation has been performed on the way.

Creating a coarser representation of the function $f(x)$ is a little bit more involved since we cannot have exactly the same function but an approximation to it:

$$\tilde{f}(x) = \tilde{c}_{\text{U}}(x) * \beta_n\left(\frac{x}{MT}\right) \quad (6.36)$$

and we have to devise a way of estimating the coefficients $\tilde{c}$ from the c coefficients. The easiest way is to look for the $\tilde{c}$ that minimize the L_2 norm of the error $\|f - \tilde{f}\|$. It can be proved [24] that the solution to this least-squares problem is

$$\tilde{c} = \frac{1}{2}\left((b_k^{2n+1})^{-1} * \downarrow_M \{h_k * b_k^{2n+1} * c_k\}\right), \quad (6.37)$$

with $\downarrow_M \{\cdot\}$ being the downsampling operator, c_k being the B-spline coefficients of the function f with sampling rate T, h_k being the sequence described in Eq. (6.34), and b_k^{2n+1} being the sequence $b_k^{2n+1} = \beta_{2n+1}(k)$. Note that $\left(b_k^{2n+1}\right)^{-1}$ is the inverse of this sequence. It can be easily understood by inverting the Z transform of the sequence b_k^{2n+1} and then performing an inverse Z transform. While b_k^{2n+1} is compactly supported, $\left(b_k^{2n+1}\right)^{-1}$ is not, and convolution with this sequence requires the design of an IIR (infinite impulse response) filter [9, 25].

6.3 Splines in Biomedical Image and Volume Registration

In this section, we show two examples of the use of splines for image and volume registration, which is one of the most challenging tasks in biomedical image processing.

The intensity-based registration can be viewed in an optimization framework in which the registration problem consists in searching for a geometric transformation of a source image/volume that gives the image/volume that best matches a target image/volume, under a chosen similarity measure. The restriction of the motion to rigid-body motion means that the distance between the points of the object is the same in the registered source and target images/volumes. Elastic registration is frequently employed in medical image analysis to combine data that describe anatomy, both because biological tissues are in general not rigid and because anatomy varies between individuals. In the intensity-based elastic registration techniques, the solution of the registration problem is the deformation field that warps the source image/volume so that the resulting image/volume best matches the target image/volume. The registration is achieved by minimizing a cost function, which represents a combination of the cost associated with the image/volume similarity and the cost associated with the smoothness of the transformation (regularization term).

Many authors have proposed to use linear combinations of B-splines placed on a regular grid to model the transformation [27–31]. The available techniques differ in the form of employed regularization term, as well as in the employed image/volume similarity metrics and the optimization method. They produce good results but have a high computational cost. The computation can be accelerated using multiscale image/volume processing, and spline pyramids provide a convenient tool for this. Moreover, spline model can be used for all computation aspects of the registration (image pyramid, transform, and the gradient of the optimization criterion) as shown in Refs [4, 31, 32].

Many examples of rigid-body image registration can be found in 3D electron microscopy [34]. Indeed, the structure of a macromolecular complex can be computed in three dimensions from a set of parallel-beam projection images of the same complex acquired in a microscope [34]. For the so-called single-particle analysis, images of the sample containing thousands of copies of the same complex are collected. Ideally, the copies have the same structure. Their orientation in the

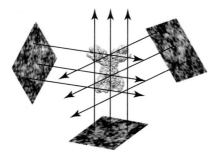

Figure 6.3 3D electron microscopy 3D electron microscopy. Experimental images corresponding to three arbitrarily chosen projection directions and a 3D model of the phosphorylase kinase 26. The arrows represent the directions of back projection in 3D space.

3D space is random and unknown, and the position of the center of each complex is unknown. These parameters have to be estimated before applying a method for 3D reconstruction. Given a first guess for the 3D model, one estimates the unknown parameters by aligning the images with the 3D model (reference model) [30] (Figure 6.3). A new reconstruction is then computed using the images and the estimated orientations and positions. The new model can be used as the reference model to resume the alignment in the next iteration of the iterative refinement of the estimated parameters [34]. It has been shown that such procedures can yield 3D models of subnanometer resolution [26].

An illustration of elastic image registration is given in Figure 6.4. One of the major difficulties in the analysis of electrophoresis 2D gels is that the gels are affected by spatial distortions due to run-time differences and dye-front deformations, which results in images that significantly differ in the content and geometry. The method proposed in Ref. [33] models the deformation field using B-splines, the advantage of which is that the model can be adapted to any continuous deformation field simply by changing the spacing between splines. The method computes quasi-invertible deformation fields so that the source image can be mapped onto the target image and vice versa, which helps the optimizer to reduce the chance of getting trapped in a local minimum and allows the simultaneous registration of any number of images.

6.4
Conclusions

In this chapter, we reviewed spline interpolation and approximation theory by presenting two spline families: tensor product splines and polyharmonic splines. Also, we presented the multiscale properties of splines. Finally, we illustrated biomedical image processing applications of splines by showing their use in image/volume registration.

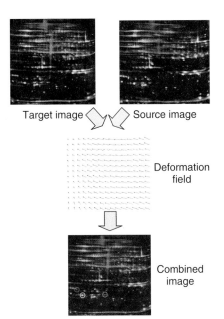

Figure 6.4 Elastic registration of 2D gels Elastic registration of two 2D protein gels (source and target) using the method proposed in [33]. The deformation field represents the continuous deformation required to convert the source into the target. The combined image shows the target image in the red channel and the warped source in the green channel. Note that there are red spots (proteins expressed in the target image and not expressed in the source image), green spots (just the opposite), and yellow spots (proteins equally expressed in both images). (Please find a color version of this figure on the color plates.)

References

1. Unser, M. (1999) Splines: a perfect fit for signal and image processing. *IEEE Signal Process. Mag.*, **16**, 22–38.
2. Unser, M. (2000) Sampling 50 years after Shannon. *Proc. IEEE*, **88**, 569–587.
3. Blu, T., Thevenaz, P., and Unser, M. (2001) MOMS: maximal-order interpolation of minimal support. *IEEE Trans. Image Process.*, **10**, 1069–1080.
4. Thevenaz, P., Ruttimann, U., and Unser, M. (1998) A pyramid approach to subpixel registration based on intensity. *IEEE Trans. Med. Imaging*, **7**, 27–41.
5. Weierstrass, K. (1884) Über die analytische Darstellbarkeit sogenannter willkürlicher Funktionen einer reellen Veränderlichen. *Sitzungsber. Königlich Preußischen Akad. Wiss. Berlin*, 633–789.
6. Whittaker, E.T. (1914) On the functions which are represented by expansion of the interpolation theory. *Proc. Roy. Soc. Edinb.*, **A35**, 181–194.
7. Shannon, C.E. (1949) Communication in the presence of noise. *Proc. Inst. Radio Eng.*, **37**, 10–21.
8. Unser, M., Aldroubi, A., and Eden, M. (1993) B-Spline signal processing: part I theory. *IEEE Trans. Signal Process.*, **41**, 821–832.
9. Unser, M., Aldroubi, A., and Eden, M. (1993) B-Spline signal processing: part II efficient design and applications. *IEEE Trans. Signal Process.*, **41**, 834–848.
10. Meijering, E. (2002) A chronology of interpolation: from ancient astronomy to modern signal and image

processing. *Proc. IEEE*, **90**, 319–342, DOI: 10.1109/5.993400.
11. Eldar, Y. and Michaeli, T. (2009) Beyond bandlimited sampling. *IEEE Signal Process. Mag.*, **26**, 48–68, DOI: 10.1109/MSP.2009.932125.
12. Thevenaz, P., Blu, T., and Unser, M. (2000) Interpolation revisited. *IEEE Trans. Med. Imaging*, **19**, 739–758.
13. Schoenberg, I.J. (1946) Contributions to the problem of approximation of equidistant data by analytic functions. *Quart. Appl. Math.*, **4**, 45–99, 112–141.
14. Piegl, L. and Tiller, W. (1997) *The NURBS Book*, Springer.
15. Duchon, J. (1977) Splines minimizing rotation-invariant semi-norms in Sobolev spaces, in *Constructive Theory of Functions of Several Variables. Lecture Notes in Mathematics*, **571** (eds. W. Schempp and K. Zeller), Springer-Verlag, pp. 85–100.
16. Iske, A. (2003) On the approximation order and numerical stability of local Lagrange interpolation by polyharmonic splines. Proceedings of 5th International Conference on Modern Developments in Multivariate Approximation, Witten-Bommerholz, Germany.
17. Beatson, R.K., Powell, M.J.D., and Tan, A.M. (2007) Fast evaluation of polyharmonic splines in three dimensions. *IMA J. Numer. Anal.*, **27**, 427–450.
18. Kybic, J. (2001) Elastic image registration using parametric deformation models. Ph.D. thesis, EPFL.
19. Van de Ville, D., Blu, T., and Unser, M. (2005) Isotropic polyharmonic B-splines: scaling functions and wavelets. *IEEE Trans. Image Process.*, **14**, 1798–1813.
20. Wahba, G. (1990) *Spline Models for Observational Data*, SIAM, Philadelphia
21. Carr, J. et al. (2001) *Reconstruction and representation of 3D objects with radial basis functions*. Proceedings of SIGGRAPH, pp. 67–76.
22. Madych, W.R. and Nelson, S.A. (1990) Polyharmonic cardinal splines. *J. Approx. Theory*, **60**, 141–156.
23. Strang, G. and Nguyen, T. (1997) *Wavelets and Filter Banks*, Wellesley-Cambridge Press.
24. Thevenaz, P. and Millet, P. (2001) Multiresolution imaging of in vivo ligand-receptor interactions. Proceedings of the SPIE International Symposium on Medical Imaging, San Diego.
25. Unser, M., Aldroubi, A., and Eden, M. (1993) The L_2 polynomial spline pyramid. *IEEE Trans. Pattern Anal. Mach. Intell.*, **15**, 364–378.
26. Venien-Bryan, C. et al. (2009) The structure of phosphorylase kinase holoenzyme at 9.9 angstroms resolution and location of the catalytic subunit and the substrate glycogen phosphorylase. *Structure*, **17**, 117–127.
27. Rueckert, D. et al. (1999) Nonrigid registration using free-form deformations: application to breast MR images. *IEEE Trans. Med. Imaging*, **18**, 712–721.
28. Studholme, C., Constable, R., and Duncan, J. (2000) Accurate alignment of functional EPI data to anatomical MRI using a physics-based distortion model. *IEEE Trans. Med. Imaging*, **19**, 1115–1127.
29. Kybic, J. et al. (2000) Unwarping of unidirectionally distorted EPI images. *IEEE Trans. Med. Imaging*, **19**, 80–93.
30. Jonic, S. et al. (2005) Spline-based image-to-volume registration for three-dimensional electron microscopy. *Ultramicroscopy*, **103**, 303–317.
31. Jonic, S. et al. (2006) An optimized spline-based registration of a 3D CT to a set of C-arm images. *Int. J. Biomed. Imaging*, 12, article ID 47197, DOI: 10.1155/IJBI/2006/47197.
32. Jonic, S. et al. (2006) Spline-based 3D-to-2D image registration for image-guided surgery and 3D electron microscopy, in *Biophotonics for Life Sciences and Medicine* (eds. M. Faupel, P. Smigielski, A. Brandenburg, and J. Fontaine), Fontis Media, Lausanne VD, Switzerland, pp. 255–273.
33. Sorzano, C. et al. (2008) Elastic image registration of 2-D gels for differential and repeatability studies. *Proteomics*, **8**, 62–65.
34. Frank, J. (1996) *Three-Dimensional Electron Microscopy of Macromolecular Assemblies*, Academic Press.

7
Wavelets

Ann Dooms and Ingrid Daubechies

7.1
Introduction

In order to digitally store or send an image, there are much better solutions than just describing its dimension in terms of pixels (*picture elements*) with their color information. The image below is made up of 2560 × 1920 pixels, each one colored with one of the $256 = 2^8$ possible gray values. This would give a total of $2560 \times 1920 \times 8 = 39.321.600$ bits to represent this image. However, the image in png format uses about half the number of bits by storing it in a more efficient way – this process is called *compression*.

When data can be reconstructed faithfully, we call the compression algorithm *lossless*. In this case, one usually exploits statistical redundancy, which will exist in natural images. *Lossy* compression, on the other hand, will introduce minor differences in the image, which do not really matter as long as they are undetectable to the human eye. One of the main ways to realize such reconstructions (see Chapter 20) is to express the image as a function and then write it as a linear combination of basic functions of some kind such that most coefficients in the expansion are small (*sparse representation*). If the basic functions are well chosen, it may be that changing the small coefficients to zero does not change the original function in a visually detectable way.

Let us explore how we could change the representation of an image of a New York scenery so that it reduces the memory requirements (we follow the recipe given in [1]). In Figure 7.1, we made two blowups of 150 × 150 pixel patches, one taken from the sky and one from the Chrysler building from which again a patch is blown up. The sky patch hardly has any difference between the pixel values, so instead of keeping track of each pixel color, we could describe it more economically by providing its size (or scale), its position in the image and an 8-bit number for the average gray value (which is, of course, almost the original value).

The 150 × 150 pixel Chrysler patch has a number of distinctive features and replacing the color of each pixel by the average gray value of the whole square would result in a patch that is not at all perceptually close to the original. On the

---

Optical and Digital Image Processing: Fundamentals and Applications, First Edition. Edited by Gabriel Cristóbal, Peter Schelkens, and Hugo Thienpont.
© 2011 Wiley-VCH Verlag GmbH & Co. KGaA. Published 2011 by Wiley-VCH Verlag GmbH & Co. KGaA.

7 Wavelets

Figure 7.1 (Successive) blowups of 150 × 150 pixel patches taken from the New York scenery.

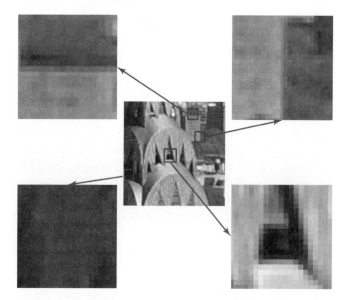

Figure 7.2 Smaller patches of the 150 × 150 pixel Chrysler patch.

other hand, it is made up of smaller patches that can be described more easily, as illustrated in Figure 7.2.

The lower left square has almost constant gray level, while the upper left and upper right squares do not, but they can be split horizontally and vertically, respectively, into two regions with almost constant gray level. The lower right square will have to be further refined to obtain such regions.

Hence instead of a huge set of pixels, all equally small, the image can be seen as a combination of regions of different sizes, each of which has more or less constant gray value. Given any subsquare of the image, it is easy to check whether it is already of this simple type by comparing it with its average gray value. If not, the region should be subdivided in four equal squares and this process is continued as long as the resulting subregions cannot be characterized by their average gray value. In some cases, it may be necessary to divide down to the pixel level, but overall, the process stops much earlier. This method is easy to implement and results in an image description using fewer bits by only storing the scale, position, and color of the necessary squares.

Moreover, this procedure can be made more efficient by starting from the other end of the scale. Instead of starting with the whole image and then subsequently dividing it, one begins at the pixel level and builds up. If the image has $2^n \times 2^n$ pixels in total, then there are $2^{n-1} \times 2^{n-1}$ subsquares of 2×2 pixels. For each such square, compute

- the average of the gray values,
- how much darker (or lighter) the average of the left half of the square is than that of the right half,
- how much darker (or lighter) the average of the top half of the square is than that of the bottom half,
- how much darker (or lighter) the average of the diagonal from top left to down right is than that of the other diagonal.

The next step is to create four new $2^{n-1} \times 2^{n-1}$ pixel images. They are obtained by replacing each 2×2 pixel square in the original by a one-pixel square colored by the four above computed values respectively, as illustrated in Figure 7.3.

The top left $2^{n-1} \times 2^{n-1}$ image contains the averages of the 2×2 squares, whereas the top right image results from vertical averaging and horizontal differencing, the bottom left image from horizontal averaging and vertical differencing, while the bottom right image contains the diagonal type of differences (all gray scales are enhanced for better viewing). Then the procedure is repeated on the top left image (and so on) (Figure 7.4).

The question is how all this can lead to compression. At each stage of the process, three species of difference numbers are accumulated, at different levels and corresponding to different positions. The total number of differences calculated is $3(1 + 2^2 + \cdots + 2^{2(n-1)}) = 2^{2n} - 1$. Together with the gray value for the whole square, this means that we end up with exactly as many numbers as gray values for the original $2^n \times 2^n$ pixel image. However, many of these differences will be very small and can just as well be dropped or put to zero leading to a new structured list of numbers which can be efficiently stored, while the reconstructed image, obtained through reversing the procedure of averaging and differencing, will have no perceptual quality loss.

This very simple image decomposition described above is the most basic example of a *wavelet decomposition*, which we now introduce in a formal mathematical way.

Figure 7.3 Four new $2^{n-1} \times 2^{n-1}$ pixel images.

Figure 7.4 A $2^{n-1} \times 2^{n-1}$ image of the 150×150 pixel Chrysler patch.

7.2
Chasing Sherlock Holmes: How to Scrutinize an Image

In the process described above of coarsely approximating the image while keeping track of the details at successive finer scales, we are basically operating in a vector space of real-valued signals – hence we need addition and scalar multiplication. The set of all possible n-dimensional complex *signals* is defined as

$$L(\mathbb{R}^n) = \{f : \mathbb{R}^n \longrightarrow \mathbb{C}\} \text{ for } n \geq 1$$

We call $L(\mathbb{R})$ the *time domain*. Often the function spaces of signals are restricted to the classical *Lebesgue spaces*

$$L^p(\mathbb{R}^n) = \left\{ f \in L(\mathbb{R}^n) \mid ||f||_p = \left(\int_{\mathbb{R}^n} |f(\mathbf{x})|^p d\mathbf{x} \right)^{1/p} < \infty \right\} \text{ where } p \geq 1$$

and

$$L^\infty(\mathbb{R}^n) = \{f \in L(\mathbb{R}^n) \mid ||f||_\infty = \sup_\mathbf{x} |f(\mathbf{x})| < \infty\}$$

The space $L^2(\mathbb{R}^n)$ is a Hilbert space for the inner product

$$\langle f, g \rangle_{L^2(\mathbb{R}^n)} = \int_{\mathbb{R}^n} \overline{f(\mathbf{x})} g(\mathbf{x}) d\mathbf{x}.$$

For $f \in L^2(\mathbb{R}^n)$ we call $||f||_2 = \langle f, f \rangle_{L^2(\mathbb{R}^n)}$ the *energy* of the signal.

An *image* can then be treated as a compactly supported function in $L^2(\mathbb{R}^2)$ which is referred to as the *spatial domain*. Through this identification with the elements of a function space, an efficient encoding of an image comes down to finding a basis of the space in which the image has a sparse representation. Moreover, the understanding or manipulation of signals in physics and engineering, in general, often requires a decomposition of their corresponding functions into basis components with specific *properties*.

It is well known that certain[1] functions belonging to Hilbert spaces can be written as a (countable) *superposition* of special (basis) functions. These functions can be chosen to represent various features and allow a signal to be *analyzed* and *synthesized*. Having f split up this way, one can carry out estimates for the pieces separately and then recombine them later. One reason for this *divide and conquer* strategy is that a typical function f tends to have many different features, for example, it may be very *spiky* and *discontinuous* in some regions, but *smooth* elsewhere or have *high frequency* in some places and *low frequency* in others, which makes it difficult to treat all of these at once. A well-chosen decomposition of the function f can isolate these features such that, for example, unwanted components (*noise*) can be separated from the useful components (*information*) in a given signal (which is known as *denoising*).

The decomposition of a function into various components is sometimes called *harmonic analysis*, whereas the set of building blocks or *atoms* is nowadays called a *dictionary*.

---

[1] Meaning, functions with suitable convergence conditions.

7.2.1
Classical Fourier Analysis

Fourier analysis is the well-known technique of harmonic analysis using trigonometric functions of various periods to represent functions at various frequencies. Every 2π-periodic square-integrable function f (i.e., $f \in L^2([-\pi, \pi])$ and extended periodically to $\mathbb{R}$) is generated by a (countable) *superposition* of integral dilations of the basic function $e^{ix} = \cos x + i \sin x$, which is a sinusoidal wave. We are talking about the *Fourier series*[2),3)] of $f \in L^2([-\pi, \pi])$

$$f = \sum_{n=-\infty}^{\infty} \frac{\alpha_n}{\sqrt{2\pi}} e^{inx}$$

where $\alpha_n = \frac{1}{\sqrt{2\pi}} \int_{-\pi}^{\pi} f(x) e^{-inx} dx$ are called the *Fourier coefficients* of f at the frequencies n. Hence the function is decomposed as an infinite linear combination of sines and cosines of various frequencies (known as the *trigonometric functions*). When n is large in absolute value, the wave e^{inx} has high frequency, while it has low frequency for $|n|$ small.

The space $L^2(\mathbb{R})$, however, is quite different from $L^2([-\pi, \pi])$ and does not contain the sinusoidal waves. If $f \in L^2(\mathbb{R})$, then recall that its *Fourier transform* is given by

$$\widehat{f}(\xi) = \lim_{n \to \infty} \frac{1}{\sqrt{2\pi}} \int_{-n}^{n} f(x) e^{-i\xi x} dx$$

(where the convergence is with respect to the norm in $L^2(\mathbb{R})$). Under suitable conditions [2], f can be recovered from $\widehat{f}$ by the *inversion formula*.

Practically, Fourier analysis is sometimes inadequate: the Fourier transform decomposes a function exactly into many components, each of which has a precise frequency, but it requires all the past as well as the future information about the signal (thus the value over the entire real line or time domain). Moreover, as the Fourier transform of a time-varying signal is a function independent of time, it does not give information on frequencies varying with time. In order to achieve time localization of a range of desired frequencies a special function $w \in L^2(\mathbb{R})$, called a *window function*, which satisfies $xw(x) \in L^2(\mathbb{R})$, is introduced into the Fourier transform. For a window function to be useful in time *and* frequency analysis, it is necessary for both w and $\widehat{w}$ to be window functions. Hence they are continuous with rapid decay in time and frequency, respectively. Thus, w is well localized in time and $\widehat{w}$ is well localized in frequency.

Multiplying a signal by a window function before taking its Fourier transform (the *windowed Fourier transform* or *(continuous) short-time Fourier transform*) has the effect of restricting the frequency information of the signal to the domain of influence of the window function and by using translates of the window function on the time axis one covers the entire time domain.

---

2) It is important to note that, in general, the above equality does not imply pointwise convergence.

3) Can be extended to more dimensions.

Hence, in this case, the dictionary is based on translations and modulations by $e^{-i\xi x}$ of the window function (a so-called *Gabor system*) and we speak of *time–frequency analysis*.

7.2.2
Forces of Nature

However, to achieve a high degree of localization in time and frequency, we need to choose a window function with sufficiently narrow time and frequency windows, but then the *Heisenberg's Uncertainty Principle* comes into play. It is an important principle with far-reaching consequences in quantum mechanics, which says that a function's feature and the feature's location cannot be both measured to an arbitrary degree of precision simultaneously. Hence it imposes a theoretical lower bound on the area of the *time–frequency window* of any window function where it can be shown that equality holds if and only if

$$w(x) = \frac{1}{2\sqrt{\pi\alpha}} e^{\frac{-x^2}{4\alpha}}$$

where $\alpha > 0$. Recall that, in this case, the windowed Fourier transform is called a *Gabor transform*. Thus Gabor transforms have the tightest time–frequency windows of all windowed Fourier transforms.

Since a sinusoid's frequency is the number of cycles per unit time, in order to capture the function's behavior, we need to zoom in to the signal to identify short-duration transients corresponding to high-frequency components (*small-scale* features) and zoom out from the signal to completely capture more gradual variations corresponding to low-frequency components (*large-scale* features). However, we cannot reduce the size of the time–frequency window beyond that of the Gabor transform. Furthermore, for any windowed Fourier transform, its time–frequency window is rigid and does not vary over time or frequency.

We, therefore, have to modify the windowed Fourier transform in a fundamentally different way to achieve *varying time* and *frequency windows*. The only way we can vary the size of the time window for different degrees of localization is by reciprocally varying the size of the frequency window at the same time, so as to keep the area of the window constant. Hence we strive for a trade-off between time and frequency localization.

This brings us seamlessly to the *wavelet transform*, an alternative and more efficient tool for analyzing functions across scales constructed using Fourier analysis. Inspired by the sinusoidal wave e^{ix} in the case of $L^2([-\pi, \pi])$, we need one single *wave* that generates a dictionary of atoms for $L^2(\mathbb{R})$ – hence the wave should decay (fast for practical purposes) to zero at $\pm\infty$. We then get a *time-scale analysis* of $L^2(\mathbb{R})$ by dilating (to obtain all stretches of the time window) and by translating (to cover the whole real line $\mathbb{R}$) the wave.

Ideally, this basis function should have a compact support in the time domain while its Fourier transform should have a compact support in the frequency domain. However, again by the famous Heisenberg's uncertainty principle, it is

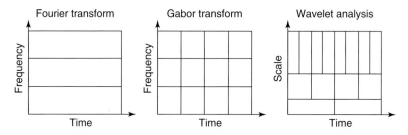

Figure 7.5 Comparison of time-frequency windows.

impossible to have a signal with finite support on the time axis which is, at the same time, bandlimited.

7.3
A Natural Evolution: The Continuous Wavelet Transform

As intuitively introduced above, we need a single *wave* function ψ to generate $L^2(\mathbb{R})$. When $\int_{-\infty}^{\infty} \psi(x)dx = 0$ (or equivalently $\widehat{\psi}(0) = 0$), then $\psi \in L^2(\mathbb{R})$ is called a *(mother) wavelet function*, as it must oscillate and thus behave as a wave. We then define *wavelets* by

$$\psi_{a,b}(x) = |a|^{-\frac{1}{2}} \psi\left(\frac{x-b}{a}\right)$$

where $a \in \mathbb{R}_0 = \mathbb{R} \setminus \{0\}$ and $b \in \mathbb{R}$. Clearly, $\psi_{a,b} \in L^2(\mathbb{R})$. We call a the *dilation parameter*, as it contracts/expands the graph of ψ, and b the *translation parameter*.

The *continuous wavelet transform* W_ψ (or *integral wavelet transform*) with respect to the wavelet function ψ is the integral transform on $L^2(\mathbb{R})$ given by

$$W_\psi(f)(a,b) = \int_{-\infty}^{\infty} f(x)\overline{\psi_{a,b}(x)}dx$$

If both the wavelet function ψ and its Fourier transform $\widehat{\psi}$ are window functions (centered around zero), then the area of the *time–frequency window* for analyzing finite energy analog signals with the help of W_ψ is constant (Figure 7.5).

Moreover, when the wavelet function ψ satisfies the *admissibility criterion*, we can recover f from $W_\psi(f)$. Obviously, it is an *overcomplete* or *redundant* representation of the signal f.[4] Indeed, instead of a one-dimensional function, one obtains a two-dimensional representation.

However, the wavelet transform of a signal does capture the localized time–frequency information of this function, unlike the Fourier transform, which sacrifices localization in one domain in order to secure it in the complementary domain. The property of time–frequency localization greatly enhances the ability

---
[4] When ψ is admissible, the wavelets form a frame (see Chapter 6 or Ref. [3]).

to study the behavior (such as smoothness, singularities, etc.) of signals as well as to change these features locally, without significantly affecting the characteristics in other regions of frequency or time.

7.4
Theory into Practice: The Discrete Wavelet Transform

In practice, it is impossible to analyze a signal for all real dilation and translation parameters. In order to obtain a discrete transform (which is not redundant), we need to extract a discrete subset of the set of functions

$$\{\psi_{a,b}\}_{(a,b)\in \mathbb{R}_0 \times \mathbb{R}}$$

which forms a *basis* of $L^2(\mathbb{R})$ and inherits the time–frequency localization property of the continuous family. Hence we need to identify an appropriate lattice of the set $\mathbb{R}_0 \times \mathbb{R}$ and select wavelets parameterized by the elements of this lattice.

In image processing, we are only interested in nonnegative frequencies, and hence we only consider positive values of the dilation parameter a. With this restriction, the admissibility criterion becomes

$$\int_0^\infty \frac{|\widehat{\psi}(\xi)|^2}{|\xi|} d\xi < \infty$$

To discretize the continuous family of wavelets $\{\psi_{a,b}\}$, we first partition the frequency domain $(0, \infty)$ into dyadic intervals by letting a take the values 2^{-j} with $j \in \mathbb{Z}$. Then we discretize the time domain at each scale separately by letting the translation parameter b take the range of values $k2^{-j}$ with $k \in \mathbb{Z}$.

Denote

$$\psi_{j,k}(x) = 2^{\frac{j}{2}} \psi(2^j x - k) \quad \text{for all} \quad j, k \in \mathbb{Z}$$

Then $\{\psi_{j,k}\}$ forms a *discrete set of wavelets* whose time–frequency windows cover the whole of the time–frequency domain. When j is small, we have wide wavelets $\psi_{j,k}$, which have very good localization in frequency, but a bad localization in time. When j is large, we have narrow wavelets, which behave in the opposite way.

The wavelet transform W_ψ of any function $f \in L^2(\mathbb{R})$, restricted to this discrete set of wavelets is given by

$$W_\psi(f)\left(\frac{1}{2^j}, \frac{k}{2^j}\right) = \int_{-\infty}^\infty f(t)\overline{\psi_{j,k}(x)} dx = \langle f, \psi_{j,k} \rangle \quad \text{for all} \quad j, k \in \mathbb{Z}$$

which is called the *discrete wavelet transform* or *DWT* of $f \in L^2(\mathbb{R})$.

When the discrete set of wavelets $\{\psi_{j,k}\}_{(j,k)\in \mathbb{Z}^2}$ forms an *orthonormal* basis of $L^2(\mathbb{R})$, we call it a *wavelet system*. Then every $f \in L^2(\mathbb{R})$ can be written as

$$f = \sum_{j\in\mathbb{Z}} \sum_{k\in\mathbb{Z}} \langle f, \psi_{j,k} \rangle \psi_{j,k}$$

which is called the *wavelet identity*. Since ψ has a zero average, each partial sum $d_j(x) = \sum_{k=-\infty}^\infty \langle f, \psi_{j,k} \rangle \psi_{j,k}$ can be interpreted as detail variations at the scale

2^{-j}. These layers of details are added at all scales to progressively improve the approximation, and ultimately recover f. We can conclude that, unlike the Fourier transform, the continuous wavelet transform is closely linked to the DWT so that we can speak of a *wavelet series* of any finite energy signal, obtained by discretizing the continuous wavelet transform.

7.5
Mallat and Meyer Digging Deeper: Multiresolution Analysis

When we want the discrete set of wavelets $\{\psi_{j,k}\}_{j,k\in\mathbb{Z}^2}$ to form a basis of $L^2(\mathbb{R})$, the choice for the wavelet ψ seems rather restrictive compared to the continuous wavelet transform. However, there is a large subclass of wavelets that arise from a structure on $L^2(\mathbb{R})$ called *multiresolution analysis* (MRA) that yields discrete families of dilations and translations that are orthonormal bases for $L^2(\mathbb{R})$[4–6]. The fundamental idea is to represent a function as a limit of successive approximations, each of which is a *smoother* or lower-resolution version of the original function. MRA is a formal approach to constructing orthogonal wavelet bases with the required behavior using a definite set of rules and procedures.

Let $f \in L^2(\mathbb{R})$. An MRA of $L^2(\mathbb{R})$ is a sequence of closed subspaces $\{V_j\}_{j\in\mathbb{Z}}$ of $L^2(\mathbb{R})$ satisfying

1) $V_j \subset V_{j+1}$ for all $j \in \mathbb{Z}$, nesting or monotonicity property
2) $f \in V_j$ if and only if $f(2\cdot) \in V_{j+1}$ for all $j \in \mathbb{Z}$, dilation property[5]
3) $\cap_{j\in\mathbb{Z}} V_j = \{0\}$, zero intersection property
4) $\cup_{j\in\mathbb{Z}} V_j$ is dense in $L^2(\mathbb{R})$, closure of the union equals $L^2(\mathbb{R})$
5) if $f \in V_0$, then $f(\cdot - k) \in V_0$ for all $k \in \mathbb{Z}$, invariance under integral translations[6]
6) $\exists \phi \in V_0$ such that $\{\phi_{0,k}\}_{k\in\mathbb{Z}}$ is an orthonormal basis[7] of V_0, where

$$\phi_{j,k}(x) = 2^{\frac{j}{2}}\phi(2^j x - k) \text{ for all } j, k \in \mathbb{Z},$$

existence of a scaling function or father wavelet.

Condition (2) implies that each of the subspaces V_j is a scaled version of the central subspace V_0 and, together with (5) and (6), we can deduce that $\{\phi_{j,k}\}_{k\in\mathbb{Z}}$ is an orthonormal basis for V_j for all $j \in \mathbb{Z}$. As the nested spaces V_j correspond to different scales, we get an effective mathematical framework for hierarchical decomposition of a signal (image) into components of different scales (resolutions).

The orthogonal complement of V_j in V_{j+1} is denoted by W_j. From (3) and (4), one can deduce that

$$L^2(\mathbb{R}) = \bigoplus_{j\in\mathbb{Z}} W_j$$

---

5) $f(2\cdot)$ denotes the function that maps x to $f(2x)$.
6) $f(\cdot - k)$ denotes the function that maps x to $f(x - k)$.
7) Can be weakened to a Riesz basis, that is, a frame of linearly independent vectors (see Chapter 6).

where the W_j are mutually orthogonal closed subspaces of $L^2(\mathbb{R})$. Now one can construct (e.g., [2]) a wavelet function ψ such that $\{\psi_{j,k}\}_{k\in\mathbb{Z}}$ is an orthonormal basis for W_j for all $j \in \mathbb{Z}$, while $\{\psi_{j,k}\}_{(j,k)\in\mathbb{Z}^2}$ is an orthonormal basis of $L^2(\mathbb{R})$.

Note that it is also possible to construct an MRA by first choosing an appropriate scaling function ϕ_0 and obtaining V_0 by taking the linear span of integer translates of ϕ_0. The other subspaces V_j can be generated as scaled versions of V_0.

As $V_0 \subset V_1$, it follows that $\phi_{0,0}$ can be expressed in terms of the orthonormal basis $\{\phi_{1,k}\}_{k\in\mathbb{Z}}$ of V_1, which is known as the *dilation equation* or *2-scale relation*

$$\phi(x) = \sum_{k\in\mathbb{Z}} \langle \phi, \phi_{1,k}\rangle \phi_{1,k}(x) = \sqrt{2} \sum_{k\in\mathbb{Z}} p_k \phi(2x-k)$$

where $p_k = \int_{\mathbb{R}} \phi(x)\overline{\phi(2x-k)}dx$. We call $\{p_k\}_{k\in\mathbb{Z}}$ the *MRA scaling sequence* or *scaling filter*. This leads to an exact formulation for the wavelet[8] that generates an orthonormal basis for $L^2(\mathbb{R})$, namely,

$$\psi(x) = \sqrt{2} \sum_{k\in\mathbb{Z}} (-1)^k \overline{p_{1-k}} \phi(2x-k)$$

We call (ϕ, ψ) a *wavelet family*.

By condition (4) above, $f \in L^2(\mathbb{R})$ can be approximated as closely as desired by a projection $f_n = P_n(f)$ onto V_n for some $n \in \mathbb{Z}$. As $V_j = V_{j-1} \oplus W_{j-1}$ for all $j \in \mathbb{Z}$, we get that

$$f_n = f_{n-1} + g_{n-1} = f_0 + g_0 + g_1 + \cdots + g_{n-2} + g_{n-1}$$

where all $f_j \in V_j$ and $g_j \in W_j$.

Consequently, instead of using infinitely many scales, the decomposition can be stopped at a given scale by representing the signal using scaling and wavelet functions. Moreover, the dilation equation comes in very handy to recursively compute the respective coefficients starting from the finest scale down to the coarser scales. Further, the coefficients can be calculated with a cascade of discrete convolutions and subsampling [3], making a swift connection with filter banks treated in Chapter 4. In fact, this wavelet analysis and synthesis can be seen as a *two-channel digital filter bank*, consisting of a *low-pass (scaling)* filter and a *high-pass (wavelet)* filter followed by decimation by factor 2. The filter bank is applied recursively on the low-pass output, which are the scaling coefficients.

The projections P_n for the wavelet ψ will correspond to an *approximation scheme* of order L (meaning that $||f - P_n(f)||_2 \le Cn^{-L}$ for all f in some class of functions) only if they can reproduce perfectly all polynomials of degree less than L. If the functions $\psi_{j,k}$ are orthogonal, then $\int \psi_{n',k}(x) P_n(f)(x) dx = 0$ whenever $n' > n$. The $\psi_{j,k}$ can thus be associated with an approximation scheme of order L only if $\int \psi_{j,k}(x) p(x) dx = 0$ for sufficiently large j and for all polynomials p of degree less than L. By scaling and translating, this reduces to the requirement

$$\int \psi(x) x^l dx = 0 \text{ for } l = 0, 1, \ldots, L-1$$

When this requirement is met, ψ is said to have L *vanishing moments*.

---

8) Not all wavelets arise from an MRA.

7.5.1
Examples

1. In 1910, Haar [7] realized that one can construct a simple piecewise constant function

$$\psi(x) = \begin{cases} 1 & 0 \leq x < \frac{1}{2} \\ -1 & \frac{1}{2} \leq x < 1 \\ 0, & \text{otherwise} \end{cases}$$

whose dilations and translations $\{\psi_{j,k}(x)\}_{(j,k)\in\mathbb{Z}^2}$ generate an orthonormal basis of $L^2(\mathbb{R})$. This wavelet naturally arises (up to normalization) from the MRA with scaling function given by the indicator function $\phi = \mathbf{1}_{[0,\,1[}$. Then all p_k are zero, except for

$$p_0 = \int_0^{\frac{1}{2}} dx = \frac{1}{2} \quad \text{and} \quad p_1 = \int_{\frac{1}{2}}^1 dx = \frac{1}{2}$$

The dilation equation becomes

$$\phi(x) = \frac{1}{\sqrt{2}}\phi(2x) + \frac{1}{\sqrt{2}}\phi(2x-1) = \frac{1}{\sqrt{2}}\phi_{1,0}(x) + \frac{1}{\sqrt{2}}\phi_{1,1}(x)$$

such that the wavelet is defined by

$$\frac{1}{\sqrt{2}}\phi(2x) - \frac{1}{\sqrt{2}}\phi(2x-1) = \frac{1}{\sqrt{2}}\phi_{1,0}(x) - \frac{1}{\sqrt{2}}\phi_{1,1}(x)$$

Let f be a row of an image, which can be seen as a piecewise constant function,

$$f : [0, 1[\longrightarrow \mathbb{R} : x \mapsto \sum_{i=0}^{2^n-1} f_i \mathbf{1}_{\left[\frac{i}{2^n}, \frac{i+1}{2^n}\right[}(x)$$

where the f_i are the respective gray values. Hence, $f \in V_n$, and thus

$$f = \sum_{k \in \mathbb{Z}} \langle f, \phi_{n,k} \rangle \phi_{n,k} = \sum_{k=0}^{2^n-1} \frac{f_k}{\sqrt{2^n}} \phi_{n,k}$$

It is now readily verified that the projections f_{n-1} and g_{n-1} of f onto V_{n-1} and $W_{n-1} = V_{n-1}^\perp$ respectively, are given by

$$f = \sum_{k=0}^{2^{n-1}-1} \frac{1}{\sqrt{2}}\left(\frac{f_{2k}}{\sqrt{2^n}} + \frac{f_{2k+1}}{\sqrt{2^n}}\right)\phi_{n-1,k} + \sum_{k=0}^{2^{n-1}-1} \frac{1}{\sqrt{2}}\left(\frac{f_{2k}}{\sqrt{2^n}} - \frac{f_{2k+1}}{\sqrt{2^n}}\right)\psi_{n-1,k}$$

Hence, the Haar wavelet system decomposes a signal into (weighted) averages and differences of two neighboring pixels; this is what we intuitively performed in the beginning of this chapter.

After iterating this process on the approximation term and by setting small differences within the result equal to zero, we obtain a sparse representation of an approximation of f. However, when removing fine scale details of the Haar wavelet transform for smooth functions, we get piecewise constant approximations, which

7.5 Mallat and Meyer Digging Deeper: Multiresolution Analysis

is far from optimal. A piecewise linear approximation would produce a smaller error in this case.

2. As B-splines satisfy the dilation equation (see Chapter 6), they allow to construct an MRA with the Riesz condition. Furthermore, in Ref. [8], it was shown that all mother wavelets can be decomposed as the convolution of a spline (responsible for a good number of appealing properties of wavelets) and a distribution.

3. While the Haar wavelets have good time localization, they are discontinuous and hence their frequency localization is bad. In 1988, Ingrid Daubechies [9] discovered a family of compactly supported wavelet families $D_N = \{(\phi_N, \psi_N) \mid N \in \mathbb{N}\}$ that become smoother as N increases.

Let us give a simplified idea of constructing an MRA with the desired properties. Compact support for ϕ necessarily implies that the MRA scaling sequence $\{p_k\}_{k \in \mathbb{Z}}$ is finite, say of length N (suppose that $p_0, p_1, \ldots, p_{N-1}$ are nonzero). Integrating the dilation equation, then gives the condition

$$\sum_{k=0}^{N-1} p_k = \sqrt{2}$$

Orthonormality of the $\phi_{0,k}$ for $k \in \mathbb{Z}$ combined with the dilation equation imposes

$$\sum_{k=0}^{N-1} p_k p_{k+2m} = \delta_{0m} \text{ for } m \in \mathbb{Z}$$

These conditions, however, do not uniquely determine the N coefficients p_k, and hence *extra requirements* can be introduced, such as vanishing moments for the associated ψ.

Let $N = 2$, then $p_0 + p_1 = \sqrt{2}$ and $p_0^2 + p_1^2 = 1$. By imposing a vanishing zeroth moment of the wavelet ψ, thus $p_1 - p_0 = 0$, constant image information can be greatly compressed.

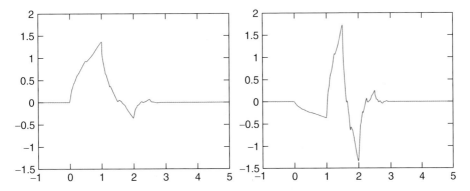

Figure 7.6 The Daubechies D_4 scaling function and wavelet function in $L^2(\mathbb{R})$. The image is obtained through the inverse transform of a unit vector in $\mathbb{R}^{1024}$. Note that the support of ϕ and ψ is [0, 3].

Let $N = 4$ and let us now impose vanishing zeroth and first moments (which annihilate linear information) of ψ. We get that $\sum_{k=0}^{3} p_k = \sqrt{2}$, $\sum_{k=0}^{3} p_k^2 = 1$, $p_0 p_2 + p_1 p_3 = 0$, $p_3 - p_2 + p_1 - p_0 = 0$, and $p_3 - 2p_2 + 3cp_1 - 4p_0 = 0$, which results for D_4 (Figure 7.6) in

$$p_0 = \frac{1+\sqrt{3}}{4\sqrt{2}}, \quad p_1 = \frac{3+\sqrt{3}}{4\sqrt{2}}, \quad p_2 = \frac{3-\sqrt{3}}{4\sqrt{2}}, \quad p_3 = \frac{1-\sqrt{3}}{4\sqrt{2}}$$

7.6
Going to Higher Dimensions: Directional Transforms

7.6.1
Separable Transforms

The DWT can be generalized to any dimension by combining several wavelet systems into a separable transform. In fact, when $\{\phi, \psi\}$ and $\{\phi', \psi'\}$ are two wavelet families, then

$$\{\psi_{j,k}^{LH}, \psi_{j,k}^{HK}, \psi_{j,k}^{HH} \mid (j,k) \in \mathbb{Z}^2\}$$

is an orthonormal basis of $L^2(\mathbb{R}^2)$, where for $\mathbf{x} = (x_1, x_2) \in \mathbb{R}^2$

$$\psi^{LH}(\mathbf{x}) = \phi(x_1)\psi'(x_2), \quad \text{low-high wavelet,}$$
$$\psi^{HL}(\mathbf{x}) = \psi(x_1)\phi'(x_2), \quad \text{high-low wavelet,}$$
$$\psi^{HH}(\mathbf{x}) = \psi(x_1)\psi'(x_2), \quad \text{high-high wavelet.}$$

Note that the role of the scaling function is carried out by

$$\phi^{LL}(\mathbf{x}) = \phi(x_1)\phi'(x_2)$$

Because of the separability, the transform can be easily implemented on an image by first applying the DWT to the rows and then to the columns (Figure 7.7). In the case of the Haar wavelet system, this process results exactly in what we obtained in Figure 7.3. The coefficients in the upper left corner form the so-called LL-sub-band and, going clockwise, we have the HL *sub-band*, HH *sub-band*, and LH *sub-band*, respectively. The transform can then be iterated on the scaling coefficients from the LL sub-band, resulting in different *levels* of LL, HL, LH, and HH sub-bands. The LL sub-band always is a coarser approximation of the input, while the other sub-bands are relatively sparse where the (few) significant numbers in these wavelet sub-bands mostly indicate edges and texture. The HL sub-band grasps vertical details, LH horizontal ones, while in the HH we find information on features with dominant orientation in the 45° and −45° directions. Clearly, features impact neighboring and scale-related wavelet coefficients.

While the wavelet transform gives a perfect compromise between spatial (or time) and frequency localization for one-dimensional signals, it is not optimal for representing images as it cannot entirely capture their geometry. Another of its shortcomings is its shift variance: a small shift of a signal can greatly disturb the wavelet coefficients pattern. Hence, one requires a multiresolution transform that yields a sparser representation for images than the wavelet transform by *looking*

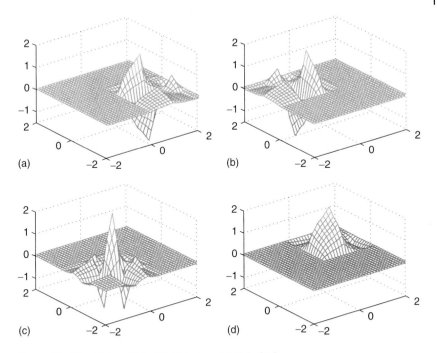

Figure 7.7 The (a) low–high, (b) high–low, (c) high–high wavelets, and (d) scaling function of the separable two-dimensional Daubechies D_4 wavelet system.

in different directions. There is, therefore, large room for improvements, and several attempts have been made in this direction both in the mathematical and engineering communities, in recent years. These include complex wavelets [10], contourlets [11], and other *directional* wavelets in the filter bank literature [12], as well as brushlets [13], ridgelets [14], curvelets [15], bandelets [16], and, recently, shearlets [17]. We focus on the popular complex wavelet constructions as well as the promising shearlets.

7.6.2
Dual-Tree Complex Wavelet Transform

The shift invariance of the magnitude of the Fourier coefficients inspired Kingsbury to create his *dual-tree complex wavelet transform* (DT-$\mathbb{C}$WT) by combining this essential property of the Fourier transform with the DWT [10], which also provides an MRA. While the Fourier representation is based on the complex-valued $e^{i\xi x}$, the DT-$\mathbb{C}$WT uses two (real-valued) wavelet systems $\{\psi_{j,k}\}_{(j,k)\in\mathbb{Z}^2}$ and $\{\psi'_{j,k}\}_{(j,k)\in\mathbb{Z}^2}$ that form a so-called *Hilbert transform pair*,[9] that is, $\psi' = \mathcal{H}\psi$ where the Hilbert

---

9) To create (approximately in a practical setting) Hilbert transform pairs, splines again prove useful.

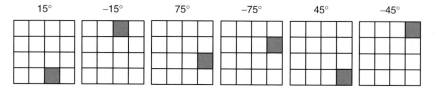

Figure 7.8 Idealized support of the Fourier spectrum of the six Kingsbury wavelets.

transform is given by

$$\mathcal{H}\psi(x) = \lim_{\varepsilon \searrow 0} \frac{1}{\pi} \int_{\varepsilon}^{\frac{1}{\varepsilon}} \frac{\psi(x-y)}{y} dy + \frac{1}{\pi} \int_{-\frac{1}{\varepsilon}}^{-\varepsilon} \frac{\psi(x-y)}{y} dy$$

These paired wavelets are combined in a complex wavelet

$$\Psi = \frac{1}{2}(\psi + i\psi')$$

that forms the *analytic*[10] counterpart of the complex sinusoids

$$e^{inx} = \cos(nx) + i\mathcal{H}(\cos(nx))$$

in Fourier analysis of $f \in L^2([-\pi, \pi])$. If we denote a_n and b_n as the (real) Fourier coefficients corresponding to the even and odd trigonometric functions respectively, then the complex coefficients are given by

$$c_n = a_n - ib_n = \langle f(x), e^{inx} \rangle$$

for which the amplitude does not change under translations of f. Owing to these observations, the DT-CWT of $f \in L^2(\mathbb{R})$ is defined by projecting the signal onto the dilated and translated copies of the analytic function Ψ.

There are two main practical constructions for the two-dimensional DT-CWT, the one proposed in Ref. [10], which has six complex wavelets oriented along the directions $\pm 15°$, $\pm 45°$, and $\pm 75°$ (Figure 7.8), and the one in Ref. [18] that can detect the principal directions $0°$, $90°$, $45°$, and $-45°$. Both transforms are approximately shift invariant.

We recall that in the two-dimensional DWT, the LH and HL sub-bands are horizontally and vertically oriented while the HH catches information from both diagonals. This is exactly where the analytic wavelet Ψ with its one-sided frequency spectrum comes to the rescue in the DT-CWT using similar separable combinations as before, but resulting in well-defined orientations depending on the choices made (Figure 7.9).

---

10) We call $g : \mathbb{R} \longrightarrow \mathbb{C}$ an *analytic signal* if there exists a real-valued signal f, such that $g = f + i\mathcal{H}f$. Hence g has no negative frequency components.

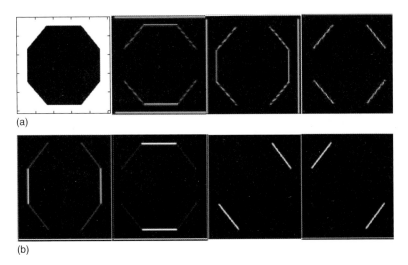

Figure 7.9 (a) Two-dimensional Daubechies D_4 wavelet decomposition of an octagon; (b) DT-CWT based on a complex spline and scheme from Ref. [18].

7.6.3
Shearlets

The *shearlet transform* [17] is a very recent newcomer in the family of directional transforms for geometric image representations that again provides an MRA.

The *continuous shearlet transform* is a generalization of the continuous wavelet transform with atoms being localized in *space, frequency,* and *orientation*. The *(mother) shearlet function* is defined in the Fourier transform domain (while satisfying appropriate admissibility conditions) as

$$\widehat{\Psi}(\Xi) = \widehat{\psi_1}(\xi_1)\widehat{\psi_2}\left(\frac{\xi_2}{\xi_1}\right)$$

where $\Xi = (\xi_1, \xi_2)$, $\widehat{\psi_1}$ the Fourier transform of a wavelet function and $\widehat{\psi_2}$ a compactly supported *bump function*, that is, $\widehat{\psi_2}(\xi) = 0$ if and only if $\xi \notin [-1, 1]$. By this condition, the shearlet function is bandlimited in a diagonal band of the frequency plane.

We define the *shearlets* by dilating, shearing, and translating the shearlet function Ψ

$$\Psi_{j,k,\mathbf{l}}(\mathbf{x}) = |\det A|^{\frac{j}{2}} \Psi(B^k A^j \mathbf{x} - \mathbf{l})$$

where $j, k \in \mathbb{R}$, $\mathbf{l} \in \mathbb{R}^2$, A and B invertible 2×2 matrices with $\det B = 1$.

The *continuous shearlet transform* S_Ψ with respect to the shearlet function Ψ is the integral transform on $L^2(\mathbb{R}^2)$ given by

$$S_\Psi(f)(j,k,\mathbf{l}) = \int_{\mathbb{R}^2} f(\mathbf{x}) \Psi_{j,k,\mathbf{l}}(\mathbf{l} - \mathbf{x}) d\mathbf{x}$$

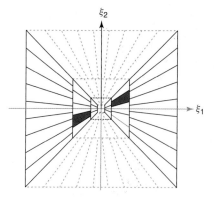

Figure 7.10 The tiling of the frequency plane induced by the shearlets [17].

For shearlet analysis, we choose A to be an anisotropic scaling matrix and B a shearing matrix given by $A = \begin{pmatrix} 4 & 0 \\ 0 & 2 \end{pmatrix}$ and $B = \begin{pmatrix} 1 & 1 \\ 0 & 1 \end{pmatrix}$. It is easily seen that by changing the scale j and shearing parameter k, arbitrary wedges of the frequency plane can be obtained (Figure 7.10).

By sampling the continuous shearlet transform on an appropriate discrete set of the scaling, shear, and translation parameters, the *discrete shearlet transform* will render a frame or even a tight frame for $L^2(\mathbb{R}^2)$. Furthermore, the atoms are associated with a generalized MRA structure, where the scaling space is not only translation invariant but also invariant under the shear operator (see *www.shearlet.org*).

7.7
Conclusion

Historically, wavelets were invented in the early 1980s by geophysicist Morlet and quantum physicist Grossmann through a formalization of the common ground in their research [9]. A few years later, Meyer made the connection with the scaling techniques from harmonics analysis and found the first smooth wavelet basis, while P. G. Lemarié and G. Battle made orthonormal wavelet bases that were piecewise polynomial. Soon after that, Mallat and Meyer invented MRA, a different approach to Meyer's construction inspired by computer vision. This framework, given by natural construction rules that embody the principle of successively finer approximations, has remained the basic principle behind the construction of many wavelet bases and redundant families.

None of the smooth wavelet bases constructed up to that point were supported inside an interval, so the algorithms to implement the transform (which were using the sub-band filtering framework without their creators knowing that it had been named and developed in electrical engineering) required, in principle, infinite filters that were impossible to implement. In practice, this meant that the infinite filters from the mathematical theory had to be truncated.

In 1987, however, Daubechies [9] managed to construct the first orthonormal wavelet basis for which ψ is smooth and supported on an interval.

Soon after this, the connection with the electrical engineering approaches was discovered and the benefit was huge within the field of image processing, in particular, in

- compression (see Chapter 20),
- restoration (see Chapter 26) – such as denoising, inpainting, and super-resolution,
- image analysis (Chapter 29) – such as segmentation and texture classification.

More exciting constructions and generalizations of wavelets followed: biorthogonal (semiorthogonal, in particular) wavelet bases, wavelet packets, multiwavelets, irregularly spaced wavelets, sophisticated multidimensional wavelet bases not derived from one-dimensional constructions, and so on.

The development of the theory benefited from all the different influences and also enriched the different fields with which wavelets are related. As the theory has matured, wavelets have become an accepted addition to the mathematical toolbox used by mathematicians, scientists, and engineers alike. They have also inspired the development of other tools that are better adapted to tasks for which wavelets are not optimal.

References

1. Daubechies, I. (2008) The influence of mathematics, in *Wavelets and Applications, The Princeton Companion to Mathematics* (eds. T. Gowers, J. Barrow-Green, and I. Leader), Princeton University Press.
2. Chui, C.K. (1992) *An Introduction to Wavelets*, Academic Press, New York.
3. Mallat, S.G. (2008) *A Signal Processing Tour of Wavelets*, Academic Press.
4. Mallat, S.G. (1988) Multiresolution representation and wavelets, Ph.D. Thesis, University of Pennsylvania, Philadelphia, PA.
5. Mallat, S.G. (1989) Multiresolution approximations and wavelet orthonormal bases of $L^2(\mathbb{R})$. Trans. Am. Math. Soc., **315**, 69–87.
6. Meyer, Y. (1986) *Ondelettes et Fonction Splines*, Seminaire EDP, Ecole Polytecnique, Paris.
7. Haar, A. (1910) Zur Theorie der orthogonalen Funktionensysteme. Math. Ann., **69**, 331–371.
8. Unser, M. and Blu, T. (2003) Wavelet theory demystified. IEEE Trans. Signal Proc., **51**, 470–483.
9. Daubechies, I. (1988) Orthonormal bases of compactly supported wavelets. Commun. Pure Appl. Math., **41**, 909–996.
10. Kingsbury, N.G. (2001) Complex wavelets for shift invariant analysis and filtering of signals. J. Appl. Comput. Harmon. Anal., **103**, 234–253.
11. Po, D.D. and Do, M.N. (2006) Directional multiscale modeling of images using the contourlet transform. IEEE Trans. Image Process., **156**, 1610–1620.
12. Antoine, J.P., Murenzi, R., and Vandergheynst, P. (1999) Directional wavelets revisited: Cauchy wavelets and symmetry detection in patterns. Appl. Comput. Harmon. Anal., **6**, 314–345.
13. Coifman, R.R. and Meyer, F.G. (1997) Brushlets: a tool for directional image analysis and image compression. Appl. Comput. Harmon. Anal., **5**, 147–187.
14. Candès, E.J. and Donoho, D.L. (1999) Ridgelets: a key to higher-dimensional

intermittency? *Phil. Trans. Royal Soc. Lond. A*, **357**, 2495–2509.
15. Candès, E.J. and Donoho, D.L. (2004) New tight frames of curvelets and optimal representations of objects with piecewise C^2 singularities. *Comm. Pure Appl. Math.*, **56**, 216–266.
16. Le Pennec, E. and Mallat, S.G. (2005) Sparse geometric image representations with bandelets. *IEEE Trans. Image Process.*, **14**, 423–438.
17. Guo, K. and Labate, D. (2007) Optimally sparse multidimensional representation using shearlets. *SIAM J. Math Anal.*, **39**, 298–318.
18. Chaudhury, K. and Unser, M. (2010) On the shiftability of dual-tree complex wavelet transforms. *IEEE Trans. Signal Process.*, **58**, 221–232.

8
Scale-Space Representations for Gray-Scale and Color Images

Iris U. Vanhamel, Ioannis Pratikakis, and Hichem Sahli

8.1
Introduction

The analysis of images depends on the scale of the objects of interest. The extent of each real-world object is limited by two scales: the inner and the outer scales. The inner scale corresponds to the resolution that expresses the pixel size and is determined by the resolution of the sampling device. The outer scale of an object corresponds to the minimum size of a window that completely contains the object, and is consequently limited by the field of view. These two scales are known in a controlled environment. However, for most contemporary image analysis problems, this is not the case. In these circumstances, it makes sense to interpret the image at all scales simultaneously. This is further advocated by the fact that the same principle is followed by the human visual front-end system. There is a need to represent an image at multiple scales: *the multiscale representation*.

Multiscale representations have been in existence since the dawn of image processing: Quad-trees are considered as one of the earliest types of multiscale representations of data. A quad-tree is a treelike representation of image data in which the image is recursively divided into smaller regions. It has been used in the classical "split-and-merge" segmentation algorithm. Pyramids also represent data at multiple scales. Here a subsampling operation is combined with a smoothing step. The latter yields reduced computational work and small memory requirements, making this representation an attractive tool for vision applications. Another form of multiscale representation is the wavelet representation. The wavelet transform is a linear operation that uses a hierarchical set of basis functions, which satisfy certain mathematical criteria and which are all translations and scalings of one another. The above-mentioned multiscale representations share the problem of lacking a firm theory that relates information of one scale to another. The introduction of *scale-space theory* as a framework for early visual operations alleviated this pitfall. This framework has been developed by the computer vision community [1–6] to handle the multiscale nature of image data. Although, Witkin [4] was the first to introduce the name *scale-space theory*, Weickert, Ishikawa, and Imiya [7] showed that the scale-space concept is more than 20 years older.

Optical and Digital Image Processing: Fundamentals and Applications, First Edition. Edited by Gabriel Cristóbal, Peter Schelkens, and Hugo Thienpont.
© 2011 Wiley-VCH Verlag GmbH & Co. KGaA. Published 2011 by Wiley-VCH Verlag GmbH & Co. KGaA.

Essentially, scale-space theory encapsulates two concepts: (i) scale-space filtering and (ii) the linking strategy or deep image structure [2, 3, 8, 9]. Scale-space filtering concerns the mechanism that embeds the signal into a one-parameter family of derived signals for which the signal content is simplified. The parameter describes the scale or resolution at which the signal is represented. The main idea is that the number of local extrema in the signal and its derivatives should decrease with scale [4]. This is quite understandable if one considers that image features can be described using differential topology. Initially, Gaussian smoothing was the only filtering process that complied with the postulated scale-space axioms. This was due to the imposition of linearity, isotropy, and homogeneity constraints, that is, uncommitted visual front-end, on the embedding process. However, these constraints are not essential [2]. Disregarding all or some of these constraints allows for the class of nonlinear scale-space filters. These filters can alleviate several inherent problems of the Gaussian scale space. The linking strategy deals with the methodology that relates signal structure at different scales. It supports the central idea of scale-space theory: important signal features persist in relatively coarse scales, even though their location may be distorted by scale-space filtering. It provides a method to track down a path in the scale-space family up to the zeroth scale, so that the signal structure can be located exactly on the original signal. Given these two concepts, scale-space theory provides a well-founded framework for dealing with signal structure at different scales. Furthermore, it can also be used for formulating multiscale feature detectors. Lindeberg et al. [3, 10] argued that the problem of selecting appropriate scales for further analysis should be addressed as well. Therefore, they introduced the notion of scale selection, which provides a mechanism for identifying interesting scale levels in the absence of a priori information about the signal's content. In the case of nonlinear scale-space filtering, the scale selection includes the identification of the optimal enhanced scale [11, 12].

This chapter is structured as follows. The first part of this chapter covers required background knowledge such as definitions, axioms, and properties, and, the expression of the scale-space filter. In the second part, several different filtering strategies are examined: (i) linear or Gaussian scale-space filtering, (ii) scale-space filtering based on the generalized heat equation, (iii) geometry-driven scale-space filtering, and (iv) scale-space filtering stemming from mathematical morphology (MM).

8.2
Background

8.2.1
Definitions

Let an image f be the mapping of the *image domain* Ω onto a *feature space* $\mathcal{F}$, $f : \Omega \mapsto \mathcal{F}$. Then we define a digital image $f \in \mathbb{R}^{N \times M}$ as the mapping of the index set $J = \{p_1, p_2, \ldots, p_N\}$ to a matrix $\{\mathbf{f}(p_1), \mathbf{f}(p_2), \ldots, \mathbf{f}(p_N)\}$, where $\mathbf{f}(p_j)$ represents

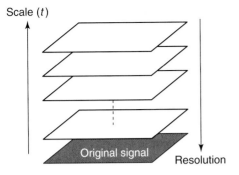

Figure 8.1 A scale-space representation of a signal is an ordered set of derived signals intended to represent the original signal at different levels of scale [3].

the feature vector $[f^{(1)}(p_j), f^{(2)}(p_j) \cdots, f^{(M)}(p_j)]$ at the point p_j. It may be noted that for a gray-scale image, $M = 1$, whereas for a color image, $M = 3$.

The *scale-space representation* of a signal is created by embedding the signal into a one-parameter family of derived signals with the scale/resolution as the parameter. The idea is that the signal content gradually simplifies. This yields a representation of the signal at a continuum of scales:

$$\mathcal{U}(f) : f(\cdot) \mapsto u(\cdot, t) \quad \text{with} \quad u(\cdot, 0) = f(\cdot) \tag{8.1}$$

where f represent the *original signal*, $\mathcal{U}$ is the embedding process (also denoted in the literature as the *multiscale analysis* [13] or the *scale-space filter* [4]), u is the scale-space family (scale-space image), and $t \in \mathbb{R}^+$ is a measure of resolution or inner scale. Figure 8.1 illustrates the scale-space representation. The transition from one scale to another is described by the *transition operator* $\mathcal{U}_{t_1,t_2}$:

$$u(\cdot, t_1) = \mathcal{U}_{t_1,t_2}(u(\cdot, t_2)) \quad \forall t_2, t_1 \in \mathbb{R}^+, t_1 \geq t_2 \tag{8.2}$$

Note that $\mathcal{U}_t$ refers to $\mathcal{U}_{t,0}$, and $\mathcal{U}_{t_1,t_1}$ is the identity operator. Furthermore, $\mathcal{U}_{0,0} = \mathcal{U}_0 = f$.

8.2.2
Axioms and Properties

In order to create a scale-space representation of a signal, the embedding $\mathcal{U}$ has to satisfy certain requirements. Several authors have postulated some fundamental hypotheses that the embedding process must satisfy in order to be a scale-space filter [1–3, 5, 13–17]. Most of these works include hypotheses that yield only one possible solution: the Gaussian scale space. In Refs [13] and [17], more general sets of fundamental of hypotheses are given along with several desirable but optional properties. All these can be categorized as architectural, stability, and morphological-requirements. With respect to these requirements, it was shown that any scale space fulfilling them can be governed by a *partial differential equation*

(PDE) with the original image as initial condition. In Ref. [17], this was extended to vector-valued images ($M \geq 1$).

8.2.2.1 Fundamental Axioms

Architectural Axioms
Axiom 1.1 *Causality (pyramidal architectural):* A coarser analysis of a signal can be deduced from a finer one without any dependence on the original signal.

$$\mathcal{U}_{0,0}(f) = f$$
$$\mathcal{U}_{t_1+t_2,0}(f) = \mathcal{U}_{t_1+t_2,t_2}\left(\mathcal{U}_{t_2,0}(f)\right) \quad \forall t_1, t_2 \in \mathbb{R}^+ \tag{8.3}$$

A strong version of the causality axiom is the semigroup property, which is included in the sequel since it is a desirable property. Note that if the semigroup property is verified, the causality axiom is also fulfilled.

Axiom 1.2 *Regularity:* The embedding process $\mathcal{U}$ should be continuous.

Axiom 1.3 *Locality:* Local spatial behavior is determined by the original signal f when the scale parameter t tends to 0.

Stability Axioms They consist of hypotheses that describe the smoothing and information reduction properties of the embedding process. In general, they state that the embedding process should not create artifacts when passing from a fine to a coarser representation. This can be achieved by using different concepts: in Ref. [13], it is achieved by the maximum principle. Other authors use concepts such as no creation of new level curves [2, 5, 18], nonenhancement of local extrema [14], decreasing number of local extrema [18], maximum–minimum principle [19, 20], positivity [1, 16], preservation of positivity [21], and Lyapunov functionals [7]. It has been demonstrated that for the Gaussian scale-space filter, these properties are closely related [7].

Axiom 1.4 *The maximum principle (comparison principle, monotony):* The filtering process should smooth the image without involving any enhancement process. For gray-scale images, it can be easily formulated: if an image f_1 is everywhere brighter than an image f_2, the filtering process should preserve this ordering. For vector-valued images, it should be verified in each image band unless an ordering has been defined. In this case the defined ordering should be preserved by filtering process. In general, no such ordering exists. Hence the maximum principle is formulated as

$$\forall t \in \mathbb{R}^+, \forall p \in \Omega, \forall i \in [1, \ldots, M]:$$
$$f_1^{(i)}(p) \leq f_2^{(i)}(p) \Rightarrow \left[\mathcal{U}_t(f_1)\right]^{(i)}(p) \leq \left[\mathcal{U}_t(f_2)\right]^{(i)}(p) \tag{8.4}$$

Axiom 1.5 *No creation of new level curves (Koenderink's formulation of causality)* is needed if one intends employing iso-intensity linking in deep image structure. It is a necessity that no new level lines are created when the scale t increases [2]. For this reason, Koenderink [2] imposed the condition that at spatial extrema with nonvanishing determinant of the Hessian, isotopes in scale space are convex upward [7], which can be expressed as follows:

$$\text{sign}(u_t) = \text{sign}(\nabla u) \tag{8.5}$$

Axiom 1.6 *Nonenhancement of local extrema* is an equivalent formulation of Koenderink's causality requirement [7]. It yields a sufficient condition for Eq. (8.5). In other words, it states that local extrema with positive or negative Hessian are not enhanced. Let x_ε be an extremum at the scale T_ε; then,

- if x_ε verifies $u_t > 0$ at T_ε, then x_ε is a minimum
- if x_ε verifies $u_t < 0$ at T_ε, then x_ε is a maximum.

Hence, for a local minimum, the eigenvalues (η_1, η_2) of the Hessian of u are positive:

$$\nabla u = \text{trace}(\text{Hessian}(u)) = \eta_1 + \eta_2 > 0 \tag{8.6}$$

which verifies Eq. 8.5. Conversely, for a local maximum,

$$\nabla u = \text{trace}(\text{Hessian}(u)) = \eta_1 + \eta_2 < 0 \tag{8.7}$$

Furthermore, this result is related to the axiom that imposes a decreasing number of local extrema, which was used in Ref. [54].

Axiom 1.7 *The maximum–minimum principle (extremum principle)* is another equivalent formulation of Koenderink's causality requirement. Along with the regularity property, it ensures that level sets can be traced back in scale. For vector-valued (color) images that have no predefined ordering relationship, it is formulated as follows:

$$\forall t \in \mathbb{R}^+, \forall p \in \Omega, \forall i \in [1, \ldots, M]:$$
$$\inf(f^{(i)}) \leq f^{(i)}(p) \leq \sup(f^{(i)}) \Rightarrow \inf(f^{(i)}) \leq [\mathcal{U}_t(f)]^{(i)}(p) \leq \sup(f^{(i)}) \tag{8.8}$$

Obviously, if an ordering relationship exists, the maximum and minimum feature vector are defined as well. Hence, the embedding process $\mathcal{U}$ should verify that

$$\forall t \in \mathbb{R}^+, \forall p \in \Omega:$$
$$\inf(f) \leq f(p) \leq \sup(f) \Rightarrow \inf(f) \leq [\mathcal{U}_t(f)](p) \leq \sup(f) \tag{8.9}$$

Axiom 1.8 *Preservation of positivity:* If the original signal is positive at all points of the image domain, the scale image should be positive at all points of the image domain as well:

$$\forall t \in \mathbb{R}^+, \forall p \in \Omega, \forall i \in [1, \ldots, M]: f^{(i)}(p) \geq 0 \Rightarrow [\mathcal{U}_t(f)]^{(i)}(p) \geq 0 \tag{8.10}$$

Axiom 1.9 *Lyapunov functionals* provide a technique to asses the stability of a system. It is a very general technique and works well for nonlinear and time-varying systems. Furthermore, Lyapunov functions are useful in proofs. However, the drawback is that proving the stability of a system with Lyapunov functions is difficult. Moreover, a failure to find the associated Lyapunov functions of a system does not imply that this system is unstable. The existence and formulation of Lyapunov functionals for a system of PDEs based on the heat equation and its generalization (variable conductance diffusion, VCD) is given in Ref. [7].

8.2.2.2 Properties

Architectural Properties

Property 1.1 *The semigroup property (recursivity) expresses a strong version of causality:*

$$\mathcal{U}_0(f) = f$$
$$\mathcal{U}_{t_1+t_2}(f) = \mathcal{U}_{t_1}(\mathcal{U}_{t_2}(f)) \quad \forall t_1, t_2 \in \mathbb{R}^+, t_1 \geq t_2 \quad (8.11)$$

It ensures that one can implement the scale-space process as a cascade smoothing that resembles certain processes of the human visual system.

Property 1.2 *Infinitesimal generator:* Given the semigroup property and regularity, functional analysis demonstrates the existence of an infinitesimal generator. In other words, there exists a limit case for the transition operator $\mathcal{U}_t$:

$$A(f) = \lim_{t \to 0} \frac{\mathcal{U}_t(f) - f}{t} \quad (8.12)$$

where A denotes the infinitesimal generator. Under these circumstances, the scale space satisfies the differential equation

$$u_t(\cdot, t) = \lim_{h \to 0^+} \frac{u(\cdot, t+h) - u(\cdot, t)}{h} = Au(\cdot, t) \quad (8.13)$$

Property 1.3 *Linearity (superposition principle)* restricts the type of scale space to the Gaussian scale space. It imposes the following:

$$\mathcal{U}_t\left(c_1^{nt} f_1 + c_2^{nt} f_2\right) = c_1^{nt} \mathcal{U}_t(f_1) + c_2^{nt} \mathcal{U}_t(f_2) \quad (8.14)$$

Property 1.4 *The convolution kernel* refers to the existence of a family of functions k_t such that

$$\mathcal{U}_t(f) = \int_\Omega k_t(p - p') f(p') dp' \quad (8.15)$$

If such a family exists, it implies that linearity and translation invariance are verified. Hence, the embedding process is the Gaussian scale-space filter. In addition, the convolution kernel has to verify the following:

- the expectation that the kernel spread the information uniformly over the image; thus for $t \to \infty$ the kernel should be flat:

$$\lim_{t \to \infty} k_t(p) = 0 \quad (8.16)$$

- the kernel is symmetrical; for example, if it is symmetrical in x, then

$$\forall p = (x, y) \in \Omega : k_t(x, y) = k_t(-x, y) \quad (8.17)$$

- separability, which states that the convolution kernel can be separated into n factors, each operating along one coordinate axis:

$$k_t(p) = k_t^{(1)}(p^{(1)}) \cdot k_t^{(2)}(p^{(2)}) \cdots k_t^{(n)}(p^{(n)}) \quad (8.18)$$

8.2.2.3 Morphological Properties

Morphological properties state that image analysis should be insensitive to shifts in the feature space and changes of position, orientation, and scale. Shape preservation states that an object has to be analyzed in the same way independent of its position and that of the perceiver: for example, translations, rotations, scale changes, and affine transformations in the spatial domain. In Ref. [2], the properties are denoted by homogeneity and isotropy, for which Koenderink pointed out that they are not essential but simplify mathematical analysis.

Property 1.5 *Feature scale invariance:* The hypothesis of scale invariance is motivated by physics: physical observations must be independent of the choice of dimensional units. In other words, there is no intrinsically preferred scale for images. For any nondecreasing real function h, the scale space should verify that

$$\mathcal{U}(h(f)) = h(\mathcal{U}(f)) \ \forall f \in \mathbb{R}^{N \times M} \tag{8.19}$$

where h is an order-preserving rearrangement of the feature values. This is a straightforward concept for gray-scale images. For higher dimensional feature spaces, it becomes more complex since an order relation cannot always be defined, for example, in incommensurable image channels. For the latter, it will have to suffice that the order relationship within each channel is preserved. A special subclass of this property is the shift invariance of the feature space, that is, h is an addition with a constant.

Property 1.6 *Average feature-value invariance:* The average value of the features over the image domain should remain constant in each image channel:

$$\int_\Omega \mathcal{U}_t(f^{(i)}) d\Omega = \int_\Omega f^{(i)} d\Omega \quad \forall t \in \mathbb{R}^+, \ \forall i \in [0, M] \tag{8.20}$$

Property 1.7 *Isometry invariance:* Isometries are bijective maps between two metric spaces that preserve distances. Hence, an isometry of the image plane is a linear transformation that preserves length. This class of isometries includes translations and rotations. Let the mapping $h : \mathbb{R}^n \mapsto \mathbb{R}^n$ be an isometry of $\mathbb{R}^n$ of the form:

$$\forall p_1, p_2 \in \Omega : d(h(p_1), h(p_2)) = d(p_1, p_2)$$

where $d(p_1, p_2)$ represents the distance between the spatial points p_1 and p_2. Let H_{iso} be the class of all isometrical transformations of $\mathbb{R}^n$; then the embedding process $\mathcal{U}$ is isometry invariant if the following is verified:

$$\mathcal{U}_t(h(f)) = h(\mathcal{U}_t(f)), \ \forall p \in \Omega, \ \forall t \in \mathbb{R}^+, \ \forall h \in H_{\text{iso}} \tag{8.21}$$

A special case of isometry invariance is translation invariance. This states that there is no a priori information about the location of the features: all points of the image domain should be treated similarly. In this case, the embedding process is independent of the spatial coordinates.

Property 1.8 *Affine invariance:* An affine transformation is any transformation that preserves colinearity and the ratios of distances. Affine transformations include geometric contraction, expansion, dilation, reflection, rotation, shear, similarity transformations, spiral similarity, translation, and any combination of these. Although affine transformations entail the class of isometries, not all of them necessarily preserve angles or lengths. Let $h : \mathbb{R}^n \mapsto \mathbb{R}^n$ be an affine transformation of $\mathbb{R}^n$ of the form:

$$\forall p \in \Omega : h(p) = h_{\text{aff}}(p) + p_{\text{cnt}} \tag{8.22}$$

where $p_{\text{cnt}} \in \mathbb{R}^n$ represents a constant vector, and h_{aff} is a linear transformation of $\mathbb{R}^n$. h is orientation preserving if $\det(h_{\text{aff}}) = 1$. Conversely if $h_{\text{aff}} = -1$, h is orientation reversing. Let H_{aff} be the class of all affine transformations of $\mathbb{R}^n$; then the embedding process $\mathcal{U}$ is affine invariant up to a rescaling of the scale parameter t if the following is verified:

$$\mathcal{U}_t(h(f)) = h(\mathcal{U}_{t'}(f)), \quad \forall p \in \Omega, \; \forall t \in \mathbb{R}^+, \; \forall h \in H_{\text{aff}} \tag{8.23}$$

Property 1.9 *Scale invariance:* The multiscale analysis should independent of the feature size.

8.2.3
PDE Equation-Based Formulation

Any embedding $\mathcal{U}$ that fulfils the fundamental axioms given above can be governed by the following PDE system:

$$\begin{cases} u_t(\cdot, t) = \Phi(\nabla^2 u(\cdot, t), \nabla u(\cdot, t), u(\cdot, t), \cdot, t) \\ u(\cdot, 0) = f(\cdot) \end{cases} \tag{8.24}$$

For the system described above, the initial condition is scale zero ($t = 0$) of the scale-space image, which yields the original signal f. In the case of discrete images, that is, images that are defined on a finite grid, a set of boundary conditions should be added. These conditions describe the PDE at the image boundary $\delta\Omega$. The von Neumann boundary conditions specify the normal derivative of the scale-space image (u_n) at the image boundary. Given that h represent any function, the von Neumann boundary conditions are defined as follows:

$$u_n(p, t) = h(p, t) \quad \forall p \in \delta\Omega, \; \forall t \in \mathbb{R}^+ \tag{8.25}$$

In the case that h is the zero function, these conditions are known as the *homogeneous von Neumann boundary conditions*.

8.2.3.1 Classification
The system of PDEs describes an *initial value problem* or *Cauchy problem*, and is employed to describe the evolution given a reasonable initial condition at $t = 0$. PDEs may be classified according to many criteria. The general theory and methods of the solution usually apply only to a given class of equations. There are six basic classification criteria:

(i) the order of the PDE, that is, the order of the highest partial derivative in the equation;
(ii) the number of independent variables;

(iii) linearity, that is, the PDE is linear if all its independent variables appear in a linear fashion.

To obtain a description for the last three criteria, assume that the PDE given in Eq. (8.24) is linear and of second order. A general form for Φ is defined as follows:

$$\Phi(\nabla^2 u(x,y,t), \nabla u(x,y,t), u(x,y,t), x, y, t)$$
$$= c_1 u_{xx}(x,y,t) + c_2 u_{yy}(x,y,t) + 2c_3 u_{xy}(x,y,t)$$
$$+ c_4 u_x(x,y,t) + c_5 u_y(x,y,t) + c_6 \qquad (8.26)$$

where c_i are coefficients that are functions of the spatial domain, feature space, and/or the scale, associating the following matrix:

$$C_{matrix} = \begin{bmatrix} c_1 & c_3 \\ c_3 & c_2 \end{bmatrix} \qquad (8.27)$$

(iv) Homogeneity: The PDE in Eq. (8.26) is homogeneous if $c_6 \equiv 0$.
(v) Types of coefficients: constant or variable.
(vi) The basic types of PDEs are (i) elliptic, (ii) hyperbolic, or (iii) parabolic. It may be noted that for general PDEs mixed types exist as well. If $\det(C_{matrix}) > 0$, the PDE is elliptic and can be used to describe steady-state phenomena. Well-known elliptic PDEs are the Laplace's and Poisson's equations. The wave equation is of the hyperbolic type, and hence $\det(C_{matrix}) < 0$. This class of PDEs is used to describe vibrating systems and wave motion. Finally the PDE for which $\det(C_{matrix}) = 0$ is obtained. This class of PDEs is parabolic and can be used to describe heat flow. A well-known parabolic PDE is the heat equation.

Common scale-space filters are described by parabolic PDEs of the second order. This was expected given the fact that this type of PDE describes heat flow and diffusion processes. Hyperbolic PDEs or mixtures between parabolic and hyperbolic PDEs can also be found.

8.2.3.2 Well-Posedness
Verifying the well-posedness of the PDE system is essential. This implies that a solution should exist, should be unique, and should continuously depend on the data. If this is not the case, regularization provides a general framework to convert ill-posed problems in early vision.

8.2.3.3 Solving the PDE
Obtaining of the representation of the original signal at a certain scale can be done analytically or by numerical approximations (Figure 8.2). For solving a PDE system analytically, it is required that there exists a Green's function or some approximation of it [22]. A *Green's function* is an integral kernel that can be used to solve an inhomogeneous differential equation with boundary conditions. It is a basic solution to a linear differential equation and can be used to calculate the solution of the system at a certain scale. The exact form of the Green's

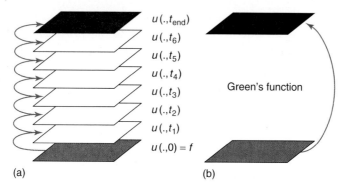

Figure 8.2 Solving the PDE of the scale-space filter. (a) Numerical approximation. (b) Analytical solution using the Green's function.

function depends on the differential equation, the definition domain, and the boundary conditions used. Let G be the Green's function associated with the linear differential equation $\phi(u(x)) = F(x)$, where $u(x)$ is an unknown function and $F(x)$ is a known nonhomogeneous term. Then the solution of this system can be obtained as follows:

$$u(x) = \phi^{-1}F(x) = \int_\Omega G(x,x')f(x') \tag{8.28}$$

One can see that the Green's function is part of the inverse operation of the linear differential equation. Since ϕ is a differential operator, it is reasonable to expect that its inverse is an integral operator. For many nonlinear systems of PDEs, the Green's function does not exist. In this case, a numerical approximation of the PDE system is required. The *numerical approximation* of PDEs is, by itself, a vast subject. In the literature, schemes have been reported using (i) finite differences [7, 23], (ii) finite elements [24, 25], (iii) multigrid methods [26, 27], (iv) approximations based on wavelets [28, 29], and (v) Monte Carlo methods. In image analysis, numerical approximations often adopt fast finite differences schemes. The finite difference is the discrete analog of the derivative. A wise use of the different finite differences methods can lead to advantageous numerical approximations. Conversely, a bad choice can lead to oscillations or instability of the approximation. The numerical approximation of PDE systems that describe parabolic initial value problems, with finite differences, can be subdivided in two major classes, namely, the *explicit time marching* scheme and the *implicit time* schemes. In the first scheme, the solution of the PDE at a certain time will depend explicitly on the solutions at previous times. Implicit schemes, such as the Crank–Nicholson method, solve a system of algebraic equations to obtain all the values for each new value of time. Both the schemes introduce a discretization error that is proportional to δt^k, with δt being the time step and k being the order of the scheme. Hence, restrictions on the time step are required in both schemes: one must be careful not to excessively increase the time step between successive solutions. Approximation is subject to approximation errors. Hence, large time steps can lead to visible and undesirable distortions of the obtained solutions. Besides the effects of approximation errors, the explicit marching scheme has another reason to restrict the time step: the stability of the approximated system. In case of the simplest explicit marching schemes, this restriction on the time step is rather severe. On the other hand, the implicit schemes may result in solving a complicated system of algebraic equations. Usually, the higher complexity occurs when the process is described on spatial domains of higher dimensions (2D or higher). As a result, semi-explicit schemes were introduced. These schemes allow for operator splitting that is fast and alleviates the complexity problem in higher dimensions. It may be noted that for both the explicit time marching and the implicit time schemes, there exist many approaches that remedy the problems of the straightforward approximation methods.

8.2.4
Variational Formulation

The PDE given in Eq. (8.24) can also be obtained using a variational framework. In this framework, the scale-space filter is obtained by minimizing a given energy functional E:

$$\min \left[E(u) \right] \tag{8.29}$$

Under general assumptions, it is required that the Euler–Lagrange derivative equals zero if u is to be a minimizer of E. This can be expressed via the steady-state solution:

$$u_t = \phi(u) \tag{8.30}$$

where ϕ denotes the Euler–Lagrange derivative. This approach has been used in image processing for quite some time. As such, several scale-space filters can be expressed as the minimization of an energy functional.

8.3
Representation

8.3.1
Gaussian Scale Space

According to Witkin [4], the term scale space refers to the multiscale representation constructed by convolution with Gaussian kernels of increasing width, or equivalently, by solving the linear homogeneous heat equation [2]:

$$\begin{cases} u_t(\cdot, t) = \Delta u(\cdot, t) & \text{in } \Omega \times \mathbb{R}^+ \\ u(\cdot, 0) = f(\cdot) \end{cases} \tag{8.31}$$

The Green's function of the system given in Eq. (8.31) is the normalized Gaussian convolution:

$$G_\sigma(\mathbf{x}) = \frac{1}{(4\pi t)^N} \exp\left[-\frac{\mathbf{x} \cdot \mathbf{x}}{4t}\right] = \frac{1}{(2\pi\sigma^2)^N} \exp\left[-\frac{\mathbf{x} \cdot \mathbf{x}}{2\sigma^2}\right], \text{ with } \mathbf{x} \in \mathbb{R}^N \quad (8.32)$$

Hence, a representation of the original signal f at a certain scale t can be obtained by convolving the original signal with a Gaussian with a kernel size σ: $u(\cdot, t) = f(\cdot) \otimes G_\sigma(\cdot)$. The extent of the simplification (filtering) is determined by the size of the kernel. From Eq. (8.32), one can derive the result $t = \frac{\sigma^2}{2}$. The Gaussian scale space was originally formulated for one-dimensional signals. The theory was extended rather fast for higher dimensional signals such as images. However, it revealed some inconsistencies with the one-dimensional case, for example, the fact that new extrema are not created is no longer verified. Although the study of nonlinear scale space for vector-valued images is fairly well established [30], the investigation of the Gaussian scale-space filter in the case of vector-valued signals is ongoing. Traditionally, each image band is convolved separately with the same filter, that is, component-wise filtering takes place. However, it has been argued that the Gaussian scale-space paradigm should be rejected when image bands are mutually dependent [31].

Gaussian scale-space theory was initially formalized for the continuous domain. A discrete version requires a rigorous discrete approximation [18]. It can be achieved by discretizing the convolution kernel (8.32) or the PDE system (8.31). For a detailed discussion on this matter the reader is referred to Ref. [18]. It may be noted that the linear scale-space filter is equivalent to minimization of the Dirichlet integral: $\min\left[E(u)\right] = \min\left[\int_\Omega \frac{|\nabla u|^2}{2} d\Omega\right]$. Figure 8.3a illustrates the linear scale-space representation for a color image.

The shortcomings of Gaussian scale-space filter are mainly due to the constraint of an uncommitted visual front-end which excludes the possibility of introducing a priori information in the filtering process. Hence image features and noise are treated similarly. Another problem of the Gaussian scale space is the dislocation of features (correspondence problem), which can cause severe difficulties during the linking process. The disadvantages of the Gaussian scale space can be resolved by using an embedding process that depends on the local image structure, that is, by employing some type of adaptive image smoothing. Image-adaptive smoothing is a well-known concept in image processing [27]. Since the work of Perona and Malik [20] such a processes can be expressed using a PDE formulation. Although their formulation has been proven to be ill posed [32], their impressive results triggered an enormous interest in PDE-based image processing and nonlinear scale-space filtering. The next section covers the Gaussian scale-space filtering and several of its nonlinear counterparts. Several filters are included that stem from the generalization of the linear heat equation, such as the Perona and Malik filter. Furthermore, we discuss geometrically driven scale spaces such as mean curvature motion and affine curvature motion. The latter is followed by MM scale spaces. There are many more filtering approaches: a well-known class of variational filters is based on the total variation (TV) norm. TV is an essential factor for the calculation of shocks in the field of fluid dynamics [7]. Osher and Rudin [33] introduced the use of TV as a tool for image restoration (see Chapter 26). They introduced the notions of TV-preserving and TV-minimizing flow. The former attempts to create a flow field away from the region's interior, while the latter falls in the class of generalization of the linear heat equation filters. Connections between wavelet shrinkage and nonlinear diffusion were unraveled in Ref. [31]. In Ref. [34], the relation between the Gabor wavelet and scale-space theory is studied. The Beltrami flow [35] is a well-known filter that treats image manifolds.

8.3.2
Variable Conductance Diffusion

VCD has been proposed as a model for low-level vision [36] and it is widely used in image processing. It is an image-adaptive filtering based on the diffusion equation (Eq. 8.33), that is, a generalized version of the heat equation, with a *variable conduction coefficient* (c) that controls the rate at which the density flows from areas with high concentration to areas of low concentration. The embedding process $\mathcal{U}_t : f(p) \mapsto u(p, t)$ with $t \in (0, T]$ under the generalized heat equation for a vector-valued image f is defined by

$$\begin{cases} u_t^{(i)} &= \text{div}\left[c\nabla u^{(i)}\right] & \text{on } \Omega \times (0, T] \\ u^{(i)}(p, 0) &= f^{(i)}(p) & \text{on } \Omega \\ \langle g(|\nabla u|)\nabla u^{(i)}, \mathbf{n}\rangle &= 0 & \text{on } \delta\Omega \times (0, T] \end{cases} \qquad (8.33)$$

If the flux ($-c\nabla u$) is parallel to the gradient, the conductance coefficient is a scalar value and referred to as the *diffusivity g*. In this case, the diffusion process is isotropic. Conversely, it is anisotropic and the conductance coefficient is matrix and is denoted by the diffusion tensor D. If the conductance coefficient is is a function of the position, the process is inhomogeneous. The diffusion process is linear if the conductance coefficient does not change with t. In the case that the conductance function depends on the spatial position, that is, we have an inhomogeneous process, its shape and argument yield a further classification.

The formulation for vector-valued images can be achieved by applying the filter to each channel separately, that is, by component-wise filtering using $c^{(i)}$. This yields a system of noncoupled PDEs. For obvious reasons, this is not advisable. In Ref. [30], Whitaker proposes a general framework for dealing with vector-valued images. It consists of a system of coupled diffusion equations for which each channel is diffused separately using a conductance function that is estimated using information from all channels. Although, this provides a general framework for dealing with vector-valued images, the literature also reports on special frameworks.

Commonly, VCD is applied on the gray scale or color input either to reduce noise (image enhancement), to obtain a piece-wise approximation of the image (segmentation), or to represent the image at different scale (scale-space filtering). However, these filters also have been used to smooth other types of information. Anisotropic diffusion filtering has been applied to enhance vector probabilities and texture estimates based on a Gabor decomposition. The basic idea remains the same: the diffusion process is steered in such a way that important features are enhanced while smoothing away noise. The latter is achieved by selecting an appropriate conductance function. The literature reports on a variety of conductance functions, which differ in shape, argument, or both. The most common type of VCD limits the feature flow according to the edges in the image. It is denoted as *edge-affected variable conductance diffusion* (EA-VCD). The argument of the conductance function represents an estimate of edgeness, and the conductance function has a form such that the diffusion process is reduced in areas that exhibit a high amount of edgeness. For EA-VCD, the conductance function (c) is often denoted as an *edge-stopping function* (g) since it stops the diffusion at edges. Typically, the edgeness is measured by the gradient magnitude. The EA-VCD can be formulated by a PDE system [7, 20, 30, 32, 37] or as an energy-minimizing process [30, 38–40]. The embedding process $\mathcal{U}_t : f(p) \mapsto u(p, t)$ with $t \in (0, T]$ under EA-VCD for a vector-valued image f is defined by the following PDE or energy functional:

$$\begin{cases} u_t^{(i)} &= \operatorname{div}\left[g(|\nabla u|)\nabla u^{(i)}\right] & \text{on } \Omega \times (0, T] \\ u^{(i)}(p, 0) &= f^{(i)}(p) & \text{on } \Omega \\ \langle g(|\nabla u|)\nabla u^{(i)}, \mathbf{n}\rangle &= 0 & \text{on } \partial\Omega \times (0, T] \end{cases} \quad (8.34)$$

$$\begin{cases} E(u) &= \int_\Omega \mathcal{G}(|\nabla u|)dp \\ u^{(i)}(p, 0) &= f^{(i)}(p) & \text{on } \Omega \end{cases} \quad (8.35)$$

where $\mathbf{n}$ denotes the normal on the image boundary $\partial\Omega$, and the last equation of Eq. (8.35) is the state of the minimization process at $t = 0$. The edge-stopping function g and the energy potential $\mathcal{G}$ verify the following criteria:

* Continuity : $g \in C^\infty(\mathbb{R})$
* Positive lower bound : $g(z) \geq 0 \ \forall z \in \mathbb{R}$
* Monotonically decreasing : $g'(z) \leq 0 \ \forall z \in \mathbb{R}$
* Positive : $\mathcal{G}(z) \geq 0 \ \forall z \in \mathbb{R}$
* Strictly increasing : $\mathcal{G}'(z) > 0 \ \forall z \in \mathbb{R}+$
* $\mathcal{G}'(0) = 0$

Since $\mathcal{G}(|\nabla u|)$ is strictly decreasing, the minimization process is equivalent to a smoothing operation. This is equivalent to the condition that g is positive. The minima of $E(|\nabla u|)$ are at some of its stationary points, which are given by the Euler–Lagrange equation and can be found by solving Eq. (8.34) for $t \mapsto \infty$ and by substituting $g(|\nabla u|) = \frac{\mathcal{G}'(|\nabla u|)}{|\nabla u|}$. The relation between these formulations often gives an additional insight into the behavior of the diffusion process. Given the variational point of view, it can be easily demonstrated that the diffusion process is ill posed if the employed energy potential does not verify certain conditions: a well-posed diffusion model can be expressed by an energy minimization process that employs a convex or linear energy potential. Well-known examples of well-posed EA-VCD diffusion models include minimizing TV flow [33] and, Charbonnier's diffusion filtering [41]. However, these models, as with the linear diffusion (Tikhonov energy potential) model, have the tendency to dislocate boundaries and to evolve away from the initial data. It is for this reason that these models are often used in conjunction with a reaction term that prohibits the diffusion from deviating too much from the original input. An alternative is the class of EA-VCD filters based on nonconvex energy potentials. These filters allow edge enhancement. Probably, the best known filter of this type is the Perona and Malik's anisotropic diffusion filter [20].

Table 8.1 gives an overview of several edge-stopping functions. Note that, in this table, x represent the regularized gradient magnitude of the vector-valued scale-space image at a given scale: $u(., t)$. The formulation of the edge-stopping function describes the diffusion process alongside an edge for scalar-valued images. The presented edge-stopping functions are classified according to their behavior across the edge. Hence, we obtain four classes, namely, ND, which indicates that there is no diffusion across the edge; FD, which implies that there are only forward diffusion processes occurring across the edges (smoothing); BFD contains edge-stopping function that yields backward diffusion (enhancement) for large-gradient magnitudes and forward diffusion (smoothing) for small-gradient magnitudes. The last class is denoted BF and entails all edge-stopping functions that result solely in backward diffusion across the edges. For the sake of completeness, the tables provide the associated energy potentials and the formulation of the diffusion amounts across the edges.

EA-VCD that belongs to class BFD and BF requires regularization. Commonly, this implies that the edge-stopping function is applied on the regularized gradient, that is, the gradient is applied on a Gaussian-smoothed image or Gaussian derivatives are used. The selection of an appropriate size for the regularization kernel often is achieved empirically. In Ref. [46], an estimate based on the median absolute deviation (MAD), [47] which is a measure of the scale or dispersion of a distribution about the median, is proposed:

$$\sigma_r(f) = 1.4826\pi\sqrt{2} \cdot \text{median}\left[|\nabla f_{\text{norm}}| - \text{median}(|\nabla f_{\text{norm}}|)|\right] \quad (8.36)$$

where $|\nabla f_{\text{norm}}|$ denotes the unit normalized gradient magnitude of the original image f. Note that the constant (1.4826) originates from the fact that the MAD of a zero-mean normal distribution with unit variance is $0.6745 = 1/1.4826$. Hence σ_n is consistent with the standard deviation for asymptotically normal distributions. Since $|\nabla f_{\text{norm}}| \in (0, 1)$, $\sigma_r(f) \in [0, 3.8253]$. The latter ensures that the regularization kernel cannot become so large that small image structures are smoothed away entirely.

Most edge-stopping functions in Table 8.1 contain a parameter that is commonly known as the *contrast parameter* (k). Its physical aspect is meaningful only in the case of BFD type filters. In these cases, it coincides with the first and only critical point in the second-order derivative of the energy potential ($k_1 = k_{\mathcal{G}'}^{(1)}$). It marks the gradient magnitude at which the amount of diffusion across the edge switches sign, that is, it switches between edge enhancement and edge blurring. This parameter is usually determined empirically. Automatic estimation may be achieved by using the argument at a percentage of the cumulative histogram of the regularized gradient.

The use of EA-VCD diffusion models has many advantages. They have been widely studied both in the continuous and in the discrete cases. In Ref. [7], the well-posedness of the Catté *et al.* regularized EA-VCD and scale-space axioms such

Table 8.1 Edge stopping functions for the edge-affected variable conductance diffusion.

Name	Type	$\mathcal{G}(x)$	$g(x) = x^{(-1)}\mathcal{G}_x(x)$
Rudin [42] (TV-minimizing flow)	ND	$\|x\|$	$\frac{1}{\|x\|}$
Green	FD	$k^2 \ln\left[\cosh\left(\frac{\|x\|}{k}\right)\right]$	$\begin{cases} 1 & \text{if } x=0 \\ \frac{k}{\|x\|}\tanh\left(\frac{\|x\|}{k}\right) & x \neq 0 \end{cases}$
Charbonnier [41, 43]	FD	$k^2\sqrt{1+\frac{x^2}{k^2}} - k^2$	$\left[1+\frac{x^2}{k^2}\right]^{-\frac{1}{2}}$
Le Clerc [20, 38]	BFD	$-k^2\exp\left[-\frac{x^2}{2k^2}\right] + k^2$	$\exp\left[-\frac{x^2}{2k^2}\right]$
Tukey bi-weight [38]	BFD	$\begin{cases} \frac{x^2}{2} - \frac{x^4}{10k^2} + \frac{x^6}{150k^4} & \text{if } \|x\| \leq \sqrt{5}k \\ \frac{5k^2}{6} & \|x\| \geq \sqrt{5}k \end{cases}$	$\begin{cases} \left[1 - \frac{x^2}{5k^2}\right]^2 & \text{if } \|x\| \leq \sqrt{5}k \\ 0 & \|x\| \geq \sqrt{5}k \end{cases}$
Weickert [7] ($C_8 = 3.31488$)	BFD	$\frac{x^2}{2} - \frac{x^2}{2}\exp\left[-C_8\frac{k^8}{x^8}\right] + \frac{\sqrt[4]{C_8 k^2}}{2}\Gamma\left(\frac{3}{4}, \frac{C_8 k^8}{x^8}\right)$	$\begin{cases} 1 & \text{if } x=0 \\ 1 - \exp\left[\frac{-C_8 k^8}{x^8}\right] & x \neq 0 \end{cases}$
Lorentzian [20, 38]	BFD	$\frac{k^2}{2}\ln\left[1+\frac{x^2}{k^2}\right]$	$\left[1+\frac{x^2}{k^2}\right]^{-1}$
Geman [44]	BD	$-4\alpha k^2\left[1+\frac{\|x\|}{k}\right]^{-1} + 4\alpha k^2$	$\frac{4\alpha k}{\|x\|}\left[1+\frac{\|x\|}{k}\right]^{-2}$
You [40] ($\aleph > 1$)	BD	$\frac{1}{\aleph}\sqrt[\aleph]{\|x\|}$	$\left[\frac{1}{\|x\|}\right]^{2-\frac{1}{\aleph}}$
Keeling I [45]	BD	$\alpha \ln\|x\|$	$\alpha\frac{1}{x^2}$
Keeling II [45]	BD	$2k\ln\|1+\frac{x}{k}\| - 2k$	$\frac{2k}{x}\left[1+\frac{x}{k}\right]^{-1}$

as the minimum–maximum principle, the nonenhancement of local extrema, some scale-space invariants, and the convergence to a steady state are proven. Furthermore the existence of Lyapunov functionals such as the Shannon entropy, generalized entropy, second-order moments, and the energy are demonstrated. These prove to be very useful in describing the information reduction of the diffusion process. In Ref. [48], the Shannon entropy was used to synchronize different EA-VCD diffusion filters for comparison, and in Ref. [11], second-order moments were taken into account for the derivation of an optimal stopping time. Another advantage is the existence of fast and simple numerical schemes that are based on finite differences. Similar to the work in Ref. [49], which deals with parallel splitting of the Navier–Stokes equations, Weickert et al. [7] proposed a fast numerical implementation for the EA-VCD that retains the properties of its continuous counterpart. This concept can be easily extended to vector-valued images. In addition, it can be shown that the numerical approximation given by

Perona and Malik [20] is a special case of Weickert's numerical scheme and that, as such, it upholds the same overall properties.

Figure 8.3(b) illustrates an EA-VCD diffusion scheme of an adapted version of the coupled diffusion for vector-valued images in conjunction with BD edge-stopping function, namely, $u_t^{(i)} = \text{div}\left[g(|\nabla u|)\frac{\nabla u^{(i)}}{|\nabla u^{(i)}|}\right]$, with $g(x) = \frac{1}{1+\frac{x^2}{k^2}}$.

8.3.3
Geometry-Driven Scale Space

Image-adaptive filtering based on curves (level sets) represents another class of scale-space filters. This type of filter deforms a given curve or image with a PDE in order to obtain the evolution of the plane curve. More precisely, the evolution

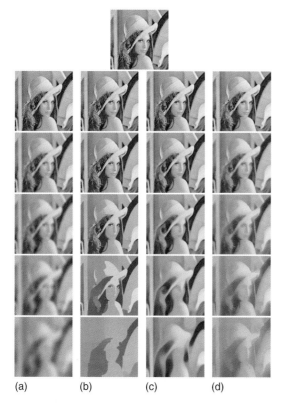

(a)   (b)   (c)   (d)

Figure 8.3 Example of scale-space representations for "Lena" color image using (a) Gaussian scale space with $t = 1.11$, 8.22, 22.35, 60.76, 165.15 (top to bottom). (b) Edge-affected variable conductance diffusion with $t = 1.11$, 60.76, 448.92, 3317.12, 24510.40 (top to bottom). (c) Mean curvature motion with $t = 1.66$, 18.30, 60.76, 201.71, 669.72 (top to bottom). (d) Levelings (Eq. 8.42) with $t = 1.66$, 18.30, 60.76, 201.71, 669.72 (top to bottom). (Please find a color version of this figure on the color plates.)

equation is governed by a second-order parabolic PDE and provides an efficient way to smooth curves that represent the contours of the objects. Deforming of curves yields the field of snakes and active contours, which is beyond the scope of this work. Henceforth, we assume that the level set function refers to an image unless specified differently. This means that the image is viewed as a collection of iso-intensity contours, each of which is to be evolved [23]. Curvature-based scale-space filters include mean curvature motion [50], affine curvature motion [51], self-snakes [37], and min–max flow [23, 52].

The evolution equation of the original signal f under curvature motion is given by the following system:

$$\begin{cases} u_t(\cdot, t) = \mathbb{F}(k(\cdot, t))|\nabla u(\cdot, t)| \\ u(\cdot, t = 0) = f(\cdot) \\ u_n = 0 \end{cases} \quad (8.37)$$

where k is the Euclidean curvature and $\mathbb{F}$ represents the speed function. This is a special case of the level set equation given by Osher and Sethian in Ref. [53], where they describe the motion of a hypersurface that propagates along its normal direction with a speed $\mathbb{F}$. The use of a curvature-dependent speed function has the advantage that the level lines will not form corners, that is, the level lines will remain smooth [23]. In the case that the speed function does not depend on the curvature, Eq. (8.37) can be related to the generalized heat as follows: $\mathbb{F}(u) = \frac{\text{div}[c\nabla u]}{|\nabla u|}$. If $\mathbb{F}(u) = \frac{\Delta u}{|\nabla u|}$, Eq. (8.37) yields the linear heat equation. It immediately becomes clear that the evolution no longer occurs only along the normal of the level lines. Instead, it is isotropic and smears sharp boundaries. A special case is the TV-minimizing flow: here the speed function is given by $\mathbb{F}(u) = \frac{k}{|\nabla u|}$. Hence, this method is also closely related to curvature-based scale-space filtering. Furthermore, curvature-based scale-space filtering is related to MM scale-space filtering. It has been demonstrated that certain curve evolutions yield the elementary multiscale dilation/erosion [7]. It may be noted that the level lines are the curves of constant gray scales. For vector-valued images, Chung et al. [54] proposed to redefine the idea of level sets using the concept of the generalized gradient [55]. The level sets are defined as the integral curves of the directions of the minimal change, which is obtained by the Di Zenzo vector–tensor gradient estimation [55]. In two-dimensional geometry, the Euclidean curvature is the amount by which a geometric object deviates from being a straight line. The curvature is a vector pointing in the direction of that circle's center, that is, it points to the interior of the curve. Obviously, the curvature at a point is closely related to the gradient at this point. The PDE formulation of the curvature is given by: $k(x, y) = \text{div}\left[\frac{\nabla u(x, y)}{|\nabla u(x, y)|}\right]$.

Mean curvature motion is considered as the standard curvature evolution equation [50]: $\mathbb{F} = k(\cdot, t)$. Essentially, it is an anisotropic variant of the linear heat equation. This can be easily demonstrated if one rewrites the linear heat equation using the gauge coordinate system: $u_t(\cdot, t) = \Delta u = u_{xx} + u_{yy} = u_{vv} + u_{ww}$. Mean curvature motion only allows diffusion along the normal of the level lines. Hence, diffusion solely occurs along the v-axis: $u_t(\cdot, t) = u_{vv} = k|\nabla u|$. For vector-valued images, there exist two possibilities: (i) component-wise filtering ($u_t^{(i)} = u_{vv}^{(i)}$) and

(ii) use of the vector-valued level lines ($u_t^{(i)} = u_{vv}^{(i)}$). Figure 8.3(c) illustrates the mean curvature filter for a color image. This filter is contrast invariant, makes nonconvex shapes convex but also blurs and dislocates the edges. In case that Eq. (8.37) is also invariant under affine transformation, then we obtain the affine curvature motion or affine shortening. In Ref. [50], it was demonstrated that $\mathcal{F} = \sqrt[3]{k(\cdot, t)}$ is the only possible equation.

Introduced by Sapiro et al. [37], self-snakes are a variant of the mean curvature motion that uses an edge-stopping function. The main goal is to prevent further shrinking of the level lines once they have reached the important image edges. For scalar images, self-snakes are governed by $u_t = |\nabla u| \text{div}\left[g(|\nabla u|)\frac{\nabla u}{|\nabla u|}\right]$. The properties and requirements of the edge-stopping function g are analogous to those given in the previous section. The latter also includes the Catté et al. regularization technique [32]. The link with the curvature can be observed by decomposing the equation into two parts. The first part corresponds to a diffusion term. Actually, it is the constraint curvature motion [50]. The second term can be viewed as a shock filter since it pushes the level lines toward valleys of high gradient, acting as Osher's shock filter [56]. For vector-valued images, one can opt for the following:

- Component-wise filtering: $u_t^{(i)} = |\nabla u^{(i)}| \text{div}\left[g(|\nabla u^{(i)}|)\frac{\nabla u^{(i)}}{|\nabla u^{(i)}|}\right]$
- Coupled system of PDEs [30]: $u_t^{(i)} = |\nabla u^{(i)}| \text{div}\left[g((|\nabla u|))\frac{\nabla u^{(i)}}{|\nabla u^{(i)}|}\right]$
- Vector-valued level lines approach: $u_t^{(i)} = g(|\nabla u|)u_{vv}^{(i)} + (\nabla g(|\nabla u|))'\nabla u^{(i)}$

The interested reader should refer to chapters 4–6 of Sethian's book [23] for more details concerning the numerical approximations and a discussion involving fast implementation methods, such as parallel algorithms and adaptive mesh refinement.

8.3.4
Mathematical Morphology Scale Space

MM has a wide range of application fields, for example, biomedical imaging, shape from shading, and edge detection, to name a few. It has also been used for scale-space filtering. The simplest form of an MM scale-space filter is based on the fundamental morphological operators, that is, dilation–erosion scale space. This scale-space filter verifies the scale-space axioms. However, MM scale-space filters may be the result of a number of different combinations of MM operations. They are generated using spatially symmetric structuring elements. Commonly, the image is filtered with a structuring element of increasing size. scale-space filtering based on MM is advantageous for quantifying shape information at different scales. The use of PDE for the multiscale MM yields a continues-scale morphology, which has a superior performance over discrete morphology in terms of isotropy and subpixel accuracy. In the mid-1990s, PDE formulations for multiscale MM started to emerge. Brockett et al. [57], modeled the basic multiscale MM operators for structuring elements, which are either compact-support convex set or concave functions. They demonstrated that convexity of the structuring element is sufficient to ensure that the scale-space filter verifies the semigroup property. Alvarez et al. [13], provide a complete scale-space modeling for multiscale erosions(dilations). They prove that the scale-space filter based on these operators verify the necessary scale-space axioms: the semigroup property, locality, regularity, and the comparison principle. Furthermore, given the morphological invariant, they provided the governing PDE for these types of scale-space filters. In Ref. [35], opening and closing by reconstruction as well as the levelings have been formulated using PDEs. In Ref. [58], PDE representations for general structuring elements were introduced. The MM filtering of vector-valued images follows two main approaches to obtain the required ordering among the feature vectors: (i) marginal ordering and (ii) lexicographical ordering. Since marginal ordering cannot guarantee the preservation of the properties of the morphological filter, it is advisable to use a method based on lexicographical ordering.

All multiscale morphological operations are generated by multiscale dilations and erosions. The multiscale erosion (dilation) is obtained by replacing the structuring element SE in the standard erosion(dilation) by a multiscale version $SE^{(t)}$. In the two-dimensional case, the above can be expressed as $SE^{(t)}(x, y, t) = t * SE(\frac{x}{t}, \frac{y}{t})$, where t is the scale parameter, and SE is the unit-scale kernel. Thus, the multiscale erosion for a two-dimensional signal f by the structuring element SE is given by $\left[f \ominus SE^{(t)}\right](x, y) = \inf_{p_i \in SE^{(t)}} \{f(x + x_i, y + y_i) - t.SE(x_i, y_i)\}$, while the multiscale dilation for a two-dimensional signal f by the structuring element SE is given by $\left[f \oplus SE^{(t)}\right](x, y) = \sup_{p_i \in SE^{(t)}} \{f(x - x_i, y - y_i) + tSE(x_i, y_i)\}$. It was demonstrated that multiscale erosion, dilation, and their combinations are governed by $u_t(\cdot, t) = |\nabla u| \phi [t, \mathcal{K}(u)]$, where ϕ is a nondecreasing function with respect to the curvature $\mathcal{K}(u)$. An example of an erosion/dilation with a flat structuring element is the elementary multiscale erosion/dilation. Here, the erosion/dilation of the original signal is achieved by a disc of radius t and origin at zero. These scale-space filters are governed by the following PDEs, respectively: $u_t = -|\nabla u|$ and $u_t = |\nabla u|$. Hence, the function $\phi(\mathcal{K}(u)) = 1$. Note that using the disk as structuring element has the advantage that the growth is isotropic. An example of a multiscale erosion/dilation PDE using a nonflat structuring element follows. Let the structuring element be of parabolic shape: $SE = -a(x^2 + y^2)$, where $a \in \mathbb{R}_0^+$ is constant; then the multiscale erosion/dilation can be expressed by $u_t = -\frac{|\nabla u|^2}{4a}$ and $u_t = -\frac{|\nabla u|^2}{4a}$. Multiscale openings and closings preserve the shape and location of vertical abrupt signal discontinuities like edges. Furthermore, the definition of scale is identical to the spatial size of geometrical objects. Levelings are a powerful class of filters that are based on the opening and closing by reconstruction. They start from a reference image consisting of several parts and a marker image inside some of these parts. The process can reconstruct entire objects with exact boundary and edge preservation. The reconstruction process is built in such a way that all smaller objects in which the marker cannot fit are removed. Hereby, using the reference image as a global constraint. The disadvantage of these filters is that they are not self-dual. As such, they treat the image foreground and background asymmetrically. The levelings, introduced by Meyer [59], alleviate this disadvantage. The levelings are robust, that is, they are strong morphological filters. Furthermore, they have several advantages. Not only do they preserve the exact position of boundaries and contours but they are also translation, rotation, and illumination invariant. Levelings can be expressed using PDEs [60], which has the additional advantage that it is closer to the true filtering than its algebraic discrete counterpart. Moreover, the last mentioned is a special case of the PDE version. One last advantage is that the PDE representation allows to stop the filtering process before convergence is reached. For a detailed description of the algebraic discrete version and the levelings, the reader is referred to Refs [35, 59–62]. Let f_{ref} be the reference image and f_{mar} be the marker image; then opening by reconstruction of f_{ref} with respect to f_{mar} using a unit scale flat structuring element can be obtained when a convergence is achieved in the following PDE system:

$$u_t = \text{sign}(f_{ref} - u)|\nabla u| = \begin{cases} |\nabla u| & \text{if } u < f_{ref} \\ 0 & \text{if } u = f_{ref} \text{ or } \nabla u = 0 \end{cases}$$
$$u(\cdot, t = 0) = f_{mar}(\cdot) \leq f_{ref}(\cdot) \tag{8.38}$$

The above describes a conditional dilation that grows the intermediate result $(u(\cdot, t))$ as long as it does not exceed the reference image. The closing by reconstruction is obtained similarly:

$$u_t = -\text{sign}(f_{ref} - u)|\nabla u| = \begin{cases} -|\nabla u| & \text{if } u > f_{ref} \\ 0 & \text{if } u = f_{ref} \text{ or } \nabla u = 0 \end{cases}$$
$$u(\cdot, t = 0) = f_{mar}(\cdot) \geq f_{ref}(\cdot) \tag{8.39}$$

A leveling combines the above by assuming that there is no specific relationship between the marker and the reference image. Thus, the leveling of an image f with respect to the marker image is obtained when the following PDE system converges:

$$u_t(\cdot, t) = \text{sign}(f(\cdot) - u(\cdot, t))|\nabla u(\cdot, t)|$$
$$u(\cdot, t = 0) = f_{mar}(\cdot) \tag{8.40}$$

The above is a switched dilation, where the switching occurs at the zero-crossing of $f - u$ and at the singularities in the gradient of u. To obtain parabolic levelings, one can use, for example, a structure element $K = -a.x^2$; then the PDE becomes

$$u_t(\cdot, t) = \frac{\text{sign}(f(\cdot) - u(\cdot, t))}{4a}|\nabla u(\cdot, t)|$$
$$u(\cdot, t = 0) = f_{mar}(\cdot) \tag{8.41}$$

Note that the numerical scheme provided in Ref. [35] is a finite difference scheme that follows the ideas presented in Ref. [56] and follows the same principles as described in the section on geometry-driven scale-space-filtering (Sec. 8.3.3). To obtain a scale space filtering based on the levelings, it was proposed to create a set of sequential levelings using a Gaussian smoothed version of the image with increasing kernel as marker [35, 60]. A more general approach and some alternatives can be found in Ref. [62]. Let $\mathcal{L}_{f_{mar}}(f)$ be the leveling of the image f

with respect to the marker image mar; then we obtain a scale-space representation by creating a series of levelings using the following strategy:

$$u(\cdot, t + \Delta t) = \mathcal{L}_{[G_{\sigma_t} \otimes f]}(u(\cdot, t))$$
$$u(\cdot, t = 0) = f \qquad (8.42)$$

Figure 8.3(d) illustrates this filter for a color image.

8.4
Conclusions

Scale-space theory provides a mathematically well-founded framework for analyzing images at multiple scales. It entails the embedding of the original image into a family of derived simplifications in which the image features are linked across the scale dimension. As a general conclusion, we may say that there are two aspects that make a multiscale representation attractive: (i) the explicit representation of the multiscale aspect of real-world images and (ii) the simplification of further processing by removing unnecessary and disturbing details. More technically, the latter motivation for using a multiscale representation also reflects the common need for smoothing as a preprocessing step to many numerical algorithms as a means of noise suppression. In multiscale image analysis tasks, the scale-space representation should be complemented with the deep image structure [2] that concerns the methodology of how to relate signal structure at different scales with a particular focus on scale selection [10, 12]; this is the first step toward the exploitation of the deep image structure.

References

1. Florack, L. (1997) *Image Structure*, Springer.
2. Koenderink, J. (1984) The structure of images. *Biol. Cybern.*, **50**, 363–370.
3. Lindeberg, T. (1994) *Scale Space Theory in Computer Vision*, Kluwer.
4. Witkin, A. (1983) Scale space filtering. *IJCAI*, **2**, 1019–1022.
5. Yuille, A. and Poggio, T. (1986) Scaling theorems for zero-crossings. *IEEE Trans. Pattern Anal. Mach. Intell.*, **8**, 15–25.
6. ter Haar Romeny, B. (2003) *Front-End Vision and Multi-Scale Image Analysis*, Kluwer Academic Publishers.
7. Weickert, J. (1998) *Anisotropic Diffusion in Image Processing*, Teubner-Verlag.
8. ter Haar Romeny, B. et al. (eds.) (1997) *Scale-Space Theory in Computer Vision*, Springer.
9. Pratikakis, I. (1998) Watershed-driven image segmentation. PhD. Vrije Universiteit Brussel.
10. Lindeberg, T. (1998) Feature detection with automatic scale selection. *Int. J. Comput. Vis.*, **30**, 77–116.
11. Mrázek, P. and Navara, M. (2003) Selection of optimal stopping time for nonlinear diffusion filtering. *Int. J. Comput. Vis.*, **52**, 189–203.
12. Vanhamel, I. et al. (2009) Scale selection for compact scale-space representation of vector-valued images. *Int. J. Comput. Vis.*, **84**, 194–204.
13. Alvarez, L. et al. (1993) Axioms and fundamental equations of image processing. *Arch. Rat. Mech. Anal.*, **123**, 199–257.

14. Sporring, J. et al. (eds.) (1997) *Gaussian Scale Space Theory*, Kluwer Academic Press.
15. Otsu, N. (1981) Mathematical studies on feature extraction in pattern recognition. PhD thesis. Umezono, sakura-mura.
16. Pauwels, E. et al. (1995) An extended class of scale-invariant and recursive scale space filters. *IEEE Trans. Pattern Anal. Mach. Intell.*, **17**, 691–701.
17. Guichard, F. (1994) Axiomatisation des analyses multi-echelles d'images et de films. PhD thesis. University of ParisIX Dauphine.
18. Lindeberg, T. (1990) Scale space for discrete signals. *IEEE Trans. Pattern Anal. Mach. Intell.*, **12**, 234–254.
19. Hummel, R. (1986) Representations based on zero crossings in scale space, in *IEEE Computer Science Conference on Computer Vision and Pattern Recognition*, 204–209.
20. Perona, P. and Malik, J. (1990) Scale space and edge detection using anisotropic diffusion. *IEEE Trans. Pattern Anal. Mach. Intell.*, **12**, 629–639.
21. Iijima, T. (1962) Basic theory on normalization of a pattern (for the case of a typical one dimensional pattern). *Bull. Electrotech. Lab.*, **26**, 368–388.
22. Fischl, B. (1997) Learning, anisotropic diffusion, nonlinear filtering and space-variant vision. PhD dissertation. Boston University.
23. Sethian, J. (1996) *Level Set Methods*, Cambridge University Press.
24. Hackbusch, W. and Sauter, S. (1997) Composite finite elements for problems containing small geometric details. Part II: implementation and numerical results. *Comput. Vis. Sci.*, **1**, 15–25.
25. Li, J. (2002) Finite element analysis and application for a nonlinear diffusion model in image denoising. *Num. Methods Partial Differential Eq.*, **18**, 649–662.
26. Acton, S. (1998) Multigrid anisotropic diffusion. *IEEE Trans. Image Process.*, **7**, 280–291.
27. Saint-Marc, P., Chen, J., and Medioni, G. (1991) Adaptive smoothing: a general tool for early vision. *IEEE Trans. Pattern Anal. Mach. Intell.*, **13**, 514–529.
28. Chambolle, A. et al. (1998) Nonlinear wavelet image processing: variational problems, compression, and noise removal through wavelet shrinkage. *IEEE Trans. Image Process.*, **7**, 319–335.
29. Steidl, G. et al. (2004) On the equivalence of soft wavelet shrinkage, total variation diffusion, total variation regularization, and sides. *SIAM J. Num. Anal.*, **42**, 686–713.
30. ter Haar Romeny, B. (ed.) (1994) *Geometry-Driven Diffusion in Computer Vision*, Springer.
31. Griffin, L. and Lillholm, M. (eds.) (2003) *Scale Space Methods in Computer Vision*, Springer.
32. Catté, F. et al. (1992) Image selective smoothing and edge detection by nonlinear diffusion. *SIAM J. Num. Anal.*, **29**, 182–193.
33. Rudin, L. and Osher, S. (1994) Total variation based image restoration with free local constraints. *Proc. IEEE Int. Conf. Image Process.*, **1**, 31–35.
34. Sgallari, F., Murli, A., and Paragios, N. (eds.) (2007) *Scale Space and Variational Methods in Computer Vision*, Springer.
35. Nielsen, M. et al. (eds.) (1999) *Scale-Space Theories in Computer Vision*, Springer.
36. Cohen, M. and Grossberg, S. (1984) Neural dynamics of brightness perception: Features, boundaries, diffusion and resonance. *Percept. Psychophys.*, **36**, 428–456.
37. Sapiro, G. (2001) *Geometric Partial Differential Equations and Image Analysis*, University Press Cambridge.
38. Black, M. et al. (1998) Robust anisotropic diffusion. *IEEE Trans. Image Process.*, **7**, 421–432.
39. Scherzer, O. and Weickert, J. (2000) Relations between regularization and diffusion filtering. *J. Math. Imaging Vision*, **12**, 43–63.
40. You, Y.-L. et al. (1996) Behavioral analysis of anisotropic diffusion in image processing. *IEEE Trans. Image Process.*, **5**, 1539–1553.
41. Charbonnier, P. (1994) Reconstruction d'image: régularisation avec prise en compte desdiscontinuités. PhD. University of Nice-Sophia Antipolis.

42. Rudin, L., Osher, S., and Fatemi, E. (1992) Nonlinear total variation based noise removal algorithms. *Phys. D*, **60**, 259–268.
43. Aubert, G. and Vese, L. (1997) A variational method in image recovery. *SIAM J. Num. Anal.*, **34**, 1948–1979.
44. Geman, D. and Reynolds, G. (1992) Constrained restoration and the recovery of discontinuities. *IEEE Trans. Pattern Anal. Mach. Intell.*, **14**, 367–383.
45. Keeling, S. and Stollberger, R. (2002) Nonlinear anisotropic diffusion filtering for multiscale edge enhancement. *Inverse Probl.*, **18**, 175–190.
46. Vanhamel, I. (2006) Vector valued nonlinear diffusion and its application to image segmentation. PhD thesis. Vrije Universiteit Brussel.
47. Hampel, F. (1974) The influence curve and its role in robust estimation. *J. Am. Stat. Assoc.*, **69**, 383–393.
48. Niessen, W. et al. (1997) Nonlinear multiscale representations for image segmentation. *Comput. Vis. Image Underst.*, **66**, 233–245.
49. Lu, T., Neittaanmaki, P., and Tai, X.-C. (1991) A parallel splitting up method and its application to Navier-Stokes equations. *Appl. Math. Lett.*, **4**, 25–29.
50. Alvarez, L., Lions, P.-L., and Morel, J.-M. (1992) Image selective smoothing and edge detection by nonlinear diffusion II. *SIAM J. Num. Anal.*, **29**, 845–866.
51. Sapiro, G. and Tannenbaum, A. (1995) Area and length preserving geometric invariant scale spaces. *IEEE Trans. Pattern Anal. Mach. Intell.*, **17**, 67–72.
52. Malladi, R. and Sethian, J. (1996) Flows under min/max curvature and mean curvature. *Graph. Models Image Process.*, **58**, 127–141.
53. Osher, S. and Sethian, J. (1988) Fronts propagating with curvature dependent speed: algorithms based on Hamilton-Jacobi formulations. *J. Comput. Phys.*, **79**, 12–49.
54. Chung, D. and Sapiro, G. (2000) On the level lines and geometry of vector-valued images. *Signal Process. Lett.*, **7**, 241–243.
55. Di Zenzo, S. (1986) A note on the gradient of a multi-image. *Comput. Vis., Graph. Image Process.*, **33**, 116–125.
56. Rudin, L. and Osher, S. (1989) *Feature-Oriented Image Enhancement with Shock Filters*, Technical, Department of Computer Science, California Institute of Technology.
57. Brockett, R. and Maragos, P. (1994) Evolution equations for continuous-scale morphological filtering. *IEEE Trans. Signal Process.*, **42**, 3377–3386.
58. Tai, X.-C. et al. (ed.) (2009) *Scale Space and Variational Methods in Computer Vision*, Springer,
59. Heijmans, H. and Roerdink, J. (eds.) (1998) *Mathematical Morphology and its Applications to Image and Signal Processing*, Kluwer.
60. Maragos, P. (2003) Algebraic and PDE approaches for lattice scale spaces. *Int. J. Comput. Vis.*, **52**, 121–137.
61. Kerckhove, M. (ed.) (2001) *Scale-Space and Morphology in Computer Vision*, Springer.
62. Meyer, F. and Maragos, P. (2000) Nonlinear scale space representation with morphological levelings. *J. Visual Commun. Image Representation*, **11**, 245–265.

9
Spatial Light Modulators (SLMs)
Philip M. Birch, Rupert Young, and Chris Chatwin

9.1
Introduction

Spatial light modulators (SLMs) are an integral part of any dynamically updatable optical system where information is to be encoded onto an optical wave front. There are numerous methods of achieving this and so this chapter reviews the common types, describes how they work, and discusses their advantages and disadvantages.

SLMs normally fall into one of two categories: phase modulators and amplitude modulators. These are discussed here, as well as important parameters including modulation range, temporal response, and wavelength range.

An SLM can be thought of as a spatially varying wavefront modulator device. Before they were invented, the traditional method was to record the modulation using photographic film. This still offers some advantages today, namely in spatial bandwidth, but it is less commonly used since unlike an SLM there is no real-time update ability. The two most common types of SLM are based around moving mirrors or manipulating the polarization state of light by the use of liquid crystals (LCs). They have many applications and are commonly found in optical processors, optical correlators, adaptive optics systems, computer-generated holograms (CGHs), displays, and beam pulse shaping, among others. Typically, these devices are electronically updatable at speeds varying from a few hertz to several megahertz.

This chapter first covers LC-SLMs starting with the nematic LC-SLMs and discusses their properties and modulation method capabilities. It then discusses ferroelectric liquid crystals (FLCs) and their properties. Multiple quantum-well devices are then covered and then the two types of mirror devices, phase and amplitude modulators, are discussed. Finally, it concludes with a discussion of the methods for achieving full complex (real and imaginary) modulation and some applications.

Optical and Digital Image Processing: Fundamentals and Applications, First Edition. Edited by Gabriel Cristóbal, Peter Schelkens, and Hugo Thienpont.
© 2011 Wiley-VCH Verlag GmbH & Co. KGaA. Published 2011 by Wiley-VCH Verlag GmbH & Co. KGaA.

9.2
Types of SLM

The phase or amplitude modulation of a wave front by an SLM can be produced by a number of physical means. The two most common ones are LC-based devices and moving mirror array devices (MMD). Other more exotic devices are based around multiple quantum-well technology, magnetostrictive, Pockles effects, and so on, although these are less common. The modulation of light that can be achieved is dependent on the type of SLM used. Some can be set up to provide multiple modulation schemes such as LC devices, while others that are more prescribed, such as mirror-based devices, can only be used for the modulation type they were designed specifically for.

9.2.1
Liquid Crystal

An LC or *mesomorphic phase* is a state of matter that exists between liquids and solids and is exhibited by some molecules. It has some of the properties of both a liquid and a crystal. The molecules are free to flow, like a liquid, but they still have at least some degree of order, like a crystal. In order to exhibit a mesomorphic phase, the molecule must have some form of anisotropic shape, either an elongated rod shape or a discoid. Depending on the properties of the particular molecule, it may also exhibit one or more classes of LC or subphases. If these phases are temperature dependent, it is known as a *thermotropic LC* and these are the most important types for SLMs.

The classification of the subphases is dependant on the amount or degree of order of the molecules within the material. If we assume that the liquid phase of matter has no long-range positional or orientational order, and the solid or crystalline phase of matter has both, then the liquid-crystal phases are somewhere in between. If the material has no (long range) positional correlation but has a degree of orientational order, the phase is known as *nematic* (see Figure 9.1). If there is some one-dimensional order in a three-dimensional material, it is known as *smectic* (i.e., the molecules are arranged into layers, but are free to move within the layer). The third case is called the *columnar phase* in which the molecules are stacked into columns giving a two-dimensional order in the three-dimensional bulk material. The nematic and smectic mesomorphic phases are the two phases that are commonly found in LC SLMs.

For the LCs with the rod-type molecules, the packing is such that the long axes approximately align with each other. The dimensionless vector **n** called the *director* is introduced to represent the preferred direction of the molecules at any point. In the *chiral nematic* phase, chirality is exhibited. The result of this is that there is a rotation of the director between layers.

In the smectic phases, the molecules are arranged to form well-defined layers that can slide over one another, giving rise to the positional and orientational ordering. The molecules are still positionally ordered along one direction or director, but

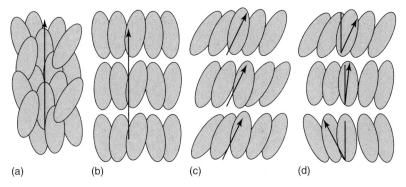

Figure 9.1 LC types. The arrow indicates the director: (a) nematic, (b) smectic A, (c) smectic C, and (d) smectic C*. Note that the angle between the plane normal and the director is fixed, so the rotation is in and out of the page as drawn.

there are many ways of packing them together; these different subphases are denoted by letters. The different phases are characterized by different types and degrees of positional and orientational order. In the smectic A phase, the molecules are oriented along the layer normal, while in the smectic C phase they are tilted away from the layer normal. Chirality can also be exhibited and this is denoted by an asterisk. For example, in smectic C* the molecules are arranged in layers, the director points to an angle other than the normal of the layers, and since the material has chirality, the angle of the director rotates around the normal of the plane, from plane to plane, (see Figure 9.1).

9.2.1.1 Nematic Liquid Crystal SLMs

Nematic devices fall into two classes: linear and twisted. Twisted nematic (TN) SLMs are commonly used in liquid crystal displays (LCDs). They are primarily designed for use as intensity modulators although phase and phase-amplitude-coupled modes can be devised (see Section 9.2.1.4). Linear nematic devices are similar but without the twist. They are commonly used as phase modulators.

In a nematic LC, both the electrical and optical properties of the material come from the rod-type shape of the molecules. The molecules have no positional ordering, but because of their shape there is some orientational ordering based on how they pack together. The anisotropic shape gives rise to two values for the dielectric permittivity depending on whether it is measured along the long axis or the two (identical) short axes (ϵ_o and ϵ_e). The dielectric anisotropy is then given by

$$\Delta\epsilon = \epsilon_e - \epsilon_o \tag{9.1}$$

An LC cell, which can be considered a single pixel of an LC-SLM, is built by sandwiching the nematic LC material between two sheets of glass. On the inside surface of the glass, an electrically conductive, but optically transmissive, electrode is placed. This is commonly made from indium tin oxide (ITO). This is connected

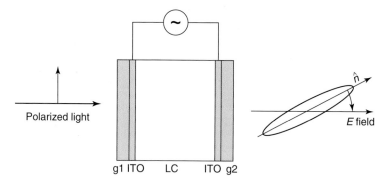

Figure 9.2 A typical nematic LC cell is shown. g1 is a glass substrate, ITO is a transparent electrode coating (indium tin oxide), LC is the liquid crystal material, and g2 is either another glass substrate or a mirror if the device is reflective. On the right is a nematic molecule showing its director, n̂, and how this becomes aligned to the electric field, E.

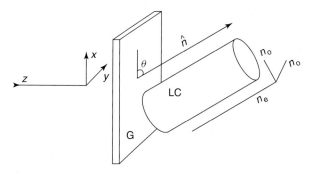

Figure 9.3 Birefringence axis. G is a glass substrate and LC is a simplified model of a liquid crystal molecule. The electric field is applied along the z-axis. This induces a rotation in the LC molecule.

to a switchable ac voltage supply as shown in Figure 9.2. Typically, the molecules are arranged such that the long axis lies in the plane of the glass (although other alignments are possible). The application of an electric field induces an electrical polarization within the molecule which manifests itself as a torque. The result is a rotation of the molecule's axis toward that of the electric field (Figures 9.2 and 9.3). The result of this is that the molecular reorientation results in a change in the birefringence, Δn, experienced by the light passing through a cell or pixel. This change clearly affects the birefringence experienced by light passing through the cell, which in turn results in the electrical controllability of the SLM. Once the electric field is removed, the molecular alignment returns to its original position. This is driven by the elastic constant of the material and its viscosity, and thus the switch-on and switch-off times of the material are different.

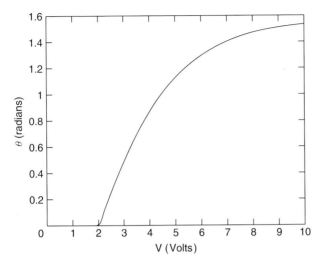

Figure 9.4 Applied voltage versus director angle for $V_c = 2$ V and $V_0 = 2$ V.

The angle of the molecular tilt is usually taken as monotonically increasing with applied voltage and can be modeled by

$$\theta = \begin{cases} 0 & V \leq V_c \\ \frac{\pi}{2} - 2\arctan\exp\left(-\frac{V-V_c}{V_0}\right) & V > V_c \end{cases} \quad (9.2)$$

and this is shown in Figure 9.4. V_c is known as the *critical voltage* and V_0 is a constant. The angle of the molecular tilt then controls the birefringence of the material and since the molecules are uniaxial this is given as

$$\frac{1}{n^2(\theta)} = \frac{\cos^2\theta}{n_e^2} + \frac{\sin^2\theta}{n_o^2} \quad (9.3)$$

Hence, taking the LC cell (or pixel), the linear nematic devices can be thought of as acting as a variable retarder. The birefringence of the cell is then

$$\Delta n(\theta) = n(\theta) - n_o \quad (9.4)$$

Note that $n(\theta)$ varies from n_e to n_o. A cell of thickness d will then have a retardance of $\Gamma(\theta) = 2\pi\Delta n(\theta)d/\lambda$ for a light of wavelength λ passing through it. The Jones matrix (see Chapter 1) for such a cell is given by

$$S(\Gamma) = \begin{bmatrix} 1 & 0 \\ 0 & \exp(-i\Gamma) \end{bmatrix} \quad (9.5)$$

9.2.1.2 Frame Rates

The molecule relaxation time of the nematic LC-SLM has typically been the limitation with regard to practical operation speed. The "on time" may be of the order of 1 or 2 microseconds but the relaxation time, or "off time," is not driven by an electric field and results in a slower time of around 30 microseconds. Apart from trying to develop new materials, two main techniques have been employed to improve on this: (i) the transient nematic effect and (ii) the dual-frequency effect.

The transient nematic effect relies on the nonlinearity of the relaxation time of the LC material [1]. Put simply, it is generally faster for the material to relax from 100 to 90%, than is it for the material to relax from 10 to 0%. By optimizing the position in the response curve, a speed improvement can generally by made.

The dual-frequency effect [2, 3] takes advantage of the fact that the dielectric anisotropy (Eq. 9.1) has an electrical frequency dependency. Nematic cells are normally driven with an ac field with a frequency of up to a few kilohertz. Above a certain frequency (the cross-over frequency) there will be a sign reversal in Eq. (9.1). The result is the direction of the induced dipole is reversed and it is now possible to drive the material off as well as on. The drive electronics are more complicated and the complex valued nature of $\Delta\epsilon$ results in absorption of the electric field, giving rise to heating effects, but there is a large improvement in the speed.

9.2.1.3 Temperature Effects

With increasing temperature, the orientational order of the molecules becomes less. This has the effect of lowering the value of Δn. This means to produce an accurately controllable device, the temperature must be either controlled or monitored and the effect calibrated out. The material, however, becomes less viscose, resulting in an improvement in the speed of the switching.

9.2.1.4 Twisted Nematic LC-SLM

The TN is perhaps the most pervasive of all types of SLM. It is widely used as an intensity modulator or LCD and is seen in television screens, computer monitors, and projectors, as well as the optical information processing laboratory. The device works by electronically controlling the rotation of the plane of polarization of the incoming light and using crossed polarization filters to modulate the signal. In nematic LC cells, the initial alignment of the director axis is controlled by the glass surface. The two glass surfaces are arranged so that these directions are at 90° to each other. Since the nematic LC tries to align with each of the planes, it results in a helical twist in the alignment of the directors through the material as shown in Figure 9.5.

The effect of the TN cell is to introduce a phase shift and to rotate the plane of polarization. However, as an electric field is applied across the cell, the molecular director aligns itself with the field (which is along the z-axis in Figure 9.3). Thus the twist is lost and the device no longer acts as a polarization rotator. The sandwiching of the device between orthogonal linear polarizers results in the device transmitting light when no field is applied and blocking light when it is applied.

There are several models using a Jones matrix formulation for TN-LC cells, but they are typically modeled by assuming that the device is built up from a set of thin birefringent wave plates, each one rotated slightly from the next. The Jones matrix is then given as follows [4]:

$$\mathbf{W_{TN}} = \exp(-i\beta)\mathbf{R}(-\alpha)\mathbf{T_{TN}}(\alpha, \beta) \qquad (9.6)$$

where β is the birefringence, α is twist angle of the whole cell, and

$$\mathbf{R}(\theta) = \begin{bmatrix} \cos(\theta) & \sin(\theta) \\ -\sin(\theta) & \cos(\theta) \end{bmatrix} \quad (9.7)$$

and

$$\mathbf{T}_{TN}(\alpha, \beta) = \begin{bmatrix} \cos(\gamma) - i\beta/\gamma \sin(\gamma) & \alpha/\gamma \sin(\gamma) \\ -\alpha/\gamma \sin(\gamma) & \cos(\gamma) + i\beta/\gamma \sin(\gamma) \end{bmatrix} \quad (9.8)$$

and

$$\gamma = \sqrt{\alpha^2 + \beta^2} \quad (9.9)$$

For display applications, the color in TN-SLM displays is produced by one of two means. An individual color filter is placed in front of each pixel – red, green, and blue. Three pixels are then required to represent one color and so the total resolution of the device is reduced. In projectors, a higher quality image is sometimes produced by using three SLMs with a single color filter and a trichromic beam splitter is used to combine the images onto a single optical path. The resolution of the device is maintained but at the expense of having three SLMs plus more complex optics. A third method of using pulsed color light or a high-speed rotating color wheel is not possible with TN-SLMs since the response rate it too slow for human persistence of vision to merge the colors together. This method is, however, used in FLC displays and MEMS (microelectromechanical systems) mirror devices such as the Texas Instruments DLP (digital light projector) technology.

Owing to the mass production of TN-SLMs, the cost of such devices is low compared to a linear nematic SLM. These are generally made by small companies serving the academic community. However, there has been considerable interest in using the TN-SLM in optical processing applications due to its low cost. A problem arises, however, that the desired change in birefringence (to generate a phase shift) with increasing the applied electric field across the cell is coupled to a rotation in the polarization of the transmitted light. The phasor diagram of the possible modulation curve of a hypothetical device is shown in Figure 9.15b. Here there is an output polarizer placed after the SLM; thus, phase and amplitude are coupled together in an undesirable way.

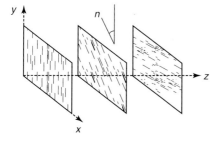

Figure 9.5 An example of how the director, **n**, twists through a twisted nematic LC-SLM. In this example, the twist angle is 90° but other arrangements are possible.

Several attempts have been made to provide solutions to this problem. The early TN-SLMs were relatively thick and sometimes it was possible to get a respectable amount of phase shift from the device before the helix became untwisted. Thus the devices were operated at a lower voltage than required to modulate intensity, so sometimes this required new drive electronics. Reference [5] shows how to set up the input and output polarizers to achieve a phase-mostly configuration.

For some applications such as in correlators, the nearest Euclidean distance on the devices response curve to the desired phase has been proposed and demonstrated [6]. A pseudorandom encoding scheme has also been used [7]. It has been demonstrated that some devices can be used in double pass mode [8]. By doing this, the effective phase shift attainable from the device is doubled and some of the polarization rotation is undone. From our own experience, we have found that we could achieve 2π phase modulation in the visible mode and only suffer a small amount of intensity variation, although this appears to be very device dependent.

Another method of achieving phase-only modulation is with polarization eigen states. It has been shown that by encoding the incident light into an elliptical polarization state, which happens to be an eigenstate of Eq. (9.6), phase-only modulation can be produced [9].

9.2.1.5 Modulation Methods

Phase-only modulation can be achieved using a linear nematic device. Phase will be modulated according to the change in optical path length as the director of the molecules rotates. There will consequently be only phase modulation along the extraordinary axis. It is therefore common to use linearly polarized light. If the input light is represented by the Jones vector (and neglecting constants outside the brackets):

$$\mathbf{V} = \begin{bmatrix} 0 \\ 1 \end{bmatrix} \tag{9.10}$$

and a linear vertical output analyzer is placed after the SLM:

$$\mathbf{P} = \begin{bmatrix} 0 & 0 \\ 0 & 1 \end{bmatrix} \tag{9.11}$$

then for a transmissive nematic SLM the modulation is

$$\mathbf{T} = \mathbf{PS}(\Gamma)\mathbf{V} \tag{9.12}$$

$$= \begin{bmatrix} 0 & 0 \\ 0 & 1 \end{bmatrix} \begin{bmatrix} 1 & 0 \\ 0 & \exp(-i\Gamma) \end{bmatrix} \begin{bmatrix} 0 \\ 1 \end{bmatrix} \tag{9.13}$$

$$= \begin{bmatrix} 0 \\ \exp(-i\Gamma) \end{bmatrix} \tag{9.14}$$

Since the modulation is achieved by a change in optical path length, the phase shift will be wavelength dependent. The polarization dependency may be a problem for some applications. It can be made polarization independent by placing two orthogonally aligned SLMs together [10] or by operating the SLM in double pass mode with a quarter wave plate between the SLM and the mirror [11].

Amplitude modulation can be achieved by rotating the axis of the input polarization and output analyzer by 45°:

$$T = \begin{bmatrix} 0.5 & 0.5 \\ 0.5 & 0.5 \end{bmatrix} \begin{bmatrix} 1 & 0 \\ 0 & \exp(-i\Gamma) \end{bmatrix} \begin{bmatrix} 0.5 \\ -0.5 \end{bmatrix} \quad (9.15)$$

$$= K \begin{bmatrix} \sin(\Gamma/2) \\ \sin(\Gamma/2) \end{bmatrix} \quad (9.16)$$

which, ignoring the static constant, K, gives an amplitude modulation as Γ changes from 0 to 2π. When the device is at the maximum transmission, it is effectively acting as a half-wave plate. The wave plate will rotate the linearly polarized light by 90°, allowing it to pass through the output polarizer. This setup is, however, wavelength dependent, so the transmission is dependent on λ.

A variation on this configuration is to place the input polarizer at 45° and place a mirror behind the SLM so that it becomes double pass [12]. This effectively doubles the retardance of the LC material and introduces a coordinate transformation [13]. The Jones matrix of the reflective SLM is then given by

$$J = \begin{bmatrix} -1 & 0 \\ 0 & -1 \end{bmatrix} S^T S \quad (9.17)$$

$$T = \begin{bmatrix} 0.5 & 0.5 \\ 0.5 & 0.5 \end{bmatrix} \begin{bmatrix} \exp(-i\Gamma) & 0 \\ 0 & \exp(i\Gamma) \end{bmatrix} \begin{bmatrix} 0.5 \\ 0.5 \end{bmatrix} \quad (9.18)$$

$$= K \begin{bmatrix} \sin(\Gamma) \\ \sin(\Gamma) \end{bmatrix} \quad (9.19)$$

where K is a constant. Since now the retardance has doubled, the amplitude actually goes from minus one to plus one instead of from zero to one. This capability can be used to remove DC biasing in CGHs and correlators [12, 14].

The operational wavelength of nematic devices is dependent on the LC and substrate materials. They are typically designed for use in the visible mode but use in the near-infrared mode has been demonstrated and we have successfully used TN devices in the ultraviolet mode (351 nm) [15].

9.2.1.6 Ferroelectric

Ferroelectric SLMs [16] typically use a smectic C* LC. Like nematic SLMs, they consist of a liquid crystal material sandwiched between two electrically conductive glass plates. The separation between the glass plates is however so small (about 1–3 µm) that it interferes with the helical arrangement of the directors that arises from the chiral nature of the molecules. This gives rise to only two possible states, that is, it is a bistable device and is sometimes referred to as surface-stabilized ferroelectric liquid crystal (SS-FLC) – see Figures 9.6 and 9.7. The state of the device is controlled by the applied electric field. Each molecule has a ferroelectric dipole aligned orthogonal to its director. Like the nematic type, it can be thought of as an electronically controllable wave plate. However, this time, the angle of the optical axis rotates rather than the retardance.

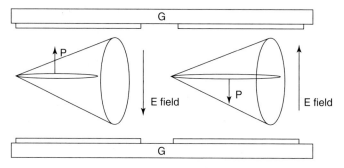

Figure 9.6 A surface-stabilized ferroelectric liquid crystal. Normally the director is free to take any angle around the cone; the addition of thin glass plates restricts this to two possible states. The glass plates, G, are arranged above and below the molecules and so the director of the molecule is in the plane of the glass. The addition of the electric field causes the molecule to take up one of two positions.

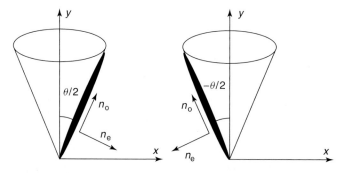

Figure 9.7 The FLC cell viewed through the glass. The two possible states of the molecules director are shown. Since the molecule is in one of two states and is birefringent, the whole cell can be thought of as a wave plate of fixed retardance but the angle of the optical axis can be switched from $-\theta/2$ to $\theta/2$.

Unlike nematic devices, the electrical polarization is not induced in the molecule but is permanent. This means that the device can be driven both on and off. The state is only dependent on the direction of the field. This makes the switching electronics much simpler than dual-frequency nematic devices. The speed of switching is also greatly increased. It is possible to drive the pixels at several kilohertz. However, the continuous application of a static electric field will cause ion migration in the LC material, resulting in damage to the device. It is therefore necessary to ac balance the field over time, that is, in the next frame the electric field for a particular pixel must be reversed. These can result in problems in an optical system. If the FLC SLM is used as an intensity modulator, an inverse image is produced every second frame which must be removed by either pulsing the light source or shuttering.

Optically, the device can be modeled using a fixed retardance Jones matrix and a rotation matrix provided by rotating the coordinate system before entering the

SLM and rotating it back after. The coordinate transformer is

$$\mathbf{R}(\theta) = \begin{bmatrix} \cos\theta & \sin\theta \\ -\sin\theta & \cos\theta \end{bmatrix} \quad (9.20)$$

So,

$$\mathbf{W}_{\text{FLC}} = \mathbf{R}(\theta)\mathbf{S}(\Gamma)\mathbf{R}(-\theta) \quad (9.21)$$

$$= \begin{bmatrix} \cos^2\theta + \exp(-i\Gamma)\sin^2\theta & \cos\theta\sin\theta(\exp(-i\Gamma)-1) \\ \cos\theta\sin\theta(\exp(-i\Gamma)-1) & \cos^2\theta + \exp(-i\Gamma)\sin^2\theta \end{bmatrix} \quad (9.22)$$

If we use a vertically polarized linear incident light source and add a horizontal linear polarizer to the output the transmitted amplitude, **T**, will be

$$\mathbf{T} = \begin{bmatrix} 1 & 0 \\ 0 & 0 \end{bmatrix} \mathbf{W}_{\text{FLC}} \begin{bmatrix} 0 \\ 1 \end{bmatrix} \quad (9.23)$$

$$= \begin{bmatrix} \cos\theta\sin\theta(\exp(-i\Gamma)-1) \\ 0 \end{bmatrix} \quad (9.24)$$

$$= K\,\text{sgn}(\theta) \quad (9.25)$$

where K is a constant and so the transmission depends on Γ and θ and is a maximum when $\theta = \pi/4$ and $\Gamma = \pi$. The problem here is that Γ is clearly wavelength dependent and so the device has to be designed and built for a fixed wavelength to maintain good transmission and θ must be $\pi/4$. This has proven to be problematic, with many devices only having a switching angle of about half of this. The ability to produce a ferroelectric material with the desired parameters is the subject of ongoing research (see Ref. [17], for example).

The FLC SLM is limited to binary modulation, but this limitation is ameliorated by its high switching speed compared to the nematic SLMs. FLCs have been used in display applications with gray levels being achieved by time multiplexing the device. Binary phase switching has been shown to be useful for adaptive optics [18, 19], correlators, and CGHs.

Multilevel phase solutions have been demonstrated in the past [20, 21]. Cascading of several FLC SLMs together, each with half of the previous SLM's phase step has been shown. Although this can produce a high-speed multiphase level SLM, the solution is expensive, bulky, and will possibly suffer from a low optical transmission due to the large number of air–glass–LC refractive index boundaries and diffraction effects.

A more recent development is the analog FLC device from Boulder Nonlinear Systems Inc [22]. This device has 512 × 512 electrically addressed pixels that have a gray level control as opposed to the binary switching in the standard SS-FLC.

9.2.1.7 Addressing Methods
Liquid crystal SLMs have two main addressing methods: optical and electrical.

EA-SLM Electrically addressed spatial light modulators (EA-SLMs) are the most commonly available ones on the market. For LC devices, there are typically three types of addressing mechanisms: direct addressing, multiplex addressing, and

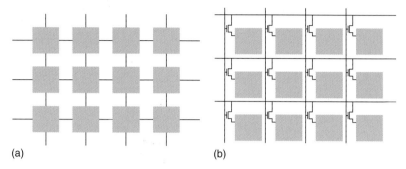

Figure 9.8 (a) Multiplexed addressing. The desired voltage V is applied half to the rows and half to the columns. This addressing mode is really only suitable for a limited number of pixels with limited gray level range. (b) The addition of a transistor (and additional circuitry) is used to overcome this limitation.

silicon backplane addressing (known as *liquid crystal on silicon* or *LCoS*) [23]. Direct address of individual pixels gives a very precise control of the driving voltage but can only be used with a limited number of pixels. Multiplex addressing, where the row and column voltage of the SLM pixels are addressed, is more commonly used today in transmissive devices. Early devices suffered from considerable cross talk, but this has been reduced by using thin-film transistor (TFT) addressing. A simple example of the addressing scheme is shown in Figure 9.8. LCoS SLMs overcome some of the addressing problems by placing each liquid crystal pixel on top of a CMOS memory structure. These can be directly addressed, removing the cross-talk problem and allowing very large pixel count SLMs (1920×1080 pixels are common). Since they are built on a silicon backplane, they are, by nature, a reflective device rather than a transmissive one.

OA-SLM Optically addressed spatial light modulators (OA-SLMs), also known as *light valves,* are less common than the electrically addressed equivalents. An example of a liquid crystal light valve is shown in Figure 9.9. Incident writing light is projected onto the left-hand surface of the device shown in Figure 9.9. This light pattern is converted into an electric field across the LC material. A method for doing this is to use a sheet of photodiode material (for example, hydrogenated amorphous silicon). The diode conducts in the presence of light and so creates a potential difference between the diode and the ITO electrode on the other side of the LC material. This electric field is then proportional to the incident light. The device is then illuminated from the other side, and this light is then modulated by the LC material. This light is then reflected off the dielectric mirror so it is a double pass device. This arrangement is unpixelated.

One of the uses of OA-SLMs is to convert incoherent light into coherent light. This can be used as the input stage of an optical processor. Another use is to convert the wavelength of light. Again, this could be used as the front end of an optical processor. In a more recent application, the OA-SLM is used to combine

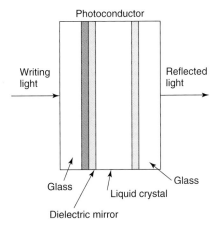

Figure 9.9 A liquid crystal light value.

several EA-SLMs to produce extremely high-resolution devices. The OA-SLM does not have the same restriction of having an electronic addressing scheme and so it can easily support very high resolutions [17] and a very high number of active pixels. Several limited resolution EA-SLMs can be combined together such that they project onto the OA-SLM and this can then be used to produce an SLM with a large number of pixels. This has been used to generate CGHs with 100 Mbit/frame [24, 25].

9.2.2
Multiple Quantum-Well SLMs

Multiple quantum-well (MQW) SLMs are solid-state structures that consist of thin layers of two alternating semiconductor materials. The semiconductors have different bandgaps and a typical material combination used in construction is layered GaAs and AlGaAs of around 10 nm thickness (see Figure 9.10). The layer with the narrowest bandgap is called the *well layer* and the wider bandgap is the *barrier layer*. The layered structure gives rise to the constraint of photon-induced excitons (electron holes) and absorption peaks that are not observed in the bulk material. The application of an electric field then has the effect of causing the peak absorption to shift in wavelength via the quantum-confined Stark effect. By utilizing this shift in absorption, it is then possible to use the device as an intensity modulator. By arranging the MQW into an array, the device becomes an SLM.

The main advantage of such a device is the high switching speed. A problem is that the contrast ratio can be quite low (sometimes only 3:1) and the devices have a strong wavelength dependency. Reference [26] describes a 2 × 128 pixel device with a contrast ratio of 335:1, a modulation rate of 15 MHz and 8 bits of gray level. Larger arrays have also been demonstrated [27, 28]. Speeds of several gigahertz are possible for single-pixel devices [29]. Optically addressed devices have been developed; see Refs [30, 31].

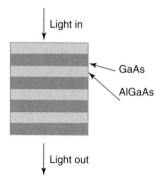

Figure 9.10 An example of an MWQ structure.

9.2.3
Mirror Devices

9.2.3.1 Amplitude Modulators

The MEMS devices marketed as a DLP by Texas Instruments has taken about 50% of the business projector market that was in the past dominated by TN-SLMs and cathode ray tube (CRT) projectors. The mass market projector application has driven down the cost while increasing the pixel count, making them attractive for other SLM applications.

The basic principle of operation is an MEMS light-switching device called a *digital micromirror device* (*DMD*) [32, 33]. It is a CMOS memory cell with a metal mirror that can reflect light in one of two directions depending on the state of the underlying memory cell. The angle is binary and can rotate $\pm 12°$. A simple optical setup is shown in Figure 9.11. Although the devices are binary in operation, they have an extremely high frame of over 30 kHz and resolutions currently up to 1920×1080 pixels [34].

A gray level response is simulated by switching the mirrors on and off very rapidly and using pulse width modulation. Color projection is achieved by either

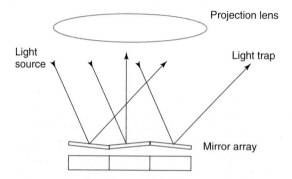

Figure 9.11 TI DMD optical setup projecting the pattern *off, on, off*.

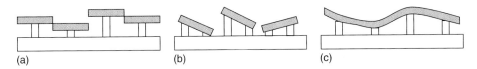

Figure 9.12 (a) Piston style phase-modulating mirror; (b) wedge mirrors; and (c) continuous faceplate mirror.

the addition of a color wheel or by combining three separate DMDs with red, green, and blue filters.

9.2.3.2 Phase Modulators

With mirror devices, phase is modulated by changing the path length that the light must travel over. This is typically achieved by mirrors that are moved in a pistonlike fashion, mirrors with wedge motion, or mirrors with a continuous deformable faceplate, as shown in Figure 9.12. Since it is an optical path-length modulation and not directly phase modulation, consideration must be given to polychromatic light modulation. For applications such as atmospheric adaptive optics, where the aberrations are caused by optical path-length shifts, the mirror provides an ideal conjugate to correct polychromatic light. (An example of a pure phase-modulating device is the FLC SLM as described in Section 9.2.1.6. A polychromatic wave front will still only have a π phase shift applied to all wavelengths; however, the transmittance will not be the same for all wavelengths.)

Many mirror devices for modulating phase are typically developed for adaptive optical correction of aberrations rather than for optical processing applications. They are generally characterized by having a high speed, large sweep, but relatively low number of degrees of freedom. There are a variety of methods used to achieve this, the major types including the following:

1) Segmented mirrors: Flat mirrors are driven by actuators (typically piezoelectric crystals with a feedback mechanism to compensate for hysteresis). These tend to be large devices with very good optical quality [35].
2) Continuous faceplate devices: Discrete actuators cause a controlled deformation in a thin flexible membrane mirror. These can be driven by piezoelectric actuators or by electrostatic forces [36, 37].
3) MEMS deformable mirrors: These can be discrete or continuous faceplate devices, and are characterized by being built using MEMS technology [38].

There are a large number of other driving mechanisms including ferrofluids [39] and bimorph mirrors.

The Fraunhofer MEMS phase former kit is a device that may be more suitable for image or data processing, since it is a pixelated phase modulator [40, 41]. The device has 200 × 240 pixels that can modulate phase with 8 bits of accuracy with a frame rate of 500 Hz. The mirrors are integrated with CMOS circuitry and can be individually addressed. The maximum stroke of the mirrors is 500 nm, which corresponds to a phase shift of twice this due to the double pass reflection mode of the device.

9.3
Fully Complex Modulation Methods

For many applications, it is highly desirable to modulate both the phase and the amplitude of light. Direct Fourier transform holograms could then be implemented to produce CGHs without having to calculate the iterative Fourier transform algorithm, amplitude correction could be performed in adaptive optics, and bandpass filtering could easily be performed in optical correlation signals. This is not possible with current SLM technology. The SLMs typically perform phase *or* amplitude modulation separately or mix them together in an undesirable way such as with the TN-SLM.

There have been several attempts to overcome this. One of the most straightforward methods is to combine two SLMs: one to modulate amplitude and one to modulate phase [42]. Amako demonstrated such an arrangement using a TN-SLM for the amplitude modulation and a linear nematic for the phase [43]. This has also been demonstrated by cascading two TN-SLMs together [44]. Although the technique works, there are some considerations.

- The TN-SLM will produce an unwanted phase shift as well as amplitude modulation. This must be compensated for using the linear nematic SLM.
- The physical size of the system is greatly increased. There are now two SLMs, polarizers, and relay optics to consider.
- The cost of the system is roughly doubled compared to a single SLM technique.
- The physical dimensions of the pixels is small (possibly 10 μm), so the alignment of the system must be very good.
- The pixel size, resolution, pitch, and aspect ratio of the two SLMs must match each other.

To do away with the need for two SLMs, macropixel methods have been demonstrated. In these methods, multiple pixels are used to encode the complex data. The minimum number of pixels required is two, but often more are used. The pixels are then combined together by either blurring the two by a spatial filter or reconstruction of the complex field in the far field. Both on-axis and off-axis techniques have been demonstrated.

Off-axis encoding is usually based around the phase detour method first demonstrated by Brown and Lohmann [45]. This method does not map well on to the square pixels of an EA-SLM but a modified version was shown by Lee using four pixels [46], and this was later reduced to three [47]. By using either an analog FLC or a linear nematic SLM, the authors have shown that this can be reduced to two pixels [12, 14]. The Lee method [46] is shown in Figure 9.13a. The complex datum is represented by four pixels. The pixel values are the positive real value, the positive imaginary value, the negative real value, and the negative imaginary value. Each of these is placed onto the SLM in such a way that, at an angle of 45°, there is a $\pi/4$ phase lag between each pixel. In the far field, these four pixels combine and reproduce the complex data point. If an SLM that is capable of producing negative and positive amplitude modulation is used, the number of pixels required

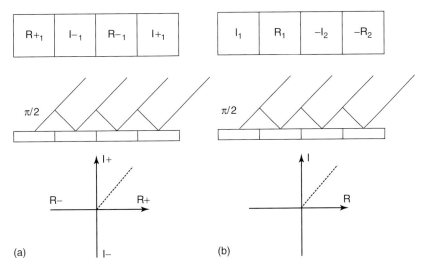

Figure 9.13 Phase detour. (a) The four pixel method; (b) two pixel method.

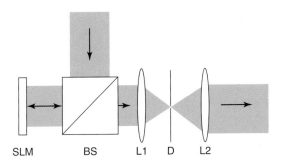

Figure 9.14 Phase detour complex modulation. BS, beam splitter; D, diaphragm; L1 and L2 are Fourier transform lenses.

is reduced to two, although every second macroblock has to be negated to maintain a continuous phase lag between them; see Figure 9.13b.

The problem with this technique is that it is only accurate at a fixed angle of 45° and the reconstruction is in the Fourier plane. By introducing a spatial filter, this limitation can be overcome [14]. A similar method, but using four pixels of a TN-SLM, has also been shown [48]. This arrangement is shown in Figure 9.14.

On-axis techniques tend to be implemented using CGHs. They typically employ a phase-modulating device such as a nematic SLM or mirror device and use this to generate some amplitude modulation in the Fresnel or Fraunhofer field. There are a variety of methods of designing these holograms, such as the iterative Fourier transform algorithm [49, 50], evolutionary algorithms [51], and direct searches [52], among others.

9.4 Applications

Matching the correct SLM to the application is critical to achieve the best results. There are too many different applications to discuss here in detail, but some of the major types are covered. For some applications, such as display, the selection process is simple. Amplitude modulation is required with typically large number of pixels and a frame rate of at least 30 Hz. Specific devices such as TNLC SLMs and the TI-DLP devices have been developed for this specific need. For other applications such as CGHs, it may be desirable to have full complex modulation. However, since this is not always practical, a compromise has to be made. These are typically realized by the use of a phase-only SLM instead. Nematic SLMs are often suitable for this. Some common applications and typical SLM solutions are shown in Table 9.1.

A summary of the phasor diagrams for most of the common modulation modes is shown in Figure 9.15. Figure 9.15a shows the characteristics of linearly polarized light modulated by a linear nematic device; here 2π modulation has been achieved. This could equally be achieved using an analog phase-modulating mirror. In practice, if 2π modulation cannot be achieved, the circle will be incomplete. A TN device often has phase coupled with some amplitude modulation. This is shown in Figure 9.15b. Here 2π modulation has not been achieved. Real devices may differ from this idealized curve but they do usually follow a spiral type pattern. Pure amplitude modulation is shown in Figure 9.15c. This is achievable by a number of means,

Table 9.1 Common applications and their commonly used SLMs.

Application	Modulation	SLM	Typical resolution and frame rate
Display	Amplitude modulation	TNLC SLM	Up to and above 1920 × 1200 pixels at 30–60 Hz
		DLP–mirror device	Up to 1920 × 1200 pixels at several kilohertz
Correlation	Usually limited to phase only	NLC	1920 × 1200 pixels at 60 Hz and higher
	Binary phase only	FLC	512 × 512 pixels at several kilohertz
Computer-generated holograms	Usually limited to phase only	NLC	1920 × 1200 pixels at 60 Hz and higher
	Binary phase only	FLC	512 × 512 pixels at several kilohertz
Adaptive optics	Phase modulation	Mirror devices	A few hundred pixels at 100 Hz
		NLC devices	A few hundred pixels at 60–500 Hz

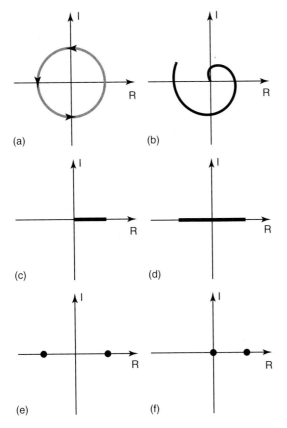

Figure 9.15 Phasor diagrams. (a) A nematic LC device with linearly polarized light; (b) a twisted nematic device; (c) achievable with an analog mirror device, linear nematic, twisted nematic or MQW-SLM; (d) achievable with a linear nematic or analog FLC; (e) SS-FLC; and (f) binary mirror device such as the TI-DLP or SS-FLC.

two examples being linear nematic SLMs and binary amplitude mirror devices (by time division multiplexing). The ability to introduce an additional π shift allows positive and negative modulation as shown in Figure 9.15c. This can be achieved using a linear nematic or analog FLC. Binary devices have a more limited phasor diagram. A typical FLC type or moving mirror with $0 - \pi$ phase modulation is shown in Figure 9.15e and the amplitude modulation equivalent is shown in Figure 9.15f.

9.5 Conclusions

SLMs provide a method of encoding information onto a light beam that can be updated in real time. This chapter has provided a review of the common types of

SLMs on the market or in development. The phasor diagram of possible modulation modes for the common SLMs is shown in Figure 9.15. Methods of achieving the various types of modulations have been discussed including phase, amplitude, and full complex modulation. The types of SLMs are typically broken down into phase modulators, such as the linear nematic LC; or intensity modulators, such as the TN-LC or the DLP mirror system. The most common methods of achieving modulation are typically based on mirrors or LC materials, although other devices such as the MQW SLM will perhaps become more common in the future owing to its very high switching speed.

References

1. Wu, S.-T. and Wu, C.-S. (1989) High-speed liquid-crystal modulators using transient nematic effect. *J. Appl. Phys.*, **65**, 527–532.
2. Bücher, H.K., Klingbiel, R.T., and VanMeter, J.P. (1974) Frequency-addressed liquid crystal field effect. *Appl. Phys. Lett.*, **25**, 186–188.
3. Kirby, A.K. and Love, G.D. (2004) Fast, large and controllable phase modulation using dual frequency liquid crystals. *Opt. Express*, **12**, 1470–1475.
4. Yariv, A. and Yeh, P. (1984) *Optical Waves in Crystals*, Wiley, New York.
5. Yamauchi, M. and Eiju, T. (1995) Optimization of twisted nematic liquid crystal panels for spatial light phase modulation. *Opt. Commun.*, **115**, 19–25.
6. Juday, R.D. (1989) Correlation with a spatial light modulator having phase and amplitude cross coupling. *Appl. Opt.*, **28**, 4865–4869.
7. Cohn, R.W. (1998) Pseudorandom encoding of complex-valued functions onto amplitude-coupled phase modulators. *J. Opt. Soc. Am. A*, **15**, 868–883.
8. Barnes, T.H. et al. (1989) Phase-only modulation using a twisted nematic liquid crystal television. *Appl. Opt.*, **28**, 4845–4852.
9. Davis, J.A., Moreno, I., and Tsai, P. (1998) Polarization eigenstates for twisted-nematic liquid-crystal displays. *Appl. Opt.*, **37**, 937–945.
10. Kelly, T.-L. and Love, G.D. (1999) White-light performance of a polarization-independent liquid-crystal phase modulator. *Appl. Opt.*, **38**, 1986–1989.
11. Love, G.D. (1993) Liquid-crystal phase modulator for unpolarized light. *Appl. Opt.*, **32**, 2222–2223.
12. Birch, P.M. et al. (2000) Two-pixel computer-generated hologram with a zero-twist nematic liquid-crystal spatial light modulator. *Opt. Lett.*, **25**, 1013–1015.
13. Goodman, J. (1996) *Introduction to Fourier Optics*, McGraw-Hill.
14. Birch, P. et al. (2001) Dynamic complex wave-front modulation with an analog spatial light modulator. *Opt. Lett.*, **26**, 920–922.
15. Chatwin, C. et al. (1998) UV microstereolithography system that uses spatial light modulator technology. *Appl. Opt.*, **37**, 7514–7522.
16. Clark, N.A. and Lagerwall, S.T. (1980) Submicrosecond bistable electro-optic switching in liquid crystals. *Appl. Phys. Lett.*, **36**, 899–901.
17. Mias, S. et al. (2005) Phase-modulating bistable optically addressed spatial light modulators using wide-switching-angle ferroelectric liquid crystal layer. *Opt. Eng.*, **44**, 014003, DOI: 10.1117/1.1828471.
18. Love, G.D. et al. (1995) Binary adaptive optics: atmospheric wave-front correction with a half-wave phase shifter. *Appl. Opt.*, **34**, 6058–6066.
19. Birch, P.M. et al. (1998) Real-time optical aberration correction with a ferroelectric liquid-crystal spatial light modulator. *Appl. Opt.*, **37**, 2164–2169.
20. Freeman, M.O., Brown, T.A., and Walba, D.M. (1992) Quantized complex

ferroelectric liquid crystal spatial light modulators. *Appl. Opt.*, **31**, 3917–3929.
21. Broomfield, S.E., Neil, M.A.A., and Paige, E.G.S. (1995) Programmable multiple-level phase modulation that uses ferroelectric liquid-crystal spatial light modulators. *Appl. Opt.*, **34**, 6652–6665.
22. http://www.bnonlinear.com (2009).
23. McKnight, D.J., Johnson, K.M., and Serati, R.A. (1994) 256 × 256 liquid-crystal-on-silicon spatial light modulator. *Appl. Opt.*, **33**, 2775–2784.
24. Stanley, M. et al. (2003) *100-Megapixel Computer-Generated Holographic Images from Active Tiling: A Dynamic and Scalable Electro-Optic Modulator System*, vol. 5005, SPIE, pp. 247–258, DOI: 10.1117/12.473866.
25. Stanley, M. et al. (2004) *3d Electronic Holography Display System using a 100-Megapixel Spatial Light Modulator*, vol. 5249, SPIE, pp. 297–308, DOI: 10.1117/12.516540.
26. Wang, Q. et al. (2005) Arrays of vertical-cavity electroabsorption modulators for parallel signal processing. *Opt. Express*, **13**, 3323–3330.
27. Worchesky, T.L. et al. (1996) Large arrays of spatial light modulators hybridized to silicon integrated circuits. *Appl. Opt.*, **35**, 1180–1186.
28. Junique, S. et al. (2005) GaAs-based multiple-quantum-well spatial light modulators fabricated by a wafer-scale process. *Appl. Opt.*, **44**, 1635–1641.
29. Liao, L. et al. (2005) High speed silicon Mach-Zehnder modulator. *Opt. Express*, **13**, 3129–3135.
30. Tayebati, P. et al. (1997) High-speed all-semiconductor optically addressed spatial light modulator. *Appl. Phys. Lett.*, **71**, 1610–1612, DOI: 10.1063/1.119993.
31. Bowman, S.R. et al. (1998) Characterization of high performance integrated optically addressed spatial light modulators. *J. Opt. Soc. Am. B*, **15**, 640–647.
32. Hornbeck, L.J. (1990) Deformable-mirror spatial light modulators, in *Spatial Light Modulators and Applications III*, Volume 1150 of Critical Reviews of Optical Science and Technology (ed U. Efron), SPIE, pp. 86–102.
33. Hornbeck, L. (1996) Digital light processing and MEMS: an overview. Advanced Applications of Lasers in Materials Processing/Broadband Optical Networks/Smart Pixels/Optical MEMs and Their Applications. IEEE/LEOS 1996 Summer Topical Meetings, pp. 7–8.
34. http://www.dlinnovations.com (2009).
35. Doel, A.P. et al. (2000) The MARTINI adaptive optics instrument. *New Astron.*, **5**, 223–233.
36. Roddier, F., Northcott, M., and Graves, J.E. (1991) A simple low-order adaptive optics system for near-infrared applications. *PASP*, **103**, 131–149.
37. Grosso, R.P. and Yellin, M. (1977) Membrane mirror as an adaptive optical element. *J. Opt. Soc. Am*, **67**, 399.
38. Vdovin, G. and Sarro, P.M. (1995) Flexible mirror micromachined in silicon. *Appl. Opt.*, **34**, 2968–2972.
39. Brousseau, D., Borra, E.F., and Thibault, S. (2007) Wavefront correction with a 37-actuator ferrofluid deformable mirror. *Opt. Express*, **15**, 18190–18199.
40. Gehner, A. et al. (2006) MEMS analog light processing-an enabling technology for adaptive optical phase control, in MEMS/MOEMS Components and Their Applications III (eds. Oliver, S., Tadigapa, S., and Henning, A.), vol. 6113, SPIE, San Jose.
41. Wildenhain, M. et al. (2008) Adaptive optical imaging correction using wavefront sensors and micro mirror arrays. Photonik International, originally published in German in Photonik 3/2007.
42. Gregory, D.A., Kirsch, J.C., and Tam, E.C. (1992) Full complex modulation using liquid-crystal televisions. *Appl. Opt.*, **31**, 163–165.
43. Jun Amako, T.S. and Miura, H. (1992) Wave-front control using liquid crystal devices. *Appl. Opt.*, **32**, 4323–4329.
44. Neto, L.G., Roberge, D., and Sheng, Y. (1996) Full-range, continuous, complex modulation by the use of two coupled-mode liquid-crystal televisions. *Appl. Opt.*, **35**, 4567–4576.
45. Brown, B.R. and Lohmann, A.W. (1966) Complex spatial filtering with binary masks. *Appl. Opt.*, **5**, 967–969.

46. Lee, W.H. (1970) Sampled Fourier transform hologram generated by computer. *Appl. Opt.*, **9**, 639–643.
47. Burckhardt, C.B. (1970) A simplification of Lee's method of generating holograms by computer. *Appl. Opt.*, **9**, 1949–1949.
48. van Putten, E.G., Vellekoop, I.M., and Mosk, A.P. (2008) Spatial amplitude and phase modulation using commercial twisted nematic LCDs. *Appl. Opt.*, **47**, 2076–2081.
49. Turunen, J. and Wyrowski, F. (1997) Introduction to Diffractive Optics, in *Diffractive Optics for Industrial and Commercial Applications*, Akademie Verlag GmbH, Berlin.
50. Gallagher, N.C. and Liu, B. (1973) Method for computing kinoforms that reduces image reconstruction error. *Appl. Opt.*, **12**, 2328–2335.
51. Birch, P. *et al.* (2000) A comparison of the iterative Fourier transform method and evolutionary algorithms for the design of diffractive optical elements. *Opt. Lasers Eng.*, **33**, 439–448, DOI: 10.1016/S0143-8166(00)00044-0.
52. Seldowitz, M.A., Allebach, J.P., and Sweeney, D.W. (1987) Synthesis of digital holograms by direct binary search. *Appl. Opt.*, **26**, 2788–2798.

10
Holographic Visualization of 3D Data
Pierre-Alexandre Blanche

10.1
Introduction

Since its early beginnings, holography has been related to imaging, though not necessarily to optical visualization. Indeed, its discovery is attributed to Dennis Gabor, who first presented it in 1947 as a theory for improving electron microscopy resolution [1]. However, we had to wait for the invention of the laser for the first holographic 3D images to be recorded. They were demonstrated nearly simultaneously in 1962 by Yuri Denisyuk in the Soviet Union [2] and by Emmett Leith and Juris Upatnieks at the University of Michigan [3]. Since then, improvements to 3D image reproduction by holography have emerged from many contributors in academia, the private sector, and even individuals who embrace holography as a hobby.

Stricto sensu, holography is the reproduction of both the intensity *and* the phase information of a given scene. Theoretically, this is the only technique to accurately display true 3D information. Occlusion, parallax, and even some amount of depth perception can be generated via many different means (some of which are discussed in this book); but only holography can provide wavefront reconstruction. This is evidenced when the picture of a hologram is taken with a camera: the focus needs to be adjusted to obtain a sharp image of the zone of interest, and one can take a picture of a crisp forefront and blurred background or vice versa by adjusting the focal point. As in real life, when objects are located at different distances, you need to change the focus of the camera or accommodate your eye strain to obtain a sharp image. This is due to the phase information contained in the real scene, as well as in the holographic image.

This unique property makes the hologram a very valuable tool for some applications and its use goes far beyond image visualization. Holograms are found in interferometry, data encoding, structure certification, quality control, vibration analysis, optical computing, and nondestructive testing, to cite only a few examples [4]. In 3D information visualization, the phase reproduction is not only a pretty effect, it provides realism to the image reproduction, and removes physiological side effects such as dizziness and eye fatigue that plague other techniques [5]. With

Optical and Digital Image Processing: Fundamentals and Applications, First Edition. Edited by Gabriel Cristóbal, Peter Schelkens, and Hugo Thienpont.
© 2011 Wiley-VCH Verlag GmbH & Co. KGaA. Published 2011 by Wiley-VCH Verlag GmbH & Co. KGaA.

holography, the eye–brain system is not fooled, since the light itself reproduces the configuration it should have had if the real object were there.

Unfortunately, what make holography so powerful for 3D reproduction makes it also extremely difficult to apply. Holograms that let you believe that the real object is there, and not just replayed, are yet to come. The reasons are numerous but certainly the first is that holography relies on light diffraction, which is dispersive. Color dispersion is the challenge of holography and scientists realized early on that they had to compromise some of the 3D information for better image reproduction. One of the most important contributions was the rainbow hologram, invented by Stephen Benton [6]. Rainbow holograms allow hologram viewing with white light, rather than with monochromatic quasi-laser light, at the expense of one-dimensional parallax instead of two. In this case, the wavefront is no longer accurately reproduced and some information is lost. This is the price to pay for easier 3D image reproduction.

Yet another constraint to accurately reproducing the phase information in holography is the need to access the original object. Indeed, to record the wavefront of the light coming from the scene of interest, holographic recording setups use interference between two laser beams. One beam carries a reference wavefront, and the other is scattered by the object; in this case, the object needs to be physically present. This is a strong limitation if the goal is visualization of 3D information that is not necessarily accessible to laser light. Indeed, computer-generated data, or 3D images gathered from other instruments cannot be used in a classical holographic setup. To overcome this problem, several solutions have been developed such as computer-generated holograms (CGHs) [7], holographic stereograms [8], and electroholography [9].

A CGH is the result of the calculation of the interference pattern that would have been engendered if the object was introduced into a holographic recording setup. Classical optics is deterministic (in opposition to quantum optics), and from a computer model of a virtual object, it is possible to calculate the interference pattern between a virtual scattered beam and a reference wavefront at a specific location. There are now many algorithms capable of performing such a calculation [10]. However, the amount of processing required to compute such a fringe pattern increases with the level of detail desired in the hologram. CGHs reproducing simple wavefronts similar to those from a lens or a mirror are routinely used to characterize precision optics [11]. CGHs reproducing basic geometrical shapes or simple objects have been shown in the past. However, one persistent limit of this technique is the reproduction of textures. Indeed, object texture is defined at the highest level of image quality and the number of rays in the ray tracing algorithm limits that precision. One can argue that it is only a matter of computational time, but when that time is longer than a human lifespan, we can definitely say we have reached a limit. Eventually, advances in computer technology will one day overcome this problem.

The second technique to generate a "hologram" without the presence of the object is integral holography. In this case, the word hologram is placed within quotes since phase information is definitely lost and cannot be retrieved. Though not producing holograms *stricto sensu*, holographic stereography uses a similar technique that is worth discussing here. Holographic stereography is derived from integral photography and integral imaging, a process that can be traced back to the works of Gabriel Lippmann in 1908 [12]. The idea is to rearrange the information contained in several 2D pictures of a scene taken from different angles, so that, when replayed, each eye of the viewer intercepts a different perspective and a 3D effect is created [8]. This is a powerful autostereoscopic technique, since any given 3D data, from satellite scenes to medical imaging passing through computer models, can be displayed to show parallax and occlusion.

Holography, for the most part, deals with still images. But it has long been realized that it can also be applied to dynamic imaging. Just as the Lumière brothers cinematograph proved that movement can be reproduced on the basis of a succession of still images, a series of holograms can be displayed to render movement. The difference between the cinematograph and its holographic counterpart is that holograms cannot be projected on a conventional diffusing screen to be seen by a large audience. Direct viewing of the holographic media is required, akin to looking directly at the film running in a projection camera. Such a technique was pioneered by Lehmann [13] and BeBitetto [14] toward the end of the 1970s; however, the small size of the film and the limited number of viewers ended any commercial aspiration based on that technique. There has been some research into screens that diffract and focus the light from the hologram into multiple viewing zones, an area still being investigated [15].

One of the last breakthroughs in holography has been the digital generation of the diffraction pattern. We mentioned earlier that, in a CGH, the interference structure is calculated. Generally, it is printed on a conventional computer printer and then photographically reduced to the size at which it diffracts the light. Now, that structure can directly be reproduced on a spatial light modulator (SLM) to diffract a reading beam and generate the holographic image. Since the SLM can be reconfigured at will, the hologram can be changed continuously. This is holographic television. Such a system has been developed by the MIT using an acousto-optic crystal [9] and by QinetiQ (United Kingdom) using an optical light modulator [16]. The limitations inherent to this technique are the same as for static CGHs, that is, the lengthy computation time and the resolution of the modulator.

10.2
Reproducing the Amplitude and the Phase

Recording good holograms is technically challenging to the point that some elevate the discipline to the rank of art. Nevertheless, basic concepts can easily be described in a few schematics. Holograms are formed by the recording of the interference between two waves. Holography is not limited to optical electromagnetic wavelengths since there also exist acoustic holograms, microwave holograms, X-ray holograms, or, as in the case of Gabor's invention, electron beam holograms [1]. For display applications, visible laser beams are used to provide the needed coherence. Without

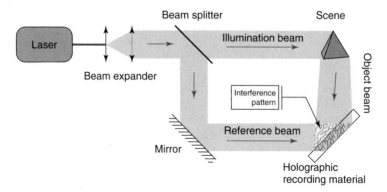

Figure 10.1 Recording a hologram. Coherent object and reference beams interfere at the holographic recording material location.

coherence, there is no interference and no pattern is recorded. One beam is called the *reference beam* and usually does not have any amplitude or phase information, being a clean beam with a flat wavefront that is used as a carrier wave. The other beam is the object beam and it is the one that contains both the amplitude and the phase (wavefront) information. When these beams cross each other, an intensity modulation pattern is created due to the resulting constructive and destructive interferences. For an extensive textbook covering the mathematical formulation of holographic imaging, see Ref. [17].

To record the hologram of a scene, the object beam is sent to the zone of interest and it is the *scattered light* that interferes with the reference beam (Figure 10.1). It has to be noted that a mean direction can be defined for the scattered object beam and so the carrier period d is strongly represented in the Fourier spectrum of the hologram.

Reading such a hologram is done by sending the reference beam alone to the holographic material, which gets diffracted to reproduce the object beam. A virtual image of the object is perceived by the observer at exactly the same place as it was during the recording (Figure 10.2). If the wavelength or the wavefront of the reading beam are changed compared to those of the reference beam used during recording, the image will be distorted at best, or not even visible in some cases. This is where pure holography reaches its limitation for 3D visualization.

As shown in Figure 10.2, when replaying a first-generation hologram, the orthoscopic (correct parallax) scene always appears behind the media. If one wants the object to appear in front of the plate, a second recording is needed using the first hologram as a master to create a *second-generation hologram*. An example of the geometry for recording a second-generation hologram is presented in Figure 10.3. In this case, the beam reading the first-generation hologram is the conjugate of the reference beam used to record it. Most of the time, it just means that the beam is sent in the opposite direction (or the hologram is flipped). A real but pseudoscopic

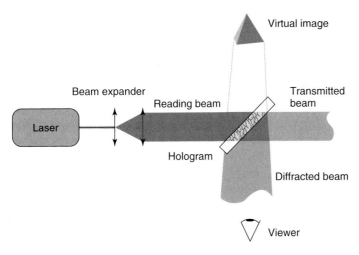

Figure 10.2 Reading a hologram. To conserve the correct parallax, the image is virtual and located behind the holographic plate.

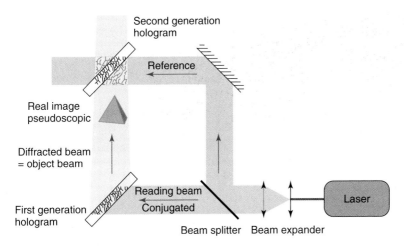

Figure 10.3 Recording a second-generation hologram. The first-generation hologram is used as a master to generate a real but pseudoscopic image.

(inverted parallax) image of the object is formed by the diffracted beam. In order to have the object seen as floating in front of the hologram, the recording plate for the second-generation hologram needs to be placed behind the real image. The reference beam can be set as for any other recording.

To reproduce the correct image of a second-generation hologram, the reading beam needs to be the conjugate of the reference beam used to record it

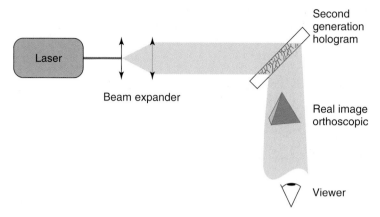

Figure 10.4 Viewing a second-generation hologram. The reading beam is conjugated according to the reference beam used to record the hologram. The image is real and orthoscopic.

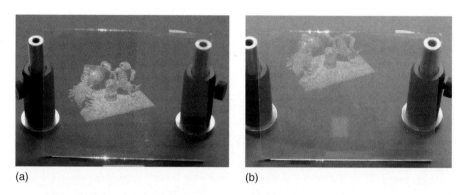

Figure 10.5 Picture of a hologram taken from different angles. The hologram is read out with a red laser light at 633 nm. (Hologram courtesy of Pierre Saint Hilaire.) (Please find a color version of this figure on the color plates.)

(Figure 10.4). So, a real, orthoscopic image is produced in front of the holographic plate (pseudoscopic imaging twice restores the correct parallax).

Pictures of the images formed by a first-generation hologram are presented in Figure 10.5. When the hologram is observed at different angles, the scene is perceived with parallax and occlusion. Holograms are sometimes referred as *windows* to another world; if the window is viewed obliquely, the scene is cropped (compare Figure 10.5a,b).

Crisp pictures of a hologram are certainly pretty but it is when something goes wrong that most of the information is gathered. Figure 10.6a shows how the same

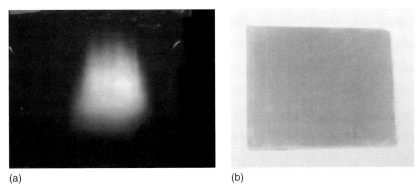

Figure 10.6 (a) Picture of the same hologram from Figure 10.5 replayed with white light. (b) The same hologram on the top of a diffusive background. (Hologram courtesy of Pierre Saint Hilaire.) (Please find a color version of this figure on the color plates.)

hologram as in Figure 10.5 appears when replayed with polychromatic (white) light. The light is dispersed in its various components, a rainbow appears, and the scene is no longer intelligible. This property is used for the manufacture of dispersive elements for spectrometers [18]. Figure 10.6b shows the hologram in front of a diffusive background, a sheet of paper in this case. This particular hologram has been recorded on a silver halide emulsion and bleached to obtain a phase hologram (see Section 11.5 for a discussion on the different types of holographic recording media).

When the hologram shown in Figure 10.5 is read out, flipped left to right or back to front in the reading beam, the diffracted beam reconstructs a real image (similar to the one in Figure 10.3) and the scene become pseudoscopic (Figure 10.7). This latter term is opposed to orthoscopic (normal) and describes an image where the parallax is inverted back to front: the background appears to shift faster than the foreground when the viewer moves. Note that when the image is sharp, the background (table and posts) is blurred and vice versa; the hologram reconstructs the wavefront and the image is actually in a different plane than the recording medium.

10.3
Different Types of Holograms

10.3.1
Transmission versus Reflection

There are two types of holograms according to the way they are recorded and replayed: transmission and reflection. Their properties are quite different, especially for their visualization (Figure 10.8).

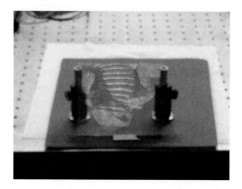 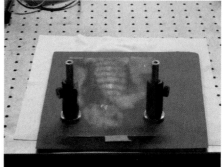

Figure 10.7 Picture of the same hologram as Figure 10.5 positioned in a different orientation; see the text for details. (Hologram courtesy of Pierre Saint Hilaire.) (Please find a color version of this figure on the color plates.)

Transmission holograms are so called because the reading beam has to pass through the recording medium and the output beam is diffracted to the opposite side of the medium from the incident beam. When recording such a hologram, the reference and the object beams are incident on the same side of the recording material. This is the kind of hologram suggested by Gabor and recorded by Leith and Upatnieks with an off-axis reference beam.

Reflection holograms act as dichroic mirrors, reflecting the reading beam back in a different direction. The surface of the "mirror" is spatially structured such that it shapes the wavefront. Reflection holograms are also wavelength selective. As its name suggests, the reading beam is "reflected" back from the surface of the hologram. In this case, since it is a diffractive element, the correct term should be "diffracted back." When recording such a hologram, the reference and object beams are incident from opposite sides of the recording material. That was the geometry used by Denisyuk, in the particular case that the reference beam is scattered back to the media by the object (see Section 10.3.2).

For a transmission hologram, the Bragg planes, that is, the elementary diffractive elements or the hologram fringes are, on average, perpendicular to the material surface. The range of orientation is between 45° and 135° relative to the hologram surface. In the case of a reflection hologram, the Bragg planes are parallel to the surface on average. The range of orientation is between 45° and −45° relative to the hologram surface. An important intermediate case arises when one of the beams propagates within the critical angle of the recording material. This is the domain of edge-lit holograms, which exhibit features of both reflection and transmission holograms. A good treatment of edge-lit holography can be found in Refs [4] and [21].

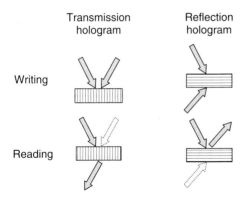

Figure 10.8 Comparison of transmission and reflection hologram geometries.

At this point, it should be noted that most large-production holograms like security holograms[1] on credit cards and commercial products are, in fact, *transmission* holograms, even though they are viewed in reflection. This is because a mirrorlike coating has been applied on the grating structure, which, nevertheless, has the properties of a transmission hologram.

As a general rule of thumb, transmission holograms are color dispersive and reflection holograms are color selective. Illuminated with white light, a transmission hologram will diffract a rainbow (Figure 10.6a) but a reflection hologram will only reflect a specific color. This very property is the basis for the first white light viewable holograms (Denisyuk holograms). As discussed later, the color reproduction strongly depends on whether the hologram is a transmission or a reflection type.

10.3.2
Denisyuk Hologram

To record a Denisyuk hologram, the object is positioned on the back of the transparent recording material. The reference beam traverses the plate and is scattered back by the scene to interfere with itself (Figure 10.9). Since there is no way to balance the intensity ratio between the object and reference beams, this type of recording works better with bright objects. Objects can also be painted according to the laser beam wavelength in order to provide better contrast.

Denisyuk holograms can be viewed with white light, owing to their wavelength selectivity. However, in this case, they have a limited depth of field and the scene appears blurry only a couple of inches behind the hologram. To minimize this effect, the object should be placed as close as possible to the holographic plate

---

1) Security holograms on high-value products are now made of reflection holograms on photopolymer material.

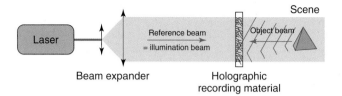

Figure 10.9 Recording setup for a Denisyuk hologram.

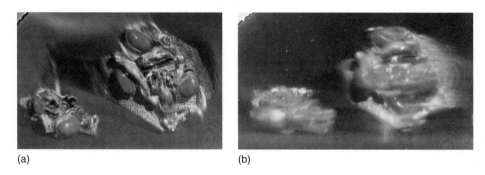

Figure 10.10 Picture of a Denisyuk hologram replayed with (a) a halogen point source and (b) fluorescent tubes. (Hologram courtesy of Arkady Bablumian.) (Please find a color version of this figure on the color plates.)

during recording, and the reading source should be as pointlike as possible (like an arc lamp or the sun). Extended spatial sources (like fluorescent tubes) reproduce a blurry image. Examples are given in Figure 10.10. Since the holographic plate is transparent for the recording process, it is generally further painted black to avoid background distraction when viewed (the black paint also acting as a protective barrier).

10.3.3
Color Reproduction

Transmission holograms are color dispersive and reflection hologram are color selective, resulting in significant differences in color reproduction. Nonetheless, the first step is the same; record three different holograms for each individual additive color: red, green, and blue. These holograms are superimposed in the recording material and will result in diffraction of the intended composite color.

Since it is not the natural color of the scene that is reproduced but a combination of red, green, and blue, the holograms are said to have pseudocolor reproduction. However, an extended color gamut can be produced to display rich, quasi-natural colors that match the sensitivity of the human eye. For better color rendition, some holographers use four or even five colors [4].

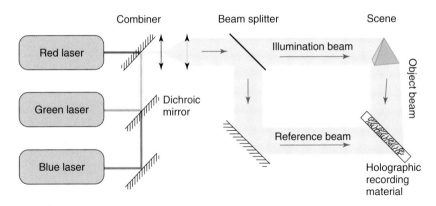

Figure 10.11 Recording a color reflection hologram.

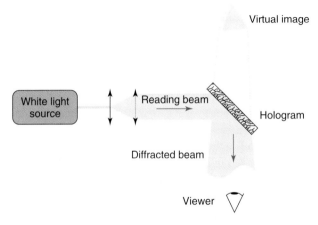

Figure 10.12 Reading a color reflection hologram.

10.3.3.1 Reflection

The scheme of a reflection hologram recording setup is presented in Figure 10.11. Three laser beams having one fundamental color (red, green, or blue) are injected into the classical holographic recording setup. Note that compared to Figure 10.1, the holographic plate has been rotated by 90° so the hologram is now a reflection type. Each beam is writing a hologram of the scene at a particular wavelength. Ideally, when replayed, each hologram will reproduce the object only at the same wavelength used to record it. So, a white light source can be used to read the hologram and reproduce the colored scene (Figure 10.12).

If the three holograms are recorded at the same time, the lasers need to be incoherent with each other so they do not interfere. If this precaution is not taken, cross-talking holograms are generated. Often the three holograms are sequentially recorded and some of the beams do come from the same continuously or discretely tunable laser source.

One of the challenges to recording a color hologram is adjusting the color balance to reproduce true white. Indeed, material sensitivity changes with wavelength and the exposure should be carefully calculated so that the diffraction efficiency of each hologram is the same. Some materials are not even sensitive enough in some parts of the spectrum to record a hologram in that color. In this case, different holographic recording plates are used and then bound together to make the color hologram. For an extended discussion on holographic materials, see Section 11.5.

Another problem that arises with reflection holograms is material shrinkage or swelling. When this happens, the distance between Bragg planes is modified and the reconstructed wavelength is changed. This effect can be use to the holographer's advantage to change the reconstruction color to better match the eye sensitivity compared to the wavelengths available from laser sources.

10.3.3.2 Transmission

The setup introduced in Figure 10.11 cannot be used to record a transmission color hologram. Indeed, transmission holograms do not reproduce the color at which they have been recorded (Figure 10.5) but disperse light in its different components (Figure 10.6a). So, three different pairs of beams need to be used to write the three holograms (Figure 10.13). The illumination beams are directed along the same direction, but the reference beams are incident at different angles to the holographic recording material. This prevents color cross talk during reconstruction. The selection of the angles depends on the dispersion properties of the hologram [19]. By having the red reference beam incident at a smaller angle and the blue reference beam incident at a larger angle, color cross talk is reduced.

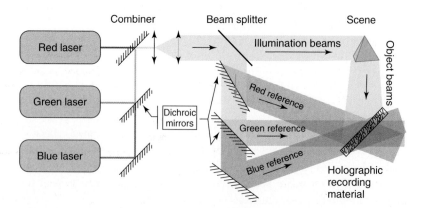

Figure 10.13 Recording a color transmission hologram. (Please find a color version of this figure on the color plates.)

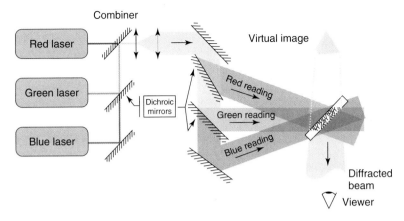

Figure 10.14 Reading a color transmission hologram. (Please find a color version of this figure on the color plates.)

The discussion we had about color balance and sequential recording for reflection hologram still applies here.

To read a color transmission hologram, three monochromatic light sources are needed (Figure 10.14). Their color and angle should match those of the reference beams used to record the hologram. If not, there will be some misalignment of the scene for the different colors, and chromatic aberrations will appear.

The use of transmission geometry for color holography looks much more complicated than reflection geometry. But it is sometimes the only solution that exists (mostly because of the material) and, in some approximations, like the holographic stereogram (see Section 10.4), it can become easier to implement. It also has to be noted that with the proper choice of recording angles (correction of the angle for wavelength different Bragg angles), full-color images can be reconstructed using white light source illumination.

One obvious advantage of transmission geometry is less sensitivity to material shrinkage or swelling than in reflection, since the Bragg plans are orthogonal to the volume of the material.

10.3.4
Phase versus Amplitude: The Diffraction Efficiency

The *diffraction efficiency of a hologram* is defined as the ratio between the intensity of the incident light and the diffracted light. In imaging, this parameter is important since it will quantify the brightness of the reconstructed image for any reading source.

The diffraction efficiency depends on the recording material properties, such as the thickness, the amplitude of the modulation, the average absorption, and the type of modulation recorded. The diffraction pattern written into the holographic material can be of two types: a refractive index modulation (phase hologram) or an absorption modulation (amplitude hologram). In some materials, both types of modulation are present, resulting in a more complicated situation.

Phase or amplitude holograms are basically the same with respect to imaging, but differ markedly with respect to the diffraction efficiency. Phase holograms can actually achieve a theoretical efficiency of 100% (all the incident light is diffracted to reproduce the object beam). However, since a portion of the light is absorbed within an amplitude hologram, the theoretical diffraction efficiency is only 3.7% [19] and therefore amplitude holograms are not widely used. This is also the reason that silver halide recording materials are further bleached to obtain a phase hologram out of what was initially an amplitude hologram (see Section 11.5).

10.3.5
Surface Relief Holograms

For transmission phase holograms, the diffractive pattern can be contained in the bulk of the material, or the surface of the material (surface relief hologram). An optical path difference exists because of the refractive index difference between air and the hologram medium. If an index-matching liquid is put on the top of such a hologram, the diffraction disappears. Reflection holograms are always volume holograms since the Bragg planes are, on average, parallel to the surface of the material.

Surface relief holograms are an important class of holograms because of the possibility to mass reproduce the structure by embossing. This has popularized holography and opened the doors of mass marketing. All of the metallic tinted holograms affixed on consumable products are of this type. The colored rainbow seen when looking at optical disks (CD, DVD, Blue ray) is also the result of the same phenomena.

The manufacturing of surface relief hologram begins with the production of a master hologram on photoresist. A layer of metal is then grown on the top of the surface by electrodeposition. The metal is removed and used to emboss a soft material like a thermoplastic polymer. This last step can be reproduced at will since the metal structure is sturdy and can be reused. To be seen in reflection, the surface relief hologram is covered with a thin layer of metal by vapor deposition and a protective layer of transparent material is overlaid (Figure 10.15). Another option is to directly emboss a previously metalized thermoplastic material. The diffraction is obtained because of the difference in optical path length that exists between light that is reflected from the peaks and that from the valleys.

Depending on the application and the desired quality of the holographic image, surface relief holograms can have two or multiple step levels, be slanted, and reproduce color.

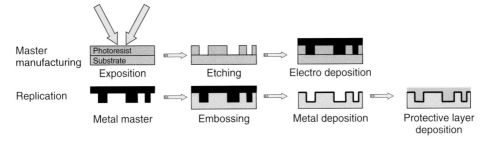

Figure 10.15 Surface relief hologram manufacturing steps.

10.3.6
Thin versus Thick Hologram

Thick holograms obey the Bragg diffraction regime, while thin holograms follow the Raman–Nath regime [20]. The major difference lies in the maximum achievable diffraction efficiency. Bragg holograms can reach 100% in the first order, while Raman–Nath can only reach 33.9% and will diffract in multiple orders.

Kogelnik first gave a thick grating condition in the coupled wave theory [19], refined later by Moharam and Young [21] and Gaylord and Moharam [22]. The thick grating condition is $d/\Lambda > 10$ where d is the thickness of the grating and Λ its period. Thin and thick conditions do not necessarily overlap and there exists an intermediate domain where the hologram behaves in between both modes.

The thickness of the holographic material alone does not define the hologram's mode of operation, but in a practical way, thin imaging holograms are only a couple of micrometers thick, whereas thick holograms are recorded on material 10 or several tens of microns thick.

10.4
Holographic Approximations

In the previous sections, we introduced different setups for recording holograms that reconstitute both the amplitude *and* the phase of an object. We also have seen the significant limitations that accompany these techniques, like the short depth of field for the Denisyuk hologram, the need to use laser beams for reading transmission holograms, or the problem of chromatic dispersion and image deformation due to material thickness variation for reflection holograms.

Moreover, for each holographic recording technique we introduced earlier (with the exception of the Bragg-selective volume reflection hologram), the scene has to be illuminated with a laser. This is a strong restriction for displaying 3D data that are not necessarily accessible to the laser, or too large to be illuminated by a reasonable amount of power. Indeed, it is not possible to use computer-generated data or information gathered by other instruments like medical or satellite imagers to make a hologram in the way we have described. The reason is that, even though the volume information is contained in that data, the phase information in the light scene has been discarded.

The only solutions to the problem of producing holograms from arbitrary 3D data is to computer generate it (with the disadvantages we have already discussed), or to use rapid prototyping machines to manufacture a tabletop size reproduction of the object. This has been done in the past, and has given good results. But the resolution, quality of the rendering, as well as speed, accessibility, and practicality of the rapid prototyping process are frequently drawbacks. The value of a hologram is also dubious when the object itself is transportable and can be replicated.

For all these reasons, holographers have realized that they have to sacrifice some of the phase features of holograms to make them more versatile – with this goal in mind, the rainbow hologram was invented.

10.4.1
Rainbow Hologram

The invention of the rainbow hologram is attributed to Stephen Benton who first described the technique in his 1969 article [6]. The rainbow hologram gives up one dimension of the parallax for the ability to view white light instead of a laser light source. The name comes from the fact that the object appears colored by the spectrum of the reading source: red at large angles, blue at lower angles.

The rainbow hologram is a second-generation transmission hologram from a master that has been recorded as presented in Figure 10.1. When recording the second generation, a slit is placed in front of the master to limit the parallax (Figure 10.16). Since humans see parallax by the horizontal separation of the eyes, the slit is usually horizontal, restraining the vertical parallax. The width of the slit will define the sharpness (due to color dispersion) of the reconstructed image.

Figure 10.17a shows the reconstruction of a rainbow hologram with a monochromatic light source. The viewer will only see the hologram when properly aligned with the image of the slit that defines a horizontal line. When a white light source is used for illuminating the rainbow hologram, the spectrum of the source is spread and the viewer can see the object in different colors moving (left–right in Figure10.17b).

It is possible to reproduce color images with the rainbow hologram technique. To do so, three holograms are recorded with the fundamental colors: red, green, and blue. Next, the rainbow hologram is recorded three times, using the same laser sources. When read out with white light, three vertically displaced spectra are reproduced and when the viewer is positioned at the correct height, he/she will view the correct colors. False colors will be viewed should he/she move his/her head up or down from this point.

10.4 Holographic Approximations | 217

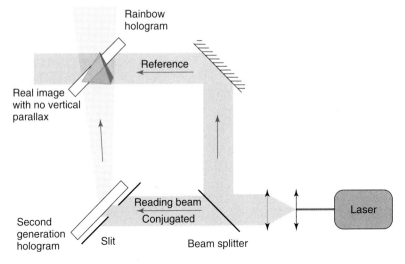

Figure 10.16 Recording a rainbow hologram.

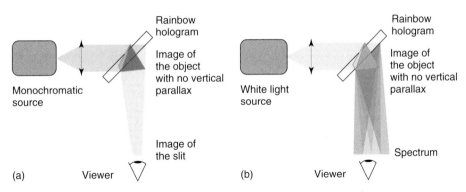

Figure 10.17 Reconstructing the image from a rainbow hologram (a) with a monochromatic source and (b) with a white light source. (Please find a color version of this figure on the color plates.)

10.4.2
Holographic Stereogram

Holographic stereograms have their roots in the works of Gabriel Lippmann on integral photography in 1908 [12], long before holography was even conceptualized. It is also interesting to note that Lippmann's photography, developed at the end of the nineteenth century [23] by reproducing colors from interference, is very close to a reflection hologram, but without reproducing the phase. Integral photography is an autostereoscopic technique that consists of taking a picture with a lens array

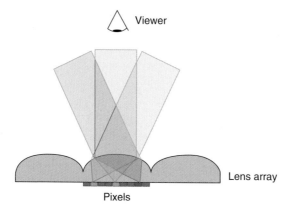

Figure 10.18 Principle of integral photography. When the viewer moves left or right, he/she views different pixels due to the lens array. (Please find a color version of this figure on the color plates.)

in front of the photographic plate and replaying the scene through that same lens array (Figure 10.18). Integral photography reproduces parallax but not the phase of a scene, so the viewer can see occlusions and experience parallax, but there is only one plane of accommodation.

Integral photography does not require any eyewear to see the 3D effect and there are multiple images reproduced. When the viewer moves in front of the display, the image changes continuously in its natural way.

Integral photography can also be implemented by taking pictures of the same scene at different positions and "processing the images" so that when played behind the correct lens array, parallax is reproduced. This technique is very powerful since, for the first time, the 3D aspect of *any kind of data* can be rendered. Indeed, if the information can be printed according to different view angles or from different positions, it can be processed according to integral photography principles; this is termed *integral imaging*.

In integral imaging, the data does not even necessarily need to be 3D. For example, it could be different images showing movement (short film) or showing a transition between totally different images (morphing). This technique is used to print 3D cards that are mass produced for advertisement and decoration.

Philips, the Netherlands-based electronics company, has even developed a 3D TV based on the same principle: the Philips WOWvx 3D TV. In spite of the great advantage that it does not require any eyewear to view the 3D effect, this 3D TV has had little commercial success and it has now been discontinued. In this application, a liquid crystal lens array was positioned in front of an HDTV screen, so that the pixels in Figure 10.18 are now dynamic, resulting in the reproduction of a moving scene. The high resolution of the HDTV screen was traded for the profit of 3D

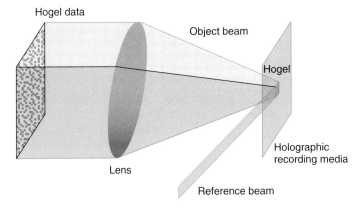

Figure 10.19 Example of a recording configuration for integral holography.

impression. The liquid crystal lens array could be turned off to reconstitute HDTV programs in 2D and full resolution.

The problem that integral imaging cannot escape is the spatial extension of the lens array that degrades the horizontal and vertical resolution of the final image. Reducing the size of the lens increases the 2D resolution but reduces the number of pixels creating the depth impression and degrading the resolution of the third dimension. Finally, the absence of the phase information limits the depth of field without blurring. Holographic stereography solves the spatial extension of the lens and the pixel resolution right away. As discussed, it is also possible to address the last issue of the depth of field to some extent by adding some phase information for a better rendering.

Figure 10.19 presents a holographic stereography recording setup. As in classical holography, a reference and an object beam interfere in a recording medium. However, there is no object on the optical table and the full hologram is recorded pixel by pixel (this holographic pixel is referred to as a *hogel* [24]). Instead of a physical object, the object beam intensity is spatially modulated by some information, or hogel data. This information comes from the same image processing as is used in integral imaging. The object beam is then focused by a lens on a small portion of the holographic recording medium (the actual hogel). The reference beam is a simple, flat wavefront. The full hologram is recorded hogel after hogel to fill the entire surface of the medium. Each time a new hogel is printed, the data are changed by using different slides or an SLM.

The hologram does not need a lenslet array in front of it to be replayed. The lens function is directly encoded as a holographic optical element in each hogel. Therefore, the hogels can be smaller and contain much more information than the combination pixel/lens array. As a result, the resolution of the 3D image is improved. The lateral size of the hogel can be submillimeter and match the eye's resolution. 3D resolution is given by the number of pixels in the hogel data, which can be millions.

A reflection or transmission geometry can be used to record autostereograms by integral holography. Color reproduction is also possible using a technique similar

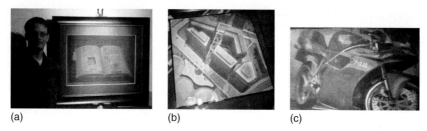

Figure 10.20 Pictures from holographic stereograms. (a) "Tractatus Holographis" from Jaques Desbiens, printed by RabbitHoles (*www.rabbitholes.com*). (b) Computer-generated building by Zebra Imaging (*www.zebraimaging.com*). (c) Ducati 748 from Geola (*www.geola.lt*). (Please find a color version of this figure on the color plates.)

to those introduced in the previous section. There exist several companies that offer to print holographic stereograms based on images provided by the client; Figure 10.20 shows some pictures of such stereograms.

As in integral imaging, integral holography does not reproduce the phase of the scene. This induces some deformation of the image for objects far away from the main image plane, as well as some eye fatigue due to the discrepancy between the eye's accommodation and the position of the objects [5]. Usually, in holography, the phase information is directly encoded by the optical setup since the wavefront of the object is carried by the object beam. This is not possible in integral holography since the wavefront of the object beam is shaped by the final lens, and different parts of the hogel data could be at different places in the scene. So, the idea is to add to the amplitude data some "fine structure" that will diffract the reading beam and reproduce some of the phase (Figure 10.21). The holographic pixel can now reproduce the wavefront and, by extension of the name hogel, has been transformed to a "wafel" [25].

The limitation in reconstructing the wavefront structure comes from the maximum frequency of the chirp function that can be encoded in the device modulating the object beam (photographic film/SLM/micromirror array). This resolution, and the reduction factor in the optical system, determines the thickness of the depth of field that it is possible to achieve. It is interesting to note that the wafel structure converges to a Fresnel hologram when the resolution is high enough.

10.5
Dynamic Holography

10.5.1
Holographic Cinematography

The move from static 3D images to dynamic holographic displays is a dream that physicists have pursued for nearly half a century now. The early attempts

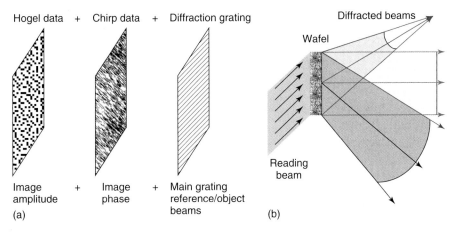

Figure 10.21 (a) The wafel is consists of the hogel data, the chirp data, and the main diffraction grating due to the interference of the object and the reference beams. (b) The diffracted beam from a wafel can have different wavefronts and intensities according to the diffracted direction.

were based on the conventional cinematographic technique, displaying successive holograms recorded on photographic strip film [13, 14]. The problem with this method is that the viewers have to look directly at the film to see the hologram. Indeed, the hologram cannot be projected on a larger screen as in 2D theaters. This strongly limits the size of the hologram and the number of viewers, which is a setback for entertainment, but does not preclude use in research-related topics, such as microscopy.

To accommodate larger viewing angles and/or more viewers, large diameter lenses, mirrors, or scattering screens have been used in front of the optics to record and display the hologram [15]. These methods rely on projection optics that redirect the light from the hologram to several view zones where the viewers are supposed to be. A rather large number of view zones can be created in this way to accommodate a significant audience. However, the holographic recording process still involves the use of a coherent light source, which strongly limits the size of the scene that can be illuminated. It is possible to expose an entire room, but not an external location. There is also a problem with generating the hologram from scenes that do not physically exist, such as computer-animated movies.

As we have discussed for static holography, the stereographic technique helps with both these points. By sacrificing the phase information, it is possible to record the movie in incoherent light with several cameras, or to generate several 2D perspectives from a computer scene. Then, by applying the integral imaging transformation, the hogels are calculated and recorded with the same kind of holographic setup presented in Figure 10.19. Each image of the holographic movie is now a holographic stereogram that reproduces the parallax (but not the phase). The same scaling methods as those explained for the holographic cinematograph can be used to enlarge the image.

10.5.2
Real-Time Integral Holography

Owing to the development process needed to reveal the hologram recorded in permanent media such as silver halide emulsions or photopolymers, the delay between recording and the playback of such a hologram is on the order of hours to days. Furthermore, the recording and the display of the hologram usually take place at different locations. So the observation of 3D images in real-time, lifelike shows, telepresence, or manipulation of the data by the end user is not possible with such technology. The solution is to use a dynamic holographic recording media that is self-developing. By doing so, new holograms can be continuously recorded into the material while the viewer is looking at the diffracted light. In this case, both recording and display happen at the same place and at the same time.

Early attempts have used photorefractive crystals as a dynamic holographic medium [26]. But the small aperture of the crystal ($<5cm^2$) was a limiting factor for the display size. It is possible to grow larger crystals but at a tremendous cost that is not practical for a comfortable display. Recent developments in photorefractive polymer materials have made it possible to manufacture a large-scale (1000 cm^2) dynamic display [27] (Figure 10.22a). Using integral holography, a refresh rate of a couple of seconds has been demonstrated as well as full parallax and color. The advantage of integral holography is the reduced amount of computer processing that is needed to calculate the hogel compared to the CGH approach. Indeed, large images with HDTV resolution can be calculated in real time with a regular desktop computer. Figure 10.22b shows a picture of a dynamic hologram recorded in a photorefractive polymer, illustrating the possibility of 3D telepresence. The subject

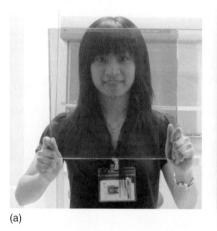 

(a) (b)

Figure 10.22 (a) A 12″ × 12″ photorefractive screen. (b) Example of 3D telepresence using a photorefractive polymer screen. (Please find a color version of this figure on the color plates.)

is filmed by 16 cameras, the hogels are calculated in real time, and then transmitted by an Ethernet link into another location where they are printed in the medium and directly observable by the viewers. Once printed, another hologram is immediately loaded and the 3D image is refreshed.

This application has some limitations due to the material sensitivity and required laser power. Indeed, with about 100 mJ cm^{-2} of power required to record the hogels in the photorefractive material, a 12″ × 12″ screen size, and a refresh rate of 24 holograms per second (video rate), the laser used to write the hologram should have a power of 2 kW. Such a pulsed laser with kilohertz repetition rate does not exist. The most powerful pulsed lasers suitable for this application have a power of about 40 W, so the material sensitivity needs to decrease by a factor 50 to achieve these goals.

10.5.3
Holographic Video

The ultimate device for 3D data visualization would be a *holographic video imaging system* where CGHs are displayed digitally at video rate. Such a system has been pursued by both the MIT9 and the UK-based company QinetiQ16. The principle is to digitally generate the computed interference pattern and refresh the hologram to represent the dynamics. With such a device, both the amplitude and the wavefront of the image are reproduced, which theoretically give the best possible rendering.

There are two parts to the system: the computer code that calculates the diffraction pattern and the optical system that reproduces it and actually diffracts the light. For the latter, MIT is using an acousto-optic crystal in which a sound wave modulates the refractive index of the material. An extended laser beam is diffracted by the crystal to form a line of the holographic image and a scanning system is used to enlarge the hologram size in the vertical direction. The QinetiQ physical device is constituted of an optically addressed SLM in which the hologram is written by a beam, modulated by a computer-addressed SLM. This allows the diffraction pattern to be downsized to the optical wavelength.

In both approaches, full-color holograms refreshed at video rate have been demonstrated. The problem is the size (or resolution) of the hologram which, at the present time, ranges only over a few square inches (or cubic inches for the 3D volume). Reproduction of textures, which are related to resolution, is also difficult and currently under investigation [28]. These limitations derive from the first part of the system: the calculation of the CGH. Indeed, both groups acknowledge that, if the optical setup is scalable, the computer-processing speed limits the size or resolution of the image due to the amount of processing power required to calculate the CGH. A direct estimation can be made of the computational load deduced from the metrics provided by QinetiQ [16]:

- The number of diffractive elements (hogels) required for a full parallax 2″ × 2″ size is 10 000.
- Each hogel consists of 1024 × 1024 pixels.
- There are 4096 diffraction-table lookup entries for the hologram computation.

- There should be 24 frames per second.
- The number of multiply accumulate (MAC) to floating point operation (flop) conversion factors is 613.

The multiplication of all those entries gives 6×10^{17} flops, nearly one quintillion operations per second. This is 300 times faster than the actual world's fastest computer, the Jaguar Cray XT5 (2×10^{15} flops). It will take another 10 years for the fastest computer to reach this computational power according to Moore's law (fps doubling every 14 months), and it will not hit desktops until the year 2046 (number of transistors double every two years). Data rate for such a device is 2^{12} bps considering 8 bits gray scale per pixel. This is another bottleneck for the real-time transmission of holographic images.

Fortunately, some approximations can be used to reduce the number of operations that need to be carried out to calculate the hologram. One of them is exploited by SeeReal, a Germany-based company [29]. Since each piece of a hologram contains the necessary information to reconstruct the entire image, SeeReal calculates only the diffraction pattern for the objects visible to the viewer. This way, only a limited region (subhologram) needs to be processed and the computation requirement is reduced by a claimed factor 10 000 allowing real-time processing for some basic images. The data transmission rate is on the order of magnitude of 10^8 bps. The downside of this technique is that the hologram is only visible in a very limited zone (about the size of the human pupil) and an eye-tracking system needs to be added to recalculate the hologram according to the viewer position in real time (solution developed by SeeReal).

10.6
Conclusion

Holography has been identified as the only way to visualize 3D scenes and data in a natural way, reproducing both the amplitude and the wavefront of a scene. 3D display is so critical that it is bound to people as tightly as their physiology; we are meant to see in 3D.

There have been many prognostics in the past announcing the blooming of full-color, full-parallax, holographic cinema or television within a couple of years. Unfortunately, this has not happened and such devices still look futuristic. But science and technology are walking hand in hand to make this dream come true and the end of the tunnel is in view. All the building blocks for a real-time holographic electronic display have been identified and laboratory devices have been demonstrated. The ultimate solution looks closer than ever.

In the mean time, some approximations have been conceded to manufacture more realistic setups that reproduce 3D at the expense of the phase information and/or one-dimension parallax. Autostereoscopic viewing that does not require the use of eyewear is of great interest and is definitely an improvement compared to other methods.

There is no doubt in the author's mind that the future will be holographic.

Acknowledgment

The author is grateful to Cory Christenson, Robert A. Norwood, Arkady Bablumian, and Pierre Saint Hilaire for their careful correction of the manuscript.

References

1. Gabor, D. (1949) Microscopy by recorded wavefronts. *Proc. R. Soc. (London)*, **197** (1051), 454–487.
2. Denisyuk, Y.N. (1962) On the reflection of optical properties of an object in a wave field of light scattered by it. *Dokl. Akad. Nauk. SSSR*, **144** (6), 1275–1278.
3. Leith, E.N. and Upatnieks, J. (1962) Reconstructed wavefronts and communication theory. *J. Opt. Soc. Am.*, **52** (10), 1123–1130.
4. Hariharan, P. (1996) *Optical Holography: Principles, Techniques, And Applications*, Cambridge Studies in Modern Optics.
5. Lambooij, M. et al. (2009) Visual Discomfort and visual fatigue of stereoscopic displays: a review. *J. Imaging Sci. Tech.*, **53** (3), 030201–030214.
6. Benton, S. (1969) Hologram reconstructions with extended incoherent sources. *J. Opt. Soc. Am.*, **59**, 1545–1546.
7. Brown, B.R. and Lohmann, A.W. (1966) Complex spatial filtering with binary masks. *Appl. Opt.*, **5** (6), 967–969.
8. DeBitetto, D.J. (1969) Holographic panoramic stereograms synthesized from white light recordings. *Appl. Opt.*, **8** (8), 1740–1741.
9. St Hilaire, P. et al. (1990) Electronic display system for computational holography. *Proc. SPIE*, **1212**, 174–182.
10. Lesem, L.B., Hirsch, P.M., and Jordan, J.A. (1969) The Kinoform: a new wavefront reconstruction device. *IBM J. Res. Dev.*, **13**, 150–155.
11. MacGovern, A.J. and Wyant, J.C. (1971) Computer generated holograms for testing optical elements. *Appl. Opt.*, **10** (3), 619–624.
12. Lippmann, G. (1908) Epreuve réversibles donnant la sensation du relief. *J. Phys.*, 4ème Série, **VII**, 821–825.
13. Caulfield, H.J. (1979) *Handbook of Optical Holography*, Academic Press, New York.
14. DeBitetto, D.J. (1970) A front-lighted 3-D holographic movie. *Appl. Opt.*, **9** (2), 498–499.
15. Leith, E.N., Brumm, D.B., and Hsiao, S.S.H. (1972) Holographic cinematography. *Appl. Opt.*, **11** (9), 2016–2023.
16. Slinger, C., Cameron, C., and Stanley, M. (2005) *Computer-Generated Holography as a Generic Display Technology*, Computer Published by the IEEE Computer Society, pp. 46–53, August 2005.
17. Benton, S. and Bove, V.M. (2008) *Holographic Imaging*, Wiley-Inter Science.
18. Barden, S.C., Arns, J.A., and Colburn, W.S. (1998) Volume-phase holographic gratings and their potential for astronomical applications. *Proc. SPIE*, **3355**, 866–876.
19. Kogelnik, H. (1969) Coupled wave theory for thick hologram gratings, *Bell Syst Tech. J.*, **48** (9), 2909–2947.
20. Raman, C.V. and Nath, N.S.N. (1936) The diffraction of light by high frequency sound waves: part I. and II. *Proc. Indian Acad. Sci. Sec. A*, **2**, 406–420.
21. Moharam, M.G. and Young, L. (1978) Criterion for Bragg and Raman-Nath diffraction regimes. *Appl. Opt.*, **17** (11), 1757–1759.
22. Gaylord, T.K. and Moharam, M.G. (1981) Thin and thick gratings: terminology clarification. *Appl. Opt.*, **20** (19), 3271–3273.
23. Lippmann, G. (1894) Sur la théorie de la photographie des couleurs simples et composées par la méthode interférentielle. *J. Phys. Thiorique appl.*, **3** (1), 97.
24. Lucente, M. (1994) Diffraction-specific fringe computation for electroholography, Ph.D. Dissertation, MIT Department of Electrical Engineering and Computer Science, Cambridge, MA.

25. Smithwick, Q.Y.J. et al. (2010) Interactive holographic stereograms with accommodation cues. *Proc. SPIE*, **7619** (03), 1–13.
26. Ketchel, B.P. et al. (1999) Three-dimensional color holographic display, *Appl. Opt.*, **38** (29).
27. Tay, S. et al. (2008) An updateable holographic 3D display. *Nature*, **451**, 694–698.
28. Smithwick, Q.Y.J. et al. (2009) Real-time shader rendering of holographic stereograms. *Proc. SPIE*, **7233** (02), 1–112.
29. Häussler, R., Schwerdtner, A., and Leister, N. (2008) Large holographic displays as an alternative to stereoscopic displays. *Proc. SPIE*, **6803** (0M), 1–9.

Further Reading

Komar, V.G. and Serov, O.B. (1989) Work on holographic cinematography in the USSR. *Proc. SPIE*, **1183**, 170.

Lee, W.H. (1970) Sampled Fourier transform hologram generated by computer. *Appl. Opt.*, **9**, 639–643.

Leseberg, D. and Bryngdahl, O. (1984) Computer-generated rainbow holograms. *Appl. Opt.*, **23**, 2441–2447.

Wyrowski, F., Hauck, R., and Bryngdahl, O. (1987) Computer-generated holography: hologram repetition and phase manipulation. *J. Opt. Soc. Am. A*, **4**, 694–698.

11
Holographic Data Storage Technology

Kevin Curtis, Lisa Dhar, and Pierre-Alexandre Blanche

11.1
Introduction

Digital data is ubiquitous in modern life. The capabilities of current storage technologies are continually being pushed by applications as diverse as the distribution of content, archiving of valuable digital assets, and downloading over high-speed networks. Current optical data storage technologies, such as the compact disk (CD), digital versatile disk (DVD), and Blu-ray disk (BD), have been widely adopted because of the ability to provide random access to data, the availability of inexpensive removable media, and the ability to rapidly replicate content (video, for example).

Traditional optical storage technologies, including CD, DVD, and BD, stream data 1 bit at a time, and record the data on the surface of the disk-shaped media. In these technologies, the data are read back by detecting changes in the reflectivity of the small marks made on the surface of the media during recording. The traditional path for increasing optical recording density is to record smaller marks, closer together. These improvements in characteristic mark sizes and track spacing have yielded storage densities for CD, DVD, and BD of approximately 0.66, 3.2, and 17 Gbit/in.^{2}, respectively. BD has decreased the size of the marks to the practical limits of far-field recording.

To further increase storage capacities, multilayer disk recording is possible [1], but signal-to-noise losses, and reduced media manufacturing yields, making it impractical to use significantly more than two layers. Considerable drive technology changes are also needed to deal with the signal-to-noise losses inherent in multiple layers. Taking all these issues into consideration, the practical limit for the storage capacity of BD is thought to be around 100 GB, with a transfer rate of 15–20 MB/s.

Alternative optical recording technologies, such as near-field [2, 3] and super-resolution methods [4, 5], aim to increase density by creating still smaller data marks. However, neither near-field nor super-resolution methods have shown compelling improvements over BD.

Another approach that produces multiple layers is two-photon recording in homogeneous media [6–8]. This method uses a first laser wavelength to record by

---

Optical and Digital Image Processing: Fundamentals and Applications, First Edition. Edited by Gabriel Cristóbal, Peter Schelkens, and Hugo Thienpont.
© 2011 Wiley-VCH Verlag GmbH & Co. KGaA. Published 2011 by Wiley-VCH Verlag GmbH & Co. KGaA.

producing a local perturbation in the absorption and fluorescence of the media, which introduces a small, localized index change through the Kramers–Kronig relationship [9]. A second wavelength is used to read out the data by stimulating an incoherent fluorescence at a different wavelength. Many layers of bits are recorded to achieve high density. Unfortunately, two-photon approaches suffer from an inherent trade-off between the cross section of the virtual or real state (sensitivity) and the lifetime of this state (transfer rate). In addition, in at least one example [6], the media is partially erased by each readout. Thus, two-photon techniques face both difficult media development and transfer rate or laser power issues.

With all other optical technologies facing obstacles to significant performance improvements, interest in holographic data storage (HDS) has dramatically increased in recent years. For example, at the 2008 Joint International Symposium on Optical Memories and Optical Data Storage held in Hawaii, nearly half of the papers were related to holographic systems, media, components, and data channels.

11.2
Holographic Data Storage Overview

Holography is in its essence a technique for recording optical wavefronts. In 1948, Hungarian scientist Dennis Gabor first conceived of the notion of "two-step optical imagery" [10]. Although Gabor conceived of this strategy to improve the resolution of the electron microscope, he first demonstrated the feasibility of the holographic approach in 1948 using photons. Holographic recording, for the most part, lay dormant as a research subject until the early 1960s when it was revived by Leith and Upatnieks [11]. This resurrection was energized by the introduction of a new coherent light source, the laser. Leith and Upatnieks pioneered the technique of off-axis holography, and recast holographic theory in a communications framework.

As mentioned above, the notion of using holography for data storage was first suggested by P. J. van Heerden in 1963. He suggested bleaching alkali halide color centers in a medium as a method of recording data-bearing holograms. Both wavelength and angle multiplexing were described. Soon after this work, others [12, 13] began to expand and refine the system and theoretical concepts needed for HDS. A large effort involving many companies in the United States in the early 1970s worked on holographic storage, mostly using photorefractive crystals as a recording medium. In the 1980s, many Japanese companies tried to develop optical disk versions of the technology. In the mid 1990s, a number of US companies again started developing holographic systems and materials as the drive components (modulators, detectors, and lasers) became commercially available. This momentum has carried over into the 2000s, with a large number of companies and universities worldwide again researching HDS.

HDS breaks through the density limitations of conventional storage technologies by going beyond two-dimensional layered approaches, to write data in three dimensions. Two methods of holographic storage have been proposed. One is recording bitwise (single bit at a time) and the other is recording and reading data pagewise (large blocks of bits at a time). In bitwise holographic storage (also called *microholographic storage*) [14–17], multiple layers of small localized holograms are recorded at the focus of two counterpropagating beams. Each of these holograms represents a single bit that is subsequently read out by monitoring the reflectance of a single focused beam. Bitwise holographic storage is appealing because the drive technology and components are similar to traditional optical storage, and because the media is homogeneous and hence easy to manufacture. However, there are several serious drawbacks. First, it is difficult to achieve fast transfer rates. Also, it requires the invention of a material that is optically nonlinear. The technique also requires a complex servo system because the two recording beams must be dynamically focused into the same volume. Finally, the multiple layers of microholograms cause distortion in the optical beams, which significantly limits the achievable density [18].

This chapter focuses on pagewise holographic storage. In this approach, data is stored as shown in Figure 11.1. Light from a single laser beam is split into two beams, the signal beam (which carries the data) and the reference beam. The hologram is formed where these two beams intersect in the recording medium. The process for encoding data onto the signal beam is accomplished by a device called a *spatial light modulator* (*SLM*) (see Chapter 9). The SLM translates the electronic data of 0s and 1s into an optical "checkerboard" pattern of light and dark pixels. This is called a *data page* and typically is $\sim 1000 \times 1000$ pixels. At the point where the reference beam and the data carrying signal beam intersect, the hologram is recorded in the light-sensitive storage medium. A chemical reaction occurs causing

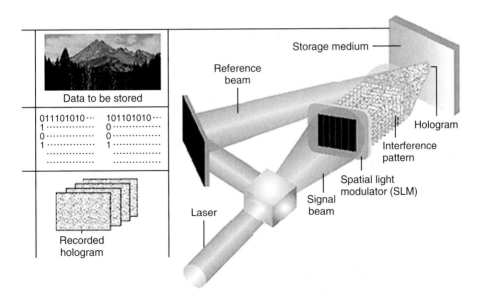

Figure 11.1 Writing data.

the hologram to be stored. By varying the reference beam angle or media position, hundreds of unique holograms are recorded (multiplexed) in the same volume of material.

Many multiplexing techniques have been proposed. Typically, the primary multiplexing method is angle multiplexing, which allows data pages to be superimposed by varying the reference beam angle. For each reference beam angle a different data page is stored. In this way, hundreds of unique holograms are recorded within the same volume of material. This group of angle-multiplexed holograms is referred to as a *book* (or sometimes, a *stack*). To achieve high capacity, fully multiplexed books may be recorded at every location on the medium, thus multiplying the 2D resolution-limited capacity by a factor of several hundred.

For data recovery, a *probe or reference beam* (which is nominally a replica of the recording reference beam) is used to illuminate the medium. The hologram diffracts the probe beam, thus reconstructing a replica of the encoded signal beam. This *diffracted beam* is then projected onto a pixelated photodetector that captures the entire data page of over one million bits at once. This parallel readout provides holography with its fast transfer rates. Figure 11.2 illustrates a basic geometry for reading holograms. The detected data page images are then processed and decoded in order to recover the stored information.

In addition to angle multiplexing, polytopic multiplexing [19] is used to achieve high density. In polytopic multiplexing, the book spacing is determined by the beam waist or signal bandwidth. Since the waists do not overlap, undesired reconstructions from neighboring books can be filtered out by introducing an aperture that passes only the desired beam waist. Writing is performed as in the nonpolytopic case, excepting that the aperture is present and the book spacing

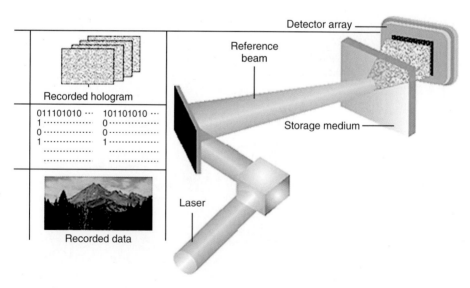

Figure 11.2 Reading data.

is determined by the polytopic criterion. Upon readout, the desired page and several of its neighbors may be simultaneously reconstructed since the probe beam must illuminate data-bearing fringes from several books at once. However, the physical aperture blocks the components diffracted from the neighboring books. This is particularly important when, as is required for high-density storage, a high numerical aperture and a thick medium are used.

11.2.1
Drive Architecture

The basic drive architecture [20, 21] is shown in Figures 11.3 and 11.4. Figure 11.3 shows how the architecture works during writing operation. The laser is a tunable 405 nm source. The laser goes through a shutter that can block the light or pass the light. This allows for the write exposure times to be varied. The light is split into two paths, a reference path and a data path, by the optical divider. The reference path goes to a galvanometer (galvo) mirror that changes the angle of the reference beam for angle multiplexing the holograms. The galvo mirror rotates to change the angle of the reflected beam. This mirror is imaged into the center of media so that the mirror changes the angle of incidence at the media without changing the location of the beam. The read galvo on the other side of the media is turned so that the through light is reflected into a beam dump.

The data path consists of a beam expander/apodizer that expands the beam to the correct size and flattens the intensity profile of the incoming Gaussian beam so that the SLM can be illuminated uniformly. After this expander, the beam passes

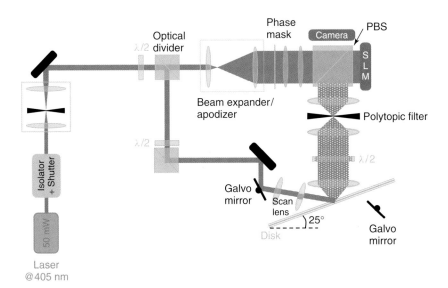

Figure 11.3 Diagram of light path during write operation.

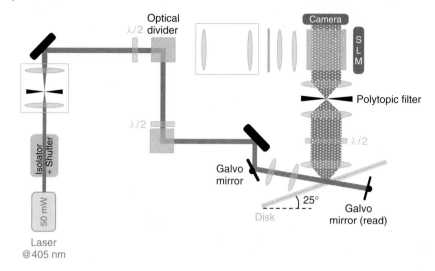

Figure 11.4 Diagram of light path during read operation.

through a phase mask to make the Fourier transform of the data beam's intensity more uniform in the media. The phase mask is imaged onto the SLM with a 4F imaging system. A 4F imaging system consists of two lenses separated by two focal lengths, with two image planes one focal length outside of the lens system (i.e., the total length is four focal lengths, hence "4F"). Since the phase function is in the image plane, it does not impact the intensity detected by the camera on recovery. The SLM encodes the light beam with the data and is then filtered by taking the Fourier transform of the SLM with a relay lens and placing a filter at the Fourier plane. This filter is the polytopic filter on the recovery path, but it removes the higher diffraction orders of the SLM and decreases and defines the hologram's size. Typically, the filter is about 1.2 Nyquist area in width. The Nyquist area represents the minimum aperture size required to resolve the SLM pixels according to the sampling theorem, and is one quarter of the area of the zeroth order of the SLM centered on the optical axis. The next relay lens converts the data page back into an image. The storage lens is the lens that is next to the media. This high NA extreme isoplanatic lens (~0.65) decreases the volume of the hologram in the media. This minimizes the overlap of the books required so as to reduce the dynamic range (M/#) requirement of the media. The reference and data beam intersect in the media and form the hologram. To angle multiplex a book of holograms in the same location, the data page on the SLM and the angle of the galvo mirror is changed to record another hologram. To access more books, the disk is rotated or translated along a radius relative to the optics.

Figure 11.4 shows the path of the light during a read operation. For read, all of the light from the laser is put into the reference path. The galvo mirror is adjusted to the correct angle to read out the desired hologram, and the disk is positioned in the correct radius and theta to read out the desired book. The reference beam passes through the media and is retro reflected by the read galvo. The read galvo consists of both a mirror and a quarter wave film. The read galvo is synchronized with the write galvo to retroreflect the light but with a 90° polarization change. The hologram is reconstructed with this retroreflected beam and the polytopic filter eliminates the cross talk from adjacent books. Owing to the polarization change, the reconstruction goes through the polarizing beam splitter (PBS) and is detected on the CMOS active pixel detector array (camera). Again, to read another page the galvo mirror is changed to another angle, and to access other books the disk is moved relative to the head.

11.2.2
Consumer Drive Implementation

In order to simplify the professional drive implementation, the monocular architecture [22–24] was proposed and over 713 Gbit/in.^{2} densities achieved. This approach reduces the drive size and complexity by putting both the data and reference beams through one objective lens, as shown in Figure 11.5a. The laser's output is collimated and circularized and then split into reference and data paths. The curved surface of the objective lens contains a thin-film angular filter for implementing the polytopic and Nyquist filtering. For reasons of both cost and size, the components are very small. The SLM and camera pixel pitches are 4 and 3 µm respectively. The objective lens has a focal length of 2.5 mm and an NA of 0.85. Angular multiplexing is accomplished by changing the mirror angle Figure 11.5b or by shifting the objective lens position Figure 11.5c. For reasons of speed, it is desirable to use a fast mirror actuator for multiplexing and move the objective lens to accomplish radial tilt correction. Again, books of holograms are

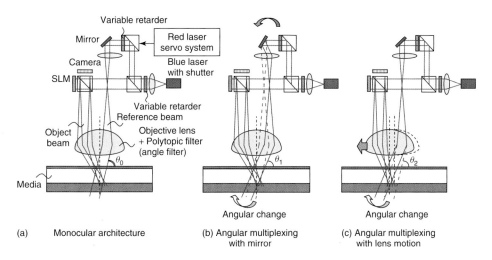

Figure 11.5 Monocular concept with multiplexing.

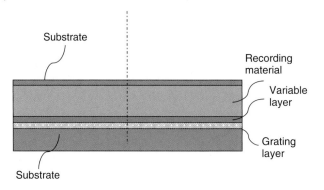

Figure 11.6 Drawing of media structure.

recorded by angle/polytopic multiplexing. This can be made compatible with the current Blu-ray products.

The media structure for a monocular is shown in Figure 11.6. The first cover substrate is 100 μm thick and has an antireflective coating. The recording material is 1.5 mm thick and records the holograms. The variable layer absorbs during recording and then transmits the light during readout. This is to prevent reflection holograms from being recorded. A polarizer with as little as 20 : 1 contrast ratio would be an excellent example of a variable layer. The polarizer can be radial symmetric polarizer formed by nanoimprint lithography [25]. On write, the polarization of both beams is set so that the polarizer absorbs the light. On readout, the polarization of the reference is changed so as to pass through the polarizer and is incident on the grating layer. This layer is a reflective grating with a period of approximately 315 nm that is chirped or blazed to reduce higher orders. This is to cause the reflection of the beam to allow for phase conjugate readout and is designed to retroreflect the center angle of the reference beam sweeps of angles. A servo pattern like the standard push–pull servo of a CD/DVD can be placed in this layer to be read by a red laser. Both the grating and the servo pattern can be stamped into the bottom substrate and then coated with metal for reflection. The bottom layer is a 1 mm polycarbonate substrate that provides the mechanical support and the attachment to the drive.

In addition, high-speed replication process has been demonstrated for holographic storage [26–28]. This allows for CD/DVD-type whole media at a time replication. Thus these products can support high-volume content distribution applications.

11.3
Tolerances and Basic Servo

The tolerances of a volume holographic system are quite different from those of traditional surface-recording methods. The basic positional tolerances are very

Table 11.1 Tolerances examined and results of measurements.

Condition	Measured tolerance (0.3 dB SNR drop)	Notes
Reference beam angle, tangential (Bragg) direction	±0.007°	Degradation due to page Bragg mismatch
Media radial offset	±45 µm	Limited by vignetting of polytopic filter
Media rotation	±0.020°	Bragg mismatch due to rotation
Media tangential offset	±20/50 µm	Limited by vignetting of polytopic filter
Reference beam angle, radial (degenerate) direction	±0.015°	Limited by NA/system geometry (SNR loss due to Bragg mismatch)
Disk tilt – Tangential direction (with reference compensation)	±0.125°	Data lens aberrations and imperfect phase conjugation
Disk defocus (media normal offset)	±50 µm	Data lens aberrations and imperfect phase conjugation
Camera defocus	±50 µm	–
Wavelength detuning	±0.3 nm	Limited by NA/system geometry
Temperature detuning	+4 °C	Same as wavelength
Wavefront for recording (safe rule of thumb)	<100 milliwave RMS	With focus limited to <2 waves. Read-only limit is much higher
Beam overlap (reference and signal)	Radial 100 µm theta > 100 µm	From DVT build and test data

large (~100× larger), but the tilt/angle tolerances are much tighter (~10×) than in traditional optical recording.

Table 11.1 below summarizes the tolerances for a drive using angle-polytopic multiplexing and the physical process that determines the tolerance.

To achieve these tolerances, a mechanical support for this disk is used in conjunction with two servo systems. The disk is contacted at the outer radius (beyond the data area) by a roller that deflects the disk vertically. This roller sets the radial disk tilt and the disk height to within specification even with standard DVD media tilt margins. Since the disk does not spin, the wear is not an issue. The first servo system is a dedicated servo that reads servo marks that are embossed in the disk substrate with a red servo laser that does not affect the blue-sensitive media. This allows theta to be determined to ~10 nm resolution and the load/unload error is less than 5 µm in radial. Thus the positional tolerances are easily handled. However, the tilt, temperature, and wavelength tolerances have to be handled.

The reference angle, tilt, media temperature, and wavelength tolerances are handled with a wobble servo [29]. Tilt and reference angle can be compensated by changing the pitch/angle of the reference beam using a galvo mirror. Temperature changes can be compensated by changing the angle and wavelength of the light to read out the data at a different temperature than recorded [30]. However, the servo signal on these parameters must come from the holograms themselves and the error signal/correction must be generated simultaneously with reading the data in order not to effect transfer rate of the drive. The wobble servo satisfies both these requirements. By reading out the pages in a book with slightly off-the-peak data page strength in reference beam angle (before the peak for even pages and after the peak for odd pages), servo signals for angle, tilt, and wavelength/temperature parameters are generated. This adjusting of the reference angle between alternate holograms is called *wobbling the angle*. By looking at the difference in signal-to-noise ratio (SNR) between even and odd pages with this wobble, an error signal for the reference angle can be generated. For example, if the difference in SNR is 0 dB then the angles are tracking perfectly (both offsets before and after the reference angle are equal). In addition to the reference beam angle error signal, the imparted wobble produces changes that can be used to determine misalignment in pitch and wavelength. This is possible because the presence of these angular misalignments causes the wobble offset to produce a shift in the best Bragg-matched region of the holographic image. The shift can be detected as a change in the position of the intensity centroid of the detected images. By looking at the centroids of the pages in the two dimensions while wobbling the reference angle, error signals for pitch and wavelength are generated. Because changes in the alignment of sequential holograms are slowly varying, and because the error signals are relatively noisy, low-gain servo compensators are required, such as the recursive least squares filter. The system recovers sequential holograms with less than 0.15 dB average SNR loss per hologram when compared to careful optimal alignment.

11.4
Data Channel Overview

In an ideal holographic system, each binary digit of user data would set the state of a single pixel on the SLM, which would then be recorded with perfect fidelity in the recording medium. On reading the holograms, the reconstructed image of each original SLM pixel would impinge on a single camera pixel, without misalignment or distortion. Each camera pixel's intensity would correspond to the original binary value of the input data: "1" for a bright pixel, and "0" for a dark one.

In practical drives, of course, none of these ideal conditions are met. The InPhase professional drive uses multiple levels of error correction codes (ECCs) to preserve the integrity of the data. Within a data page, a low-density parity check (LDPC) code provides powerful error correction capabilities for instances where a page is only partially recoverable. A second-level Reed–Solomon code [31] provides chapter-level protection, allowing the data to be recovered even when whole pages cannot be recovered. Finally, a third-level "Anthology" code (also Reed–Solomon) allows the data to be recovered even when entire chapters cannot be recovered with the inner two levels. In addition to this ECC, the data must be detected even if misaligned and with the incorrect magnification. Thus the 2D data page (image of the data) can be detected on the 2D camera without alignment. The discussion below briefly covers the oversampling method of detection.

In our present method [32], the page overall location is determined, blocks of known bits called *reserve blocks* that are uniformly positioned throughout the page are used to find the local misalignment, and using the known bits, the data pixels are resampled. The overall page location is found by using large sync marks at the top of the page. Once the sync marks are found, the relative location for the reserve blocks is known. The reserve blocks are blocks of 16 pixels with very good autocorrelation functions that allow the local misalignment to be accurately measured to better than 1/10th of the pixel. The reserve blocks are separated by 64 data pixels in both x and y directions on an approximately 1000×1000 pixel page. With the local misalignment known, the filter can resample the data pixels to correct for misalignment and magnification errors. However, since the 4×4 detector window may have an arbitrary fractional alignment to the data pixel image, a single set of coefficients will not perform this task optimally. Therefore, a different set of coefficients was generated for each possible fractional alignment. The algorithm utilizes coefficients optimized for fractional alignments, δ_X, δ_Y, in increments of 5% of the detector element spacing, resulting in a table of size $21 \times 21 = 441$ different 16-coefficient sets. Coefficient sets were derived over 2048 different binary states of the neighborhood for each fractional alignment case. The coefficients were optimized to produce a minimum mean square error estimate of the *state* (1 or 0) of the central data pixel image, thus effectively removing much of the intersymbol interference (ISI) and maximizing the SNR.

The oversampling ratio itself is of primary importance – how much larger the SLM pixels are than the detector pixels. Figure 11.7 shows the effect of the *oversampling ratio* – defined as the data pixel image spacing divided by the detector element spacing – on SNR for three local alignment cases. For these simulations, ISI was established by an aperture of width 1.08 Nyquist of the data pattern, pixel area fill factors for both the detector and the SLM were set to 90%, and coherent pseudorandom noise with power 1/20th of an "on" pixel power was added. Figure 11.7 shows that with a ratio of only 4/3, linear oversampling can achieve performance that is as good as Nyquist sampling (2 linear) regardless of misalignment.

11.5
Materials for Holography

When starting to record holograms, the most important decision that determines the kind of hologram, the technique used, the setup, the laser source, and so on, is the choice of holographic recording media. HDS is no exception. We first introduce the holographic recording material in a general context before coming back to the specificities of data storage.

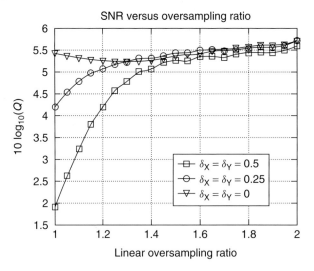

Figure 11.7 Simulated effects of oversampling ratio on SNR for misaligned pages.

Historically, the predominating material was the photographic emulsion. However, manufacturers stopped their production during the 1990s, since this medium was mainly intended for X-ray medical imaging (holography does not constitute a large market). Photopolymers are now a better alternative since several suppliers exist. Dichromated gelatin is a latent technology that is still in use, but due to the limited time available between the preparation of the gelatin and its use, holographers have to prepare their own mixture. In addition to the following three major categories, there exist some more exotic materials that we briefly introduce – photothermoplastic, photochromic, and photorefractive materials.

11.5.1
Silver Halide Photographic Emulsion

The difference between the holographic photographic emulsion and the classical black-and-white photographic emulsion is the size of the silver halide grains. It is on the order of magnitude of a micrometer (10^{-6} m) for the photographic emulsion, but needs to be much smaller for holography since the resolution of the "image" is now about 1000 lines per millimeter. Typical holographic emulsions have grains on the order of magnitude of nanometers (10^{-9} m).

Since the sensitivity of the emulsion is directly proportional to the size of the grain, holographic emulsions are less sensitive to light than their photographic counterpart. This is still one of the most sensitive media for holography, and only requires a fraction of joules per square centimeter of exposure down to millijoules per square centimeter.

Other than the grain-size discrepancy, photographic and holographic emulsions share the same chemistry. The material requires wet processing for the latent image to be revealed and fixed. If no additional steps are taken, the initial hologram has an amplitude structure and the diffraction efficiency can only be of a few percent (see Section 10.3.4). To increase the diffraction efficiency, the hologram needs to be "bleached," which means that an index modulation is generated from the gray-scale image. This happens through yet another wet process, where the silver particles are rehalogenated (Ag → AgH). In this case, the diffraction efficiency can reach several tens of percent.

Silver halide emulsions are sensitive throughout the entire visible spectrum, permitting the use of this material for color holography. Dye sensitizers can even be added to the emulsion to improve the sensitivity in some regions of the spectrum. Nevertheless, each emulsion is generally intended for a particular wavelength region (green, blue, or red) and different emulsions should be used to achieve a full-color hologram.

Silver halide emulsions have benefited from the many years of research and development in the photographic industry, and are still a material of choice for holographic imagery because of their many advantages. One of the last holographic emulsion manufacturers is Russian-based Slavich [33].

11.5.2
Photopolymers

Photopolymers consist of a photoreactive, polymerizable system dispersed within a host. Figure 11.8 shows the steps in the formation of the refractive index modulation in response to an interference pattern. During holographic recording, the optical interference pattern (step 1 in Figure 11.8) initiates a pattern of polymerization in the photoreactive system – polymerization is induced in the light intensity maxima of the interference pattern (steps 2–3) while no polymerization occurs in the nulls. This patterned polymerization sets up a concentration gradient in the unreacted species. Unpolymerized species diffuse (steps 4–6) from the nulls to the maxima of the interference pattern to equalize its concentration in the recording area (steps 6–7), creating a refractive index modulation set up by the difference between the refractive indices of the photosensitive component and the host material (steps 8–9) [34–36].

The polymerization that occurs during the hologram recording process results in changes in the dimensions and the bulk refractive index of the material [37]. Changes such as these, if not controlled, can distort the imaging process, degrade the fidelity with which holographic data can be recovered, and ultimately limit the density of data the material can support. The design of photopolymer media for holographic storage applications must therefore balance the advantages of photopolymers (photosensitivity and large refractive index modulations) against the impact of the changes that accompany the polymerization.

Many photopolymer materials based on various chemistries have been developed in the past. Some require wet processing for the phase structure to be revealed and fixed (Polaroid DMP128® [38]), some only require a thermal post treatment (DuPont Omnidex® [39]), and others need uniform light exposure (Bayer Bayfol-HX® [40]).

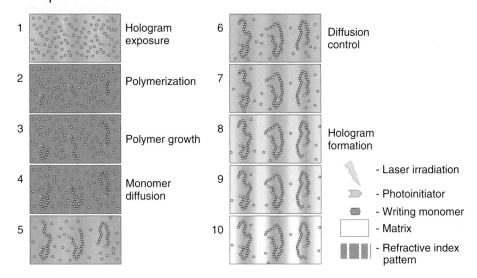

Figure 11.8 Hologram-recording mechanism for photopolymers.

Photopolymers are widely used for security holograms and this is one of the reasons why some companies (like DuPont) no longer distribute their products to private individuals. An alternative to these cutting-edge photopolymers is the use of optical glue. Optical glues are based on the same photo polymerization principle, but are not optimized for holography, which means that their sensitivity is not tuned properly. But that does not prevent enthusiasts from using them as holographic recording materials.

The key advantage of photopolymers is their ease of use due to the simpler postprocessing compared to silver halide emulsion. On the other hand, they are several orders of magnitude less sensitive ($10-10^4$ J cm^{-2}), and their spectral response is biased toward the blue and the UV. Red sensitization is the major advantage of the commercial products, allowing the materials to be used with all wavelengths (panchromatic).

11.5.3
Dichromated Gelatin

Gelatin can be viewed as a *natural* photopolymer. When sensitized by dichromate $(NH_4)_2Cr_2O_7$, chain cross-linking happens due to the formation of Cr^{3+} ions that bind to the carboxylate groups of neighboring chains. This hardens the gelatin and locally changes its index of refraction.

To reveal or amplify the index modulation, the dichromated gelatin must undergo wet processing that involves reducing the dichromate, swelling the gelatin, and then hardening it by immersion in different baths of heated water and organic solvent (usually isopropyl alcohol) [41]. A proper development process can induce an index modulation larger than 10% of the mean index, which is comparatively huge. One of the advantages of this material is that it can be reprocessed; if the quality of the final hologram does not please the author, the wet processing can be redone to further enhance the index modulation or change the level of swelling/shrinkage to modify the color of the hologram.

Dichromated gelatin material is very well adapted to thick volume-phase holograms. The resolution is one of the best with the potential to write 10 000 lines per millimeter. Unfortunately, this material is plagued with several serious problems. The first is its sensitivity; hundreds of millijoules per square centimeter are required during the exposure. This means that a fairly large and expensive laser is needed for any decent size hologram. The wavelength sensitivity is also limited to the blue with some extension to the green (five times less sensitive at 532 nm than at 488 nm). Red can be reproduced for color holograms but it involves shrinkage of the material since the hologram cannot be recorded directly with a red laser. Plates have a limited shelf life since reduction of Cr^{6+} ions to Cr^{3+} occurs even in the dark (though longer life can be obtained by freezing the plate). Lastly, the gelatin is sensitive to humidity and is hygroscopic. Thus, the hologram must be encapsulated to be protected from ambient moisture and bacterial degradation.

11.5.4
Miscellaneous Recording Media

In addition to the three major materials we just introduced, there exist many other holographic recording media. In fact, any material where an index modulation or an amplitude modulation can be photoinduced can be used. The material most well adapted to the desired application must be found.

11.5.4.1 Photothermoplastics

We have already mentioned photoresist materials for surface relief holograms (see Section 10.3.5). Another class of surface relief materials is photothermoplastics. One of the unique properties of these materials is that they are reusable. The process is illustrated in Figure 11.9. Electrical charges are deposited on the top of a thermoplastic film laid down on the top of a photoconductor. When illuminated, charges go through the photoconductor and pair with the charges on

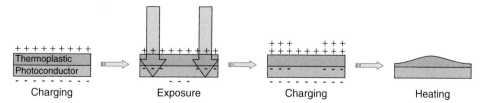

Figure 11.9 Recording procedure in a photothermoplastic material.

the thermoplastic. This creates electrostatic compression due to charge attraction. To further enhance that pressure, additional charging can be applied. When heated, the thermoplastic softens and its thickness is modulated by the electrostatic pressure. A surface relief hologram is thus revealed. To erase this grating, the thermoplastic polymer is uniformly charged and heated up so that pressure flattens the surface.

11.5.4.2 Photochromics

A photochromic material undergoes a change in its absorption coefficient upon illumination. Some doped glasses as well as some dye-doped polymers react this way. Obviously, this kind of material generates amplitude holograms that have limited diffraction efficiency (4%). Some have the advantage of not requiring any development process and they are also reusable since the absorption change is not permanent. Bacteriorhodopsin is even metastable and goes through a cycle of transformation according to the illumination wavelength. This material has also been useful for data storage.

11.5.4.3 Photorefractive

The photorefractive effect is presented in Figure 11.10: it begins by the light-induced generation of pairs of electrical charges inside the material [42]. Charge carriers have different mobility according to their sign. Therefore, the primary charge carrier moves into the material (various mechanisms could be involved here), and gets trapped in the dark regions. A local space charge field is generated due to the spatial redistribution of charge. It is this latter electric field that changes the material's index of refraction through the electro-optic effect (once again, several mechanisms can be involved). The blueprint of the photorefractive effect is the phase shift between the illumination pattern and the resulting index modulation, since it is not the charge location that matters but rather the gradient: the electrical field.

There exist two main classes of photorefractive materials: inorganic crystals and polymers. For holographic imaging, crystals have the disadvantage of being small due to the difficulty of their growth. Polymers, on the other hand, can be spread out on large substrates to give fairly large devices.

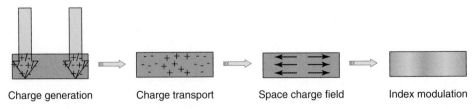

Figure 11.10 The photorefractive effect.

11.6
Material for Data Storage

One of the major challenges in the history of HDS has been the development of a recording material that could fulfill the promise of this technology. Holographic storage requires a material that can rapidly image (capture) the complex optical interference patterns generated during the writing step such that the imaged patterns (i) are generated with adequately high contrast; (ii) are preserved (both dimensionally and in their contrast) during subsequent, spatially colocated writing; (iii) remain unchanged by readout of the holograms; (iv) are robust to environmental effects; (v) can survive over long periods (many years) of time; and (vi) can be inexpensively manufactured in high volume.

Early efforts [43–45] in HDS sought to demonstrate the basic capabilities of the approach – the ability to record and recover data pages with high fidelity and to multiplex to achieve storage density. Because of this initial focus on the feasibility of the technology, these efforts mostly used photorefractive crystals which were well known [9]. While valuable for basic demonstrations of the technology, photorefractive crystals suffered from limited refractive index contrast, slow writing sensitivity, and unwanted volatility where the readout of holograms led to erasure. Attention turned to alternatives such as photochromic/thermoplastic [46, 47], photoaddressable [48], and photopolymer materials where, similarly, optical interference patterns could be imaged as refractive index modulations in thick materials. While other materials such as photographic emulsions can be used to record holograms, they cannot be made thick and therefore have little Bragg selectivity.

Table 11.2 provides a basic comparison of some of the materials that have been used for holographic storage. The postprocessing column implies that some process step (typically a heating or baking process) is required to make the holograms develop or to make them nonvolatile (not erase during readout).

Table 11.2 Basic comparison of materials used for holographic storage.

	Dynamic range (M/#)	Photo sensitivity	Dimensional stability (bulk index/ shrinkage)	Thickness (can make at least 1 mm)	Optical quality (scatter, flatness)	Non volatile	Media postprocessing	Media cost
Photorefractives	X	X	✓	✓	✓		X	XX
Photochromatics	X	XX	✓	✓	-		X	-
Photo-addressibles	X	XX	✓	XX	-		✓	✓
Photopolymers	✓	✓	-	✓	✓		✓	✓

"X" means the parameter is not good enough; "XX" means the parameter is very bad (approximately off desired value by factor of 100); "✓" means good; and "-" means parameter is okay.

Photopolymers [49–51] quickly became the leading candidates as, in addition to their imaging properties, they exhibited high contrast and recording speeds and were nonvolatile. While almost all photopolymers are for write once applications (permanent holograms), they can be made rewritable [52]. The first photopolymer systems to be used were borrowed from the holographic display market where small numbers of holograms were recorded in thin (typically ≤ 10 μm) materials [53, 54]. Again, the additional capabilities of the materials advanced and further established the viability of holographic storage with demonstrations of tens of bits per micron squared [55, 56]. However, because these materials were optimized for display applications, which are relatively insensitive to dimensional changes, optical quality, and scatter, further development of this materials platform was needed to meet the requirements for HDS.

The recording transfer rate depends on the photosensitivity of the material – how well it uses light to image the optical interference patterns. The material's photosensitivity is a measure of how much energy is required to achieve an index of refraction perturbation. The hologram's index perturbation and material thickness determines the hologram's diffraction efficiency. The diffraction efficiency of a hologram is how much of the readout light is diffracted by the hologram to reconstruct the stored data page (signal). For fast read transfer rates, high diffraction efficiency is needed so that the required energy reaches the detector with short integration times. High photosensitivity results in rapid recording rates in that short (fast) exposures from a low-power diode-based laser can generate holograms of sufficient strength to have a usable SNR. The diffraction efficiency or signal strength of each hologram determines the SNR since for most materials and systems the dominant noise source is media scattering. This again highlights the importance of the material's refractive index contrast, thickness, and photosensitivity for a commercially viable system.

The material's M/# is a measure of the achievable diffraction efficiency that a material can support at a given density. The diffraction efficiency of multiplexed holograms scales with the M/# of the material. The M/# scales linearly with the material's maximum index change and the thickness of the media. M/# was precisely derived for photorefractive materials and is now used for all materials defined by the equation,

$$\eta = \frac{M/\#^2}{N^2} \tag{11.1}$$

In the above equation, η is the diffraction efficiency of the holograms, N is the number of holograms multiplexed in the same volume, and M/# is the constant that relates the two quantities. M/# is critical because as holograms are multiplexed (storage density is increased) the diffraction efficiency goes down as $1/N^2$ the number of superimposed holograms. Thus to achieve high density, good SNR, and fast readout rates requires materials with high M/#'s. A material's dynamic range is a function of its M/# and its scatter level. A material with a large M/# (30–60) and low scatter floor is said to have a large dynamic range and should therefore be capable of supporting holograms at high density with good SNR.

The properties of the material such as its dimensional stability, optical quality, scatter level, linearity, and volatility affect the fidelity of the recording and readout process. The dimensional stability of the material is reflected in the resistance of the material to shrinkage or expansion during the recording process and thermal changes. Scatter level of the media, in many cases, determines the required diffraction efficiency that the holograms must be recorded to in order to be sufficiently above the noise level. Since scatter is a coherent noise source, it is particularly important in these systems.

For commercial systems, the manufacturing process and usage must also fit the needs of the application. For example, archival storage demands long (>50 years) lifetimes (i.e., data retention after recording). Adequate shelf life is also required. (The media has to last for several years after being manufactured but before being used to record data. It also has to survive shipping transients before recording.) Thermal or solvent-based post-recording processes (common for display materials) are not viable practices in the data storage marketplace. Finally, the manufacturing must be inexpensive. This requires the use of inexpensive plastic substrates, and fast Takt times for high volume manufacturing.

Photopolymers can meet the requirement for holographic storage and have become the primary focus of development worldwide.

There have been two main approaches to designing photopolymer systems for HDS systems. One approach, pioneered by Polaroid and then later Aprilis, Inc., uses a low-shrinkage chemistry known as a *cationic ring-opening polymerization (CROP)* to control the dimensional changes that occur during holographic recording [57]. These materials consist of a photoacid generator, sensitizing dye, CROP monomers, and binder [58]. The CROP monomers typically have a functionality of greater than 2 and are based on cyclohexeneoxides. In this strategy, a preimaging exposure is used to increase the viscosity of the material to prepare for holographic recording. The CROP imaging chemistry is also tuned for high photosensitivity.

The other approach, introduced by Bell Laboratories, Lucent Technologies, and InPhase Technologies (spun out of Bell Labs), is known as the *two-chemistry approach* [59]. The material is composed of two independently polymerizable systems – one system reacts to form a three-dimensional cross-linked polymer matrix in the presence of the second photopolymerizable monomer system. The photopolymerizable monomer is the imaging component, as it reacts during holographic recording. This strategy produces high-performance recording media as a result of several important attributes. The matrix is formed *in situ*, which allows thick and optically flat formats to be formed. The three-dimensional cross-linked nature of the polymer matrix creates a mechanically robust and stable medium. The matrix and photopolymerizable monomer system are chosen to be compatible in order to yield media with low levels of light scatter. The independence of the matrix and monomer systems avoids cross-reactions between the two that can dilute the refractive index contrast due to premature polymerization of the imaging component.

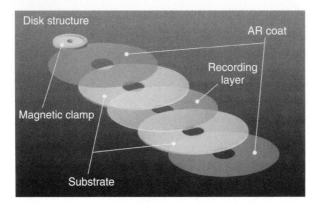

Figure 11.11 Schematic of media construction.

11.7
Media for Holographic Data Storage

A schematic of the Tapestry® media for the professional archival drive is shown in Figure 11.11. The active recording layer (the photopolymer material) is sandwiched between two plastic substrates. Both substrates are coated on their outer surfaces with antireflection (AR) thin films to minimize reflection losses of the incident blue laser light during both recording and readout. (The coating is a nonmetallic and transparent dielectric stack.) A tracking pattern is molded onto the outer surface of bottom substrate and read out through reflection of a servo laser at 850 nm. A metal hub is attached for magnetically clamping the spindle of the drive, and the outer edge of the disk and the center hole area are sealed against moisture.

The two substrates are bonded together using Zerowave® bonding process [60]. This process results in media with flatness of $\lambda/5$ over the disk measured in transmitted wavefront even when using inexpensive plastic substrates that are themselves not flat. This flatness improves SNR and decreases the use of the servo system. The optical quality survives archival testing at very high temperatures (80–100 °C).

The Tapestry media and drive are designed for archival storage of digital assets for use in libraries and jukeboxes. Because of this, the media is housed within a magneto-optical-style cartridge for compatibility with the robotic handling of existing library systems. (In addition, the cartridge protects the unrecorded disk against exposure to ambient light.)

Figure 11.12 is a photograph of the media and the cartridge.

11.8
Conclusions

The commercialization of holographic storage has come a long way. Working systems have now been demonstrated and are near commercial availability. Materials

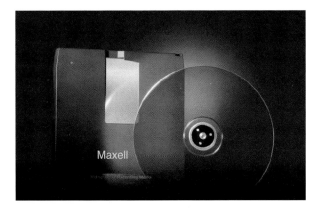

Figure 11.12 Picture of media and cartridge.

for both data storage and display have also improved with particular emphasize on polymer-based materials. The long-term capacity of holographic systems can allow tens of terabytes on optical-disk-sized media. For a more detailed description, see Ref. [61].

References

1. Mitsumori, A. *et al.* (2008) Multi-layer 400 GB optical disk, Poster presented at the Joint International Symposium on Optical Memories and Optical Data Storage, Paper MB01, July 2008, Waikoloa, HI.
2. Bruls, D. *et al.* (2006) Practical and robust near field optical recording systems, Poster presented at the International Symposium on Optical Memories Takamatsu, Paper Mo-C-01, October 2006, Japan.
3. van den Eerenbeemd, J.M.A. *et al.* (2006) Towards a multi-layer near field recording system, dual layer recording results. Poster presented at the International Symposium on Optical Memories Takamatsu, Paper Tu-F-03, October 2006, Japan.
4. Kim, J. *et al.* (2006) The error rate improvement of Super-RENS Disc. Poster presented at the International Symposium on Optical Memories Takamatsu, Paper Mo-B-01, October 2006, Japan.
5. Tominaga, J. and Nakano, T. (2005) *Optical Near-Field Recording – Science and Technology*, Springer-Verlag, New York.
6. Walker, E.P. *et al.* (2008) Terabyte recorded in a two-photon 3D disc. Poster presented at the Joint International Symposium on Optical Memories and Optical Data Storage, Paper MB01, July 2008, Waikoloa, HI.
7. Shipway, A.N. *et al.* (2005) A new media for two-photon volumetric data recording and playback. Poster presented at the Joint International Symposium on Optical Memories and Optical Data Storage, paper MC6, July 2005, Honolulu, HI.
8. Akselrod, M. *et al.* (2007) Progress in bit-wise volumetric optical storage using alumina-based materials. Poster presented at the Optical Data Storage Conference, Paper MA2, Portland, OR.
9. Yariv, A. (1985) *Optical Electronics*, Holt, Rinehart, and Winston, Inc., Orlando, FL.
10. Gabor, D. (1948) A new microscope principle. *Nature*, **161**, 777.
11. Leith, E.N. and Upatnieks, J. (1962) Reconstructed wavefronts and communication theory. *J Opt. Soc. Am.*, **52**, 1123–1128.

12. Fleisher, H. et al. (1965) Optical and Electro-optical information processing. *An Optically Accessed Memory Using the Lippmann Process for Information Storage*, MIT Press.
13. Leith, E.N. et al. (1966) Holographic data storage in three-dimensional media. *Appl. Opt.*, **5**, 1303–1311.
14. Orlic, S. et al. (2008) Microholographic data storage towards dynamic disk recording, Poster presented at the Joint International Symposium on Optical Memories and Optical Data Storage, Paper MB05, July 2008, Waikoloa, HI.
15. McLeod, R.R. et al. (2005) Micro-holographic multi-layer optical disk data storage. Poster presented at the Joint International Symposium on Optical Memories and Optical Data Storage, Paper MB03, July 2005, Honolulu, HI.
16. Horigome, T. et al. (2005) Drive system for micro-reflector recording employing blue laser diode. Poster presented at the International Symposium on Optical Memories Takamatsu, Paper Mo-D-02, October 2005, Japan.
17. Lawrence, B.L. (2008) Micro-holographic storage and threshold holographic recording materials. Poster presented at the Joint International Symposium on Optical Memories and Optical Data Storage, paper MB03, July 2008, Waikoloa, HI.
18. McLeod, R.R. (2009) Impact of phase aberrations on three-dimensional optical data storage in homogeneous media. *J. Opt. Soc. Am. B*, **26**, 308–317.
19. Anderson, K. and Curtis, K. (2004) Polytopic multiplexing. *Opt. Lett.*, **29**, 1402–1404.
20. Curtis, K. and Wilson, W.L. (2006) Architecture and Function of InPhase's Holographic Drive. Poster presented at the Invited Talk, Asia-Pacific Data Storage Conference, Lakeshore Hotel, August 28–30, 2006, Hsinchu, Taiwan.
21. Anderson, K. et al. (2006) High speed holographic data storage at 500Gb/in2. *SMPTE Motion Imaging J.*, 200–203.
22. Hoskins, A. et al. (2008) Monocular architecture. *Jap. J. Appl. Phys.*, **47**, 5912–5914.
23. Hoskins, A. et al. (2007) Monocular architecture. Poster presented at the International Workshop on Holographic Memory, International Workshop on Holographic Memories, October 26–28, 2007, Penang, Malaysia.
24. Shimada, K. et al. (2009) High density recording using Monocular architecture for 500GB consumer system. Poster presented at the Optical Data Storage Conference, Florida.
25. Ahn, S.W. et al. (2005) Fabrication of a 50 nm half-pitch wire grid polarizer using nano-imprint lithography. *Nanotechnology*, **16**, 1874–1877.
26. Chuang, E. et al. (2007) Demonstration of holographic ROM mastering, replication, and playback with a compact reader, Poster presented at the International Workshop on Holographic Memory, October 26–28, Penang, Malaysia.
27. Chuang, E. et al. (2008) Demonstration of holographic read-only-memory mastering, replication, and playback with a compact reader. *Jpn. J. Appl. Phys.*, **47**, 5909–5911.
28. Chuang, E. et al. (2006) Consumer holographic ROM reader with mastering and replication technology. *Opt. Lett.*, **31**, 1050–1052.
29. Ayres, M. et al. (2008) Wobble alignment for angularly multiplexed holograms. ODS/IDSOM Hawaii, Paper ThC01.
30. Hoskins, A. et al. (2006) Temperature compensation strategy for holographic storage. ODS, Paper WC4, Montreal Canada April 23-26.
31. Marchant, A. (1990) *Optical Recording: A Technical Overview*, Addison-Wesley.
32. Ayres M. et al. Image over-sampling for page-oriented optical data storage. *Appl. Opt.*, **45**, 2459–2464.
33. http://www.slavich.ru/.
34. Zhao, G. and Mouroulis, P. (1994) Diffusion model of hologram formation in dry photopolymer materials. *J. Mod. Opt.*, **41**, 1929–1939.
35. Colburn, W.S. and Haines, K.A. (1971) Volume hologram formation in photopolymer materials. *Appl. Opt.*, **10**, 636–641.

References

36. Bloom, A. et al. (1974) Organic recording medium for volume-phase holography. *Appl. Phys. Lett.*, **24**, 612–614.
37. Dhar, L. et al. (1996) Temperature-induced changes in photopolymer volume holograms. *Appl. Phys. Lett.*, **73**, 1337–1339.
38. Guenther, B.D. and Hay, W. (1986) Holographic properties of DMP-128 photopolymers, Final Report, 15 November 1985 – 31 December. Duke University, Durham, NC. Defense Technical Information Center OAI-PMH Repository (United States), AD Number: ADA172825.
39. Gambogi, W.J., Weber, A.M., and Trout, T.J. (1993) Advances and applications of DuPont holographic photopolymers. *Proc. SPIE*, **2043**.
40. Bruder, F.-K., Deuber, F., Facke, T., Hagen, R., Honel, D., Jurberg, D., Kogure, M., Rolle, T., and Weiser, M.-S. (2009) Full-Color self-processing holographic photopolymers with high sensitivity in red – the first class of instant holographic photopolymers. *J. Photopolym. Sci. Technol.*, **22** (2), 257–260.
41. Chang, B.J. and Leonard, C.D. (1979) Dichromated gelatin for the fabrication of holographic optical elements. *Appl. Opt.*, **18** (14), 2407–2417.
42. Günter, P. and Huignard, J.-P. (eds.) (2007) *Photorefractive Materials and Their Applications*, Springer Series in Optical Sciences, Vol. 1–3, Springer.
43. Chen, F.S. (1968) Holographic storage in lithium niobate. *Appl. Opt.*, **13**, 223–225.
44. d'Auria, L. et al. (1974) Experimental holographic read-write memory using 3D storage. *Appl. Opt.*, **13**, 808–818.
45. Staebler, D.L. et al. (1975) Multiple storage and erasure of fixed holograms in Fe-doped LiNbO3. *Appl. Phys. Lett.*, **26**, 182–184.
46. Lorincz, E. et al. (2000) Rewritable holographic memory card system. Poster presented at the Optical Data Storage Conference, Whistler, British Columbia, Canada.
47. Dubois, M. et al. (2005) Characterization of micro-holograms recorded in a thermoplastic medium for three-dimension optical data storage. *Opt. Lett.*, **30**, 1947.
48. Eickmans, J. et al. (1999) Photoaddressable polymers: a new class of materials for optical data storage and holographic memories. *Jap. J. Appl. Phys.*, **38**, 1835–1836.
49. Curtis, K. and Psaltis, D. (1992) Recording of multiple holograms in photopolymer films. *Appl. Opt.*, 7425–7428.
50. Curtis, K. and Psaltis, D. (1994) Characteristics of DuPont photopolymer films for 3-D holographic storage. *Appl. Opt.*, **33**, 5396–5399.
51. Rhee, U.S. et al. (1993) Characteristics of the DuPont photopolymer for angularly multiplexed page oriented holographic memories. *Opt. Eng.*, **32**, 1839–1847.
52. Trentler, T. et al. (2004) Blue-sensitivity rewriteable holographic media, Poster presented at the Optical Data Storage Conference, Monterey, CA.
53. Hay, W.C. and Guenther, B.D. (1988) Characterization of Polaroid's DMP-128 holographic recording photopolymer. *SPIE, Holographic Opt.: Des. Appl.*, **883**, 102–105.
54. Booth, B.L. (1975) Photopolymer material for holography. *Appl. Opt.*, **14**, 593–601.
55. Curtis, K. et al. (1994) High density holographic storage in thin films. OSA Topical Meeting on Optical Data Storage, 10 (OSA Technical Digest Series), Paper TuA4-1, OSA, Washington, DC, p. 43.
56. Pu, A. et al. (1994) Storage density of peristrophic multiplexing. OSA Annual Meeting, 16 (OSA Technical Digest Series), Paper MO5, OSA, Washington, DC, p. 58.
57. Waldman, D.A. et al. (1997) Volume shrinkage in slant fringe gratings of a cationic ring-opening holographic recording material. *J. Imaging Sci. Tech.*, **41**, 497–514.
58. Waldmann, D.A. et al. (2003) CROP holographic storage media for optical data storage at greater than 100 bits/μm^2. *Proc. SPIE*, **5216**, 10.

59. Dhar, L. et al. (1999) Recording media that exhibit high dynamic range for digital holographic data storage. *Opt. Lett.*, **24**, 487–489.
60. (a) Campbell, S. et al. Method for fabricating a multilayer optical article. US Patent 5,932,045; (b) Dhar, L. et al. Method and apparatus for multilayer optical articles. US Patent 7,112,359.
61. Curtis, K. et al. (2010) *Holographic Storage: from Theory to Practical Systems*, John Wiley & Sons, Inc.

12
Phase-Space Rotators and their Applications in Optics

José A. Rodrigo, Tatiana Alieva, and Martin J. Bastiaans

12.1
Introduction

Light propagation through paraxial optical systems (also known as $ABCD$ systems), involving lenses, mirrors, prisms, free-space propagation intervals, gradient-index fibers, and so on, is described by the linear canonical integral transformation. It is associated with affine operations of the signal's phase-space representation (e.g., the Wigner distribution), which provides a description of the signal in terms of position and (spatial) frequency simultaneously. Scaling, shearing, and rotation are examples of affine phase-space operations with important applications in optics and signal processing. The well-known Fourier transformation, for instance, produces a $\pi/2$ rotation of the signal's phase-space representation in the position–frequency plane. Several relevant optical information processing operations, such as convolution and correlation, are directly related to the Fourier transformation.

Rotations in the position–frequency plane through arbitrary angles are associated with the fractional Fourier transformation, which is a generalization of the ordinary Fourier transformation. The fractional Fourier transformation, introduced in optics almost two decades ago [1, 2], opened new perspectives in optical information processing, with attractive applications for beam characterization, phase retrieval, filtering, encryption, and so on. The cascade of a fractional Fourier transformer (FT) together with image rotators describes all possible phase-space rotators.

Paraxial optical systems are easily described in the framework of the ray transformation matrix, which we use for the analysis of phase-space rotators. This approach also permits one to design optical systems for these phase-space operations by using basic elements such as cylindrical lenses and free-space intervals. The implementation of a lens by means of a spatial light modulator leads to a programmable optical setup, which is able to perform phase-space rotations almost at real time – a property that is required for various applications.

The chapter starts with an introduction of the representation of two-dimensional signals in phase space. Then we recall the matrix formalism associated with paraxial optical systems, and, in particular, for phase-space rotators. It allows to establish four basic uniparametric phase-space rotations, from which other attractive operations can be derived. Finally, their optical implementations and main applications are studied.

12.2
Signal Representation in Phase Space: The Wigner Distribution

The Wigner distribution [3] was originally introduced in quantum mechanics as a distribution function that permitted a description of mechanical phenomena in a phase space. After its introduction in optics [4–8], it has been found to be of great use for the description of partially coherent fields.

While the mechanical phase space is associated with position and momentum of particles, the phase space in optics is connected to geometrical optics, where the propagation of optical rays is considered and where we are interested in the position and the direction of an optical ray. In phase space, the Wigner distribution thus represents an optical field in terms of a ray picture, and this representation is independent of whether the light is partially or completely coherent.

A description by means of a Wigner distribution is particularly useful when the optical signals and systems can be described by quadratic-phase functions, that is, when we are in the realm of paraxial approximations. Although formulated in Fourier-optical terms, the Wigner distribution forms a link between such diverse fields as geometrical optics, wave optics, matrix optics, and radiometry, treated in the paraxial approximation.

The reader who is interested in phase-space optics has a number of very useful references available (see, for example, Refs [9–11]).

12.2.1
Description of Partially Coherent Light

We consider the general case of partially coherent light, where the optical field is described by a stochastic process $\tilde{f}(\mathbf{x}, t)$; the two-dimensional column vector $\mathbf{x} = [x, y]^t$ represents the transverse coordinates x and y in a plane $z =$ constant. To describe partially coherent light, we start with the mutual coherence function [12–15]

$$\tilde{\Gamma}(\mathbf{x}_1, \mathbf{x}_2, t_1, t_2) = E\{\tilde{f}(\mathbf{x}_1, t_1)\tilde{f}^*(\mathbf{x}_2, t_2)\} =: \tilde{\Gamma}(\mathbf{x}_1, \mathbf{x}_2, t_1 - t_2) \tag{12.1}$$

where we have assumed that the stochastic process is temporally stationary. After Fourier transforming the mutual coherence function $\tilde{\Gamma}(\mathbf{x}_1, \mathbf{x}_2, \tau)$ with respect to the time difference τ, we get the mutual power spectrum:

$$\Gamma(\mathbf{x}_1, \mathbf{x}_2, \nu) = \int \tilde{\Gamma}(\mathbf{x}_1, \mathbf{x}_2, \tau) \exp(i2\pi\nu\tau) \, d\tau =: \Gamma(\mathbf{x}_1, \mathbf{x}_2) \tag{12.2}$$

Note that we omit the temporal frequency coordinate ν, because in the sequel we restrict ourselves to monochromatic light; in this case, the mutual power spectrum might as well be called the mutual intensity – the term that we use from now on.

12.2 Signal Representation in Phase Space: The Wigner Distribution

We recall that in the completely coherent case, $\Gamma(x_1, x_2)$ takes the product form $f(x_1) f^*(x_2)$.

Important basic examples of coherent signals, as they appear in a plane $z =$ constant, are as follows:

1) an impulse at position x_0, $f(x) = \delta(x - x_0)$. In optical terms, the impulse corresponds to a point source of light;
2) the crossing with that plane of a plane wave with spatial frequency p_0, $f(x) = \exp(i2\pi p_0^t x)$. The plane-wave example shows us how we should interpret the spatial-frequency vector p_0. We assume that the wavelength of the light equals λ, in which case the length of the wave vector k equals $2\pi/\lambda$. If we express the wave vector in the form $k = [k_x, k_y, k_z]^t$, then $2\pi p_0 = 2\pi [p_x, p_y]^t = [k_x, k_y]^t$ is simply the transversal part of k, that is, its projection onto the plane $z =$ constant. Furthermore, if the angle between the wave vector k and the z axis equals θ, then the length of the spatial-frequency vector p_0 equals $\sin\theta/\lambda$;
3) the crossing with that plane of a generalized spherical wave (with quadratic phase, in the paraxial approximation), the curvature of which is described by the real symmetric 2×2 matrix $H = H^t$: $f(x) = \exp(i\pi x^t H x) = \exp[i\pi(h_{xx}x^2 + 2h_{xy}xy + h_{yy}y^2)]$. The simplest case, a spherical wave $\exp[i\pi h(x^2 + y^2)]$, arises for $h_{xx} = h_{yy} = h$ and $h_{xy} = 0$, for which we have the same curvature in all directions; lines in the (x, y) plane with equal phase, determined by the equation $h(x^2 + y^2) =$ constant, then take the form of a circle. In general, we have to deal with the equation $h_{xx}x^2 + 2h_{xy}xy + h_{yy}y^2 =$ constant, which leads to an ellipse (of which the circle is a particular case) or a hyperbola, depending on the entries of the curvature matrix H.

Basic examples of partially coherent signals are as follows:

4) completely incoherent light with "intensity" $I(x)$, $\Gamma(x_1, x_2) = I(x_1) \delta(x_1 - x_2)$. Note that $I(x)$ is a nonnegative function;
5) spatially stationary light, $\Gamma(x_1, x_2) = s(x_1 - x_2)$. We remark that its intensity distribution is constant: $\Gamma(x, x) = s(0)$. We see later that the spatial Fourier transform of $s(x)$, $\hat{s}(p) = \int s(x) \exp(-i2\pi p^t x) \, dx$, is a nonnegative function.

12.2.2
Wigner Distribution

From the mutual intensity $\Gamma(x_1, x_2)$, we define the Wigner distribution $W(x, p)$ as its spatial Fourier transform with respect to the coordinate difference $x' = x_1 - x_2$:

$$W(x, p) = \int \Gamma(x + x'/2, x - x'/2) \exp(-i2\pi p^t x') \, dx' \qquad (12.3)$$

Note that the Wigner distribution is real, which is a direct consequence from the Hermitian symmetry of the mutual intensity, $\Gamma(x_1, x_2) = \Gamma^*(x_2, x_1)$. Moreover, the mutual intensity can be reconstructed from the Wigner distribution simply by applying an inverse Fourier transformation. From the many nice properties of the Wigner distribution we explicitly mention the important property of space and frequency shift covariance: if $W(x, p)$ is the Wigner distribution that corresponds to $\Gamma(x_1, x_2)$, then $W(x - x_0, p - p_0)$ is the Wigner distribution that corresponds to the space- and frequency-shifted version $\Gamma(x_1 - x_0, x_2 - x_0)\exp[i2\pi p_0^t(x_1 - x_2)]$. Additional properties of the Wigner distribution can be found elsewhere (see, for instance, Refs [9,16–24] and the many references cited therein).

Since the Wigner distribution is more or less "in the middle" between the pure space description, that is, in terms of (x_1, x_2), and the pure spatial-frequency description, that is, in terms of (p_1, p_2), we might as well define it in terms of the spatial Fourier transform $\hat{\Gamma}(p_1, p_2)$ of $\Gamma(x_1, x_2)$:

$$\hat{\Gamma}(p_1, p_2) = \iint \Gamma(x_1, x_2) \exp[-i2\pi(p_1^t x_1 - p_2^t x_2)] \, dx_1 \, dx_2 \tag{12.4}$$

the definition that is very much like Eq. (12.3):

$$W(x, p) = \int \overline{\Gamma}(p + p'/2, p - p'/2) \exp(i2\pi x^t p') \, dp' \tag{12.5}$$

There is also another function that is somehow "in the middle" between the pure space and the pure spatial-frequency description, viz., the Fourier transform of the mutual intensity with respect to the coordinate mean $x = (x_1 + x_2)/2$:

$$\int \Gamma(x + x'/2, x - x'/2) \exp(-i2\pi x^t x') \, dx =: A(x', p') \tag{12.6}$$

This Fourier transform is known as the *ambiguity function* [25] and was introduced in optics by Papoulis [26]. We immediately notice a Fourier transform relationship between the Wigner distribution and the ambiguity function:

$$A(x', p') = \iint W(x, p) \exp[-i2\pi(x^t p' - p^t x')] \, dx \, dp \tag{12.7}$$

This Fourier transform relationship implies that properties for the Wigner distribution have their counterparts for the ambiguity function and vice versa: moments for the Wigner distribution become derivatives for the ambiguity function, convolutions in the "Wigner domain" become products in the "ambiguity domain," and so on. In this chapter, we concentrate on the Wigner distribution.

Let us return to our five basic examples. The space behavior $f(x)$ or $\Gamma(x_1, x_2)$, the spatial-frequency behavior $\hat{f}(p)$ or $\hat{\Gamma}(p_1, p_2)$, and the Wigner distribution $W(x, p)$ of (i) a point source, (ii) a plane wave, (iii) a generalized spherical wave, (iv) an incoherent light field, and (v) a spatially stationary light field have been represented in Table 12.1.

We remark the clear physical interpretations of the Wigner distributions.

1) The Wigner distribution of a point source $f(x) = \delta(x - x_0)$ reads $W(x, p) = \delta(x - x_0)$, and we observe that all the light originates from one point $x = x_0$ and propagates uniformly in all directions p.
2) Its dual, a plane wave $f(x) = \exp(i2\pi p_0^t x)$, also expressible in the frequency domain as $\hat{f}(p) = \int f(x) \exp(-i2\pi p^t x) \, dx = \delta(p - p_0)$, has as its Wigner distribution $W(x, p) = \delta(p - p_0)$, and we observe that for all positions x the light propagates in only one direction, that is p_0.

Table 12.1 Wigner distribution of some basic examples: (i) point source, (ii) plane wave, (iii) generalized spherical wave, (iv) incoherent light, and (v) spatially stationary light.

	$f(x)$	$\hat{f}(p)$	$W(x, p)$
(i)	$\delta(x - x_0)$	$\exp(-i2\pi x_0^t p)$	$\delta(x - x_0)$
(ii)	$\exp(i2\pi p_0^t x)$	$\delta(p - p_0)$	$\delta(p - p_0)$
(iii)	$\exp(i\pi x^t H x)$	$[\det(-iH)]^{-1/2} \exp(-i\pi p^t H^{-1} p)$	$\delta(p - Hx)$
	$\Gamma(x_1, x_2)$	$\hat{\Gamma}(p_1, p_2)$	$W(x, p)$
(iv)	$I(x_1)\, \delta(x_1 - x_2)$	$\hat{I}(p_1 - p_2)$	$I(x)$
(v)	$s(x_1 - x_2)$	$\hat{s}(p_1)\, \delta(p_1 - p_2)$	$\hat{s}(p)$

3) The Wigner distribution of the generalized spherical wave $f(x) = \exp(i\pi x^t H x)$ takes the simple form $W(x, p) = \delta(p - Hx)$, and we conclude that at any point x only one frequency $p = Hx$ manifests itself. This corresponds exactly to the ray picture of a spherical wave.

4) Incoherent light, $\Gamma(x + x'/2, x - x'/2) = I(x)\, \delta(x')$, yields the Wigner distribution $W(x, p) = Ix$. Note that it is a function of the space variable x only, and that it does not depend on the frequency variable p: the light radiates equally in all directions, with intensity profile $I(x) \geq 0$.

5) Spatially stationary light, $\Gamma(x + x'/2, x - x'/2) = s(x')$, is dual to incoherent light: its frequency behavior is similar to the space behavior of incoherent light and vice versa, and $\hat{s}(p)$, its intensity function in the frequency domain, is nonnegative. The duality between incoherent light and spatially stationary light is, in fact, the Van Cittert–Zernike theorem. The Wigner distribution of spatially stationary light reads as $W(x, p) = \hat{s}(p)$; note that it is a function of the frequency variable p only, and that it does not depend on the space variable x. It thus has the same form as the Wigner distribution of incoherent light, except that it is rotated through $90°$ in the space-frequency domain. The same observation can be made for the point source and the plane wave; see examples (1) and (2), which are also each other's duals.

In the next section, we consider the transformation of the Wigner distribution during its propagation through paraxial optical systems.

12.3
Matrix Formalism for the Description of Phase-Space Rotations

The propagation of monochromatic light through first-order optical systems is described by a linear canonical integral transformation, represented by the operator $\mathcal{R}^T$, and parametrized by the well-known real 4×4 ray transformation matrix T [27] (see also Chapter 1), which relates the position x_i and direction p_i of an incoming ray to the position x_0 and direction p_0 of the outgoing ray:

$$\begin{bmatrix} x_0 \\ p_0 \end{bmatrix} = \begin{bmatrix} A & B \\ C & D \end{bmatrix} \begin{bmatrix} x_i \\ p_i \end{bmatrix} = T \begin{bmatrix} x_i \\ p_i \end{bmatrix} \tag{12.8}$$

Here, we use normalized coordinates $xs^{-1/2} =: x$ and $ps^{1/2} =: p$, where $s^{1/2}$ has the dimension of length, resulting in dimensionless submatrices B and C.

The evolution of the complex field amplitude $f(x)$ during its propagation through a first-order optical system is thus described by

$$f_0(x_0) = \mathcal{R}^T[f_i(x_i)](x_0) = \int f_i(x_i) \, K_T(x_i, x_0) \, dx_i \tag{12.9}$$

where, for $\det B \neq 0$, the kernel $K_T(x_i, x_0)$ is given by [28]

$$K_T(x_i, x_0) = (\det iB)^{-1/2}$$
$$\times \exp\left[i\pi \left(x_i^t B^{-1} A x_i - 2 x_i^t B^{-1} x_0 + x_0^t D B^{-1} x_0\right)\right] \tag{12.10}$$

while for the singular case $B = 0$, corresponding to generalized imaging, it reduces to

$$K_T(x_i, x_0) = |\det A|^{-1/2} \exp\left(i\pi x_0^t CA^{-1} x_0\right) \delta\left(x_i - A^{-1} x_0\right) \tag{12.11}$$

For partially coherent light, we have to describe its propagation through a first-order optical system in terms of the mutual intensity; instead of Eq. (12.9), we then get the expression

$$\Gamma_0(x_{0,1}, x_{0,2}) = \iint \Gamma_i(x_{i,1}, x_{i,2}) \, K_T(x_{i,1}, x_{0,1}) \, K_T^*(x_{i,2}, x_{0,2}) \, dx_{i,1} \, dx_{i,2} \tag{12.12}$$

In terms of Wigner distributions, the input–output relationship for a first-order optical system takes the elegant form

$$W_0(Ax + Bp, Cx + Dp) = W_i(x, p) \tag{12.13}$$

which is independent of a possible degeneracy of the submatrix B and which holds for completely coherent as well as partially coherent light.

The ray transformation matrix T is symplectic:

$$T^{-1} = \begin{bmatrix} A & B \\ C & D \end{bmatrix}^{-1} = \begin{bmatrix} D^t & -B^t \\ -C^t & A^t \end{bmatrix} = \begin{bmatrix} 0 & I \\ -I & 0 \end{bmatrix} T^t \begin{bmatrix} 0 & -I \\ I & 0 \end{bmatrix} \tag{12.14}$$

We recall that, owing to this symplecticity, the 4×4 matrix T is described by only 10 free parameters.

On the basis of the modified Iwasawa decomposition [29–31], any ray transformation matrix T can be written as a product of three such matrices,

$$\begin{bmatrix} A & B \\ C & D \end{bmatrix} = \begin{bmatrix} I & 0 \\ -G & I \end{bmatrix} \begin{bmatrix} S & 0 \\ 0 & S^{-1} \end{bmatrix} \begin{bmatrix} X & Y \\ -Y & X \end{bmatrix} =: T_L T_S T_0 \tag{12.15}$$

where

$$\begin{aligned} G &= -\left(C A^t + D B^t\right)\left(A A^t + B B^t\right)^{-1} &= G^t \\ S &= \left(A A^t + B B^t\right)^{1/2} &= S^t \\ X + iY &= \left(A A^t + B B^t\right)^{-1/2} (A + iB) &= \left(X^t - iY^t\right)^{-1} \end{aligned} \tag{12.16}$$

The first matrix in the decomposition, T_L, corresponds to a generalized lens, which performs an anamorphic quadratic-phase modulation. The scaling operation, described by the second matrix T_S, is also well understood. The last matrix – which is orthosymplectic, that is, both orthogonal $(T_0^t = T_0^{-1})$ and symplectic (see Eq. (12.14)) – is a general expression for a variety of attractive transformations [31–33], corresponding to rotations in phase space. These phase-space rotators can be elegantly described by the unitary matrix $U = X + iY = (U^*)^{-1}$. It is easy to verify that for a phase-space rotator the general input–output relation (12.8) can be replaced by the shorter version $(x - ip)_o = U\,(x - ip)_i$.

Let us formulate an important property for the projection of the Wigner distribution onto a plane in the four-dimensional phase space. Note that the projection onto the x plane results in the intensity of the signal: $\int W(x, p)\,dp = \Gamma(x, x)$. If $W_o(x, p)$ is the Wigner distribution that arises after a linear canonical transformation, then $\int W_o(x, p)\,dp = \Gamma_o(x, x)$ can be written as

$$\Gamma_o(x_0, x_0) = \iint W_o(x', p')\,\delta(x' - x_0)\,dx'\,dp'$$
$$= \iint W_i(D^t x' - B^t p', -C^t x' + A^t p')\,\delta(x' - x_0)\,dx'\,dp'$$
$$= \iint W_i(x, p)\,\delta(Ax + Bp - x_0)\,dx\,dp \quad (12.17)$$

where we have substituted W_o from Eq. (12.13). If the ray transformation matrix is orthogonal, T_0, the latter expression corresponds to a projection of the original Wigner distribution $W_i(x, p)$ onto the x_0 plane instead of the x plane, where $Ax + Bp = x_0$ and $Cx + Dp = p_0 = 0$. Indeed, for an orthogonal $A\,B\,C\,D$ matrix, the orthogonality of $[x^t, 0^t]$ and $[0^t, p^t]$ implies the orthogonality of $[x^t D, -x^t C]$ and $[-p^t B, p^t A]$. As an example, when $A = D = 0$ and $B = -C = I$, which is associated with the ordinary Fourier transformation, we get the projection $\iint W_i(x, p)\,\delta(p - x_0)\,dx\,dp = \int W_i(x, x_0)\,dx = \hat{\Gamma}_i(x_0, x_0)$, which is the intensity of the Fourier transform.

12.4
Basic Phase-Space Rotators for Two-Dimensional Signals

The ray transformation matrix of a phase-space rotator takes the form

$$T_0 = \begin{bmatrix} X & Y \\ -Y & X \end{bmatrix} \quad (12.18)$$

and has only four free parameters according to the symplecticity conditions

$$XY^t = YX^t, \quad X^t Y = Y^t X, \quad XX^t + YY^t = I, \quad (12.19)$$

which follow directly from Eq. (12.14). There are four basic uniparametric phase-space rotators: the symmetric fractional FT $\mathcal{F}(\gamma, \gamma)$ and the antisymmetric fractional FT $\mathcal{F}(\gamma, -\gamma)$, associated with the unitary matrices $U_f(\gamma, \gamma)$ and

$U_f(\gamma, -\gamma)$, respectively, which are special cases of the separable fractional FT $\mathcal{F}(\gamma_x, \gamma_y)$ with diagonal matrix

$$U_f(\gamma_x, \gamma_y) = \begin{bmatrix} \exp(i\gamma_x) & 0 \\ 0 & \exp(i\gamma_y) \end{bmatrix} \qquad (12.20)$$

and the image rotator $\mathcal{R}(\alpha)$ and the gyrator $\mathcal{G}(\beta)$, associated with the matrices

$$U_r(\alpha) = \begin{bmatrix} \cos\alpha & \sin\alpha \\ -\sin\alpha & \cos\alpha \end{bmatrix} \quad \text{and} \quad U_g(\beta) = \begin{bmatrix} \cos\beta & i\sin\beta \\ i\sin\beta & \cos\beta \end{bmatrix} \qquad (12.21)$$

respectively. The general separable fractional FT $\mathcal{F}(\gamma_x, \gamma_y)$ can always be represented as a cascade of symmetric and antisymmetric FTs, $\mathcal{F}(\gamma_s, \gamma_s)$ and $\mathcal{F}(\gamma_a, -\gamma_a)$, where $\gamma_x = \gamma_s + \gamma_a$ and $\gamma_y = \gamma_s - \gamma_a$. The separable fractional FT $\mathcal{F}(\gamma_x, \gamma_y)$ describes rotations in the (x, p_x) and (y, p_y) planes through angles γ_x and γ_y, respectively. The image rotator $\mathcal{R}(\alpha)$ is associated with an anticlockwise rotation in the (x, y) and (p_x, p_y) planes through an angle α, and the gyrator $\mathcal{G}(\beta)$ corresponds to a rotation in the twisted (x, p_y) and (y, p_x) planes through an angle β. All four basic uniparametric phase-space rotators are additive with respect to their parameters. Combinations of the basic phase-space rotators may also be additive in some of their parameters. We mention, in particular, the cascade $\mathcal{R}(-\alpha)\,\mathcal{F}(\gamma_x, \gamma_y)\,\mathcal{R}(\alpha)$, which corresponds to a rotated separable fractional FT and which is additive with respect to γ_x and γ_y when the rotation angle α is fixed. Its unitary matrix $U(\alpha, \gamma_x, \gamma_y)$ follows readily from $U_r(-\alpha)\,U_f(\gamma_x, \gamma_y)\,U_r(\alpha)$ and takes the form

$$\begin{bmatrix} \cos^2\alpha \exp(i\gamma_x) + \sin^2\alpha \exp(i\gamma_y) & \sin\alpha\cos\alpha\,[\exp(i\gamma_x) - \exp(i\gamma_y)] \\ \sin\alpha\cos\alpha\,[\exp(i\gamma_x) - \exp(i\gamma_y)] & \sin^2\alpha \exp(i\gamma_x) + \cos^2\alpha \exp(i\gamma_y) \end{bmatrix} \qquad (12.22)$$

Any basic phase-space rotator can always be realized as a cascade of the others. As an example, we mention the cascade $\mathcal{R}(-\pi/4)\,\mathcal{F}(\gamma, -\gamma)\,\mathcal{R}(\pi/4)$, which leads to the gyrator $\mathcal{G}(\gamma)$. To realize an arbitrary phase-space rotator, we thus have many possibilities. We mention, in particular, $\mathcal{F}(\gamma/2, -\gamma/2)\,\mathcal{R}(\varphi)\,\mathcal{F}(-\gamma/2, \gamma/2)\,\mathcal{F}(\gamma_x, \gamma_y)$, which follows directly from the unitary matrix U, if we represent it as

$$U = \begin{bmatrix} \exp(i\gamma_x)\cos\alpha & \exp[i(\gamma_y + \gamma)]\sin\alpha \\ -\exp[i(\gamma_x - \gamma)]\sin\alpha & \exp(i\gamma_y)\cos\alpha \end{bmatrix} \qquad (12.23)$$

Another useful combination with which any phase-space rotator can be realized is $\mathcal{R}(\beta)\,\mathcal{F}(\gamma_x, \gamma_y)\,\mathcal{R}(\alpha)$, i.e., a separable fractional FT embedded in between two image rotators [34]. Some combinations that lead to attractive phase-space rotators are $\mathcal{R}(\alpha)\,\mathcal{F}(\pi, 0)$, a cascade of a rotator and an image reflector (along the x-axis), and $\mathcal{R}(\pi/2)\,\mathcal{F}(\pi + \gamma_1, \gamma_2)$, which lead to the unitary matrices

$$U_1(\alpha) = \begin{bmatrix} -\cos\alpha & \sin\alpha \\ \sin\alpha & \cos\alpha \end{bmatrix} \quad \text{and} \quad U_2(\gamma_1, \gamma_2) = \begin{bmatrix} 0 & \exp(i\gamma_y) \\ \exp(i\gamma_x) & 0 \end{bmatrix} \qquad (12.24)$$

respectively. Note that the latter two phase-space rotators are not additive with respect to their parameters.

12.4 Basic Phase-Space Rotators for Two-Dimensional Signals

The kernel $K_f^{\gamma_x,\gamma_y}(\mathbf{x}_i, \mathbf{x}_0)$ of the fractional Fourier transformation is separable for orthogonal coordinates: $K_f^{\gamma_x,\gamma_y}(\mathbf{x}_i, \mathbf{x}_0) = K_f^{\gamma_x}(x_i, x_0) \, K_f^{\gamma_y}(y_i, y_0)$, where

$$K_f^{\gamma_x}(x_i, x_0) = (i \sin \gamma_x)^{-1/2} \exp\left[i\pi \frac{(x_0^2 + x_i^2)\cos\gamma_x - 2x_i x_0}{\sin \gamma_x}\right] \quad (12.25)$$

with $\gamma_x \neq n\pi$. For $\gamma_x = \pi/2$, we have the normal Fourier transformation kernel, except for the phase factor $i^{-1/2}$. To compensate for that and to get a transformation that has the common 2π periodicity with respect to γ_x, we have to consider a slightly modified kernel, where an additional phase factor $\exp(i\gamma_x/2)$ has been added. The separable fractional FT then takes the form

$$F^{\gamma_x,\gamma_y}(\mathbf{x}_0) = \exp[i(\gamma_x + \gamma_y)/2] \int f(\mathbf{x}_i) \, K_f^{\gamma_x,\gamma_y}(\mathbf{x}_i, \mathbf{x}_0) \, d\mathbf{x}_i \quad (12.26)$$

with $\gamma_x, \gamma_y \neq n\pi$. For $\gamma_x = \gamma_y = 0$, the fractional Fourier transformation corresponds to the identity transformation $F^{0,0}(\mathbf{x}) = f(\mathbf{x})$; in general, for $\gamma_{x,y} = n_x, y\pi$, we get $F^{n_x\pi, n_y\pi}(x, y) = f[(-1)^{n_x}x, (-1)^{n_y}y]$, and coordinate reversion (image reflection) along one or both axes occurs for odd values of $n_{x,y}$. The properties of the fractional Fourier transformation have been studied in detail in Ref. [35].

We remark that adding the phase factor $\exp[i(\gamma_x + \gamma_y)/2]$ ensures the recovery of the ordinary Fourier transformation and its inverse for $\gamma_{x,y} = \pm\pi/2$. Nevertheless, except for some particular cases, this issue in the kernel definition is not important for most of the applications involving phase-space rotators. Besides, the matrix formalism avoids such a difference in the kernel definition of the fractional Fourier transformation.

In the case of the gyrator operation, Eq. (12.9) takes the explicit form:

$$F^{\gamma}(x_0, y_0) = \frac{1}{|\sin \gamma|} \iint f_i(x_i, y_i)$$
$$\times \exp\left[i2\pi \frac{(x_0 y_0 + x_i y_i)\cos\gamma - (x_i y_0 + x_0 y_i)}{\sin\gamma}\right] dx_i \, dy_i \quad (12.27)$$

with $\gamma \neq n\pi$. For $\gamma = 0$, it corresponds to the identity transformation, $F^0(\mathbf{x}) = f(\mathbf{x})$; in general, for $\gamma = n\pi$, we get $F^{n\pi}(\mathbf{x}) = f[(-1)^n(\mathbf{x})]$, and coordinate reversion (image reflection) along both axes occurs for odd values of n. For $\gamma = \pi/2$ it reduces to the antisymmetric fractional Fourier transformation $\mathcal{F}(\pi/2, -\pi/2)$, followed by an additional rotation $\mathcal{R}(-\pi/2)$ of the output coordinates through an angle $-\pi/2$; indeed, $U_g(\pi/2) = U_r(-\pi/2) \, U_f(\pi/2, -\pi/2)$. We could as well use the relationship $U_g(\pi/2) = U_r(-\pi/4) \, U_f(\pi/2, -\pi/2) \, U_r(-\pi/4)$, corresponding to $\mathcal{G}(\gamma) = \mathcal{R}(-\pi/4) \, \mathcal{F}(\gamma, -\gamma) \, \mathcal{R}(\pi/4)$ with $\gamma = \pi/2$, and use rotations of both the input and output coordinates through an angle $\pi/4$ and $-\pi/4$, respectively.

While the kernel of the fractional FT is given as a product of plane and elliptic waves, the gyrator kernel is given as a product of plane and hyperbolic waves. Its main properties and applications have been analyzed in detail recently [36–38].

In the next section, we present an optical setup with which it is possible to realize an arbitrary phase-space rotator that can be decomposed as $\mathcal{R}(-\alpha) \, \mathcal{F}(\gamma_x, \gamma_y) \, \mathcal{R}(\alpha)$ and for which the unitary matrix $U(\alpha, \gamma_x, \gamma_y) = U_r(-\alpha) \, U_f(\gamma_x, \gamma_y)$

$U_r(\alpha) = X + iY$ takes the form (12.22). The general formula (12.10) for the kernel in this case reduces to

$$K^{\alpha,\gamma_x,\gamma_y}(x_i, x_0) = (\det iY)^{-1/2}$$
$$\times \exp\left[i\pi \left(x_i^t Y^{-1} X x_i - 2x_i^t Y^{-1} x_0 + x_0^t X Y^{-1} x_0\right)\right] \quad (12.28)$$

with

$$Y^{-1} = \begin{bmatrix} \dfrac{\cos^2 \alpha}{\sin \gamma_x} + \dfrac{\sin^2 \alpha}{\sin \gamma_y} & \dfrac{\sin 2\alpha}{2 \sin \gamma_x} - \dfrac{\sin 2\alpha}{2 \sin \gamma_y} \\ \dfrac{\sin 2\alpha}{2 \sin \gamma_x} - \dfrac{\sin 2\alpha}{2 \sin \gamma_y} & \dfrac{\sin^2 \alpha}{\sin \gamma_x} + \dfrac{\cos^2 \alpha}{\sin \gamma_y} \end{bmatrix} \quad (12.29)$$

$$Y^{-1}X = XY^{-1}$$
$$= \begin{bmatrix} \dfrac{\cos \gamma_x \cos^2 \alpha}{\sin \gamma_x} + \dfrac{\cos \gamma_y \sin^2 \alpha}{\sin \gamma_y} & \dfrac{\cos \gamma_x \sin 2\alpha}{2 \sin \gamma_x} - \dfrac{\cos \gamma_y \sin 2\alpha}{2 \sin \gamma_y} \\ \dfrac{\cos \gamma_x \sin 2\alpha}{2 \sin \gamma_x} - \dfrac{\cos \gamma_y \sin 2\alpha}{2 \sin \gamma_y} & \dfrac{\cos \gamma_x \sin^2 \alpha}{\sin \gamma_x} + \dfrac{\cos \gamma_y \cos^2 \alpha}{\sin \gamma_y} \end{bmatrix}$$
$$(12.30)$$

and with $\det iY = -\sin \gamma_x \sin \gamma_y$. In particular, we obtain the antisymmetric fractional FT $\mathcal{F}(\gamma, -\gamma)$ (for $\alpha = 0$ and $\gamma_x = -\gamma_y = \gamma$) and the gyrator $\mathcal{G}(\beta)$ (for $\alpha = \pi/4$ and $\gamma_x = -\gamma_y = \beta$).

12.5
Optical System Design for Phase-Space Rotators and their Experimental Implementations

The fractional FT is a phase-space rotator that has attracted major attention in recent years. Different systems for the optical implementation of this operation have been proposed (see, for example, Refs [39–41] and the references cited therein). Some of them are based on the application of glass lenses, while in the others spatial light modulators (SLMs) are used for lens implementation. In many applications, phase-space rotators with tunable transformation angles are required. Since a variation of the distance between optical elements usually leads to system misalignment, it is preferable to fix their positions. In such systems, a transformation parameter can be changed only by lens power variation. Changing lenses, however, is also inappropriate for the same reason of misalignment and because of the large number of required items. Nevertheless, for certain phase-space rotators, a change in the transformation parameter can be achieved by a mere rotation of the cylindrical lenses around the optical axis. Another possibility is to use digital lenses implemented by SLMs. Recent developments of SLM technology (e.g., a pixel size of 8 μm and a dynamic range of 256 levels) allow one to achieve an appropriate phase modulation for a large range of focal distances (see Chapter 9).

12.5 Optical System Design for Phase-Space Rotators and their Experimental Implementations

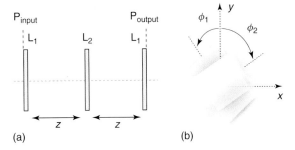

Figure 12.1 (a) Optical setup scheme for the phase-space rotator $\mathcal{R}(-\alpha)\,\mathcal{F}(\gamma_x,\gamma_y)\,\mathcal{R}(\alpha)$. Note that the separable fractional FT is obtained when $\alpha = 0$. P_{input} and P_{output} are the input and output planes. Free-space intervals z are fixed. (b) Each of the generalized lenses L_1 and L_2 is an assembled set of two cylindrical lenses, located at rotation angles ϕ_1 and ϕ_2, respectively.

Moreover, the application of SLMs may provide an almost real-time variation in the transformation parameters, which is required for some tasks.

It has been shown [42, 43] that a minimum of three lenses (here we are speaking about generalized lenses, to be discussed below) are needed to construct an optical setup with fixed lens positions that can perform a separable fractional Fourier transformation of tunable order. It is only in this flexible setup that a change in the fractional angle does not cause an additional scaling and/or phase factor – the effects that occur for other proposed systems [39–41].

Specifically, this fractional Fourier transformation system consists of three generalized lenses with a fixed distance z between them, as displayed in Figure 12.1a. The first and the last lenses are identical ($L_3 = L_1$). Each generalized lens L_j ($j = 1, 2$) is an assembled set of two crossed cylindrical lenses, active in the two orthogonal directions x and y, with phase modulation functions $\exp[-i\pi g_x^{(j)} z\,(s/\lambda z)\,(x^2/s)]$ and $\exp[-i\pi g_y^{(j)} z\,(s/\lambda z)\,(y^2/s)]$, respectively, where we still have the possibility of choosing a proper normalization parameter s; the lens powers $g_{x,y}^{(j)}$ are given by

$$g_{x,y}^{(1)} z = 1 - (\lambda z/s) \cot(\gamma_{x,y}/2)$$
$$g_{x,y}^{(2)} z = 2 - (s/\lambda z) \sin \gamma_{x,y}$$
(12.31)

With reference to Figure 12.2b, the cylindrical lenses are thus oriented such that $\phi_1^{(1,2)} = 0$ and $\phi_2^{(1,2)} = \pi/2$, where the angles are measured counterclockwise and $\phi = 0$ corresponds to the y axis.

For the normalization parameter $s = 2\lambda z$, we get the operation curves $g_{x,y}^{(1)} z = 1 - \cot(\gamma_{x,y}/2)/2$ and $g_{x,y}^{(2)} z = 2 - 2 \sin \gamma_{x,y}$, which are depicted in Figure 12.2. Although $\gamma_{x,y}$ may take any value, $\gamma_{x,y} \in (0, 2\pi)$, we restrict it to the interval $\gamma_{x,y} \in [\pi/2, 3\pi/2]$ that corresponds to convergent lenses, in which case the second operation curve remains invertible; at the same time, the first operation curve remains bounded. This interval is sufficient for a large list of fractional FT applications. Nevertheless, the entire interval $\gamma_{x,y} \in (0, 2\pi)$ can be covered, if necessary, thanks to the relation $F^{\gamma_x+\pi,\gamma_y+\pi}(\mathbf{x}) = F^{\gamma_x,\gamma_y}(-\mathbf{x})$.

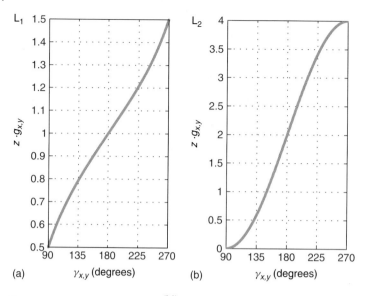

Figure 12.2 Operation curves $z \cdot g_{x,y}^{(1,2)}$ for the generalized lenses L_1 and L_2 are displayed in (a) and (b), respectively. Normalization parameter is $s = 2\lambda z$.

Note that the phase-space rotator $R(-\alpha)\,\mathcal{F}(\gamma_x, \gamma_y)\,R(\alpha)$, with unitary matrix described by the expression (12.22), can be easily achieved by rotating the above-mentioned fractional Fourier transform system through an angle α around the optical axis. In other words (see also Figure 12.1b), the cylindrical lenses are now oriented according to $\phi_1^{(1,2)} = \alpha$ and $\phi_2^{(1,2)} = \alpha + \pi/2$. Thus the phase modulation function associated with each generalized lens L_j ($j = 1, 2$) takes the form

$$\Psi^{(j)}(x, y) = \exp\left[-i\pi g_x^{(j)} z \frac{s}{\lambda z}\left(\frac{x\cos\alpha - y\sin\alpha}{\sqrt{s}}\right)^2\right]$$

$$\times \exp\left[-i\pi g_y^{(j)} z \frac{s}{\lambda z}\left(\frac{y\cos\alpha + x\sin\alpha}{\sqrt{s}}\right)^2\right] \quad (12.32)$$

This optical configuration permits one to perform various attractive operations. For example, for $\alpha = 0$, as described above, we recover the fractional Fourier transform setup associated with Eq. (12.25), whereas for $\alpha = \pi/4$ and $\gamma_x = -\gamma_y = \gamma$, we obtain the gyrator operation Eq. (12.27). It has been tacitly assumed that the generalized lenses in the above-mentioned setup are implemented by SLMs, which allows a fast modification of the transformation angles α, γ_x, and γ_y. The feasibility of such a setup for the case of fractional Fourier transformation has been demonstrated experimentally [43].

We remark that in the special case $\gamma_x = -\gamma_y = \gamma$, the corresponding setup can also be built using glass cylindrical lenses (fixed power). This subclass of phase-space rotators includes the gyrator and the antisymmetric fractional FT. Each generalized lens L_j ($j = 1, 2$) of such a setup consists again of two cylindrical lenses, but now with fixed power $g_{x,y}^{(j)} = j/z$ and no longer crossing at a fixed angle of $\pi/2$. The transversal axes of the cylindrical lenses form angles $\phi_1^{(j)} = \varphi^{(j)} + \alpha + \pi/4$ and $\phi_2^{(j)} = -\varphi^{(j)} + \alpha - \pi/4$ with the y axis; note that the two cylindrical lenses cross at an angle $\phi_1^{(j)} - \phi_2^{(j)} = 2\varphi^{(j)} + \pi/2$. The angles $\varphi^{(1,2)}$ follow from $\sin 2\varphi^{(1)} = (\lambda z/s)\cot(\gamma/2)$ and $2\sin 2\varphi^{(2)} = (s/\lambda z)\sin\gamma$. From the requirement $|\cot(\gamma/2)| \leq 1$, we conclude that the angle interval $\gamma \in [\pi/2, 3\pi/2]$ is covered again if $\lambda z/s = 1$. This scheme (with normalization parameter $s = \lambda z$) has been used for the experimental realization of the gyrator reported in Ref. [38].

Many applications such as beam characterization, phase retrieval, chirp detection, phase-space tomography, and so on, demand the acquisition of fractional Fourier power spectra for various angles, associated with intensity distributions. For this purpose, the third generalized lens ($L_3 = L_1$) is not required. The optical setup for the measurement of fractional Fourier power spectra is displayed in Figure 12.3. Two reflective SLMs (SLM 1 and SLM 2) operating in the phase-only modulation regime have been used for lens implementation. The input signal is projected on SLM 1 where the phase $\arg[\Psi^{(1)}(x, y)]$ associated with the first generalized lens is addressed. At a distance corresponding to the optical path z, SLM 2 is located, which implements the second generalized lens. Finally, the intensity distribution of the output signal is registered by a CCD camera (see Figure 12.3). This setup allows one to change the fractional order at almost real time and to store the measured fractional Fourier power spectra as a video file [43]. The same setup, but with generalized lenses given by Eq. (12.32), can be used for the measurement of the power spectra corresponding to the rest of the phase-space rotators described by Eq. (12.22).

Whereas for the above-mentioned phase-space rotators, flexible optical schemes that contain only three generalized lenses exist, four generalized lenses are needed

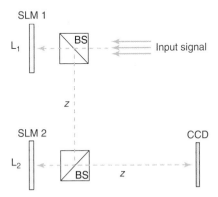

Figure 12.3 Experimental setup for the measurement of fractional Fourier power spectra. The generalized lenses L_1 and L_2 are addressed into SLM 1 and SLM 2, respectively. The SLMs are set at reflection geometry operating in phase-only modulation. The intensity distribution, associated with the corresponding Wigner distribution projection, is registered by a CCD camera.

for the realization of a tunable image rotator. The details for the construction of this setup can be found in Ref. [42].

12.6
Applications of Phase-Space Rotators in Optics

Phase-space rotators are used for beam characterization, system analysis, optical information processing, and so on. For instance, the measurement of fractional Fourier power spectra permits one to reconstruct the Wigner distribution of fields without interferometric or iterative techniques (see Chapter 35). Here, we discuss the applications of these operations for mode conversion, phase retrieval, and chirp detection.

Gaussian beams

$$\text{HG}_{m,n}(\mathbf{x}; w) = 2^{1/2} \frac{H_m\left(\sqrt{2\pi}\, x/w\right) H_n\left(\sqrt{2\pi}\, y/w\right)}{\sqrt{2^m m!\, w} \sqrt{2^n n!\, w}} \exp\left(-\pi \frac{r^2}{w^2}\right) \quad (12.33)$$

and

$$\text{LG}_{p,l}^\pm(\mathbf{x}; w) = w^{-1} \left(\frac{p!}{(p+l)!}\right)^{1/2} \left(\sqrt{2\pi}\, \frac{x \pm iy}{w}\right)^l L_p^l\left(2\pi \frac{r^2}{w^2}\right) \exp\left(-\pi \frac{r^2}{w^2}\right) \quad (12.34)$$

modulated by Hermite H_m, or Laguerre L_p^l polynomials, where p and l are the radial and azimuthal indices, $r^2 = x^2 + y^2$, and w is the beam waist [37, 38], play an important role in optics. These modes arise naturally from laser resonators [44] or can be generated by using computer-generated holograms. Both kinds of beams are eigenfunctions of the symmetric fractional Fourier transformation. This guarantees that their intensity distributions do not change, except for a scaling, during beam propagation in a homogeneous medium. The modes $\text{HG}_{m,n}$ can be transformed into $\text{LG}_{p,l}$ and vice versa, where $p = \min\{m, n\}$ and $l = |m - n|$, under the gyrator operation with angle $(2k+1)\pi/4$.

In general, other stable modes can be generated by using a phase-space rotator described by the unitary matrix

$$U(\gamma_1, \gamma_2) = \frac{1}{\sqrt{2}} \begin{bmatrix} \exp(i\gamma_1) & \pm i \exp(i\gamma_2) \\ \pm i \exp(i\gamma_1) & \exp(i\gamma_2) \end{bmatrix} \quad (12.35)$$

which at angles $\gamma_1 = \gamma_2 = 0$ reduces to the gyrator matrix $U_g(\pi/4)$ (see Ref. [45]). Indeed, the gyrator works as a tunable mode converter, which can be realized by a simple and cheap optical setup based on glass cylindrical lenses. Note that the modes obtained from the HG one, under this operation, are described by the same Wigner distribution but rotated in phase space.

The action of the gyrator transformation on the HG mode leads to the appearance of the longitudinal orbital angular momentum projection of the beam, the value of which depends on the parameter of the transformation: $l\hbar \sin 2\gamma$ per photon [46]. The transformation of the $\text{LG}_{3,2}^+$ mode under a gyrator operation with angles

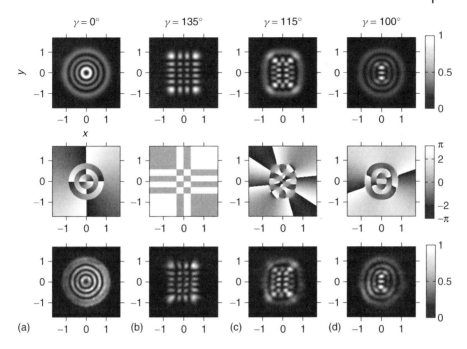

Figure 12.4 Numerical simulation of the optical gyrator setup for $LG_{3,2}^+$ input mode (a): the first and second rows correspond to intensity and phase distribution, respectively. Experimental results are displayed in the third row (a)–(d). Units in x and y axis are given in millimeters.

$\gamma = 135°$, $115°$, and $100°$ is demonstrated in Figure 12.4. The first and the second rows (intensity and phase distribution) correspond to numerical simulations associated with the optical gyrator setup. Here, we have chosen a beam waist that coincides with the normalized distance: $w = (2\lambda z)^{1/2}$, where $z = 50$ cm and $\lambda = 532$ nm. The corresponding experimental results, displayed in the third row of Figure 12.4, show a good agreement with the theoretical predictions.

Another application of phase-space rotators is related to phase retrieval. Since in optics only intensity distributions can be measured directly, several algorithms have been proposed for phase recovery from intensity measurements. The measurement of Wigner distribution projections, associated with the signal's intensity distribution and its power spectrum, permits one to reconstruct the signal phase distribution up to a constant. The Gerchberg–Saxton algorithm exploits this fact by using an iterative procedure and several constraint parameters [47]. Its efficiency has been demonstrated in many different applications ranging from beam characterization to holography. It was originally based on the Fourier transform operation and then extended to the Fresnel diffraction domain as well as to the symmetric fractional Fourier transformation [48]. In contrast to the well-established Gerchberg–Saxton algorithms based on the Fourier transform operation, the algorithms based on fractional Fourier transformation or Fresnel diffraction use more than one constraint image. This fact leads to significant improvements and a faster convergence in the phase reconstruction. Nevertheless, there exist signals for which the phase distributions cannot be retrieved from these input data. A representative and important example corresponds to the LG modes. This issue arises from the fact that the LG beam is an eigenfunction of the symmetric fractional Fourier transformation and the Fresnel transformation (in a mild sense, i.e., except for scaling and quadratic phase), as well as from the fact that the corresponding intensity distributions satisfy $|LG_{p,l}^+|^2 = |LG_{p,l}^-|^2$. Therefore, within this context, the $LG_{p,l}^+$ beam cannot be distinguished from the $LG_{p,l}^-$ one. In order to solve this problem, an astigmatic operation such as an antisymmetric fractional Fourier transformation or a gyrator operation has to be used.

As an example, we consider this phase retrieval problem by applying an antisymmetric fractional FT. In this case, the input signal corresponds to the $LG_{4,1}^+$ mode (see Figure 12.5, $\gamma = 0°$) for which the phase distribution has to be retrieved by using, for example, the intensity distributions associated with the antisymmetric fractional Fourier power spectra for the angles $\gamma = 225°$, $248°$, and $259°$. The first and the second rows in Figure 12.5b–d display the intensity distributions obtained by numerical simulation and measured experimentally, respectively, corresponding to the fractional Fourier transform setup previously discussed [43].

The phase recovery process can be summarized as follows. First, the fractional Fourier transform of a complex signal with random phase distribution, often used as an initial phase estimation, and amplitude corresponding to the squared root of the known intensity (in our case $|LG_{4,1}^+|$) is calculated. Then, the output amplitude distribution is replaced by a constraint image, which corresponds to the squared root of the measured power spectrum, and the inverse fractional Fourier transformation of this signal is performed. The phase of the result is used in the next iteration loop. After a sufficient number of iterations, the retrieved complex

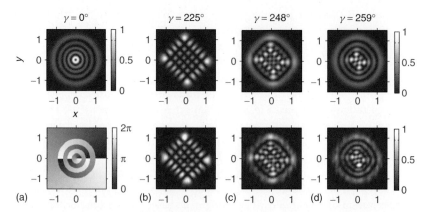

Figure 12.5 Transformation of the $LG_{4,1}^+$ input mode (a) under antisymmetric fractional Fourier transformation. Numerical simulation results are displayed in the first row (b–d), whereas the experimental results are shown in the second row (b–d) for the transformation angles $\gamma = 225°$, $248°$, and $259°$. Units in x and y axis are given in millimeters.

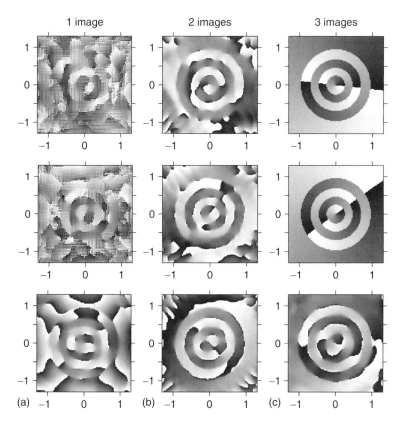

Figure 12.6 Retrieved phase distribution corresponding to one, two and three constraint images associated with the antisymmetric fractional Fourier transformation at angles $\gamma = 225°$, $248°$ and $259°$, respectively. The simulations were performed for $LG_{4,1}^+$ and $LG_{4,1}^-$ input signals (first and second rows, respectively). The third row displays the retrieved phase calculated from the experimental results shown in Figure 12.5b–d, corresponding to $LG_{4,1}^+$ input signal.

field coincides with the original one ($LG_{4,1}^+$). The algorithm convergence speeds ups when we consider more than one constraint image. The reader can find further details in Refs [48, 49] and the references cited therein.

The phase distributions for the $LG_{4,1}^+$ and the $LG_{4,1}^-$ beams, retrieved after 45 iterations by using one, two, and three simulated constraint images (amplitudes corresponding to the antisymmetric Fourier spectra), are displayed in Figure 12.6, first and second row, respectively.

Phase retrieval results obtained from the experimental data associated with the $LG_{4,1}^+$ transformation are displayed in the third row in Figure 12.6. We underline that the wavefront distortion arising from the optical setup is also included in such retrieved phase (Figure 12.6c, third row). This phase retrieval approach can also be based on other astigmatic operations, as it has been demonstrated in Ref. [49] for the case of the gyrator operation.

Phase-space rotators can also be used for detection and processing of chirplike waves, which are present in medical and industrial signals and images, optical beams, and so on. The application of an ordinary FT, a fractional FT, and a gyrator permits the detection of plane, elliptic, and hyperbolic waves, respectively. These signals act as noise and can be extracted by means of filtering in the appropriate phase-space domains.

A generalized chirp wave corresponds to the point-spread function associated with a phase-space rotator, which is described by unitary matrix U and is thus given as

$$R^U[\delta(x_i - v)](x_0) = (\det iY)^{-1/2}$$
$$\times \exp\left[i\pi \left(v^t Y^{-1} X v - 2v^t Y^{-1} x_0 + x_0^t X Y^{-1} x_0\right)\right] \quad (12.36)$$

where $v = [v_x, v_y]^t$ denotes the position of the Dirac δ function in the input plane. Therefore, a chirp function can be easily localized as a Dirac δ function $\delta(x_i - v)$ by applying the inverse transformation parametrized by U^{-1}. In the case of a hyperbolic chirp, $f(x, y) = \exp\left[i2\pi c (x - v_x)(y - v_y)\right]$, the gyrator operation has to be applied. Indeed, when $c = \pm \cot \gamma$, a Dirac δ function is obtained:

$$R^{U_g(\mp\gamma)}[f(x_i, y_i)](x_0) = \delta(x_0 \pm v \cos \gamma) \exp\left(\pm i\pi v_x v_y \sin 2\gamma\right) \quad (12.37)$$

Thus, by analyzing the overall gyrator spectrum, we can detect a local maximum associated with a Dirac δ function for the correct angle γ of the transformation. As an example, we consider several gyrator power spectra corresponding to the input signal displayed in Figure 12.7 ($\gamma = 0°$). In this case, the input signal is a linear superposition of three hyperbolic waves, which are detected at the angles $\gamma = 15°$, $25°$, and $35°$ (Figure 12.7). This procedure can also be applied for image denoising as demonstrated in Ref. [36].

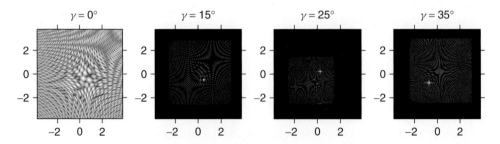

Figure 12.7 Intensity distribution at $\gamma = 0°$ corresponds to an input signal given as a superposition of three hyperbolic waves. These chirp functions included in the input signal are detected when a gyrator operation is performed at the angles $\gamma = 15°$, $25°$, and $35°$.

12.7
Conclusions

It has been shown that the matrix formalism associated with first-order optical systems is a powerful framework that permits, in particular, the description of basic transformations that correspond to rotations in phase space. On the basis of this approach, it is possible to reveal the basic phase-space rotators and design flexible optical setups for their implementations, and to establish new methods for signal characterization and information processing. We have studied these relevant setups as well as their experimental implementations and main applications.

Acknowledgments

The financial support of the Spanish Ministry of Science and Innovation under projects TEC2005–02180 and TEC2008–04105 and of the Santander–Complutense project PR–34/07–15914 is acknowledged. José A. Rodrigo gratefully acknowledges a "Juan de la Cierva" grant. Martin J. Bastiaans appreciates the hospitality at Universidad Complutense de Madrid.

References

1. Mendlovic, D. and Ozaktas, H.M. (1993) Fractional Fourier transform and their optical implementation. *J. Opt. Soc. Am. A*, **10**, 1875–1881.
2. Lohmann, A.W. (1993) Image rotation, Wigner rotation, and the fractional Fourier transform. *J. Opt. Soc. Am. A*, **10**, 2181–2186.
3. Wigner, E. (1932) On the quantum correction for thermodynamic equilibrium. *Phys. Rev.*, **40**, 749–759.
4. Bastiaans, M.J. (1978) The Wigner distribution function applied to optical signals and systems. *Opt. Commun.*, **25**, 26–30.
5. Bastiaans, M.J. (1979) Wigner distribution function and its application to first-order optics. *J. Opt. Soc. Am.*, **69**, 1710–1716.
6. Bastiaans, M.J. (1986) Application of the Wigner distribution function to partially coherent light. *J. Opt. Soc. Am. A*, **3**, 1227–1238.
7. Dolin, L.S. (1964) Beam description of weakly-inhomogeneous wave fields. *Izv. Vyssh. Uchebn. Zaved. Radiofiz.*, **7**, 559–563.
8. Walther, A. (1968) Radiometry and coherence. *J. Opt. Soc. Am.*, **58**, 1256–1259.
9. Bastiaans, M.J. (2008) Applications of the Wigner distribution to partially coherent light beam, in *Advances in Information Optics and Photonics*, (eds. A.T. Friberg and R. Dandliker), SPIE, Bellingham, WA, pp. 27–56.
10. Testorf, M., Hennelly, B., and Ojeda-Castañeda, J. (2009) *Phase Space Optics: Fundamentals and Applications*, McGraw-Hill, New York.
11. Testorf, M.E., Ojeda-Castañeda, J., and Lohmann, A.W. (2006) *Selected Papers on Phase-Space Optics, SPIE Milestone Series*, vol. MS 181, SPIE, Bellingham, WA.
12. Bastiaans, M.J. (1977) A frequency-domain treatment of partial coherence. *Opt. Acta*, **24**, 261–274.
13. Papoulis, A. (1968) *Systems and Transforms with Applications in Optics*, McGraw-Hill, New York.
14. Wolf, E. (1954) A macroscopic theory of interference and diffraction of light from finite sources. I. Fields with a narrow

spectral range. *Proc. R. Soc. Lond. Ser. A*, **225**, 96–111.
15. Wolf, E. (1955) A macroscopic theory of interference and diffraction of light from finite sources. II. Fields with a spectral range of arbitrary width. *Proc. R. Soc. Lond. Ser. A*, **230**, 246–265.
16. Boashash, B. (2003) Time-frequency signal analysis and processing: a comprehensive reference, in *Particular Part 1: Introduction to the Concepts of TFSAP*, (ed. B. Boashash), Elsevier, Oxford, UK.
17. Claasen, T.A.C.M. and Mecklenbräuker, W.F.G. (1980) The Wigner distribution a tool for time-frequency signal analysis; Part 1: continuous-time signals. *Philips J. Res.*, **35**, 217–250.
18. Cohen, L. (1989) Time-frequency distributions a review. *Proc. IEEE*, **77**, 941–981.
19. Cohen, L. (1995) *Time-Frequency Analysis*, Prentice Hall, Englewood Cliffs, NJ.
20. Dragoman, D. (1997) The Wigner distribution function in optics and optoelectronics, in *Progress in Optics*, vol. 37 (ed. E. Wolf), North-Holland, Amsterdam, pp. 1–56.
21. Hlawatsch, F. and Boudreaux-Bartels, G.F. (1992) Linear and quadratic time-frequency signal representations. *IEEE Signal Process. Mag.*, **92**, 21–67.
22. Lee, H.W. (1995) Theory and applications of the quantum phase-space distribution functions. *Phys. Rep.*, **259**, 147–211.
23. Mecklenbrauker, W. and Hlawatsch, F. (1997) *The Wigner Distribution - Theory and Applications in Signal Processing*, Elsevier Science, Amsterdam.
24. Torre, A. (2005) *Linear Ray and Wave Optics in Phase Space*, Elsevier, Amsterdam.
25. Woodward, P.M. (1953) *Probability and Information Theory with Applications to Radar*, Pergamon, London. Chapter 7.
26. Papoulis, A. (1974) Ambiguity function in Fourier optics. *J. Opt. Soc. Am.*, **64**, 779–788.
27. Luneburg, R.K. (1966) *Mathematical Theory of Optics*, University of California Press, Berkeley and Los Angeles, CA.
28. Collins, S.A. Jr. (1970) Lens-system diffraction integral written in terms of matrix optics. *J. Opt. Soc. Am.*, **60**, 1168–1177.
29. Sundar, K., Mukunda, N., and Simon, R. (1995) Coherent-mode decomposition of general anisotropic Gaussian Schell-model beams. *J. Opt. Soc. Am. A*, **12**, 560–569.
30. Simon, R. and Mukunda, M. (1998) Iwasawa decomposition in first-order optics: universal treatment of shape-invariant propagation for coherent and partially coherent beams. *J. Opt. Soc. Am. A*, **15**, 2146–2155.
31. Wolf, K.B. (2004) *Geometric Optics on Phase Space*, Springer-Verlag, Berlin.
32. Simon, R. and Wolf, K.B. (2000) Structure of the set of paraxial optical systems. *J. Opt. Soc. Am. A*, **17**, 342–355.
33. Simon, R. and Wolf, K.B. (2000) Fractional Fourier transforms in two dimensions. *J. Opt. Soc. Am. A*, **17**, 2368–2381.
34. Alieva, T. and Bastiaans, M.J. (2005) Alternative representation of the linear canonical integral transform. *Opt. Lett.*, **30**, 3302–3304.
35. Ozaktas, H.M., Zalevsky, Z., and Kutay, M.A. (2001) *The Fractional Fourier Transform with Applications in Optics and Signal Processing*, Wiley, New York.
36. Rodrigo, J.A., Alieva, T., and Calvo, M.L. (2007) Applications of gyrator transform for image processing. *Opt. Commun.*, **278**, 279–284.
37. Rodrigo, J.A., Alieva, T., and Calvo, M.L. (2007) Gyrator transform: properties and applications. *Opt. Express*, **15**, 2190–2203.
38. Rodrigo, J.A., Alieva, T., and Calvo, M.L. (2007) Experimental implementation of the gyrator transform. *J. Opt. Soc. Am. A*, **24**, 3135–3139.
39. Malyutin, A.A. (2006) Tunable Fourier transformer of the fractional order. *Quantum Electron.*, **36**, 79–83.
40. Moreno, I., Ferreira, C., and Sánchez-López, M.M. (2006) Ray matrix analysis of anamorphic fractional Fourier systems. *J. Opt. A: Pure Appl. Opt.*, **8**, 427–435.
41. Sahin, A., Ozaktas, H.M., and Mendlovic, D. (1998) Optical implementations of two-dimensional fractional

Fourier transforms and linear canonical transforms with arbitrary parameters. *Appl. Opt.*, **37**, 2130–2141.
42. Rodrigo, J.A., Alieva, T., and Calvo, M.L. (2006) Optical system design for orthosymplectic transformations in phase space. *J. Opt. Soc. Am. A*, **23**, 2494–2500.
43. Rodrigo, J.A., Alieva, T., and Calvo, M.L. (2009) Programmable two-dimensional optical fractional Fourier processor. *Opt. Express*, **17**, 4976–4983.
44. Siegman, A.E. (1986) *Lasers*, University Science Books, Mill Valley, CA.
45. Bastiaans, M.J. and Alieva, T. (2005) Propagation law for the generating function of Hermite-Gaussian-type modes in first-order optical systems. *Opt. Express*, **13**, 1107–1112.
46. Alieva, T. and Bastiaans, M.J. (2007) Orthonormal mode sets for the two-dimensional fractional Fourier transformation. *Opt. Lett.*, **32**, 1226–1228.
47. Gerchberg, R.W. and Saxton, W.O. (1972) A practical algorithm for the determination of phase from image and diffraction plane pictures. *Optik*, **35**, 237–246.
48. Zalevsky, Z., Mendlovic, D., and Dorsch, R.G. (1996) Gerchberg-Saxton algorithm applied in the fractional Fourier or the Fresnel domain. *Opt. Lett.*, **21**, 842–844.
49. Rodrigo, J.A., Duadi, H., Alieva, T., and Zalevsky, Z. (2010) Multi-stage phase retrieval algorithm based upon the gyrator transform. *Opt. Express*, **18**, 1510–1520.

13
Microscopic Imaging

Gloria Bueno, Oscar Déniz, Roberto González-Morales, Juan Vidal, and Jesús Salido

13.1
Introduction

Microscopic imaging provides invaluable research support for important fields such as medicine, biology, material sciences, forensics, and so on. The underlying ability in all forms of microscopy is that of obtaining images from which useful high-level information can be extracted. This ability depends on a multitude of factors such as optics, illumination, image processing, segmentation, computational power, and so on. In microscopy, some of these generic factors require specific hardware, techniques, and algorithms.

In the past decades, we have witnessed a significant growth in the application of microscopy for investigations in a wide variety of disciplines. Digital imaging and analysis have enabled microscopists to acquire an enormous amount of useful information. Microscopy is now routinely used in modern research centers, criminal investigations, hospitals, and so on.

Enlargements of 5 or 10 times can be obtained with a single lens, though that is not enough for cell imaging, for example. The first compound (two-lens) microscopes were introduced by Galileo in 1610 and Kepler in 1611. These rudimentary microscopes enabled scientists to see, for the first time, microorganisms, cells, bacteria, blood cells, and so on. Until the eighteenth century, however, microscopes had a basic structural defect: aberrations. The basic functionality of a microscope is to make light rays converge at a given point. Lenses made of a single piece of glass cannot focus all the light rays in one point. Thus, there were spherical aberrations (rays that traverse the periphery of the lens do not converge with those that traverse the center of the lens) as well as chromatic aberrations (different frequencies – colors – converge at different distances to the lens).

Achromatic lenses, introduced by Amici in 1810, allowed to correct aberrations. Abbe developed the theory of microscopic imaging and resolution [1]. In 1893, Köhler proposed a method to provide uniform sample or specimen illumination. Zernike developed phase contrast microscopy in 1935. Differential interference contrast (DIC) microscopy was introduced by Nomarski in 1955 [1]. Over the years, several other advancements were introduced such as binocular microscopes, stereomicroscopes, and so on. Some of the techniques involved are explained later.

The most recent breakthrough in microscopy has been the introduction of digital imaging, which greatly enhances the process of extracting information about the specimen from a microscopic image. Digital imaging is not only a natural evolution but it has also become essential for the analysis and interpretation of modern microscopic images. For these reasons, digital imaging, and, in general, "computer-aided microscopy" are covered more extensively in this chapter. For more details on the history of microscopy, see Ref. [2].

13.2
Image Formation: Basic Concepts

This section introduces the reader to a number of basic concepts that are used throughout the chapter. It is only intended to be a review; for a more extensive treatment the reader can see Refs [1, 3–5].

13.2.1
Types of Image

Castleman [5] described an appealing approach for studying microscopic imaging in which four types of images are dealt with. First, there is the *optical image*, which is the image created by the optical components of the microscope and projected onto the image sensor. This image contains useful information about the specimen, although it is an imperfect representation: it is the 2D projection of a 3D object; it is subject to distortion and noise. Second, there is a *continuous image*, which can be represented as a 2D continuous function that represents light intensity at a given spatial position. Third, the *digital image* is produced by digitization of the continuous optical image, which produces a matrix of numeric data. This process can introduce additional noise and severely affect the quality of the captured image. Fourth, there is the *displayed image*, which is essentially a conversion from digital to optical, since the human eye cannot view an image in digital form.

13.2.2
Image Formation in the Optical Microscope

The optical image is created when light from a source passes through the specimen and then through the magnifying objective, which is essentially a convex lens. This light is projected on the so-called focal point. The focal distance is the distance between the lens and the plane in which the focal point lies. The focal point lies inside the microscope tube. Light goes past the focal point inside the microscope tube. A magnified, inverted, image of the specimen is projected on the so-called intermediate image plane, at the eyepiece diaphragm. The eyepiece lens further magnifies the image and produces the version that the eye sees.

The magnification in the intermediate plane depends on the distance between the specimen and the objective. This image is further magnified by a factor depending on the focal length of the eyepiece.

13.2.3
Light

Light is the probe that can be used to determine the structure of objects. We examine here only those aspects of light that will be most useful to the posterior sections. The kinds of light most frequently encountered in microscopy are as follows:

- **Monochromatic**: waves having the same frequency (same color).
- **Polarized**: waves in which the vibrations occur in a single plane.
- **Coherent**: waves of a given wavelength that maintain the same phase relationship while traveling through space (laser light is coherent).
- **Collimated**: waves having coaxial paths of propagation (without convergence or divergence).

The eye–brain system perceives differences in light intensity and frequency (color), although it does not see differences in phase or state of polarization. As discussed later, such properties can convey useful information about the specimen, too.

Light interacts with matter in a number of different ways as follows:

- **Transmission/absorption**: Transparent substances have 100% transmission. Microscope objectives are designed to transmit light in specific visible wavelengths such as ultraviolet or infrared. Fluorescence filters are also designed to absorb or transmit specific wavelengths.
- **Reflection**: Mirrored surfaces reflect light. Mirrors are used inside the microscope to deflect light to various imaging devices.
- **Refraction**: Light rays entering substances with different refractive index are bent according to Snell's law. Lenses and objectives focus light by refraction.
- **Fluorescence**: Some substances absorb light of a specific wavelength and reemit part of that energy as light of longer wavelength.

13.2.4
Resolution

The spot in the image plane produced by a point in the focal plane is the so-called point spread function (PSF). The intensity distribution associated with the PSF is called the *Airy disk pattern*. The resolution of a microscope, which indicates its ability to reproduce in the image small structures present in the specimen, is the minimum distance by which two point sources must be separated in order for them to be recognized as separate. When light from the various points of a specimen passes through the objective and is reconstituted as the image, the various points of the specimen appear in the image as small Airy disk patterns (not points). The

PSF's overlap as the points get closer together, and thus one would have to consider how much contrast is needed to recognize the two objects as distinct. There are, however, two distances commonly used for comparing the resolution power of optical systems: the Abbe distance and the Rayleigh distance. Before presenting the corresponding equations, however, we need to introduce another important concept: numerical aperture (NA).

The NA represents the range of angles at which the lens of an objective (see below) can accept or emit light rays. Objectives with lower angular aperture can include only a narrower cone of light as compared to objectives with higher angular aperture. Dry objectives have, in practice, an NA of not more than 0.95, while oil immersion objectives can yield an NA of 1.4.

The importance of the NA in microscopy stems from the fact that it also represents the resolving power of the lens. The Abbe and Rayleigh distances mentioned above are defined as

$$r_{Abbe} = \frac{\lambda}{2NA} = 0.5 \left(\frac{\lambda}{NA}\right) \tag{13.1}$$

$$r_{Rayleigh} = 0.61 \left(\frac{\lambda}{2NA}\right) \tag{13.2}$$

r is the smallest distance between two points lying close together in the specimen but which are still seen as separate points in the image. λ is the wavelength of the light used. As NA increases, r becomes smaller, the distance between adjacent points becomes smaller and hence resolution is better.

13.3
Components of a Microscopic Imaging System

The basic components of a modern optical upright microscope are depicted in Figure 13.1. Note that there exists also the inverted microscope, where the objectives are located beneath the state and the condenser is above it.

- **Objective**: The objective is probably the most important component of a microscope. Its function is to collect diffracted light from the specimen and produce a magnified image. Manufacturers commonly identify objectives by their magnification and NA. Typical magnification values are 4×, 5×, 10×, 20×, 40×, 50×, and 100×. Objectives are commonly parfocal, that is, they can be changed and the specimen image will stay in focus.
- **Eyepiece**: This piece holds one or two lenses that bring the images to focus for the eye. Some microscopes allow fitting a camera (see below).
- **Camera**: CCD or CMOS cameras are present in some microscopes to capture images. In fluorescence microscopy, large exposure times are needed and so cooled cameras are common.
- **Adjustment knobs**: These knobs allow to adjust the focus. In motorized microscopes (see below), there may be knobs or joysticks for controlling the stage position.

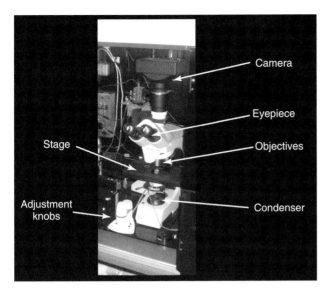

Figure 13.1 Basic components of an optical microscope.

- **Specimen holder or stage**: This is the platform where the specimen lies. The platform has a hole in the center through which light passes. Specimens are typically mounted on glass slides. Some slides include a bar code on their sides to identify the specimen. Some microscope models include a motorized slide feeder.
- **Illumination**: Normally situated below the condenser, it can range from a simple incandescent lamp to an optic-fiber-guided high-power external laser illuminator.
- **Condenser**: The condenser is located between the illuminator and the specimen and it serves to concentrate the light.
- **Computer**: Modern microscopes (or their cameras) are attached to computers that normally have large storage space and processing power. Other useful options are image processing accelerating hardware, autofocusing hardware, large format or dual monitors, 3D mouse, fast network connection, and so on.

Some microscopes are motorized with a stage that allows to position the slide in the x, y (and sometimes z) axis. This is particularly useful for automated scanning of the slide at different magnifications. Other common motorized functions are objective and filter change.

13.4
Types of Microscopy

Proper illumination of the specimen is crucial in achieving high-quality images in microscopy. As mentioned above, an advanced procedure for microscopic illumination was first introduced in 1893 by August Köhler as a method of providing

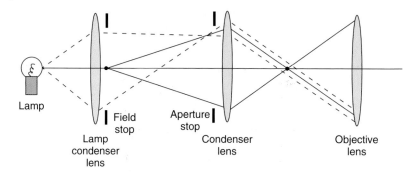

Figure 13.2 Köhler illumination is achieved when the light from a small source is focused by a well-corrected lamp condenser lens (which was introduced by Köhler himself) onto the aperture stop that is in front focal plane of the substage condenser lens. This method of illumination enables a small, high-intensity light source to be used because the back surface of the lamp condenser acts as a uniformly illuminated, large area source.

optimum specimen illumination [6]. The illumination system of the microscope, when adjusted for proper Köhler illumination (Figure 13.2), must satisfy several requirements. The illuminated area of the specimen plane must not be larger than the field of view for any given objective/eyepiece combination. Also, the light must be of uniform intensity (in practice it is not uniform but one can use postprocessing techniques, such as, for example, the top-hat morphological operation, to correct this) and the NA may vary from a maximum (equal to that of the objective) to a lesser value that will depend upon the optical characteristics of the specimen. Moreover, according to the type of object or specimen to be visualized, different techniques must be used. These techniques are oriented to improve specimen contrast or visualize specific object's molecules that fluoresce in the presence of excitatory light. Basically the most important techniques are bright-field microscopy, phase contrast microscopy, dark contrast microscopy, DIC microscopy, fluorescence microscopy, confocal microscopy, and electron microscopy. These techniques are briefly described in the following.

13.4.1
Bright-Field Microscopy

Bright-field microscopy is the simplest of all the optical microscopic illumination techniques. The illumination used is transmitted white light, and the specimen is illuminated from below toward the condenser and observed from above through the objective lens and the eyepiece.

The advantage of bright-field microscopy is its simplicity of setup and the fact that no sample preparation is required, allowing viewing of live cells. Bright-field microscopy has, however, low apparent optical resolution due to the blur of out of focus material. Another drawback lies in distinguishing details in unstained samples and transparent objects. Although transparent objects induce phase shifts between interacting beams of light due to scattering and diffraction, they remain nearly invisible, because the eye cannot detect differences in phase. In order to better distinguish these transparent objects, two illumination techniques are used: phase contrast and dark-field microscopy.

13.4.2
Phase Contrast Microscopy

Phase contrast microscopy transforms differences in the relative phase of object waves to amplitude differences in the image. As light travels through a medium, interaction with this medium causes its amplitude and phase to change in a manner that depends on the properties of the medium. Changes in amplitude give rise to familiar absorption of light, which gives rise to colors when it is wavelength dependent. Although the phase variations introduced by the sample are preserved by the microscope, this information is lost in the process that measures the light. In order to make phase variations observable, it is necessary to combine the light passing through the sample with a reference so that the resulting interference reveals the phase structure of the sample. The difference in densities and composition within the imaged objects, however, often give rise to changes in the phase of light passing through them, and hence they are sometimes called *"phase objects."* Use of the phase-contrast technique makes these structures visible and allows their study with the specimen still alive.

When specimens are examined in positive contrast mode, the conventional mode of viewing, objects with a higher refractive index and therefore higher density, amplitude, and optical path length than the surrounding medium appear dark and vice versa. Thus high-contrast images may be obtained. To avoid confusion regarding bright and dark contrast in phase-contrast images, it is useful to reconsider the term *optical path difference*, which is the product of refractive index and object thickness, and is related to the relative phase shift between object and background waves.

However, the size and orientation of asymmetric objects also affect intensity and contrast. Further, there are optical artifacts present in every phase-contrast image. Phase-contrast images show also characteristic patterns of contrast in which the observed intensity does not correspond directly with the optical path difference of the object. These patterns are sometimes referred to as *phase artifacts* or *distortions*. They are *phase halos*, which always surround *phase objects* and *shade-off*, frequently observed on large and extended objects. These patterns should be recognized as a natural result of the optical system. Interpreting phase contrast images requires care. Thus, a small object with a high refractive index and a large object with a lower refractive index can show the same optical path difference and yet appear to the eye to have the same intensity.

It was Fritz Zernike [7–8] in the 1930s who created an optical design that could transform differences in phase to differences in amplitude. Zernike realized both that it is necessary to interfere with a reference beam, and maximize the contrast achieved with the technique, and it is necessary to introduce a phase shift to this reference so that the no-phase-change condition gives rise to completely destructive interference. This phase-contrast technique proved to be such advancement in microscopy that Zernike was awarded the Nobel Prize (physics) in 1953.

13.4.3
Dark Contrast Microscopy

In dark-field microscopy, image formation is based solely on diffracted wave components. It requires blocking out the central light rays that ordinarily pass through or around the specimen and allowing only oblique rays to illuminate the specimen. Therefore, it works on the principle of illuminating the specimen with light that will not be collected by the objective lens. Thus, the nondiffracted rays are removed altogether so that the image is only composed of diffracted wave components. Dark-field is used to enhance the contrast in unstained specimens, since the component of nondiffracted background light is very large in them, resulting in bright, low-contrast images in which details are poorly visible. The unstained specimens with dark-field illumination appear as brightly illuminated objects on a dark background.

Oblique light rays emanating from a dark-field condenser strike the specimen from every azimuth and are diffracted, reflected, and refracted into the microscope objective. If no specimen is present and the NA of the condenser is greater than that of the objective, the oblique rays cross and all such rays will miss entering the objective because of the obliquity. The field of view will appear dark. To learn more about how to create a dark-field image, readers are referred to Ref. [3].

Dark-field image may first appear to be a negative of the bright-field image, although different effects are visible in each. In bright-field microscopy, features are visible where either a shadow is cast on the surface by the incident light, or a part of the surface is less reflective. Raised features that are too smooth to cast shadows will not appear in bright-field images, but the light that reflects off the sides of the feature will be visible in the dark-field images. The appearance of a dark-field image is similar to one of self-luminous or fluorescent objects on a dark background. However, details in dark-field images are broader and less distinct compared to other imaging modes such as phase contrast. Further, if the NA of the objective selected is too restricted, many diffracted waves are also eliminated, resulting in a loss of definition of fine details in the specimen but they are almost entirely free of artifacts.

The main limitation of dark-field microscopy is the low light levels seen in the final image. This means the sample must be very strongly illuminated, which can cause damage to the sample. One of the advantages of dark-field optics is that it allows the detection of weak diffracted light signals, and may be the method of choice for viewing fine structural details. Very small specimens are all easily seen in well-adjusted dark-field optics. Dark-field optics is also a very simple yet effective technique and is well suited for uses involving live and unstained biological samples. Considering the simplicity of the setup, the quality of images obtained from this technique is very good.

13.4.4
Differential Interference Contrast (DIC) Microscopy

In the continuing search for better image contrast, Nomarski developed a method (DIC microscopy) for converting optical gradients into intensity differences. A crucial element in DIC microscopes is the Wollaston prism, which is constructed by cementing together two quartz wedges whose polarization directions are perpendicular. DIC microscopes act by converting differences in the refractive index into differences in amplitude (intensity). A polarized light ray (Figure 13.3) is split into two components by the Wollaston prism, generating two perpendicularly polarized rays that travel in parallel and are separated by a small distance. When these two parallel rays encounter two different structures (with different refraction indices), their phase will change differentially. Another prism recombines the two rays. If the phase shift of the two rays is the same, nothing happens. If the phase shift is different for the two rays, the polarization plane of the resulting ray will be different from the original. The recombined rays then pass through an analyzer (another polarizing filter orientated perpendicularly to the illumination polarizer). Polarizer filters all but the altered rays, thus generating a high contrast image of the object.

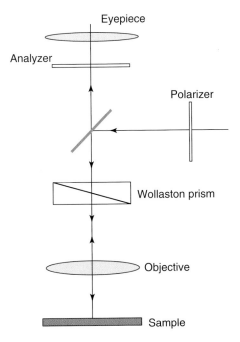

Figure 13.3 Basic operation of a DIC microscope.

13.4.5
Fluorescence Microscopy

Fluorescence microscopy [10, 11], currently a fast growing area of research, is based on the fluorescence of the specimen itself or in its fluorescence when treated with specific fluorescent chemicals (fluorochromes). The fluorescence microscope emits light that illuminates the specimen. This excitation light causes fluorescence, that is, the specimen emits radiation of a larger wavelength. The basic task of the fluorescence microscopy is then to separate the excitation radiation from the fluorescent light (Figure 13.4).

The light source of a fluorescence microscope is usually of much higher intensity than in other modalities (usually a mercury or xenon burner). The excitation filter passes only the desired excitation wavelengths. The dichroic mirror separates the two light paths: the excitation light is reflected onto the ocular, while the fluorescence light passes through it. The emission filter passes only the specific wavelengths of the fluorescence. These three basic components are usually found inside the so-called filter cube.

Fluorescence microscopy provides a level of specificity that is not available in other modalities. Cells and cell structures, for example, can be robustly identified amid nonfluorescent tissue. Even single molecules can be located under certain circumstances. A typical problem in fluorescence microscopy is photobleaching: the permanent loss of fluorescence in the fluorochrome by light exposure. In time-lapse captures, fluorescence light might fade abruptly after a few seconds of exposure.

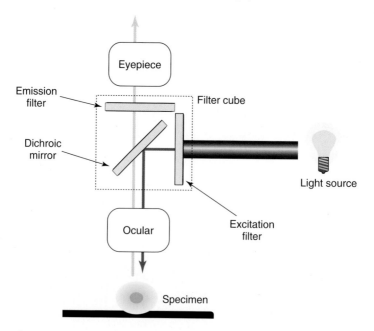

Figure 13.4 Basic operation of a fluorescence microscope.

13.4.6
Confocal Microscopy

In conventional wide-field microscopy, the entire specimen is illuminated, and the formed image can be viewed directly. Note, however, that typically some parts of the specimen will be out of focus. In confocal microscopy [12, 13], illumination consists of a beam of light that scans the specimen. Each beam is focused on the particular point of the specimen onto which it is applied. The formed image is much sharper than in conventional microscopy. In fact, confocal microscopy lies between wide-field light microscopy and electron microscopy in terms of resolution. Confocal microscopy is particularly useful for thick specimens, and it also allows obtaining of 3D structure by focusing on different depth layers.

Figure 13.5 shows the basic operation of the confocal microscope. The key element is a pinhole, which blocks out-of-focus light. A second pinhole may be used to product the light beam for illumination.

As fewer photons are collected, large exposure times are usually needed, which translates into slow frame rates. This can be alleviated by high-intensity light sources like lasers. In confocal microscopy a computer is essential for sequential light collection. The scan is usually achieved by a motorized stage.

13.4.7
Electron Microscopy

The electron microscope uses electrons to illuminate the specimen. There are two types of electron microscopes: the scanning electron microscope and the scanning transmission electron microscope. Just as light is refracted and focused by an optical

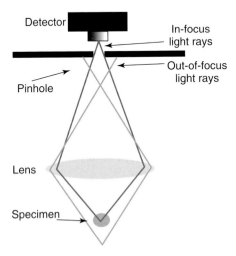

Figure 13.5 Basic operation of a confocal microscope.

lens, electron beams can be directed and focused by magnetic or electric fields. The spatial resolution of the electron microscope is limited by the wavelength associated with electrons, which is roughly 100 000 times smaller than optical wavelengths. Thus, the resolution is, in theory, 0.10 nm, which allows to visualize specific molecules and obtain high-resolution images of subcellular structures.

The main disadvantages of the electron microscope are its technical complexity (cathode sources and vacuum chambers included) and high cost. Besides, specimens must be coated with a conducting metal (typically gold) in order to accelerate the electrons onto them.

13.5
Digital Image Processing in Microscopy

Microscope image processing has been going on since the 1960s when it was realized that some of the techniques of image acquisition and analysis, first developed for television, could also be applied to images obtained through the microscope. One of the earliest projects in this domain was the cytophotometric data conversion (CYDAC) project [14], where image analysis was done off-line. Owing to advances in computer technology, the use of digital images in microscopy for storage, analysis, or processing prior to display and interpretation have dramatically increased [3, 5, 15, 16].

In order to obtain a digital image, an optical image must be digitalized. The digital image is an array of integers obtained by sampling and quantizing the optical image. The design or analysis of an image processing algorithm must take into account both the continuous and the digital image. Although usually the processing is only done on the digital image, the continuous image may be obtained by interpolation. In practice, digitization and interpolation cannot be done without loss of information and the introduction of noise and distortion. However, digitization and interpolation must both be done in a way that preserves the image content and that does not introduce noise or distortion.

Many processing techniques and algorithms may be applied to a digital image to enhance, detect, and extract information about the specimen from a microscopic image. It is not in the scope of this chapter to discuss all possible methods, but to just to mention some of the most important and popular processing techniques based on the authors' experience in microscope imaging. This experience is mainly for biomedical applications where there is a need for objective image processing methods for diagnosis and interpretation [17]. For more details the reader can see Refs [5, 18–20].

13.5.1
Image Preprocessing

The first processing techniques to be considered are those procedures applied to correct some of the defects in the acquired images that may be present due to imperfect detectors, noncorrect focus, nonuniform illumination, and so on. The most popular preprocessing methods to be applied on microscopic images are (i) *illumination correction*, (ii) *autofocusing*, and (iii) *mosaic generation*.

1) **Illumination correction**: One of the problems with microscopic images is the spatially varying contrast due to nonuniform illumination. This problem can have many sources: irregular surfaces, nonperfect alignment of the condenser lens system, faulty reference voltages, or contaminated apertures. It can be minimized by a careful setup of the imaging conditions, or if it cannot be eliminated, it can be assumed to be constant over some period of time. This assumption allows correction by image processing. Usually, it is possible to acquire a background image obtained with a reference surface without tissue sample. This image can then be used to correct the subsequent images by *arithmetic operations*, that is, by subtracting or dividing (in case of linear camera or sensor) the background image and subsequent correction of the gray level range to the original data. Two other commonly used methods to correct the illumination are *fitting a background* function and *rank or morphological leveling*. By selecting a number of points in the image, a set of values at different positions can be obtained. These can be used to perform least-squares fitting of a function that approximates the background and can be subtracted or divided similar to the previous approach. When the surface is irregular and cannot be fit to a polynomial function, rank leveling based on neighborhood operations is used. The basic idea is to compare each pixel with its neighbors or combine the pixel values in some small region similar to a grayscale erosion or dilation depending on the dark or light background, respectively.

2) **Autofocusing**: It is essential to obtain a correct focus of the specimen to be digitalized. Although autofocusing is a long-standing topic in the literature and a variety of focus algorithms have been proposed, the selection of an appropriate focus algorithm for specific experimental microscopy conditions remains ad hoc and time consuming. Automatic focusing methods fall into two main categories: passive and active systems. Passive methods estimate the position of the lens by finding the position where image sharpness is maximum. Active methods are based on emitting a wave in order to estimate the distance to the object of interest and consequently adjusting the lens position. They are widely used in microscopy. The most commons algorithms can be classified into four groups: (i) *derivative* – based on gradient filters and wavelet operations; (ii) *statistical* – based on variance, autocorrelation, and standard deviation analysis; (iii) *histogram based* – based on gray level range and entropy analysis; and (iv) *threshold based*. A good review of these techniques can be found in Ref. [20].

3) **Mosaic generation**: In microscopy, a large number of fields of view are required to capture a complete slide. These fields of view or tiles must be assembled in such a way that a globally consistent virtual slide is formed. This is known as *mosaic generation*. To solve this problem, rigid registration method based on global alignment algorithms have been used [21]. Figure 13.6 shows an example of the so-called simultaneous contrast illusion.

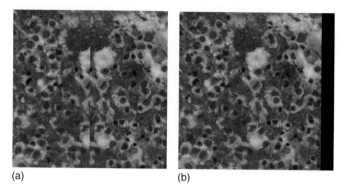

Figure 13.6 Example of mosaicking a cytology digitalized at 20×. (a) Tiling two tiles and (b) mosaicking two tiles.

13.5.2
Image Enhancement

Image enhancement is the process of increasing the contrast and quality of an image for better visualization of certain features and to facilitate further image analysis while keeping the information content. The limited capacity of optical imaging instruments and the noise inherent in optical imaging make image enhancement a requirement for many microscopic image processing applications. See, for example, Figure 13.7 where noise, nonuniform illumination, and color distortion may be observed due to the digitalization for the same sample of a prostate biopsy digitalized at different zooms: (a) 4×, (b) 10×, (c) 20×, and (d) 40×.

The design of an enhancement algorithm should take full advantage of the multidimensional and multispectral information included in the microscopic image due to its acquisition at different focal planes and time intervals, as well as different spectral channels. There are many image enhancement or contrast algorithms [22]. These algorithms can be classified into two categories: spatial domain methods and frequency or transform domain methods. They can be applied to both gray-level and color images by performing the operations on the different color channels according to the color models used [23].

The spatial domain methods include operations carried out on a whole image or on a local region. The most popular methods include histogram equalization for contrast enhancement, image averaging, and nonlinear filtering such as median filtering, spatial bandpass filtering, directional and steerable filtering for noise reduction, and image subtraction for illumination and white balance correction.

Frequency domain methods manipulate image information in transform domains, such as Fourier and wavelet domains. Often, salient image features can be more easily isolated and extracted in the transform domain than in the spatial domain. For example, one can amplify certain coefficients in the Fourier domain

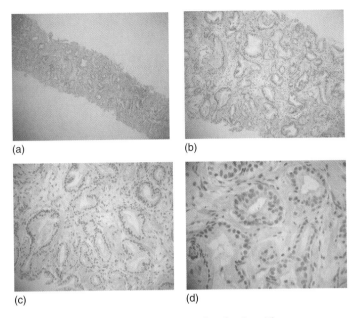

Figure 13.7 Sample of a prostate biopsy digitalized at different zooms. Microscope image digitalized at (a) 4×, (b) 10×, (c) 20×, and (d) 40×.

and then recover the image information in the spatial domain. These techniques are usually used for noise reduction and edge enhancement.

13.5.3
Segmentation

Image segmentation is the process that detects and partitions a digital image into disjoint regions consisting of connected sets of pixels, each of which typically corresponds to either an object or a region of interest (ROI) or the background. It is an important process for further image analysis, such as object measurement and classification.

Though human vision may identify ROIs and shapes without too much effort, optical illusions can occur and the segmentation output may differ from objective reality in terms of depth and motion perception, color and brightness constancies, and feature perception [24]. Figure 13.8 shows an example of the so-called simultaneous contrast illusion, where there are just four gray levels marked on the image, but the human eye would recognize more than four. However, a digital processing may properly identify all gray levels but it takes longer to partition the ROIs.

There are also many image segmentation techniques that are application oriented. Most of the algorithms have been developed for 2D gray-level images but they may be extended to color, multispectral, and 3D images.

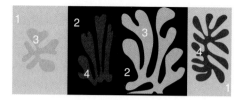

Figure 13.8 Image segmentation and effect of optical illusion.

Segmentation algorithms may be classified into two complementary categories: region-based methods and edge-based methods. Region-based techniques partition the image into homogeneous sets of pixels according to the similarity of image properties such as intensity, texture, energy, and so on. Each pixel in the image is labeled and assigned to a particular region or object. Edge-based techniques establish object boundaries by detecting discontinuities associated with image properties. The combination of these methods can lead to improved segmentation performance.

The most popular region-based segmentation methods are as follows: (i) *Thresholding* is a method in which pixels are grouped and classified either into the object of interest or background according to their value with respect to the threshold, where the threshold corresponds to an image property. Thus all pixels at or above the threshold are assigned to the object and all pixels below the threshold are assigned to the background. The selection of the threshold value is crucial and can significantly affect the resulting boundaries and areas of segmented objects. (ii) *Region growing* is a bottom-up method that exploits spatial context by combining pixels and adjacent regions into larger regions, within which the pixels have similar properties. (iii) *Splitting* is a top-down method that breaks the whole image into disjoint regions within which the pixels have similar properties. (iv) *Watershed* is a morphological method based on a flooding simulation where the image is a topographic surface. The goal is to produce the watershed lines or dams on this surface by flooding the image from the minima and building a dam when water from different minima is about to merge. These dams correspond to object contours [25].

For edge-based segmentation the most popular methods include the following: (i) *Gradient based* method detects edges by looking for the pixels with large gradient magnitude. (ii) *Laplacian based* method finds the edges by searching for zero-crossings in the second derivative of the image. (iii) *Hough transform* can fit a parameterized boundary function to a scattered set of edge points. The Hough transform can detect object boundaries of regular shape, such as lines, circles, and ellipses in an image by recognizing evidence in their transformed parameter space. It is particularly useful when the input image is noisy and the data points are sparse. (iv) *Active contours (ACs) models or snakes* can be used to refine boundaries that have been found by other methods. The contour is parameterized by a set of connected points that form a close curve, which can move so as to minimize an energy function. Properties of the image such as intensity and gradient contribute to the energy of the curve or AC, as do constraints on the continuity and curvature

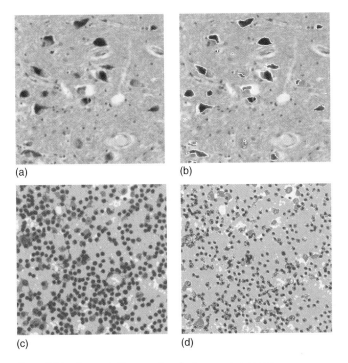

Figure 13.9 Segmentation results of microscopic images: (a) Brain autopsy at 40×, (b) segmentation of mesencephalic cells, (c) lung cytology at 40×, and (d) segmentation of different structures; (a) and (c) original digital images, (b) contour-based segmentation, and (d) region-based segmentation.

of the contour itself. In this way, the snake contour can move in a continuous manner, ensuring continuity and smoothness until it fits the object boundary [26].

Real-world applications in digital microscopy often pose very challenging segmentation problems. For example, one of the main problems dealing with microscopic image for anatomical pathology applications is the huge size of the images. The size of these images usually goes from 1 to 10 GB. In order to process these images, parallel processing and grid-based methods must be used [27]. Figure 13.9 shows results of the segmentation processing edge-based and region-based methods for different tissue types.

13.5.4 Classification

Once the ROI or objects in the image have been segmented and their meaningful features extracted, a classification system identifies them by labeling or assigning each of them to one of several previously defined categories or classes, such as benign or malign, different tissue structures, and so on. A suitable classification system and a sufficient number of training samples are needed for a successful classification. The classification system includes (i) the evaluation of classification performance in terms of accuracy, reproducibility, robustness, and the ability to fully use the information content of the data and (ii) the classification accuracy assessment to take into account sources of errors. In addition to errors from the classification itself, other sources of errors, interpretation errors, and poor quality of training or test samples all affect the accuracy of classification.

There are several methods available for classification [28]. These methods can be divided into different categories attending to different criteria; the most common ones are as follows:

1) Whether statistical parameters such as mean vector and covariance matrix are used or not.
   a. **Parametric**: They assumed a Gaussian or a particular functional form for the probability distribution function (pdf) of the data and statistical parameters generated from the training samples. The most common parametric classifier is the *maximum likelihood*. A parametric classifier requires less training data since it is used only to estimate the parameters. A well-trained parametric classifier can be quite effective at multiclass, multifeature classification, even in the presence of noise.
   b. **Nonparametric**: No statistical parameters are needed to distinguish image classes. The functional form of the pdf is not given or it is known to be non-Gaussian, and one must estimate the pdfs directly from the training data. They usually require considerably more training data than the parametric classifiers. Among the most commonly used nonparametric classification approaches are *artificial neural networks (ANNs), decision trees, support vector machines*, and *expert systems*.
2) Whether training samples are used or not.
   a. **Supervised classification**: The spectral information given by the training samples are used to train the classifier and classify the data. The most common supervised methods are *maximum likelihood, minimum distance, ANN*, and *decision tree classifier*. The ANN classifier has the advantage that it is not necessary to know the statistics of the features. However, it is difficult to prove optimality or to predict error rates.
   b. **Nonsupervised classification**: This is based on the statistical information inherent in the image. No prior definitions of the classes are used. The most commonly used nonsupervised methods are *ISODATA* and *K-means clustering*.

All these classification methods may be also divided according to other criteria:
3) whether just the pixel information is used;
4) whether spatial and contextual information is used.

The type and number of features selected for classification are also very important to increase classifier performance. Ideally a small number of highly discriminating uncorrelated features should be used. Increasing the number of features increases the dimensionality and hence increases the requirements for training set and test set size. There are well-known procedures for reducing the large number of features down to a smaller number without losing their discriminating power. Principal component analysis (PCA) and linear discriminant analysis (LDA), also known as *Fisher discriminant analysis* (*FDA*), are among the best-known methods. Both generate a new set of features, each of which is a linear combination of the original features. PCA uncorrelates the new features and maximizes their variance, whereas LDA maximizes the ability of the features to discriminate among the classes. In both the cases, the new features are ranked so that one can select only a few of the most useful ones, thereby reducing the number of features.

PCA is also used as an unsupervised spectral unmixing method. Spectral unmixing isolates the light from one stain or fluorophore to a single color channel, for specimen analysis in bright-field and fluorescence microscopy, respectively. Spectral unmixing algorithms assume that the measured signal at each pixel is a linear combination of the overlapping spectra at that point. This assumption may be done just when the absorption and fluorophore concentrations and stain are low; in other cases, appropriate correction terms are necessary [5]. In recent years, spectral imaging has found utility in microscopy applications ranging from general cell visualization [29], and with cells of variously colored fluorescent proteins [30].

The spectral unmixing algorithms may be based on supervised or unsupervised classification methods, depending on the requirement of prior knowledge of the fluorophore or stain spectra. The most popular methods for unsupervised spectral unmixing are PCA, independent component analysis (ICA), and nonnegative matrix factorization (NMF).

PCA relies on a statistical analysis of the whole dataset (i.e., spectra of all the pixels in the image) to find meaningful explanations for the similarities and differences in the data. PCA creates a linear transformation of an *n*-dimensional space that minimizes the variance along each new axis while maximizing the variance between the new axes. A dataset may be reduced in dimensionality by eliminating low-variance components. To use PCA for spectral unmixing, we must assume that the gray-level values have a normal distribution (i.e., Gaussian pdf). ICA also extracts linearly independent signals from images but this is done by means of using all possible projections of the data to realize independence across all statistical orders. The differences between PCA, ICA, and NMF arise from different constraints imposed on the set of sensitivities of each color channel to the spectrum of each fluorophore and the signal concentrations. Both PCA and ICA allow any values for these concentrations but NMF does not allow negative values. In addition, NMF does not impose the assumption of signals being statistically independent, and this is more realistic for most applications.

Classification sometimes is combined with segmentation into one step similar to the case of multispectral pixel.

13.6
Conclusions

Microscopic imaging provides invaluable research support for important fields such as medicine, biology, material sciences, forensics, and so on. The underlying ability in all forms of microscopy is that of obtaining images from which useful high-level information can be extracted. This ability depends on a multitude of factors such as optics, illumination, image processing, segmentation, computational power, and so on. In microscopy, some of these generic factors require specific hardware, techniques, and algorithms. This chapter has described, from a tutorial perspective, the fundamentals of microscopic imaging, with an emphasis on showing those specificities. Besides presenting established results and techniques from a global perspective, it also has outlined current lines of work in the field.

Acknowledgments

This work was partially funded by the Spanish Research Ministry and Junta de Comunidades de Castilla-La Mancha through projects DPI2008-06071 and PAI08-0283-9663 and PII2I09-0043-3364.

References

1. Davidson, M.W. and Abramowitz, M. (2002) Optical microscopy, in *Encyclopedia of Imaging Science and Technology*, vol. 2 (ed. J. Hornak), Wiley, pp. 1106–1141.
2. Croft, W.J. (2006) *Under the Microscope: A Brief History of Microscopy*, World Scientific Publishing Pvt. Ltd., Singapore.
3. Murphy, D.B. (2001) *Fundamentals of Light Microscopy and Electronic*, 1st edn, Wiley-Liss.
4. Heath, J.P. (2005) *Dictionary of Microscopy*, John Wiley & Sons, Inc.
5. Wu, Q., Merchant, F., and Castleman, K. (2008) *Microscope Image Processing*, Academic Press.
6. Kapitza, H.G. (1994) *Microscopy from the Very Beginning* (ed. S. Lichtenberg), Carl Zeiss, Oberkochen.
7. Zernike, F. (1942) Phase-contrast, a new method for microscopic observation of transparent objects. Part I and Part II. *Physica*, 9, 686–698, 974–986.
8. Zernike, F. (1955) How I discovered phase contrast. *Science*, 121, 345–349.
9. Bennett, A. et al. (1951) *Phase Microscopy: Principles and Applications*, John Wiley & Sons, Inc., New York.
10. Herman, B. and Tanke, H.J. (1998) *Fluorescence Microscopy*, Springer.
11. Dai, X. (2009) *Fluorescence Microscopy Techniques: Concepts and Applications in Biological and Chemical Systems*, VDM Verlag Dr. Müller.
12. Pawley, J.B. (2006) *Handbook of Biological Confocal Microscopy*, 3rd edn, Springer, Berlin.
13. Hibbs, A.R. (2004) *Confocal Microscopy for Biologists*, Springer.
14. Mendelsohn, M.L. et al. (1968) Digital transformation and computer analysis of microscopic images. *Adv. Opti. Electron Microsc.*, 2, 77–150.
15. Sluder, G. and Wolf, D.E. (2003) *Digital Microscopy*, 2nd edn, Academic Press.
16. Inoue, S. and Spring, K.R. (1997) *Video Microscopy*, 2nd edn, Springer.
17. García, M. et al. (2006) Critical comparison of 31 commercially available digital slide systems in pathology. *Int. J. Surg. Pathol.*, 14, 1–21.

18. Russ, J.C. (2006) *The Image Processing Handbook*, 5th edn, CRC.
19. Russ, J.C. and Russ, J.C. (2007) *Introduction to Image Processing and Analysis*, CRC.
20. Sun, Y., Duthaler, S., and Nelson, B.J. (2004) Autofocusing in computer microscopy: selecting the optimal focus algorithm. *Microsc. Res. Tech.*, **65**, 139–149.
21. Steckhan, D. *et al.* (2008) Efficient large scale image stitching for virtual microscopy. Poster presented at the 30th Annual International IEEE EMBS Conference, pp. 4021–4023.
22. Gonzalez, R.C. and Woods, R.E. (2007) *Digital Image Processing*, 3rd edn, Prentice Hall.
23. Koschan, A. and Abidi, M. (2008) *Digital Colour Image Processing*, 3rd edn, Wiley-Interscience.
24. Changizi, M.A. (2008) Perceiving the present and a systematization of illusions. *Cogn. Sci.*, **32**, 459–503.
25. Soille, P. (2004) *Morphological Image Analysis: Principles and Applications*, 2nd edn, Springer.
26. Kass, K., Witkin, A., and Terzopoulos, D. (1988) Snakes: active contour models. *Int. J. Comput. Vis.*, 321–331.
27. Bueno, G. *et al.* (2009) Image processing methods and architectures in diagnostic pathology. *Folia Histochem. Cytobiol.*, **47** (4), 691–697.
28. Duda, R.O., Hart, P.E., and Stork, D.H. (2000) *Pattern Classification*, 2nd edn, Wiley-Interscience.
29. Maiti, D., Sennoune, S., and Martinez-Zaguilan, R. (2003) Proton gradients in human breast cancer cells determined by confocal, multiphoton, and spectral imaging microscopy, *FASEB J.*, **17** (4), A467.
30. Haraguchi, T. (2002) Spectral imaging fluorescence spectroscopy. *Genes Cells*, **7** (9), 881–887.

14
Adaptive Optics in Microscopy
Martin J. Booth

14.1
Introduction

High-resolution optical microscopes are widely used as imaging tools across many scientific disciplines. Various imaging modes and contrast mechanisms have been developed, ranging from conventional bright field microscopes to super-resolving fluorescence microscopes with resolutions well below the diffraction limit (Chapter 13). These microscopes have found particular application in imaging of biological specimens owing to their abilities in imaging biological structure and function.

The performance of these microscopes is often compromised by aberrations that lead to a reduction in the image resolution and contrast. These aberrations may arise from imperfections in the optical system or may be introduced by the physical properties of the specimen. The problems caused by aberrations can be overcome using adaptive optics, whereby aberrations are corrected using a dynamic element, such as a deformable mirror [1]. This technology was originally conceived for the compensation of the aberrating effects of the atmosphere and was first developed for military and astronomical telescopes. Adaptive optics systems have also been introduced for other applications such as laser beam shaping, optical communications, data storage, ophthalmology, and microscopy [2].

The development of adaptive optics for microscopy has required various departures from the traditional adaptive optics schemes that are used in astronomy [3]. In this chapter, we explain the sources of aberrations in microscopy, their effects, and their correction with adaptive optics. Different methods of wavefront sensing, indirect aberration measurement, and aberration correction devices are discussed. The aim of this chapter is not to provide a complete overview of either adaptive optics or high-resolution microscopy, but rather to cover the issues encountered when combining these two technologies. For more details of the parent technologies, the reader is directed to the mentioned texts.

Optical and Digital Image Processing: Fundamentals and Applications, First Edition. Edited by Gabriel Cristóbal, Peter Schelkens, and Hugo Thienpont.
© 2011 Wiley-VCH Verlag GmbH & Co. KGaA. Published 2011 by Wiley-VCH Verlag GmbH & Co. KGaA.

14.2
Aberrations in Microscopy

Aberrations are deviations of the optical system from its ideal form and they may be characterized in a number of ways, such as ray deviation in geometrical optics or wavefront distortion in wave optics. The theory of aberrations has been studied extensively by many authors (a notable recent source is Ref. [4]). In this section, we review the aspects of aberration theory that are most important for understanding the issues of adaptive optics in microscopy.

14.2.1
Definition of Aberrations

For the purpose of understanding the operation of an adaptive optical system, it is best to think of aberrations in terms of distortions of an optical wavefront. A collimated beam consists of flat wavefronts that maintain their shape on propagation through a uniform medium. The wavefront may become distorted if it is reflected by a nonflat surface as different portions of the beam travel through different path lengths (Figure 14.1). Alternatively, spatial variations in the medium's refractive index cause parts of the beam to propagate at differing speeds, leading to aberration of the wavefront. If the refractive index is denoted as n, then the local speed of propagation is c/n, where c is the velocity of light *in vacuo*. This aberration can, therefore, be quantified in terms of optical path length (OPL), which is equivalent to the distance traveled by a wave had it been propagating in free space at a velocity c. The OPL can be calculated as

$$\text{OPL} = \int_C n(\mathbf{r})\, d\mathbf{r} \tag{14.1}$$

where C is the path of propagation and $\mathbf{r}$ is the spatial coordinate vector. C represents the path of a ray through the system, which, in general, will be nonlinear due to refraction caused by variations in $n(\mathbf{r})$. However, if the variation in refractive index

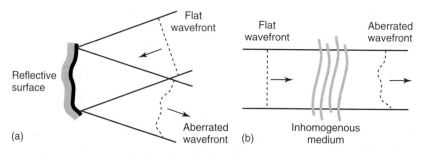

Figure 14.1 Wavefront aberrations: (a) aberrations introduced by reflection from a nonplanar surface and (b) aberrations caused by propagation through a nonuniform refractive index distribution.

is small, then the path C can be approximated by a straight line. Alternatively, the aberration can be described in terms of the phase ϕ of the wavefront given by

$$\phi = \frac{2\pi}{\lambda} \text{OPL} \tag{14.2}$$

where λ is the wavelength. Aberrations are usually modeled as a function in an entrance or exit pupil of the optical system. In most microscope systems, the pupils are circular, so it is useful to express the aberration in a normalized form, defined over a unit radius circle. The complex pupil function is defined as

$$\begin{aligned} P(r,\theta) &= A(r,\theta) \exp[i\phi(r,\theta)], \quad 0 \le r \le 1 \\ &= 0, \quad r > 1 \end{aligned} \tag{14.3}$$

where $A(r, \theta)$ represents amplitude variations, $\phi(r, \theta)$ is the phase aberration, and $i = \sqrt{-1}$. Amplitude variations frequently occur in optical systems due to absorption or reflection losses, but in practical microscope systems $A(r, \theta)$ can often be assumed to be constant. This is particularly the case in aberrated systems as phase variations usually have a much larger effect on performance than amplitude variations.

14.2.2
Representation of Aberrations

It is usual to express aberrations as a series of basis functions or aberration modes, for example,

$$\phi(r,\theta) = \sum_i a_i \psi_i(r,\theta) \tag{14.4}$$

where the functions $\psi_i(r, \theta)$ are an appropriate set of basis modes and a_i are the modal coefficients. If the series is rapidly convergent, then the aberration can be represented in a concise manner using only a small number of coefficients. Representing aberrations in this way can simplify the design, control, and characterization of adaptive optics. The choice of modes for a particular application is often influenced by some aspect of the system, such as the deformation modes of a deformable mirror or the statistics of the induced aberrations. Otherwise, the modal representation may be chosen through mathematical convenience. For example, Zernike polynomials are often used for systems with circular apertures as they form a complete, orthogonal set of functions defined over a unit circle (Figure 14.2) [5]. Other orthogonal sets of modes have also been derived for use in specific microscope systems – this is discussed in Section 14.5.2.

It is useful to consider the different effects that aberration modes have on an imaging system. This can be illustrated using the Zernike modes, as shown in Figure 14.2. The piston mode is simply a constant phase offset and, in most

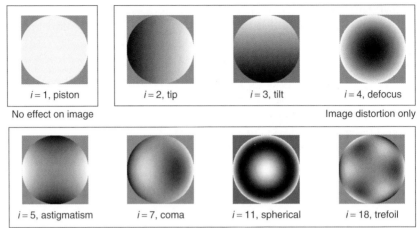

Figure 14.2 Aberration modes based upon Zernike circle polynomials. Examples of low-order modes are given along with their mode index *i* and the mode name. The aberration modes are grouped according to their influence on the imaging process (see text for details).

microscopes, has no effect on the imaging process.[1] The tip and tilt modes consist of linear phase gradients in orthogonal directions and correspond to lateral translations of an image. Similarly, the defocus mode is equivalent to a refocusing of the imaging system. Therefore, the three modes tip, tilt, and defocus represent image translations or distortions in three dimensions. They do not, however, affect the resolution or contrast of the image. The higher order modes, such as those in the lower part of Figure 14.2, lead to distortion of the focal spot and consequently have a detrimental effect on image quality.

14.2.3
Effects of Aberrations in Microscopes

When a collimated beam is focused by an objective lens, the wavefronts take the ideal form of a spherical cap that propagates through the focusing cone toward the focal point. In an aberration-free system, the resulting focus has a finite size determined by the diffraction properties of the light. To a first approximation, the focal spot has a diameter given by

$$d = \frac{\lambda}{2\mathrm{NA}} \tag{14.5}$$

---

1) As it does not usually affect the imaging process, this piston component is frequently neglected in analysis. As throughout this chapter, the aberrations are usually assumed to have zero mean phase and, hence, zero piston component.

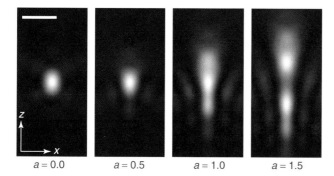

a = 0.0   a = 0.5   a = 1.0   a = 1.5

Figure 14.3 Effects of spherical aberration on the focal spot. The images show simulated focal intensities with different values of a, the amplitude in radians rms of Zernike mode 11. The wavelength was 633 nm and the NA was 0.9. The optic axis is in the direction labeled z. The image brightness has been normalized in each case. The relative maximum intensity decreases from the left image to the right image as 1, 0.78, 0.34, and 0.25. Scale bar 1 μm.

where the numerical aperture (NA) of the objective lens is given by

$$\text{NA} = n \sin \alpha \tag{14.6}$$

in which n is the refractive index of the medium into which the light is focused and α is the angle of the marginal rays to the optic axis. The same approximation leads to an expression for the width of the focal spot along the optic axis as

$$w = \frac{n\lambda}{(\text{NA})^2} \tag{14.7}$$

These expressions for d and w represent the diffraction-limited size of the focal spot for a given wavelength and NA, but these can only be achieved if the focused wavefronts maintain their ideal form. If refractive index variations are present in the propagation medium, they induce aberrations in the spherical wavefronts. This, in turn, leads to a reduction in intensity and a broadening of the focal spot, so that its size is larger than that predicted by Eqs (14.5) and (14.7). The effect is illustrated in Figure 14.3.

The performance of a microscope depends upon its ability to accurately reproduce object features in the image. In an aberration-free system, the imaging properties are determined by a diffraction-limited point spread function (PSF), whose properties are intrinsically linked to the focal spot described above. Just as aberrations distort the focal spot, in an imaging system they cause the PSF to be distorted, leading to image blurring and reduction of contrast. In some microscopes, particularly high-resolution systems such as confocal or multiphoton microscopes, aberrations also cause a reduction in imaging efficiency, leading to a lower detected signal level. These effects conspire to further reduce the image quality. Figure 14.4 illustrates the detrimental effects of aberrations on the microscope image quality and shows the benefit of their correction through adaptive optics.

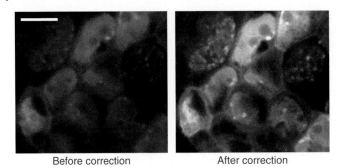

Before correction After correction

Figure 14.4 Correction of specimen-induced aberrations in an adaptive two-photon fluorescence microscope. The signal level, contrast, and resolution are affected by the aberrations, but are restored using aberration correction [6]. The specimen was a mouse embryo labeled with the fluorescent markers DAPI and GFP imaged with a 1.2 NA water immersion lens at $\lambda = 850$ nm. Scale bar 20 μm.

14.2.4
Sources of Aberrations in Microscopes

It is common for microscope specimens to have spatial nonuniformities in refractive index that lead to significant aberrations. One common source of such aberrations is a planar mismatch in refractive index, such as that between the microscope coverslip and the specimen mounting medium, which introduces spherical aberration whose magnitude is proportional to the focusing depth [7]. Biological specimens, in particular, also exhibit variations in refractive index that arise from the three-dimensional nature of cells and tissue structures [8]. In general, these aberrations become greater in magnitude and more complex in form as the focusing depth is increased.

As with all optical systems, microscopes can also suffer from aberrations due to imperfections in the optical components. In practice, no system can be totally free from aberrations and so systems are designed to maintain aberrations below a particular tolerance for a given set of imaging conditions, such as wavelength, magnification, and field of view. Significant aberrations can be introduced if a microscope is used outside its design specifications, for example, at the incorrect wavelength or at a different temperature (see Chapter 11 of Ref. [9]).

14.2.5
Effects of the Numerical Aperture

Aberrations can affect all optical microscopes, but the effects are most significant for those employing high NA objective lenses to provide the finest resolution. The choice of the objective lens can have a significant effect on the aberrations encountered in a microscope.

The resolution of a microscope ultimately depends upon the focal spot size. It is clear from Eqs 14.5 to 14.7 that the focal spot size can be reduced only by using a shorter wavelength or larger NA. In many applications, there is little freedom to reduce the wavelength as it is determined by factors such as the available light sources or the transmission properties of the optical elements. Higher resolution is, therefore, obtained by increasing the NA of the objective lens. According to Eq. (14.6), there are two parameters that can be altered to achieve this: (i) increasing the refractive index n and (ii) increasing the aperture angle α. The refractive index can be changed by using an immersion medium, such as water or oil, but the index at the focus is frequently determined by the nature of the specimen. For example, most biological specimens are predominantly water based and so should be imaged using a water immersion lens. If instead a lens designed for use with a different index medium is employed, it will suffer from spherical aberration due to the index mismatch.

As the wavelength and immersion medium are determined by other factors, the only remaining parameter that we can adjust is α. Consequently, high NA lenses have large focusing angles – for example, a 0.95 NA dry objective has $\alpha = 72°$. This focusing geometry leads to high susceptibility to aberrations if refractive index variations are present in the specimen, as the marginal rays traverse a much longer optical path within the specimen than the paraxial rays. The effects of this are illustrated in Figure 14.5, where the aberrations induced when focusing through a refractive index mismatch for a low NA and high NA lens are compared. The light propagates into a medium of higher refractive index. Refraction caused by the mismatch causes a refocusing effect, whereby the apparent focus depth is different from the nominal focus depth; this difference increases with NA. Another consequence is the introduction of spherical aberration, which is more significant with the high NA lens. While aberrations are also present in the low NA system, they are much smaller in magnitude than those in the high NA system. The phase functions in Figure 14.5 were obtained using the expressions derived in Ref. [10].

As the problems caused by aberrations are usually most severe when high NA objective lenses are used, most adaptive optics systems have been developed for use in the highest resolution microscopes. A range of such high-resolution microscope techniques is discussed in Section 14.4.

14.3
Principles of Adaptive Optics

Adaptive optics systems enable the dynamic correction of aberrations through the reconfiguration of an adaptive optical element, for example, a deformable mirror. They were originally introduced to compensate for the optical effects of atmospheric turbulence on telescope imaging [1]. While adaptive systems can be implemented in various forms, the essential components are the same: (i) a method of aberration measurement, (ii) an adaptive element for aberration correction, and (iii) a control system. The overall operation of adaptive optics is based upon the principle of phase conjugation: the goal is that the correction element introduces an equal

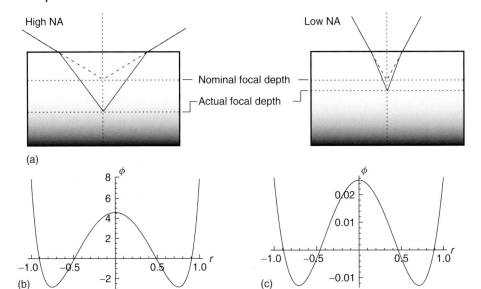

Figure 14.5 Effects of numerical aperture on specimen-induced aberrations. (a) Schematic illustration of the focusing geometry in high and low NA systems showing the focal shift. The marginal rays are shown and are incident from the objective lens, which is situated above the specimen. (b) The pupil phase ϕ in radians as a function of pupil radius r for a high NA dry lens focusing into glass with $n = 1.5$, NA= 0.9, wavelength 633 nm, and a 20 μm nominal focus depth. (c) The equivalent phase function for a 0.3 NA lens.

but opposite phase aberration to that present in the optical system.[2] As the sum of these two aberrations is zero, diffraction-limited operation should be restored. In practice, total compensation is rarely achieved due to the limitations of the correction element in creating the required aberration or due to the accuracy of aberration measurement. Instead, systems are usually designed to maintain the correction error below an acceptable tolerance. This tolerance can be quantified in terms of maximum acceptable phase error or path length error. Equivalently, this can be quantified using the Strehl ratio – the ratio of maximum focal intensity with aberration to the aberration-free focal intensity [5]. For small aberrations, the Strehl ratio S is related to root-mean-square (rms) phase error ϕ_{rms} in radians by

$$S \approx 1 - \phi_{rms}^2 \qquad (14.8)$$

An alternative, empirically determined relationship provides a better approximation for larger aberrations in most practical situations [4]:

2) For monochromatic light, this goal can be relaxed; the requirement is rather that the correction phase be equal and opposite, modulo 2π. This is important as the use of wrapped phase functions can effectively extend the range of some correction elements.

$$S \approx \exp\left(-\phi_{rms}^2\right) \qquad (14.9)$$

A typical target specification for an adaptive microscope system would be $\phi_{rms} = 0.3$ rad, corresponding to an OPL error of $\lambda/20$ or Strehl ratio $S \approx 0.9$.

14.3.1
Methods for Aberration Measurement

The most direct way to measure aberrations is to use a wavefront sensor. Typical configurations of an adaptive optics system incorporating a wavefront sensor are shown in Figure 14.6. As it is difficult to measure optical phase directly, a wavefront sensor always incorporates a method of converting phase variations to intensity

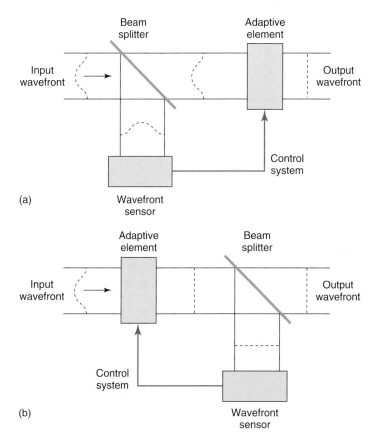

Figure 14.6 Schematic diagrams of conventional adaptive optics systems, where part of the main beam is split off and coupled into the wavefront sensor. (a) Open-loop operation, where the wavefront sensor measures the input aberration directly. (b) Closed-loop operation, where the wavefront sensor measures the error in the corrected wavefront.

variations, which can be measured. One common method is to use interferometry, in which phase variations are encoded as distortions in the fringes of an interference pattern [11]. Alternatively, aberrations are often measured using a Shack–Hartman wavefront sensor (SHWS). In this sensor, the wavefront is passed through an array of microlenses, essentially splitting up the wavefront into component zones [1]. Each microlens creates a focal spot on a camera, the position of which depends upon the local phase gradient in the corresponding wavefront zone. By measuring the position of each spot, the whole wavefront shape can be reconstructed. Direct wavefront sensing is widely used in adaptive optics, but it is not straightforward to incorporate such a sensor into an adaptive microscope. The issues are discussed in Section 14.5.

Indirect wavefront measurement systems do not employ a wavefront sensor, but rather determine the required aberration correction through an optimization procedure. In these "sensorless" systems, a metric related to the image quality, such as total intensity, is either maximized or minimized, as appropriate. A sequence of test aberrations is introduced by the adaptive element and the corresponding values of the optimization metric are obtained. These values are then used in an estimation algorithm that predicts the required correction aberration. This procedure is summarized in the flow diagram of Figure 14.7. An example of a sensorless measurement process is shown in Figure 14.8. Adaptive systems using

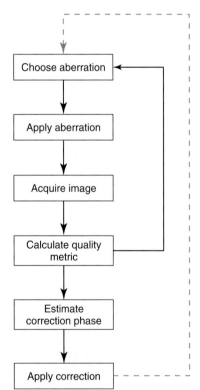

Figure 14.7 Flow chart of the indirect aberration measurement scheme for an image-based sensorless adaptive optics system. A sequence of images is acquired, each with a different aberration applied. For each image, the quality metric is calculated and from this set of measurements the correction phase is estimated and then applied. The dashed line represents an optional repeat of the process, if full correction is not completed during the first cycle.

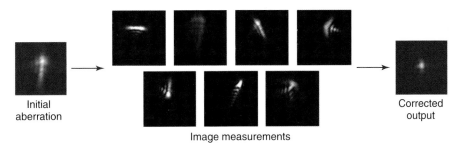

Figure 14.8 Example of an indirect aberration measurement process using images of a focused laser spot. A sequence of predetermined aberrations was applied using an adaptive element to produce the seven spot images. The optimization metric was calculated for each image as a weighed sum of image pixel values and the correction aberration was determined from these metric values [12].

indirect aberration measurement can be simpler than the direct sensing equivalent as only the adaptive element needs to be included. This method may, therefore, be useful when adaptive optics is incorporated into an existing optical system. It may also be useful in applications where there are problems using conventional wavefront sensing, for example, when it is impossible to access the aberrated wavefront. This is the case in some microscopes, where the necessary wavefront exists only in the vicinity of the focus.

14.3.2
Aberration Correction Devices

The most common adaptive correction elements are deformable mirrors or liquid-crystal-based spatial light modulators. Both sets of devices use the same basic principle: aberrations are introduced by altering the OPL traveled by a wavefront when it traverses the adaptive element. However, there are considerable differences in these devices, which must be taken into account when designing an adaptive optics system.

The shape of a deformable mirror can be changed by applying appropriate control signals to an array of actuators that exert forces on the mirror surface. Local displacement of the surface changes the distance traveled by the wavefront reflected off the mirror, introducing an additional phase variation across the wavefront. Most deformable mirrors are actuated using electrostatic or electromagnetic forces or by piezoelectric devices through direct actuation or in bimorph form. Many devices use a continuous mirror faceplate, although versions with segmented surfaces are also available. Deformable mirrors have high optical efficiency and are wavelength and polarization independent. The fastest mirrors can be reconfigured at rates of a few kilohertz. The deformable mirrors that have been used for adaptive microscopy typically have a few tens of actuators. Other mirrors being developed for astronomy have thousands of actuators.

Liquid crystal devices can also be used to produce or correct aberrations. Whereas a deformable mirror alters the OPL by changing the distance a wavefront has traveled, a liquid crystal spatial light modulator (LCSLM) instead changes the effective refractive index of the medium through which the wavefront propagates (Chapter 9). Different liquid crystal technologies have been used, the most common being nematic liquid crystals. Many of these LCSLMs are adapted from devices developed for the display industry and so are relatively inexpensive compared to deformable mirrors and are driven by standard video outputs. The operation of these devices is typically polarization and wavelength dependent and is limited to speeds of a few tens of hertz. An advantage of this technology is that LCSLMs have a large number of pixels and so are capable of producing more complex phase patterns than a deformable mirror. Light efficiency can, however, be low in comparison with deformable mirrors.

Various factors influence the choice of adaptive element for a particular application. These include light efficiency, number of actuators (or pixels), correction range, and speed. In some areas of microscopy, light efficiency is of paramount importance. For example, in biological microscopy of live specimens, it is highly desirable to minimize the illumination intensity in order to reduce phototoxic effects. Furthermore, in fluorescence imaging, fluorophore densities should be maintained at the lowest possible level to minimize perturbation of the specimen. These factors conspire to ensure that the detected light levels are very low, and high detection efficiency becomes essential. For this reason, mainly deformable mirrors have been used in adaptive microscopes. LCSLMs are poorly suited for use with unpolarized, broadband light, such as fluorescence. Some microscope configurations are, however, compatible with LCSLMs, particularly when aberration correction is only required in the illumination path (such configurations are discussed in Section 14.4). Although losses may occur in this path, they are readily compensated by increasing the illumination power, while maintaining acceptable power levels in the specimen.

The required number of mirror actuators or LCSLM pixels is related to the complexity of the aberrations in the system. If a deformable mirror has more actuators, it will generally be able to compensate more complex aberrations. Most implementations of adaptive microscopes have achieved significant improvement in image quality using only a few low-order aberrations modes, requiring mirrors with a few tens of actuators. Deformable mirrors are often specified using the maximum throw of the mirror surface – this is typically a few micrometers to a few tens of micrometers. While this quantity gives some indication of the correction range, a larger value does not necessarily mean a better mirror. The maximum throw might only be obtained for one specific mirror shape and does not provide information about the achievable ranges for individual aberration modes. The suitability of a deformable mirror for a particular application depends upon these factors and the nature of the system aberrations. LCSLMs generally have large numbers of pixels and are, thus, able to produce many aberration modes with high complexity. They can also produce large amplitude aberrations using phase wrapping – that is, using jumps of 2π radians to wrap phase large variations into the range $0-2\pi$ radians – although this approach can only be used for narrow band illumination.

14.4
Aberration Correction in High-Resolution Optical Microscopy

A wide range of high-resolution microscopy methods have been developed based around different optical configurations and contrast mechanisms (Chapter 13). In all of these microscopes, the image resolution and contrast are affected by aberrations, although the exact effects and the benefit of adaptive correction are different in each case. In this section, we briefly review these methods and discuss how adaptive correction systems can be incorporated in each case.

14.4.1
Microscope Configurations

One can think of a microscope as consisting of two parts: an illumination path and a detection (or imaging) path. In *dia*-illumination microscopes, these paths are totally separate using different condenser and objective lenses (as illustrated in Figure 14.9a). In *epi*-illumination microscopes, illumination and detection are performed via the same objective lens (Figure 14.9b). There are important points related to these configurations that influence the design and implementation of adaptive optical microscopes. Aberrations can clearly be induced in both the illumination and detection paths, so one could conclude that correction would be required in both paths. However, depending upon the contrast mechanism and imaging configuration used, the aberrations in a particular path may have no effect on the final image quality. A well-known example of this is the conventional microscope, where condenser aberrations have no effect on the overall imaging properties of the microscope. It follows that aberration correction may only be required in a single

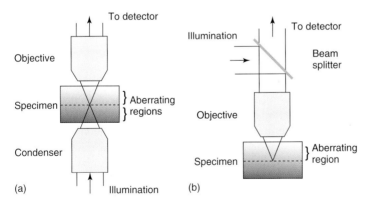

Figure 14.9 Diagram of the focusing configurations in (a) *dia*-illumination microscopes and (b) *epi*-illumination microscopes. (a) Specimen regions below the focal plane contribute to illumination path aberrations, whereas regions above the focal plane lead to detection path aberrations. (b) The illumination and detection paths are affected by the same specimen region.

pass, through either the illumination or detection system. Adaptive microscopes can, therefore, be categorized as single or dual pass, depending upon whether correction is required in one or both of the illumination and detection paths. Dual pass systems may require two separate correction elements, although in many cases it is possible to use a single adaptive correction element if the induced aberrations are the same in both paths [13].

High-resolution optical microscopy methods can also be categorized into two groups: point-scanning and widefield microscopies. In widefield microscopes, a whole two-dimensional image is acquired simultaneously. Conventional microscopes are the most common example of this type. In point-scanning microscopes, the image is acquired in a point-by-point manner. These two types of microscopes offer different challenges in designing adaptive optical systems; these are discussed in the following sections.

14.4.2
Point-Scanning Microscopes

Scanning optical microscopes are widely used for high-resolution imaging, mainly because certain implementations provide three-dimensional resolution with optical sectioning and are, thus, particularly useful for imaging the volume structures of biological specimens.[3] In these microscopes, illumination is provided by a laser that is focused by an objective lens into the specimen. The light emitted from the specimen is collected, usually through the same objective lens, and its intensity is measured by a single photodetector. The focal spot is scanned through the specimen in a raster pattern and the image is acquired in a point-by-point manner. The resulting data are stored and rendered as images in a computer.

The most common example of this type is the confocal microscope, which can be operated in reflection or fluorescence mode (Chapter 13). Three-dimensional resolution is achieved by the placement of a pinhole in front of the photodetector. In a reflection mode confocal microscope, the illumination is scattered by objects not only in the focal region but also throughout the focusing cone. In fluorescence mode, emission is generated not only in the focus but also in out-of-focus regions. The pinhole ensures that mainly light from the focal region falls upon the detector and light from out-of-focus planes is obscured. It is critical in the confocal microscope that both the illumination and detection paths are diffraction limited. This ensures (i) that the illuminating focal spot is as small as possible and (ii) that the focus is perfectly imaged on to the detector pinhole. Therefore, in an adaptive confocal microscope, aberration correction must be included in both paths. This dual pass adaptive system can usually be implemented using a single deformable

---

3) Optical sectioning refers to the ability of the microscope to exclude light from outside the focal plane. This permits, for example, discrimination of the axial position of a planar specimen. A degree of three-dimensional resolution can be obtained in conventional microscopes, but they are not able to perform axial discrimination of planar specimens.

mirror, if the path length aberrations are the same for both the illumination and the emission light. This is the case if there is no significant dispersion in the specimen or chromatic aberration in the optics.

Several other point-scanning microscope modalities have been introduced, including two-photon excitation fluorescence (TPEF) microscopy, second harmonic generation (SHG) and third harmonic generation (THG) microscopies, and coherent anti-Stokes–Raman (CARS) microscopy [9]. Rather than using confocal pinholes to provide three-dimensional resolution, these methods rely upon nonlinear optical processes to confine the emission to the focal spot. This effect can be illustrated by comparing the TPEF microscope with a confocal "single-photon" fluorescence microscope. In the confocal microscope, fluorescence emission is proportional to the illumination intensity, so significant emission is generated throughout the focusing cone. In the TPEF microscope, fluorescence emission is proportional to the square of the illumination intensity, so out-of-focus fluorescence is significantly reduced and emission is only generated in significant amounts in the focal spot. A pinhole is not required to obtain three-dimensional resolution, so most TPEF microscopes use large area detectors to maximize signal collection. Although they rely upon other physical processes, nonlinear imaging modalities such as SHG, THG, and CARS exhibit similar resolution properties. When using large area detectors, the fidelity of imaging in the detection path is unimportant, so the effects of any aberrations in this path are negated. It follows that single pass adaptive optics is appropriate for these microscopes as aberration correction need only be implemented in the illumination path. Adaptive optics systems have been successfully combined with several point-scanning microscope systems including confocal [13], TPEF [6, 14, 15], harmonic generation [16, 17], and CARS [18]. Example images of aberration correction in an adaptive THG microscope are shown in Figure 14.10.

14.4.3
Widefield Microscopes

In widefield microscopes, a whole two-dimensional image is acquired simultaneously using a wide area detector array, such as a charge-coupled device (CCD) camera. This is, of course, the principle behind all conventional microscopes since the first demonstrations in the sixteenth century. Although the principle of a conventional widefield microscope is old, it is still the central engine behind many modern high-resolution imaging techniques. In conventional microscopes, widefield illumination is provided using either transmission optics or, in the case of reflection or fluorescence modes, via the objective lens in an *epi* configuration. In either case, the image quality depends only on the optics of the detection path and is independent of the fidelity of the illumination path. Aberration correction is, therefore, necessary only in the detection path and a single pass adaptive optics system will suffice.

A number of techniques have been developed for providing optically sectioned, three-dimensionally resolved images from a widefield microscope. They all rely

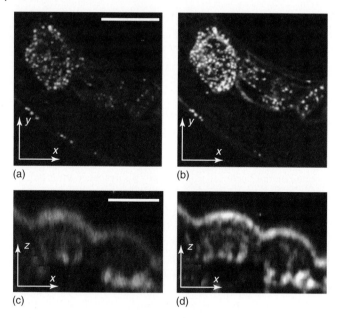

Figure 14.10 Adaptive aberration correction in a THG microscope [16]. (a,b) Lateral plane (*xy*) images of a *Caenorhabditis elegans* specimen before (left) and after (right) aberration correction. Scale bar 50 μm. (b,c) Axial plane (*xz*) images of live mouse embryos. Scale bar 10 μm.

upon the use of nonuniform patterns to illuminate the specimen and a combination of optical and/or computational techniques to separate in-focus light from out-of-focus contributions. One class of widefield sectioning microscopes is best considered as a parallel form of a point-scanning microscope. For example, in Nipkow-disk-based confocal microscopes, an array of focal spots is scanned across the specimen. Each spot has a conjugate detector pinhole that, as in the confocal microscope, admits light only from the focal plane. The light passing through the pinholes is imaged by a camera, which with suitable averaging provides a two-dimensional sectioned image of the specimen (see Chapter 10 of Ref. [9]). As with the confocal microscope, aberrations can affect this microscope in both the illumination and detection path and a dual pass adaptive system is required.

Multispot parallel versions of the TPEF microscope have also been demonstrated. In the simplest case, a microlens array on a rotating disk provides multiple spot illumination, much like the Nipkow disk confocal microscopes [19]. As with the majority of TPEF microscopes, these systems operate without confocal pinholes – the fluorescence emission is imaged directly onto the camera. It is clear that aberrations in the illumination path will affect the quality of the focal spots used to illuminate the specimen, just as they affect the point-scanning TPEF microscope. The effect of aberrations in the detection path is more subtle. In the previous section, it was explained that detection path aberrations have no effect on TPEF microscope image quality when a large area detector is used. However, in this case, the quality of the final image depends upon the fidelity of imaging the fluorescence emission onto the camera. Although all of the collected emission from a focal spot falls upon the camera, it may be detected by the wrong pixels if the spot image is blurred by aberrations. This indicates that dual pass aberration correction is required.

Other widefield sectioning microscopes use illumination patterns to demarcate the focal plane of the microscope. In the simplest form, a sinusoidal spatial variation in intensity is projected into the specimen. The pattern appears clearly in the focus, but rapidly becomes weaker as one moves away from the focal plane. A raw image of a specimen illuminated in this way consists of a patterned in-focus section superimposed upon out-of-focus blur. The in-focus section can be retrieved using a combination of optical and computational techniques. In one system, this is achieved by translating the sinusoidal pattern and processing the resulting sequence of images [20]. Other methods use correlation optical filtering before acquiring the images [21, 22]. In all of these systems, aberrations affect both the illumination and detection paths, so dual pass adaptive systems are required. Adaptive optics has been successfully implemented in widefield microscopes both in conventional imaging [23, 24] and with optical sectioning [20].

The different microscope systems mentioned above and their categorizations are summarized in Table 14.1. The groupings indicate common properties between each microscope that influence the design and implementation of adaptive aberration correction. This summary covers the majority of microscope architectures. Some advanced microscopy methods, particularly super-resolution methods, have been omitted for the sake of brevity. However, the approaches discussed here could, in principle, also be extended to these systems.

Table 14.1 Classification of common microscope methods with regard to the implementation of adaptive optics.

	Point scanning	Widefield
Single pass (illumination only)	Two-photon fluorescence SHG, THG CARS	
Single pass (detection only)		Conventional
Dual pass	Confocal fluorescence Confocal reflection Two-photon fluorescence (with pinhole)	Structured illumination Nipkow disk confocal Disk-based sectioning Programmable array microscopy Multispot two photon

14.5
Aberration Measurement and Wavefront Sensing

Aberration measurement is a significant challenge in adaptive microscopy. The standard methods developed for other adaptive optics applications are not easily adapted to the more complicated optical systems of the microscope. In this section, we discuss these issues and explain some of the methods that can be employed.

14.5.1
Direct Wavefront Sensing in Microscopy

Conventional wavefront sensors, such the Shack–Hartman sensor, are widely used in adaptive optics, but their use in microscope systems has been much more limited. A major reason for this is the difficulty of incorporating this method into an adaptive microscope. The sensor requires a well-defined wavefront in the pupil of the system; this could be provided, for example, by a distant pointlike emitter, such as a guide star in an astronomical telescope. Equivalently, in the microscope this well-formed wavefront could be produced by a pointlike emitter in the focus. In general, however, the light emitted from the focal region arises from the two- or three-dimensional structure of the specimen, which produces a superposition of wavefronts in the pupil. The effects of this superposition depend upon the coherence of the emitted light: If coherent, for example when using back reflected laser illumination in a point-scanning microscope, then the various contributions will interfere, causing phase and amplitude variations in the pupil; this will lead to ambiguous wavefront sensor readings. Potential ambiguities are illustrated in Figure 14.11. The illumination light is focused through the specimen, which introduces a phase aberration $\Phi(r, \theta)$; this leads to distortion of the focal spot. If the specimen is a pointlike scatterer, then it is excited by the illumination field and reemits spherical wavefronts. These wavefronts pass back through the specimen, acquiring the aberration $\Phi(r, \theta)$ in the process, and are collected by the objective lens. In this case, the aberration $\Psi(r, \theta)$ measured by the sensor is the aberration induced in the emission path, that is, $\Psi(r, \theta) = \Phi(r, \theta)$, which is by good fortune also the same as that induced in the illumination path. The measurement in this case would clearly be suitable for controlling a correction element. However, if the specimen has a planar, mirrorlike structure in the focal region, a different aberration will be measured (Figure 14.11). The illumination path aberration is again $\Phi(r, \theta)$. A ray passing through the illumination pupil on one side of the optic axis reflects off the mirror and passes through the diametrically opposing point in the detection pupil. It follows that the aberration induced by the specimen in the detection path is the spatial inversion of that from the illumination path. Therefore, we denote the detection path aberration by $\tilde{\Phi}(r, \theta)$, where the tilde indicates spatial inversion about the optic axis: $\tilde{\Phi}(r, \theta) = \Phi(r, \theta + \pi)$. It is instructive here to express the specimen aberration as $\Phi(r, \theta) = \Phi_e(r, \theta) + \Phi_o(r, \theta)$, where Φ_e and Φ_o represent the even and odd components of Φ. The measured aberration $\Psi(r, \theta)$

14.5 Aberration Measurement and Wavefront Sensing

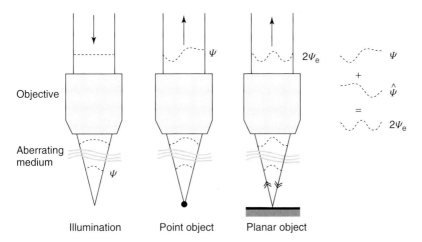

Figure 14.11 The effects of specimen structure on wavefront measurements in dual path microscopes. An aberration Ψ is induced in the illumination wavefronts. As the point object scatters the light from a point in the focus, the phase variations arising in the illumination path are lost. Only the emission path aberration Ψ is measured. For a planar reflector, the illumination wavefront is spatially inverted on reflection before acquiring further aberration in the detection path. The measured aberration is $2\Psi_e$, twice the even component of Ψ.

is the sum of the illumination and emission path aberrations:

$$\Psi = \Phi + \tilde{\Phi} = \Phi_e + \Phi_o + \tilde{\Phi}_e + \tilde{\Phi}_o \tag{14.10}$$

For clarity, we have dropped the explicit dependence of these function on the coordinates (r, θ). However, since $\tilde{\Phi}_e = \Phi_e$ and $\tilde{\Phi}_o = -\Phi_o$, the measured aberration can be written as

$$\Psi = 2\Phi_e \tag{14.11}$$

and we see that the wavefront sensor measures twice the even component of the aberration, but is not sensitive to the odd component. However, for the pointlike specimen, we obtained

$$\Psi = \Phi = \Phi_e + \Phi_o \tag{14.12}$$

which shows that both even and odd components are detected. There is an obvious ambiguity between these two measurements. Similar problems occur for specimens with more complex three-dimensional structure due to the coherent superposition of wavefronts emanating from different parts of the specimen.

The three-dimensional nature of the specimen leads to another problem: the total amount of light scattered from out-of-focus planes can be far more intense than the light from the focal region, which is required for the aberration measurement. This contrast can be improved by including a spatial filter in an intermediate image plane between the microscope objective and the sensor. This filter performs a similar function to the pinhole in a confocal microscope by admitting light predominantly from the focal region. In addition to this optical sectioning property, the pinhole has a further effect of smoothing the wavefront. The filter would normally be larger than that used in a confocal microscope to avoid oversmoothing and loss of aberration information. An alternative, albeit more complex, method of excluding out-of-focus light uses coherence gating [25]. In this method, the coherence properties of the illumination are used to ensure that only light from the focal region contributes to the aberration measurement. The optical sectioning effect is determined by the coherence length of the source.

To some degree, the wavefront sensing problem is mitigated in a fluorescence microscope as fluorescence emission is an incoherent process. While the level of fluorescence depends upon the intensity of the focal spot and, hence, the aberrations in the illumination path, only the emission path aberrations can be measured by the sensor. However, in a three-dimensional specimen, fluorescence can be generated throughout the illumination cone and all of this light falls upon the sensor. It is possible, therefore, that the sensor becomes swamped with out-of-focus emission that obscures the desired in-focus light. As in the reflection mode microscope, for direct wavefront sensing to be effective, a method for eliminating the effects of out-of-focus emission is needed. Again, the use of a pinhole filter may be beneficial.

In two-photon microscopy, fluorescence emission is already confined to the focal spot, so it is, in principle, possible to use this emission as the wavefront source. This approach is limited by the very low light levels that are typically produced by these microscopes. Emission is also confined to the focus in other nonlinear methods, such as SHG, THG, or CARS microscopy. However, the signal generation process in these microscopes is coherent, so interaction between the phase aberrations in the illumination path and the detection path may still play a role, in a similar manner to the reflection mode microscope.

14.5.2
Indirect Wavefront Sensing

The complexities of using direct wavefront sensing, as detailed in the preceding discussion, mean that most implementations of adaptive microscopes so far have employed indirect wavefront sensing. In these systems, the initial aberration is measured and corrected through a sequential method. A sequence of images is acquired for which different aberrations are intentionally introduced by the adaptive element. For each of the acquired images, a quality metric M is calculated; this may be, for example, the total image intensity or image sharpness. From this set of measurements of M, an estimate of the initial aberration is made and the corresponding correction is introduced using the adaptive element. Such a process is illustrated in Figure 14.8. This aberration correction procedure is equivalent to a mathematical optimization problem, where the goal is to find the maximum or the minimum value of a function. Various approaches based upon heuristic optimization algorithms can be considered to solve this problem. For example, an exhaustive search of all possible adaptive element configurations would guarantee the best solution, but it is not likely to be a practical approach due to the excessive number of required measurements. Various model-free stochastic algorithms ranging from random searches to genetic algorithms have been used [26, 27]; these provide improved aberration correction but frequently require a large number of measurements. In most situations, however, it is not necessary to resort to stochastic algorithms as efficient, model-based, deterministic methods can be derived.

It is desirable to use an algorithm that minimizes the number of measurements. This has obvious benefits in terms of speed of operation and helps mitigate problems that may arise due to specimen motion between measurements. The need for fewer measurements also means lower cumulative specimen exposure, which leads to reduced photobleaching of fluorescence and reduced phototoxicity in live specimens. The number of measurements can be minimized by using a model-based algorithm, which takes into account the functional form of the optimization metric. In many practically relevant situations, the metric has a well-defined maximum with paraboloidal form. If the aberration is expressed as a series of N basis modes, such as Zernike polynomials,

$$\phi(r,\theta) = \sum_{i=1}^{N} a_i \psi_i(r,\theta) \tag{14.13}$$

then the metric has the general parabolic form:

$$M \approx q \left(1 - \sum_{k=1}^{N} \sum_{l=1}^{N} \gamma_{kl} a_k a_l \right) \tag{14.14}$$

where q and γ_{kl} are constants. This function has a clearly defined maximum, but in this formulation each term depends upon the coefficients of two aberration modes. For the purposes of optimization, a more convenient from of Eq. (14.14) would have terms each of which contains only one coefficient. This can be obtained by using a new set of aberration modes $\xi_i(r, \theta)$ chosen specifically to have the desired properties, giving a metric with the form

$$M \approx q \left(1 - \sum_{i=1}^{N} \beta_i b_i^2 \right) \tag{14.15}$$

where β_i are constants and b_i are the coefficients of the new modes [20]. This transformation is useful as the different modal coefficients appear separately in the expansion of M. This means that each mode can be optimized separately from the others using a one-variable quadratic maximization algorithm without explicit knowledge of the other variables. This property arises because the modes satisfy a form of orthogonality relationship that is specific to the type of microscope and choice of metric. One approach for designing such a scheme is to choose a metric and set of modes with known properties so that the metric has the required form. This approach has been used in some sensorless adaptive systems, for example, using Lukosz modes with a metric derived from the low-frequency content of an image [23]. Alternatively, it has been shown that any set of basis modes (e.g., Zernike polynomials) can be transformed into new set of modes for which the metric has the form of Eq. (14.15) [20].

In practice, maximization is implemented by intentionally applying predetermined aberrations, or biases, into the system using the adaptive element. First, a positive bias $+x\xi_i$ of the measured mode ξ_i is added, where x is a suitably chosen amplitude, and the metric value M_+ is calculated. Then, a negative bias of the same mode $-x\xi_i$ is applied, and the metric value M_- is found. Additionally, the metric value M_0 is calculated when no aberration is applied. The correction aberration is then found through parabolic maximization as

$$a_{\text{corr}} = \frac{x(M_+ - M_-)}{2M_+ - 4M_0 + 2M_-} \qquad (14.16)$$

The measurement and correction cycle is then repeated for each of the N modes of interest. This process of measuring a single mode is illustrated in Figure 14.12 using an image-based sensorless adaptive transmission microscope with a holographic scatterer plate as the specimen [23]. In this case, Lukosz aberration modes were used in conjunction with a metric calculated from the low spatial frequency content of the image.

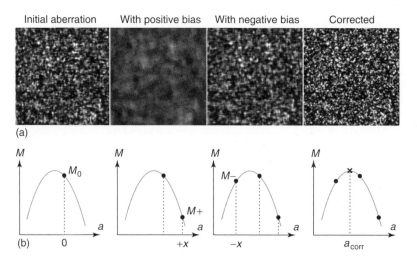

Figure 14.12 Indirect wavefront measurement and correction of a single aberration mode. (a) Transmission microscope images. (b) The quadratic maximization process. Three images are taken, each with a different amount of the mode applied, and the corresponding values of M are calculated. The correction aberration is predicted by quadratic interpolation between these three values.

In Eq. (14.16), three measurements of M are required for each mode. However, the value of M_0 is the same for each mode, as the measurement is taken with zero applied aberration bias. Hence, the N modal coefficients b_i can be obtained from a total of $2N + 1$ metric measurements. Similarly, $2N + 1$ evaluations are required to solve for the unknowns in Eq. (14.15); in addition to the N modal coefficients, there are $N + 1$ unknown variables: q and the N values of β_i, all of which depend upon the image structure. Therefore, it is important to note that this method allows for the determination of the coefficients b_i independently of the effects of specimen structure.

The quadratic approximation used in Eq. (14.15) is valid for small values of the coefficients b_i. For large aberrations, the approximation is not reliable. This places an upper limit on the size of aberrations for which a given sensorless scheme of this type can operate. For some systems, it is possible to derive different combinations of modes and metrics that are valid for different aberration amplitude ranges [12, 23]. This permits a sequence of coarse and finer correction cycles in order to gradually correct large aberrations.

In these sensorless systems, it is important to exclude aberration modes that have no (or a weak) effect on the optimization metric. In most microscope systems, the first four Zernike modes, piston, tip, tilt, and defocus, should be excluded: piston is a constant phase shift, so it has no effect on the image; tip and tilt result in an image shift, so have they no effect on image-based metrics such as total intensity;[4] a similar argument applies to defocus in a three-dimensional imaging system where an averaged intensity over a volume is used.[5] It is also important to exclude the tip, tilt, and defocus modes to avoid image repositioning or refocusing by the adaptive system. This might occur, for example, if a bright object was situated close to a dimmer-imaged region – a system using an intensity-based metric and including these three modes would tend to move the imaged region to the brighter object.

14.6
Control Strategies for Adaptive Microscopy

In the previous sections, we have considered the choice of aberration correction and measurement methods for adaptive microscopes. The adaptive optics system is completed by a control system that takes the wavefront measurement, processes the information, and uses it to control the adaptive correction element. The way in which aberrations arise in microscope systems leads to important differences in the requirements of adaptive microscopes and other adaptive optics systems. Some of these factors are discussed in this section.

---

4) If bright objects are located only at the periphery of the image, it is possible that the tip or tilt mode could cause them to move out of the imaged region. In this case, the intensity metric could be affected.
5) The Zernike defocus mode, which is quadratic in the radial coordinate, only approximately models the refocusing effect in high NA systems. Strictly speaking, one should instead use a high NA defocus function, which includes higher order radial terms.

14.6.1
Choice of Signal for Wavefront Sensing

The close correspondence between the measured aberration and the aberration that causes image degradation is an important factor in the design of the adaptive optics system. Problems can arise when the wavelength of the measurement light is different to the illumination, as chromatic aberrations may be present in the optics, or dispersion in the specimen. This could be a significant problem in multiphoton microscopies, where the emission wavelength is a fraction of the excitation wavelength. To avoid this problem, it is desirable to measure the aberration at the excitation wavelength. This could be implemented by using illumination light scattered from the sample as the source for a wavefront sensor. However, this measurement would be subject to the ambiguities discussed in Section 14.5.1. Despite this potential limitation, there would be significant benefit in using backscattered light for sensing in biological microscopy, as the necessary illumination powers would be much lower than those required for, say, fluorescence microscopy. The detrimental effects of photobleaching and phototoxicity could, therefore, be significantly reduced during the aberration measurement process.

Any wavefront sensing method, whether direct or indirect, requires sufficient light levels in order to provide a reliable measurement. The light emitted from the specimen must, therefore, be bright enough in the region of measurement to exceed the background noise level. In some parts of the specimen, those corresponding to dark image regions, this will not be achieved and wavefront measurement will be impossible. It is possible, however, that these apparently dark regions contain dim features that would be visible if aberration correction were implemented. An effective control strategy for an adaptive microscope should overcome this limitation. One way to tackle this is to derive a scheme that uses local averages – this is based on the hypothesis that aberrations vary slowly across a field of view, so the aberrations required in dark regions can be interpolated from those measured in the surrounding bright regions. This approach is inherent in the image-based indirect sensing schemes outlined in Section 14.5.2.

A further advantage of indirect wavefront sensing is that the optimized metric is directly related to the image quality. When using direct sensing, the measured wavefront is coupled out of the optical train at a point remote from the object and the image sensor. It is possible, therefore, that the measured wavefront contains aberrations different from those that determine the image formation. These noncommon path errors are introduced as the wavefronts do not follow exactly the same path through the system. This problem is avoided in sensorless systems as the correction is determined directly from the image quality metric.

14.6.2
Aberration Dynamics

One important factor to consider in designing a control strategy is the way in which aberrations vary over time. High-resolution optical microscopes are often used to image biological structures and processes in fixed or live specimens. A fixed specimen is clearly stationary, so light passing through the specimen on a particular path will always suffer the same aberration. Even when imaging fast biological processes in live specimens, aberrations usually vary slowly with time. This is because most aberrations arise from the large-scale structure of a specimen through which the light is focused – for example, the overall arrangement of cells in thick tissue – rather than the small-scale rapid changes that are being imaged. Therefore, aberration compensation in microscopes can often be achieved through semistatic correction – the adaptive element can be set with a single correction aberration, which is maintained while imaging a particular region. The correction needs to be modified only when a different specimen region is imaged. The temporal bandwidth required of an adaptive element in such a system is much lower than that required for, say, compensation of atmospheric turbulence in a telescope.

In a system with a single adaptive element in the pupil plane, this semistatic method can provide the same aberration correction for the whole imaged region. Rapid changes in aberration correction may, however, be required in point-scanning microscopes if aberrations change significantly across the image field. These field-dependent aberrations are discussed in the next section.

14.6.3
Field-Dependent Aberrations

The form of the induced aberrations depends upon the path the light takes through the specimen. The light associated with one point in an image propagates through a different part of the specimen as compared to the light associated with a nearby point. If the specimen is reasonably homogeneous and the field of view is not too large, then the aberrations will be similar across the whole field. In this situation, a single setting of the adaptive element would be sufficient to completely correct the aberrations and produce a diffraction-limited image. However, if aberrations change significantly across the field of view, a single correction aberration cannot completely compensate for the whole field.[6] This phenomenon has different consequences in point-scanning and widefield microscopes.

In widefield microscopes, the full image field is acquired simultaneously, so in a system using a single adaptive element it is not possible to obtain complete correction of the whole image. Full aberration correction could be achieved if it is limited to a subregion of the field of view. Alternatively, partial correction of the whole field is possible. In this case, partial correction would mean that the image quality is improved on average, but not optimally in every location. This could mean that no single image point is fully corrected and, moreover, that the aberrations could be made worse in certain parts of the image.

6) In astronomy, this field-dependent aberration variation is referred to as *anisoplanatism*.

In principle, this problem can be overcome in a point-scanning system as only a single point need be imaged at any one time. The adaptive element can, therefore, be reconfigured to change the aberration correction for each point in the scan. In practice, this approach is limited by the response time of the adaptive element, which is typically orders of magnitude longer than the pixel dwell time of the microscope. This limitation is slightly mitigated if aberrations vary slowly across the field of view, as the bandwidth of the adaptive element can then be much lower than the pixel frequency.

The problem of field-dependent aberrations could potentially be solved using multiconjugate adaptive optics (MCAO), where multiple adaptive elements are used. In astronomy, MCAO was originally proposed to compensate simultaneously for multiple aberrating layers in the atmosphere. Each of the adaptive elements is designed to be in a plane conjugate to each aberrating layer. The same principle can be applied to the three-dimensional structure of microscope specimens, which can be approximated by a stack of aberrating layers [28]. The collection of adaptive elements forms an inverse specimen that unwraps any aberrations suffered by light emanating from any part of the specimen. An MCAO system would permit full correction of field varying aberrations in both widefield and point-scanning microscopes. Adaptive point-scanning microscopes could, therefore, be implemented using semistatic correction, avoiding the pixel rate limitations mentioned earlier. The additional complexity of these systems means that practical implementation of such MCAO in microscopes would require significant development of aberration correction, measurement, and control schemes.

14.7
Conclusion

Aberrations can be a significant problem in high-resolution optical microscopy. Even if the greatest care is taken to minimize aberrations in the design of the microscope, the specimen forms part of the optical system that is outside of the designer's control. Refractive index variations in the specimen induce aberrations that degrade the imaging properties of the microscope, leading to reductions in resolution, contrast, and signal levels. The incorporation of adaptive optics into these systems can enhance the performance by removing the problems caused by aberrations. Adaptive optics can restore the optimal imaging properties of the microscope, ensuring that they are maintained even when focusing deep into thick specimens. The compensation of aberrations can have other beneficial consequences including faster imaging, reduced photobleaching, and lower phototoxicity.

The combination of these technologies does, however, present numerous challenges. The conventional models for adaptive optics systems are not easily translated to high-resolution microscopes. The optical layouts of these microscopes are often much more complicated than the telescope systems for which adaptive optics was first developed. Conventional methods of wavefront sensing are not ideally suited to use in microscopes and the current capabilities of adaptive elements place restrictions on the aberration correction process. However, adaptive correction has been successfully implemented in a number of microscope architectures and clear improvements in imaging quality have been demonstrated. There is plenty of scope for further development of this technology and numerous application areas that could benefit. In particular, the development of adaptive microscopes will further enable the use of high-resolution optical imaging in more challenging situations, such as imaging deeper into living specimens.

Acknowledgments

The two-photon microscope images in Figure 14.4 were obtained by Dr. D. Débarre. The third harmonic images of Figure 14.10 were obtained by Dr. A. Jesacher and Dr. A. Thayil. This work was funded by the Engineering and Physical Sciences Research Council and the Biotechnology and Biological Sciences Research Council.

References

1. Tyson, R.K. (1991) *Principles of Adaptive Optics*, Academic Press, London.
2. Dainty, J.C. (ed.) (2007) *Adaptive Optics for Industry and Medicine*, World Scientific, London.
3. Booth, M.J. (2007) Adaptive optics in microscopy. *Philos. Transact. A Math. Phys. Eng. Sci.*, **365**, 2829–2843.
4. Mahajan, V.N. (2001) *Optical Imaging and Aberrations, Part II. Wave Diffraction Optics*, SPIE, Bellingham, WA.
5. Born, M. and Wolf, E. (1983) *Principles of Optics*, 6th edn, Pergamon Press, Oxford.
6. Débarre, D. et al. (2009) Image-based adaptive optics for two-photon microscopy. *Opt. Lett.*, **34**, 2495–2497.
7. Gibson, S.F. and Lanni, F. (1992) Experimental test of an analytical model of aberration in an oil-immersion objective lens used in three-dimensional light microscopy. *J. Opt. Soc. Am. A Opt. Image Sci. Vis.*, **9**, 154–166.
8. Schwertner, M., Booth, M.J., and Wilson, T. (2004) Characterizing specimen induced aberrations for high NA adaptive optical microscopy. *Opt. Exp.*, **12**, 6540–6552.
9. Pawley, J.B. (2006) *Handbook of Biological Confocal Microscopy*, 3rd edn, Springer, New York.
10. Booth, M.J., Neil, M.A.A., and Wilson, T. (1998) Aberration correction for confocal imaging in refractive-index-mismatched media. *J. Microsc.*, **192**, 90–98.
11. Hariharan, P. (2003) *Optical Interferometry*, Academic Press, London.
12. Booth, M.J. (2007) Wavefront sensorless adaptive optics for large aberrations. *Opt. Lett.*, **32**, 5–7.
13. Booth, M.J. et al. (2002) Adaptive aberration correction in a confocal microscope. *Proc. Natl. Acad. Sci. U.S.A.*, **99**, 5788–5792.
14. Marsh, P.N., Burns, D., and Girkin, J.M. (2003) Practical implementation of adaptive optics in multiphoton microscopy. *Opt. Exp.*, **11**, 1123–1130.
15. Rueckel, M., Mack-Bucher, J.A., and Denk, W. (2006) Adaptive wavefront correction in two-photon microscopy using coherence-gated wavefront sensing. *Proc. Natl. Acad. Sci. U.S.A.*, **103**, 17137–17142.
16. Jesacher, A. et al. (2009) Adaptive harmonic generation microscopy of mammalian embryos. *Opt. Lett.*, **34**, 3154–3156.

17. Olivier, N., Débarre, D., and Beaurepaire, E. (2009) Dynamic aberration correction for multiharmonic microscopy. *Opt. Lett.*, **34**, 3145–3147.
18. Wright, A.J. *et al.* (2007) Adaptive optics for enhanced signal in CARS microscopy. *Opt. Exp.*, **15**, 18209–18219.
19. Bewersdorf, J., Pick, R., and Hell, S.W. (1998) Multifocal multiphoton microscopy. *Opt. Lett.*, **23**, 655–657.
20. Débarre, D. *et al.* (2008) Adaptive optics for structured illumination microscopy. *Opt. Exp.*, **16**, 9290–9305.
21. Wilson, T. *et al.* (1996) Confocal microscopy by aperture correlation. *Opt. Lett.*, **21**, 1879–1881.
22. Hanley, Q.S. *et al.* (1999) An optical sectioning programmable array microscope implemented with a digital micromirror device. *J. Microsc.*, **196**, 317–331.
23. Débarre, D., Booth, M.J., and Wilson, T. (2007) Image based adaptive optics through optimisation of low spatial frequencies. *Opt. Exp.*, **15**, 8176–8190.
24. Kner, P. *et al.* (2010) High-resolution wide-field microscopy with adaptive optics for spherical aberration correction and motionless focusing. *J. Microsc.*, **237**, 136–147.
25. Feierabend, M., Rückel, M., and Denk, W. (2004) Coherence-gated wave-front sensing in strongly scattering samples. *Opt. Lett.*, **29**, 2255–2257.
26. Sherman, L. *et al.* (2002) Adaptive correction of depth-induced aberrations in multiphoton scanning microscopy using a deformable mirror. *J. Microsc.*, **206**, 65–71.
27. Wright, A.J. *et al.* (2005) Exploration of the optimisation algorithms used in the implementation of adaptive optics in confocal and multiphoton microscopy. *Microsc. Res. Tech.*, **67**, 36–44.
28. Kam, Z. *et al.* (2007) Modelling the application of adaptive optics to wide-field microscope live imaging. *J. Microsc.*, **226**, 33–42.

15
Aperture Synthesis and Astronomical Image Formation
Anna Scaife

15.1
Introduction

The vastness of space is filled with a myriad of different physical processes. For many years, our understanding of these processes was limited to the very nearest objects, visible to us in optical light. It was only in 1932, when Karl Jansky's discovery of radio "noise" from the galactic plane became the first observational evidence for radio emission from astronomical bodies, that the radio universe began to be realized. Even then, it was not until the advent of the Second World War, when the use of radar technology became important, that radio studies of the sun and meteors became a major topic of research. At that time, these observations faced a serious instrumental resolution problem. Just as for optical observations, the resolving power of a radio telescope is limited by the diameter of the telescope measured in wavelengths. Since radio wavelengths (meters to centimeters) are so much greater than those of optical light (nanometers), the earliest radio telescopes were limited to beamwidths of several degrees. This problem was overcome in 1946, when Ryle and Vonberg [1] published the first observational data from a radio interferometer. Their measurements allowed the activity of sunspots to be distinguished from the galactic radio background for the first time and provided information at a resolution that would have otherwise required a very large aerial structure to be built.

This new measurement technique was a revolution in radio astronomy. It enabled observations to be made at previously inconceivable resolutions. The science quickly began to evolve. Ryle and Vonbergs' first interferometer had produced a pattern of fringes from the sun as it was scanned across the sky; these fringes were superimposed on the smoothly varying emission from the galactic plane. The next step in the evolution of radio interferometry was to remove even this background radiation. In 1952, Ryle [2] introduced the concept of phase switching into radio interferometry. By the simple process of applying half a wavelength of pathlength to one of the antennae in a pair, he removed the smooth background emission leaving only the fringe oscillations.

Optical and Digital Image Processing: Fundamentals and Applications, First Edition. Edited by Gabriel Cristóbal, Peter Schelkens, and Hugo Thienpont.
© 2011 Wiley-VCH Verlag GmbH & Co. KGaA. Published 2011 by Wiley-VCH Verlag GmbH & Co. KGaA.

Through the 1950s and 1960s, there were a large number of advances in radio interferometry, but it was in 1962 that the very important step of Earth rotation synthesis imaging was introduced [3]. The concept was demonstrated with the ambride One-Mile Radio Telescope. In 1956, the One-Mile Telescope published maps of the strong radio sources Cassiopeia A and Cygnus A, which contained a degree of structural detail incomparable with other current observations [4]. Aperture synthesis had changed the face of radio astronomy, and it was for this pioneering advance that Ryle was awarded the Nobel Prize in 1974.

With the advent of synthesis arrays, radio interferometric mapping became increasingly important and, in addition to the arrays constructed in Cambridge, a number of large synthesis telescopes were built around the world. Perhaps, the most notable were the Westerbork Synthesis Radio Telescope in the Netherlands [5] and the Very Large Array in New Mexico [6]. The technique itself was pushed further with the development of very long baseline interferometry (VLBI). VLBI was introduced to achieve sub-arcsecond resolution at radio wavelengths using antenna separations of hundreds to thousands of kilometers. This was achieved by recording data from individual antennae and combining them later, with no real-time communication link.

The technology of radio interferometry is still evolving and future plans include the construction of the square kilometer array (SKA) telescope. With a square kilometer of collecting area at its disposal, more than 30 times that of any existing array, the SKA will provide a huge jump forward in sensitivity. It will incorporate an interferometric array of phased aperture arrays giving it an instantaneous field of view of several degrees, while baselines of several thousand kilometers will allow it to retain sub-arcsecond resolution.

15.2
Image Formation from Optical Telescopes

The first recorded images from optical telescopes were projected directly onto photographic emulsions and saved as glass or paper plates. One of the most frequently used optical surveys of this type was the Palomar Observatory Sky Survey (POSS; [7]). This survey was carried out between 1949 and 1957 using the 48 in. Samuel Oschin Telescope at the Palomar Observatory (Figure 15.1), and eventually covered the entire Southern Hemisphere in 935 plates each of 14 in. at each frequency. It was later followed by the POSS-II survey [8], which covered the Northern Hemisphere in 894 plates. Modern optical telescopes have, in general, had their observers' eyepieces replaced by computerized sensors. These sensors, such as charge-coupled devices (CCDs), produce much higher quality photometry at greater sensitivity than the original photographic plates. However, due to the discrete number of pixels available when using such devices, it is often necessary to undersample the image in order to achieve a reasonably large field of view. While this presents a problem for ground-based telescopes, it was also a significant issue for the wide-field cameras on the Hubble Space Telescope (HST), where a single

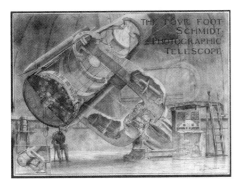

Figure 15.1 A 1941 drawing of the 48 in. Samuel Oschin Telescope at the Palomar Observatory by Russel Porter.

pixel of the CCDs was the same width as the full width half maximum (FWHM) of the optical point spread function (PSF). This undersamples the received signal by a factor of $\simeq 2.5$ in Nyquist terms.

In order to correct for this undersampling, optical systems employ a technique known as dithering, where the position of the detector is shifted by an offset less than the width of an individual pixel. These dithered observations are used to restore the image using a variety of linear reconstruction techniques. Common among these are the interlacing and shift-and-add techniques. The first of these, "interlacing," places the pixels from the individual dither positions in alternating pixels in the output image according to their relative alignment. The second, "shift-and-add," simply shifts the dither images to their appropriate position and adds them onto a subsampled image. Neither of these techniques are ideal, the latter causing significant additional correlation of the noise between pixels.

Consequently, an improvement to these methods is the most widely used image restoration technique. The *Drizzle* algorithm [9] allows the user to shrink the pixels in the constituent drizzle images before placing them onto the subsampled pixels of the output image, rotating them according to the coordinate transform between the Cartesian plane of the CCD and the astronomical image, and compensating for the optical distortions introduced by the camera. The shrunken input pixels are averaged onto the output pixels in a weighted fashion, where the weight is proportional to the area of overlap between the input and output pixel. Drizzle presents a number of improvements over the standard interlacing and shift-and-add methods, although both these techniques represent limiting cases of the algorithm. Importantly, Drizzle preserves the photometry of the input image, which is advantageous for postimaging analysis. In addition, the linear weighting scheme is equivalent to inverse variance weighting and results in optimal signal to noise.

15.3
Single-Aperture Radio Telescopes

In practical terms, the simplest form of radio telescope is a dipole antenna. Antennae such as these have been used many times historically to explore the radio universe and are still in regular use today. In spite of this, most people who are asked to visualize a radio telescope will instead picture a dish. For this reason, we employ the dish as our example in what follows, and indeed the principles described can be easily extended to the dipole.

Unlike the dipole, a radio dish does not receive power from all parts of the sky. The exact reception pattern, $A(\theta, \phi)$, of an individual dish will be the Fourier transform of its *aperture distribution*, the distribution of excitation currents as a function of distance from the center of the dish. The reception pattern of an antenna is called the *beam*. It follows from the Fourier relationship between the beam and the aperture distribution that the *larger* a telescope dish, the *smaller* is its beam. Hence, if very fine detail is required in an image of the sky, it is necessary to employ a very large radio dish to make the measurements. For the resolutions required by modern radio astronomy, the required dish sizes are often impractical.

The sensitivity of a radio telescope is also related to the size of its dish. The power density received by an antenna from a source of flux density S is $SA_{\text{eff}}/2$, where A_{eff} is the effective area of the dish. The effective area is some fractional multiple of the true dish area, usually between 0.6 and 0.7, and arises as a consequence of the imperfect conduction on the dish surface.

In practice, the degree of accuracy to which flux density can be measured is governed by the noise introduced to the antenna mainly through its amplification. This noise can be reduced by (i) increasing the bandwidth, B, of the signal being passed through the system and (ii) increasing the integration time of the signal. The first of these measures reduces the correlation in the signal, which scales as $1/B$, and consequently increases the number of independent measurements. The second measure again increases the number of independent measurements by a factor τ, where τ is the integration time, and so the relative uncertainty in a measurement of antenna temperature, T_A, is also inversely proportional to the square root of the integration time. The antenna temperature has contributions not only from the source but also from the receiver itself, the atmosphere, and also contributions from any side-lobe structure in the beam.

Single-dish radio telescopes suffer from a much poorer rejection of systematics than interferometric telescopes. A typical single-dish measurement employs a beam-switching strategy to implement a differencing scheme. This beam switching is usually done by tilting a subreflector within the main dish so that the source under observation is observed in an interleaved fashion with a comparatively empty region of sky. This type of on–off observing is done in order to eliminate time-varying sky and instrumental effects.

If a larger region of sky is to be observed, then, rather than a single point, a raster scan technique is used. In this case, the region is observed twice in orthogonal directions. The differences in the resulting images are then minimized, often using a least-squares method. This technique is called *basket weaving* and the ends of the scan lengths in both directions are used as base-level references.

15.4
Aperture Synthesis

To achieve the sub-arcsecond resolution in the radio band of the electromagnetic spectrum, as has been achieved by optical telescopes such as the Palomar 48 in. Schmidt telescope, would require prohibitively large radio dishes. This problem of building larger and larger radio dishes in order to improve both the resolution and sensitivity of radio-astronomical observations was overcome by the introduction of interferometry [1]. In its original form, radio interferometry was only suitable for making one-dimensional tracks across a source and it was the advent of *aperture synthesis* [3] that truly revolutionized science. This new technique used the rotation of the Earth itself in order to synthesize apertures on a scale that could never be realized in practice. The arrival of aperture synthesis redefined the observable radio universe and revealed the inner workings of many previously poorly understood objects. This chapter is by no means an exhaustive description of the intricacies of interferometry and aperture synthesis, but is rather an overview of some of the most fundamental aspects and commonly used techniques.

15.4.1
Principles of Earth Rotation Aperture Synthesis

The simplest introduction to radio interferometry is to consider the instantaneous response to a point source from a two-element interferometer. When the distance between the two constituent antennae is not very large and the source may be considered to lie in the far field of the instrument, this may be approached using simple geometric representations. In this case, the two antennae, separated by distance D, known as the *baseline*, will receive a planar wavefront from the source S (Figure 15.2). The wavefront will necessarily be delayed in reaching one of the antennae by a time $\tau_g = (D/c)\sin\theta$, where θ is the angle of the source from the normal to the plane of the antennae. With the interferometer operating as shown in Figure 15.2, the output of the multiplication of the two signals V_1 and V_2 at a frequency ν will be

$$F = 2\sin(2\pi\nu t)\sin(2\pi\nu(t - \tau_g)) \tag{15.1}$$

The multiplied signals are then integrated over a defined time period. The combination of multiplication and integration is a correlation, and the combined voltage multiplier and integrator system is known as the *correlator*. Since the variation of $\nu\tau_g$ will, in general, be far smaller than νt, this multiplication may be approximated by

$$F = \cos(2\pi\nu\tau_g) = \cos\left(\frac{2\pi D\sin\theta}{\lambda}\right)$$

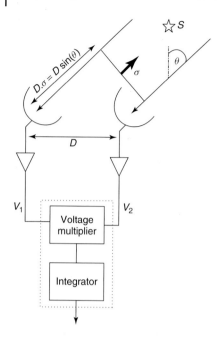

Figure 15.2 A simple two-element interferometer showing the path difference between the two antennae and the correlator that consists of a voltage multiplier and integrator.

In practice, an instrumental time delay τ_i is inserted into the back end of one antenna before multiplication in order to compensate for τ_g toward a defined position in the sky. This predefined position is known as the *phase center* and is generally, although not necessarily, aligned with the peak of the aperture illumination function. In interferometry, the aperture illumination function is known as the *primary beam*, for reasons that will become apparent. This single instrumental delay, τ_i, is only appropriate for one position in the sky; a source offset from the phase center will see a time delay $\tau = \tau_g - \tau_i$.

If the integrator has a time constant $2T \gg \Delta\nu^{-1}$, the output from the correlator will be the autocorrelation function

$$R(\tau) = \frac{1}{2T} \int_{-T}^{T} V(t)V(t-\tau)dt \tag{15.2}$$

This signal will, of course, be bandlimited by the amplifiers in the telescope system and so, using the Wiener–Khinchin relation,[1]

$$R(\tau) = \epsilon(\tau)\cos(2\pi\nu_0\tau) \tag{15.3}$$

where $\epsilon(\tau)$ is an envelope function determined by the bandpass known as the *delay pattern* and ν_0 is the center of the frequency passband.

---

1) The power spectrum of a deterministic signal is the Fourier transform of the autocorrelation of that signal: $|H(\nu)|^2 = \int_{-\infty}^{\infty} R(\tau)e^{-i2\pi\nu\tau}d\tau$, and conversely $R(\tau) = \int_{-\infty}^{\infty} |H(\nu)|^2 e^{i2\pi\nu\tau}d\nu$.

The rotation of the Earth will cause the position of sources in the sky to be constantly changing with time and it is therefore necessary to constantly update and adjust the instrumental delay, which is often referred to as *path compensation*. If this path compensation is designed such that $\tau_g|_{\theta_0} - \tau_i = 0$, then for a source offset from the phase center, θ_0, by a small amount $\Delta\theta$, the fringe rate will be

$$\cos(2\pi\nu_0\tau) = \cos\left[2\pi\nu_0\left(\frac{D}{c}\sin(\theta_0 - \Delta\theta) - \tau_i\right)\right]$$
$$\approx \cos\left[2\pi\sin(\Delta\theta)\nu_0(D/c)\cos(\theta_0)\right] \quad (15.4)$$

The argument of this function is written in such a way as to emphasize the contribution from $\nu_0(D/c)\cos(\theta_0)$. This quantity represents the length of the baseline as projected onto the plane normal to the direction to the phase center. It is measured in wavelengths and is interpreted as the *spatial frequency*, u. Consequently, we can rewrite Eq. (15.4) as

$$\cos(2\pi\nu_0\tau) = \cos(2\pi u l) \quad (15.5)$$

where $l = \sin(\Delta\theta) \approx \Delta\theta$.

The overall response of the interferometer can therefore be written as

$$R(l) = \int_S \cos\left[2\pi u(l - l')\right] A(l')\epsilon(l')I(l')dl' \quad (15.6)$$

We can now extend the idea of a two-element interferometer to a two-dimensional synthesis array. Accordingly, we need to exchange our one-dimensional notation with a corresponding 2D geometry. For example, our phase center $\theta_0 \to \mathbf{s}_0$. Momentarily neglecting the shape of the bandpass response, we may rewrite the response of the telescope correlator as

$$R(\mathbf{D}_\lambda, \mathbf{s}_0) = \Delta\nu \int_{4\pi} A(\sigma)I(\sigma)\cos\left[2\pi\mathbf{D}_\lambda\cdot(\mathbf{s}_0 + \sigma)\right]d\Omega \quad (15.7)$$

If we define a complex *visibility*, V, as

$$V = |V|e^{i\phi_V} = \int_{4\pi} A(\sigma)I(\sigma)e^{-i2\pi\mathbf{D}_\lambda\cdot\sigma}d\Omega \quad (15.8)$$

then we are able to express Eq. (15.7) as

$$R(\mathbf{D}_\lambda, \mathbf{s}_0) = \Delta\nu\{\cos[2\pi\mathbf{D}_\lambda\cdot\mathbf{s}_0]\Re\{V\} - \sin[2\pi\mathbf{D}_\lambda\cdot\mathbf{s}_0]\Im\{V\}\}$$
$$= A_0\Delta\nu|V|\cos[2\pi\mathbf{D}_\lambda\cdot\mathbf{s}_0 - \phi_V] \quad (15.9)$$

The coordinates of σ are generally given as (l, m), where l and m are direction cosines away from the phase center; $\mathbf{D}_\lambda$ is the baseline vector (u, v, w) projected onto the plane normal to the direction of the phase center. We might therefore reexpress Eq. (15.8) as

$$V(u, v, w) = \int_{-\infty}^{\infty}\int_{-\infty}^{\infty} A(l, m)I(l, m)e^{-i2\pi\left[ul+vm+w(\sqrt{1-l^2-m^2}-1)\right]}$$
$$\frac{dl\,dm}{\sqrt{1-l^2-m^2}} \quad (15.10)$$

and it is this equation that is most often used to express visibility.

The w term in the exponent of Eq. (15.10) is often neglected since, for a restricted range of l and m, as is often the case due to the limited nature of $A(l, m)$, this term becomes negligible. In these circumstances, the visibility equation reduces to a two-dimensional Fourier transform and can be reversed to recover the sky intensity distribution from the measured visibilities:

$$I'(l, m) = \int_{-\infty}^{\infty} \int_{-\infty}^{\infty} V(u, v) e^{i2\pi(ul+vm)} du dv \qquad (15.11)$$

where $I'(l, m)$ is the true sky intensity, $I(l, m)$, multiplied by the primary beam and the normalization factor $1/\sqrt{1 - l^2 - m^2}$.

The plane normal to the direction of observation $\mathbf{s}_0$ is an important quantity in aperture synthesis. Known as the uv plane because of the (u, v) coordinate system, it is equivalent to the two-dimensional Fourier plane. A baseline of projected length $|\mathbf{u}|$ on this plane will measure that Fourier component of the sky intensity distribution $\tilde{I}(\mathbf{u})$.

In the specific case where the baseline lies exactly east–west, the rotation of the Earth will cause the baseline vector $\mathbf{u}$ to rotate as a function of time and describe a circle of radius $|\mathbf{u}|$ in the uv plane. For a perfectly east–west baseline, this circle will remain in the plane as the baseline has no component parallel to the Earth's rotation axis, and it was this property that was exploited by the earliest Earth rotation aperture synthesis telescopes. By changing the length of the baseline, or by using an array of several different baselines all spaced along an east–west line, it is possible to use the rotation of the Earth to synthesize an aperture much larger than it is practical to engineer (Figure 15.3).

Unlike a single-dish telescope, which measures sky intensity directly, an interferometer measures the Fourier components of a sky intensity distribution that correspond to the length of its projected baselines. A full synthesis observation requires a 12-hour track to complete the circular synthesized aperture. This requirement being time consuming, very soon telescopes and algorithms that could recover images from incomplete syntheses were designed.

15.4.2
Receiving System Response

Modern interferometers are rarely east–west and consequently their projected baselines will not describe circles in the uv plane. During an observation toward a particular direction, non-east–west baselines will instead describe ellipses in the uv plane. A baseline is measured in units of wavelength in order to relate it simply to an angular scale, $\Delta\theta \approx D_\lambda^{-1}$. During a synthesis observation the length of the projected baseline will change deterministically as a function of the astronomical pointing direction and local sidereal time. Since the sky signal is intrinsically real, the complex amplitude, $V(\mathbf{u})$, received by a baseline, will obey the relationship

$$V(-\mathbf{u}) = V^*(\mathbf{u}) \qquad (15.12)$$

where V^* denotes the complex conjugate of V.

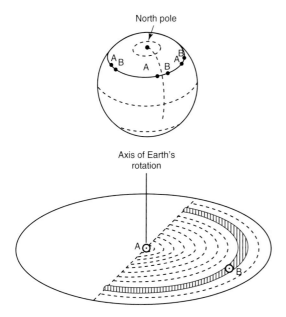

Figure 15.3 Geometric description of Earth rotation aperture synthesis [3].

An array of N antennae will provide $N(N-1)/2$ baselines, each describing a separate arc in the uv-plane. This filling of the uv-plane synthesizes an aperture of the dimension of the longest baseline, hence the name *aperture synthesis*.

Unlike a true filled aperture, however, it is not possible to measure the zero-spacing flux at the uv point $(0, 0)$. This is because the dishes measure only correlated emission and can necessarily be separated only by distances greater than the dish size. A consequence of this is that the total intensity of the sky being measured cannot be found from synthesis measurements, which will always have a total measured flux of zero. A further consequence of the incomplete filling in uv space is that the recovered sky is convolved with the Fourier transform of the pattern of the uv loci described by the baselines. These loci in the uv plane may be thought of as a weighting, $W(u, v)$, and their Fourier transform is known as the *synthesized* beam. This distinguishes it from the aperture illumination function, known as the *primary* beam. The relationship between the measured visibilities and the recovered sky intensity can be seen easily using the convolution theorem. Since the true sky intensity distribution $I(l, m)$ is multiplied by the primary beam and its Fourier transform $\tilde{I}(u, v)$ is sampled in uv space,

$$[I(l, m) \times A(l, m)] * \tilde{W}(l, m) = \left[\tilde{I}(u, v) * \tilde{A}(u, v)\right] \times W(u, v) \qquad (15.13)$$

In the case of an interferometer in which not all the aperture illumination functions of the separate antennae may be exactly the same, the primary beam is found as their autocorrelation. In a single-dish-filled aperture telescope, the primary and synthesized beams are the same, referred to only as the beam.

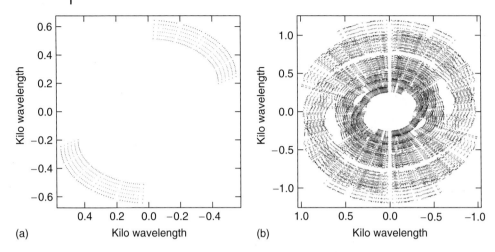

Figure 15.4 (a) The *uv* coverage of a single baseline in the AMI array. The six frequency channels are shown with the lowest frequency channel in red and the highest frequency channel in dark blue. (b) The *uv* coverage of the whole AMI array. (Please find a color version of this figure on the color plates.)

To illustrate this mathematical description, we use the example of the Arcminute Microkelvin Imager (AMI) Telescope operated by the Astrophysics group at the Cavendish Laboratory (Cambridge, UK). This telescope is an aperture synthesis array with 10 3.7 m dishes operating in six frequency bands from 14.3 to 17.9 GHz. With dishes of this size, and operating at these frequencies, the primary beam have an FWHM of approximately 20 arcminutes. The synthesized beam is approximately 2 arcminutes. Since the size of both the primary beam and the synthesized beam are a function of wavelength they vary with frequency. In the case of the synthesized beam, we can see that this is an effect of the coverage in the *uv* plane at each frequency. Figure 15.4a illustrates the *uv* coverage for a single baseline in the array. The separate frequency channels are color coded to illustrate their different coverages, and the conjugate symmetry described in Eq. (15.12) can be seen by the presence of two symmetric ellipses in the *uv* plane. The effect of periodic time samples being removed for calibration is also evident in this figure. The shape of the synthesized beam will depend on the distribution of the *uv* coverage and the synthesized beam that corresponds to the coverage shown in Figure 15.4b is shown in Figure 15.5.

Conversely, since an interferometer only measures discrete Fourier components of the sky signal and does not completely sample all Fourier space it will not measure sky flux on all scales. Where this becomes an issue, it is referred to as *flux loss*. It is particularly troublesome when observing astronomical objects that are extended with respect to the resolution of telescope. If flux is lost purely on the largest scales due to the finite separation of the dishes then it may be corrected for quite simply in an approximate sense. If the flux is lost on scales intermediate to

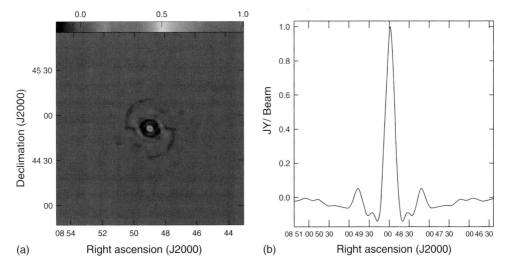

Figure 15.5 (a) A map of the synthesized beam resulting from the uv coverage shown in Figure 15.4(b). (b) A horizontal profile through the synthesized beam. (Please find a color version of this figure on the color plates.)

the smallest and largest measured scales, as may happen where the spacing of the dishes in an array is not well distributed for sampling the Fourier representation of the sky, it is not possible to correct without a priori knowledge of the intensity distribution. These losses may, however, be used constructively. For example, should one wish to measure the emission from pointlike objects embedded in a large-scale noisy background, one might construct an interferometer with only very long baselines. Such an interferometer will be insensitive to the large-scale fluctuations and will only measure the small-scale emission from the point sources. Building interferometers for specific purposes in this way is a form of direct filtering in Fourier space.

15.5
Image Formation

15.5.1
Derivation of Intensity from Visibility

Since the data measured by an interferometer are, to a suitable approximation, a collection of Fourier components, the measured sky intensity may be recovered through the use of Fourier transforms. We may consider the Fourier transform of the sky intensity distribution illuminated by the primary beam as $V(u, v) \iff I(l, m)A(l, m)$. Our measured visibilities are then given by

$$V_{\text{measured}} = W(u, v)w(u, v)V(u, v) \tag{15.14}$$

where we have applied some user-defined weighting, $w(u, v)$. This weighting function allows manipulation of the form of the synthesized beam, $b_0(l, m)$, which is now defined as

$$b_0(l, m) \iff W(u, v)w(u, v) \tag{15.15}$$

A single-dish radio telescope will return the power from the sky in a single direction, at a resolution determined by the size of its dish. This power must then be combined in a scanning strategy to form a map of the sky. In a single pointing, an interferometer returns a field of view greater than its resolution, $A(l, m) \gg b_0(l, m)$, which contains structures on the scales of its constituent baselines. This makes interferometers much more suitable for conducting large surveys of the sky than single-dish telescopes. The fact that they can instantaneously image areas much larger than their resolution proves advantageous in terms of survey speed, and the collecting area of many small dishes will quickly overtake the maximum collecting area available to a single dish without experiencing major engineering and gravitational deformation problems.

For obvious reasons, the best signal-to-noise ratio in our recovered map will be achieved by using natural weighting and setting the weights to the reciprocal of the variance for each measured $W(u, v)$. In practice, this leads to a poorly synthesized beam shape and therefore it is common practice to use *normal* weighting where the weights are factored by the inverse of the local area density of uv points. Since the uv points will necessarily be denser at shorter spacings, this has the effect of weighting up points with larger values of $|\mathbf{u}|$. This makes the weight of the data more uniform across the uv plane and the weighted distribution will approximate a uniformly filled annulus. Since the synthesized beam is the Fourier transform of the $W(u, v)w(u, v)$ distribution, this becomes a Bessel function.

If instead of a direct Fourier transform a discrete fast Fourier transformation is used, then a couple of further complications are introduced. The first of these is the necessity of evaluating the V_{measured} on a regular grid; and the second is the problem of aliasing. Following Ref. [10], the ouput of the gridding process can be described as

$$\frac{w(u, v)}{\Delta u \Delta v} {}^2\mathrm{III}\left(\frac{u}{\Delta u}, \frac{v}{\Delta v}\right) \{C(u, v) * [W(u, v)V(u, v)]\} \tag{15.16}$$

The convolution kernel, $C(u, v)$, ensures a smooth visibility distribution, which is resampled on a grid of cell size $\Delta u, \Delta v$. ${}^2\mathrm{III}$ is the two-dimensional shah function [11].

The choice of gridding kernel, $C(u, v)$, is vital to the quality of the recovered sky intensity distribution. It has a twofold purpose: to both smooth the uv distribution for resampling and reduce the contamination from aliasing. The latter is of considerable importance and therefore the Fourier transform of an optimal gridding kernel, $\tilde{C}(l, m)$, should be maximally concentrated within $A(l, m)$, the primary beam. This is equivalent to maximizing the quantity [12]

$$R = \frac{\int \int_A |\tilde{C}(l, m)|^2 dl dm}{\int_{-\infty}^{\infty} \int_{-\infty}^{\infty} |\tilde{C}(l, m)|^2 dl dm}. \tag{15.17}$$

The function that maximizes Eq. (15.17) is the spheroidal function and the most commonly used gridding kernel is the prolate spheroidal wavefunction (PSWF), a special case of the spheroidal function. A particularly nice quality of these functions is that they are their own Fourier transform.

The full measurement equation for an interferometer, Eq. (15.10), is often approximated to its two-dimensional form shown in Eq. (15.11). This approximation is made when the third term in the exponent, $w(\sqrt{1-l^2-m^2})$, is small: $\ll 1$. In the case of many, if not most, current datasets, this approximation is quite valid. However, when the w term approaches or exceeds unity, this will no longer be the case. Increasingly, for modern and planned future experiments, the w term is becoming important. As can be seen from Eq. (15.10), the effect of the w term is to introduce a phase shift differing per baseline; a second effect is that the morphology of a particular structure in the sky will differ as seen by differing baselines. The three-dimensional nature of the (u, v, w) coordinate system suggests that an obvious solution to this problem may be to simply use a 3D Fourier transform, rather than a 2D transform, in l, m, and n. Although this approach can be used, in practice, it is often not for several reasons: the 3D Fourier transform is significantly slower than its 2D equivalent, and with data cubes of order $>10^3$, its use is unfeasible; for a large field of view the recovered cube is then largely empty; depending on the antenna configuration, it is generally the case that the third dimension is heavily undersampled.

A far better method is the w-projection [13]. This method re-expresses Eq. (15.10) as a multiplication:

$$V(u,v,w) = \int\int \frac{I(l,m)}{\sqrt{1-l^2-m^2}} G(l,m,w) e^{-i2\pi(ul+vm)} dl dm \tag{15.18}$$

where

$$G(l,m,w) = e^{-i2\pi[w(\sqrt{1-l^2-m^2})]} \tag{15.19}$$

Since we are making a multiplication in the map plane, the convolution theorem states that we must be making a convolution in the uv plane. The convolution kernel for this will be

$$\tilde{G}(u,v,w) = \frac{\sin\left(\frac{\pi(u^2+v^2)}{2w}\right)}{2w} \tag{15.20}$$

The net effect of this convolution is to project the data from its true point (u, v, w) to $(u, v, w = 0)$ from whence it can be transformed using the standard 2D Fourier transform.

15.5.1.1 Full-Sky Imaging

For very large fields of view observed with fine resolution, it has been proposed that the visibility equation, Eq. (15.10), might be reformulated to encompass a spherical or wavelet basis and consequently avoid the tangent plane approximation all together. The visibility equation may be rewritten as

$$V(\mathbf{u}) = \sum_{lm} (A^l \cdot I^l)_{lm} \int_{S^2} e^{-i2\pi \mathbf{u} \cdot \hat{\mathbf{s}}^l} Y_{lm}(\hat{\mathbf{s}}^l) d\Omega(\hat{\mathbf{s}}^l) \tag{15.21}$$

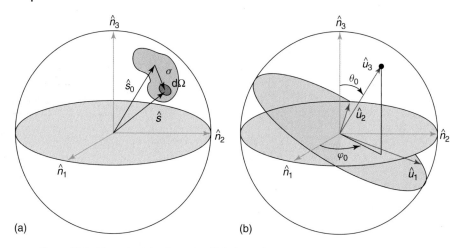

Figure 15.6 The celestial sphere. (a) Global coordinate system for an observation with phase and beam center in the direction $\hat{s}_0$; (b) Local coordinate system showing the rotation R_0 which maps from the global to the local coordinate system.

where $(A^l \cdot I^l)_{lm}$ is the beam-modulated sky intensity and a rotation matrix $R_0 \equiv R(\phi_0, \theta_0, 0)$ has been used to transform to a local frame denoted $\hat{s}^l$ via $\hat{s}^l = R_0^{-1}\hat{s}^n$ (Figure 15.6). Noting the addition formula for spherical harmonics and the Jacobi–Anger expansion of a plane wave, this equation reduces to

$$V(\mathbf{u}) = 4\pi \sum_{lm} (-i)^l j_l(2\pi||\mathbf{u}||) Y_{lm}(\hat{u})(A^l \cdot I^l)_{lm} \tag{15.22}$$

[14, 15]. Computing visibilities in this manner allows for arbitrarily wide fields of view and beam sizes to be considered. However, although the forward problem of obtaining visibilities from a sky intensity is possible, the recovery of the beam-modulated sky intensity becomes more difficult. Theoretically, one may reverse Eq. (15.22) to obtain

$$\int_{S^2} V(\mathbf{u}) Y_{lm}^*(\hat{u}) d\Omega(\hat{u}) = 4\pi(-i)^l j_l(2\pi||\mathbf{u}||)(A^l \cdot I^l)_{lm} \tag{15.23}$$

However, in practice, this would require full sampling of the visibility function in $\mathbb{R}^3$, which is rarely, if ever, feasible in interferometry.

A better approach to full-sky imaging is to use a basis with compact support such as the spherical Haar wavelet basis (see Chapter 7). In this context, the visibility equation may be formulated as

$$V(\mathbf{u}) = \sum_{k=0}^{N_{J_0}-1} \lambda_{J_0,k} \eta_{J_0,k}(\mathbf{u}) + \sum_{j=J_0}^{J-1} \sum_{k=0}^{N_j-1} \sum_{m=0}^{2} \gamma_{j,k}^m \delta_{j,k}^m(\mathbf{u}) \tag{15.24}$$

(see Ref. [15]). Here $\lambda_{J_0,k}$ and $\gamma_{j,k}^m$ are the scaling and wavelet coefficients of the beam-modulated sky intensity and $\eta_{J_0,k}$ and $\delta_{j,k}^m$ are the scaling and wavelet coefficients of the plane wave $e^{-i2\pi u \cdot \hat{s}^l}$, respectively. The inverse problem involves inverting the linear system given by this equation; the system is overdetermined and may be solved in a least-squares sense. A further advantage of the wavelet basis is that the beam-modulated sky intensity distributions will contain localized high-frequency content, making their wavelet representation very sparse. In addition to true sparsity, where the wavelet coefficients are identically zero, there will be many coefficients approaching zero. This enables a degree of thresholding to be performed by setting these coefficients also identically to zero without introducing significant errors. These benefits outweigh the lack of analytic methods for computation in the wavelet method, which are, of course, possible in the case of the spherical basis where the expansion of the plane wave can be performed analytically.

Full-sky interferometry solutions have been developed in principle, but are rarely used in practice. Current telescopes are well served by flat-sky approximations; however, the next generation of interferometers will have very large fields of view and will require these methods to be more fully explored.

15.5.2
Deconvolution Techniques

The incomplete filling of the aperture plane caused by the discrete number of baselines in an interferometer causes a ringing effect in the map plane when the visibilities are Fourier transformed to recover the sky intensity. A simple example of this would be the east–west Earth rotation array: when the array has completed one full synthesis, 12 hours, the aperture plane will be almost exactly filled between the scales corresponding to the shortest and longest baselines available. This is well described by a filled annulus in the aperture plane. The Fourier transform of an annulus centered on 0,0 is the difference of two Bessel functions – another Bessel function. Since the aperture plane distribution has been *multiplied* by the annulus, the sky plane distribution will be *convolved* will the Bessel function. The Bessel function has both positive and negative values and, consequently, wherever there is positive flux in the measured sky, the recovered sky will also contain negative flux. The synthesized beam Bessel function is often referred to as the *dirty* beam of the telescope and the ringing associated with its positive peak is called *side-lobe* structure.

It is often preferable to remove this convolution from the recovered sky. This is done not only for aesthetic reasons but also for more practical purposes: the negative and indeed positive side lobes of a bright source will often act to obscure further emission in the area of interest. The method of this deconvolution is an ongoing topic of research. The more commonly used and established methods with a few modern extensions are outlined here.

15.5.2.1 The CLEAN Algorithm

The most powerful, most commonly used, and most widely debated deconvolution algorithm in radio astronomy is the CLEAN algorithm. For low dynamic range deconvolution, the CLEAN algorithm is a reasonably fast and hugely reliable matching pursuit method. In its simplest incarnation, CLEAN will search a user-defined area of recovered sky and iteratively subtract a dirty beam, or its multiple, from the map maxima until the residual map is consistent with noise. At every position that a dirty beam has been subtracted, a *clean* beam will be placed on a CLEANed sky map. This clean beam will usually take the form of a smooth Gaussian modeled on the central region of the dirty-beam Bessel function,[2] but may alternatively be defined by the user. This form of deconvolution explicitly assumes that the sky intensity can be represented entirely as a combination of δ-functions and although it implicitly assumes that this representation is sparse, there is no explicit regularization term to enforce this. The first implementation of this algorithm was made by Högbom [16] in a simple three-step process where a dirty beam multiplied by a factor γ, known as the loop gain and typically in the range 0.01–0.1, was subtracted from the point of maximum amplitude in the map. The position and amplitude of this subtraction was then stored as a δ-function in a model sky. This step was then repeated until the residual map was consistent with noise. At this point, the model sky was convolved with a *clean* beam and added to the residuals of the subtracted map.

It is important to convolve the model sky with a clean beam in order to properly interpret the measurements that have been made. A telescope is not sensitive to scales smaller than its resolution, that is, the width of the dirty-beam main lobe. To consider the finer-resolution model sky as a true interpretation of the measurements is inherently flawed. However, the Gaussian clean beam is itself an approximation as it introduces a taper in the visibility plane. Note that the CLEAN algorithm must be applied to the data without correction for the primary beam as the two functions are nonassociative.

The original CLEAN algorithm was improved by Clark in 1980 [17]. This new procedure used the known analytic response to a point source of an interferometer in the uv plane. This CLEAN method uses a series of minor cycles to roughly deconvolve a small area using the Högbom method, allowing the components to be identified quickly, followed by a major cycle where the identified components are removed precisely using their analytic form in Fourier space. This method was found to be up to an order of magnitude faster than the original Högbom clean. A variation on this method is the Cotton–Schwab CLEAN [18] where, in the major cycle, the Fourier components of the identified sources are removed directly from the ungridded visibilities. This approach avoids aliasing problems and allows for

---

2) A note on nomenclature: it is the case that both the clean and dirty beams can be referred to as the synthesized beam. It is usually assumed that the difference will be obvious from context.

sources outside the field of view of the primary pointing to be removed using information from another.

The usefulness of the CLEAN deconvolution method is apparent from its widespread use in radio astronomy. However, there are a number of pros and cons to the procedure that have been discussed at length in the literature. The nonlinear nature of CLEAN and its propensity for misuse often cause some concern. The most common problem with maps deconvolved using CLEAN is the presence of CLEAN artifacts, a regular pattern of "lumps" seen where there is large-scale smooth emission present. These occur as a result of CLEAN misidentifying the side lobes of previously subtracted components as new maxima.

A further question mark over the correct use of CLEAN is the convergence condition. Schwarz [19] showed that CLEAN would converge under three conditions: (i) the beam matrix must be symmetric, (ii) the beam matrix must be positive definite or positive semidefinite, and (iii) the dirty map must be in the range of the dirty beam. All three of these requirements are fulfilled by standard datasets, for which it is typical to run CLEAN until the largest clean component remaining is some (small, ≤ 3) multiple of the theoretical thermal noise. Schwarz [19, 20] showed that if the number of CLEAN components did not exceed the number of independent visibilities in the data, then CLEAN will converge to a solution that is the least-squares fit of the Fourier transform of the delta function components to the measured visibility data.

A further known problem with CLEAN is the effect of having a pointlike source located between the two pixels in the recovered image. Although it is possible to center pointlike emission from a single source on a grid cell, this approach becomes problematic in fields that contain a large number of sources. The CLEAN algorithm will approximate such structures using delta functions centered on the nearest pixel centers. This has the effect of introducing artifacts into the data and reducing the dynamic range, here defined as the ratio of the peak flux density to the root mean square noise level, of the recovered image. Cotton and Uson [21] showed that without any correction this could cause the dynamic range to be limited to 10^4.

There are a number of modern extensions to the CLEAN algorithm using both multiple scales of clean component apart from the standard delta function (e.g., MS-CLEAN, ASP-CLEAN), and exploiting the increased coverage of the uv-plane from multiple frequency data by introducing a frequency dependence to the clean components (e.g., MF-CLEAN).

15.5.3
Maximum Entropy Deconvolution (The Bayesian Radio Astronomer)

John Skilling once wrote that Bayesian probability calculus was *"the unique calculus of rationality."* It is unsurprising therefore that Bayesian deconvolution methods, such as the widely used maximum entropy method (MEM), are heavily utilized. It is also satisfying that their principles are mathematically expressible more readily than those of the CLEAN algorithm.

A single point source at position $\mathbf{x_0}$ with a flux density S_0 will contribute a visibility, $V(\mathbf{u})$:

$$V(\mathbf{u}) = S_0 e^{-i2\pi \mathbf{u} \cdot \mathbf{x_0}} + N(\mathbf{u}) \qquad (15.25)$$

where $N(\mathbf{u})$ is the complex Gaussian noise vector belonging to that visibility with variance $\sigma(\mathbf{u})$. If we have a set of visibilities, $\{V_i\}$ belonging to $\{u_i\}$, then we may express the likelihood function as

$$\text{pr}(\{V_i\}|S_0, \mathbf{x_0}, \{\sigma_i\}) = \prod_i \frac{1}{2\pi\sigma_i^2} e^{\frac{-|V_i - S_0 e^{i2\pi \mathbf{u} \cdot \mathbf{x_0}}|^2}{2\sigma_i^2}} \qquad (15.26)$$

It is immediately obvious from this form that we may maximize the likelihood by minimizing the exponent of the exponential, which itself is the constraint statistic χ^2.[3] Under the necessary assumption of a complex multiscale sky intensity, we can immediately recognize that simply minimizing the χ^2 will not produce a unique solution. Bayesian probability theory states that

$$\text{pr}(S_0, \mathbf{x_0}|\{V_i\}) \propto \text{pr}(S_0, \mathbf{x_0})\text{pr}(\{V_i\}|S_0, \mathbf{x_0}) \qquad (15.27)$$

where $\text{pr}(S_0, \mathbf{x_0})$ is the prior on the flux, which we may assume is uniform and, in the case of S_0, positive. If we regard our image as a positive array, $\{I_i\}$, we may define an entropy E,

$$E = -\sum_i^N p_i \log p_i, \qquad p_i = I_i / \sum I \qquad (15.28)$$

Using the entropy of a dataset automatically invokes the positivity constraint and the use of a smooth function such as Eq. (15.28) ensures that this constraint is met more easily than requiring each $I_i \geq 0$ separately. Equation (15.28) is not the only form of entropy used in the context of image reconstruction; however, this form yields most easily to Bayesian reasoning. Suggested first by Frieden [22], a more extensive list of its virtues above other forms of entropy was compiled by Bryan and Skilling [23]. The entropy alone is also not enough to recover an image, instead the two constraint statistics, χ^2 and E, are both used. This is done by maximizing E subject to χ^2 being less than some user-defined maximum, $\chi^2_{\max}$. Since the surfaces of entropy are convex and of χ^2 are elliptical, this will produce a *unique* solution. The problem is better formalized as a Lagrangian function,

$$Q = E - \lambda \chi^2 \qquad (15.29)$$

where λ is a Lagrange multiplier. Since, in interferometry, the total flux, $\sum I$, is important, the Lagrange function can be modified to include this constraint. The maximum entropy solution, $\{I_i\}$, is then found by maximizing the Lagrange function:

$$Q = E - \mu \sum I - \lambda \chi^2 \qquad (15.30)$$

3) $\chi^2 = \sum_i \frac{(\text{data}_i - \text{model}_i)^2}{\sigma_i^2}$

Since the MEM method is essentially model based, it may return a result that contains features smaller than the resolution of the telescope. This is allowed because there is evidence for these features in the data. When the structure is reconstructed on a scale smaller than the practical resolution it is known as *super-resolution* (see Chapter 27). This kind of reconstruction is feasible to a moderate degree only when the data has sufficiently high signal to noise and the magnitude of that noise is well known. It is the case that an underestimation of the noise will often produce spurious super-resolution in the recovered map due to features in the noise being interpreted as true signals. The principle of super-resolution has been shown to be sound in many cases using observationally higher resolution images of known sources for comparison; however, it is the case that super-resolution features found in an MEM map will not necessarily correspond to those in the sky and it should be used with care.

Since visibility datasets $\{V_i\}$ are often much larger than their corresponding images $\{I_j\}$, it is useful to approximate the χ^2 term in the Lagrange function by the misfit in the image plane. Notably, Cornwell and Evans [24] implemented the approximation

$$\chi^{2\prime} = \sum_i \left(\sum_j B_{i,j} \cdot I_j - D_i \right)^2 \tag{15.31}$$

in the widely used Astronomical Image Processing System (AIPS). In this method, $\{B_i\}$ represents the dirty beam and $\{D_i\}$ the dirty image. A direct calculation of χ^2 in the map plane is complicated by both the residual errors in the weights from the visibility gridding step and also the correlation of $\{D_i\}$. Cornwell and Evans [24] maximize the Lagrange function with the preceding form of the χ^2 term numerically using an adaptation of a quasi-Newton–Raphson nonlinear least-squares algorithm.

Another approach is to sample the maximum entropy posterior distribution in order to maximize the Lagrange function. Markov chain Monte Carlo (MCMC) methods have been used extensively to perform this type of analysis. These methods are very powerful in the case of a unimodal, linearly degenerate or nondegenerate posterior. When a posterior may be multimodal or possess long curving degeneracies, then a much more efficient way of evaluating the posterior is to use the nested sampling technique [25]. Unlike standard MCMC techniques such as simulated annealing, which recover the Bayesian evidence as a by-product of the posterior distribution, nested sampling evaluates the Bayesian evidence directly and recovers the posterior as a by-product. The nested sampling method is implemented in the BayeSys [26] application where the image reconstruction is performed using a collection of flux components called atoms. These atoms can not only be utilized in much the same way as the delta function approach of CLEAN but may also be supplied with additional prior information in the form of distribution functions.

15.5.4
Compressed Sensing

Compressed sensing (CS) (also compressive sensing; [27]) relies on the principle that a signal is sparse or compressible in *some basis*. By definition, this is true if its expansion contains only a small number of nonzero, or indeed significant coefficients. If this is the case, then CS demonstrates that the signal need not be sampled at the Nyquist rate but instead requires a much smaller number of measurements to be accurately known. In addition, the sampling must satisfy a restricted isometry property (RIP), which conditions the corresponding sensing for further reconstruction. Strikingly, many random sampling constructions respect the RIP with very high probability (see Chapter 23). For instance, the sensing obtained by a small number of randomly sampled measurements from a basis incoherent with the sparsity basis will almost always ensure this property when the number of measurements is bigger than a few multiples of the image sparsity level. In the case of radio interferometry [28], one might imagine the example of a small number of visibility measurements in Fourier space made by observing a sparse sky intensity function, such as a distribution of pointlike sources. For instance, let us we imagine we have a generic sky intensity $\mathbf{I}$ that in reality will be continuous, but here we consider it as Nyquist sampled, and we suppose that this intensity function is sparse or compressible in a basis Ψ such that the decomposition

$$\mathbf{I} = \Psi \alpha \tag{15.32}$$

has only a small number, K, of nonzero or significant coefficients. If we then measure this sky intensity in the sensing basis Φ to obtain a set of measurements $\mathbf{V}$

$$\mathbf{V} = \Theta \alpha + \mathbf{n} \text{ with } \Theta = \Phi \Psi \tag{15.33}$$

where $\mathbf{n}$ is some independent and randomly distributed noise. We can then formulate the compressive sensing basis pursuit denoise (CS-BPDN) problem:

$$\min_{\alpha'} ||\alpha'||_1 \text{ subject to } || \mathbf{y} - \Theta \alpha'||_2 \leq \epsilon \tag{15.34}$$

The L_1 norm on the basis coefficients is a way of reflecting sparsity in our basis. From a Bayesian perspective, the L_1 norm may be seen as the negative logarithm of a Laplacian prior distribution on each independent component of α. Including this kind of constraint explicitly assumes sparsity. Conversely, the L_2 norm can be considered as the negative logarithm of a Gaussian, a function that is much less highly peaked and tailed than the Laplacian. The quantity ϵ must be set so that the pure image is a feasible solution of the constraint in the BPDN problem. Various estimators for ϵ exist according to the distribution of $\mathbf{n}$ in the sensing model (Eq. 15.33). The sky intensity itself may then be recovered by the simple operation $\mathbf{I}^* = \Psi \alpha^*$, where α^* is the set of coefficients that minimize Eq. (15.34).

This description is by no means an exhaustive description of CS and for a more complete treatment we refer the reader to Chapter 23.

15.6
Conclusions

This chapter has attempted to introduce the reader to the fundamentals of astronomical image formation, with an emphasis on the technique of aperture synthesis for high-resolution radio astronomy. The techniques used for this type of science are in a constant state of evolution, a situation made necessary by the increasingly sophisticated nature of the telescopes being constructed as radio astronomers strive for a more complete understanding of the Universe.

Although the heavens have been observed for over 400 years using optical telescopes, it is easy to forget that it has been less than 100 years since the construction of the first radio telescope. Since then, the contributions of radio astronomy to modern living, direct and indirect, are numerous: from the development of precision radar techniques to mobile phone location and medical imaging techniques to wireless internet standards. Although many early advances have now been superseded it is inconceivable that the new generation of radio telescopes, such as the SKA, will not bring their own unique improvements.

References

1. Ryle, M. and Vonberg, D.D. (1946) Solar radiation on 175 Mc./s. *Nature*, **158**, 339–340.
2. Ryle, M. (1952) A new radio interferometer and its application to the observation of weak radio stars. *Proc. R. Soc. Lond. Ser. A*, **211**, 351–375.
3. Ryle, M. (1962) The New Cambridge radio telescope. *Nature*, **194**, 517–518.
4. Ryle, M., Elsmore, B., and Neville, A.C. (1965) High-resolution observations of the radio sources in Cygnus and Cassiopeia. *Nature*, **205**, 1259–1262.
5. Baars, J.W.M. et al. (1973) The synthesis radio telescope at Westerbork. *IEEE Proc.*, **61**, 1258–1266.
6. Thompson, A.R. et al. (1980) The very large array. *APJS*, **44**, 151–167.
7. Minkowski, R.L. and Abell, G.O. (1968) The National Geographic Society-Palomar observatory sky survey, in *Basic Astronomical Data: Stars and stellar systems*, (ed. K.A. Strand), University of Chicago Press, Chicago, p. 481.
8. Reid, I.N. et al. (1991) The second Palomar sky survey. *PASP*, **103**, 661–674.
9. Fruchter, A.S. and Hook, R.N. (2002) Drizzle: a method for the linear reconstruction of undersampled images. *PASP*, **114**, 144–152.
10. Thompson, A.R., Moran, J.M., and Swenson, G.W. Jr. (2001) *Interferometry and Synthesis in Radio Astronomy*, 2nd edn, Wiley, New York.
11. Bracewell, R.N. (1956) Two-dimensional aerial smoothing in radio astronomy. *Aust. J. Phys.*, **9**, 297.
12. Schwab, F.R. (1984) Relaxing the isoplanatism assumption in self-calibration; applications to low-frequency radio interferometry. *Astron. J.*, **89**, 1076–1081.
13. Cornwell, T.J., Golap, J., and Bhatnagar, S. (2003) W projection: a new algorithm for non-coplanar baselines. EVLA Memo Series, 67.
14. Ng, K. (2001) Complex visibilities of cosmic microwave background anisotropies. *P. Rev. D*, **63**, 123001. arXiv:astro-ph/0009277.
15. McEwen, J.D. and Scaife, A.M.M. (2008) Simulating full-sky interferometric observations. *Mon. Not. R. Astron. Soc.*, **389**, 1163–1178. 0803.2165.

16. Högbom, J.A. (1974) Aperture synthesis with a non-regular distribution of interferometer baselines. *Astrophys. J. Suppl. Ser.*, **15**, 417.
17. Clark, B.G. (1980) An efficient implementation of the algorithm 'CLEAN'. *Astron. Astrophys.*, **89**, 377.
18. Schwab, F.R. (1984) Optimal gridding of visibility data in radio interferometry, in *Indirect Imaging. Measurement and Processing for Indirect Imaging* (ed. J.A. Roberts), Cambridge University Press, Cambridge, England, p. 333.
19. Schwarz, U.J. (1978) Mathematical-statistical description of the iterative beam removing technique (Method CLEAN). *Astron. Astrophys.*, **65**, 345.
20. Schwarz, U.J. (1979) The method clean – use, misuse and variations (invited Paper), in *IAU Colloq. 49: Image Formation from Coherence Functions in Astronomy*, Astrophysics and Space Science Library, vol. 76 (ed. C. vanSchooneveld), p. 261.
21. Cotton, W.D. and Uson, J.M. (2007) Image pixelization and dynamic range. *EVLA Memo. Ser.*, **114**.
22. Frieden, B.R. (1972) Restoring with maximum likelihood and maximum entropy. *J. Opt. Soc. Am. (1917-1983)*, **62**, 511.
23. Bryan, R.K. and Skilling, J. (1980) Deconvolution by maximum entropy, as illustrated by application to the jet of M87. *Mon. Not. R. Astron. Soc.*, **191**, 69–79.
24. Cornwell, T.J. and Evans, K.F. (1985) A simple maximum entropy deconvolution algorithm. *Astron. Astrophys.*, **143**, 77–83.
25. Skilling, J. (2006) Nested sampling for Bayesian computations. Proceedings Valencia/ISBA 8th World Meeting on Bayesian Statistics.
26. Skilling, J. (2004) *BayeSys and MassInf*. Tech. report. Maximum Entropy Data Consultants Ltd.
27. Candés, E.J. (2006) Compressive sensing, in *Proceedings of the International Congress of Mathematicians*, vol. 3 (eds. M. Sanz-Sole, J. Soria, and J.L. Verdera), European Mathematical Society, Madrid.
28. Wiaux, Y. *et al.* (2009) Compressed sensing imaging techniques for radio interferometry. *Mon. Not. R. Astron. Soc.*, **395**, 1733–1742. 0812.4933.

16
Display and Projection

Tom Kimpe, Patrick Candry, and Peter Janssens

16.1
Introduction

This chapter discusses a widespread application of photonics: display systems. The two main families of display systems – direct view displays and projection displays – are described.

Direct view displays are displays that are looked at directly by the user. With direct view displays, the image is presented to the unaided eye of the user without the help of any additional optics or magnification. The size of the image produced by the display is equal to the actual size viewed by the user.

Projection displays make use of a magnification lens to magnify and project an image onto a projection screen. The user looks at the magnified image instead of directly at the image-forming element or modulator.

The aim of this chapter is to provide a good understanding of the basic technological concepts of direct view and projection display systems. This is done by describing the state of the art, focusing on some important ongoing trends and new types of displays, and finally explaining some specific applications in greater detail such as presentation and education, simulation, and medical imaging.

16.2
Direct View Displays

16.2.1
Working Principle

For direct view displays, the screen and the spatial light modulator (see Chapter 9) have the same components. Similar to projectors, the spatial light modulator consists of a two-dimensional array of pixels. Unlike projectors, each of these pixels of a color display typically consists of three subpixels; one for each primary color (red, green, and blue). By modulating the light intensity of each of the subpixels, the image is created on the screen.

Optical and Digital Image Processing: Fundamentals and Applications, First Edition. Edited by Gabriel Cristóbal, Peter Schelkens, and Hugo Thienpont.
© 2011 Wiley-VCH Verlag GmbH & Co. KGaA. Published 2011 by Wiley-VCH Verlag GmbH & Co. KGaA.

There are three different classes of direct view displays. In transmissive displays, the light source is placed behind the spatial light modulator, while for a reflective display, environmental light (e.g., from the sun or from office lighting) is reflected by the screen. In the case of an emissive display, the light is emitted by the pixels themselves.

16.2.2
Transmissive Displays

In a transmissive display, the image content is built by controlling the amount of light that is transmitted through the display. Computer monitors and most flat screen televisions are transmissive liquid crystal displays (LCDs). They consist of a liquid crystal (LC) layer stacked between two polarizers and two alignment layers (see Figure 16.1 and Chapter 9). The light source of the display, which is placed behind this stack of layers, is called the *backlight unit*. In order to obtain a color display, each pixel is divided into three subpixels, each having a different color filter (red, green, and blue).

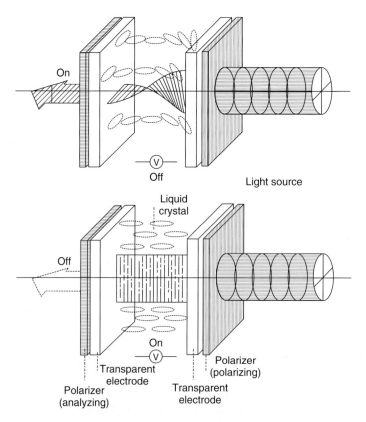

Figure 16.1 Working principle of an LCD pixel [1].

The LC layer consists of elongated and asymmetric organic molecules. At room temperature, the LC molecules form a phase between a solid state and a liquid. Although they can move with respect to each other, they tend to align themselves in a specific order. This global ordering of the molecules leads to macroscopic effects on the propagation of light through the LC layer. In particular, they are able to rotate the polarization of light.

The most common implementation of LC displays, called *twisted nematic* (TN), is shown in Figure 16.1. The ordering of the LC molecules is determined through two crossed alignment layers. The structure of the alignment layer forces the first layer of molecules to align in the desired orientation. All other molecules will align in such a way that a smooth transition in the orientation of the molecules over the complete LC material is obtained. The polarization of the light traveling through the LC layer is now forced to rotate; thus it can pass through the second polarizer, which is oriented at 90° with respect to the first polarizer. In this state, the light is transmitted and the subpixel appears bright.

The LC layer is placed between transparent electrodes made from indium tin oxide (ITO). When a voltage is applied over the LC layer, the molecules tend to line up with the electric field due to their asymmetric dielectric structure. The polarization of the light traveling through the LC layer is now no longer rotated and the subpixel appears dark.

The most common type of backlight is based on cold cathode fluorescent lamps (CCFLs), which are very efficient in generating light. It consists of a tube-shaped recipient filled with mercury vapor and two electrodes. The walls of the tube are coated with a white phosphor. The mercury vapor is excited by the electrons released by the cathode and traveling through the lamp. By de-excitation of the mercury atoms, UV photons are released, which are converted into white light by the phosphor. Nowadays, these CCFLs are increasingly being replaced by white or RGB light-emitting diodes (LEDs). LEDs are advantageous, as they allow local dimming of the display. This not only leads to higher contrast ratios but also reduces the energy consumption of the display. RGB LEDs have the additional advantage of an increased color gamut, but they are more expensive compared to white LEDs.

LCDs can be made very light and thin: the thickness of the display is mainly determined by the type of backlight that is used and the weight is much lower than comparable cathode ray tube (CRT) displays. They can also be produced in large sizes and high resolutions.

However, LCDs have a number of disadvantages, such as a small viewing angle, slow response time, limited contrast ratio, and poor color reproduction. The viewing angle is the maximum angle at which a display can be viewed with acceptable image quality. For nonorthogonal viewing of TN LCDs, colors and contrast degrade. The response time is determined by the time needed by the LC molecules to rotate to the desired position, which is of the order of a few milliseconds, and which limits the refresh rate of the panel. The slow response time also causes motion blur: fast-moving objects appear to be blurred. One of the solutions for motion blur is to flash the backlight when the LC is settled in the desired state. The contrast ratio is

16 Display and Projection

determined by the brightest and the darkest state of the panel. Even in the off state, some light leaks through the display, leading to the relatively low contrast ratio.

There are other types of LCDs (e.g., vertical alignment and in-plane switching panels), which offer better-viewing angles and black levels, but these are more expensive and have slower response times.

16.2.3
Emissive Displays

Examples of emissive displays are CRT monitors, plasma displays, LED displays, and organic light-emitting diode (OLED) displays. The main principle of all these displays is the same: by varying the current through each of the pixels, the amount of light produced by the pixels can be controlled to create the desired image.

16.2.3.1 CRT Display

For a long time, CRT [2] was the dominant display technology. A CRT display consists of a vacuum tube (Figure 16.2), which is made out of glass. One side of this vacuum tube is flat and coated with a matrix of phosphors at the inside. In total, there are three different types of phosphors used to produce color images. The three cathodes at the other side of the vacuum tube eject electrons, which are accelerated by an electrostatic potential of up to a few kilovolts; these electron beams are scanned over the phosphors line by line by the steering yokes. The energy of the electrons is converted into light by the phosphors and the brightness is controlled by a modulation of the current in the electron beam. A perforated metal plate serves as a mask that prevents, for example, the "red" electron beam from hitting a green or a blue phosphor.

CRT displays produce vivid colors, and they have a fast response time and a wide viewing angle. They also offer good contrast ratios due to the excellent black level and the high brightness of the phosphors.

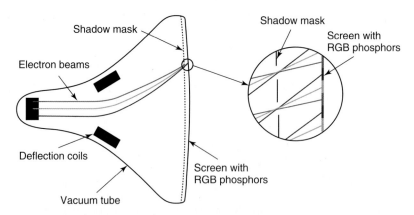

Figure 16.2 Working principle of a CRT display. (Please find a color version of this figure on the color plates).

However, they get large and heavy for large screen diagonals, because the vacuum tube has to be larger and also longer, making the display thicker. Therefore, CRT displays are impractical for large screens. Also the power consumption of a CRT display is higher than a similar-sized LCD display. Another disadvantage is burn-in: the phosphors lose their luminance with use – the more a phosphor has been used the less efficient it becomes. This leads to a reduced efficiency, but more importantly, ghost images can appear when stationary images are shown on the display.

16.2.3.2 Plasma Display

A plasma is a partially ionized gas: it is a state of matter, where a part of the atoms have lost one or more of their electrons. These free charges (both electrons and remaining ions) make the plasma conductive. In an electric field, the electrons and ions move in opposite directions and the collisions excite the gas, releasing more electrons and ions in the plasma. Collisions can also lead to the excitation of an electron to a higher energy level. When this electron returns to its original state, light is emitted (usually in the UV band of the spectrum).

In a plasma display, each subpixel is a little vessel, containing noble gases such as neon and xenon. By applying a sufficiently high voltage (typically above 300 V) over the electrodes of the pixels, the gas is ionized and plasma is formed. The UV photons emitted by the plasma are converted into visible light by the phosphor coating of the vessel (Figure 16.3). The intensity of the pixel is determined by the current through the electrodes. Phosphors similar to those in CRT displays are used to produce the three colors. This means that plasma displays offer the advantages of CRT displays (vivid colors, large viewing angle, fast response time, excellent contrast ratio). Unlike CRT displays, plasma screens can be made thin and large. The main disadvantages of plasma displays are the burn-in of the phosphors and the large power consumption compared to similar-sized LCDs.

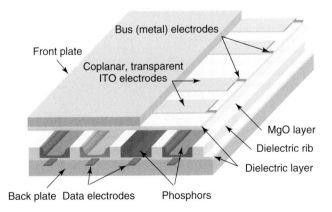

Figure 16.3 Working principle of a plasma display [3].

Figure 16.4 LED wall: each pixel consists of a red, green, and blue LED. (Please find a color version of this figure on the color plates.)

16.2.3.3 LED Display

LEDs are semiconductor junctions that generate light when current passes through them. The light originates from the recombination of holes and free electrons. The wavelength of the emitted light depends on the band gap of the semiconductor and the amount of light can be controlled by the current through the diode. LEDs are available in different of colors, including red, green, and blue.

The smallest building block of an LED display is 1 pixel, which consists of a red, green, and blue LED (Figure 16.4). By combining the pixels into a matrix structure, large displays can be built. This has the advantage that there is no limitation on the size and the shape of LED displays, so they can be used for very creative displays. It is difficult to mount LED pixels very close to each other: typical distances between the pixels range from a few millimeters up to a centimeter. Therefore, these displays are meant to be viewed from relatively large distances. The main advantages of LED displays are the low power consumption per square meter, because LEDs are very efficient light sources, and the long lifetime of the LEDs. In addition, an LED display offers a wide color gamut and fast response times.

16.2.3.4 OLED Display

An OLED is a thin-layer semiconductor structure made of organic molecules. When a current is applied, light is emitted by the organic layer. An OLED consists of a stack of layers. Electrons are injected in the electron transport layer by the cathode and guided to the emissive layer. At the other end of the stack, electrons are extracted from the hole-transport layer at the anode and the created holes move toward the emissive layer. At the boundary of the emissive and the conductive layer, holes and electrons are recombined creating light.

As the only function of the substrate is to support the other layers, it can be made very thin and even flexible. Usually, the substrate is coated with a reflective layer and the light not only leaves the OLED at the side of the cathode as shown in Figure 16.5 but transparent OLED displays are also possible. The cathode is made of the transparent material ITO. The color of the OLED depends on the type of

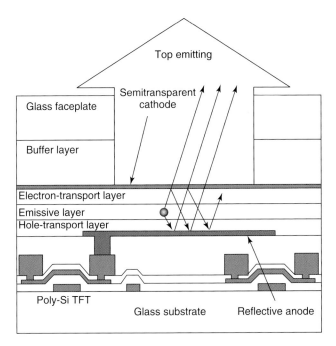

Figure 16.5 Structure of a top-emitting OLED [4].

organic molecule in the emissive layer. By grouping red, green, and blue subpixels into a large matrix, OLED displays are created. The intensity of the light is changed by current modulation for active matrix OLEDs and by pulse-width modulation for passive matrix OLEDs.

OLED displays have many advantages. They can be very thin, light, and flexible. Compared to LCD displays, where a lot of light is wasted by the color filters and the polarizers, OLED displays only generate the light when and where it is needed: OLED displays are very efficient devices and they reach very high contrast ratios. Another important advantage with respect to LCD is the very wide viewing angle. In addition, OLED displays have the potential of being very cheap to produce, as they can be made by inkjet printing.

One of the main disadvantages of OLED displays is burn-in: pixels that are used more frequently will have a lower light output than other pixels. This is similar to the burn-in of CRT and plasma displays. Another disadvantage of OLEDs is that they are damaged by the humidity in the air; so a very good sealing of the display is required.

16.2.4
Reflective Displays

In a reflective display, the light of the environment is used and there is no need to produce light; as such, these displays consume less power. The fact that the display does not generate its own light improves the reading comfort in a bright environment (e.g., sunlight), because the brightness of the display adapts itself to the brightness of the environment. On the other hand, they are very difficult to read in dark environments and it is also difficult to obtain good color reproduction. There are different types of reflective displays. Here, we limit ourselves to reflective LCDs and electronic paper.

16.2.4.1 Reflective LCD

Reflective LCDs are not much different from transmissive LCDs. For a reflective LCD, one of the glass substrates is replaced by a mirror. There is also no need for a second polarizer. The light from the environment passes through the first polarizer and the LC layer. Then it is reflected by the mirror and passes again through the LC layer and the polarizer. If the polarization of the light is rotated by passing twice through the layer structure, the pixel appears dark. These displays offer the same properties as transmissive LCDs, but due to the low efficiency of LCDs they are relatively dark: they appear gray, rather than white.

16.2.4.2 Electronic Paper (e-Paper)

Another example of a reflective display technology is electronic paper or e-paper. Several types of e-paper exist, of which electrophoretic is the most common.

An example of such an electrophoretic display is e-ink, which is used in electronic book readers (EBRs). In this type of display, small cells or microcapsules are placed between two electrodes. Each cell contains a clear fluid and black and white particles, which have an opposite charge. When the bottom electrodes are positively charged, the white particles move to the top of the cell and the black pigments move to the bottom. In this way the pixel will reflect light and appear white (Figure 16.6).

The main advantage of electronic paper, apart from the reading comfort, is the low power consumption–power is required only to refresh a page. Refreshing

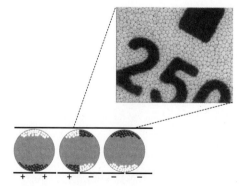

Figure 16.6 Schematic side view and microscopic image of the e-ink electrophoretic display [5].

pages, however, is a slow process making these displays unsuited for video content.

16.3
Projection Displays

Nowadays, projection displays are based on microdisplays (MDs) (two-dimensional spatial light modulators). The MDs are illuminated by a high-intensity light source and imaged on a front or rear screen. Many different MD technologies, light source technologies, and projector architectures have been developed and proposed. This short overview addresses the most important among the current technologies and some recent innovations.

16.3.1
Basic Concepts and Key Components

Three MD technologies are currently in use: high-temperature polysilicon (HTPS) LC, digital light processing (DLP), and liquid crystal on silicon (LCoS).

HTPS and LCoS are both LC MDs. The modulation of light intensity is based on polarization rotation. Therefore, polarized light and light-polarizing components are necessary to obtain light intensity modulation. With active matrix addressing of the individual pixels, it is possible to realize high-resolution MDs. An active matrix display is an analog dynamic memory with optical readout. HTPS MDs operate in transmissive mode, LCoS MDs in reflective mode. Thin-film transistors (polysilicon semiconductor) are applied in the HTPS MDs. A CMOS backplane is applied for the reflective LCoS MDs. The TN LC film with organic alignment layers is the traditional LC technology for HTPS MDs. Recently, vertically aligned nematic (VAN) LC with inorganic alignment layers have also been introduced. Current LCoS panels have VAN LC with inorganic alignment layers. LCoS MDs with the highest resolution have 8192×4320 pixels with a 1.75" diagonal panel size [6].

DLP technology is based on the digital micromirror device (DMD) (see Chapter 9).This is a matrix of electrostatically controlled aluminum mirrors acting as fast light switches. Incident light can be reflected in one of the two directions corresponding to the on state and the off state. Gray scales are made with pulse-width modulation.

The MD aperture must be uniformly illuminated by the illumination optics. In most cases, the illumination optics is based on a fly's eye integrator or on a rod integrator. In the case of a fly's eye integrator, it is the first integrator plate illuminated by a collimated light beam from a parabolic reflector (Figure 16.7). The first integrator plate has lenslets with the same aperture ratio as the MD. Each lenslet of the first integrator plate makes an image of the lamp arc on the corresponding lenslet of the second integrator plate. Each lenslet of the first integrator plate is imaged on the MD by the corresponding lenslet of the second

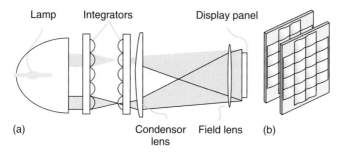

Figure 16.7 Fly's eye integrator [7].

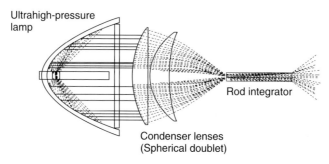

Figure 16.8 Rod integrator [8].

integrator plate and the condenser lens. The light distribution on the lenslets of the first integrator plates are all added on the MD aperture and the nonuniformities of light distributions on the individual lenslets of the first integrator plate are canceled out. In the case of the rod integrator, the multiple reflections inside the rod integrator result in a uniform light distribution at the exit aperture of the rod integrator. The exit aperture of the rod integrator has the same aspect ratio as the MD. The exit aperture of the rod integrator is imaged on the MD aperture (Figure 16.8). Fly's eye and rod integrators have high collection efficiency and give a uniformly illuminated MD aperture.

The light is produced by a high-intensity light source. The light sources are usually high-intensity gas discharge lamps: ultrahigh pressure (UHP) lamps, or xenon lamps. Recently, high-brightness red, green, and blue LEDs have also been used, and red, green, and blue lasers are under development.

Xenon lamps are DC-current high-intensity discharge lamps and are applied in the high light output projectors. The operating voltage is in the 20–30 V range and the power range is from hundreds of watts to several kilowatts. These short-arc lamps have a high luminance. The average luminance is around 1.7 Mcd m^{-2} in the case of a 2.2 kW xenon lamp. The efficacy is around 35 lm W^{-1} and the colorimetry is very good. Lamp life is limited to around 1000 h and end of life is caused by lumen output degradation or flicker.

The UHP lamp technology was introduced in 1995 by Philips. This is a high-pressure mercury vapor discharge lamp technology with a good matching (mainly the short-arc length of ~1.3 mm and the high efficacy of ~60 lm W^{-1}) with the requirements for a projection display light source. The luminance of the arc is around 1 Gcd mm^{-2}. UHP lamps operate with an AC current; this means that each electrode alternately functions as a cathode and an anode. These arc lamps are commercially available in a power range of about 80–350 W. The spectral composition of the light depends strongly on the mercury pressure. The lifetime of UHP lamps is >10 kh for ~100 W types and around 1500–2000 h for higher power types.

LEDs, based on AlGaInP/GaAs for red (~625 nm) and GaInN/GaN for green (~525 nm) and blue (~460 nm), are solid-state surface emitters with attractive properties for projection displays: long lifetime (~50 kh); saturated colors for the red, green, and blue (typical full width at half maximum $\Delta\lambda$ of ~25 nm). Moreover, fast on and off switching (~1 µs) is possible, the driving voltage is low (<5 V), and RGB LEDs do not emit UV or IR radiation. Although the luminance (~30 Mcd mm^{-2} for high-brightness green and red LEDs) has increased considerably over the last 10 years, it is still low compared with high-intensity gas discharge lamps (>1 Gcd m^{-2}). LED-based projectors have therefore a relative low light output (10–500 lm), but this technology allows a strong miniaturization of projectors.

Lasers can also be used as a light source for projection instead of lamps or LEDs. Lasers offer several advantages with respect to lamps and LEDs. The luminance is very high and the lifetime of RGB lasers can reach values higher than 20 000 h. Owing to the narrow spectrum of lasers ($\Delta\lambda$ < 1 nm), more saturated primaries are obtained, leading to a much larger color gamut. Finally, laser projectors will potentially consume less power because lasers offer a more efficient way to produce light. Wall plug efficiencies of 10–25% have been published. Unfortunately, lasers cause granular interference patterns on the screen due to the coherent nature of laser light. This phenomenon is called *speckle*, and is not easily solved. Lasers are also more expensive than lamps and could give rise to eye safety issues, especially in the case of high-brightness projectors.

The étendue is a fundamental characteristic of an optical system and in particular for a projection system. It is a measure of the light flux gathering capability of an optical system and a pure geometrical quantity independent of the flux content. The optical throughput of a projector is limited by the aperture with the smallest étendue, which is usually the MD. An MD with an area A (mm^2) combined with a projection lens with an f-number F_{no} has an étendue E_S (in mm^{2}sr), called the *system étendue*:

$$E_S = \pi \cdot A \cdot \sin^2 \theta_{1/2} = \frac{\pi \cdot A}{4 \cdot F_{no}^2}$$

This approximation is valid for a flat surface with a uniform divergence angle $\theta_{1/2}$ over that surface [9]. For example, the étendue value of a 0.67″ DMD in combination with an F/2.5 projection lens is ~15 mm^2 sr.

A light source is also characterized by its étendue. The étendue of a light source can never decrease; it can only increase by, for instance, aberrations, scattering, or diffraction in the optical system. Light will be lost each time the light passes an optical component with a lower étendue than the previous étendue-limiting component (geometrical losses). Ideally, the system étendue of the optical system has to match the light source étendue.

The étendue of a lamp–reflector combination depends not only on the arc size but also on the dimensions of the reflector [9]. The étendue of lamp–reflector combinations are usually much larger than the system étendue; the optical system is then étendue limited. Geometrical losses will occur in étendue-limited systems. The following formula approximates the collection efficiency ε_{coll} of a UHP lamp with arc length d (in mm) in an optical system with a system étendue E_S [10]:

$$\varepsilon_{coll} = \frac{2}{\pi} \cdot \tan^{-1}\left(\frac{E_S}{3.8 \cdot d^2 + 0.9 \cdot d + 0.8}\right)$$

The étendue E_{LED} of high-brightness LEDs is large compared to the system étendue of MDs, and also large geometrical losses will occur. An LED is a flat surface emitter with a uniform divergence angle $\theta_{1/2}$ over the light-emitting surface [7]:

$$E_{LED} = n^2 \cdot \pi \cdot \sin^2\theta_{1/2} \cdot A_{LED}$$

For example, an LED with $A_{LED} = 4$ mm² emitting into silicone with a refractive index (n) of 1.51 and a half angle $\theta_{1/2}$ of 90° has an étendue of 28.7 mm²sr. We have to compare this with, for instance, the system étendue of 0.67" 1080p DMD with F/2.5 projection lens ($\theta_{1/2} = 11.5°$), which has an étendue of 14.7 mm²sr.

The étendue of a laser is much lower than the étendue of a lamp or an LED. For example, a solid-state laser (single mode) with optical power of 500 mW green light (532 nm) has a beam diameter of 80 μm ($1/e^2$ value) and a divergence angle $\theta_{1/2} < 0.29°$.

The light output Φ (in lumen) of a projection systems is a function of the system étendue E_S, the average light source luminance L (in lm mm^{-2}sr), and the system luminance transfer efficiency η:

$$\Phi = L \cdot E_S \cdot \eta$$

The efficiency of a projection system is a function of the system étendue, the effectiveness of the light source spectrum, the light source luminance, and the efficacy of the light source (lm W^{-1}) [11].

16.3.2
Projector Architectures

Five projector architectures are described. The three-panel transmissive technology is based on the HTPS MDs usually in combination with UHP lamps. For the three-panel reflective type, the important MD technologies are DLP and LCoS; they are usually applied in combination with xenon or UHP light sources. Single-panel systems are also based on DLP or LCoS MDs and use UHP or LED light sources.

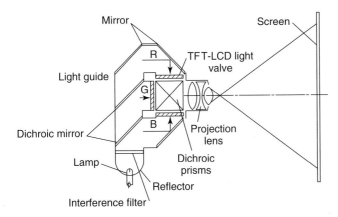

Figure 16.9 Three-panel HTPS architecture [12].

Laser illumination can not only be combined with any architecture for lamps or LEDs but also with image-scanning architectures.

16.3.2.1 Three-Panel Transmissive

This basic architecture for HTPS MD projectors was already proposed in 1986 [12] (Figure 16.9). The light from the white light source is separated into red, green, and blue beams by two dichroic mirrors. Three HTPS MDs modulate the red, green, and blue light beams, respectively. The dichroic prism assembly (X-cube) recombines the three light beams. A single projection lens images the three panels on the screen.

Many improvements have been developed since the advent of this important projection technology. TN mode LC with organic alignment layers is used for these panels; recently, VAN LCs in combination with inorganic alignment layers have also been applied for higher contrast ratio and longer life. The transmission of the HTPS MDs increased significantly by increasing the aperture ratio from ~47 to >70%. Microlens arrays have also been introduced to increase the transmission of the HTPS MDs. Panel and pixel sizes decreased also at an impressive rate; 8.5 μm pixels have been realized on 0.7″ diagonal panels with 1920 × 1080 pixels.

The invention of the polarization convertor was also very important for the success of this projection technology [13].

16.3.2.2 Three-Panel Reflective

An important class of projectors is based on three reflective MDs. The applied MD technologies are either DMD or LCoS panels. High resolution, for example, 4096 × 2160 on 1.55 in. LCoS, and high light output level (e.g., 30 000 lm with DLP), are possible with these technologies. Xenon short-arc lamps are used for the high light output projectors.

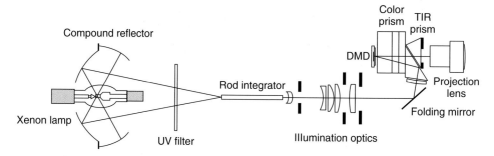

Figure 16.10 Optical architecture three-chip DLP projector [14].

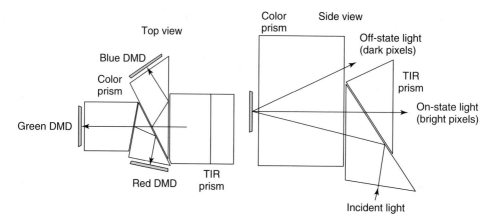

Figure 16.11 Top view and side view of the color and TIR prisms [14].

Figure 16.10 shows a typical three DMD system. A compound elliptical and spherical reflector is applied in this case to optimize light collection. The illumination system has a rod integrator. Moreover, a prism system (color prism and total internal reflection prism) separates the white light beam into red, green, and blue beams and recombines the three beams again after modulation by the DMDs. One projection lens images the three DMDs on the screen (Figure 16.11).

An LCoS-based configuration is shown in Figure 16.12. A fly's eye integrator is used in this case and two dichroic mirrors are applied to make the red, green, and blue beams. Wire-grid polarizing beam splitters are applied for high contrast ratio. A dichroic prism is used to recombine the three colors. One projection lens images the three LCoS panels on the screen.

16.3.2.3 One-Panel Reflective DLP with UHP Lamp

The light from the arc lamp is focused to the entrance of the rod integrator. The exit aperture of the rod integrator is imaged on the DMD by the illumination optics.

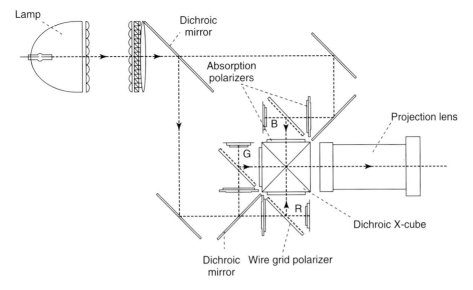

Figure 16.12 Wire-grid and X-cube LCoS projection architecture [15].

At the air gap of the TIR prism, the light is reflected to the DMD. In the on state of the DMD mirror, the light is reflected back and goes through the projection lens. In the off state of the DMD mirror, the light propagates, but not through the projection lens. Color images are generated by the color-field sequential technique. The light is focused on the dielectric color filters of the color wheel that rotates synchronously with the video signal. The red, green, and blue data of the video signal is displayed on the DMDs when, the red, green, and blue filter segments of the color wheel are illuminated, respectively (Figure 16.13).

16.3.2.4 One-Panel Reflective LCoS with High-Brightness LEDs

This is the architecture of a so-called pico projector. A single-panel color-field sequential is possible with fast MDs as DMDs and ferro-electric-liquid-crystal-on-silicon (FLCoS) MDs [16]. LED-based color-field sequential projectors are possible without a rotating color wheel and higher field rates can be applied to reduce color breakup.

The light from the red, green, and blue LEDs is combined by the dichroic color combiner. A fly's eye integrator with a polarizing conversion system images the LED light beam via the polarizing beam splitter on the FLC LCoS MD. One projection lens images the MD on the screen (Figure 16.14).

16.3.2.5 Three-Panel Grating Light Valve Projector

Lasers can be used in combination with the traditional 2D imagers such as DLP and LCoS and the small étendue of the lasers allows them to be used in laser scanning projectors. One option is to scan the laser beam itself over the surface of

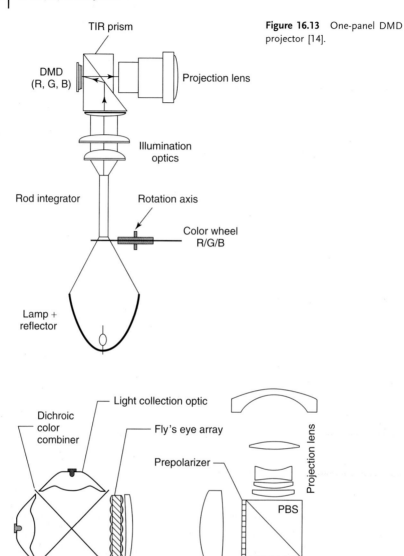

Figure 16.13 One-panel DMD projector [14].

Figure 16.14 Ferroelectric LCoS (FLCoS) architecture [16]. (Please find a color version of this figure on the color plates.)

the screen. This requires a laser source that can be modulated very fast. Projectors using this approach can be made very small (picoprojectors). In another approach, a modulated line of light is scanned over the screen [17].

The laser source of each of the primaries is focused into a homogeneously illuminated line on a one-dimensional imager, called a *grating light valve* (GLV).

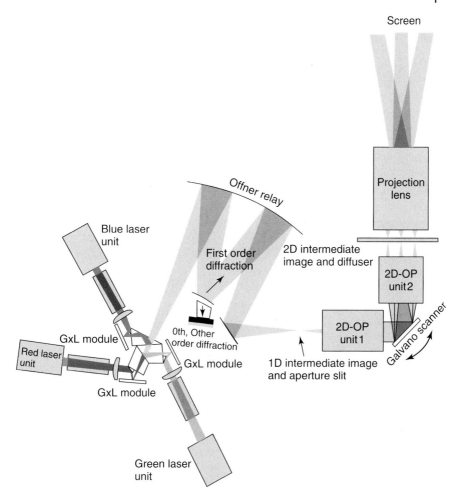

Figure 16.15 Grating light valve laser projector [17].

This imager consists of a large number of ribbons, perpendicular to the line of laser light. For each pixel typically four ribbons are required; two of these can move, while the other two are fixed. By moving down the two ribbons, light will be diffracted into the first order and by means of relay optics (Offner relay in Figure 16.15) the light in the zeroth order is blocked. The light in the first order is transmitted toward a galvano scanner, which sweeps the modulated line of light over the screen. As such, a two-dimensional image is constructed. The scanning occurs at a very high speed to avoid flicker. In the approach depicted in the figure, a two-dimensional intermediate image is created on a diffuser to reduce speckle. The image formed on this diffuser is then projected on the screen by the projection lens.

16.4
Applications

16.4.1
Medical Imaging Displays

16.4.1.1 Medical LCD Displays

At the end of the 1990s, LCDs were already widely used for consumer applications such as desktop monitors or laptop displays. But for medical imaging applications, CRTs displays were still the standard because they offered superior quality compared to LCDs. Around the year 2000, LCD panels started having specifications that came close to specifications of medical CRTs. However, LCD technology still had several important drawbacks compared to CRT displays. First of all, LCDs suffer from severe nonuniformity and spatial noise [18]. In comparison with CRT displays, there is a lot more variation in the pixel-to-pixel behavior of LCDs. There was a serious concern in the medical imaging community that this pixel noise could obscure or interfere with subtle features of a medical image. Secondly, a typical LCD display contains several defective pixels. In modalities such as mammography, radiologists sometimes are looking for features of only a few pixels in size. A high number of pixel defects therefore was unacceptable. Finally, the behavior of LCDs is much more complex compared to medical CRT displays. One well-known example is that LCDs suffer from viewing angle dependency [19]. Also, the less-predictable and less-fluent transfer curve of an LCD required more complex calibration algorithms. These disadvantages created a lot of hesitation in the medical imaging community and slowed down adoption of medical LCDs. For high-demand modalities such as mammography, these drawbacks were just not acceptable and needed to be truly solved before the medical community was ready to adopt LCDs [20].

16.4.1.2 Calibration and Quality Assurance of Medical Display Systems

One important difference between traditional LCDs and medical display systems is the fact that medical displays are being calibrated to the Digital Imaging and Communications in Medicine (DICOM) gray-scale standard display function (GSDF) standard [21] to guarantee sufficient and consistent image quality. In most countries, this calibration is required for display systems that are being used for primary diagnosis [22]. Why this calibration is needed and how it is being performed is explained in the next section.

Every display system has a specific native curve that describes the luminance behavior (cd m^{-2}) of the display in function of the digital drive level (DDL) or gray level. Without any display calibration, the perceptual distance (perceived contrast) between consecutive gray levels would not be constant [23]. Some gray levels could be very close to each other and therefore would not be perceived as being different anymore. Other gray levels could be too far apart, resulting in banding artifacts.

To overcome these problems, display systems are being calibrated to DICOM GSDF. The GSDF describes the luminance (cd m^{-2}) as a function of just noticeable

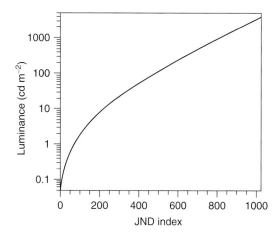

Figure 16.16 DICOM gray-scale standard display function.

difference (JND) index (Figure 16.16) [24]. In the ideal situation, the difference between every pair of consecutive gray levels should be larger than 1 JND (such that the difference is visible). Moreover, the difference between every pair of consecutive gray levels should also be constant so that the perceived contrast is same as well.

Calibrating the display to the DICOM GSDF curve is done to achieve these objectives. The native curve of the display is measured (typically at the center of the display and for on-axis viewing) with a luminance sensor. On the basis of the measured curve, a lookup table (calibration lookup table) is calculated and configured. This lookup table is configured such that the resulting luminance curve as measured on the display, will match the DICOM GSDF curve as good as possible (Figure 16.17).

In general, there exist two types of tests to verify compliance to DICOM GSDF: visual inspection tests and physical measurements. In the category of the visual inspection tests, three test patterns are broadly used. The Society of Motion Picture and Television Engineers (SMPTE) pattern originates from the time where mainly CRT displays were used in medical imaging. A first goal of the SMPTE pattern is to check spatial resolution and aliasing characteristics of the display system. A second goal of the SMPTE pattern is to check whether the luminance and contrast settings are in the correct range. One has to be aware that the SMPTE pattern is used today mainly for historical reasons but it does not provide sufficient guarantees that a display is well calibrated.

The Briggs pattern is a contrast-detail pattern. A monitor is well calibrated if all the low-contrast squares are visible with equal perceived contrast. If all the low-contrast squares are visible, then it means that all gray-scale transitions of the display are above 1 JND. If, moreover, the perceived contrast for each of the squares is also equal, then the monitor is also perceptually linearized, which means that each of the gray-scale transitions has the same step size in JND space.

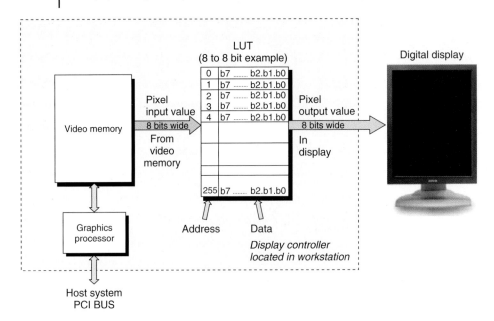

Figure 16.17 Calibration lookup table of a medical display.

The American Association of Physicists in Medicine (AAPM) Task Group 18 [22] is a national task force consisting of medical imaging experts and organizational affiliates dealing with performance evaluation of electronic display devices. They have made available recommendations for quality assurance checks of monitors. Their recommendations provide guidelines to practicing medical physicists and engineers for in-field performance evaluation of electronic display devices intended for medical use and include the well-known TG18 test pattern. This test pattern combines both spatial resolution and contrast-detail tests.

16.4.2
Other Applications

Projection displays are used in a wide range of applications. Presentation, education, and conference rooms represent a big segment in which single-panel DLP and HTPS projectors are mainly used. Another important segment is the large venue market: auditoria, rental business, religious institutions (large churches), and digital cinema. These applications require high-brightness projectors (>5000 lm). Therefore, three-panel DLP and LCoS find use in this category. The home theater is a special segment, which constitutes a high-end application where a high contrast ratio (>10 000 : 1) is required and LCoS is mainly used. Another important category is that of high-end professional applications: control rooms (e.g., power plants, telecommunication, broadcast, and military applications), simulation and training (e.g., aircraft, ships, trains, and cars), and virtual reality (e.g., oil and gas industry, automotive design, and scientific research). All these technologies are applied in this category.

Digital cinema is an important application to which the three-panel DMD architecture is very well suited. A typical example is the Barco DP-3000 2k digital cinema projector, where three 1.2″ diagonal 2048 × 1080 pixel DMDs in combination with one 7 kW xenon high-intensity discharge lamp are used. This projector is used in movie theaters with screen widths up to 30 m. Colorimetry and contrast ratio are very important specifications and all the characteristics of this projector comply with the Digital Cinema Initiative (DCI) standard.

Simulation and training is another demanding application. A simulator for jet plane pilot training, for instance, requires a higher resolution (e.g., >30 MP) than the highest achievable resolution of a single projector. Therefore, setups of multiple projectors with good edge blending are developed to fulfill the resolution requirements. Contrast ratio and dynamic range are extremely important characteristics of a simulation projector.

Dynamic range is defined as the ratio between the maximum and the minimum luminance value (as measured by means of a physical measurement) that a display can produce. Typical dynamic range values for LCDs are in the range 500 : 1 up to 2000 : 1. Some LCDs offer the possibility to dynamically change the backlight value when a scene becomes, on average, darker or brighter. In this way, dynamic range values of up to 1 000 000 : 1 can be obtained (even though the simultaneous dynamic range that can be obtained still is limited to typically 2000 : 1).

One can question what the value is of having a dynamic range of 1 000 000 : 1. After all, there is a limit to the dynamic range [25] that the human eye can simultaneously look at. A contrast of 4000 : 1 is approximately what the eye can see simultaneously and when fully adapted to the average luminance of the scene. This limitation of the eye can easily be experienced on a sunny summer day. When one is outside in the sun, the luminance can be 10^6 cd m^{-2} or higher. Still the eye is able to see perfectly in this environment. However, when moving indoors (typical ambient light of 100–200 cd m^{-2}), it will become clear that the eye requires many seconds or even minutes before truly being adapted to the new average luminance level. During this adaptation period, it is very difficult to perceive low-contrast details. By means of this adaptation process, the eye can handle a dynamic range of more than ten orders of magnitude.

Nighttime training requires luminance levels of 10^{-5} cd m^{-2}; for daytime training, luminance levels of >60 cd m^{-2} are necessary. For this application, the projection screens are usually not flat surfaces. Spherical screens are often applied because the complete field of view of the pilot must be filled. Electronic geometrical predistortion (warping) is necessary to compensate the distortion when the surface of the screen is not flat. For these applications, fast-moving objects must be reconstructed by projection displays with high resolution, which also requires displays with a narrow temporal aperture. A possible solution is a shutter system synchronized with MD scanning. The SIM7 projector is specifically developed for this purpose and has 3 MP resolution (2048 × 1536 pixels per LCoS panel), a dimmer and adjustable apertures for high dynamic range, three fast-switching

LCoS panels with 10 000 : 1 contrast ratio, and a shutter system to eliminate the perceived image blur caused by the pixel hold time.

Usually, virtual reality applications also require a very high pixel count in combination with relative high brightness and contrast ratio. Therefore, multiple projector setups with edge blending are necessary, often in rear projection configurations. Immersive display systems and stereoscopic images are also required in certain virtual reality applications. The Galaxy NW-12 projector is a WUXGA (1920 × 1200) three-panel DMD projector, specifically developed for virtual reality application and with multiple stereoscopic capabilities. One particular stereo imaging capability is the active wavelength multiplexing, using interference filters.

16.5
Conclusion

This chapter explained the basic concepts behind photonics technology – display systems. The two main families of display systems have been described – direct view displays and projection displays. For each of these families, the basic concepts and current state of the art have been discussed. Specific attention was paid to describe ongoing trends such as the gradual move toward LED and lasers as light source for projection displays and transmissive direct view display systems. A preview of potential future types of display systems was given as well with the description of, for example, OLED displays, laser projectors, and electronic paper displays.

Finally, the chapter concluded with some selected applications such as education and simulation. A detailed description was given of the use of display systems for the high-end medical imaging market.

References

1. O'Mara, W. (1993) *Liquid Crystal Flat Panel Displays: Manufacturing Science and Technology*, Van Nostrand Reinhold.
2. Compton, K. (2003) *Image Performance in CRT Displays*, SPIE Press Book.
3. Boeuf, J.P. (2003) Plasma display panels: physics, recent developments and key issues. *J. Phys. D: Appl. Phys.*, **36**, 53–79.
4. Sasaoka, T. (2001) A 13.0-inch AM-OLED display with top emitting structure and adaptive current mode programmed pixel circuit (TAC). *SID Symp. Dig.*, **32**, 384–387.
5. Crawford, G.P. (2005) *Flexible Flat Panel Displays*, John Wiley & Sons, Inc.
6. JVC (2009) JVC Develops World's First Full-Coverage, High Resolution Super Hi-Vision Projector. Press release, May 12 2009.
7. Harbers, G. et al. (2007) Performance of high power light emitting diodes in display illumination applications. *IEEE J. Display Technol.*, **3** (2), 98–109.
8. Sekiguchi, A. et al. (2004) Étendue-density homogenization lens optimized for high-pressure mercury lamps. *J. SID*, **12** (1), 105–111.
9. Stupp, E. and Brennesholtz, M. (1998) *Projection Displays*, John Wiley & Sons, Inc.
10. Derra, G. et al. (2005) UHP lamp systems for projection applications. *J. Phys. D: Appl. Phys.*, **38**, 2995–3010.

11. Stroomer, M. (1989) CRT and LCD projection TV: a comparison. SPIE Projection Display Technology, Systems and Applications, Vol. 1081, pp. 136–143.
12. Morozumi S. et al. (1986) LCD full-color video projector, *SID Symp. Digest Tech. Pap.*, **17**, 375–376.
13. Imai, M. et al. (1992) A polarization convertor for high-brightness liquid crystal light valve projector. Japan Display '92, pp. 235–238.
14. Meuret, Y. (2004) De contrast- en kleurenproblematiek van digitale projectoren met een grote helderheid. Ph.D. dissertation, University Ghent.
15. Robinson, M.G. et al. (2006) Three-panel LCoS projection systems. *J. SID*, **14** (3), 303–310.
16. Darmon, D. et al. (2008) LED-Illuminated pico projector architectures, *SID Symp. Digest Tech. Pap.*, **39**, 1070–1073.
17. Kikuchi, H. et al. (2009) High-pixel-rate grating-light-valve laser projector, *J. SID*, **17** (3), 263–269.
18. Kimpe, T. et al. (2005) Solution for nonuniformities and spatial noise in medical LCD displays by using pixel-based correction. *J. Digital Imaging*, **18**, 209–218.
19. Badano, A. et al. (2003) Angular dependence of the luminance and contrast in medical monochrome liquid crystal displays. *Med. Phys.*, **30**, 2602.
20. Bacher, K. et al. (2006) Image quality performance of liquid crystal display systems: influence of display resolution, magnification and window settings on contrast-detail detection. *Eur. J. Radiol.*, **58**, 471–479.
21. National Electrical Manufacturers Association (1998) Digital imaging and communications in medicine (DICOM), supplement 28: Grayscale standard display function. *http://medical.nema.org/dicom/final/sup28 ft.pdf* (accessed January 1998).
22. Samei, E. et al. (2005) Assessment of display performance for medical imaging systems: executive summary of AAPM TG18 report. *Med. Phys.*, **32**, 1205.
23. Kimpe, T. and Tuytschaever, T. (2007) Increasing the number of gray shades in medical display systems; how much is enough? *J. Digit. Imaging*, **20**, 422–432.
24. Barten, P. (1992) Physical model for the contrast sensitivity of the human eye. *Proc. SPIE*, **1666**, 57.
25. Brainard, D., Pelli, D., and Robson, T. (2002) *Encyclopedia of Imaging Science and Technology*, Wiley, pp. 172–188.

17
3D Displays

Janusz Konrad

17.1
Introduction

Humanity has been fascinated with virtual reproduction of the three-dimensional (3D) world for centuries. After all, effortless, 3D perception of surroundings is a natural ability for about 90% of the population, namely, those who benefit from *binocular stereopsis*, that is, the capacity to fuse two viewpoints captured by the eyes. Seeking a "being there" experience, we have developed many visual devices that are capable of mimicking the appearance of a 3D scene. Such devices use mechanical, optical, or electro-optical means to simulate a field of light that, in turn, produces a different projection on the retinas of the viewer's two eyes. The brain interprets the disparity between these light patterns as depth.

One should note that retinal disparity is not the sole cue that human vision exploits to perceive depth; 2D cues, such as perspective, occlusion, shading, relative object size, and motion parallax, are important for depth perception as well. However, these cues are auxiliary and cannot invoke a full 3D perception without retinal disparity. Clearly, of interest are displays that deliver binocular stereopsis or even motion parallax, that is, a change in binocular stereopsis induced by viewer motion in front of the display.

The numerous 3D display technologies developed to date offer many benefits but are also plagued by certain deficiencies. From the early days of stereoscope (mid-19th century), through parallax stereograms (early twentieth century), and polarized 3D movies (1960s), to quality 3D films by IMAX™ (1980s and 1990s) and electronic 3D (today), advances in materials, electronics, and optics have led to significant improvements in 3D image quality, visual comfort, and benefit/cost ratio. Owing to these improvements, 3D cinema today has a solid footing in the entertainment market, with 3D blockbusters being released monthly. In specialized applications, such as image-guided surgery, remote robot guidance, or battlefield reconnaissance, 3D hardware is becoming a commodity item. Furthermore, 3D home entertainment is no more an elusive mirage, but more a clear target within our reach, and at its forefront is 3D video gaming with over two dozen video games available today.

Optical and Digital Image Processing: Fundamentals and Applications, First Edition. Edited by Gabriel Cristóbal, Peter Schelkens, and Hugo Thienpont.
© 2011 Wiley-VCH Verlag GmbH & Co. KGaA. Published 2011 by Wiley-VCH Verlag GmbH & Co. KGaA.

A wider adoption of 3D displays in the near future is likely to be driven by the ongoing digital multimedia revolution. While 3D visualization in the past relied on custom components and technologies, 3D display devices of today are well positioned to take advantage of the all-digital processing chain, from capture, through processing and editing, to compression and transmission. Today, 3D digital image processing can be seamlessly added to the existing pipeline without reinventing other steps. It is this all-digital infrastructure that is at the core of the recent resurgence of 3D cinema and that is becoming a foundation for the 3D home theater in the future.

Yet challenges remain in the quest for an effortless "being there" experience. One challenge is in the development of new display technologies to allow high-quality, high-resolution 3D experience without glasses. The second challenge is in the area of compression for 3D displays, as they require massive amounts of data. A significant progress has been made on this front by the adoption of the multiview coding (MVC) extension of the advanced video coding (AVC) standard. The third challenge, which we focus on here, is in the development of effective and efficient signal processing methods to overcome 3D displays' deficiencies and enhance 3D viewer experience.

In this chapter, we present an overview of today's main planar electronic 3D display technologies from a signal processing perspective[1]. We briefly describe the underlying physics and discuss deficiencies of *planar stereoscopic displays*, the simplest and most common systems that provide binocular stereopsis, and of *planar multiview displays* that provide motion parallax and are expected to dominate 3D home theater market in the future. Then, we focus on the role of signal processing in addressing display deficiencies, and highlight signal processing challenges in the quest for the ultimate 3D experience.

17.2
Planar Stereoscopic Displays

The earliest stereoscopic 3D display, the *stereoscope*, was developed in the mid-19th century. Two drawings were placed in front of viewers' eyes with a mechanical barrier to channel each drawing to one eye. The drawings were later replaced by black-and-white pictures that were captured by two closely spaced cameras, and lenses were added to enhance the visual experience. Despite the extreme popularity of stereoscopes during the Victorian era, 3D photography has never rivaled the popularity of its "flat" counterpart.

The above principle of delivering two images through separate optical channels to viewer's eyes is shared by stereoscope's modern cousins, head-mounted displays (HMDs). Instead of two photographs, two small screens are placed directly in front

---

1) For a discussion of holographic and volumetric 3D displays used in professional applications, the reader is referred to a recent review article [1].

of the eyes, thus avoiding image multiplexing. We do not consider such systems here, but concentrate on the delivery based on image multiplexing with or without glasses.

17.2.1
Stereoscopic Displays with Glasses

Stereoscopic viewing with glasses is based on one of three approaches: spectral filtering, light polarization, and temporal shuttering. One filter or shutter for each eye ensures a separate delivery of either the left or the right channel. Glasses-based systems have the advantage of providing an entire audience with a 3D experience at reasonable cost.

17.2.1.1 Spectral Filtering
The simplest form of 3D reproduction by means of spectral filtering is the *anaglyph* method in which left and right images are jointly displayed in complementary or nearly complementary colors and separated by correspondingly colored glasses. It was developed in the mid-nineteenth century for line drawings and later extended to photographs. Until today, it has remained the most printer-compatible 3D system. It became quite popular with the advent of motion pictures in the early twentieth century, but was subsumed by polarized systems by the 1950s.

Ideally, one would like to divide the visible light spectrum (370–730 nm) into two complementary bands (colors), such as 370–550 nm and 550–730 nm, by means of "brick-wall" optical filters. In practice, however, typical transmission characteristics of the filters have a gradual roll-off. For example, Figure 17.1 shows

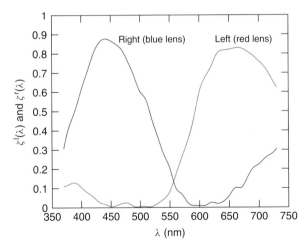

Figure 17.1 Spectral transmission functions of anaglyph Roscolux filters: $\zeta^l(\lambda)$ of orange-red (#25), $\zeta^r(\lambda)$ of brilliant-blue (#69).

transmission curves for practical red and blue lenses. Usually, the left-image data are placed in the red channel of a color photograph or electronic display, and the right-image data are placed in the blue channel. Viewed through red–blue glasses, the image invokes a perception of depth. Although there exists some flexibility in composing the green channel, this strongly depends on the spectral characteristics of the glasses. For example, for red–blue glasses shown in Figure 17.1, placing the right-image data instead of the left-image data in the green channel (approximately 500–550 nm) improves depth perception. This is due to the fact that blue–green mixtures fall in the passband of the right spectacle unlike the red–green mixtures that do not match the left lens well.

The most important benefit of anaglyph stereo is that it can be used on essentially any device that is capable of reproducing color, whether electronic, film, or paper. In addition, with simple, easy-to-manufacture glasses it is the least expensive technique for mass 3D visualization. Its main deficiencies, however, include visual discomfort due to different spectral content presented to each eye (color rivalry) and optical cross talk between channels (poor spectral separation of colored lenses). Anaglyphs also have very limited color reproduction capability.

Although the interchannel cross talk and color gamut can be, to a degree, controlled by a careful selection of color lenses *vis-à-vis* screen phosphors or print dyes, the range of possible improvements is rather small. This is primarily due to a limited selection of spectral lens characteristics and rather rudimentary algorithms for anaglyph preparation. Significant improvements, however, are possible by employing signal processing methods to optimize anaglyph images with respect to reproducible color gamut given spectral absorption curves of the lenses, spectral density functions of the display primaries, and colorimetric characteristics of the human visual system. A practical solution of this problem is described in Section 17.4.1.

The main deficiency of the anaglyph method stems from limited color capture by each eye (either 370–550 nm or 550–730 nm). This can be mitigated through multiband spectral filtering, where the visible light spectrum is divided into two complementary *sets of wavelength intervals*. When those sets are uniformly spread out across the light spectrum (Figure 17.2), a vastly improved color reproduction in each eye is possible. Recently, Infitec GmbH, Germany, developed a pair of triband filters (coarse comb filters) [2] whose spectral responses are complementary but include a sufficient composition of wavelengths to warrant good color reproduction in each eye.

For 3D visualization, two projectors are equipped with complementary filters, while viewers wear corresponding glasses. With suitable processing, colors are represented in terms of spectral characteristics of the left and right filters independently, that is, in terms of left-lens primaries $R^l B^l G^l$ and right-lens primaries $R^r G^r B^r$ (Figure 17.2). If this color representation is accurate, no visual discomfort results, despite the different color primaries used for the left and right eyes, because the human eye cannot distinguish different compositions of the same color. Although filters with ideal spectral curves (Figure 17.2) cannot be realized in practice, filters

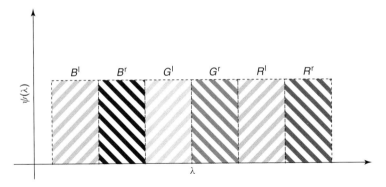

Figure 17.2 Spectral characteristics of ideal multiband filters illustrating the principle of operation of displays using Infitec glasses.

with sufficiently accurate responses to minimize optical cross talk while assuring spectral complementarity can be manufactured today.

The multiband filters are new to the commercial market and are far more expensive than traditional anaglyphic or polarized glasses. The majority of multiband filter systems require two projectors, although recently Barco introduced a single-projector solution where a single filter switches its spectral characteristics from frame to frame, thus permitting time-sequential stereo. Whether based on one or two projectors, the system uses standard projection screen unlike systems based on polarization (nondepolarizing screen is needed). To date, stereoscopic 3D projection with Infitec glasses (Dolby® 3D Digital Cinema) has captured a portion of the 3D movie theater market.

The perceived 3D image quality in a system based on multiband filters greatly depends on the accuracy of color representation *vis-à-vis* filter spectral functions, and the complementarity of these functions between the left and right channels (amount of cross talk). While the first issue is a standard problem in colorimetry, the latter can be addressed by techniques similar to that discussed in Section 17.4.2.

17.2.1.2 Light Polarization

Light polarization is another 3D display method permitting full-color reproduction. In a typical arrangement, two projectors are equipped with differently polarized filters. For example, if linear polarization is used, one projector is equipped with a horizontally oriented polarizing filter, while the other carries a vertical polarizing filter. A special silver screen is necessary to ensure that polarization is maintained during projection. When a viewer uses correspondingly polarized glasses, each eye captures light from (ideally) one view only, producing a sensation of depth.

The ability of a polarized system to render a full-color gamut greatly reduces the visual discomfort typical of anaglyphic systems. In real-world systems, however, cross talk between left and right views is inevitable. View separation using linear polarizers is both imperfect and sensitive to rotation: as orientation of the glasses departs from that of projector polarizers, for example, because of viewer head motion, the optical cross talk between channels increases. These effects can be largely mitigated by employing circularly polarized filters, but even then some optical cross talk remains because of insufficient extinction properties of the filters, and light depolarization on the screen.

Although light polarization is easier to implement with projectors (the filter is small as it only needs to cover the projector lens), it has also been implemented using LCD screens since they produce polarized light. Planar's StereoMirror™ display [3] offers a desktop solution that is compatible with passive polarized glasses by combining two LCDs and a beamsplitter that flips the polarization of one of the panels.

Yet another issue in dual-projector/dual-display systems is the alignment of the two projected images. Unless carefully controlled, image misalignment may cause significant eye strain. This is less of an issue in dual-display systems, like the one from Planar, because of the flatness of the LCD panels used and short projection distances, but it can be a problem in dual-projector systems. The problem can be eliminated altogether if static polarizers in front of left and right projectors are replaced by a dynamic polarization modulator in front of single projector. By interleaving left and right images in a video sequence, and switching the modulator polarity in synchronism with vertical blanking, the projected left and right images are differently polarized and can be easily separated by viewer-worn glasses. However, a projector/display capable of doubled refresh rate is needed. This is not a problem for CRT- or DLP-based projectors, but LCD projectors are unsuitable owing to unacceptable ghosting resulting from slow refresh rates[2]. Two examples of polarity modulation technology are ZScreen® developed by Stereographics Corp. (currently RealD Inc.) that alternates circular polarization of consecutive images and μPol™ (micropol) developed by Vrex Inc. which alternates linear polarization of consecutive lines of an image. To date, stereoscopic 3D projection based on ZScreen® modulator and polarized glasses has been deployed commercially in hundreds of electronic cinemas (RealD theaters) capturing majority of this market.

Although the dynamic modulator solution does away with left/right image alignment issues, light output and cross talk remain of concern. While the light absorption in a filter or modulator can be compensated for by increased brightness of a projector, the cross talk problem necessitates signal processing solutions, such as that described in Section 17.4.2; by preprocessing the left and right images, one can largely suppress the perceived cross talk. Such a solution is applicable to systems using circular polarization since cross talk is largely rotation invariant.

---

2) Although the newer LCD flat panels improve refresh rates, and plasma flat panels are relatively fast, one would have to manufacture 50 in. polarity modulators which is difficult and cost prohibitive.

In systems with linear light polarization, the cross talk varies with viewer head rotation and the problem is rather intractable.

17.2.1.3 Light Shuttering

In order to resolve image alignment and, to a degree, light output issues, shuttered glasses (e.g., liquid-crystal shutters (LCS)) were introduced in the late 20th century. Instead of light polarization or spectral filtering, they block light by means of fast switching lenses that, working in synchronism with the screen, become opaque when the unintended view is rendered on the screen, and transparent when the intended view is displayed.

CRT displays in combination with LCS have long offered high-quality stereoscopic visualization on the desktop. Such systems are characterized by full CRT spatial resolution, full CRT color gamut, and no flicker for CRTs with refresh rates of at least 120 Hz. However, this approach is expensive since each viewer needs a pair of costly LCS glasses. Moreover, similarly to 3D systems using polarization modulator and polarized glasses, the CRT/LCS combination suffers from optical cross talk. The source of the cross talk is different however, with the main factors being screen phosphor persistence, LCS extinction characteristics, and LCS timing errors [4].

The CRT/LCS cross talk is approximately shift invariant, because CRT phosphor persistence and LCS light extinction characteristics are quite uniform spatially (although the latter slightly changes toward shutter periphery). This approximate shift invariance permits application of signal processing algorithms to predistort the original left and right images in order to compensate for the cross talk added during display [5, 6]. Details of such an approach are discussed in Section 17.4.2.

The widespread replacement of CRTs by slower response LCD panels has halted the march toward "stereo on the desktop"; the slow refresh rate of LCD panels stemming from hold times causes significant cross talk. Although various methods, such as black frame insertion, motion-adaptive frame interpolation, and backlight modulation, have been developed recently to increase the refresh rates to 120 Hz and even 240 Hz, LCD screens remain unsuitable for 3D due to "scanlike" image update[3], preventing flicker-free, ghost-free 3D experience [7]. Recently, Viewsonic and Samsung have released 3D LCD monitors and NVIDIA LCS glasses, but it is unclear whether the above issues have been overcome. Also, Panasonic and Samsung have announced 3D plasma panels using LCS glasses, but it remains to be seen whether the phosphors-induced cross talk can be sufficiently controlled in those units.

17.2.2
Autostereoscopic Displays (without glasses)

The need to use glasses for viewing stereoscopic content has long been identified as a major inconvenience due to physical viewer discomfort, the cost of glasses (especially of the LCS variety), and potential for image misalignment in two-projector

---

[3] Image update in a typical LCD screen takes place line by line sequentially from the top of the screen down in such a way that at any given time instant lines from two consecutive frames are displayed simultaneously.

systems. Moreover, in scenarios where both 3D screens and 2D screens, which are not time synchronized with the 3D ones, are present in the field of view of LCS glasses, the shutters optically interfere with the nonsynchronized displays, resulting in screen flicker and, therefore, additional viewer discomfort. Similarly, polarizing glasses may interfere with viewing conventional 2D LCD monitors because of the inherent polarization of light produced by these monitors.

As an alternative to systems using glasses, a number of techniques have been proposed since the early 1900s [8] that apply spatial multiplexing of left and right images in combination with a light-directing mechanism that presents those views separately to the viewer's eyes. These so-called *autostereoscopic* techniques have gained a wide recognition with the arrival of electronic, pixel-addressable displays, such as LCD and plasma flat panels, to which they are directly applicable [9, 10]. The main multiplexing techniques today permitting separate channeling of each view to one eye are *parallax barrier* and *microlenses* (also called *lenticulars*). Note that lenticulars have also been used for image capture [11].

In parallax-barrier displays, an opaque layer (e.g., a very thin sheet of aluminum or inked screen) with narrow, regularly spaced slits is placed very close to a pixel-addressable screen (Figure 17.3a). Note that for a given slit, each eye's viewing angle is different and also that each slit acts horizontally as an approximate pin-hole projector. If the slits are precisely aligned with pixel columns, and the overall geometry is carefully adjusted (pixel pitch, slit width and pitch, and barrier distance from the screen surface), a viewer sitting at a prescribed distance from the screen will see different sets of pixel columns with each eye, corresponding to the left and right views.

In lenticular displays, a sheet of narrow, thin cylindrical microlenses is typically attached to a pixel-addressable screen at approximately one focal length, so that light passing through lenses focuses the underlying panel (Figure 17.3b) to either infinity or the intended viewing distance. Again, for a given microlens the viewing angles of the eyes are different and each microlens focuses light horizontally. If the microlenses are precisely aligned with pixel columns, and if pixel pitch, microlens pitch, and focal length are carefully selected, a viewer sitting at a prescribed distance will also see different sets of pixel columns with each eye.

The locations in front of a screen from which a consistent 3D perception is experienced, that is, the left eye sees pixels with left-image data and the right eye sees pixels with right-image data, is called a *viewing zone*. The size of the viewing zones is related to the geometry of the display mentioned above.

When viewer's eyes straddle two different viewing zones in a two-view display, the left and right views presented are reversed and the viewer sees a *pseudoscopic*, or depth-inverted, image. The physical adjacency of correct (orthoscopic) and inverted (pseudoscopic) view zones can lead to visual confusion and discomfort for the viewer. Both parallax-barrier and lenticular technologies suffer from this shortcoming, which can be considered a type of optical cross talk since one slit or lens can produce an image using its neighbor's data.

One method of reducing cross talk in parallax-barrier displays is to reduce the width of each slit aperture. This approach has the disadvantage of reducing the

17.2 Planar Stereoscopic Displays | 377

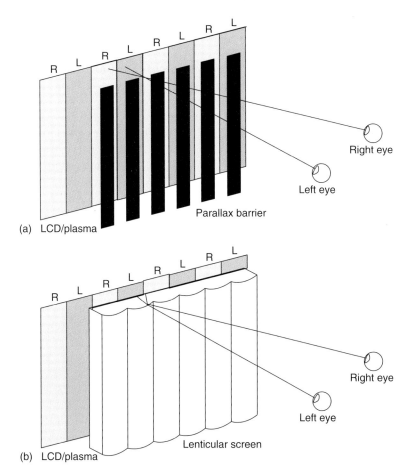

Figure 17.3 Schematic illustration of how (a) parallax-barrier and (b) lenticular displays work.

display's light efficiency. Cross talk can also be minimized by introducing guard bands with no image information in between the view data for each slit/microlens. With guard bands, the image would black out for one or both eyes rather than display neighbor's data, but only at the expense of reducing viewing zone size. In electronic displays, guard bands are often impractical because of the structure of the underlying pixel array. In practice, display systems are designed to balance cross talk, size of viewing zones, and light output requirements, although the "sweet spot" of an autostereoscopic display, that is, location where the image appears depth-correct, tends to be fairly narrow.

A few additional remarks are in order here. CRT displays cannot be used with parallax-barrier or lenticular technology, because a precise geometric relationship between pixels and slits/microlenses is required; light patterns on a CRT screen suffer from unacceptable jitter. While parallax-barrier systems are easier to manufacture than lenticulars, their main deficiency is a reduced light output due to the opaque barrier.

Furthermore, all autostereoscopic displays based on spatial multiplexing suffer from the loss of horizontal resolution delivered to each eye; one-half of the pixels in each horizontal row is delivered to the left eye while the other half is delivered to the right eye. In order to spatially multiplex left and right images on the screen, both images need to be horizontally subsampled by a factor of 2. Clearly, such subsampling necessitates prior half-band horizontal low-pass filtering in order to avoid aliasing. Such filtering is a special case of a more general prefiltering employed in spatially multiplexed multiview displays that are discussed in Section 17.4.3.

17.3
Planar Multiview Displays

Stereoscopic displays, while relatively simple and inexpensive, cannot accurately recreate a true 3D experience. In glasses-based systems, the same two views are presented regardless of the viewing position, which can lead to mental confusion about the shape of the object, and in autostereoscopic systems, an apparent depth inversion of the scene may take place.

The stereoscopic displays described thus far also fail to provide the important depth cue of motion parallax, that is, a change of viewpoint due to viewer motion. Motion parallax is an important complement to stereopsis, particularly when a scene is relatively static and the viewer is free to inspect it from a variety of positions. Motion parallax can provide more meaningful depth information over a larger range of working distances between the viewer and the object than can stereopsis, and it provides depth cues to a larger portion of the population (including those with deficiencies in binocular vision).

Several display technologies provide motion parallax in addition to stereoscopic depth by displaying a range of viewpoints of a scene to the viewer. We refer to displays of this type as *multiview displays*. This section describes two main displays of this type: active multiview systems, where a viewer's position is used to calculate and present the appropriate images, and passive multiview displays, where several viewpoints of the scene are projected simultaneously into a viewing zone through optical means.

17.3.1
Active Multiview 3D Displays

In order to display motion parallax, a device must present view information to an observer that is specific to his or her location. *Active* multiview displays provide motion parallax by tracking the viewer location, generating two appropriate views for the viewer's eyes, and using an optical mechanism, such as any of those described in Section 17.2 to present those views to the viewer [9]. An active multiview display, at least ideally, presents the same stereo and parallax cues as would a true 3D object.

As view rendering depends in this case on head position, *irregular, temporal view multiplexing* takes place; the degree of irregularity depends on the motion pattern of viewer's head. Furthermore, the selection of suitable view for on-screen rendering can be thought of as subsampling of this view's full motion sequence, and thus a temporal anti-alias filter should be employed. However, because of movement-induced irregularity, or time-variance, the design of an optimal filter is nontrivial. Moreover, even if one were able to find an optimal filter, another significant deficiency would remain; because of the tracking employed, the computed view is correct only for the viewer being tracked. Essentially, active multiview 3D displays are single-user displays.

17.3.2
Passive Multiview 3D Displays

The inability to support multiple viewers is a major weakness of active multiview 3D displays. An alternative is to display all views simultaneously by means of spatial multiplexing [12]. If each view (light from the corresponding pixels) is directed into a distinct viewing zone in front of the 3D display, then lateral viewer motion across zones induces motion parallax. The viewer must, however, rest at a specified distance from the screen, as otherwise the effect is largely lost. Since all views are displayed at the same time, and no head tracking is used, multiple viewers can experience 3D sensation that is consistent with their respective positions in front of the display. Support for multiple viewers, motion parallax, and elimination of glasses have proved very appealing and a number of parallax-barrier and lenticular multiview 3D displays have been successfully launched commercially in the last decade [13, 14]. Multiview displays that require no glasses are often referred to as *automultiscopic* displays.

The technologies used to implement autostereoscopic displays, namely, parallax barriers and lenticulars (Figure 17.3), are also the most common methods for implementing automultiscopic displays. For example, Figure 17.4a schematically shows a cross section of a single row of pixels in a seven-view lenticular display. If the slits or lenses are parallel to pixel columns (i.e., are vertical), the display is capable of displaying horizontal parallax only, thus providing depth information to a viewer with strictly vertical head orientation and moving from side to side. While full-parallax passive multiview displays have been developed, they are currently poorly suited to electronic implementation because of insufficient pixel density of plasma and LCD panels.

A serious deficiency of multiview displays with vertical lenses is horizontal resolution loss. For example, if each lens covers N columns of subpixels (R,G,B), then at any position the viewer can see only one out of N columns. This results in a severe imbalance between vertical and horizontal image resolutions perceived.

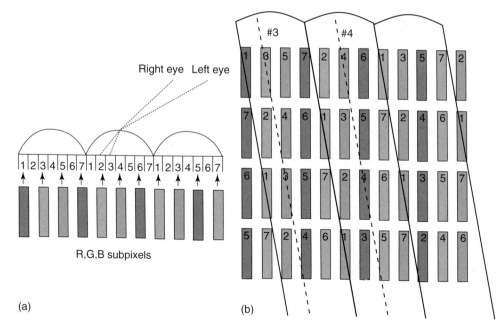

Figure 17.4 Illustration of the principle of operation of lenticular seven-view display: (a) single-row cross section with marked locations where views 1–7 are perceived; and (b) impact of lenticule slant on the spatial distribution of same-view pixels (a dashed line passes through subpixels that are visible from one location in front of the screen; 1–7 indicate the views from which R, G, or B value is rendered). (Please find a color version of this figure on the color plates.)

In order to correct this imbalance and also to minimize vertical banding due to the presence of a black mask between subpixel columns[4], a slight slant of the lenses was introduced as shown in Figure 17.4b. The slant angle has to be carefully selected in order to minimize Moiré patterns resulting from interference of regular structure of the lenses and screen raster [12]. This tilt causes same-view subpixels to be distributed more uniformly on the screen, as shown in Figure 17.6, as opposed to the case without tilt where all single-view subpixels are vertically aligned but spaced far apart horizontally. Unfortunately, while in the nontilted case the layout of single-view subpixels is regular, it is not the case in displays with tilted lenses. Since in each case the extraction of single-view subpixels amounts to spatial subsampling, a suitable prefiltering must be employed to avoid aliasing. While this is straightforward for nontilted lenticules, it is more complicated when a tilt is involved. In Section 17.4.3, we address issues related to the characterization of such aliasing as well as the design of appropriate anti-alias filters.

---

4) This is the so-called "picket-fence" effect resulting from periodic visibility of a black mask present between subpixel columns as the viewer moves in front of a lenticular screen.

17.4
Signal Processing for 3D Displays

As pointed out earlier, both projection-based 3D systems and direct-view 3D systems suffer from various shortcomings. Although some of these shortcomings are inherent to the display itself and can only be overcome by its redesign (e.g., reduced brightness in parallax-barrier displays), others can be, at least partially, corrected by suitable signal processing. As 3D visualization attempts to enter massive consumer markets, curing 3D display ills is critical. Any resurgence of 3D in the past has always resulted in a pullback for various reasons, be it a limited 3D content, excessive "in your face" 3D effects, or display deficiencies (e.g., cross talk, flicker, and image quality). Today, with the all-digital 3D chain many of these deficiencies can be addressed by signal processing. As for 3D content, it seems that Hollywood has taken notice. However, will content producers stay away from "poking" viewers' faces in 3D?

We describe below examples of signal processing techniques that help improve the quality of 3D images that are reproduced by various stereoscopic and multiview displays.

17.4.1
Enhancement of 3D Anaglyph Visualization

The perceived 3D image quality in anaglyphic viewing can be improved by means of signal processing as proposed by Dubois [15]. The main idea is to optimize the anaglyph image so that when displayed on a screen with known spectral characteristics and viewed through lenses with known absorption functions, the image is perceived as close as possible to the original stereo pair viewed without glasses.

Let $\{I_i^l, I_i^r\}$, $i = 1, 2, 3$, be a stereoscopic image pair with i denoting one of RGB color components. A pixel at location x is described by two RGB triplets, expressed as vectors: $I^l[x] = [I_1^l[x], I_2^l[x], I_3^l[x]]^T$ and $I^r[x] = [I_1^r[x], I_2^r[x], I_3^r[x]]^T$. Note that $I^l[x]$ and $I^r[x]$ are triplets of tristimulus values expressed with respect to some primary system that may be gamma-corrected, so that they can be directly displayed on a standard CRT monitor.

If $s_i(\lambda)$, $i = 1, 2, 3$, are spectral density functions of the RGB display phosphors, and $\bar{p}_k(\lambda)$, $k = 1, 2, 3$, are color-matching functions for selected primary colors, then the left- and right-image colors perceived at x can be expressed as follows:

$$\widetilde{I}_k^l[x] = \sum_{j=1}^{3} c_{kj} I_j^l[x], \quad \widetilde{I}_k^r[x] = \sum_{j=1}^{3} c_{kj} I_j^r[x], \quad c_{kj} = \int \bar{p}_k(\lambda) s_j(\lambda) d\lambda$$

for $k, j = 1, 2, 3$. In vector notation, we can write: $\widetilde{I}^l[x] = C I^l[x]$ and $\widetilde{I}^r[x] = C I^r[x]$, where $[C]_{kj} = c_{kj}$. This is a simple transformation between two sets of primary colors. For a specific monitor, this results in a fixed 3×3 matrix C [15]. Then, at each x we have $\widetilde{I}[x] = [(\widetilde{I}^l[x])^T (\widetilde{I}^r[x])^T]^T = C_2 I[x]$, where $I[x] = [(I^l[x])^T (I^r[x])^T]^T$ and the 6×6 matrix C_2 has two matrices C on the diagonal and zeros elsewhere.

Let I_j^a, $j = 1, 2, 3$, be an anaglyph image to be displayed on the same monitor but viewed through glasses with colored lenses with spectral transmission functions $\zeta^l(\lambda)$ and $\zeta^r(\lambda)$ (Figure 17.1). A similar transformation of color coordinates takes place again except for the presence of color filters. The two color transformation matrices are now as follows:

$$[A^l]_{kj} = \int \bar{p}_k(\lambda) s_j(\lambda) \zeta^l(\lambda) d\lambda, \quad [A^r]_{kj} = \int \bar{p}_k(\lambda) s_j(\lambda) \zeta^r(\lambda) d\lambda$$

and the new color coordinates are $\widetilde{J}^l[x] = A^l I^a[x]$ and $\widetilde{J}^r[x] = A^r I^a[x]$, where $I^a[x] = [I_1^a[x], I_2^a[x], I_3^a[x]]^T$. Combining the left and right color coordinates into one vector leads to a simple transformation $\widetilde{J}[x] = R I^a[x]$, where $R^T = [(A^l)^T (A^r)^T]$ is a 6×3 matrix.

The goal is now as follows. Given a stereo pair $\{I^l[x], I^r[x]\}$ with $0 \leq I_j^l[x], I_j^r[x] \leq 1$ for $j = 1, 2, 3$, find an anaglyphic image $I^a[x]$ with $0 \leq I_j^a[x] \leq 1$ for $j = 1, 2, 3$, such that the image $\widetilde{J}$ perceived through the glasses is as similar as possible to the input stereo pair $\widetilde{I}$. Dubois proposed a metric to numerically compute the subjective difference between a stereo pair and anaglyphic image based on certain heuristics, and developed a projection-based optimization method. The images produced by this method result in a wider color gamut perceived by viewers compared to those produced by rudimentary anaglyph preparation methods used to date.

17.4.2
Ghosting Suppression in Polarized and Shuttered 3D Displays

The optical cross talk encountered in many 3D screens is perceived by a viewer as double edges or "ghosts" at high-contrast features misaligned between left and right images because of disparity. In systems using polarized glasses, the cross talk is due to imperfect light extinction in the glasses and light depolarization on the projection screen. In systems using liquid-crystal shutters, the cross talk is caused by phosphor persistence of the CRT monitor or projector (primarily green)[5], imperfect light extinction of LCS in the opaque state (light leakage), and by timing errors of LCS (opening/closing too early or too late). In systems using a single DLP projector with LCS glasses, phosphor persistence is not an issue and only extinction characteristics and timing errors of the shutters play a role. Although in each system improvements may be possible by a careful selection of components, manipulation of image contrast, and adjustment of disparity magnitude, the degree of potential improvement is quite limited.

---

5) Since 3D displays based on plasma flat panels and LCS glasses are currently under development, it is yet unclear how serious an issue will phosphor persistence be in such systems.

A significant reduction of the perceived cross talk, even its complete elimination under certain conditions, is possible by employing signal processing. The basic idea is to create "anti-cross talk," that is, predistort images so that upon display ghosting is largely suppressed [5]. Recently, a computationally efficient algorithm that accounts for specific screen and LCS characteristics has been developed [6]. The algorithm is based on a simple cross talk model:

$$J_i^l = I_i^l + \phi_i(I_i^r, I_i^l), \quad J_i^r = I_i^r + \phi_i(I_i^l, I_i^r), \quad i = 1, 2, 3 \quad (17.1)$$

where J_i^l, J_i^r are RGB components of images that are perceived by the left and right eyes, respectively, I_i^l and I_i^r are RGB components driving the monitor, and ϕ_i's are *cross talk functions* for the three color channels. The cross talk functions ϕ_i quantify the amount of cross talk seen by an eye in terms of unintended *and* intended stimuli. They are dependent on the particular display/glasses combination used and need to be quantified; examples of functions for a specific CRT monitor and LCS glasses can be found in Ref. [6]. The above cross talk model ignores the point spread functions of the monitor and glasses.

If the mapping in Eq. (17.1), which transforms (I_i^l, I_i^r) into (J_i^l, J_i^r) for $i = 1, 2, 3$, is denoted by T with the domain $D(T)$ and range $R(T)$, then the task is to find the inverse mapping T^{-1} that transforms (I_i^l, I_i^r), images we would like to see, into cross-talk-biased images (G_i^l, G_i^r) that drive the monitor, that is, find (G^l, G^r) satisfying

$$I_i^l = G_i^l + \phi_i(G_i^r, G_i^l), \quad I_i^r = G_i^r + \phi_i(G_i^l, G_i^r), \quad i = 1, 2, 3 \quad (17.2)$$

For given cross talk functions ϕ_i, this mapping can be computed off-line and stored in a 400 kB look-up table for 8-bit color components ($256 \times 256 \times 3 \times 2$) and 6.3 MB table for 10-bit components, both easily handled by modern graphics cards.

Since for $\mathcal{D}(T) = [0, 255] \times [0, 255]$, $\mathcal{R}(T)$ is only a subset of $[0, 255] \times [0, 255]$, the inverse mapping T^{-1} operating on $[0, 255] \times [0, 255]$ may result in negative tristimulus values that cannot be displayed. The algorithm tries to "carve out" intensity notches (Figure 17.5) that will get filled with the unintended light, but in dark parts of an image this is not possible. Two solutions preventing negative tristimulus values are linear mapping of RGB components, at the cost of reduced image contrast, and black-level saturation, which leads to loss of detail in dark image areas ("black crush") – both applied prior to T^{-1}. Since neither solution is acceptable, a compromise between cross talk reduction and image quality degradation is usually sought [6]. This method has been successfully used in LCS-based systems, but it is equally applicable to polarization-based systems using CRT monitors, CRT projectors, or DLP projectors. Also, the recently announced 3D displays using plasma panels and LCS glasses may benefit from this approach as they are likely to suffer from cross talk as well.

An example of application of the above algorithm is shown in Figure 17.5. Note the reduced image intensity in areas corresponding to bright features present in the other image; the unintended light due to optical cross talk fills these "holes" during display, leading to cross talk compensation.

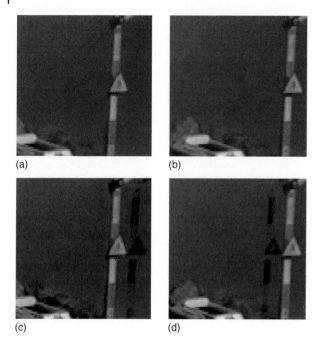

Figure 17.5 Example of cross talk compensation: (a) left, and (b) right original images (luminance only), and (c,d) the same images after cross talk compensation using the algorithm described in Ref. [6] with luminance clipping below the amplitude of 30. (From Konrad et al. [6]. Reproduced with permission of the Institute of Electrical and Electronics Engineers © 2000 IEEE.)

17.4.3
Anti-Alias Filtering for Multiview Autostereoscopic Displays

When preparing an image for passive multiview display, data from several views must be multiplexed together. Clearly, this involves spatial-view subsampling as only a subset of RGB subpixels is extracted from each view (Figure 17.4b). For example, the locations of red subpixels that are activated by one view only are shown in Figure 17.6 as circles. Ideally, all these subpixels would be seen by one eye only. Note that these subpixels are not regularly spaced; hence, their locations cannot be described by a 2D lattice [16]. A similar observation can be made for green and blue subpixels. Thus, it is not obvious as to how to quantify the aliasing resulting from such an irregular subsampling. Two approaches have been developed to date.

In the first approach [17], the irregular subsampling layout $\mathcal{V} \subset R^2$ (circles in Figure 17.6) is *approximated in spatial domain* by a lattice or union of cosets [16], denoted as $\Psi \subset R^2$. If Γ is the orthonormal lattice of the screen raster (dots), then $\mathcal{V} \subset \Gamma \subset R^2$. The goal is to find Ψ such that, in some sense, it best approximates the set $\mathcal{V}$. One possibility is to minimize the distance between sets Ψ and $\mathcal{V}$.

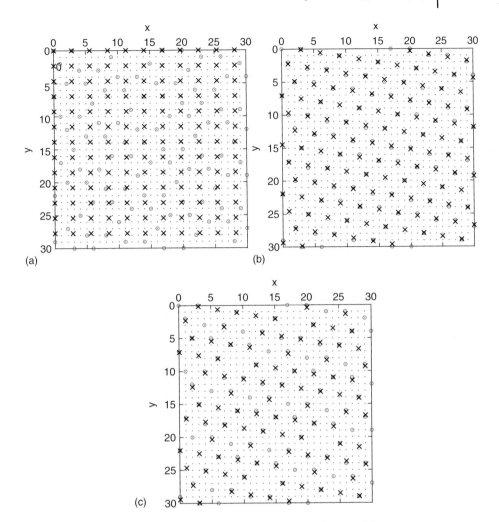

Figure 17.6 Approximation of single-view subpixels using (a) orthogonal lattice; (b) nonorthogonal lattice; and (c) union of 21 cosets. Dots denote the orthonormal RGB screen raster, circles (O) denote red subpixels of this raster that are activated when rendering one view ($\mathcal{V}$), while crosses (×) show locations from model Ψ. (From Konrad and Halle [1]. Reproduced with permission of the Institute of Electrical and Electronics Engineers © 2007 IEEE.) (Please find a color version of this figure on the color plates.)

Let $d(\mathbf{x}, \mathcal{A})$ be the distance from point $\mathbf{x}$ to a discrete set of points $\mathcal{A}$, that is, $d(\mathbf{x}, \mathcal{A}) = \min_{\mathbf{y} \in \mathcal{A}} \|\mathbf{x} - \mathbf{y}\|$. Then, one can find an optimal approximation by minimizing the mutual distance between two point sets Ψ and $\mathcal{V}$ as follows:

$$\min_{\eta} \beta_{\mathcal{V}} \sum_{\mathbf{x} \in \mathcal{V}} d(\mathbf{x}, \Psi) + \beta_{\Psi} \sum_{\mathbf{x} \in \Psi} d(\mathbf{x}, \mathcal{V}) \tag{17.3}$$

where η is a vector of parameters describing the sampling structure Ψ. For a 2D lattice η is a 3-vector, and thus the minimization in Eq. (17.3) can be accomplished by hierarchical exhaustive search over a discrete state space. As the weights $\beta_\mathcal{V}$ and β_Ψ are adjusted, different solutions result. If $\beta_\Psi = 0$, a very dense (in the limit, infinitely dense) Ψ would result, while for $\beta_\mathcal{V} = 0$, a single-point $\Psi \subset \mathcal{V}$ would be found optimal. Instead of a combination of both distances, one could use either of them under constrained minimization (e.g., a constraint on the density of Ψ).

Applied to a number of layouts $\mathcal{V}$, the above method has been shown to be effective in identifying regular approximations of $\mathcal{V}$, from orthogonal-lattice approximations (quite inaccurate), through nonorthogonal-lattice approximations (more accurate), to union-of-cosets approximations (most accurate), as shown in Figure 17.6. Note a progressively improving alignment between single-view irregular points in $\mathcal{V}$ (circles) and regular points in Ψ (crosses). Having identified a regular-structure approximation, it is relatively straightforward to identify the passband of suitable low-pass anti-alias prefilters [16] that must precede view multiplexing.

In an alternative approach, the irregularity is *approximated in frequency domain* [18]. The main idea is based on the observation that in 1D a sequence of unit impulses (Kronecker deltas) $g[n]$ has the discrete-time Fourier transform (DTFT) in the form of a periodic impulse train (sum of delayed Dirac delta functions). Therefore, the subsampling of a signal $x[n]$, by multiplying it with the sequence $g[n]$, results in the convolution of their DTFTs: $\mathcal{F}\{x[n]\} * \mathcal{F}\{g[n]\}$. Clearly, the spectrum $\mathcal{F}\{x[n]\}$ is going to be replicated at locations of the periodic impulse train $\mathcal{F}\{g[n]\}$. A similar relationship holds in 2D with respect to bisequences and 2D periodic impulse trains.

The above relationships also hold if $g[n, m]$ is a bisequence of *irregularly spaced* unit impulses, that is, defined on $\mathcal{V}$. Then, $\mathcal{F}\{g[n, m]\}$ is a 2D train of Dirac delta functions located on a reciprocal lattice Λ^* of the least dense lattice Λ such that $\mathcal{V} \subset \Lambda$ [19]. In addition, impulses in this train are scaled differently at different frequency locations [19]. The consequence of this is that the signal spectrum is replicated with different gains at different frequency locations, and replications with smaller gain factors may have less of an impact on aliasing than those with large gain factors. Although to completely prevent aliasing one could consider the worst case scenario, that is, limiting the signal spectrum so that the closest off-origin spectral replications do not "leak" into the baseband Voronoi cell, such a design would be too restrictive. Alternatively, a controlled degree of aliasing may be permitted, but then either the actual signal spectrum or its model must be known. Good results were obtained for separable Markov-1 image model [18]; an anti-alias filter's passband boundary was found by identifying frequencies at which the ratio of baseband spectrum magnitude to that of its closest spectral replication (with suitable gain) was equal to or exceeded 1.

Figure 17.7 shows contour plots of the magnitude response of anti-alias filters based on the spatial- and frequency-domain approximations discussed above. Note the rotated hexagonal passband for the model based on nonorthogonal lattice, and the diamond-shaped passband, with horizontal and vertical extensions, for

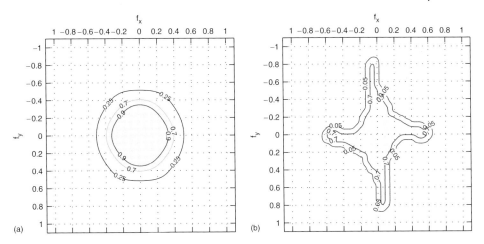

Figure 17.7 Contour plots of optimized magnitude response of an anti-alias filter: (a) based on nonorthogonal lattice in spatial domain; (b) based on irregular sampling structure followed by Markov-1 spectrum modeling. The desired magnitude response passband is shaded. Both frequency axes are normalized to the Nyquist frequency. (From Konrad and Halle [1]. Reproduced with permission of the Institute of Electrical and Electronics Engineers © 2007 IEEE.)

the frequency-domain approximation. Applied in practice, a prefiltering based on either specification results in effective suppression of aliasing artifacts, although prefiltering based on specifications from Figure 17.7b preserves horizontal and vertical detail of 3D images better.

The benefit of approximation in frequency domain is that anti-alias filtering can be adapted to specific spectra, whether of a particular image or model. In principle, given sufficient computing power (already available in high-end graphics cards), one can imagine anti-alias filtering adapted to each displayed image.

17.4.4
Luminance/Color Balancing for Stereo Pairs

Stereoscopic filmmakers are very careful to use stereo cameras with as similar parameters, such as focus, luminance, and color, as possible. Although human visual system can deal with small differences in focus, luminance, or color between the views observed by the two eyes, the viewer will experience visual discomfort if the views are severely mismatched. Furthermore, any such mismatch adversely affects correspondence estimation with severe consequences for intermediate view interpolation (Section 17.4.5) and 3D data compression. For example, color or luminance bias in one of the images may lead to incorrect correspondences and thus distorted intermediate views, and also to increased transmission rates of 3D data as compression algorithms are sensitive to such drifts.

In the following discussion, we describe a simple method that helps to deal with global luminance and color mismatch issues [20]. The underlying assumption in this method is that the mismatch is due to unequal camera parameters (lighting conditions are considered the same for the two viewpoints), and therefore it is global and can be modeled by a simple, linear transformation applied to the whole image. The transformation parameters are found by requiring that sample means and variances of color components be identical in both images after the transformation.

The method can be implemented as follows. Given the RGB stereo pair $\{I_i^l, I_i^r\}$, $i = 1, 2, 3$, suppose that I_i^l undergoes the following linear transformation:

$$\hat{I}_i^l[x] = a_i I_i^l[x] + b_i, \quad i = 1, 2, 3, \tag{17.4}$$

where a_i and b_i are such parameters that

$$\hat{\mu}_i^l = \mu_i^r, \quad (\hat{\sigma}_i^l)^2 = (\sigma_i^r)^2, \quad i = 1, 2, 3 \tag{17.5}$$

with $\hat{\mu}_i^l, \mu_i^r$ being the means and $(\hat{\sigma}_i^l)^2, (\sigma_i^r)^2$ being the variances of $\hat{I}_i^l$ and I_i^r, respectively. From Eq. (17.4) it follows that

$$\hat{\mu}_i^l = a_i \mu_i^l + b_i, \quad (\hat{\sigma}_i^l)^2 = a_i^2 (\sigma_i^l)^2 \tag{17.6}$$

Solving Eqs (17.5) and (17.6) for the parameters a_i and b_i, one obtains the following:

$$a_i = \frac{\sigma_i^r}{\sigma_i^l}, \quad b_i = \mu_i^r - \frac{\sigma_i^r}{\sigma_i^l} \mu_i^l$$

Clearly, balancing the color between two mismatched views amounts to the calculation of means and variances of two images (to obtain the transformation parameters), and then one multiplication (gain a_i) and one addition (offset b_i) per color component per pixel. Although this model is quite crude, it is surprisingly effective when viewpoint mismatch is caused by camera miscalibration. Figure 17.8 shows an example of the application of this approach; the original images I^l, I^r have been captured by identical cameras but with different exposure and shutter-speed settings. Clearly, the original left view has a deeper orange tint compared to the right view. After transformation, the orange tint is reduced, and the colors are much better matched between the views.

(a) I^l (b) I^r (c) $\hat{I}^l$

Figure 17.8 Example of color balancing in a stereo pair captured by identical cameras but with different exposure parameters: (a) original left view I^l; (b) original right view I^r; and (c) left view after linear transformation from Eq. (17.4). (Please find a color version of this figure on the color plates.)

17.4.5
Intermediate View Interpolation

In order that a 3D display deliver realistic, natural experience to the viewer, it must render the correct perspective for each viewer position or, in other words, permit motion parallax. Although a head tracker could identify viewer head position and a stereo camera could be moved accordingly, this would not work with prerecorded data. An alternative would be to employ a camera array to capture many views simultaneously [21] and select suitable views in accordance with viewer motion in front of the 3D screen. However, the design, synchronization, and maintenance of such an array are cumbersome and the cost is quite prohibitive; except for experimental installations such arrays are not commonly used. Yet another alternative is to use a small number of cameras (Figure 17.9) and render virtual (missing) views based on the available ones. One class of such methods falls into the category of *image-based rendering* that relies on the available views to generate a 3D model and then render virtual views of the scene [22]. Although quite versatile, such methods are computationally demanding and imprecise

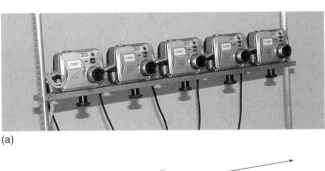

(a)

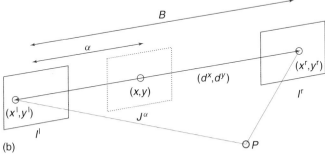

(b)

Figure 17.9 (a) *PENTACAM*, a linear, 5-camera array developed at Boston University as a testbed for intermediate view interpolation research (middle cameras are used to capture ground-truth intermediate views for quantitative comparison of view interpolation algorithms); (b) images captured by leftmost and rightmost cameras, I^l and I^r, respectively, are shown with homologous points (projections of 3D point P). If virtual camera were present at position α, (x, y) would be the coordinates of the homologous point in virtual view J^α (dashed lines).

for complex 3D scenes. However, video capture for multiview 3D displays is constrained in terms of camera baseline, thus resulting in limited-range disparities, and in terms of camera convergence angle (no more than a few degrees), thus limiting the geometric distortions (keystoning). These constraints facilitate various simplifying assumptions that allow virtual view computation without explicit 3D modeling. Such methods are commonly known as *intermediate view interpolation* or *intermediate view reconstruction* [23–25] and are discussed below.

For the sake of this chapter, we assume parallel camera geometry so that the virtual cameras are located on the line passing through the optical centers of the real cameras and all have parallel optical axes (Figure 17.9b). We further assume that the stereo camera baseline B is normalized to unity and the relative distance of a virtual camera from the left camera is α (i.e., for the left-camera position $\alpha = 0$ and for the right-camera position $\alpha = 1$). The goal now is as follows: given the left image I^l and right image I^r compute an intermediate image J^α that would have been captured by a virtual camera. Usually, this requires two steps: correspondence estimation between captured views and interpolation of luminance/color from the corresponding points.

The goal of correspondence estimation is to identify the so-called *homologous points* in the left and right images, that is, points that are projections of the same 3D point onto imaging planes of the left and right cameras, as depicted in Figure 17.9b. Clearly, homologous points are photometrically similar (luminance and color are alike), and can be related to each other by a 2D vector called *disparity*. Traditionally, correspondence estimation is performed in the coordinate system of either the left or right image, that is, either for each pixel in the left image its homolog is sought in the right image or *vice versa*. Such a problem has been extensively researched to date and its detailed analysis is beyond the scope of this chapter. We refer the reader to the literature for further reading on correspondence estimation based on block matching [23], optical flow [26], dynamic programming [22], and graph cuts [27]. This is not an exhaustive list but it is rich enough to give the reader a good understanding of issues involved.

However, the correspondence problem in view interpolation is slightly different. Let the left and right images, I^l and I^r, respectively, be sampled on 2D lattice Λ [16], that is, $I^l[\mathbf{x}]$, $I^r[\mathbf{x}]$, $\mathbf{x} = (x, y)^T \in \Lambda$. Also, let $\mathbf{d} = (d^x, d^y)^T$ be a disparity vector resulting from correspondence estimation (Figure 17.9b). What traditional correspondence estimation seeks is to find a left disparity field $\mathbf{d}^l[\mathbf{x}]$, $\mathbf{x} \in \Lambda$ such that $I^l[\mathbf{x}] \approx I^r[\mathbf{x} + \mathbf{d}^l[\mathbf{x}]]$ or a right disparity field $\mathbf{d}^r[\mathbf{x}]$ such that $I^r[\mathbf{x}] \approx I^l[\mathbf{x} + \mathbf{d}^r[\mathbf{x}]]$ for all $\mathbf{x} \in \Lambda$. Having computed either disparity field, it is still unclear as to how to reconstruct the intermediate image J^α for all $\mathbf{x} \in \Lambda$.

Since the camera geometry is assumed parallel or almost parallel, the homologous point in virtual view J^α (Figure 17.9b) is located on the same scan line as both $\mathbf{x}^l$ and $\mathbf{x}^r$ or very close to it [22]. Consider a scenario such that for each pixel $\mathbf{x}^l$ in the left image its homolog $\mathbf{x}^r$ is found in the right image, namely, $\mathbf{x}^r = \mathbf{x}^l + \mathbf{d}[\mathbf{x}^l]$. Then, the homologous point in the intermediate image J^α is located at $\mathbf{x} = \mathbf{x}^l + \alpha \mathbf{d}[\mathbf{x}^l]$ and the corresponding luminance and color are defined by $J^\alpha[\mathbf{x}]$. Since $\mathbf{x}$ is not necessarily in Λ, because of the fact that it depends on α and $\mathbf{d}$, the set of photometric reconstructions $J^\alpha[x^l + \alpha d[x^l]]$, $x^l \in \Lambda$ is *irregularly spaced*. In order to recover J^α on lattice Λ, an additional, fairly complex step of interpolating regularly spaced samples from irregularly spaced ones is needed.

This additional complexity can be avoided by reformulating the correspondence problem so as to solve for disparities defined on Λ in the intermediate image instead, that is, one may seek d's such that

$$I^l[x - \alpha d[x]] \approx I^r[x + (1 - \alpha)d[x]].$$

This can be thought of as pivoting; the disparity vector d passes through location $x \in \Lambda$ in J^α while its ends are free to move in I^l and I^r. In order to find d for all pixels in J^α, some methods, such as block matching and optical flow, can be easily adapted [24, 28], whereas other methods may require more extensive modifications.

Having found a disparity field, the interpolation of an intermediate image should be ideally as simple as copying either the corresponding left or right pixel to the intermediate image J^α. However, owing to noise, illumination effects, and various distortions (e.g., aliasing) the matching pixels, even if truly homologous, are usually not identical. This is even more true when correspondence is performed at block level. Perhaps the simplest option is to reconstruct J^α by means of disparity-compensated linear interpolation with a simple two-coefficient kernel:

$$J^\alpha[x] = \xi^l \widetilde{I}^l[x - \alpha d[x]] + \xi^r \widetilde{I}^r[x + (1 - \alpha)d[x]], \qquad \xi^l + \xi^r = 1 \qquad (17.7)$$

at all $x \in \Lambda$ in the intermediate image J^α. Note that $\widetilde{I}$ denotes an interpolated luminance/color since, in general, $x - \alpha d[x]$ and $x + (1 - \alpha)d[x]$ are not in Λ. This is a simple interpolation of irregularly spaced samples from regularly spaced ones that can be effectively and efficiently computed, e.g., by bicubic interpolation. The constraint on the sum of the coefficients ($\xi^l + \xi^r = 1$) assures a unit gain of the filter. In general, good results have been obtained with the following selection of the coefficients:

$$\xi^l = 1 - \alpha, \qquad \xi^r = \alpha$$

which assures high weighting of an image close to the intermediate position and low weighting of a distant image. This implements low-pass filtering that suppresses noise in the intermediate image.

Figure 17.10 shows an example of intermediate view interpolation based on optical flow correspondence [28] and linear interpolation from Eq. (17.7) for a pair of images captured by stereo camera with a small baseline. Note the overall high fidelity of interpolation, which is especially notable in the magnified sections of the image; the tree behind one of the columns is correctly reconstructed as being only partially occluded (Figure 17.10b).

The interpolation from Eq. (17.7) works well if disparity is precisely estimated; otherwise the reconstructed images suffer from edge blurring due to averaging of misaligned edges. An alternative to this is disparity-compensated nonlinear interpolation [24] where pixels in J^α are optimally selected from *either* the left or the right image. Furthermore, occlusion areas, where a specific scene feature is visible

Figure 17.10 Example of intermediate view interpolation for images captured by a stereo camera with 3-inch baseline using disparity estimation based on an optical flow model [28] and simple linear interpolation from Eq. (17.7): (a) original left view; (b) reconstructed intermediate view; and (c) original right view. On the right are magnified windows from the middle of the image, highlighting the accuracy of interpolation (e.g., partially occluded tree). (Please find a color version of this figure on the color plates.)

in only one view, pose a significant challenge for view interpolation algorithms. Recently, several methods have been developed that account for occlusions in view interpolation [25–27].

Intermediate view interpolation algorithms can replace cumbersome camera arrays, thus providing data input for multiview displays. They can be also used as an efficient prediction tool in multiview compression [29]. Yet another interesting application is parallax adjustment. Among viewers there exists a very significant variation in sensitivity to stereoscopic cues [30]; while some viewers have no problem with the fusion of a particular stereo image, others may be unable to do so or may feel eye strain. The problem may be further compounded by the presence of excessive 3D cues in the viewed images. Although viewers may experience no fusion problems with properly acquired stereoscopic images, that is, with moderate parallax, significant difficulties may arise when parallax is excessive. For images acquired by a trained operator of a stereoscopic camera, this should occur infrequently. However, if the images are acquired under conditions optimized for one viewing environment (e.g., TV screen) and are then presented in another environment (e.g., large-screen cinema), viewer discomfort may occur. In order to minimize viewer discomfort, the amount of parallax can be adjusted by means of intermediate view interpolation (i.e., by reducing the virtual baseline). This amounts to equipping a future 3D screen with a "3D-ness" or "depthness" knob similar to the "brightness" and "contrast" adjustments used today.

17.5 Conclusions

A survey chapter such as this can only give the reader a glimpse into the field of 3D displays. To date, 3D displays have experienced ups and downs in popularity, with very few types being consistently successful in some specialized areas. Beyond these fads and niches, however, 3D has never before threatened the dominance of "flat" imaging.

Today, the digital content revolution has changed everything. The CCD has replaced film, the desktop computer has become a video editing station, the graphics technology keeps breaking boundaries, and digital imaging has become ubiquitous. The digital bitstream is one of the most important foundations for the reemergence of 3D, as it facilitates 3D data compression, transmission, and storage in ways unthinkable in the analog world before. Furthermore, it facilitates digital signal processing of 3D content that can help address many deficiencies of 3D displays.

New signal processing algorithms, new 3D display devices, and new ways to create 3D content, are, together, rapidly curing 3D's longtime ills and ushering in a new era of ubiquitous 3D. We believe that the early successes we are seeing today in 3D will only expand as public interest grows, as more commercial applications succeed, as new algorithms and devices are developed, and as computation becomes ever cheaper.

The future enabled by the combination of 3D signal processing, digital content, and new display technologies looks very promising. Already today, stereoscopic endoscopes allow doctors to more effectively screen colon cancer, while audiences in a packed theater attempt to dodge a sword in a digital 3D movie. We can only imagine, or perhaps invent, the kinds of applications that will emerge when complete 3D display systems become widely available.

Acknowledgments

The author would like to acknowledge a contribution to the results presented here from his past students, especially Anthony Mancini, Bertrand Lacotte, Peter McNerney, Philippe Agniel, Ashish Jain, and Serdar Ince. He would also like to thank Prof. Eric Dubois of the University of Ottawa, Canada, for providing Figure 17.1 and Prof. Prakash Ishwar of Boston University for reading this chapter and providing valuable feedback. The author would also like to acknowledge the support from NSF under Grant ECS-0219224 and Boston University under the SPRInG Award.

References

1. Konrad, J. and Halle, M. (2007) 3-D displays and signal processing: an answer to 3-D ills? *IEEE Signal Process. Mag.*, **24**, 97–111.
2. Jorke, H. and Fritz, M. (2003) Infitec- A new stereoscopic visualization tool by wavelength multiplexing imaging. Proceedings Electronic Displays.
3. Fergason, J. *et al.* (2005) An innovative beamsplitter-based stereoscopic/3D display design. Proceedings SPIE Stereoscopic Displays and Virtual Reality Systems, vol. 5664, pp. 488–494. DOI: 10.1117/12.588363.
4. Bos, P. (1991) Time sequential stereoscopic displays: the contribution of phosphor persistence to the ''ghost'' image intensity, in *Proceedings ITEC'91 Annual Conference, three-Dimensional Image Techniques* (ed. H. Kusaka), pp. 603–606.
5. Lipscomb, J. and Wooten, W. (1994) Reducing crosstalk between stereoscopic views. Proceedings SPIE Stereoscopic Displays and Virtual Reality Systems, vol. 2177, pp. 92–96.
6. Konrad, J., Lacotte, B., and Dubois, E. (2000) Cancellation of image crosstalk in time-sequential displays of stereoscopic video. *IEEE Trans. Image Process.*, **9**, 897–908.
7. Woods, A. and Sehic, A. (2009) The compatibility of LCD TVs with time-sequential stereoscopic 3D visualization. Proceedings SPIE Stereoscopic Displays and Applications, vol. 7237.
8. Okoshi, T. (1976) *Three Dimensional Imaging Techniques*, Academic Press.
9. Sexton, I. and Surman, P. (1999) Stereoscopic and autostereoscopic display systems. *IEEE Signal Process. Mag.*, **16**, 85–99.
10. Dodgson, N. (2005) Autostereoscopic 3-D displays. *IEEE Comput.*, **38**, 31–36.
11. Adelson, E. and Wang, J. (1992) Single lens stereo with a plenoptic camera. *IEEE Trans. Pattern Anal. Mach. Intell.*, **14**, 99–106.
12. Van Berkel, C., Parker, D.W., and Franklin, A.R. (1996) Multiview 3D-LCD. Proceedings SPIE Stereoscopic Displays and Virtual Reality Systems, vol. 2653, pp. 32–39.
13. Lipton, L. and Feldman, M. (2002) A new stereoscopic display technology: The SynthaGram. Proceedings SPIE

Stereoscopic Displays and Virtual Reality Systems, vol. 4660, pp. 229–235.
14. Schmidt, A. and Grasnick, A. (2002) Multi-viewpoint autostereoscopic displays from 4D-Vision. Proceedings SPIE Stereoscopic Displays and Virtual Reality Systems, vol. 4660, pp. 212–221.
15. Dubois, E. (2001) A projection method to generate anaglyph stereo images. Proceedings IEEE International Conference on Acoustics, Speech, and Signal Processing, vol. 3, pp. 1661–1664.
16. Dubois, E. (1985) The sampling and reconstruction of time-varying imagery with application in video systems. *Proc. IEEE*, **73**, 502–522.
17. Konrad, J. and Agniel, P. (2006) Subsampling models and anti-alias filters for 3-D automultiscopic displays. *IEEE Trans. Image Process.*, **15**, 128–140.
18. Jain, A. and Konrad, J. (2007) Crosstalk in automultiscopic 3-D displays: Blessing in disguise? Proceedings SPIE Stereoscopic Displays and Virtual Reality Systems, vol. 6490, pp. 12.1–12.12.
19. Jain, A. (2006) Crosstalk-aware design of anti-alias filters for 3-D automultiscopic displays. Master's thesis. Boston University.
20. Franich, R. and ter Horst, R. (1996) Balance compensation for stereoscopic image sequences, ISO/IEC JTC1/SC29/WG11 - MPEG96.
21. Wilburn, B. *et al.* (2004) High speed video using a dense camera array. Proceedings IEEE Conference Computer Vision Pattern Recognition, pp. 28–41.
22. Kang, S., Li, Y. and Tong, X. (2007) *Image Based Rendering*, Now Publishers.
23. Izquierdo, E. (1997) Stereo matching for enhanced telepresence in three-dimensional videocommunications. *IEEE Trans. Circuits Syst. Video Technol.*, **7**, 629–643.
24. Mansouri, A.-R. and Konrad, J. (2000) Bayesian winner-take-all reconstruction of intermediate views from stereoscopic images. *IEEE Trans. Image Process.*, **9**, 1710–1722.
25. Ince, S. and Konrad, J. (2008) Occlusion-aware view interpolation. *EURASIP J. Image Video Process.* DOI: 10.1155/2008/803231.
26. Ince, S. and Konrad, J. (2008) Occlusion-aware optical flow estimation. *IEEE Trans. Image Process.*, **17**, 1443–1451.
27. Kolmogorov, V. and Zabih, R. (2002) Multi-camera scene reconstruction via graph cuts. Proceedings European Conferecne Computer Vision, pp. 82–96.
28. Konrad, J. (1999) Enhancement of viewer comfort in stereoscopic viewing: parallax adjustment. Proceedings SPIE Stereoscopic Displays and Virtual Reality Systems, vol. 3639, pp. 179–190.
29. Ince, S. *et al.* (2007) Depth estimation for view synthesis in multiview video coding. Proceedings 3DTV Conferecne, Kos Island, Greece.
30. Tam, W. and Stelmach, L. (1995) Stereo depth perception in a sample of young television viewers. International Workshop on Stereoscopic and 3D Imaging, Santorini, Greece, pp. 149–156.

18
Linking Analog and Digital Image Processing
Leonid P. Yaroslavsky

18.1
Introduction

Informational optics is the branch of optics that deals with optical information processing, imaging, and optical measurements. The history of informational optics from ancient magnifying glasses through Galileo's telescope and van Leeuwenhoek's microscope to the huge diversity of modern optical imaging and measuring devices is a history of creating and perfecting optical imaging and measuring devices and mastering new bands of waves and new kinds of radiation. A revolutionary stage in the evolution of informational optics was the marriage of optics and electronics in the second half of the twentieth century. Nowadays, informational optics is reaching its maturity. It has gone digital. Digital computers and signal processors are becoming an inherent integral part of optical imaging and measuring devices. Here are only a few examples. Digital cameras are replacing analog photographic cameras and completely eliminating photographic alchemy; a new generation of computerized synthetic aperture telescopes and optical interferometers (see Chapter 14) is helping astronomers study stars in hundred times finer detail than is possible with classical analog optics; computer-controlled adaptive optics and digital video processing allow perfect sharp imaging through turbulent media with super-resolution exceeding the resolving power of the sensors (see Chapter 27); nowadays, optical and holographic interferometers widely make use of digital processors to achieve higher accuracy, versatility, and informational throughput; computerized tomographic synthesis and digital image processing methods have revolutionized medical imaging and nondestructive testing industry.

There are three new major qualities that have been brought into optical systems by digital computers and processors. These are as follows:

- The most substantial advantage of digital computers as compared to analog electronic and optical information processing devices is their flexibility and adaptability: no hardware modifications are necessary to reprogram digital computers to solving different tasks. This feature also makes digital computers an

---

Optical and Digital Image Processing: Fundamentals and Applications, First Edition. Edited by Gabriel Cristóbal, Peter Schelkens, and Hugo Thienpont.
© 2011 Wiley-VCH Verlag GmbH & Co. KGaA. Published 2011 by Wiley-VCH Verlag GmbH & Co. KGaA.

ideal vehicle for processing optical signals since they can adapt rapidly and easily to varying signals, tasks, and end-user requirements.
- Digital computers integrated into optical information processing systems enable them to perform any operations needed in addition to elementwise and integral signal transformations, such as spatial and temporal Fourier analysis, signal convolution, and correlation that are characteristic for analog optics. This removes the major limitation of analog optical information processing.
- Acquiring and processing quantitative data contained in optical signals and connecting optical systems to other informational systems and networks is most natural when data are handled in a digital form. In the same way as money is the general equivalent in economics, digital signals are the general equivalent in information handling. Digital signals that represent optical ones are, so to say, purified information carried by optical signals deprived of their physical carrier. Digital signals, thanks to their universal nature, are also the ideal means for integrating different information systems.

As always, there is a trade-off between good and bad features. The fundamental limitation of signal processing in computers is the speed of computations. What optics does in parallel and with the speed of light, computers perform as a sequence of very simple logical operations with binary digits, which is fundamentally slower whatever the speed of these operations is. Obviously, the optimal design of image processing systems requires appropriate combination of analog and digital processing using the advantages and taking into consideration the limitations of both.

The marriage of analog electro-optical and digital processing requires appropriate linking analog and digital signals and transforms. In this chapter, we address several aspects of this fundamental issue specifically for digital imaging. In Section 18.2, we outline basic principles and review methods of digital representation of images and imaging transforms, such as convolution and Fourier integral transforms. In Section 18.3, we discuss applying these principles to the optimization of methods for building, in computers, continuous image models for image resampling, image recovery and super-resolution from sparse data, differentiating, and integrating. In Section 18.4, digital-to-analog conversion in digital holography is illustrated by the results of analysis of its influence on image reconstruction of physical parameters of devices used for recording computer-generated hologram.

18.2
How Should One Build Discrete Representation of Images and Transforms?

In informational optics, physical reality can be considered a continuum, whereas computers have only a finite number of states. How can one represent physical reality of optical signals and transforms in computers? The answer to this question is discretization.

18.2.1
Signal Discretization

Signal discretization is converting continuous signals into discrete ones represented by a finite set of numbers. In principle, there might be many different ways of signal discretization. However, our technological tradition and all technical devices that are used at present for such a conversion implement a method that can mathematically be modeled as signal approximation by its expansion over a finite set of basis functions:

$$a(x) \approx \sum_{k=0}^{N-1} \alpha_k \varphi_k^{(r)}(x); \quad \alpha_k = \int_X a(x) \varphi_k^{(d)}(x) dx \quad (18.1)$$

where $a(x)$ is the image signal as a function of spatial coordinate x, $\{\varphi_k^{(d)}(x)\}$, and $\{\varphi_k^{(r)}(x)\}$ are sets of basis functions used for signal discretization and their reconstruction from the discrete representation, correspondingly, k is the basis function index, and N is the number of discrete representation coefficients $\{\alpha_k\}$ of the signal.

For different bases, signal approximation accuracy for a given N might be different. Obviously, the discretization bases that provide better approximation accuracy for a given N are preferable. However, the accuracy of signal recovery from its discrete representation is not the only issue in selecting discretization and reconstruction basis functions. Another issue is that of complexity of generating and implementing discretization and reconstruction basis functions and computing the signal representation coefficients.

In principle, discretization and reconstruction basis functions can be implemented in a form of prefabricated templates in the image discretization and display devices (or, in digital processing, in computer memory). However, in practice, the number N of the required template functions is most frequently very large (for instance, on the order of $10^6 \div 10^7$ for modern digital cameras). This is why signal discretization and reconstruction basis functions are usually generated (in hardware or software) by means of certain modifications of a unique "mother" function. Three families of such basis functions are best known: shift (convolutional) basis functions, scale (multiplicative) basis functions, and wavelets.

Shift (convolutional) basis functions, usually called *sampling functions*, implement the simplest method for generating basis functions from one "mother" function, that of spatial translation (coordinate shift) of a "mother" function $\varphi_0^{(s)}(x)$, such that

$$\varphi_k^{(s)}(x) = \varphi_0^{(s)}[x - (k + u^{(s)}) \Delta x]; \quad \varphi_k^{(r)}(x) = \varphi_0^{(r)}[x - (k + u^{(r)}) \Delta x] \quad (18.2)$$

and signal representation coefficients are samples of signal convolution with sampling basis functions:

$$\alpha_k = \int_{-\infty}^{\infty} a(\xi) \varphi_0^{(s)}(x - \xi) d\xi \bigg|_{x=(k+u)\Delta x} \quad (18.3)$$

where Δx is an elementary shift interval called the *sampling interval* and $(u^{(s)}, u^{(r)})$ are analogous (not necessarily integer) shift parameters, which, together with the sample index k, determine positions of sampling points in the signal coordinate system.

If the number of signal samples is unlimited, the best signal approximation, in terms of the minimum of mean square approximation error, is achieved when sinc-functions

$$\operatorname{sinc}\left\{\pi[x-(k+u)\Delta x]/\Delta x\right\} = \frac{\sin\left\{\pi[x-(k+u)\Delta x]/\Delta x\right\}}{\pi[x-(k+u)\Delta x]/\Delta x} \qquad (18.4)$$

are used as sampling and reconstruction basis functions. The signal mean square reconstruction error is, in this case, minimal possible for shift basis functions and it is equal to the energy of signal Fourier spectrum components outside the frequency interval $[-1/2\Delta x, 1/2\Delta x]$ called the sampling *baseband*. In the idealistic case of bandlimited functions, whose spectrum vanishes outside this interval, the approximation error is zero. These statements constitute the meaning of the classical sampling theorem.

For real-life signals that are not bandlimited and are sampled with a finite number of samples using sampling bases other than physically not realizable sinc-functions, the approximation error is never zero. It is caused by distortions of signal spectra within the baseband $[-1/2\Delta x, 1/2\Delta x]$ and by aliasing signal spectral components outside the baseband [1].

In 2D and higher dimensions, the signal translation parameter is a vector rather than a scalar. Therefore, in higher dimensions, in addition to the amount of translation, there is an additional degree of freedom for generating basis functions – the selection of directions, in which translations are implemented. The most simple and widespread is the rectangular sampling grid, which corresponds to rectangular signal basebands, although hexagonal sampling grids or sampling grids in skewed coordinates may be, in some cases, more appropriate in terms of minimization of the signal approximation error.

Scale (multiplicative) basis functions are built by means of scaling argument of a "mother" function proportionally to the function index k:

$$\varphi_k(x) = \varphi_0(kx) \qquad (18.5)$$

Such scaling basis functions are also called *Fourier kernels*. Typical examples are sinusoidal functions $\{\cos(2\pi kx/X)\}$ and $\{\sin(2\pi kx/X)\}$, where X is a finite interval, within which signals are approximated. Signal expansions over sets of sinusoidal basis functions are versions of the Fourier series expansion on finite intervals. Mathematical treatment of the Fourier series expansion is much simplified if pairs of cosine and sine functions are replaced by complex exponential functions $\{\exp(i2\pi kx/X) = \cos(i2\pi kx/X) + i\sin(i2\pi kx/X)\}$, $i = \sqrt{-1}$. With them, the Fourier series expansion takes the form

$$a(x) = \sum_k \alpha_k \exp(i2\pi kx/X), \text{ with } \alpha_k = \frac{1}{X}\int_{-X/2}^{X/2} a(x)\exp(-i2\pi kx/X)dx \qquad (18.6)$$

The basis of exponential functions $\{\exp(i2\pi kx/X)\}$ can be treated in another way as well – as having been generated by means of multiplying the "mother" function:

$$\exp(i2\pi kx/X) = \prod_{l=1}^{k} \exp(i2\pi x/X) = \exp\left(i\pi \sum_{l=1}^{k} 2x/X\right) \qquad (18.7)$$

This is why such bases are also called *multiplicative* bases. Another example of family of basis functions built on this principle is the family of Walsh functions [1].

A distinctive feature of shift (convolution) bases is that, for them, signal representation coefficients depend on signal values in the vicinity of the corresponding sampling point, and therefore they carry local information about signals. In contrast to them, signal discrete representation for scale (multiplicative) bases is "global" because signal representation coefficients depend on the entire signal they represent. It is frequently useful to have a combination of these two features in the signal discrete representation. This is achieved with *wavelet basis functions* built using both shifting and scaling of "mother" functions. At present, numerous versions of wavelet bases are known (see Chapter 7 for a more detailed description of wavelets).

18.2.2
Imaging Transforms in the Mirror of Digital Computers

The most frequently used mathematical models of image formation in optical and holographic systems are convolutional and Fourier integral transforms. For digital processing, one has to find their appropriate representation in computers. At first glance, obtaining discrete representation of analog transforms is a trivial task: one has to only replace integrals by integral sums of integrand values in sampling points. However, this solution that originates from traditions of classical numerical mathematics, based on signal Taylor expansion approximation, disregards physical characteristics of signal sampling and reconstruction devices and, owing to this, cannot properly treat the problem of accuracy of the discrete representation. The rigorous approach to discrete representation of analog signal transforms is based on the following consistency and mutual correspondence principles [1]:

- Discrete representation of signal transformations should parallel that of signals.
- Discrete transformation corresponds to its analog prototype if both act to transform identical input signals into identical output signals.
- Digital processors incorporated into optical information systems should be regarded and treated together with signal discretization and reconstruction devices as integrated analog units and should be specified and characterized in terms of equivalent analog transformations.

Tables 18.1–18.3 summarize discrete representations of such most widely used imaging transforms as the convolution and Fourier transforms. These representations are built for different sampling conditions on the assumption that analog signals are discretized using sampling basis functions. Detailed derivations can be found in Refs [1–3].

Table 18.1 Convolution integral and digital filters (in 1D denotations).

The convolution integral of a signal $a(x)$ with shift invariant kernel $h(x)$	$b(x) = \int_{-\infty}^{\infty} a(\xi)h(x-\xi)d\xi$
Digital filter for samples $\{a_k\}$ and $\{b_k\}$ of input and output signals	$b_k = \sum_{n=0}^{N_h-1} h_n a_{k-n}$
	N_h is the number of nonzero samples of h_n
Discrete impulse response $\{h_n\}$ of the digital filter for input and output signal sampling bases $\phi_0^{(s)}(x)$ and $\varphi_0^{(r)}(x)$ and sampling interval Δx	$h_n = \int_{-\infty}^{\infty} \int_{-\infty}^{\infty} h[x-\xi-n\Delta x]\varphi^{(r)}(\xi)\phi^{(s)}(x)dxd\xi$

As one can see in Tables 18.2 and 18.3, taking into consideration different possible sampling conditions leads to extension of the canonical discrete Fourier transform (DFT) to shifted DFT, discrete cosine transform (DCT), scaled DFT, and rotated DFT. Shifted, scaled, and rotated DFTs contain analog shift, scale, and rotation angle parameters that determine the geometrical position of the image, its spectrum sampling grids in analog image and its spectrum coordinate systems. Their presence enables efficient computational means for image perfect and arbitrary translation, rescaling with arbitrary scale factor and rotation, which is particularly useful in template matching with subpixel accuracy, in numerical reconstruction of color holograms, and in many other applications [3, 4].

DCT is a special case of shifted DFT with shift parameters (1/2, 0) of virtual signals, obtained from original ones by means of padding them with their mirror reflection from their borders. Such a symmetrization eliminates signal discontinuities at its borders, which otherwise appear due to periodical replication of the signal that corresponds to working in the domain of DFTs. This radically improves the energy compaction property of the transform. DCT is well known for its application in image compression. Not less important is its application for boundary effect free fast digital convolution [1, 3].

Similar representations that maintain correspondence with analog transforms can be derived for the integral Fresnel transform [3] used for the numerical reconstruction of holograms and synthesis of computer-generated holograms (CGHs).

18.2.3
Characterization of Discrete Transforms in Terms of Equivalent Analog Transforms

Characterization of discrete transforms in terms of equivalent continuous transforms is necessary for proper design and performance analysis of digital image processing systems. In this section, we address this issue for digital filters and DFTs.

Table 18.2 Discrete representations of 1D integral Fourier transform.

1D direct and inverse integral Fourier transforms of a signal a(x)

$$\alpha(f) = \int_{-\infty}^{\infty} a(x) \exp(i2\pi fx) \, dx \qquad a(x) = \int_{-\infty}^{\infty} \alpha(f) \exp(-i2\pi fx) \, df$$

Direct and inverse canonical discrete Fourier transforms (DFTs)

$$\alpha_r = \frac{1}{\sqrt{N}} \sum_{k=0}^{N-1} a_k \exp\left(i2\pi \frac{kr}{N}\right) \qquad a_k = \frac{1}{\sqrt{N}} \sum_{r=0}^{N-1} \alpha_r \exp\left(-i2\pi \frac{kr}{N}\right)$$

Sampling conditions:
- Signal and signal sampling device coordinate systems are identical as also those of the signal spectrum and the assumed signal spectrum sampling device.
- Signal samples $\{a_k\}$ as well as samples $\{\alpha_r\}$ of its Fourier spectrum are positioned in such a way that samples with indices $k = 0$ and $r = 0$ are taken in signal and spectrum coordinates at points $x = 0$ and $f = 0$, respectively.
- Signal and its Fourier spectrum sampling intervals Δx and Δf satisfy the "cardinal" sampling relationship $\Delta x = 1/N\Delta f$ associated with the signal baseband $N\Delta f$.

Direct and inverse shifted DFTs (SDFT(u, v)):

$$\alpha_r^{u,v} = \frac{1}{\sqrt{N}} \sum_{k=0}^{N-1} a_k \exp\left[i2\pi \frac{(k+u)(r+v)}{N}\right] \qquad a_k = \frac{1}{\sqrt{N}} \sum_{k=0}^{N-1} \alpha_r^{u,v} \exp\left[-i2\pi \frac{(k+u)(r+v)}{N}\right]$$

Sampling conditions:
- Signal and signal sampling device coordinate systems as well as, correspondingly, those of signal spectrum and of the assumed signal spectrum discretization device, are shifted with respect to each other in such a way that signal sample $\{a_0\}$ and, correspondingly, sample $\{\alpha_0\}$ of its Fourier spectrum are taken in signal and spectrum coordinates at points $x = -u\Delta x$ and $f = -v\Delta f$.
- Signal "cardinal" sampling: $\Delta x = 1/N\Delta f$.

Direct and inverse discrete cosine transform (DCT):

$$\alpha_r = \frac{2}{\sqrt{2N}} \sum_{k=0}^{N-1} a_k \cos\left(\pi \frac{k+1/2}{N} r\right) \qquad a_k = \frac{1}{\sqrt{2N}} \left[\alpha_0 + 2 \sum_{r=1}^{N-1} \alpha_r \cos\left(\pi \frac{k+1/2}{N} r\right)\right]$$

Sampling conditions:
- Special case of SDFT for sampling grid shift parameters: $u = 1/2; v = 0$.
- Analog signal of final length is, before sampling, artificially padded with its mirror copy to form a symmetrical sampled signal of double the length: $\{a_k = a_{2N-1-k}\}$.

Direct and inverse scaled shifted DFTs [ScSDFT(u,v;σ)]

$$\alpha_r^\sigma = \sum_{k=0}^{N-1} a_k \exp\left[i2\pi \frac{(k+u)(r+v)}{\sigma N}\right]$$

Sampling conditions:
- Sampling rate is σ times the cardinal rate: $\Delta x = 1/\sigma N\Delta f$
- Sampling shift parameters: $u, v \neq 0$

18.2.3.1 Point Spread Function and Frequency Response of a Continuous Filter Equivalent to a Given Digital Filter

A linear filter is fully characterized by its *point spread function* (PSF), or by its Fourier transform, the filter *frequency response*. Given the discrete PSF $\{h_n\}$ of a digital filter, one can, in accordance to the above formulated consistency and mutual correspondence principles, find the overall PSF $h_{eq}(x, \xi)$ of an analog filter

Table 18.3 Discrete representations of 2D integral Fourier transform.

2D direct and inverse integral Fourier transforms of a signal $a(x,y)$

$$\alpha(f,p) = \int_{-\infty}^{\infty}\int_{-\infty}^{\infty} a(x,y)\exp[i2\pi(fx+py)]dxdy \qquad a(x,y) = \int_{-\infty}^{\infty}\int_{-\infty}^{\infty} \alpha(f,p)\exp[-i2\pi(fx+py)]df\,dp$$

2D separable direct and inverse canonical DFTs

$$\alpha_{k,l} = \frac{1}{\sqrt{N_1 N_2}}\sum_{k=0}^{N_1-1}\sum_{l=0}^{N_2-1}\alpha_{r,s}\exp\left[-i2\pi\left(\frac{kr}{N_1}+\frac{ls}{N_2}\right)\right] \qquad a_{k,l} = \frac{1}{\sqrt{N_1 N_2}}\sum_{k=0}^{N_1-1}\sum_{l=0}^{N_2-1}\alpha_{r,s}\exp\left[-i2\pi\left(\frac{kr}{N_1}+\frac{ls}{N_2}\right)\right]$$

Sampling conditions:
- Sampling in a rectangular sampling grid with cardinal sampling rates $\Delta x = 1/N_1\Delta f_x$, $\Delta y = 1/N_2\Delta f_y$.
- Zero sampling grid shift parameters.

Scaled shifted DFTs

$$\sigma_1,\sigma_2 \alpha_{r,s}^{u,v/p,q} = \sum_{k=0}^{N_1-1}\sum_{l=0}^{N_2-1} a_{k,l}\exp\left\{i2\pi\left[\frac{(k+u)(r+v)}{\sigma_1 N_1}+\frac{(l+p)(s+q)}{\sigma_2 N_2}\right]\right\}$$

Sampling conditions:
- Sampling in a rectangular sampling grid. Sampling rates $\Delta x = 1/\sigma_1 N_1 \Delta f_x$; $\Delta y = 1/\sigma_2 N_2 \Delta f_y$.
- Nonzero sampling grid shift parameters (u,v) and (p,q).

$$\sigma\alpha_{r,s}^{\theta} \propto \sum_{k=0}^{N-1}\sum_{l=0}^{N-1} a_{k,l}\exp\left[i2\pi\left(\frac{\tilde{r}k+\tilde{s}l}{\sigma N}\cos\theta - \frac{\tilde{s}k-\tilde{r}l}{\sigma N}\sin\theta\right)\right] \quad \begin{array}{l}\tilde{k}=k+u;\ \tilde{r}=r+v;\\ \tilde{l}=l+p;\ \tilde{s}=s+q\end{array}$$

Rotated and scaled DFTs

$$\sigma\alpha_{r,s}^{\theta} \propto \sum_{k=0}^{N-1}\sum_{l=0}^{N-1} a_{k,l}\exp\left[i2\pi\left(\frac{\tilde{r}k+\tilde{s}l}{\sigma N}\cos\theta - \frac{\tilde{s}k-\tilde{r}l}{\sigma N}\sin\theta\right)\right]; \quad \begin{array}{l}\tilde{k}=k+u;\ \tilde{r}=r+v;\\ \tilde{l}=l+p;\ \tilde{s}=s+q\end{array}$$

Sampling conditions:
- Sampling in a rectangular sampling grid in a rotated coordinate system
$$\begin{bmatrix}x^\theta\\y^\theta\end{bmatrix} = \begin{bmatrix}\cos\theta & \sin\theta\\ -\sin\theta & \cos\theta\end{bmatrix}\begin{bmatrix}x\\y\end{bmatrix} \text{ with } \theta \text{ as a rotation angle}$$
- Sampling rates $\Delta x = 1/\sigma N\Delta f_x$; $\Delta y = 1/\sigma N\Delta f_y$; nonzero sampling grid shift parameters (u,v) and (p,q)

---

equivalent to a given digital filter as [1, 3]

$$h_{eq}(x,\xi) = \sum_{k=0}^{N_b-1}\sum_{m=0}^{N_h-1} h_n \varphi^{(r)}(x-n\Delta x)\varphi^{(s)}[\xi - (k-n)\Delta x] \qquad (18.8)$$

where N_b is the number of samples of the filter output signal $\{b_k\}$ involved in reconstruction, from its discrete representation, of analog output signal $b(x)$ of the equivalent analog filter, N_h is the number of nonzero samples of the digital filter PSF, and Δx is the signal sampling interval [1].

It is more convenient to design digital filters and to characterize and analyze their equivalent continuous filters through the overall frequency response $H_{eq}(f,p)$ of the filter found as the Fourier transform of its overall PSF:

$$H_{eq}(f,p) = \int_{-\infty}^{\infty}\int_{-\infty}^{\infty} h_{eq}(x,\xi)\exp[i2\pi(fx-p\xi)]\,dx\,d\xi$$

$$= \left[\sum_{k=0}^{N_b-1}\exp[i2\pi(f-p)k\Delta x]\right]\int_{-\infty}^{\infty}\varphi^{(r)}(x)\exp(i2\pi fx)dx$$

$$\times \int_{-\infty}^{\infty}\varphi^{(s)}(\xi)\exp(-i2\pi p\xi)d\xi \times \left[\sum_{m=0}^{N_h-1}h_m\exp(i2\pi pm\Delta x)\right] \quad (18.9)$$

This expression contains four multiplicands:

$$SV(f,p) = \sum_{k=0}^{N_b-1}\exp[i2\pi(f-p)k\Delta x] \qquad (18.10a)$$

$$\Phi^{(r)}(f) = \int_{-\infty}^{\infty}\varphi^{(r)}(x)\exp(i2\pi fx)dx \qquad (18.10b)$$

$$\Phi^{(s)}(p) = \int_{-\infty}^{\infty}\varphi^{(s)}(x)\exp(i2\pi px)dx \qquad (18.10c)$$

$$CFrR(p) = \sum_{n=0}^{N_h-1}h_n\exp(i2\pi pn\Delta x) \qquad (18.10d)$$

The term $SV(f,p)$ reflects the fact that the digital filters defined in Table 18.1 and obtained as a discrete representation of the convolution integral with shift invariant kernel are shift variant because a finite number N_b of samples $\{b_k\}$ of the filter output signal is involved in the reconstruction of the filter analog output signal $b(x)$. This discrepancy vanishes in the limit when $N_b \to \infty$.

The terms $\Phi^{(s)}(.)$ and $\Phi^{(r)}(.)$ are frequency responses of signal sampling and reconstruction devices, respectively.

The term $CFrR(p)$ is the continuous frequency response of the digital filter. Being a Fourier series, it is a periodical function in the frequency domain. If the reconstruction and sampling basis functions of the signal are ideal sinc-functions, the sampling and reconstruction devices act as ideal low-pass filters that remove all but one of its periods. In reality, this is not the case, and therefore one should anticipate aliasing effects in the convolution results similar to those in signal reconstruction from its sampled representation.

It is shown [1, 3] that continuous frequency response of the digital filter $CFrR(p)$ can be represented via coefficients $\{\eta_r\}$ of DFT of the filter discrete PSF $\{h_n\}$:

$$\eta_r = \frac{1}{\sqrt{N}}\sum_{n=0}^{N-1}h_n\exp\left[i2\pi\frac{n-(N-1)/2}{N}r\right] \qquad (18.11)$$

as

$$CFrR(p) \propto \sum_{r=0}^{N-1} \eta_r \, \text{sincd}\,[N; \pi(pN\Delta x - r)] \tag{18.12}$$

where

$$\text{sincd}(N; x) = \frac{\sin x}{N \sin(x/N)} \tag{18.13}$$

is the discrete sinc-function and N is the number of samples of the filter input signal.

These results lead to the following important conclusions:

- Given signal sampling and reconstruction devices and the number N_b of samples used for reconstruction of analog signals from their samples, the overall frequency response $H_{eq}(f, p)$ of the digital filter is fully defined by DFT coefficients $\{\eta_r\}$ of the discrete PSF $\{h_n\}$ of the filter.
- These coefficients are its samples taken within the baseband with a sampling interval $1/N\Delta x$.

These conclusions constitute the base for the design of digital filters with a desired analog frequency response.

This has been illustrated in Section 18.3.

18.2.3.2 Point Spread Function of the Discrete Fourier Analysis

An immediate application of DFTs is discrete Fourier spectral analysis performed by applying an appropriate version of DFTs to signal samples. In this process, the computer, together with the signal sampling device, plays the role of a sampling device for the Fourier spectra of the signal; spectral samples obtained from signal samples are regarded as samples of Fourier spectra of the analog signal.

The ultimate characteristic of sampling devices is their PSF. For spectral analysis, the PSF $h_{DFA}(f, r)$ of spectral sampling, which defines the resolving power of the discrete Fourier analysis (DFA), can be found from the relationship:

$$\alpha_r = \int_{-\infty}^{\infty} \alpha(f) h_{DFA}(f, r) df \tag{18.14}$$

where $\alpha(f)$ is the Fourier spectrum of signal $a(x)$ and $\{\alpha_r\}$ are its samples.

Replacing in Eq. (18.14) samples $\{\alpha_r\}$ of the Fourier spectrum of the analog signal by their expression as scaled DFTs of signal samples and linking the latter with the signal and its Fourier spectrum, one can obtain the PSF of the discrete spectrum analysis defined by the equation [1, 3]:

$$h_{DFA}(f, r) = N \, \text{sinc}\,d\left[N; \left(\frac{r}{\sigma} - fN\Delta x\right)\right] \Phi_s(f) \tag{18.15}$$

where σ is a sampling scale parameter, which, together with the number N of signal samples, links the signal sampling interval Δx with that of Δf of its spectrum.

Equation (18.15) has a clear physical interpretation:

- Discrete spectral analysis can be treated as the sampling, with a discrete sinc-function, of the spectrum of an analog signal, masked by the sampling device frequency response $\Phi_s(f)$.
- When $\sigma = 1$, spectrum samples are taken with a cardinal sampling interval $\Delta f_{card} = 1/N\Delta x$. When $\sigma > 1$, the signal spectrum is "oversampled"; its samples are taken with sampling interval $\Delta f_{card}/\sigma < \Delta f_{card}$ and are sincd-interpolated versions of the "cardinal" samples that correspond to the case $\sigma = 1$. When $\sigma < 1$, the spectrum is "undersampled."

Figure 18.1a,b illustrate the resolving power of the DFA. Figure 18.1a shows spectra of continuous sinusoidal signals with frequencies 64/256 and 65/256 (in fraction of the width of the signal baseband) and sampled to 256 samples as well as the spectrum of the sum of these two signals. Figure 18.1b shows the same for signals with normalized frequencies 64/256 and 65.5/256. One can conclude from these figures that, in DFA, sinusoidal signals can be reliably resolved if the difference of their frequencies is larger than about $1.5/N\Delta x$.

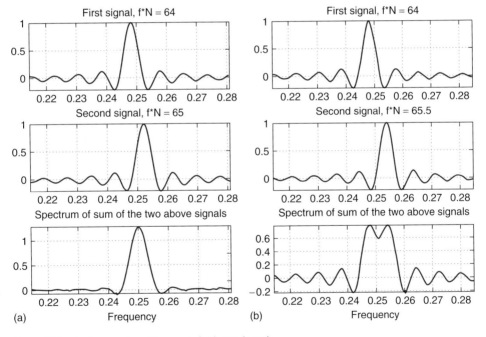

Figure 18.1 Analog spectra of two periodical signals with close frequencies and their mixture computed by means of the discrete Fourier analysis: (a) signals with frequencies 64/256 and 65/256 are not resolved; (b) signals with frequencies 64/256 and 65.5/256 are resolved in the mixture.

18.3
Building Continuous Image Models

When working with sampled images in computers, one frequently needs to return to their analog originals. Typical applications that require reconstruction of analog image models are image resampling, image restoration from sparse samples, and differentiation and integration, to name a few. In this section, we illustrate the use of the above concepts for these applications.

18.3.1
Discrete Sinc Interpolation: The Gold Standard for Image Resampling

Image resampling assumes reconstruction of a continuous approximation of the original nonsampled image by means of interpolation of available image samples to obtain samples in between the given ones. In some applications, for instance in computer graphics and print art, simple numerical interpolation methods, such as nearest neighbor or linear (bilinear) interpolations, can provide satisfactory results. In applications, such as optical metrology, that are more demanding in terms of the interpolation accuracy, higher order spline interpolation methods are frequently recommended. However, all these methods are not perfectly accurate and introduce signal distortions in addition to those caused by the primary image sampling.

The discrete signal interpolation method that is capable, within limitations defined by the given finite number of signal samples, of securing reconstruction of analog images without adding any additional interpolation errors is discrete sinc interpolation [4].

A perfect resampling filter can be designed as a discrete representation of the ideal analog shift operator with frequency response $H(f) = \exp(i2\pi \delta x f)$, where δx is the analog signal shift in signal coordinate x and f is the frequency. According to Eq. (18.12), samples $\{\eta_r = \exp(i2\pi r\delta x/N\Delta x)\}$ of this frequency response in the baseband $N\Delta x$, defined by the sampling interval Δx and the number of signal samples N, are examples of DFT of the filter discrete PSF $\{h_n\}$. These samples define the discrete sinc-interpolator with discrete PSF $\{h_n = \text{sinc d}\{N, \pi[n - (N-1)/2 - \delta x/\Delta x]\}\}$ [4].

Discrete sinc interpolation, by definition, preserves samples of the spectrum of the analog signal in its baseband. All the other convolutional interpolation methods with PSF, other than the discrete sinc-function distort signal spectrum, therefore introduce interpolation error. This means that discrete sinc interpolation can be regarded the "gold standard" of discrete signal interpolation. Discrete sinc interpolation is also competitive with other less-accurate interpolation methods in terms of the computational complexity, thanks to its implementation through fast Fourier or DCT transforms [4]. The latter implementation is the most recommended because it is virtually completely free of oscillation artifacts at signal borders characteristic for implementation through FFT.

Perfect interpolation capability of the discrete sinc interpolation in comparison to other numerical interpolation methods is illustrated in Figure 18.2, which shows

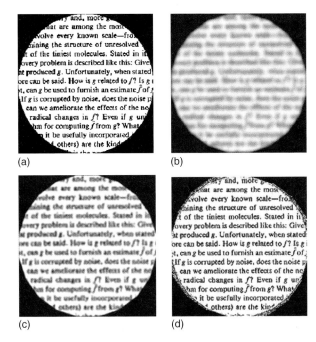

Figure 18.2 Discrete sinc interpolation versus conventional interpolation methods: results of rotation of the (a) test image through 3 × 360° in 60 steps of 18°; (b) bilinear interpolation ; (c) bicubic interpolation; and (d) discrete sinc interpolation.

the results of rotation of a test image (Figure 18.2a) through 3 × 360° in 60 equal steps, using two most popular methods of image resampling, bilinear and bicubic ones (Figure 18.2b,c), and that of discrete-sinc interpolation (Figure 18.2d) [4].

Images shown in the figure clearly show that, after 60 rotations though 18° each, bilinear and bicubic interpolation methods completely destroy the readability of the text, while discrete sinc-interpolated rotated image is virtually unchanged and is not distinguishable from the original one. Further comparison data can be found in Ref. [4].

Bilinear and bicubic interpolations are spline interpolations of the first and second order. The higher the spline order, the higher is the interpolation accuracy, and the higher the computational complexity of spline interpolation. Note that the higher the spline order, the closer their PSFs approximate the discrete sinc-function.

18.3.1.1 Signal Recovery from Sparse or Nonuniformly Sampled Data

In many applications, sampled image data are collected in an irregular fashion or are partly lost or unavailable. In these cases, it is required to restore missing data and to convert irregularly sampled signals into regularly sampled ones. This problem can be treated and solved as a least square approximation task in a framework of the discrete sampling theorem for "bandlimited" discrete signals that have a limited number of nonzero transform coefficients in a certain transform domain [5]. We illustrate possible applications of this signal recovery technique through two examples.

One of the attractive potential applications is image super-resolution from multiple digital video frames with chaotic pixel displacements due to atmospheric turbulence, camera instability, or similar random factors [6]. By means of elastic registration of the sequence of frames of the same scene, one can determine, for each image frame and with subpixel accuracy, pixel displacements caused by random acquisition factors. Using these data, a synthetic fused image can be generated by placing pixels from all available turbulent video frames in their proper positions on the correspondingly denser sampling grid according to their found displacements. In this process, some pixel positions on the denser sampling grid will remain unoccupied, when a limited number of image frames is fused. These missing pixels can then be recovered assuming that the high-resolution image is bandlimited in a domain of certain transform. In many applications, including the discussed one, using DCT as such transform is advisable owing to the excellent energy compaction capability of DCT.

The super-resolution from multiple chaotically sampled video frames is illustrated in Figure 18.3, which shows one low-resolution frame (a); an image fused from 15 frames (b); and a result of iterative restoration (c) achieved after 50 iterations [6]. Image band limitation was set in this experiment twice the baseband of raw low-resolution images.

Yet another application that the discussed sparse data recovery technique can find in computed tomography, where it frequently happens that a substantial part of slices, which surrounds the studied body slice, is a priori known to be an empty field. This means that slice projections (sinograms) are "bandlimited" functions in the domain of the Radon transform. Therefore, whatever number of projections is available, a certain number of additional projections, commensurable, according to the discrete sampling theorem, with the size of the slice empty zone, can be obtained and the corresponding resolution increase in the reconstructed images

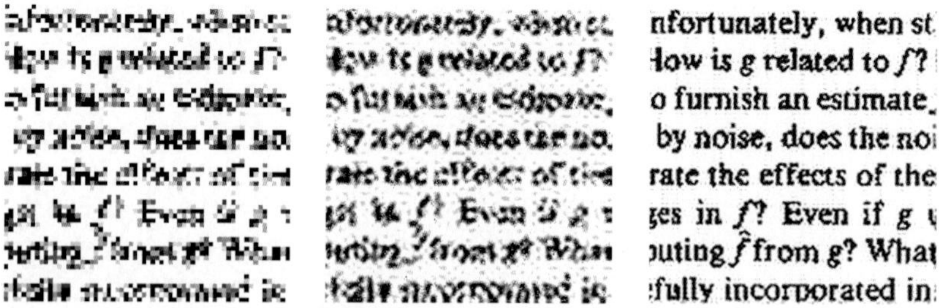

Figure 18.3 Iterative image interpolation in the super-resolution process: (a) a low-resolution frame; (b) an image fused by the elastic image registration from 15 frames; and (c) a result of bandlimited restoration of image (b).

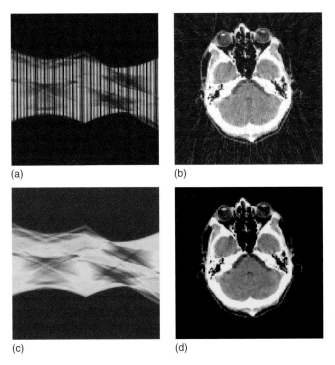

Figure 18.4 Super-resolution in computed tomography: (a) a set of initial projections supplemented with the same number of presumably lost projections (shown dark) to double the number of projections; initial guesses of the supplemented projections are set to zero; (b) image reconstructed from initially available projections; (c) a result of restoration of missing projections; and (d) an image reconstructed from the restored double set of projections.

can be achieved. Figure 18.4 illustrates such super-resolution by means of recovery of a missing 50% of the projections achieved using the fact that, by segmentation of the restored image, it was found that the outer 55% of the image area was empty.

18.3.2
Image Numerical Differentiation and Integration

Signal numerical differentiation and integration are operations that are defined for continuous functions and require measuring infinitesimal increments or decrements of signals and their arguments. Therefore, numerical computing signal derivatives and integrals assume one or another method of building analog models of signals specified by their samples through explicit or implicit interpolation between available signal samples.

Because differentiation and integration are shift invariant linear operations, methods of computing signal derivatives and integrals from their samples can be conveniently designed, implemented, and compared through their frequency responses.

Frequency responses of ideal analog differentiation and integration filters are, correspondingly $H_{\text{diff}} = -i2\pi f$ and $H_{\text{int}} = i/2\pi f$, where f is the signal frequency. Therefore, according to Eq. (18.12), samples of frequency responses of perfect digital differentiating and integrating filters must be set correspondingly to $\{\eta_r^{\text{diff}} = -i2\pi r/N\}$ and $\{\eta_r^{\text{integr}} = iN/2\pi r\}$. We refer to these filters as the *differentiation ramp filter* and the *DFT-based integration filter*, respectively. Being defined in the frequency domain, these filters can be efficiently implemented by means of FFT-type fast transforms. In order to diminish boundary effects, filtering should be carried out using the DCT-based convolution algorithm [3, 4]. One can show that such numerical differentiation and integration imply the discrete sinc interpolation of signals.

In classical numerical mathematics, a common approach to numerical computing of signal derivatives and integrals is based on a Taylor series signal expansion. This approach results in algorithms implemented through discrete convolution of the signal in the signal domain. The following simplest differentiating kernels $\{h_n^{\text{diff}}\}$ of two and five samples are recommended in manuals on numerical methods, such as [7]: $\{h_n^{\text{diff}(D1)}\} = \{-1, 1\}$ and $\{h_n^{\text{diff}(D2)}\} = \{-1/12, 8/12, 0, -8/12, 1/12\}$, which we refer to as numerical differentiating methods D1 and D2, respectively.

The best known numerical integration methods are the Newton–Cotes quadrature rules [7]. The three first rules are the trapezoidal, the Simpson, and the 3/8-Simpson ones. Another frequently used alternative is cubic spline integration [8]. In these integration methods, a linear, a quadratic, a cubic, and a cubic spline interpolation, respectively, are assumed between the sampled data.

Frequency responses of the above-described perfect and conventional numerical differentiation and integration methods are presented for comparison in Figures 18.5 and 18.6, respectively. One can see from these figures that conventional numerical differentiation and integration methods entail certain, and sometimes very substantial, distortions of spectral contents of the signal at high frequencies. In particular, numerical differentiation methods D1 and D2 attenuate signals at high frequencies. Among integration methods, Simpson and 3/8-Simpson integration methods, being slightly more accurate than the trapezoidal method at middle frequencies, tend to generate substantial artifacts if signals contain higher frequencies. Frequency response of the 3/8-Simpson rule tends to infinity at a frequency that is two-third of the maximum frequency, and frequency response of the simple Simpson rule has almost the same tendency for the maximal frequency in the baseband. This means that round-off computation errors and noise that might be present in the input data will be overamplified by Simpson and 3/8-Simpson integration filters in these frequencies.

Figure 18.7 presents results of the comparison of differentiation errors for different numerical differentiation methods applied to pseudorandom test signals of different bandwidths [4]. They convincingly evidence that D1 and D2 methods

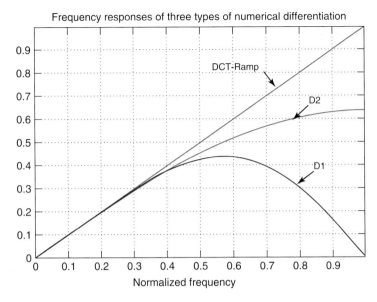

Figure 18.5 Absolute values of frequency responses of three described numerical differentiation filters: D1, D2, and ramp filter implemented using DCT.

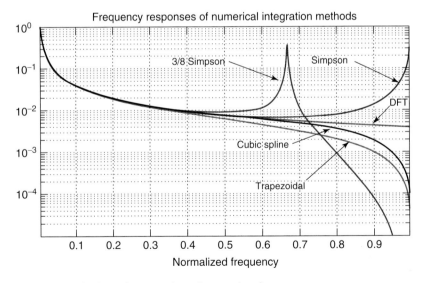

Figure 18.6 Absolute values, in a logarithmic scale, of frequency responses of described numerical integration filters: trapezoidal, Simpson, 3/8 Simpson, cubic spline, and DFT-based ones.

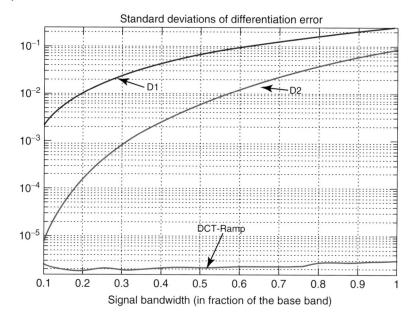

Figure 18.7 Standard deviations of differentiation errors, in fraction of standard deviation of the derivative, averaged over 100 realizations of pseudorandom test signals for D1, D2, and DCT-ramp differentiation methods as a function of test signal bandwidth (in fractions of the baseband defined by the signal sampling rate).

provide reasonably good differentiation accuracy only for signals with bandwidth less than 0.1–0.2 of the baseband, which means that a substantial signal oversampling is required in order to maintain acceptable differentiation error. But even for such signals, standard deviation for the normalized error is much lower for the DCT-ramp method. For signals with broader bandwidth, this method outperforms other methods, in terms of the differentiation accuracy, by at least 3–4 orders of magnitude.

Similar comparison and conclusions can be made for different numerical integration methods [4, 9].

18.4
Digital-to-Analog Conversion in Digital Holography. Case Study: Reconstruction of Kinoform

CGHs have numerous applications as spatial filters for optical information processing, as diffractive wavefront correcting elements for large optical objectives and telescope mirrors, beam-forming elements (for instance, for laser tweezers), laser focusers, deflectors, beam splitters, and multiplicators, and for information display. Being generated in computers as sets of numbers, CGHs are then recorded on a physical medium to be used in analog optical setups. Methods of this digital-to-analog conversion have their impact on optical characteristics of CGHs. In this section we demonstrate how this impact can be studied by means of an example of optical reconstruction of kinoform.

Kinoform is one of the simplest hologram-encoding methods for phase-modulating spatial light modulators (SLMs). In this method, amplitude components of samples of the computed holograms are forcibly set to a constant and only their phase components are preserved and used for recording on phase-modulating SLM. These samples of kinoform are recorded, one by one, on resolution cells of the SLM arranged in a rectangular grid, as illustrated in Figure 18.8.

The resulting recorded hologram can be represented mathematically as

$$\Gamma(\xi,\eta) = w(\xi,\eta) \sum_r \sum_s \Gamma_{r,s} h_{\text{rec}}(\xi - \xi_0 - r\Delta\xi, \eta - \eta_0 - s\Delta\eta) \quad (18.16)$$

where $\{\Gamma_{r,s}\}$ are samples of the computed hologram with $\{r,s\}$ as 2D sample indices; (ξ,η) are physical coordinates on the hologram-recording SLM; $(\Delta\xi, \Delta\eta)$ are sampling intervals of the rectangular sampling grid along coordinates; (ξ,η), (ξ_0, η_0) are coordinates of the first sample of the hologram in the reconstruction setup; $h_{\text{rec}}(\xi,\eta)$ are hologram-recording device aperture functions, and $w(\xi,\eta)$ is a hologram window function that defines the physical dimensions of the recorded hologram: $0 \leqslant w(\xi,\eta) \leqslant 1$, when (ξ,η) belongs to the hologram area and $w(\xi,\eta) = 0$, otherwise.

At the reconstruction stage, this hologram is subjected to analog optical transformation in an optical setup to reconstruct an image. Assume that the computer-generated kinoform is designed for image reconstruction in the far diffraction zone or in the focal plane of a lens, which corresponds to an optical Fourier transform. In this case, the reconstruction image can be represented as

$$A_{\text{rcstr}}(x,y) = \int_{-\infty}^{\infty} \int_{-\infty}^{\infty} \Gamma(\xi,\eta) \exp\left(-i2\pi \frac{x\xi + y\eta}{\lambda Z}\right) d\xi d\eta \quad (18.17)$$

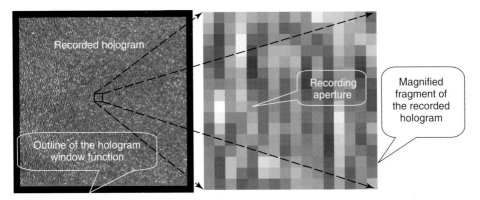

Figure 18.8 Definitions related to recorded physical computer-generated holograms.

where λ is the wavelength of reconstructing coherent illumination and Z is the hologram-to-object distance or the focal length of the lens.

Samples $\{\Gamma_{r,s}\}$ of the kinoform are linked with samples of the input image $\{A_{k,l}^{(kf)}\}$ through, in general, shifted DFT:

$$\Gamma_{r,s} = \sum_{k=0}^{N_1-1} \sum_{l=0}^{N_2-1} A_{k,l}^{(kf)} \exp\left\{i2\pi\left[\frac{(k+u)(r+p)}{N_1} + \frac{(l+v)(s+q)}{N_2}\right]\right\} \quad (18.18)$$

where $\{k,l\}$ are pixel indices, $\{N_1, N_2\}$ are input image dimensions, and (u, v) and (p, q) are shift parameters associated with shifts of samples of the input image with respect to its coordinate system and shifts of kinoform samples with respect to the coordinate system of the optical setup. Note that the input image $\{A_{k,l}^{(kf)}\}$ should be, before the synthesis of kinoform, appropriately preprocessed to secure minimal distortions of the reconstructed image caused by neglecting amplitude components of samples of computed hologram (for details see, for instance, Ref. [2]).

Substituting Eq. (18.16) in Eq. (18.17) and replacing, in the former, samples of kinoform $\{\Gamma_{r,s}\}$ by their expression (Eq. (18.18)) through input image samples $\{A_{k,l}^{(kf)}\}$, one can obtain, after completing corresponding integrations and summations, that the image reconstructed by the kinoform in the optical setup, is described, for shift parameters $p = q = 0$ and $\xi_0 = \eta_0 = 0$, by the following expression [2]:

$$A_{\text{rcstr}}(x, y) = \sum_{do_x=-\infty}^{\infty} \sum_{do_y=-\infty}^{\infty} \sum_{k=0}^{N_1-1} \sum_{l=0}^{N_2-1}$$

$$\times \overline{H}_{\text{rec}}\left[\left(\frac{k+u}{N_1} - do_x\right)\frac{\lambda Z}{\Delta \xi}, \left(\frac{l+v}{N_2} - do_y\right)\frac{\lambda Z}{\Delta \eta}\right] A_{k,l}^{(kf)}$$

$$\times W\left(x - \frac{k+u}{N_1}\frac{\lambda Z}{\Delta \xi} - do_x\frac{\lambda Z}{\Delta \xi}; y - \frac{l+v}{N_2}\frac{\lambda Z}{\Delta \eta} - do_y\frac{\lambda Z}{\Delta \eta}\right) \quad (18.19)$$

where

$$W(\bar{\xi}, \bar{\eta}) = \int_{-\infty}^{\infty}\int_{-\infty}^{\infty} w(\xi, \eta) \exp\left(-i2\pi \frac{\xi\bar{\xi} + \eta\bar{\eta}}{\lambda Z}\right) d\xi\, d\eta \quad (18.20)$$

is the Fourier transform of the hologram window function $w(\xi, \eta)$ and

$$H_{\text{rec}}(\bar{\xi}, \bar{\eta}) = \int_{-\infty}^{\infty}\int_{-\infty}^{\infty} h_{\text{rec}}(\xi, \eta) \exp\left[-i2\pi \frac{\xi\bar{\xi} + \eta\bar{\eta}}{\lambda Z}\right] d\xi\, d\eta \quad (18.21)$$

is the Fourier transform of the hologram-recording aperture $h_{\text{rec}}(\xi, \eta)$, or frequency response of the hologram-recording device.

Equation (18.19) has a clear physical interpretation illustrated in Figure 18.9:

- The object wavefront is reconstructed in a number of diffraction orders $\{do_x, do_y\}$.

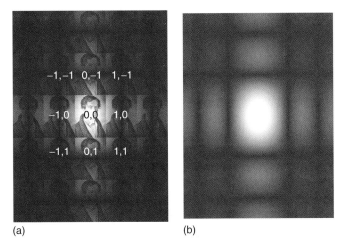

Figure 18.9 (a) Computer simulation of optical reconstruction of computer-generated kinoform; numbers are diffraction order indices $\{do_x, do_y\}$. (b) Reconstruction masking function $H_{rec}(x, y)$ for a rectangular hologram-recording aperture of size $\Delta\xi \times \Delta\eta$.

- In each particular diffraction order (do_x, do_y), the reconstructed wavefront is a result of the interpolation of samples $\left\{A_{k,l}^{(kf)}\right\}$ of the object wavefront used for synthesis of kinoform, with an interpolation kernel $W(x, y)$, which is the Fourier transform of the recorded hologram window function $w(\xi, \eta)$; samples of this object wavefront are weighted by samples of the frequency response $H_{rec}(x, y)$ of the hologram-recording device.

This is how physical parameters of the SLM used for recording computer-generated kinoform and those of the optical setup used for reconstructing images affect reconstructed images.

18.5 Conclusion

Optimal design of image processing systems is always hybrid with appropriately selected sharing of functions between analog, optical, and digital components of the system. This requires characterization and design of digital image processing in terms of the hybrid processing. In this article, we addressed the following aspects of this problem:

- discrete representation of analog convolution and Fourier transforms for sampled signals, which maintains mutual correspondence with analog and discrete transform;
- characterization of digital filtering and discrete Fourier analysis in terms of corresponding analog operations;

- building continuous image models from their sampled representation on examples of image resampling with perfect discrete sinc interpolation, image recovery from sparse data, and signal differentiation and integration;
- digital-to-analog conversion in computer-generated holography (with an example of optical image reconstruction of computer-generated kinoforms).

References

1. Yaroslavsky, L. (2004) *Digital Holography and Digital Image Processing*, Kluwer Academic Publishers, Boston.
2. Yaroslavsky, L. (2009) *Introduction to Digital Holography*, Bentham E-book Series, Digital Signal Processing in Experimental Research, Vol. 1 (eds. L. Yaroslavsky and J. Astola), Bentham Science Publishers Ltd., ISSN: 1879-4432, 2009, eISBN: 978-1-60805-079-6.
3. Yaroslavsky, L. (2007) Discrete transforms, fast algorithms and point spread functions of numerical reconstruction of digitally recorded holograms, in *Advances in Signal Transforms: Theory and Applications*, EURASIP Book Series on Signal Processing and Communications (eds. J. Astola and L. Yaroslavsky), Hindawi, pp. 93–142.
4. Yaroslavsky, L. (2007) Fast discrete sinc-interpolation: a gold standard for image resampling, in *Advances in Signal transforms: Theory and Applications*, EURASIP Book Series on Signal Processing and Communications (eds. J. Astola and L. Yaroslavsky), Hindawi, pp. 337–405.
5. Yaroslavsky, L.P., Shabat, G., Salomon, B.G., Ideses, I.A., and Fishbain, B. (2009) Nonuniform sampling, image recovery from sparse data and the discrete sampling theorem. *J. Opt. Soc. Am. A*, **26**, 566–575.
6. Yaroslavsky, L.P., Fishbain, B., Shabat, G., and Ideses, I. (2007) Super-resolution in turbulent videos: making profit from damage. *Opt. Lett.*, **32** (20), 3038–3040.
7. Press, W.H., Flannery, B.P., Teukolsky, S.A., and Vetterling, W.T. (1987) *Numerical Recipes. The Art of Scientific Computing*, Cambridge University Press, Cambridge.
8. Mathew, J.H. and Fink, K.D. (1999) *Numerical Methods Using MATLAB*, Prentice Hall.
9. Yaroslavsky, L.P., Moreno, A., and Campos, J. (2005) Frequency responses and resolving power of numerical integration of sampled data. *Opt. Express*, **13** (8), 2892–2905.

19
Visual Perception and Quality Assessment
Anush K. Moorthy, Zhou Wang, and Alan C. Bovik

19.1
Introduction

"Quality" according to the international standards organization (ISO) is *the degree to which a set of inherent characteristics of a product fulfills customer requirements* [1]. Even though this definition seems relatively straightforward at first, introspection leads one to the conclusion that the ambiguity inherent in the definition makes the quality assessment task highly subjective and, hence, difficult to model. Indeed, over the years, researchers in the field of visual quality assessment have found that judging the quality of an image or a video is a challenging task. The highly subjective nature of the task, coupled with the human visual systems' peculiarities, makes this an interesting problem to study and in this chapter we attempt to do just that.

This chapter is concerned with the algorithmic evaluation of quality of an image or a video, which is referred to as objective quality assessment. What makes this task difficult is the fact that the measure of quality produced by the algorithm should match up to that produced by a human assessor. To obtain a statistically relevant measure of what a human thinks the quality of an image or video is, a set of images or videos are shown to a group of human observers who are asked to rate the quality on a particular scale. The mean rating for an image or a video is referred to as the mean opinion score (MOS) and is representative of the perceptual quality of that visual stimulus. Such an assessment of quality is referred to as subjective quality assessment. To gauge the performance of an objective algorithm, the scores produced by the algorithm are correlated with MOS; a higher correlation is indicative of better performance. In this chapter, we focus on a subset of image quality assessment/video quality assessment algorithms (IQA/VQA), which are referred to as full-reference (FR) algorithms. In these algorithms, the original, pristine stimulus is available along with the stimulus whose quality is to be assessed. The FR IQA/VQA algorithm accepts the pristine reference stimulus and its distorted version as input and produces a score that is representative of the visual quality of the distorted stimulus [2].

One of the primary questions that arises when we talk of visual quality assessment is as follows: "why not use mean square error (MSE) for this purpose?" An MSE

Optical and Digital Image Processing: Fundamentals and Applications, First Edition. Edited by Gabriel Cristóbal, Peter Schelkens, and Hugo Thienpont.
© 2011 Wiley-VCH Verlag GmbH & Co. KGaA. Published 2011 by Wiley-VCH Verlag GmbH & Co. KGaA.

between two N-dimensional vectors x and y is defined as

$$\text{MSE} = \frac{1}{N} \sum_{i=1}^{N} (x_i - y_i)^2 \qquad (19.1)$$

A low MSE value indicates that the two vectors are similar. Generally, in order to follow a convention where a higher value indicates greater similarity, the peak-signal-to-noise ratio (PSNR) is used. PSNR is defined as

$$\text{PSNR} = 10 \log_{10} \left(\frac{L^2}{\text{MSE}} \right) \qquad (19.2)$$

where L is the dynamic range of the pixel values (e.g., $L = 255$ for grayscale images). Through this chapter, we use MSE and PSNR interchangeably.

Let us now discuss the valid question of why one should not use MSE for visual quality assessment. MSE has, after all, several elegant properties and is a prime candidate of choice to measure the deviation of one signal from another. How is visual quality assessment different from this task? The major difference for visual quality assessment (as for audio quality assessment) is the ultimate receiver. For images and videos, the ultimate receiver is the human observer. Immaterial of whether there exists a difference between the stimuli under test, the difference is not perceptually significant as long as the human is unable to observe the difference. This begs the question, are not all differences equally significant for a human? The answer is an emphatic no! As vision researchers have observed, the human visual system (HVS) is replete with peculiarities. The properties of the HVS–as discussed in the next section–govern the perceivability of the distortions and, hence, an algorithm that seeks to evaluate visual quality must be tuned to human perception. MSE, as many researchers have argued, is not tuned to human perception and, hence, does not make for a good visual quality assessment algorithm [3, 4].

We begin this chapter with a short description of how visual stimulus is processed by the human. We then describe various FR IQA/VQA algorithms. Our discussion then moves on to how one assesses the performance of an IQA/VQA algorithm and we describe some standard databases that are used for this task. Finally, we conclude this chapter with a discussion of possible future research directions in the field of quality assessment.

19.2
The Human Visual System

The first stage of the HVS is the eye, where the visual stimulus passes through the optics of the eye and then on to the photoreceptors at the back of the eye. Even though the eye exhibits some peculiarities including lens aberrations, these are generally not modeled in HVS-based IQA/VQA algorithms. The optics of the eye are band limited and act as a low-pass filter; hence, some HVS-based IQA/VQA systems model this using a point-spread function (PSF).

The photoreceptors are classified as rods, which are responsible for vision under scotopic conditions, and cones, which are responsible for vision under photopic conditions. Cones are also responsible for encoding color information. The distribution of rods and cones in the eye is not uniform. In fact, the number of photoreceptors is high at a region called the *fovea* and falls off as one moves away from the fovea [5]. This is important for IQA/VQA because of the fact that the human does not assimilate the entire visual stimulus at the same "resolution." The part of the stimulus that is imaged on the fovea has the highest resolution and regions that are imaged farther away have lower resolution. To assimilate the entire stimulus, the human scans the image using a set of fixations, followed by rapid eye movements called *saccades*. Little or no information is gathered during a saccade. This implies that for visual stimuli, certain regions may be of greater importance than others [6].

The information from the photoreceptors is then processed by the retinal ganglion cells. The ganglion cells are an interesting area of study and many researchers have devoted their energies toward such research. However, we do not dwell upon the ganglion cells here; the interested reader is referred to Ref. [5] for a thorough explanation. The information from the ganglion cells is passed onto the lateral geniculate nucleus (LGN), which has been hypothesized to act as a "relay" station [5, 7]. The LGN is the first location along the visual pathway where the information from the left and the right eye merges. The LGN receives not only the feed-forward information from the retinal cells but also feedback information from the next stage of processing–the primary visual cortex (area V1) [7]. The amount of feedback received leads one to believe that the LGN may not be *just* a relay station in the visual pathway [7]. Further, recent discoveries show that the LGN may perform certain normalization computations [8]. Further research in understanding the LGN and its functioning may be important for visual quality assessment algorithm design.

Moving along, the information from the LGN is projected onto area V1 or the primary visual cortex. Area V1 is hypothesized to encompass two types of cells–simple cells and complex cells. Simple cells are known to be tuned to different orientations, scales, and frequencies. This tuning of cells can be regarded as the HVS performing a scale-space-orientation decomposition of the visual stimulus (Chapter 8). This is the rationale behind many HVS-based IQA/VQA systems performing a waveletlike decomposition of the visual stimulus. Complex cells are currently modeled as receiving inputs from a set of simple cells [5, 7]. Complex cells are known to be direction selective. Even though V1 is connected to many other regions, one region of interest is area V5/MT, which is hypothesized to play a key role in motion processing [9, 10]. Area MT along with its neighboring area MST is attributed with computing motion estimates. It is not surprising that motion processing is essential, since it allows us to perform many important tasks, including depth perception, tracking of moving objects, and so on. Humans are extremely good at judging velocities of approaching objects and in discriminating opponent velocities [5, 11]. A significant amount of neural activity is devoted to motion processing. Given that the HVS is sensitive to motion, it is imperative that objective measures of video quality take motion into consideration.

Even though we have made progress in understanding the HVS, there is still a lot to be done. Indeed, some researchers have claimed that we have understood only a significantly small portion of the primary visual cortex [12]. Each of the above-mentioned areas is an active field of research. Interested readers are directed to Refs [5] and [7] for good overviews and Ref. [13] for an engineering perspective toward understanding and analyzing the HVS.

Now that we have looked at the basics of human visual processing, let us list out some peculiarities of the HVS. These are relevant for IQA/VQA, since many of these peculiarities govern the discernibility of distortions and, hence, of quality.

Light adaptation refers to the property that the HVS response depends much more upon the difference in the intensity between the object and the background than upon the actual luminance levels. This allows the human to see over a very large range of intensities.

Contrast sensitivity functions (CSFs) model the decreasing sensitivity of the HVS with increasing spatial frequencies. The HVS also exhibits varying sensitivity to temporal frequencies. Generally, most models for QA assume that the spatial and temporal responses are approximately separable. A thorough modeling would involve a spatiotemporal CSF [14].

Masking refers to the property of the HVS in which the presence of a "strong" stimulus renders the weaker stimulus imperceptible. Types of masking include texture masking–where certain distortions are masked in the presence of strong texture; contrast masking–where regions with larger contrast mask regions with lower contrast; and temporal masking–where the presence of a temporal discontinuity masks the presence of some distortions.

After having described the HVS briefly, we now discuss some visual quality assessment algorithms. We broadly classify QA algorithms as (i) those based on the HVS, (ii) those that utilize a feature-based approach, and (iii) structural and information-theoretic approaches. For VQA, we also describe algorithms that utilize motion information explicitly–motion-modeling-based approaches. While the HVS-based approaches seem the best way to evaluate visual quality, our limited understanding of the HVS leads to poor HVS models, which in turn do not function well as QA algorithms. Feature-based approaches employ heuristics and extracted features are generally only tenuously related to the HVS. Structural and information-theoretic measures, on the other hand, utilize an approach based on natural scene statistics (NSS) [15], which are hypothesized to be the inverse problem to that of modeling the HVS [16]. For VQA, explicit incorporation of motion is of prime importance and motion-modeling-based approaches do just that.

19.3
Human-Visual-System-Based Models

HVS-based models for IQA/VQA generally follow a series of operations akin to those hypothesized to occur along the visual pathway in humans. The first major component of these models is a linear decomposition of the stimulus over multiple scales and orientations. Contrast sensitivity is parameterized by a CSF. Generally, the spatial CSF is modeled using a low-pass filter (since the HVS is not as sensitive to higher frequencies) and the temporal CSF is modeled using bandpass filters. Parameters for the filters are estimated from psychovisual experiments. It is generally far more easier to model the spatial and temporal CSFs separately instead of modeling a spatiotemporal CSF [17]. The spatial and temporal responses of the HVS are approximately separable [5]. Masking is another HVS property that is taken into account for IQA. A good overview of HVS-based models for IQA/VQA can be found in Ref. [18].

19.3.1
Visual Difference Predictor (VDP)

Visual difference predictor (VDP) first applies a point nonlinearity to the images to model the fact that visual sensitivity and perception of lightness are nonlinear functions of luminance, followed by a CSF [19]. A modified version of the cortex transform [20] is then used to model the initial stage of the human *detection mechanisms*, followed by masking. To account for the fact that the probability of detection increases with an increase in stimulus contrast, VDP then applies a psychometric function, followed by a probability summation.

19.3.2
Visual Discrimination Model (VDM)

The Sarnoff visual discrimination model (VDM), which was later modified to the popular Sarnoff JND metric for video [21], was proposed by Lubin [22]. A PSF is first applied to the images, followed by a modeling of the retinal cone sampling. A Laplacian pyramid performs a decomposition of the signal and a contrast energy measure is computed. This measure is then processed through a masking function and a just-noticeable-difference (JND) distance measure is computed to produce the quality index.

19.3.3
Teo and Heeger Model

Teo and Heeger proposed a model for IQA based on the HVS [23]. The model performs a linear decomposition of the reference and test images using a hex-quadrature mirror filter (QMF), and then squares each coefficient at the output. Contrast normalization is accomplished by computing:

$$R^\theta = k \frac{(A^\theta)^2}{\sum_\phi (A^\phi)^2 + \sigma^2} \tag{19.3}$$

where A^θ is a coefficient at the output of the linear transform at orientation θ and k and σ^2 are the scaling and saturation constants. ϕ sums over all the possible orientations of the linear transform, thus performing a normalization. The final error measure is then the vector distance between the responses of the test and reference images.

19.3.4
Visual Signal-to-Noise Ratio (VSNR)

Proposed by researchers at Cornell, visual signal-to-noise ratio (VSNR), which aims to evaluate the effect of *suprathreshold* distortion, uses parameters for the HVS model derived from experiments where the stimulus was an actual image as against sinusoidal gratings or Gabor patches [24]. Many arguments that support the use of natural images/videos for estimating HVS parameters are enlisted in Ref. [13]. VSNR first computes a difference image from the reference and distorted images. This difference image is then subjected to a discrete wavelet transform. Within each sub-band, VSNR then computes the visibility of distortions, by comparing the contrast of the distortion to the detection threshold, and then computes the RMS contrast of the error signal (d_{pc}). Finally, using a strategy inspired from what is termed as *global precedence* in the HVS, VSNR computes a global precedence preserving contrast (d_{gp}). The final index is a linear combination of d_{pc} and d_{gp}.

19.3.5
Digital Video Quality Metric (DVQ)

Watson *et al.* proposed the digital video quality metric (DVQ), which evaluates visual quality in the discrete cosine transform (DCT) domain [25]. We note that even though the algorithm is labeled as a '*metric*', it is not a metric in the true sense of the word. We continue its usage in this chapter for this and other metrics; however, appropriate use of terminology for IQA/VQA metrics is essential (see Ref. [26] for relevant discussion). DVQ metric evaluates the quality in the YOZ opponent color space [27]. We note that this space is unusual in the sense that most researchers operate in the YUV color space. However, the authors propose arguments for their choice [25]. An 8×8 block DCT is then performed on the reference and test videos. The ratio of the DCT (AC) amplitude to the DC amplitude is computed as the local contrast, which is then converted into JNDs using thresholds derived from a small human study, followed by contrast masking. The error scores (which can be viewed as quality scores) are then computed using a Minkowski formulation.

19.3.6
Moving Picture Quality Metric (MPQM)

The metric proposed by Van den Branden Lambrecht [28] uses Gabor filters for spatial decomposition, two temporal mechanisms, and a spatiotemporal CSF. It also models a simple intrachannel contrast masking. One difference in MPQM is the use of segmentation to identify regions–uniform areas, contours, and textures–within an image and the error scores in each of these regions is pooled separately. An elementary evaluation of the metric was done to demonstrate its performance.

19.3.7
Scalable Wavelet-Based Distortion Metric for VQA

This VQA algorithm filters the reference and test videos using a low-pass filter [29]. The Haar wavelet transform is then used to perform a spatial frequency decomposition. A subset of these coefficients is selected for distortion measurement. This model utilizes a CSF for weighting as well as masking. The reason this metric differs from other HVS-based metrics is that the parameters used for the CSF and masking function are derived from human responses to natural videos as against sinusoidal gratings as is generally the case (as in VSNR [24]). The algorithm then computes a quality score for the distorted video using the differences in the responses from the reference and distorted videos.

19.4
Feature-Based Models

Feature-based models generally extract features from images or videos that are deemed to be of importance in visual quality assessment. For example, some algorithms extract the presence of edges and the relative edge strength using simple edge filters or a Fourier analysis. Some of the models extract elementary motion features for VQA. The primary argument against such models is the fact that the extracted features may not be correlated with the HVS. Some other arguments include the use of unmotivated thresholds and pooling strategies. However, as we see later, some of these models perform well in terms of correlation with human perception.

19.4.1
A Distortion Measure Based on Human Visual Sensitivity

Karunasekera and Kingsbury filter the difference (error) image (computed from the reference and distorted images) using direction filters—vertical, horizontal, and two diagonal orientations to form oriented edge images [30]. The outputs of each of these orientations are processed independently, and then pooled to produce a distortion measure. Masking computation based on an activity measure and brightness is undertaken. A nonlinearity is then applied to obtain the directional error. This model proposed in Ref. [30] is an extension of the authors' previous blocking measure proposed in Ref. [31] and incorporates ringing and blurring measures using the above-described process.

19.4.2
Singular Value Decomposition and Quality

Singular value decomposition (SVD) is a well-known tool from linear algebra which has been used for a host of multimedia applications. In Ref. [32], the authors use

SVD for IQA. SVD is applied on 8 × 8 blocks in the reference and test images and then the distortion per block is computed as $D_i = \sqrt{\sum_{i=1}^{n}(s_i - \widehat{s_i})}$, where s_i and $\widehat{s_i}$ are the singular values for block i from the reference and test images. The final quality score is computed as follows: $M - SVD = \sum_{i \in all\_blocks} |D_i - D_{mid}|$, where D_{mid} represents the median of the block distortions. Even though the authors claim that the algorithm performs well, its relation to the HVS is unclear as is the significance of the SVD for IQA.

19.4.3
Curvature-Based Image Quality Assessment

The IQA index, proposed in Ref. [33], first uses the discrete wavelet transform to decompose the reference and test images. In each sub-band, mean surface curvature maps are obtained as follows:

$$H = \frac{I_{uu} + I_{vv} + I_{uu}I_v^2 + I_{vv}I_u^2 - 2I_u I_v I_{uv}}{2(1 + I_u^2 + I_v^2)^{3/2}} \tag{19.4}$$

where I_{uu}, I_{vv}, I_u, and I_v are the partial derivatives of the image I. The correlation coefficient between the curvatures of the original and distorted images is then evaluated. These correlation coefficients are then collapsed across the sub-bands to produce a quality score.

19.4.4
Perceptual Video Quality Metric (PVQM)

Proposed by Swisscom/KPN research, the perceptual video quality metric (PVQM) extracts three features from the videos under consideration [34]. *Edginess* is essentially indicated by a difference between the (dialated) edges of the reference and distorted frames computed using an approximation to the local gradient. The edginess indicator is supposed to reflect the loss in spatial detail. The *temporal indicator* for a frame is the correlation coefficient between adjacent frames subtracted from 1. Finally, a *chrominance indicator* based on color saturation is computed. Each of these indicators is pooled separately across the video and then a weighted sum of these pooled values is utilized as the quality measure. PVQM utilizes a large number of thresholds for each of the indicators as well as for the final pooling. Some of these thresholds are claimed to be based on psychovisual evaluation.

19.4.5
Video Quality Metric (VQM)

Pinson and Wolf proposed the VQM [35], which was the top performer in the video quality experts group (VQEG) phase-II studies [36]. Owing to its performance, VQM has also been standardized by the American National Standards Institute and the international telecommunications union (ITU) has included VQM as a normative measure for digital cable television systems [37]. VQM, which was trained on the

VQEG phase-I dataset [38], first performs a spatiotemporal alignment of the videos followed by gain and offset correction. This is followed by the extraction of a set of features that are thresholded. The final quality score is computed as a weighted sum of these features. The computed features include a feature to describe the loss of spatial detail; one to describe a shift in the orientation of the edges; one to describe the spread of chrominance components; as well as one to describe severe color impairments. VQM also includes elementary motion information in the form of the difference between frames and a quality improvement feature that accounts for improvements arising from (for example) sharpness operations. The quality score ranges from 0 to 1.

19.4.6
Temporal Variations of Spatial-Distortion-Based VQA

Ninassi *et al.* proposed a VQA index recently [39]. They model temporal distortions such as mosquito noise, flickering, and jerkiness as an evolution of spatial distortions over time. A spatiotemporal tube consisting of a spatiotemporal chunk of the video computed from motion vector information is created, which is then evaluated for its spatial distortion. The spatial distortion is computed using the WQA quality index [40]. A temporal filtering of the spatial distortion is then undertaken, followed by a measurement of the temporal variation of the distortion. The quality scores are then pooled across the video to produce the final quality index.

19.4.7
Temporal Trajectory Aware Quality Measure

One of the few algorithms that utilizes motion information is the one proposed by Barkowsky *et al.*–the Tetra VQM [41]. Information from a block-based motion estimation algorithm for each (heuristically determined) shot [42] is utilized for temporal trajectory evaluation of the distorted video. This information is logged in a temporal information buffer, which is followed by a temporal visibility mask. Spatial distortion is evaluated by MSE–the authors claim that this is for reducing the complexity of the algorithm. A spatial–temporal distortion map is then created. The pooling stage first applies a mask to reflect human foveation and then a temporal summation is performed. The proposed algorithm also models the frame rate, pauses, and skips. All of these indicators are then combined to form a quality score for the video.

19.5
Structural and Information-Theoretic Models

A structural approach for IQA was proposed by Wang and Bovik [43]. This approach was later modified for VQA [44]. These models are based on the premise that the HVS extracts (and is, hence, sensitive to) structural information in the stimulus.

Loss of structural information is, hence, related to perceptual loss of quality. Information-theoretic models utilize NSS in order to quantify loss of information in the wavelet domain. Recent research indicates how these two metrics are closely related to each other and to mutual masking hypothesized to occur in the HVS [45].

19.5.1
Single-Scale Structural Similarity Index (SS-SSIM)

For two image patches drawn from the same location of the reference and distorted images, $\mathbf{x} = \{x_i | i = 1, 2, \ldots, N\}$, and $\mathbf{y} = \{y_i | i = 1, 2, \ldots, N\}$, respectively, single-scale structural similarity index SS-SSIM computes three terms, namely, luminance, contrast, and structure as follows [43]:

$$l(\mathbf{x}, \mathbf{y}) = \frac{2\mu_x \mu_y + C_1}{\mu_x^2 + \mu_y^2 + C_1} \tag{19.5}$$

$$c(\mathbf{x}, \mathbf{y}) = \frac{2\sigma_x \sigma_y + C_2}{\sigma_x^2 + \sigma_y^2 + C_2} \tag{19.6}$$

$$s(\mathbf{x}, \mathbf{y}) = \frac{\sigma_{xy} + C_3}{\sigma_x \sigma_y + C_3} \tag{19.7}$$

where C_1, C_2, and C_3 are small constants. The constants C_1, C_2, and C_3 ($C_3 = C_2/2$) are included to prevent instabilities from arising when the denominator tends to zero. μ_x, μ_y, σ_x^2, σ_y^2, and σ_{xy} are the means of $\mathbf{x}$, $\mathbf{y}$, the variances of $\mathbf{x}$, $\mathbf{y}$, and the covariance between $\mathbf{x}$ and $\mathbf{y}$, respectively, computed using a sliding window approach. The window used is a 11×11 circular-symmetric Gaussian weighting function $w = \{w_i | i = 1, 2, \ldots, N\}$, with standard deviation of 1.5 samples, normalized to sum to unity ($\sum_{i=1}^{N} w_i = 1$).

Finally, the SSIM index between signal $\mathbf{x}$ and $\mathbf{y}$ is defined as follows:

$$\text{SSIM}(\mathbf{x}, \mathbf{y}) = \frac{(2\mu_x \mu_y + C_1)(2\sigma_{xy} + C_2)}{(\mu_x^2 + \mu_y^2 + C_1)(\sigma_x^2 + \sigma_y^2 + C_2)} \tag{19.8}$$

This index produces a map of quality scores having the same dimensions as that of the image. Generally, the mean of the scores is utilized as the quality index for the image. The SSIM index is an extension of the universal quality index (UQI) [46], which is the SSIM index with $C_1 = C_2 = 0$.

19.5.2
Multiscale Structural Similarity Index (MS-SSIM)

Images are naturally multiscales. Further, the perception of image quality depends upon a host of scale-related factors. To evaluate image quality at multiple resolutions, in [47], the multiscale structural similarity index (MS-SSIM) index was proposed.

In MS-SSIM, quality assessment is accomplished over multiple scales of the reference and distorted image patches (the signals defined as $\mathbf{x}$ and $\mathbf{y}$ in the previous discussion on SS-SSIM) by iteratively low-pass filtering and downsampling the signals by a factor of 2. The original image scale is indexed as 1, the first downsampled version is indexed as 2, and so on. The highest scale M is obtained after $M-1$ iterations.

At each scale j, the contrast comparison (Eq. (19.6)) and the structure comparison (Eq. (19.7)) terms are calculated and denoted as $c_j(\mathbf{x},\mathbf{y})$ and $s_j(\mathbf{x},\mathbf{y})$, respectively. The luminance comparison (Eq. (19.5)) term is computed only at scale M and is denoted as $l_M(\mathbf{x},\mathbf{y})$. The overall SSIM evaluation is obtained by combining the measurement over scales:

$$\text{SSIM}(\mathbf{x},\mathbf{y}) = [l_M(\mathbf{x},\mathbf{y})]^{\alpha_M} \cdot \prod_{j=1}^{M} [c_j(\mathbf{x},\mathbf{y})]^{\beta_j} \cdot [s_j(\mathbf{x},\mathbf{y})]^{\gamma_j} \quad (19.9)$$

The highest scale used here is $M=5$.

The exponents α_j, β_j, and γ_j are selected such that $\alpha_j = \beta_j = \gamma_j$, and $\sum_{j=1}^{M} \gamma_j = 1$.

19.5.3
SSIM Variants

SS-SSIM proposed in [43] was extended to the complex wavelet domain in Ref. [48] and the proposed index was labeled as the complex wavelet structural similarity index (CW-SSIM). CW-SSIM is computed as

$$S(c_x, c_y) = \frac{2|\sum_{i=1}^{N} c_{x,i} c_{y,i}^*| + K}{\sum_{i=1}^{N} |c_{x,i}|^2 + \sum_{i=1}^{N} |c_{y,i}|^2 + K} \quad (19.10)$$

where $c_x = \{c_{x,i} | i = 1, \ldots, N\}$ and $c_y = \{c_{y,i} | i = 1, \ldots, N\}$ are two sets of complex wavelet coefficients drawn from the same location in the reference and test images. CW-SSIM has been used for face recognition [49] and for a host of other applications [50].

Other variants of SSIM that modify the pooling strategy from the mean to fixation-based pooling [51], percentile pooling [6], and information-content-based pooling [52] have also been proposed. Modifications of SSIM include a gradient-based approach [53] and another technique based on pooling three perceptually important parameters [54]. In [55], a classification based on the type of region is undertaken, after applying SSIM on the images. A weighted sum of the SSIM scores from each of the regions is combined to produce a score for the image.

19.5.4
Visual Information Fidelity (VIF)

It is known that when images are filtered using oriented bandpass filters (e.g., a wavelet transform), the distributions of the resulting (marginal) coefficients are highly peaked around zeros and possess heavy tails [15]. Such statistical descriptions of natural scenes are labeled as natural scene statistics (NSS) and this has been an active area of research. Visual information fidelity (VIF) [56] utilizes the Gaussian scale mixture (GSM) model for wavelet NSS [57]. VIF first performs a scale-space-orientation wavelet decomposition using the steerable pyramid [58] and models each sub-band in the source as $C = S \cdot U$, where S is a random field (RF) of scalars and U is a Gaussian vector RF. The distortion model is $D = GC + v$, where G is a scalar gain field and v is additive Gaussian noise RF. VIF then assumes that the distorted and source images pass through the HVS and the HVS uncertainty is modeled as *visual noise*: N and N' for the source and distorted image, respectively, where N and N' are zero-mean uncorrelated multivariate Gaussians. It then computes $E = C + N$ and $F = D + N'$. The VIF criterion is then evaluated as

$$\text{VIF} = \frac{\sum_{j \in \text{allsub-bands}} I(C^j; F^j | s^j)}{\sum_{j \in \text{allsub-bands}} I(C^j; E^j | s^j)}$$

where $I(X; Y|Z)$ is the conditional mutual information between X and Y, conditioned on Z; s^j is a realization of S^j for a particular image; and the index j runs through all the sub-bands in the decomposed image.

19.5.5
Structural Similarity for VQA

SS-SSIM defined for images was applied on a frame-by-frame basis on videos for VQA and was shown to perform well [44]. Realizing the importance of motion information, the authors [44] also utilized a simple motion-based weighting strategy that was used to weight the spatial SSIM scores. Video SSIM, the resulting algorithm, was shown to perform well [44].

19.5.6
Video VIF

VIF was extended to VQA in Ref. [59]. The authors justified the use of VIF for video by first motivating the GSM model for the spatiotemporal NSS. The model for video VIF then is essentially the same as that for VIF with the exception being the application of the model to the spatiotemporal domain as opposed to the spatial domain.

19.6
Motion-Modeling-Based Algorithms

As described before, the areas MT and MST in the HVS are responsible for motion processing. Given that the HVS is sensitive to motion, it is imperative that objective measures of quality take motion into consideration.

An observant reader would have observed that the models for VQA described so far were essentially IQA algorithms applied on a frame-by-frame basis. Some of these algorithms utilized motion information; however, the incorporation was ad hoc. In this section, we describe some recent VQA algorithms that incorporate motion information. We believe that the importance of spatiotemporal quality assessment as against a spatial-only technique for VQA cannot be understated.

19.6.1
Speed-Weighted Structural Similarity Index (SW-SSIM)

Speed-weighted structural similarity index (SW-SSIM) first computes SS-SSIM at each pixel location in the video using SS-SSIM in the spatial domain [60]. Motion estimates are then obtained using Black and Anandan's optical flow computation algorithm [61]. Using a histogram-based approach for each frame, a global motion vector for that frame is identified. Relative motion is then extracted as the difference between the absolute motion vectors (obtained from optical flow) and the computed global motion vectors. Then, a weight for each pixel is computed. This weight is a function of the computed relative and global motion and the stimulus contrast. The weight so obtained is then used to weight the SS-SSIM scores. The weighted scores are then pooled across the video and normalized to produce the quality index for the video. The described weighting scheme was inspired by the experiments into human visual speed perception done by Stocker and Simoncelli [62].

19.6.2
Motion-Based Video Integrity Evaluation (MOVIE)

Motion-based video integrity evaluation (MOVIE) first decomposes the reference and test videos using a multiscale spatiotemporal Gabor filter bank [63]. Spatial quality assessment is conducted using a technique similar to MS-SSIM. A modified version of the Jepson and Fleet algorithm for optical flow [64] is used to produce motion estimates. The same set of Gabor filters is utilized for optical flow computation and quality assessment. Translational motion in the spatiotemporal domain manifests itself as a plane in the frequency domain [65]. MOVIE assigns positive excitatory weights to the response of those filters that lie close to the spectra plane defined by the computed motion vectors and negative inhibitory weights to those filter responses that lie farther away from the spectral plane. Such weighting results in a strong response if the motion in the test and reference video coincide and a weak response is produced if the test video has motion deviant from that in the reference video. The mean-squared error between the responses from the test and reference filter banks then provides the temporal quality estimate. The final MOVIE index is the product of the spatial and temporal quality indices – a technique inspired from the spatial and temporal separability of the HVS.

We have described a handful of algorithms in this chapter. There exist many other algorithms for visual quality assessment that are not covered here. The reader is referred to Refs [2, 18, 66–69] for reviews of other such approaches to FR QA. Further, we have covered only FR QA algorithms – partly due to the maturity of this field. Algorithms that do not require a reference stimulus for QA are called no-reference (NR) algorithms [2]. Some examples of NR IQA algorithms are included in Refs [70–75]. Some NR VQA algorithms can be found in Refs [76–78].

There also exists another class of algorithms, reduced reference (RR) algorithms, in which the distorted stimulus contains some additional information about the pristine reference [2]. Recent RR QA algorithms can be found in Ref. [79–81].

Having described visual quality assessment algorithms, let us now move on to analyzing how performance of these algorithms can be computed.

19.7
Performance Evaluation and Validation

Now that we have gone through the nuts and bolts of a set of algorithms, the question of evaluating the performance of the algorithm follows. A demonstration that the ranking of a handful of videos/images produced by the algorithm scores matches human perception alone cannot be an accurate measure of algorithm performance. To compare algorithm performance, a common testing ground that is publicly available must be used. This testing ground, which we refer to as a *dataset*, must contain a large number of images/videos that have undergone all possible kinds of distortions. For example, for IQA, distortions should include Gaussian noise, blur, distortion due to packet loss, distortion due to compression, and so on. Further, each image in the dataset must be rated by a sizable number of human observers in order to produce the subjective quality score for the image/video. This subjective assessment of quality is what forms the basis of performance evaluation for IQA/VQA algorithms. The ITU has listed procedures and recommendations on how such subjective assessment is to be conducted [82]. After a group of human observers have rated the stimuli in the dataset, an MOS that is the mean of the ratings given by the observers (which is computed after subject rejection [82]) is computed for each of the stimuli. The MOS is representative of the perceived quality of the stimulus. To evaluate algorithm performance, each stimulus in the dataset is evaluated by the algorithm and receives an objective score. The objective and subjective scores are then correlated using statistical measures of correlation. The higher the correlation, the better the performance of the algorithm.

Traditional measures of correlation that have been used to evaluate performance of visual quality assessment algorithms include the following: Spearman's rank ordered correlation coefficient (SROCC), linear (Pearson's) correlation coefficient (LCC), root-mean-squared error (RMSE), and outlier ratio (OR) [38]. SROCC is a measure of the prediction monotonicity and can directly be computed using algorithmic scores and MOS. OR is a measure of the prediction consistency. LCC and RMSE measure the prediction accuracy and are generally computed after transforming the algorithmic scores using a logistic function. This is because LCC assumes that the data under test are linearly related and tries to quantify this linear relationship. However, algorithmic quality scores may be nonlinearly related to subjective MOS. After transforming the objective algorithmic scores using the logistic, we eliminate this nonlinearity, which allows us to evaluate LCC and RMSE. The logistic functions generally used are those proposed in Refs [38] and [83]. The parameters of the logistic function are those that provide the best fit between the objective and subjective scores. In general, an algorithm with higher values (close to 1) for SROCC and LCC and lower values (close to 0) for RMSE and OR is considered to be a good visual quality assessment algorithm. One final statistical measure is that of statistical significance. This measure indicates whether, given a particular dataset, the obtained differences in correlation between algorithms is statistically significant. The F-statistic and ANOVA [84] have been utilized in the past for this purpose [38, 83].

Even though the above-described procedure is one that is currently adopted, there have been efforts directed at improving the performance evaluation of visual quality assessment algorithms. For example, a recently proposed approach by Wang and Simoncelli calls for the use of a technique called *maximum differentiation competition (MAD)* [85]. Images and video sequences live in a high-dimensional signal space, where the dimension is equal to the number of pixels. However, only hundreds or at most thousands of images can be tested in a practical subjective test, and thus their distribution is extremely sparse in the space of all possible images. Examining only a single sample from each orthant of an N-dimensional space would require a total of 2^N samples, an unimaginably large number for an image signal space with dimensionality on the order of thousands to millions. In Ref. [85], the authors propose an efficient methodology—MAD, where test images were synthesized to optimally distinguish competing perceptual quality models. Although this approach cannot fully prove the validity of a quality model, it offers an optimal means of selecting test images that are most likely to falsify it. As such, it provides a useful complementary approach to the traditional performance evaluation method. Other approaches based on maximum likelihood difference scaling (MLDS) [86] have been proposed as well [87, 88].

Now that we have an idea of the general procedure to evaluate algorithm performance, let us discuss the databases that are used for these purposes. For IQA, one the most popular databases, which is currently the de facto standard for all IQA algorithms, is the Laboratory for Image and Video Engineering (LIVE) IQA database [83]. The LIVE image database was created by researchers at the University of Texas at Austin and consists of 29 reference images. Each reference image was distorted using five different distortion processes (compression, noise, etc.) and with a varying level of severity for each distortion type. A subjective study was conducted as outlined earlier and each image was viewed by approximately 29 subjects. A host of leading IQA algorithms were then evaluated for their performance in Ref. [83]. MS-SSIM and VIF were shown to have correlated extremely well with human perception [83]. To demonstrate how the structural and information-theoretic approaches compare to the often-criticized PSNR, in Table 19.1 we (partially) reproduce the SROCC values between these algorithms and DMOS on the LIVE IQA database. The reader is referred to Ref. [83] for other correlation measures and statistical analysis for these and other algorithms.

Other IQA datasets include the one from researchers at Cornell [89], the IVC dataset [90], the TAMPERE image dataset [91], and the one from researchers at Toyama University [92]. Readers interested in a package that encompasses some of the discussed algorithms are referred to Ref. [93].

Table 19.1 Spearman's rank ordered correlation coefficient (SROCC) on the LIVE IQA database.

	J2k#1	J2k#2	JPG#1	JPG#2	WN	Gblur	FF
PSNR	0.9263	0.8549	0.8779	0.7708	0.9854	0.7823	0.8907
MS-SSIM	0.9645	0.9648	0.9702	0.9454	0.9805	0.9519	0.9395
VIF	0.9721	0.9719	0.9699	0.9439	0.9282	0.9706	0.9649

J2k, JPEG2000 compression; JPG, JPEG compression; WN, white noise; Gblur, Gaussian blur; and FF, Rayleigh fast-fading channel.

For VQA, the largest known publicly available dataset is that from the VQEG and is labeled as the VQEG FRTV phase-I dataset[1] [38]. Even though the VQEG has conducted other studies, none of the data has been made publicly available [36, 94]. The VQEG dataset consists of a total of 320 distorted videos created by distorting 20 reference video sequences. Even though the VQEG dataset has been widely used, it is not without its flaws. The VQEG dataset (and the associated study) was specifically aimed at television and, hence, contains interlaced videos. Deinterlacing algorithms used before VQA may add distortions of their own, thereby possibly reducing the correlation of the algorithmic scores with subjective DMOS. Further, the dataset consists of nonnatural videos, and many VQA algorithms that rely on the assumption of natural scenes face a distinct disadvantage[2] Again, the perceptual separation of the dataset is such that humans and algorithms have difficulty in making consistent judgments. It is to alleviate many such problems and to provide for a common publicly available testing bed for future VQA algorithms that the researchers at LIVE have developed the LIVE VQA and LIVE wireless VQA databases [95, 96]. These databases have recently been made publicly available at no cost to researchers in order to further the field of VQA.

Before we conclude this chapter, we would like to stress on the importance of utilizing a common publicly available test set for evaluating IQA and VQA algorithms. Even though a host of new IQA/VQA algorithms have been proposed, comparing algorithmic performance makes no sense if one is unable to see relative algorithmic performance. The LIVE image dataset and the associated scores for algorithms, for example, allow for an objective comparison of IQA algorithms. The publicly available dataset must encompass a range of distortions and perceptual distortion levels so as to accurately test the algorithm. Reporting results on small and/or private test sets fails to prove the performance of the proposed algorithm.

1) We refer to this as the VQEG dataset henceforth.
2) This is not only for those algorithms that explicitly utilize NSS models, such as VIF, but also for models such as SSIM that have been shown to be equivalent to NSS-based models [45, 64].

19.8 Conclusion

In this chapter, we discussed HVS briefly and studied some HVS-based algorithms for IQA/VQA. We then looked at some feature-based approaches and moved on to describe structural and information-theoretic measures for IQA/VQA. We then stressed the importance of motion modeling for VQA and described recent algorithms that incorporate motion information. This was followed by a description of performance evaluation techniques for IQA/VQA and relevant databases for this purpose.

The area of visual quality assessment has been an active area of research for a long period of time. With the advent of structural approaches (and its variants) for IQA, the field of FR IQA seems to have found a winner [83]. This is mainly because the simplicity of the structural approach, coupled with its performance, makes it an algorithm of choice in practical systems. Further simplifications of the index make it even more attractive [97]. Even though authors have tried to improve the performance of these approaches, the improvements have been minimal and in most cases do not justify the additional complexity [6, 51, 52]. What remains unanswered at this stage is the relationship between the structural approach and the HVS. Even though some researchers have demonstrated a link [45], further research is required for understanding the index.

FR VQA is a far tougher problem to solve, since incorporation of motion information and temporal quality assessment are fields still in the nascent stage. Even though the structural approach does well on the VQEG dataset, improvements can be achieved by using appropriate motion-modeling-based techniques [60].

The field of RR/NR QA is one that has tremendous potential for research. RR/NR algorithms are particularly useful in applications where quality evaluation is of utmost importance, but the reference stimulus is unavailable–for example, in systems that monitor quality of service (QoS) at the end user. Further, FR/NR/RR algorithms that not only perform well but are computationally efficient need to be developed for real-time quality monitoring systems.

Having reviewed some algorithms in this chapter, we hope that the reader has obtained a general idea of approaches that are utilized in the design and analysis of visual quality assessment systems and that his/her interest in this broad field, which involves researchers from a host of areas ranging from engineering to psychology, has been piqued.

References

1. International Standards Organization (ISO). http://www.iso.org/iso/home.htm.
2. Wang, Z. and Bovik, A.C. (2006) *Modern Image Quality Assessment*, Morgan & Claypool Publishers.
3. Girod, B. (1993) What's wrong with mean-squared error? in *Digital Images and Human Vision* (ed. A.B. Watson), MIT Press Cambridge, MA, USA pp. 207–220.
4. Wang, Z. and Bovik, A.C. (2009) Mean squared error: Love it or leave it? - a new look at fidelity measures. *IEEE Signal Process. Mag.*, **26**, 98–117.

5. Sekuler, R. and Blake, R. (1988) *Perception*, Random House USA Inc.
6. Moorthy, A.K. and Bovik, A.C. (2009) Visual importance pooling for image quality assessment. *IEEE J. Sel. Top. Signal Process.*, **3**, 193–201.
7. Hubel, D., Wensveen, J., and Wick, B. (1988) Eye, brain, and vision.
8. Mante, V., Bonin, V., and Carandini, M. (2008) Functional mechanisms shaping lateral geniculate responses to artificial and natural stimuli. *Neuron*, **58**, 625–638.
9. Born, R. and Bradley, D. (2005) Structure and function of visual area MT. *Annu. Rev. Neurosci.*, **28**, 157–189.
10. Rust, N.C. et al. (2006) How MT cells analyze the motion of visual patterns. *Nat. Neurosci.*, **9**, 1421–1431.
11. Wandell, B. (1995) *Foundations of vision*, Sinauer Associates.
12. Olshausen, B. and Field, D. (2005) How close are we to understanding V1? *Neural Comput.*, **17**, 1665–1699.
13. Carandini, M. et al. (2005) Do we know what the early visual system does? *J. Neurosci.*, **25**, 10577–10597.
14. Daly, S. (1998) Engineering observations from spatiovelocity and spatiotemporal visual models. Proceedings of SPIE 3299, San Jose, CA, pp. 180–191.
15. Simoncelli, E. and Olshausen, B. (2001) Natural image statistics and neural representation. *Ann. Rev. Neurosci.*, **24**, 1193–1216.
16. Sheikh, H.R., Bovik, A.C., and De Veciana, G. (2005) An information fidelity criterion for image quality assessment using natural scene statistics. *IEEE Trans. Image process.*, **14**, 2117–2128.
17. Kelly, D.H. (1983) Spatiotemporal variation of chromatic and achromatic contrast thresholds. *J. Opt. Soc. Am.*, **73** (6), 742–750.
18. Nadenau, M. et al. (2000) Human vision models for perceptually optimized image processing-a review. Proceedings of the IEEE 2000.
19. Daly, S. (1992) Visible differences predictor: an algorithm for the assessment of image fidelity. Proceedings of SPIE, 1666, San Jose, CA, p. 2.
20. Watson, A.B. (1987) The cortex transform- Rapid computation of simulated neural images. *Comput. Vis. Graph. Image Process.*, **39**, 311–327.
21. Lubin, J. and Fibush, D. (1997) Sarnoff JND vision model. T1A1, **5**, 97–612.
22. Lubin, J. (1995) A visual discrimination model for imaging system design and evaluation, in *Vision Models for Target Detection and Recognition: In Memory of Arthur Menendez*, ed. E. Pelip, World Scientific, p. 245.
23. Teo, P. and Heeger, D. (1994) Perceptual image distortion. *SID International Symposium Digest of Technical Papers*, **25**, 209–209.
24. Chandler, D.M. and Hemami, S.S. (2007) VSNR: a wavelet-based visual signal-to-noise ratio for natural images. *IEEE Trans. Image Process.*, **16**, 2284–2298.
25. Watson, A., Hu, J., and McGowan, J. III. (2001) Digital video quality metric based on human vision. *J. Electron. Imaging*, **10**, 20.
26. Bovik, A. (2009) Meditations on visual quality. IEEE COMPSOC E-LETTER, Technology Advances.
27. Peterson, H., Ahumada, A. Jr, and Watson, A. An improved detection model for DCT coefficient quantization, Human Vision, Visual Processing, and Digital Display IV, San Jose, CA, pp. 191–201.
28. Van den Branden Lambrecht, C. and Verscheure, O. (1996) Perceptual quality measure using a spatiotemporal model of the human visual system. Proceedings of SPIE, San Jose, CA, pp. 450–461.
29. Masry, M., Hemami, S., and Sermadevi, Y. (2006) A scalable wavelet-based video distortion metric and applications. *IEEE Trans. circuits Syst. Video Technol.*, **16**, 260–273.
30. Karunasekera, S. and Kingsbury, N. (1994) A distortion measure for image artifacts based on human visual sensitivity. IEEE International Conference on Acoustics, Speech, and Signal Processing, 1994. Adelaide, South Australia, ICASSP-94.
31. Karunasekera, S. and Kingsbury, N. (1995) A distortion measure for blocking

artifacts in images based on human visual sensitivity. *IEEE Trans. Image process.*, **4**, 713–724.

32. Shnayderman, A., Gusev, A., and Eskicioglu, A. (2003) A multidimensional image quality measure using singular value decomposition, 5294, San Antonio, TX, pp. 82–92.

33. Yao, S. *et al.* (2007) Image quality measure using curvature similarity. IEEE International Conference on Image Processing, San Antonio, TX, pp. 437–440.

34. Hekstra, A. *et al.* (2002) PVQM-a perceptual video quality measure. *Signal Process Image Commun*, **17**, 781–798.

35. Pinson, M.H. and Wolf, S. (2004) A new standardized method for objectively measuring video quality. *IEEE Trans. Broadcast.*, 312–313.

36. (VQEG), V.Q.E.G. (2003) Final report from the video quality experts group on the validation of objective quality metrics for video quality assessment phase II, http://www.its.bldrdoc.gov/vqeg/projects/frtv_phaseII.

37. (2004) Objective perceptual video quality measurement techniques for digital cable television in the presence of a full reference, International Telecommunications Union (ITU), Std. ITU-T Rec. J. 144.

38. Video Quality Experts Group (VQEG). (2000) Final report from the video quality experts group on the validation of objective quality metrics for video quality assessment phase I, http://www.its.bldrdoc.gov/vqeg/projects/frtv_phaseI.

39. Ninassi, A. *et al.* (2009) Considering temporal variations of spatial visual distortions in video quality assessment. *IEEE J. Sel. Top. Signal Process.*, **3**, 253–265.

40. Ninassi, A. *et al.* (2008) On the performance of human visual system based image quality assessment metric using wavelet domain. Proceedings SPIE Human Vision and Electronic Imaging XIII, San Jose, CA, 6806.

41. Barkowsky, M. *et al.* (2009) Temporal trajectory aware video quality measure. *IEEE J. Sel. Top. Signal Process.*, **3**, 266–279.

42. Cotsaces, C., Nikolaidis, N., and Pitas, I. (2006) Video shot detection and condensed representation. a review. *IEEE Signal Process. Mag.*, **23**, 28–37.

43. Wang, Z. *et al.* (2004) Image quality assessment: From error measurement to structural similarity. *IEEE Signal Process. Lett.*, **13**, 600–612.

44. Wang, Z., Lu, L., and Bovik, A.C. (2004) Video quality assessment based on structural distortion measurement. *Signal Process. Image commun.*, 121–132.

45. Seshadrinathan, K. and Bovik, A.C. (2008) Unifying analysis of full reference image quality assessment. 15th IEEE International Conference on Image Processing, 2008. ICIP 2008, San Diego, CA, pp. 1200–1203.

46. Wang, Z. and Bovik, A.C. (2002) A universal image quality index. *IEEE Signal Process. Lett.*, **9**, 81–84.

47. Wang, Z., Simoncelli, E.P., and Bovik, A.C. (2003) Multi-scale structural similarity for image quality assessment. Proceedings IEEE Asilomar Conference on Signals, Systems, and Computers, (Asilomar) Pacific Grove, CA.

48. Wang, Z. and Simoncelli, E. (2005) Translation insensitive image similarity in complex wavelet domain. IEEE International Conference on Acoustics, Speech, and Signal Processing, 2005. Proceedings (ICASSP'05), 2.

49. Gupta, S., Markey, M.K., and Bovik, A.C. (2007) Advances and challenges in 3D and 2D+ 3D human face recognition, *Pattern Recognition in Biology*, Nova Science Publishers.

50. Sampat, M.P. *et al.* (2011) Complex wavelet structural similarity: a new image similarity index. *IEEE Trans. Image Process.*, **18** (11), 2385–2401.

51. Moorthy, A.K. and Bovik, A.C. (2009) Perceptually significant spatial pooling techniques for image quality assessment. Human Vision and Electronic Imaging XIV. Proceedings of the SPIE, San Jose, CA, 7240.

52. Wang, Z. and Shang, X. (2006) Spatial pooling strategies for perceptual image quality assessment. IEEE International Conference on Image Processing.

53. Chen, G., Yang, C., and Xie, S. (2006) Gradient-based structural similarity for

image quality assessment. IEEE International Conference on Image Processing, San Jose, CA, pp. 2929–2932.

54. Li, C. and Bovik, A. (2009) Three-component weighted structural similarity index. Proceedings of SPIE, 7242, p. 72420Q.

55. Gao, X., Wang, T., and Li, J. (2005) A content-based image quality metric, *Lecture Notes in Computer Science*, Springer Verlag, LNCS 3642, 231.

56. Sheikh, H.R. and Bovik, A.C. (2006) Image information and visual quality. *IEEE Trans. Image Process.*, **15**, 430–444.

57. Wainwright, M. and Simoncelli, E. (2000) Scale mixtures of Gaussians and the statistics of natural images. *Adv. Neural Inf. Process. Syst.*, **12**, 855–861.

58. Simoncelli, E. et al. (1992) Shiftable multiscale transforms. *IEEE Trans. Inf. Theory*, **38**, 587–607.

59. Sheikh, H.R. and Bovik, A.C. (2005) A visual information fidelity approach to video quality assessment. First International Workshop on Video Processing and Quality Metrics for Conusmer Electronics, Scottsdale, AZ.

60. Wang, Z. and Li, Q. (2007) Video quality assessment using a statistical model of human visual speed perception. *J. Opt. Soc. Am.*, **24**, B61–B69.

61. Black, M.J. and Anandan, P. (1996) The robust estimation of multiple motions: Parametric and piecewise-smooth flow fields. *Comput. Vis. Image Unders.*, **63**, 75–104.

62. Stocker, A. and Simoncelli, E. (2006) Noise characteristics and prior expectations in human visual speed perception. *Nat. Neurosci.*, **9**, 578–585.

63. Seshadrinathan, K. (2008) Video quality assessment based on motion models. PhD thesis. The University of Texas at Austin.

64. Fleet, D. and Jepson, A. (1990) Computation of component image velocity from local phase information. *Int. J. Comput. Vis.*, **5**, 77–104.

65. Watson, A.B. and Ahumada, A.J. (1985) Model of human visual-motion sensing. *J. Opt. Soc. Am.*, **A2**, 322–342.

66. Ahumada, A. Jr. (1993) Computational image quality metrics: a review. *SID Digest*, **24**, 305–308.

67. Wang, Z., Sheikh, H., and Bovik, A. (2003) Objective video quality assessment, *The Handbook of Video Databases: Design and Applications*, CRC Press, pp. 1041–1078.

68. Seshadrinathan, K. et al. (2009) *Image Quality Assessment in the Essential Guide to Image Processing*, Academic Press, Chapter 20.

69. Seshadrinathan, K. and Bovik, A.C. (2009) *Video Quality Assessment in The Essential Guide to Video Processing*, Academic Press, Chapter 14.

70. Wang, Z., Bovik, A., and Evans, B. (2000) Blind measurement of blocking artifacts in images, *Proceedings IEEE International Conference on Image Processing*, vol. 3, Vancouver, Canada, Citeseer, pp. 981–984.

71. Li, X. et al. (2002) Blind image quality assessment. International Conference on Image Processing, New York.

72. Marziliano, P. et al. (2002) A no-reference perceptual blur metric. Proceedings of the International Conference on Image Processing, vol. 3, Rochester, NY, pp. 57–60.

73. Gastaldo, P. et al. (2005) Objective quality assessment of displayed images by using neural networks. *Signal Process. Image Commun.*, **20**, 643–661.

74. Sheikh, H.R., Bovik, A.C., and Cormack, L.K. (2005) No-reference quality assessment using natural scene statistics: JPEG2000. *IEEE Trans. Image Process.*, **14**, 1918–1927.

75. Gabarda, S. and Cristobal, G. (2007) Blind image quality assessment through anisotropy. *J. Opt. Soc. Am.*, **A24**, B42–B51.

76. Winkler, S., Sharma, A., and McNally, D. (2001) Perceptual video quality and blockiness metrics for multimedia streaming applications. Proceedings of the International Symposium on Wireless Personal Multimedia Communications, pp. 547–552.

77. Farias, M. and Mitra, S. (2005) No-reference video quality metric based on artifact measurements. IEEE International Conference on Image Processing, vol. 3, pp. 141–144.

78. Pastrana-Vidal, R. and Gicquel, J. (2007) A no-reference video quality metric

based on a human assessment model. Third International Workshop on Video Processing and Quality Metrics for Consumer Electronics VPQM, Scottsdale, AZ, vol. 7, pp. 25–26.
79. Wang, Z. et al. (2006) Quality-aware images. IEEE Trans. Image Process., **15**, 1680–1689.
80. Hiremath, B., Li, Q., and Wang, Z. (2007) Quality-aware video, 3.
81. Li, Q. and Wang, Z. (2009) Reduced-reference image quality assessment using divisive normalization-based image representation. IEEE J. Sel. Top. Signal Process., **3**, 202–211.
82. ITU-R BT.500-11. (2002) Methodology for the subjective assessment of the quality of television pictures. International Telecommunication Union, Geneva, Switzerland.
83. Sheikh, H.R., Sabir, M.F., and Bovik, A.C. (2006) A statistical evaluation of recent full reference image quality assessment algorithms. IEEE Trans. Image Process., **15**, 3440–3451.
84. Sheskin, D. (2004) *Handbook of Parametric and Nonparametric Statistical Procedures*, CRC Press.
85. Wang, Z. and Simoncelli, E.P. (2008) Maximum differentiation (MAD) competition: a methodology for comparing computational models of perceptual quantities. J. Vis., **8**, 1–13.
86. Maloney, L. and Yang, J. (2003) Maximum likelihood difference scaling. J. Vis., **3**, 573–585.
87. Charrier, C., Maloney, L.T., and Knoblauch, H.C.K. (2007) Maximum likelihood difference scaling of image quality in compression-degraded images. J. Opt. Soc. Am., **24**, 3418–3426.
88. Charrier, C. et al. (2010) Comparison of image quality assessment algorithms on compressed images. SPIE Conference on Image Quality and System Performance, San Jose, CA.
89. Chandler, D.M. and Hemami, S.S. (2007) A57 database, http://foulard.ece.cornell.edu/dmc27/vsnr/vsnr.html.
90. Le Callet, P. and Autrusseau, F. (2005) Subjective quality assessment IRCCYN/IVC database, http://www.irccyn.ec-nantes.fr/ivcdb/.
91. Ponomarenko, N. et al. (2008) Tampere image database, http://www.ponomarenko.info/tid2008.htm.
92. Toyama image database, http://mict.eng.u-toyama.ac.jp/mict/index2.html.
93. Gaubatz, M. MeTriX MuX visual quality assessment package, http://foulard.ece.cornell.edu/gaubatz/metrix_mux/.
94. Video Quality Experts Group (VQEG) (2008) Final report of video quality experts group multimedia phase I validation test, TD 923, ITU Study Group 9.
95. Seshadrinathan, K. et al. LIVE video quality assessment database, http://live.ece.utexas.edu/research/quality/live_video.html.
96. Moorthy, A.K. and Bovik, A. LIVE wireless video quality assessment database, http://live.ece.utexas.edu/research/quality/live_wireless_video.html.
97. Rouse, D. and Hemami, S. (2008) Understanding and simplifying the structural similarity metric. 15th IEEE International Conference on Image Processing, 2008. San Diego, CA, ICIP 2008, pp. 1188–1191.

20
Digital Image and Video Compression
Joeri Barbarien, Adrian Munteanu, and Peter Schelkens

20.1
Introduction

For a long time, data compression was primarily intended to counteract the bandwidth limitations faced by computing equipment when handling internal and external data transfer and storage. Improvements in network and storage technology have gradually reduced these bandwidth limitations. Unfortunately, data volumes have continued to increase at the same time because of augmented resolutions, dynamic ranges, and so on. As a consequence, communication and storage bandwidth have increased insufficiently to inhibit the need for efficient compression. At the same time, owing to the heterogeneity of communication channels and sensing, storage, and display devices, compression systems have become subject to new functionality requirements [1] that have primarily emerged to adapt to these heterogeneous operating circumstances of quality, resolution and temporal scalability, intelligent image transfer scheduling, random access, error-resilient transmission, and so on.

This chapter intends to introduce the basic concepts of image and video coding technology and to provide future-proof guidelines for the design of compression systems operating on visual data. The chapter begins by discussing the individual components of a coding system and ends by providing some insight into a number of widely employed coding technologies, but it does not intend to be comprehensive.

20.2
Typical Architecture

An image or video encoder is typically composed of five main components as illustrated in Figure 20.1. For video material, processing is performed on a frame-by-frame basis. In the first processing step, the image or video frame X, which, in the general case, can be considered to have n spatial dimensions and s spectral components, is subjected to a spectral transformation C that decorrelates the spectral image components, yielding the spectrally decomposed

---

Optical and Digital Image Processing: Fundamentals and Applications, First Edition. Edited by Gabriel Cristóbal, Peter Schelkens, and Hugo Thienpont.
© 2011 Wiley-VCH Verlag GmbH & Co. KGaA. Published 2011 by Wiley-VCH Verlag GmbH & Co. KGaA.

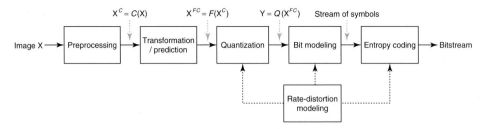

Figure 20.1 The general architecture of an image encoding chain.

image $\mathbf{X}^C = C(\mathbf{X})$. Typically, the decorrelation is performed using a classical color transform (e.g., RGB to YCbCr), a more advanced transform (e.g., Karhunen–Loève transform), or a prediction technique. Next, the spectrally decomposed image $\mathbf{X}^C$ is subjected to a spatiotemporal transformation or prediction F (the temporal component is only relevant in the case of video sequences), which reduces the spatiotemporal correlation between neighboring pixel values and concentrates the energy of the signal in a few coefficients of the output image $\mathbf{X}^{FC} = F(\mathbf{X}^C)$.

Next, the quantization step Q projects the real-valued transform coefficients in the transformed image $\mathbf{X}^{FC}$ into a discrete set of values, each representing a good (in some distortion sense) approximation of the original coefficients. Thereafter, the quantized image coefficients $\mathbf{Y} = Q(\mathbf{X}^{FC})$ are mapped into a set of symbols by the bit-modeling module, and the resulting stream of symbols is subsequently losslessly compressed by the entropy encoding module to produce the final coded bit stream. The rate-distortion modeling component typically steers the quantization, bit-modeling, and entropy encoding steps in order to meet the rate or quality constraints under lossy encoding conditions. The decoder follows exactly the same steps as the encoder, but performed in the inverse order.

20.3
Data Prediction and Transformation

20.3.1
Removing Data Redundancy

In uncompressed digital image and video materials, each pixel value is represented by a fixed number of bits per (spectral) component (usually between 8 and 16). This representation typically contains a significant amount of redundant information. One of the reasons for this is that the uncompressed representation does not take the statistical distribution of the data into account. In general, the process of exploiting the probability distribution of the input symbols to generate a compact binary representation is called *entropy coding* and is discussed further in Section 20.5. In its simplest form, entropy coding assigns a binary code to the set of possible input symbols in such a way that the most probable symbol gets assigned the shortest code word. However, when representing a sequence of input symbols having a long-tailed, quasi-uniform probability distribution, the compacting ability of entropy coding will be significantly lower than that when compressing a sequence of input symbols with a strongly peaked probability distribution. Unfortunately, the pixel values in natural image and video data tend to exhibit long-tailed probability distributions. For a more efficient compression of image and video data, higher order redundancy, stemming from the fact that in the uncompressed representation neighboring pixel values are spatially, temporally, and spectrally correlated, needs to be removed. Two approaches tend to be followed to exploit the correlation between neighboring pixel values: data prediction and data transformation. Both approaches are based on the same principle, which is to map the original pixel values, having a long-tailed probability distribution, to a new set of symbols having a more peaked probability distribution, thereby enabling more efficient entropy coding.

20.3.2
Spatial Prediction

Spatial prediction schemes predict each pixel value based on one or a set of previously encoded spatially neighboring pixel values and return the prediction error, that is, the difference between the actual pixel value and the predicted one. If the pixel values are highly correlated, the prediction will be good and the probability distribution of the prediction error will be strongly peaked around zero, enabling a compact representation using entropy coding. In a very simple example of spatial prediction, the pixels are visited in raster scan order and each pixel value $p(x, y), 0 \leq x < W, 0 \leq y < H$, with W and H, respectively, denoting the image's width and height in pixels, is predicted by $p'(x, y)$, which is defined as

$$p'(x, y) = \begin{cases} p(x-1, y), & x > 0 \\ 0, & x = 0 \end{cases} \tag{20.1}$$

This essentially means that, with the exception of the pixel values at the left image border, each pixel value is predicted from the previously coded, left neighboring pixel value. The effectiveness of this approach is illustrated in Figure 20.2. The histogram of the pixel values in the "Lena" image is clearly long tailed, while the

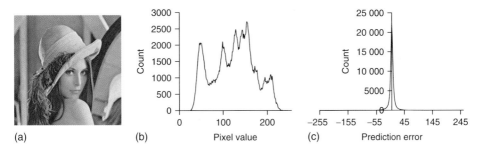

Figure 20.2 (a) The gray-scale "Lena" image, (b) the histogram of its pixel values, and (c) the histogram of the prediction errors, after spatial prediction using Eq. (20.4).

histogram of the prediction errors generated using this simple predictor is strongly peaked around zero, enabling far more efficient entropy coding. This illustrates the effect of data prediction.

The spatial predictor presented in Eq. (20.1) is a very rudimentary illustration of the concept. Practical image coding techniques such as Joint Bi-level Image Experts Group (JBIG) [2], PNG [3], and JPEG-LS [4] use much more advanced spatial predictors.

20.3.3
Spatial Transforms

A second approach to exploiting the spatial correlation among neighboring pixel values is to apply a spatial transform. Either the spatial transforms can be applied to the entire image (global transform) or the image can be divided into nonoverlapping rectangular patches (blocks), which are then transformed independently (block-based transforms). Given an image or image patch of dimension $W \times H$, a spatial transform can be defined as an operation that decomposes the original image or image patch into a linear combination of $W \times H$ different basis images/functions $b_{i,j}(x,y), 0 \leq i < W, 0 \leq j < H, i,j \in \mathbb{Z}$ (as shown in Eq. (20.2)):

$$p(x,y) = \sum_{j=0}^{j=H-1} \sum_{i=0}^{i=W-1} c(i,j) \cdot b_{i,j}(x,y) \tag{20.2}$$

and, as a result, returns the factors $c(i,j)$ in this linear combination, the so-called transform coefficients. When the basis images $b_{i,j}(x,y)$ are known at the decoder, the original pixel values $p(x,y)$ can be reconstructed based upon the transform coefficients $c(i,j)$ using Eq. (20.2). Popular image transforms for image and video coding are the discrete cosine transform (DCT, Chapter 4) and the discrete wavelet transform (DWT, Chapter 7).

The primary reason for using a spatial transform such as the DCT or the DWT in image and video compression systems is the fact that the probability distribution of the transform coefficients rather than that of the original pixel values is typically more suited to entropy coding. While the pixel values typically exhibit a long-tailed probability distribution, the high-frequency DCT and DWT coefficients show a strongly peaked zero-mean Laplacian probability distribution, allowing for a more compact representation (Figure 20.3).

Secondly, both the DCT and the DWT perform frequency decomposition. The spatial high-frequency information, which is less important for the perceptual quality, is thus isolated in a distinct set of transform coefficients. These coefficients can be more coarsely quantized to save bit rate without jeopardizing the visual quality, as the human visual system is known to be less sensitive to high-frequency spatial variations. Thirdly, transforms allow one to straightforwardly support resolution scalability and progressive decoding. For instance, in the wavelet decomposition of an image or video frame, the LL sub-band of each decomposition level represents an approximated version of the original frame with a lower resolution and can

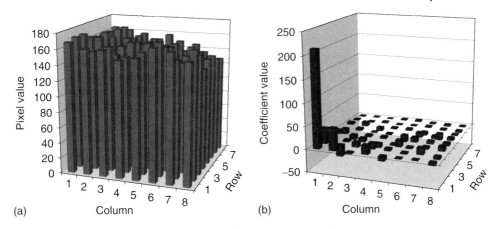

Figure 20.3 Values for a block of 8 × 8 pixels (a) and their DCT transform (b).

be independently reconstructed from the detail sub-bands of the same or lower decomposition levels (see Chapter 7).

20.3.4
Color/Spectral and Multiple-Component Prediction and Transforms

In their original, uncompressed form, color images are typically represented in the RGB color space. Unfortunately, this format exhibits a high correlation between the three spectral bands and is, therefore, often transformed to a luminance–chrominance representation, like $Y'C_bC_r$ [5]. In addition, for most image and video coding applications, the chrominance planes are further subsampled (typically with a factor 2 in both spatial dimensions) to reduce the amount of data [5]. The latter procedure is generally accepted by the coding community because of the significantly lower sensitivity of the human visual system to chrominance information compared to luminance information. For multispectral or multicomponent datasets, other transforms (DCT, DWT, and Karhunen–Loève transform) or prediction techniques can be applied along the component axis to optimally decorrelate the data [1].

20.3.5
Temporal Redundancy Removal by Motion Estimation

20.3.5.1 Motion-Compensated Prediction

When compressing video sequences, another important source of redundancy needs to be eliminated: successive frames in a sequence are typically similar and, therefore, highly correlated. The most straightforward approach to exploit this correlation is to predict the pixel values in the current frame based on the values of the pixels at the same positions in the preceding frames and to code only the prediction errors. However, realizing that the differences between successive frames in a video sequence typically stem from either camera motion

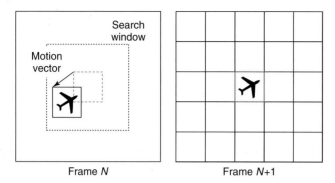

Figure 20.4 Block-based motion estimation.

or object motion in the scene, we can find a better solution in the form of motion-compensated prediction.

Given a reference frame, $p_{ref}(x, y)$, which has been encoded previously (in the simplest case, this is the previous frame in the sequence), and the frame $p(x, y)$ that is being predicted, motion-compensated prediction consists of two steps: a motion estimation (ME) step and a motion compensation (MC) step. The first step tries to estimate the object and camera motions between $p_{ref}(x, y)$ and $p(x, y)$, while the second step uses the derived motion information to predict $p(x, y)$ from $p_{ref}(x, y)$, compensating for the motion between the frames. Both steps of the motion-compensated prediction assume a model for the motion between the frames. Although more complicated models have been proposed in the literature, the most commonly used is the simple block-based motion model. This model assumes that objects are composed of rectangular/square blocks and that camera and objects move in a purely translational, rigid fashion between successive frames. Despite the model's simplicity and obvious flaws, motion-compensated prediction based on this model has been included in all video compression standards since H.261[1] [6].

In practice, block-based ME is performed as follows: the luminance (luma) component of the current frame, which is being predicted, is divided into square, nonoverlapping blocks of size $N \times N$. For each of these blocks, the encoder finds the best matching block in the luminance component of the reference frame. The best matching block is determined as the block with the minimal absolute difference from the block in the current frame within a certain search window SW (Figure 20.4).

Denoting the block in the current frame by $B_{k,l}$, with the top-left pixel located at position $(N \cdot k, N \cdot l)$, and assuming, for simplicity, that the image's width and height, W and H, are exact multiples of N so that $0 \leqslant k < W/N$ and $0 \leqslant l < H/N$,

---

1) Except for intraframe video coding standards such as Motion JPEG, SMPTE VC-3 (Avid DNxHD), and Motion JPEG 2000, which do not use motion-compensated prediction.

the procedure can be more rigorously described by the following equation:

$$\left(v_x^{k,l}, v_y^{k,l}\right) = \underset{(dx,dy)\in SW}{\arg\min} \left(\sum_{j=N\cdot l}^{j=N\cdot l+N-1} \sum_{i=N\cdot k}^{i=N\cdot k+N-1} |p(i,j) - p_{ref}(i+dx, j+dy)| \right)$$
(20.3)

For each $B_{k,l}$, the result of the ME procedure is a motion vector $\left(v_x^{k,l}, v_y^{k,l}\right)$, which is the difference between the position of the block $B_{k,l}$ in the current frame and that of its best matching block in the reference frame. Note that the motion vectors of all blocks in the frame must be signaled in the bit stream to allow proper decoding. Motion vectors are typically encoded using motion-vector prediction followed by entropy coding of the prediction errors.

After the ME is completed, MC of the reference frame is performed. On the basis of the motion vectors $\left(v_x^{k,l}, v_y^{k,l}\right)$ derived by ME using Eq. (20.3), the motion-compensated prediction $p'(x, y)$ is determined by

$$p'(N\cdot k+i, N\cdot l+j) = p_{ref}\left(N\cdot k + v_x^{k,l} + i, N\cdot l + v_y^{k,l} + j\right)$$
$$i,j \in [0, N); 0 \leq k < W/N; 0 \leq l < H/N \qquad (20.4)$$

The prediction $p'(x, y)$ is subsequently subtracted from the current frame $p(x, y)$, resulting in a prediction-error frame, which is thereafter further encoded. At the decoder side, the motion vectors are extracted from the bit stream and the motion-compensated prediction can be executed in the same way, provided the same reference frame is used.

As discussed earlier, the assumptions made by the block-based motion model are often invalid in practice, so that block-based ME fails to detect the true motion between the predicted frame and the reference frame. However, because of the block matching procedure, the selected block from the reference frame is still likely to produce a reasonably good approximation of the considered block.

20.3.5.2 Improvements over the Basic Approach

The basic scheme of block-based motion-compensated prediction, as detailed in the previous section, is used in H.261 [6]. In more recent standards, several advances have been made to improve prediction performance including subpixel accuracy, bidirectional prediction and multiple reference frames, variable block sizes, and multihypothesis prediction.

Subpixel Accuracy Equation (20.3) implicitly assumes that objects or the camera moves over an integer number of pixels between the reference frame and the predicted frame. However, in practice, movement can also occur over a fraction of the pixel distance. To alleviate this problem, subpixel-accurate motion-compensated prediction was proposed. Subpixel accuracy implies that blocks in the reference frame, which are determined by fractional motion-vectors, can be used to form the prediction of a block. In order to generate these blocks, the reference frame needs to be interpolated by means of an interpolation filter, which derives the pixel values at fractional positions based on the neighboring pixel values at integer positions.

Earlier standards such as MPEG-2 [7] limit ME accuracy to half the pixel distance and use simple bilinear interpolation. H.264/AVC [8, 9], the state of the art in video compression, allows 1/4-pixel accuracy and employs a 6-tap Wiener interpolation filter to derive the half-pixel position values and bilinear interpolation to derive the quarter-pixel position values.

Bidirectional Prediction and Multiple Reference Frames Another improvement to the original design is the addition of bidirectional prediction. The idea is as follows: if the current frame is similar to the previous frame in the sequence, it is probably also similar to the next frame in the sequence.

By changing the order in which the frames are coded from the order in which they are displayed (Figure 20.5), a frame can be predicted either from the past (previous frame) or from the future (next frame). Moreover, a block can also be predicted simultaneously from the past and the future by the weighted averaging of the two best matching blocks, which can provide a better prediction. The idea of bidirectional prediction was later generalized to allow prediction from multiple reference frames, which can either be from the future or from the past. H.264/AVC [8, 9], for instance, is typically operated with up to four reference frames to select from at any given time.

Variable Block Sizes The assumption that objects are composed of square blocks of a certain fixed size is often false. Using a single size of blocks can, therefore, reduce prediction performance. To alleviate this problem, variable block-size motion-compensated prediction can be used [8, 9] (Figure 20.6).

Multihypothesis Variable Block-Size Motion-Compensated Prediction with Multiple Reference Frames and Quarter-Sample Accuracy There are a number of choices

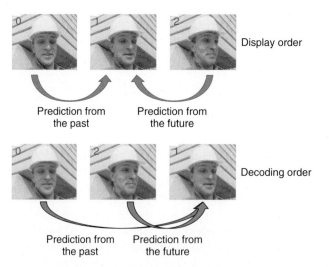

Figure 20.5 Bidirectional motion-compensated prediction with two reference frames.

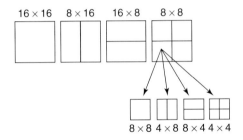

Figure 20.6 Variable block-size motion-compensated prediction in H.264/AVC.

that can be made when performing block-based motion-compensated prediction. Somehow, a choice between these different prediction modes has to be made for each top-level block. This is usually done through a rate-distortion optimal mode decision. This process associates a bit rate cost R_i and a resulting distortion D_i with each prediction mode pm_i. The distortion D_i is calculated as the square error between the original top-level block and the prediction of that block using the considered prediction mode pm_i. The bitrate cost R_i is the sum of the bit rate needed to signal the prediction mode and the bit rate needed to code the prediction error, given a fixed quantization parameter. On the basis of R_i and D_i, a global cost functional $J_i = D_i + \lambda R_i$ is calculated and the prediction mode pm_i for which J_i is minimal is chosen. Usually, the Lagrangian multiplier λ is chosen such that R_i plays a more important part in the cost calculation for lower target bit rates.

20.4 Quantization

20.4.1 Principle

Quantization techniques can be roughly classified into two main types, that is, scalar quantization, in which the quantizer operates on individual pixels, and vector quantization, in which the quantizer operates on groups of pixels. There is a huge literature available on this subject, and an excellent tutorial on quantization can be found in Ref. [10]. The view given in this section is limited only to scalar quantization, as this is the most widely used approach.

Scalar quantization is a mapping from $\mathbb{R}$ to a discrete set of values $\{0, 1, ..., N-1\}$. Let X be a random variable with probability density function (pdf) $p_X(x)$, and let $Q = \{y_i, C_i; i = \overline{0 \ldots N-1}\}$ be a scalar quantizer with output levels (code words) y_i and partition cells $C_i = [x_i, x_{i+1})$. Quantizing X with the scalar quantizer Q implies that if X is in some cell C_i, that is, $x_i \leqslant X < x_{i+1}$, then $Q(X) = i$. The inverse quantizer Q^{-1} produces the output y_i:

$$Q^{-1}(i) = y_i \tag{20.5}$$

which is an approximation of X, according to

$$Q^{-1}(Q(X)) = y_i \tag{20.6}$$

The probability density of the output $p_Y(i)$, $i = 0...N-1$ depends on the cumulative density of the input, and is given by

$$p_Y(i) = \text{prob}(Y = y_i) = \int_{x_i}^{x_{i+1}} p_X(x)\,dx \qquad (20.7)$$

The nth order distortion associated with the quantization process is given by

$$D_n = \sum_{i=0}^{N-1} \int_{x_i}^{x_{i+1}} |x - y_i|^n p_X(x)\,dx \qquad (20.8)$$

20.4.2
Lloyd–Max Quantizers

Quantizers that satisfy the necessary conditions for minimizing the distortion D_n subject to the constraint on the size N of the code are known as *Lloyd–Max quantizers* [10]. In general, it can be shown that for these quantizers the optimal output levels y_i must satisfy the following equation:

$$\int_{x_i}^{y_i} |x - y_i|^{n-1} p_X(x)\,dx = \int_{y_i}^{x_{i+1}} |x - y_i|^{n-1} p_X(x)\,dx \qquad (20.9)$$

For $n = 1$ this is equivalent to

$$\int_{x_i}^{y_i} p_X(x)\,dx = \int_{y_i}^{x_{i+1}} p_X(x)\,dx = \frac{1}{2}\int_{x_i}^{x_{i+1}} p_X(x)\,dx \qquad (20.10)$$

that is, y_i should be the median point of the partition cell C_i. If the mean square error (MSE) criterion is chosen to measure the distortion (i.e., $n = 2$ in Eq. (20.8)), then setting the partial derivatives with respect to any x_k equal to zero in Eq. (20.8) yields $(x_k - y_{k-1})^2 p_X(x_k) - (x_k - y_k)^2 p_X(x_k) = 0$. Solving for x_k we obtain

$$x_k = \frac{y_{k-1} + y_k}{2}, k = 1, 2, ..., N-1 \qquad (20.11)$$

Also, differentiating with respect to y_k yields

$$y_k = \frac{\int_{x_k}^{x_{k+1}} x p_X(x)\,dx}{\int_{x_k}^{x_{k+1}} p_X(x)\,dx}, k = 0, 1, ..., N-1 \qquad (20.12)$$

Equation (20.11) implies that the start points x_k of the partition cells C_k, $1 \leq k < N$ are the median points between the two corresponding output levels y_{k-1} and y_k. Also, Eq. (20.12) is equivalent to $y_k = E[X|X \in C_k]$, that is, the output levels y_k should be the centroids of C_k.

Equations (20.11) and (20.12) are the *necessary* conditions for an optimal scalar quantizer minimizing the MSE. However, these conditions are not *sufficient*, and examples of suboptimal quantizers satisfying Eqs (20.11) and (20.12) can be constructed [10]. However, optimality is assured for all sources having log-concave pdfs (i.e., $\log p_X(x)$ is a concave function [11]). Uniform, Laplacian, and Gaussian distributions satisfy this property, and, hence, their corresponding Lloyd–Max quantizers are optimal in D_2 sense.

However, in compression, both rate and distortion should be considered together. Usually, variable-length entropy coding, such as Huffman [12] and arithmetic coding [13] (see Section 20.5), is applied to the indices produced by the scalar quantizer/bit-modeling (Figure 20.1). Although optimal entropy-constrained scalar quantizers can be constructed, it has been shown that, at high rates (i.e., large $H(Y)$) and for smooth pdfs, the simplest quantizer, that is, optimal or close to the optimum in D_2 sense, is the uniform quantizer [10]. For this quantizer, the partition cells are uniformly distributed, that is, $x_{i+1} - x_i = \Delta$ for any $i = 0...N-1$. In particular, the uniform quantizer is optimal if the input $p_X(x)$ is a Laplacian or an exponential pdf, and close to the optimal for other distributions [10].

At low rates, the optimal entropy-constrained scalar quantizer is no longer uniform. However, a uniform quantizer with centroid code words (see Eq. (20.8)) is very nearly optimal [14]. Furthermore, we must notice that for symmetric pdfs, the rate-distortion behavior at low rates can be improved by widening the partition cell located around zero, that is, by using "deadzone uniform scalar quantizers" [11]. We may conclude that, since, for example, the wavelet coefficients in the detail sub-bands are commonly modeled as having (generalized) Laplacian distributions, the entropy-coded quantizer that is optimal or nearly optimal in mean square sense for a broad range of rates is the uniform quantizer. The main problem with this quantizer is that it cannot be used for embedded coding, and different quantization strategies have to be adopted.

20.4.3
Embedded Quantization

In embedded quantization [11], the partition cells of higher rate quantizers are embedded in the partition cells of all the lower rate, coarser quantizers. A particularly important instance of the family of embedded scalar quantizers discussed above is the family of embedded deadzone scalar quantizers. For this family, X is quantized to

$$i_p = Q_p(X) = \begin{cases} \text{sign}(X) \cdot \left\lfloor \frac{|X|}{2^p \Delta} + \frac{\xi}{2^p} \right\rfloor & \text{if } \frac{|X|}{2^p \Delta} + \frac{\xi}{2^p} > 0 \\ 0 & \text{otherwise} \end{cases} \quad (20.13)$$

where $\lfloor a \rfloor$ denotes the integer part of a, $\xi < 1$ determines the width of the deadzone, $\Delta > 0$, and $p \in \mathbb{Z}_+$.

An example of the partition cells and quantizer indices that is obtained for successive approximation quantization (SAQ), a form of embedded quantization used in JPEG 2000 (Joint Photographic Experts Group), which essentially corresponds to coding the coefficients in a bitplane-by-bitplane fashion, is given in Figure 20.7, in which s denotes the sign bit.

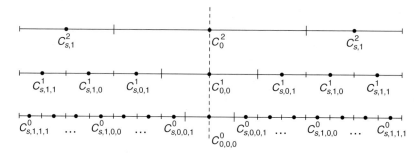

Figure 20.7 Embedded deadzone scalar quantizer for $\xi = 0$ and $\delta = 1/2$.

20.5 Entropy Coding

20.5.1 Huffman Coding

Entropy coding is a lossless coding step that is present in any compression system and converts the previously generated symbols (after prediction/transformation and/or quantization) into a set of variable-length code words such that (i) the code word length maximally approximates the information content of each symbol and (ii) the average code word length or entropy of the signal can be reached. The generated code words are additionally subjected to the condition that they should have unique prefix attributes such that they can be unambiguously interpreted by the decoder despite their variable length. A very popular algorithm that allows for the construction of such variable-length code words is Huffman coding [12].

The Huffman algorithm starts with the building of a binary tree. This procedure consists of two steps. First, the symbols are arranged according to increasing probability. s_1 is the symbol with the highest probability and s_N the symbol with the lowest probability. In the particular case illustrated in Figure 20.8, considering a source with alphabet $S = \{0, 1, \text{EOF}\}$ with associated symbol probabilities $p = \{0.7, 0.2, 0.1\}$, this yields $s_1 = $ "0," $s_2 = $ "1," $s_3 = $ "EOF," $N = 3$. These symbols $s_i, 1 \leq i \leq N$ represent the leaves of the tree. We now choose the two symbols

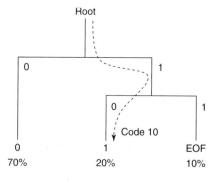

Figure 20.8 Construction of Huffman codewords for the alphabet {0; 1; EOF}.

with the lowest probability, s_{N-1} and s_N (i.e., "1" and "EOF"), and combine them together to form one symbol that we shall call T_{N-1}. This symbol gets a probability that is equal to the sum of the probabilities of s_{N-1} and s_N, T_{N-1} is a new junction in the tree with s_{N-1} and s_N as its children. Bit 0 is assigned to junction s_{N-1} and bit 1 to s_N. We now have a series of $N-1$ symbols $s_1 \ldots s_{N-2}, T_{N-1}$. These two steps are repeated until only one symbol remains. This is the *root* of the tree. By traversing the tree from the root to one of the leaves we obtain the code for the symbol associated with that leaf. In the simple case illustrated in Figure 20.8, the code words for the symbols "0," "1," and "EOF" are respectively "0," "10," and "11." Suppose that we want to encode the sequence 0-1-1-0-0-0-EOF; this results in a sequence of 10 bits (i.e., 0-10-10-0-0-0-11), while with fixed-length coding at least 2 bits per symbol are required or precisely 14 bits are required to encode the sequence.

It is important to realize that the constructed code table has to be available at both the encoder and decoder sides, either by transmission of the code table together with the encoded data or by synchronized construction at both the encoder and decoder sides. The latter case is known as *adaptive Huffman coding*, where the encoder and decoder use the same probabilities at start (e.g., predefined by standard or equal probabilities for each symbol) and where the symbol probabilities are monitored and, if significant deviations from the supposed probability distribution are observed, new Huffman tables are constructed.

Recalculating the Huffman tables at regular intervals adds a significant amount of computational complexity. This issue can be solved by using structural codes, which assume a model for the probability distribution of the input symbols and which can adapt to changes in the parameters of this model in a structured way, that is, for a given symbol and a given parameter set characterizing the probability distribution, the correct code word can be immediately inferred, without having to recalculate an entire code word table. Structural codes used in image and video compression standards include Golomb codes (JPEG-LS [4]) and Exponential-Golomb codes (H.264/AVC [8, 9]).

However, the main drawback of both Huffman and structural codes is their integer-length code words, while the information content per symbol is often not an integer value. Moreover, with these codes the smallest code word size is 1, while symbols with a very high probability may have a significantly smaller optimal length.

20.5.2
Arithmetic Coding

These problems of suboptimality in combination with the complexity of the adaptive scheme led to a search for another form of entropy coding that does not have these shortcomings: arithmetic coding [13]. To initialize the encoding process, we define an interval between 0x000 ($= L^{(1)}$) and 0xFFFF ($= H^{(1)}$) and divide this interval into subintervals i, associated with the corresponding symbols s_i in the alphabet S, choosing the subinterval sizes to be proportional to the probability of the associated symbols $p(s_i)$. With each subinterval i, we can associate its upper and lower limits H_i and L_i.

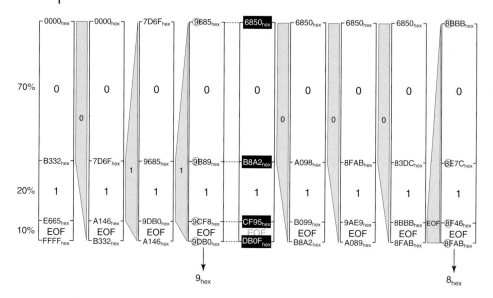

Figure 20.9 Arithmetic encoding of the symbol sequence 0-1-1-0-0-0-EOF.

To illustrate the concept, we again consider the alphabet $S = \{0, 1, \text{EOF}\}$ with equal associated symbol probabilities, $p = \{0.7, 0.2, 0.1\}$, and encode the same message, 0-1-1-0-0-0-EOF (Figure 20.9). Since the first symbol to be encoded is a "0," we select its associated subinterval and take its upper and lower limits $H^{(1)}_{"0"} = \text{0xB332}$ and $L^{(1)}_{"0"} = \text{0x0000}$ as the new upper and lower limits of the global interval. Next, the new global interval is again subdivided into subintervals having sizes proportional to the associated symbol probabilities. Since the next symbol to be encoded is "1," the selected subinterval has the upper limit $H^{(2)}_{"1"} = \text{0xA146}$ and the lower limit $L^{(2)}_{"1"} = \text{0x7D6F}$; for the next symbol "1," this becomes $H^{(3)}_{"1"} = \text{0x9DB0}$ and $L^{(3)}_{"1"} = \text{0x9685}$. We notice that the subintervals are continuously shrinking – a process that unavoidably leads to reaching the precision of the integer arithmetic. To avoid this, we shift – when the upper and lower interval limits are equal – the most significant digit (4 bits) out and transfer it to the output buffer of the encoder. In this particular case, 0x9 is shifted out. The remaining digits are left-shifted, shifting in 0x0 for the lower bound and 0xF for the upper bound, yielding new four-digit hexadecimal numbers for both bounds of the global interval. Next, the encoding process can be continued as before, until the "EOF" symbol is encoded, which results in 0x8 being shifted out. If the decoder receives 0x98 or the bit sequence 10011000, it can – based on the reconstruction of the successive subintervals and assuming that it has the starting probabilities at its disposal – reconstruct the whole sequence. Note that only 8 bits were required for encoding this whole sequence of symbols, 2 bits less than those required in the case of Huffman coding.

The superiority of arithmetic coding becomes even clearer with a second example. We work with the alphabet of two symbols, "0" and "EOF" with

$p(\text{"0"}) = 16\,382/16\,383$ and $p(\text{"EOF"}) = 1/16\,383$. A message that consists of a series of 100 000 zeros, followed by "EOF," has to be encoded. With arithmetic coding, the encoded message is 3 bytes long. Huffman coding requires 12 501 bytes to encode the same message.

Another advantage of arithmetic coding versus Huffman encoding is that the modeling of the probabilities and the code generation are separated. Because of this, making the arithmetic coder adaptive requires only a small adjustment, since probabilities can be updated on the fly, while the complexity of a Huffman coder strongly increases if we want to work adaptively. Finally, it is important to mention that nowadays context-based adaptive variable-length coding (e.g., Huffman) and arithmetic coding are popular. These techniques allow for switching between different sets of probabilities, depending on the context of the information to be encoded [11]. It can be theoretically proven that this approach delivers coding results that are better than or at least equal to those provided by single-model implementations.

20.6
Image and Volumetric Coding

20.6.1
Generic Image Coding

For compressing natural still image content – though alternative formats such as GIF and PNG exist – the market is completely dominated by standards issued by the JPEG. This committee has a long tradition in the development of still image coding standards. JPEG is a joint working group of the International Standardization Organization (ISO) and the International Electrotechnical Commission (IEC). It resides under JTC1, which is the ISO/IEC Joint Technical Committee for Information Technology. More specifically, the JPEG committee is Working Group 1 (WG1) – coding of still pictures – of JTC 1's subcommittee 29 (SC29) – coding of audio, picture, multimedia, and hypermedia information. The word "Joint" in JPEG, however, does not refer to the joint efforts of ISO and IEC, but it refers to the fact that the JPEG activities are a result of an additional collaboration with the International Telecommunication Union (ITU). This committee has defined the JPEG, JPEG-LS, JPEG 2000, and JPEG XR standards for the lossless and/or lossy encoding of image data sets. In parallel, the WG1 committee also launched other initiatives in the context of bilevel image coding and lossless coding of photographic images. The JBIG of WG1 delivered the JBIG-1 (ISO/IEC 11544:1993 | ITU-T Rec. T.82:1993) and JBIG-2 standards (ISO/IEC 14492:2001 | ITU-T Rec. T.88: 2000) that offer solutions for the progressive, lossy/lossless encoding of bi-level and halftone images. The following sections address two main standards, JPEG and JPEG 2000, which are based on two different classes of technology. Note that the decision to elaborate on these particular standards does not mean that the codecs that are not addressed are less valuable. For example, JPEG-LS is currently the best performing technology available for lossless coding of still images.

20.6.2
JPEG

The JPEG standard, formally denoted as ISO/IEC IS 10918-1 | ITU-T Recommendation T.81, was published in 1994 as a result of a process that started in 1986 [15]. Although this standard is generally considered as a single specification, in reality it is composed of four separate parts ranging from the core coding technology to file format specifications. Without any doubt, it can be stated that JPEG has been one of the most successful multimedia standards defined so far.

JPEG is a lossy compression method for continuous-tone color or gray-scale still images. JPEG features many compression parameters, allowing one to properly tune the codec to the application needs. In the first coding step, JPEG transforms color images from RGB into a $Y'C_bC_r$ (luminance/chrominance) color space. Subsequently, the chrominance part is subsampled (4 : 2 : 2 or 4 : 2 : 0) as the human eye is less sensitive to chrominance information (see Chapter 19). Next, the pixels of each component are organized into groups of 8×8 samples, called *data units*, and each data unit is compressed separately. A DCT is then applied to each data unit, resulting in 64 frequency components. Each of the 64 frequency components in a data unit is quantized by a frequency-specific quantization factor determined by a standardized quantization matrix, the values of which are linearly scaled with a quality factor determining the global reconstruction quality and, as a consequence, also the achieved compression ratio. Thereafter, all the 64 quantized frequency coefficients of each data unit are encoded using a combination of run-length encoding (RLE) and Huffman coding. An arithmetic coding variant known as the *QM coder* can optionally be used instead of Huffman coding.

20.6.3
JPEG 2000

In 1997, inspired by the promises of wavelet technology and the need for more advanced functionalities, WG1 launched a new standardization process that resulted in a new standard in 2000: JPEG 2000 (ISO/IEC 15444-1 | ITU-T Rec. T.800) [1, 11]. JPEG 2000 provides functionalities such as lossy-to-lossless progressive coding, quality scalability, resolution scalability, region-of-interest coding, and random code stream access, to mention a few. As with JPEG, JPEG 2000 is a collection of a number of standard parts, which together shape the complete standard. Meanwhile, JPEG 2000 has known vast adoption across a wide range of applications. One of the most successful adoptions is probably the acceptance by the Digital Cinema Initiatives (DCIs). JPEG 2000 was selected as the encoding technology for motion pictures for use in movie theaters.

Part 1 of the standard specifies the core coding technology and fixes the code stream specification. JPEG 2000 is an embedded wavelet-based image coding technology that adheres to the "compress once, decompress many ways" paradigm. This standard can be used as a lossy-to-lossless method to compress continuous-tone images (with multiple components). In order to cope with very large image datasets, the codec allows one to spatially tile the image datasets into manageable sizes for intended application and the underlying processing architecture. Each tile is subsequently independently encoded by utilizing wavelet technology and embedded block coding by optimized truncation (EBCOT) as bit-modeling, entropy coding, and rate-distortion optimization framework. Before applying the wavelet transform, the codec delivers support for lossy and lossless color transforms to spectrally decorrelate the dataset. Thereafter, each component is wavelet transformed. Two wavelet kernels are supported: the 5×3 integer wavelet kernel that can be configured to support both lossless and lossy coding and the 9×7 floating point wavelet kernel for optimal performance in the case of lossy or near-lossless compression. The first quantization stage is subsequently applied to the resulting coefficients, typically employing different quantization step sizes for the different sub-bands.

Subsequently, the EBCOT module activates its first layer, that is, tier 1, the low-level embedded block coding layer. Each sub-band is divided into small fixed-sized code blocks C_i, typically configured with a size ranging from 32×32 to 64×64 (dyadically sized). The coefficients in each block are quantized using SAQ, representing the coefficients in a bit plane by bit plane fashion, and the resulting binary symbols are coded using context-based adaptive arithmetic coding, yielding an embedded bit stream. In the tier 2 step, that is, the layer formation, the final embedded code stream is generated. The embedded bit streams B_i resulting from each code block C_i in tier 1 are truncated in chunks corresponding to the supported quality layers and recombined in such a way that the desired quality and resolution scalability functionalities are supported while assuring optimized rate-distortion behavior. The rate-distortion modeling and optimization are carried out by utilizing Lagrangian rate-distortion optimization.

20.7
Video Coding

The most successful video coding schemes that have been proposed in the past two decades [6–9, 16, 17] are the result of standardization efforts by ITU-T and ISO/IEC JTC-1 SC29 WG11 (MPEG). These video codecs share a similar architecture that combines motion-compensated prediction with a block-based DCT, scalar quantization, and Huffman or arithmetic entropy coding.

To illustrate the basic interaction between these building blocks, the workings of H.261 [6], the first globally accepted video coding standard, are briefly explained in Section 20.7.1. All video coding standards defined after H.261 largely operate in the same way but provide increased compression performance, primarily through the use of more complex entropy coding schemes and improved block-based motion-compensated prediction, supporting multiple reference frames, subpixel-accurate motion vectors, and variable block sizes [6–9, 16, 17]. To illustrate the evolution of video coding technology since H.261, we also describe the operation of H.264/AVC, the current state of the art in video coding technology, in Section 20.7.2.

20.7.1
H.261

The general setup of an H.261 encoder is shown in Figure 20.10. Each frame is first divided into macroblocks (MBs) consisting of a 16 × 16 block of luma samples and two collocated 8 × 8 blocks of chroma samples. The encoding of each frame proceeds on an MB-by-MB basis. The MBs in the first frame of the sequence are all intracoded, which means that they are coded without any reference to other frames in the sequence. When encoding these MBs, switches S_1–S_3 in Figure 20.10 are left open. The MBs are divided into 8 × 8 blocks of samples, which are thereafter independently transformed using a block-based DCT. The transform coefficients are subsequently quantized by dividing each coefficient by a quantization step Δ and rounding the result down to an integer value. The quantization step size is frequency dependent. In this way, the higher frequency information is more coarsely represented. In the last step, the quantized coefficients are coded using Huffman and run-length entropy coding.

Starting from the quantized coefficients, each MB is reconstructed by performing the inverse quantization and transform steps. The resulting MBs form the reconstructed frame, which is stored in the frame buffer for use as a reference in the motion-compensated prediction of MBs from the succeeding frame. In this way, the reference frame used for MC at the encoder side is exactly the same as the one used at the decoder side, which prevents accumulation of mismatch errors. Because the reconstructed frame is used as a reference for future motion-compensated prediction, the H.261 encoder is referred to as a *closed-loop video encoder*.

The MBs in the succeding frames of the sequence can be either intracoded, as described above, or predicted by block-based motion-compensated prediction from the reference frame stored in the frame buffer, which is a reconstructed version of the preceding frame in the sequence. In the latter case, switches S_1–S_3 in Figure 20.10 are closed. ME is first performed on the MB's 16 × 16 luma samples. The process finds the best matching 16 × 16 block of luma samples in

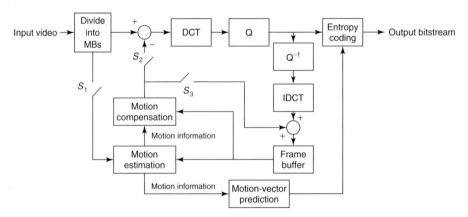

Figure 20.10 Structure of a typical H.261 encoder.

the reference frame, given a search window of [−15, 15] samples horizontally and vertically around the position of the predicted MB. The resulting motion vector is encoded using motion-vector prediction from the motion vector of the preceding block followed by entropy coding of the prediction errors using Huffman coding. The MB's motion-compensated prediction, generated by MC from the reference frame by using the derived motion vector, is subsequently subtracted from the original MB, resulting in a prediction error, which is coded in a similar way as an intracoded MB. Starting from the quantized transform coefficients generated by this process, the MB's prediction error is reconstructed and added to the MB's motion-compensated prediction. The resulting decoded MB is stored in the frame buffer as a part of the succeeding reference frame.

20.7.2
H.264/AVC

Note that the structure of the H.264/AVC encoder (Figure 20.11) is very similar to that of the H.261 encoder. Nonetheless, the first difference between both encoders is the option to reorder the frames prior to the start of the coding process to allow bidirectional prediction as explained in Section 20.3.5.2. In H.264/AVC, each frame is again encoded on a MB-by-MB basis. All MBs, even the intracoded ones, are first predicted. Intracoded MBs are spatially predicted based on the pixels of the already encoded and reconstructed MBs of the same frame, whereby the prediction proceeds along one of a set of predefined directions (horizontal, vertical, etc.). Intercoded blocks are predicted by multihypothesis, block-based, motion-compensated prediction with quarter-pixel accuracy, using variable block sizes and multiple reference frames. A block can be predicted from a single block in one of the reference frames or by weighted averaging of two blocks, each residing in one of the reference frames. For both intra- and intercoded MBs, the

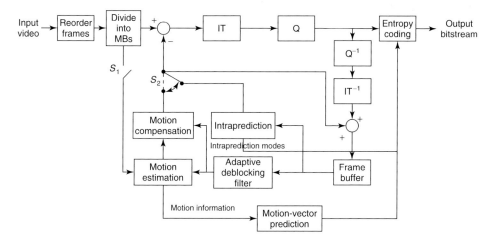

Figure 20.11 Structure of a typical H.264/AVC encoder.

prediction error is divided into smaller blocks that are transformed by applying a 4×4 or 8×8 sample integer transform [18], which is an integer approximation of the DCT. After the transform, the resulting coefficients can be quantized using either uniform scalar quantization or frequency-weighted scalar quantization. The quantized transform coefficients, the prediction modes, and motion information can be entropy coded using either context-adaptive variable length coding (CAVLC), which is based on Huffman coding and run-length coding, or context-based adaptive binary arithmetic coding (CABAC).

Similar to H.261, H.264/AVC enforces closed-loop prediction. To this end, the quantized transform coefficients are decoded by performing the inverse quantization and transform, yielding the reconstructed prediction error. This error is subsequently added to the MB's prediction (intra or inter), and the resulting decoded MB is placed in a frame stored in the reference frame buffer. Before using this frame as a reference for motion-compensated prediction, an adaptive deblocking filter is applied [19]. This filter reduces blocking artifacts, which are essentially visibly annoying discontinuities at the edges of the transform blocks and motion partitions that tend to appear when large quantization step sizes are used. By suppressing the blocking artifacts, the subjective quality at the decoder side is improved and a better reference for motion-compensated prediction is generated.

20.8
Conclusions

This chapter intended to provide a generic overview of a compression system for image and video data. The different steps of compression systems were addressed: spectral decorrelation, spatial and temporal decorrelation, quantization, entropy coding, and finally rate-distortion optimization. A few examples of practical image and video coding solutions have also been provided. The reader should consider this chapter as a starting point for further reading, for which the references are provided at the end.

Acknowledgments

This research was funded by the Fund for Scientific Research Flanders via the postdoctoral fellowships of Adrian Munteanu and Peter Schelkens, and research projects G039107 and G014610N.

References

1. Schelkens, P., Skodras, A., and Ebrahimi, T. (2009) *The JPEG 2000 Suite*, John Wiley & Sons, Ltd, Chichester.

2. ISO/IEC and ITU-T (1993) ISO/IEC 11544 | ITU-T recommendation T.82 | ISO/IEC JTC-1 SC29 WG1 and ITU-T. *Information Technology – Coded*

Representation of Picture and Audio Information – Progressive Bi-Level Image Compression.
3. ISO/IEC (2003) ISO/IEC 15948 | ISO/IEC JTC 1 SC 24. *Information Technology – Computer Graphics and Image Processing – Portable Network Graphics (PNG): Functional Specification*.
4. Weinberger, M.J., Seroussi, G., and Sapiro, G. (2000) The LOCO-1 lossless image compression algorithm: principles and standardization into JPEG-LS. *IEEE Trans. Image Process.*, **9** (8), 1309–1324.
5. Poynton, C.A. (2007) *Digital Video and HDTV: Algorithms and Interfaces*, 1st edn, Morgan Kaufmann/Elsevier Science, San Fransisco, CA.
6. ITU-T (1990) ITU-T Recommendation H.261. *Video Codec for Audiovisual Services at p x 64 kbit/s, ITU-T*.
7. ISO/IEC and ITU-T (1995) ISO/IEC 13818-2 | ITU-T recommendation H.262 | ISO/IEC JTC1/SC29/WG11 and ITU-T. *Information Technology – Generic Coding of Moving Pictures and Associated Audio Information – Part 2: Video*.
8. Richardson, I.E.G. (2003) *H.264 and MPEG-4 Video Compression*, John Wiley & Sons, Ltd, Chichester.
9. Wiegand, T. *et al.* (2003) Overview of the H.264/AVC video coding standard. *IEEE Trans. Circuits Syst. Video Technol.*, **13** (7), 560–576.
10. Gersho, A. and Gray, R.M. (1992) *Vector Quantization and Signal Compression*, Kluwer Academic Press/Springer.
11. Taubman, D. and Marcellin, M.W. (2001) *JPEG2000 – Image Compression: Fundamentals, Standards and Practice*, Kluwer Academic Publishers, Hingham, MA.
12. Huffman, D.A. (1952) A method for the construction of minimum-redundancy codes. *Proc. IRE* **40** (9), 1098–1101.
13. Witten, I.H., Neal, R.M., and Cleary, J.G. (1987) Arithmetic coding for data compression. *Commun. ACM*, **30** (5), 520–540.
14. Farvardin, N. and Modestino, J.W. (1984) Optimum quantizer performance for a class of non-Gaussian memoryless sources. *IEEE Trans. Inf. Theory*, **IT-30**, 485–497.
15. Pennebaker, W.B. and Mitchell, J.L. (1993) *JPEG Still Image Data Compression Standard*, Van Nostrand Reinhold, New York.
16. ITU-T (1995) ITU-T Recommendation H.263. *Video Coding for Low Bit Rate Communication, ITU-T*.
17. Pereira, F. and Ebrahimi, T. (2003) *The MPEG-4 Book*, Prentice Hall.
18. Wien, M. (2003) Variable block-size transforms for H.264/AVC. *IEEE Trans. Circuits Syst. Video Technol.*, **13** (7), 604–613.
19. List, P. *et al.* (2003) Adaptive deblocking filter. *IEEE Trans. Circuits Syst. Video Technol*, **13** (7), 614–619.

21
Optical Compression Scheme to Simultaneously Multiplex and Encode Images

Ayman Alfalou, Ali Mansour, Marwa Elbouz, and Christian Brosseau

21.1
Introduction

In the last two decades, the development of secure transmission systems has become a priority for many research and engineering groups. Extensive studies on the application of coherent optical methods to real-time communication and image transmission have been carried out. This is especially true when a large amount of information needs to be processed, for example, high-resolution images. In actual transceiver systems, optical images should be converted into digital form in order to transmit, store, compress, and/or encrypt them. This process requires important computing times, or image quality reduction. To avoid such problems, an optical compression of the images may be appropriate. The main advantage of optical methods, compared to digital ones, lies in their capability to provide massively parallel operations in a 2D space. Approaches that consist in processing the data closer to the sensors to pick out the useful information before storage or transmission are a topic that has generated great interest from both academic and industrial perspectives (see Chapters 22 and 23). This fact has been well developed in the Applied Optics issue on Specific Tasks Sensing (*Appl. Opt.* 45, (2006)) as well as in the tutorial article "Optical Image Compression and Encryption Methods" published in *Adv. Opt. Photon.* 1, (2009). On the basis of the latter, it appears that the complexity of optical methods increases if color images and video sequences are considered, or if the encryption rate of an image must be increased. For the purpose of simplifying the complexity of optical systems, hybrid optical–digital systems, for which processing steps are carried out using digital systems, can be proposed. In this case, the complexity of the whole system can be reduced with a reasonable speed. The major contribution of the current chapter consists in reporting on optical filtering operations which can be used to simultaneously compress and encrypt target images. We begin this chapter with a brief introduction of several optical compression methods. Then, in the third section, we develop an optical compression method based on the fusion of Fourier planes of several target images. To optimize this fusion in our first optical

Optical and Digital Image Processing: Fundamentals and Applications, First Edition. Edited by Gabriel Cristóbal, Peter Schelkens, and Hugo Thienpont.
© 2011 Wiley-VCH Verlag GmbH & Co. KGaA. Published 2011 by Wiley-VCH Verlag GmbH & Co. KGaA.

compression method, an RMS duration criterion has been applied to segmented Fourier planes. Next, we detail two methods that are implemented separately: the Joint Photographic Experts Group (JPEG) and the Joint Photographic Experts Group 2000 (JPEG2000) methods. A comparison between the digital and optical versions is also performed. A simultaneous optical compression and encryption method based on a biometric key and discrete cosine transform (DCT) is presented. We show that this simultaneous optical compression/encryption can be easily performed in optical correlation techniques. Finally, our conclusions are presented. A glossary is proposed to summarize the various algorithms explored in this chapter.

21.2
Optical Image Compression Methods: Background

Since the 1990s, many researchers have been involved in the design, implementation, and development of optical image processing systems. In this chapter, a section dealing with optical image compression is given first, with special attention given to our earlier works. These concern the implementation of a real-time image processing, such as the frequency filtering based on coherent optics using all-optical or mixed optical–digital systems.

Thanks to their parallelism, optical systems can be considered as an appropriate solution to the above-mentioned problems. However, this solution requires powerful electro-optical interfaces. This issue required major effort toward algorithmic development. Digital versus optical systems has been a major issue in the last two decades. On the one hand, digital systems benefited from the huge development of modern electronic and computing devices. On the other hand, optical systems were improved thanks to the appearance of powerful electronic–optical interfaces such as the spatial light modulators (SLMs) (see Chapter 9). But each approach has some specific drawbacks. A relevant compromise solution that takes advantage of both techniques can be found by developing hybrid digital–optical systems.

Signal processing techniques have been widely developed to improve the detection and the performance of radar, sonar, and wireless communication. Very recently, many researchers were involved in solving the pattern recognition problems by using optical correlation techniques (see Chapter 30) [1–5]. In decision theory, the correlation is widely used and is related to matched filters. Owing to some limitations (distortion, scale, and rotation effects), the correlation cannot be used to identify complex images. The detection capability has been improved by using new techniques such as composite filters [6]. Composite filters optimize cost functions, which can improve the detection and reduce the probability of false alarm or misdetection. Composite filters can be considered as advanced filters that can recognize a form in a complex image. This concept can be generalized to develop special correlation filters, the purpose of which is to optimize a criterion in order to improve the detection and to achieve the desired final decision. In general, these off-line filters can shorten the postprocessing process. In fact, the matched filter is appropriate to extract features from the scene. Postprocessing is required to merge these features in order to take some decision. If some knowledge about the pattern recognition is specified in the filter, less postprocessing is needed. This concept can be realized thanks to a segmented filter [7, 8]. This filter consists in grouping together several correlation filters. For this purpose, the Fourier plane of filtered image should be divided into several zones. We mention here that each zone should contain information originating only from one correlation filter.

Segmented filters allow optical multiplexing (fusion) of different classes of objects which should be identified by using a specific phase function for each class. In this case, the fusion of different reference objects can be done in the spectrum plane, that is, the Fourier plane. For this purpose, a power clustering criterion based on frequency partition has to be optimized. In addition, the specific optical implementation constraints should also be considered. Thanks to previous works on multiclass recognition decision, we obtained a precise knowledge of the way the information needs to be treated in the spectral plane. Hence, an optimized segmented filter to simultaneously incorporate many references was proposed by considering specific classification and selection of the spectral information belonging to each reference or that shared by several references. Such filters can be used in various optical image processing techniques, for example, correlation of color images, three-dimensional object (3D object) recognition, encryption, and compression [5, 7–9]. Such methods permit one to preserve pertinent information and to reduce the redundancy that results in image compression. The first optical compression scheme based on this approach has been proposed. To improve this work, various clustering and selection criteria were suggested [9].

Recently, many works on optical compression appeared in the literature [10]. Among them, the bandwidth-compression scheme for 2D images developed by Smith and Barrett [11] is noticeable. Their compression technique use Radon transforming, filtering, thresholding, and quantification steps. The Radon transform is an integral transform of a 2D function over a straight line that allows us to handle the 2D Fourier transform (FT) of a 2D image without performing 2D operations. Three important features of the Radon transform that justify its use for FT compression techniques are as follows:

1) The coding process can be performed by using well-known ID devices.
2) The large dynamic range of FT components can be significantly reduced by the filtering operation.
3) Only one line through the center of the 2D FT can be examined at any time and adaptively compressed.

Darakis and Soraghan [12] described a method for compressing interference patterns based on off-axis (phase-shifting) digital holography by using the phase-shifting interferometry digital holography (PSIDH) hologram compression scheme. According to Ref. [12], the compression in the reconstructed plane can minimize the normalized root-mean-square (NRMS) error better than any other technique. To achieve 3D object recognition using the PSIDH technique, Naughton and coworkers [13] suggested a digital hologram compression technique based on Fourier-domain processing. The principle of this operation consists in recording the digital holograms with an optical system based on a Mach–Zehnder interferometer. These authors showed that the compression performances, based on the correlation peak amplitude between the original and reconstructed images, are rather poor. To improve this performance, they proposed new lossy compression schemes, that is, a resized hologram. Firstly, a resampling technique is performed, which is followed by a quantization that is directly applied to the complex-valued holographic pixels. Finally, a selective removal of the discrete FT coefficients is done. In Ref. [13], the authors demonstrated that digital holograms are very sensitive to resampling due to the speckle effects, which can be minimized thanks to a high degree of median filtering. Wijaya and coworkers [14] developed advanced correlation filters to perform illumination-tolerant face verification of compressed test images at low bit rates using the JPEG2000 wavelet compression standard.

Ding and coworkers [15] proposed and validated an optical image recognition system based on volume holography and wavelet packet compression methods. The volume holographic associative storage in a photo-refractive crystal has some special properties that are adapted to develop an optical correlation system for image recognition. The wavelet packet method was introduced in the optical system to reduce the number of stored images. By selecting the best wavelet packet basis, a set of best eigen-images, stored in the crystal as reference images for the recognition step, are extracted from a large number of training images. Naughton and coworkers [16] suggested and validated a new scheme of multiple-client compression and transmission of 3D objects. They used the PSIDH technique for recording multiple views of a 3D object (Fresnel fields). The digital Fresnel fields have dimensions of (2028 × 2044) pixels with 8 bytes of amplitude information and 8 bytes of phase information for each pixel. An Internet-based Fresnel field compression was applied to reliably and accurately measure the interaction between compression and transmission times. This client–server application and associated compression algorithms were implemented in Java™, allowing them to develop a platform-independent environment for experimentation over the Internet. The authors of Ref. [17] proposed another compression technique based on FT and temporal fringe pattern storage that can simultaneously retrieve the intensity and the deformation phase. In Ref. [18], Yeom and coworkers proposed a new method to compress a full-color 3D integral image based on adapted video compression MPEG-2. In the following, the advantages and drawbacks of our compression algorithm, based on filtering in Fourier plane, are described.

21.3
Compression and Multiplexing: Information Fusion by Segmentation in the Spectral Plane

Compression of an image can be defined as the selection of necessary relevant information for a given application. To achieve a good compression ratio and a good quality of compressed images, the relevant information must be selected carefully. In Ref. [9], a new technique for optical image compression based on

21.3 Compression and Multiplexing: Information Fusion by Segmentation in the Spectral Plane

image phase spectrum was suggested. This technique consists in an algebraic multiplication of the image spectrum with specific segmented amplitude mask (SAM) in the Fourier domain. SAM can mainly locate redundant information areas in a given spectrum. To filter out irrelevant information, several selection and segmentation/fusion criteria can be used. The main idea of this approach consists in merging the spectrum of the target images into one image and transmitting it. The obtained image should contain all necessary and pertinent information to retrieve the original compressed images. To merge the spectrum and obtain a good compressed image, the first step is to divide the Fourier plane of each target image into different areas. Next, the pertinent spectral information relative to every image is merged. A discussion of a couple of features pertinent to this segmentation–fusion procedure is in order.

First, a good segmentation criterion should be chosen. The second feature deals with human vision, since human eyes are less sensitive to small details (Figure 21.1a–b). Our approach can be divided into two steps. First, the FT of every target image is calculated. Second, a phase mask is added to the obtained spectrum of each image (Figure 21.1a). These masks allow us to retrieve the original images from the mixed and the compressed final spectrum images [9] (Figure 21.1b). The fusion criterion is described in the next section. Once the mixing and compression

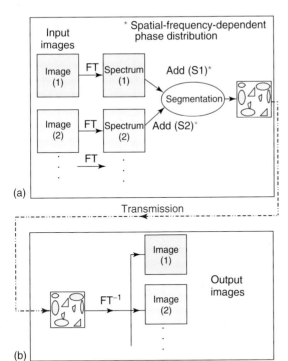

Figure 21.1 Principle of the spectral domain based compression method using a segmented criterion: (a) multiplexing and (b) demultiplexing.

procedures are achieved, the obtained compressed and mixed spectrum images can be stored or transmitted. Original images can be easily obtained by applying an inverse Fourier transform (IFT) to the received spectrum compressed images (Figure 21.1b).

As was previously mentioned, this approach is based on the decomposition of the spectral planes of target images, that is, the images that should be stored or transmitted, in different areas. This procedure can be performed using a powerful segmentation criterion: pertinent spectral information related to each image can be selected, which helps to reconstruct good images at the receiver point. The selected criterion is a key point of our approach and is described in Figure 21.1a.

Let us consider L target images. For a given position (u, v) in the Fourier plane, this technique compares the pixels of all spectra located at this position. It is worth noting that the spectra are normalized to a common fixed energy E_0 ($E_0 = 1$). The winning pixel among the L pixels of the respective spectra has the largest intensity [9]. The pixel at position (u, v) of the spectrum k is retained if Eq. (21.1) is satisfied:

$$E_{uv}^k \geq \begin{cases} E_{uv}^0 \\ E_{uv}^1 \\ \cdot \\ \cdot \\ \cdot \\ E_{uv}^{L-1} \end{cases} \quad l = 0, \ldots, L-1 \tag{21.1}$$

where E_{uv}^k denotes the energy of pixel (u, v) of the image k and L is the number of images. The intensity criterion suffers from two drawbacks: (i) all nonwinning pixels (losers), which generally cover large zones, are ignored even if the intensities of some losers are close to those of the winners; (ii) it neglects the importance of the phase during the selection. In some instances [9], the phase information is often much more important than the intensity information, and the phase variation can cause diffusing as well as focusing behaviors. To overcome these drawbacks and to take the phase information in our segmentation into account, the real part criterion can be used. This criterion consists in comparing the real part of the pixels rather than their intensities E_{uv}^k. It has the advantage of taking both the amplitude and phase information into account:

$$(A_{uv}^k \cos(\varphi_{uv}^k))^2 \geq \begin{cases} (A_{uv}^0 \cos(\varphi_{uv}^0))^2 \\ (A_{uv}^1 \cos(\varphi_{uv}^1))^2 \\ \cdot \\ \cdot \\ (A_{uv}^{L-1} \cos(\varphi_{uv}^{L-1}))^2 \end{cases} \tag{21.2}$$

where L is the number of images and $A_{uv}^k \cos(\varphi_{uv})$ represents the real part of image k at the pixel (u, v). This criterion gives better results than energetic ones. However, it could not retrieve images of good quality, for example, see Figure 21.2c. In fact, by selecting only winner pixels, it can generate isolated pixels, that is, a pixel of an

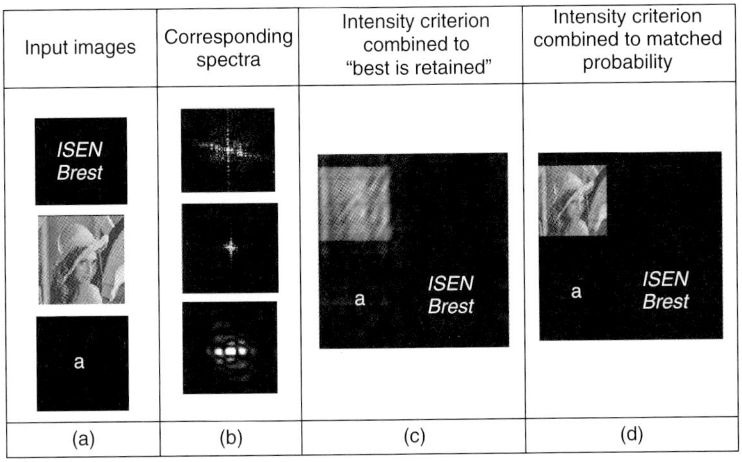

Input images	Corresponding spectra	Intensity criterion combined to "best is retained"	Intensity criterion combined to matched probability
ISEN Brest ... a		a ISEN Brest	a ISEN Brest
(a)	(b)	(c)	(d)

Figure 21.2 (a) Three input images: binary and gray-scale images, (b) amplitude spectra, (c) reconstructed images using the intensity criterion, (d) reconstructed images using the intensity criterion combined to matched probability [9].

image surrounded by pixels from different other images, and it can favor an image with respect to other images by affecting most of the pixels in this image.

To solve this problem, a probability function was added to our criterion [9]. For this purpose, many probability models or stochastic functions can be considered. In our study [9] we found that the most appropriate function was the matched probability function. The idea is to match the probability to the pixel's importance with respect to a selected criterion. The matched probabilistic selection is based on the pixel classification according to its importance. However, unlike the fixed probability-based selection, the matched probability model does not neglect the importance of any spectral pixel compared to its counterpart at the same location in the other spectra. When the difference between the first and second pixels (with respect to the chosen criterion) increases, the probability of retaining the second pixel decreases. If this difference is very small, both pixels can be likely selected. This technique is expected to lead to a significant improvement in the quality of the reconstructed image (Figure 21.2d).

Unfortunately, the approach based on the matched probability model cannot correctly handle images that have a similar spectrum as in a video sequence. To improve this approach, it was suggested to reorganize the spectrum of target images to avoid the overlapping of energetic areas [19]. This reorganization is done by shifting the center of the spectrum (Figure 21.3a). It was shown that the selection criterion can achieve better results once this operation is done. The shifting of the center should be done with meticulous care to optimize the spectrum of the final compressed image. By applying this approach to a video sequence, good experimental results have been obtained (Figure 21.3b). To optimize the reorganization, an RMS duration criterion can be used (Eq. (21.3)). Thanks to the RMS duration criterion, the minimum width of every spectrum can be estimated

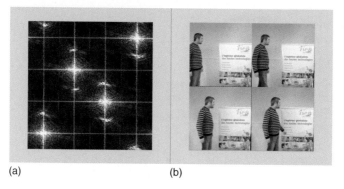

Figure 21.3 (a) Shifted spectrum, (b) the outcome of our modified approach based on energetic selection criterion and shifted spectrum [19].

by using the minimum distance to avoid overlapping of the important spectral areas. The RMS criterion chosen is based on that of Papoulis [20]. We first consider the extent of a 2D signal $I(x,y)$ with a spectrum $S_I(u,v) = A(u,v)\exp(i\psi(u,v))$. For this purpose, the RMS duration can be approximated by a 2D integral of the second moment of the signal [20] and the estimated distance can be approximated by

$$\left(\frac{\Delta}{2}\right)^2 = \int_{-\infty}^{+\infty}\int_{-\infty}^{+\infty}(u^2+v^2)\left|S_I(u,v)\right|^2 du\, dv \tag{21.3}$$

where $S_I(u,v)$ is the spectrum of the input image [21].

We showed that the shifting procedure along with the energetic selection approach can achieve good compression–decompression results. However, the shifting should be done carefully: this requires a relevant estimation and an adequate shifting of the various spectra. The procedure can hardly be implemented on all optical systems. We are currently working on the design of a hybrid optical–digital system to achieve this adaptive shifting selection procedure for which the segmentation step can be realized in the digital part of our system. In order to minimize the processing time and the required memory, the number of bits/pixels must be reduced. Considering a fixed number of bits/pixels is not relevant to our application since all pixels would have the same importance. To solve this issue, we proposed a variable number of bits/pixels (between 2 and 8 bits/pixels) for every frequency in the spectrum plane. This number is increased according to the importance of the selected frequency. In order to estimate the importance of the selected frequency, the RMS duration criterion given in Eq. (21.3) is used. Our preliminary experimental studies corroborate the performance of this optimization [22].

21.4
Optical Compression of Color Images by Using JPEG and JPEG2000 Standards

Another compression scheme based on the JPEG method [23] was suggested. This scheme can be divided into two parts:

1) Selecting the pertinent information, that is, a DCT is performed to reduce the redundancy of a target image.
2) Encoding, that is, grouping the pertinent information of an image to reduce the size of the encoded image.

It is well known that the first step requires most of processing time. Alkholidi et al. [23] proposed an optical scheme to achieve the first step by JPEG compression. A hybrid optical–digital system for achieving the JPEG compression method was realized (Figure 21.4). The input of this hybrid system is the original target image that has to be compressed. The DCT part of the JPEG compression is done using an all-optical scheme. Next, the optical quantization of the output of the DCT is done according to a specific criterion. During the quantization process, it is this criterion that determines the importance of the information and the compression ratio. Then, the output image is the compressed image, which can be stored or transmitted. Finally, to retrieve the original image, the process is inverted, that is, dequantization is followed by an optical inverse DCT.

Two advantages are as follows: first, FT can be easily performed using a convergent lens. Second, DCT is very similar to FT. As mentioned before, the DCT part of the JPEG compression system is performed using an all-optical scheme [24]. The DCT of an image $I(N,N)$ is obtained from the FT of the image which is realized using a simple convergent lens. The suggested scheme is illustrated in Figure 21.4a. First, the input image is duplicated in a special manner. Second, the FFT of the duplicated image is performed using a convergent lens. The output spectrum is multiplied by a hologram mask. The pixels of the outcome of the later multiplication are again multiplied by $\exp\{-j2\pi(mp+nq/2N)\}$, where N stands for the image size and (u,v) denotes the spectral coordinates. In the second

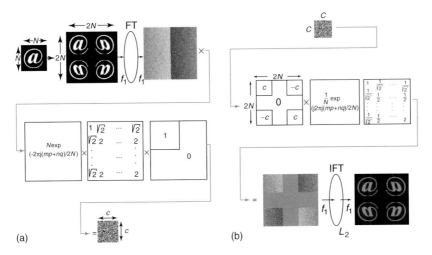

Figure 21.4 (a) Synoptic diagram of the proposed step implementing the DCT optically; (b) synoptic diagram of the optical JPEG decompression [23].

stage, the DCT is quantized and filtered in order to reduce the redundancy and eliminate the insignificant frequencies. The quantization step consists in dividing the DCT output image by a given matrix. This procedure can be realized optically by multiplying the DCT output image with a hologram in the Fourier domain. Next, the obtained image is multiplied by a low-pass filter (LPF), whose size c is the size of the compressed image which depends on the compression ratio T_c and on the image size N, that is, $T_c = 100 \left(1 - \frac{c \times c}{N \times N}\right)$. The optical setup is described in Refs [23, 24]. The compressed DCT image of size c can be stored or transmitted. At the receiver side, this image is decompressed (Figure 21.4b). For this purpose, its spectrum ($2N \times 2N$) is first extracted. Then, this spectrum and a specified hologram are multiplied on a pixel-by-pixel basis. This hologram is the inverse of the one that was previously used in the compression procedure (Figure 21.4a). After dequantization, an IFT can retrieve the duplicated image I. Now, making use of a simple mask of size ($N \times N$), the original image is retrieved.

21.4.1
Optical JPEG Implementation Results

To test our decompression system, a compressed image was generated numerically using a digital JPEG compression model proposed in [23]. This image has been decompressed using the optical system described in Refs [23, 24]. Using this system, various images (binary, gray-scale, and color) have been tested.

At this point, an illustration of this scheme is useful. In this experiment, a binary image of the λ symbol, which has been compressed using different compression ratios $T_c = 10\%, 20\%, 30\%, 40\%, 50\%,$ and 60%, is considered. Figure 21.5 shows the results obtained with a binary compressed spectrum. We notice that if the

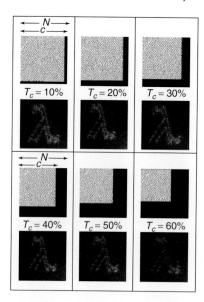

Figure 21.5 JPEG compression and optical decompression using various compression ratios: $T_c = 100 \left(1 - \frac{c \times c}{N \times N}\right)$, here c (resp. N) stands for the size of the compressed (resp. original) image [23].

compression ratio is increased, the amount of stored or transmitted information as well as the quality of the decompressed image is reduced. To consider color images, another approach based on the decomposition of a color image into its fundamental color components (red, green, and blue (RGB) or YUV) was proposed. Results of simulations conducted in our laboratory show that the above binary system can handle color images by processing its fundamental color components separately.

21.4.2
Optical and Digital JPEG Comparison

To validate the good performance of our adapted JPEG compression method (optical version), we compared it with the original digital JPEG version. For this purpose, we simulated (using Matlab) these two JPEG versions when displaying the Barbara image (a gray-scale image shown in Figure 21.6) at the input plane of each version [25]. To perform this comparison, we calculated the PSNR (i.e., peak signal-to-noise ratio defined in Eq. (21.4)) with respect to the compression ratio. The compression ratio has been adjusted in each version. Thus, in the optical version the compression ratio is calculated according to "c," the number of frequency pixels stored in the spectral domain (Figure 21.4). In the digital case, the *compression ratio* is defined as the ratio between the compressed image size and the original image size. Results of comparison between the two versions are shown in Figure 21.6. The dark gray solid line represents the PSNR values of the optical version. The PSNR values of the digital version are plotted in light gray solid line. Figure 21.6 shows a good performance of our optical approach. Indeed,

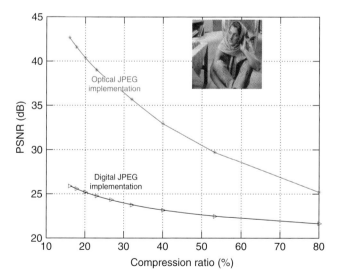

Figure 21.6 The PSNR values obtained with the optical JPEG version (a light gray solid line) and the digital JPEG version (a dark gray solid line) [25].

the optical JPEG version achieves good reconstructed quality. In addition, it can be implemented to provide a real-time system, thanks to the parallelism of optical systems:

$$\text{PSNR}\left(I_{\text{orig}}, I_{\text{dec}}\right) = 10 \log_{10}\left(\frac{p^2}{\text{MSE}\left(I_{\text{orig}}, I_{\text{dec}}\right)}\right) \text{dB} \qquad (21.4)$$

where I_{orig} denotes the original image (without compression), I_{dec} is the compressed–decompressed image (at the output of the system), and MSE is the mean square error between the two images.

21.4.3
Optical JPEG2000 Implementation

In a recent work [26], a new optical compression–decompression system based on wavelet transform (WT) and inspired by JPEG2000 was suggested and tested. Simulation results show that this optical JPEG2000 compressor–decompressor scheme leads to high-quality images (see Ref. [26]). It should also be noted that an adaptation of this technique to color images has been proposed in Ref. [26].

21.5
New Simultaneous Compression and Encryption Approach Based on a Biometric Key and DCT

In Ref. [10], we have discussed major optical compression and encryption methods that are described in the archival literature. We have also pointed out the problem of carrying out these two operations separately [9–28]. It is well known that the compression part can affect the results of the encryption part. Therefore, we developed an algorithm that simultaneously performs the compression and encryption operations, that is, taking the specification of each operation into account. In this section, a new approach, which is based on the optical DCT system, described in Section 21.4, and frequency filtering, is described. This approach can be used to simultaneously compress and encrypt any image [29, 30]. The main advantage of this method, compared to other methods, lies in the dependence between the compression and encryption stages: both stages are linked and optimized simultaneously. Other optical encryption methods are described in Chapter 33.

The optical DCT is used to merge and select relevant information in order to propose an optical system of compression and encryption [29, 30]. A filter should be applied at the output of the DCT module. Figure 21.7 shows the principle of this scheme. First, a DCT (Figure 21.7a and b) is applied to many target images. It is a well-known fact that the output of a DCT is an $(N \times N)$ spectrum, in which the significant frequency components are gathered at the left top corner of the frequency domain. For this reason, we apply a filter of size $c \ll N$, where N is the size of the original image in pixels. The filtered spectra of the target images can be easily merged together into one resulting spectrum by using a simple addition.

21.5 New Simultaneous Compression and Encryption Approach Based on a Biometric Key and DCT

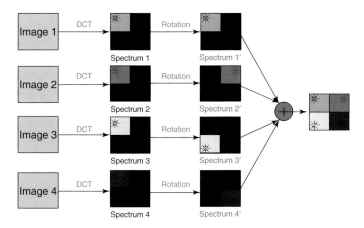

Figure 21.7 Synoptic diagram of the proposed compression and encryption approach: (a) Input image; (b) DCT(s) of the input images; (c) DCT(s) of the input images with rotation; (d) compressed and encrypted spectra [29]. (Please find a color version of this figure on the color plates.)

The resulting spectrum contains all information necessary to retrieve the original images after decompression. With an increasing number of target images, this technique may lead to saturation owing to the fact that all filtered spectra should be merged in the same area of spectrum plane.

To solve the issue, a global spectrum optimization is conducted. As the DCT concentrates the spectrum in the left top corner side of the spectrum plane, the idea of this global optimization is to use the whole spectrum plane. Thus, more filtered spectra can be inserted in the free zones of the spectrum plane. While nondestructively regrouping these various spectra in only one plane, these images are compressed and multiplexed. The number of multiplexed images depends on the minimum required size of their spectra. It is necessary to find a compromise between the desired quality and the size of filtered spectrum. It should be noticed that this operation changes the frequency distribution significantly, which results in the encryption of the mixed and compressed images.

Using the above-described scheme, encrypted images can be easily cracked. This encouraged us to modify this scheme and enhance its encoding capacity. In order to ensure a good level of encryption against any hacking attempt, we propose to change the rotation of the frequencies of the various DCT images, Figure 21.7c, before gathering them. The resulting spectrum is more similar to an FT spectrum than to a DCT spectrum. In this case anyone who does not have the necessary information on the number of spectra and their distribution in the spectral plane will not be able to decrypt this information easily (Figure 21.7d). Without rotating the image spectra, the shape of the DCT spectra is very characteristic and easily detectable thanks to the amplitude that drops gradually while moving away from the high-left corner toward the low-right corner of the spectrum.

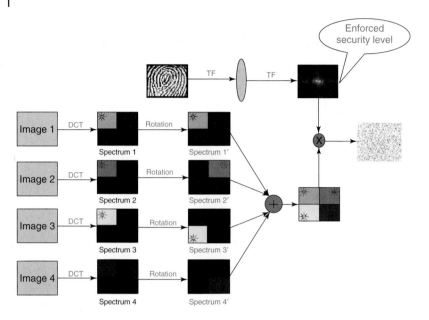

Figure 21.8 Synoptic diagram of the enhanced compression and encryption system using two security levels: (a) Input image; (b) DCT(s) of the input images; (c) DCT(s) of the input images with rotation; (d) compressed and encrypted spectra using one security level; (e) compressed and encrypted spectra using two security levels [29]. (Please find a color version of this figure on the color plates.)

However, the encryption level of the algorithm is still not satisfactory. To deal with this issue, we propose to multiply the multiplexed spectra (Figure 21.8a, b and c) with a random mask or with the spectrum of a corresponding fingerprint (Figure 21.8d). This multiplication aims at changing the spectral distribution of the DCT (Figure 21.8e) and enhances the encryption capacity of our algorithm.

After applying this method and transmitting compressed and encrypted information (Figure 21.9a), the extraction of received images is realized by inversing the various steps performed at the transmitter. First, the received spectrum is multiplied by the inverse of the random mask or the fingerprint spectrum of the authorized corresponding person (Figure 21.9b). Next, the spectrum is separated from the original filtered spectra (Figure 21.9c) and the various inverse rotations are done. Finally, the DCT is inverted in order to obtain the original images (Figure 21.9d).

Many simulations have been performed to test the performance of our approach with two security levels. To simplify the discussion and clarify our ideas, let us consider a simple video sequence composed of two images (Figure 21.10a). After applying our proposed compression and encryption enforced scheme using a random encryption key (Figure 21.8), good results are obtained (Figure 21.10b). In fact, this compressed and encrypted image shows that we succeeded in simultaneously carrying out the compression and encryption operations of the two images presented in Figure 21.10a. Figure 21.10c shows the output images that are

21.5 New Simultaneous Compression and Encryption Approach Based on a Biometric Key and DCT

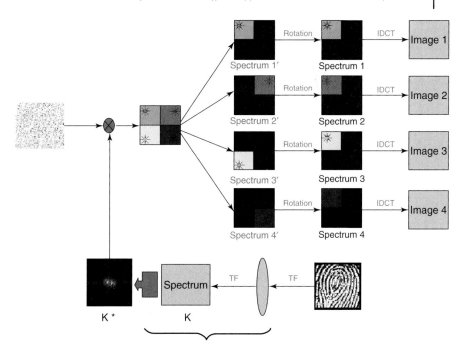

Figure 21.9 Synoptic diagram describing the system of decompression and decryption using two security levels: (a) Compressed and encrypted spectra using two security levels; (b) decompressed and encrypted spectra using one security level; (c) DCT(s) of the input images with rotation; (d) output image. (Please find a color version of this figure on the color plates.)

decrypted and decompressed using the technique presented in Figure 21.9. These reconstructed images demonstrate the validity of this compression–decompression algorithm. Decompressed images having good quality are obtained. To measure the quality of retrieving images, the MSE was suggested as a performance index. The MSE measures the difference between the original target image, the output of our system, and the intermediate images as

$$\text{MSE} = \frac{1}{N \times M} \sum_{i=1}^{N} \sum_{j=1}^{M} |I_{\text{rebuilt}}(i,j) - I_O(i,j)|^2 \qquad (21.5)$$

In the example shown in Figure 21.10, we obtained very low values, that is, MSE1 = 0.0035 and MSE2 = 0.0027. Encouraged by these results, we tested our approach in a correlation situation [30]. In many security or face recognition applications, a huge amount of secure data should be processed, for example, at airport security check points. It is obvious that in such applications, the data should be compressed and encrypted before transmitting them. Standard compression methods, for example, JPEG, can affect the shape of the images, thus reducing the outcome of the correlation stage. In addition, to encrypt the compress data, we should introduce another encryption algorithm that increases the volume of transmitted data without improving the quality of decompressed images. This

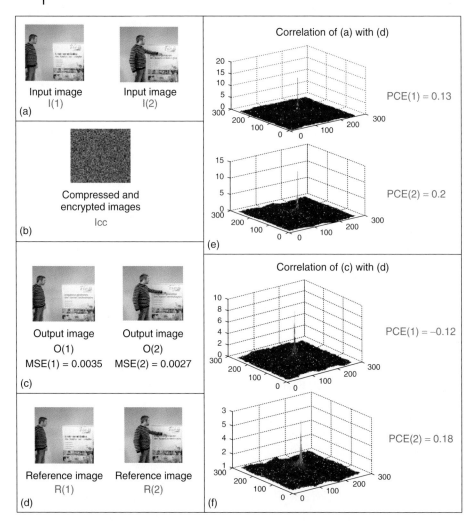

Figure 21.10 (a) Input images, (b) compressed and encrypted image, (c) reconstructed (i.e., decrypted and decompressed) ones, (d) references, (e) correlation results among original target images and the segmented filter, and (f) correlation results among reconstructed images and the same segmented filter [30].

latter problem can be solved using our approach. A recognition system based on an optical multidecision correlator is now proposed [7].

This multidecision correlator has two major advantages: it has a reconfigurable architecture and it uses the segmented composite filter, previously discussed in Refs [7, 8]. This filter divides the Fourier plane into several domains and assigns each one to one reference. Figure 21.11 illustrates the principle of such optical correlator multidecision by using SLMs that display the segmented composite

21.5 New Simultaneous Compression and Encryption Approach Based on a Biometric Key and DCT

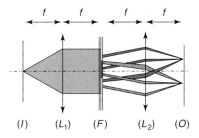

(I)   (L₁)   (F)   (L₂)   (O)

Figure 21.11 Principle of a multidecision correlator using a segmented composite POF filter: *I* denotes the input plane, L_1 is the first convergent lens used to perform the first Fourier transform, *F* is the correlator Fourier plane, that is, the filter plane, the second convergent lens is presented by L_2, and *O* is the output plane, that is, the correlation plane [7].

phase-only filters (SCPOFs) in the Fourier plane. It is worth noting that the phase filters are required in practical optical setups with optimized energy for a multidecision correlation.

Possibilities of parallelism in suitably designed filters, namely, the segmented composite filters, significantly increase the processing capacity. In addition, an optimal use of the space–bandwidth product available in the correlator can be reached, that is, by assigning one channel to every correlation. According to Ref. [7], the local saturation phenomenon in the Fourier plane is much more important in classical composite filters than in segmented filters. This is due to the implementation of composite filters, which can locally add spectral information coming from different references. On the other hand, the structure of the segmented composite filter cancels out this phenomenon by segmenting the Fourier plane in several areas, each of them corresponding to the spectrum of one image. An optimal segmentation is strictly related to the optimization of a criterion. In Ref. [7], an energetic criterion that is insensitive to phase information, Eq. (21.6), was proposed. By neglecting the phase information, some important information could be lost. To overcome this issue, the following optimization of the following criterion was suggested [30]. In the Fourier plane, we assign the pixel (i, j) to the class labeled k if and only if

$$\frac{E_{ij}^k \cos(\phi_{ij}^k)}{\sum_{i=0}^{N}\sum_{j=0}^{N} E_{ij}^k} \geq \begin{cases} \frac{E_{ij}^0 \cos(\phi_{ij}^0)}{\sum_{i=0}^{N}\sum_{j=0}^{N} E_{ij}^0} \\ \frac{E_{ij}^1 \cos(\phi_{ij}^1)}{\sum_{i=0}^{N}\sum_{j=0}^{N} E_{ij}^1} \\ \vdots \\ \frac{E_{ij}^{L-1} \cos(\phi_{ij}^{L-1})}{\sum_{i=0}^{N}\sum_{j=0}^{N} E_{ij}^{L-1}} \end{cases} \quad (21.6)$$

where L stands for the reference number, N is the size of the filter plane, E_{ij}^k and ϕ_{ij}^k denote, respectively, the spectral energy and the phase of class k at pixel location (i,j). To relate this pixel to a reference, Eq. (21.6) should be satisfied. The separation at the output plane is achieved by adding a spatial-frequency-dependent phase distribution to every class.

To validate this approach, a pattern recognition application using the two images presented in Figure 21.10a was considered. Several numerical simulations have been conducted to test the robustness of the proposed technique as well as to show that the proposed simultaneous compression and encryption do not affect the correlation's decision at the receiver. Let us consider two noncompressed bitmap images (I(1) and I(2)) as shown in Figure 21.10a. Using these two images, we generate one compressed and encrypted image (Icc) shown in Figure 21.10b.

At the receiver, the decryption and decompression of image Icc are realized, which generates two other images (both decrypted and decompressed) designated as O(1) and O(2) (Figure 21.10c). Good-quality images with small values of MSE have been obtained. Thanks to the good performance of our approach, the correlator's decision does not show any deterioration in this approach, in contrast to the JPEG compression. Original images I(1) and I(2) are correlated by using a segmented filter (Figure 21.10d). The correlation results are shown in Figure 21.10e. We mention here that the same previous filter should be used at the receiver's side to correlate O(1) and O(2) (Figure 21.10f). To evaluate the performance of the correlation stage, that is, recognition and/or discrimination, several criteria have been proposed in the literature [31]. Here, we use the peak to correlation energy (PCE) defined as the energy of the peak correlation normalized to the total energy of the correlation plane:

$$\text{PCE} = \frac{\sum_{i,j}^{M} E_{\text{peak}}(i,j)}{\sum_{i,j}^{N} E_{\text{correlation\_plane}}(i,j)} \quad (21.7)$$

where M denotes the size of the peak correlation spot and N is the size of the correlation plane. By comparing the correlation planes and PCEs (Figures 21.10e,f), we can notice that there is no significant drop in the PCE values and the correlation peaks are always very sharp and well centered. This enables us to validate our approach and to show that the proposed technique of simultaneous compression and encryption does not deteriorate the performances of the decisions of the correlator using a multicorrelation segmented filter.

21.6
Conclusions

Optical compression methods are expected to play a major role because of their real-time processing capability. This chapter contains a brief description of the state of the art of this topic. Compression methods based on spectral filtering, which can be optically implemented, were first described. The first compression method is based on spectral multiplexing to merge the spectra of target images into one resulting spectrum. This method can give good performance results, especially when the spectra of target images have been optimized. By using an RMS criterion to avoid spectra overlapping, this optimization is done by shifting the different spectra. An optical compression method using a DCT has also been proposed and discussed. This method is based on a hybrid system for which the selection of the relevant information in a JPEG standard has been implemented optically. We were able to perform the DCT thanks to a convergent lens. Another hybrid optical–digital method has been proposed to achieve the wavelet transformation, which can be considered as the basis of JPEG2000. We presented only preliminary results showing the performance of this new method. Further work is needed to conduct real experiments. To enhance the transmission of encrypted images, simultaneous compression–encryption approaches were described. Using an optical DCT system, this approach can compress and encrypt several target images. To improve the security level of this method, a random mask has been introduced to hide the spectral information needed to decompress and retrieve original images. Our experimental studies showed that this method can be a good match for an optical correlation. In closing, it is important to realize that simultaneously realized compression and encryption approaches are attractive because there is no need to use two independent operations, and also because these two stages are closely related to each other. Thus, optimization constraints can be considered in a global manner. Another compression–encryption method that can process multiple target images and at the same time achieve an optimal compression ratio was suggested [27]. Experimental results show that this method can improve the compression ratio of the standard and well-known double random key system (double random phase (DRP)) [28]. This new approach uses a double encryption layer. The first layer is based on performing multiple FT and using several random keys. The second layer is based on the use of the classical DRP system.

References

1. Vander Lugt, A. (1964) Signal detection by complex spatial filtering. *IEEE Trans. Info. Theory*, **IT-10**, 139–145.
2. Goodman, J.W. (1968) *Introduction to Fourier Optics*, McGraw-Hill, New York.
3. Horner, J.L. and Gianino, P.D. (1984) Phase-only matched filtering. *Appl. Opt.*, **23**, 812–817.
4. Alfalou, A. (1999) Implementation of optical multichannel correlators: application to pattern recognition, PhD thesis, Université de Rennes 1, Rennes (France).
5. Alfalou, A. and Brosseau, C. (2010) Understanding correlation techniques for face recognition: from basics to applications, in *Face Recognition*, IN-TECH, 354–380, ISBN-978-953-307-060-5.
6. Kumar, B.V.K.V. (1992) Tutorial survey of composite filter designs for optical correlators. *Appl. Opt.*, **31**, 4773–4801.
7. Alfalou, A., Keryer, G., and de Bougrenet de la Tocnaye, J.L. (1999) Optical implementation of segmented composite filtering. *Appl. Opt.*, **38**, 6129–6136.
8. Alfalou, A., Elbouz, M., and Hamam, H. (2005) Segmented phase-only filter binarized with a new error diffusion approach. *J. Opt. A: Pure Appl. Opt.*, **7**, 183–191.
9. Soualmi, S., Alfalou, A., and Hamam, H. (2007) Optical image

compression based on segmentation of the Fourier plane: new approaches and critical analysis. *J. Opt. A: Pure Appl. Opt.*, **9**, 73–80.
10. Alfalou, A. and Brosseau, C. (2009) Optical image compression and encryption methods. *OSA: Adv. Opt. Photonics*, **1** (3), 589–636.
11. Smith, W.E. and Barrett, H.H. (1983) Radon transform and bandwidth compression. *Opt. Lett.*, **8**, 395–397.
12. Darakis, E. and Soraghan, J.J. (2007) Reconstruction domain compression of phase-shifting digital holograms. *Appl. Opt.*, **46**, 351–356.
13. Naughton, T.J. et al. (2002) Compression of digital holograms for three-dimensional object reconstruction and recognition. *Appl. Opt.*, **41**, 4124–4131.
14. Wijaya, S.L., Savvides, M., and Vijaya Kumar, B.V.K. (2005) Illumination-tolerant face verification of low-bit-rate JPEG2000 wavelet images with advanced correlation filters for handheld devices. *Appl. Opt.*, **44**, 655–665.
15. Ding, L. et al. (2003) Wavelet packet compression for volume holographic image recognition. *Opt. Commun.*, **216**, 105–113.
16. Naughton, T.J., McDonald, J.B., and Javidi, B. (2003) Efficient compression of Fresnel fields for Internet transmission of three-dimensional images. *Appl. Opt.*, **42**, 4758–4764.
17. Ng, T.W. and Ang, K.T. (2005) Fourier-transform method of data compression and temporal fringe pattern analysis. *Appl. Opt.*, **44**, 7043–7049.
18. Yeom, S., Stern, A., and Javidi, B. (2004) Compression of 3D color integral images. *Opt. Express*, **12**, 1632–1642.
19. Cottour, A., Alfalou, A., and Hamam, H. (2008) Optical video image compression: a multiplexing method based on the spectral fusion of information, *IEEE-Conference on Information and Communication Technologies: from Theory to Applications*, IEEE, pp. 1–6.
20. Papoulis, A. (1962) *The Fourier Integral and its Applications*, McGraw-Hill, New York.
21. Alfalou, A. and Brosseau, G. (2010) Exploiting RMS time-frequency structure for multiple image optical compressions and encryption. *Opt. Lett.*, **35**, 1914–1916.
22. Alfalou, A., Elboy, M., and Kerya, G. (2010) New spectral image compression method based on a optimal phase coding and the RMS duration principle. *J. Opt.*, **12**, 115403–115415.
23. Alkholidi, A., Alfalou, A., and Hamam, H. (2007) A new approach for optical colored image compression using the JPEG standards. *Signal Process.*, **87**, 569–583.
24. Alfalou, A. and Alkholidi, A. (2005) Implementation of an all-optical image compression architecture based on Fourier transform which will be the core principle in the realisation of DCT. *Proc. SPIE*, **5823**, 183–190.
25. Alkholidi, A., Alfalou, A., and Hamam, H. (2007) Comparison between the optical and digital JPEG image compression standard. *The 4th International Conference SETIT 2007: Sciences of Electronic, Technologies of Information and Telecommunications, March 25–29*, ISBN: 978-9973-61-475-9.
26. Alkholidi, A. et al. (2008) Real-time optical 2D wavelet transform based on the JPEG2000 standards. *Eur. Phys. J. Appl. Phys.*, **44**, 261–272.
27. Alfalou, A. and Mansour, A. (2009) A new double random phase encryption scheme to multiplex & simultaneous encode multiple images, *Appl. Opt.*, **48** (4), 5933–5947.
28. Refregier, P. and Javidi, B. (1995) Optical image encryption based on input plane and Fourier plane random encoding. *Opt. Lett.*, **20**, 767–769.
29. Loussert, A. et al. (2008) Enhanced System for image's compression and encryption by addition of biometric characteristics. *Int. J. Softw. Eng. Appl.*, **2**, 111–118.
30. Alfalou, A. et al. (2009) A new simultaneous compression and

encryption method for images suitable to optical correlation. Proc. SPIE Em. Seam. Def. (Berlin) (ed. C. Lewis), 74860J, 1–8.

31. Horner, J.L. (1992) Metrics for assessing pattern-recognition performance. *Appl. Opt.*, **31**, 165.

22
Compressive Optical Imaging: Architectures and Algorithms

Roummel F. Marcia, Rebecca M. Willett, and Zachary T. Harmany

22.1
Introduction

Many traditional optical sensors are designed to collect directly interpretable and intuitive measurements. For instance, a standard digital camera directly measures the intensity of a scene at different spatial locations to form a pixel array. Recent advances in the fields of image reconstruction, inverse problems, and compressive sensing (CS) [1, 2] indicate, however, that substantial performance gains may be possible in many contexts via less direct measurements combined with computational methods. In particular, CS allows for the extraction of high-resolution images from relatively small focal plane arrays (FPAs). CS is described in detail in Chapter 23. The basic idea of CS theory is that when the image of interest is very sparse or highly compressible in some basis (i.e., most basis coefficients are small or zero valued), relatively few well-chosen observations suffice to reconstruct the most significant nonzero components. In particular, judicious selection of the type of image transformation introduced by measurement systems may significantly improve our ability to extract high-quality images from a limited number of measurements. By designing optical sensors to collect measurements of a scene according to CS theory, we can use sophisticated computational methods to infer critical scene structure and content.

These ideas are particularly relevant to imaging applications in which it is useful or even crucial to keep the FPA of an optical system relatively small. For example, in low-light settings where sensitive detectors are costly, smaller FPAs translate directly to less expensive systems. Smaller FPAs also make systems lighter and, thus, more portable. In addition, imaging systems with fewer photodetectors consume less power and, therefore, require fewer battery charges. Finally, smaller cameras can fit into tighter spaces for unobtrusive surveillance. An important goal in the design of many imaging systems, then, is to extract as much information as possible from a small number of detector array measurements.

While recent progress in the exploitation of CS theory is highly encouraging, there are several key issues in the context of optical systems that must be addressed:

Optical and Digital Image Processing: Fundamentals and Applications, First Edition. Edited by Gabriel Cristóbal, Peter Schelkens, and Hugo Thienpont.
© 2011 Wiley-VCH Verlag GmbH & Co. KGaA. Published 2011 by Wiley-VCH Verlag GmbH & Co. KGaA.

- In most real-world sensing systems, we face *physical constraints* on the nature of the scene (i.e., photon intensity is always nonnegative) and the nature of the measurements, which can be collected in hardware (i.e., photons can be accumulated, redirected, or filtered, but not "subtracted" in conventional settings).
- A typical CS assumption is that each measurement collected by a sensor is the projection of the image of interest onto a different random vector. It is not clear how *practical optical systems* are best built within this context. In many settings, collecting these projections would result in either a physically very large and cumbersome system or significant noise when sequences of measurements are collected over a limited period of time.
- Typically, it is not possible to wait hours or even minutes for an iterative reconstruction routine to produce a single image; rather, algorithms must be able to operate effectively under *stringent time constraints* and produce image estimates that satisfy the above-mentioned physical constraints.

In this chapter, we explore (i) the potential of several different physically realizable optical systems based on CS principles and (ii) associated fast numerical reconstruction algorithms for overcoming the challenges described above. Specifically, for a fixed size FPA, we describe how compressive measurements combined with sophisticated optimization algorithms can significantly increase image quality and/or resolution.

22.1.1
Organization of the Chapter

The chapter is organized as follows. We describe the compressive sensing problem and formulate it mathematically in Section 22.2. Section 22.3 describes architectures for compressive image acquisition. In particular, we focus on coded aperture imaging and how CS theory can be used for constructing coded aperture masks that can easily be implemented for improving image reconstruction resolution. Section 22.4 discusses algorithms for recovering compressively sensed images while incorporating physical constraints inherent to optical imaging. In addition, we describe a new recovery algorithm that can exploit structure in sparsity patterns. We present experimental results and provide analysis in Section 22.5, discuss the impact of noise and quantization errors in Section 22.6, and offer concluding remarks in Section 22.7.

22.2
Compressive Sensing

To understand the intuition behind the CS framework, consider a $\sqrt{n} \times \sqrt{n}$ image G, which can be written as a length-n column vector **g** and represented in terms of a basis expansion with n coefficients:

$$\mathbf{g} = \sum_{i=1}^{n} \theta_i \mathbf{w}_i$$

where $\mathbf{w}_i$ is the ith basis vector and θ_i is the corresponding coefficient. In many settings, the basis $\mathbf{W} \triangleq [\mathbf{w}_1, \ldots, \mathbf{w}_n]$ can be chosen so that only $k \ll n$ coefficients have significant magnitude, that is, many of the θ_i's are zero or very small for large classes of images; we then say that $\boldsymbol{\theta} \triangleq [\theta_1, \ldots, \theta_n]^t$ is *sparse* or *compressible*. In such cases, it is clear that if we knew *which* $k\,\theta_i$'s were significant, we would ideally just measure these k coefficients directly, which results in fewer measurements to obtain an accurate rpesentation of $\mathbf{g}$. Of course, in general, we do not know a priori which coefficients are significant. The key insight of CS is that, with slightly more than k well-chosen measurements, we can determine which θ_i's are significant and accurately estimate their values.

Furthermore, fast algorithms that exploit the *sparsity* of $\boldsymbol{\theta}$ make this recovery computationally feasible. Sparsity has long been recognized as a highly useful metric in a variety of inverse problems, but much of the underlying theoretical support was lacking. However, more recent theoretical studies have provided strong justification for the use of sparsity constraints and quantified the accuracy of sparse solutions to these underdetermined systems [1, 3].

The problem of estimating the image $\mathbf{G}^{\text{true}} \in \mathbb{R}_+^{\sqrt{n} \times \sqrt{n}}$, where n is the total number of image pixels or voxels, can be formulated mathematically as an inverse problem, where the data collected by an imaging or measurement system are represented as

$$\mathbf{y} = \mathbf{A}\mathbf{g}^{\text{true}} + \mathbf{n} \tag{22.1}$$

where $\mathbf{g}^{\text{true}} \in \mathbb{R}_+^n$ is the image $\mathbf{G}^{\text{true}}$ stored as a length-n column vector of pixel intensities, $\mathbf{A} \in \mathbb{R}_+^{m \times n}$ linearly projects the scene onto an m-dimensional set of observations, $\mathbf{n} \in \mathbb{R}^m$ is noise associated with the physics of the sensor, and $\mathbf{y} \in \mathbb{R}_+^m$ is the observed data. (Typically $\mathbf{n}$ is assumed bounded or bounded with high probability to ensure that $\mathbf{y}$, which is proportional to photon intensities, is nonnegative.) CS addresses the problem of solving for $\mathbf{g}^{\text{true}}$ when the number of unknowns, n, is much larger than the number of observations, m. In general, this is an ill-posed problem as there are an infinite number of candidate solutions for $\mathbf{g}^{\text{true}}$; nevertheless, CS theory provides a set of conditions that, if satisfied, assure an accurate estimation of $\mathbf{g}^{\text{true}}$. We first presuppose that $\mathbf{g}^{\text{true}}$ is sparse or compressible in a basis $\mathbf{W}$. Then given $\mathbf{W}$, we require that $\mathbf{A}$ in conjunction with $\mathbf{W}$ satisfies a technical condition called the *restricted isometry property* (RIP) [1]. More specifically, we say that $\mathbf{AW}$ satisfies the RIP of order s if there exists a constant $\delta_s \in (0, 1)$ for which

$$(1 - \delta_s)\|\boldsymbol{\theta}\|_2^2 \leq \|\mathbf{AW}\boldsymbol{\theta}\|_2^2 \leq (1 + \delta_s)\|\boldsymbol{\theta}\|_2^2 \tag{22.2}$$

holds for all s-sparse $\boldsymbol{\theta} \in \mathbb{R}^n$. In other words, the energy contained in the projected image, $\mathbf{AW}\boldsymbol{\theta}$, is close to the energy contained in the original image, $\boldsymbol{\theta}$. While the RIP cannot be verified for an arbitrary given observation matrix and basis, it has been shown that observation matrices $\mathbf{A}$ drawn independently from many probability distributions satisfy the RIP of order s with high probability for any orthogonal basis $\mathbf{W}$ when $m \geq Cs\log(n/s)$ for some constant C [1]. Although generating CS matrices using this procedure is simple in software, building physical systems to measure $\mathbf{AW}\theta$ can be notoriously difficult.

While the system of equations in Eq. (22.1) can be grossly underdetermined, CS theory suggests that selecting the *sparsest* solution to this system of equations will yield a highly accurate solution, subject to $\theta^{\text{true}} = \mathbf{W}^t \mathbf{g}^{\text{true}}$ being sufficiently sparse and $\mathbf{A}$ satisfying the RIP of order $2k$, where k is the sparsity of $\mathbf{g}^{\text{true}}$ [1]. This problem can be posed as the optimization problem

$$\theta^{\text{est}} = \arg\min_{\theta} \|\theta\|_0 \quad \text{subject to} \quad \|\mathbf{y} - \mathbf{AW}\theta\|_2 \leq \epsilon \quad (22.3\text{a})$$

$$\mathbf{g}^{\text{est}} = \mathbf{W}\theta^{\text{est}} \quad (22.3\text{b})$$

where the ℓ_0 quasinorm $\|\cdot\|_0$ denotes the number of nonzero elements in the argument and $\epsilon \geq 0$ is a bound on the noise. In general, ℓ_0 minimization is NP-hard, since it is combinatorial in nature and its computational complexity grows exponentially with the size of the image [4]. A common alternative is to *relax* this norm to the ℓ_1 norm [2, 5, 6], and instead solve

$$\theta^{\text{est}} = \arg\min_{\theta} \|\theta\|_1 \quad \text{subject to} \quad \|\mathbf{y} - \mathbf{AW}\theta\|_2 \leq \epsilon \quad (22.4\text{a})$$

$$\mathbf{g}^{\text{est}} = \mathbf{W}\theta^{\text{est}} \quad (22.4\text{b})$$

One of the key insights of CS is that in the noiseless case, that is, when the bound ϵ is set to zero, the solution to this relaxed problem is *equivalent* to the solution of the intractable problem in Eq. (22.3) if the matrix product $\mathbf{AW}$ satisfies the RIP 22.2. We discuss equivalent formulations of Eq. (22.4) and various ways of solving them in Section 22.4.

22.3
Architectures for Compressive Image Acquisition

Developing practical optical systems to exploit CS theory is a significant challenge being explored by investigators in the signal processing, optics, astronomy, and coding theory communities. In addition to implicitly placing hard constraints on the nature of the measurements that can be collected, such as nonnegativity of both the projection vectors and the measurements, practical CS imaging systems must also be robust and reasonably sized. Neifeld and Ke [7] describe three general optical architectures for compressive imaging: (i) sequential, where measurements are taken one at a time, (ii) parallel, where multiple measurements taken simultaneously using a fixed mask, and (iii) photon-sharing, where beam-splitters and micromirror arrays are used to collect measurements. Here, we describe some optical hardware with these architectures that have been recently considered in literature.

Rice single-pixel camera: The Rice single-pixel camera [8] is a camera architecture that uses only a single detector element to image a scene. As shown in Figure 22.1,

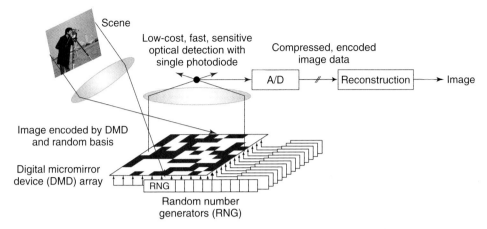

Figure 22.1 A block diagram of the Rice single-pixel camera. Light from the scene to be imaged is reflected off a digital micromirror device (DMD) array whose mirror pattern orientations are generated (pseudo)randomly. Each pattern produces a single measurement. From a collection of measurements, an estimate of the scene is constructed using compressive sensing (CS) algorithms.

a digital micromirror array is used to represent a pseudorandom binary array, and the scene of interest is then projected onto that array before the aggregate intensity of the projection is measured with a single detector. Since the individual orientations of the mirrors in the micromirror array can be altered very rapidly, a series of different pseudorandom projections can be measured successively in relatively small time. The original image is then reconstructed from the resulting observations using CS reconstruction techniques such as those described in Section 22.4. One of the chief benefits of this architecture is that any binary projection matrix can readily be implemented in this system, so that existing CS theory can be directly applied to the measurements. While this "single-pixel" imaging system demonstrates a successful implementation of CS principles, the setup limits the temporal resolution of a dynamically changing scene. Although we can rapidly collect many scenes sequentially at lower exposure, this increases the amount of noise per measurement, thus diminishing its potential for video imaging applications.

Spectral imagers: In spectral imaging settings, a vector containing spectral information about the materials being imaged is assigned to each spatial pixel location, forming a three-dimensional data cube (i.e., two spatial dimensions and one spectral dimension). However, trade-offs between spectral and spatial resolution limit the performance of modern spectral imagers, especially in photon-limited settings where the small number of photons must be apportioned between the voxels in the data cube, resulting in low signal-to-noise ratio (SNR) per voxel. In Ref. [9], investigators propose innovative, real-time spectral imagers, where each pixel measurement is the coded projection of the spectrum in the corresponding spatial location in the data cube. This was implemented using two dispersive elements separated by binary-coded masks. In numerical experiments, a $256 \times 256 \times 15$ spectral image was accurately recovered from a 256×256 observation using an appropriate reconstruction approach. In related work, objects are illuminated by light sources with tunable spectra using spatial light modulators to facilitate compressive spectral image acquisition [10].

Task-specific information compressive imagers: Compressive optical architectures have been considered for target detection systems, which use knowledge of the target's features to reduce the number of measurements required from a more conventional snapshot camera. In particular, investigators [11] note that effective target detection can be performed using a relatively small subset of coefficients of principal components, independent components, generalized matched filters, or generalized Fisher discriminant representations; as a result, directly measuring these coefficients can be more efficient than collecting direct pixel-wise observations of the entire scene and then computing these coefficients in software. The authors further describe ways to optimize the amount of photon energy allotted to each representation element (e.g., each principal component measured) in order to maximize target detection performance. While the proposed architectures are finely tuned to specific target detection tasks and cannot be used for some general-purpose imaging tasks, they, nevertheless, carefully consider the role of physical constraints introduced by optical elements and photon noise.

Coded aperture imagers: Investigators [12, 13] propose practical implementations of CS ideas using coded apertures, demonstrating that if the coded apertures are designed using a pseudorandom construction, then the resulting observation model satisfies the RIP. This approach is described in detail in the next section. This parallel architecture is highly suitable for practical and implementable compressive imaging since it provides a snapshot image (i.e., all m measurements are collected simultaneously) and does not require complex, and potentially large, imaging apparatuses.

22.3.1
Coded Apertures

Coded apertures were first developed to increase the amount of light hitting a detector in an optical system without sacrificing resolution (by, say, increasing the diameter of an opening in a pinhole camera). The basic idea is to use a mask, that is, an opaque rectangular plate with a specified pattern of openings, that allows significantly brighter observations with higher SNR than those from conventional pinhole cameras [14]. These masks encode the image before detection, and the original image is recovered from the distorted observation in postprocessing using an appropriate reconstruction algorithm. The mask pattern introduces a more complicated point-spread function than that associated with a pinhole aperture, and this pattern is exploited to reconstruct high-quality image estimates (Figure 22.2). These multiplexing techniques are particularly popular in astronomical [15] and medical [16] applications because of their efficacy at

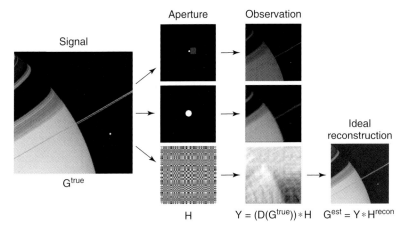

Figure 22.2 An illustration of aperture imaging. An image $\mathbf{G}^{true}$ obtained through a pinhole camera (top) results in dark observations due to the low amount of light allowed through the opening. The aperture can be enlarged (middle) to obtain brighter observations, but this yields blurry observations due to edge effects. Coded apertures, such as MURA patterns $\mathbf{H}^{MURA}$ (bottom), allow substantially more light. The observation $\mathbf{Y}$ is the convolution between a downsampling of the true image (represented by $\mathcal{D}(\mathbf{G}^{true})$) and $\mathbf{H}^{MURA}$. Although the observation has little resemblance to the original image, a highly accurate estimate $\mathbf{G}^{est}$ can be obtained in a postprocessing step involving convolution with a decoding pattern $\mathbf{H}^{recon}$.

wavelengths where lenses cannot be used, but recent work has also demonstrated their utility for collecting both high-resolution images and object depth information simultaneously [17].

Seminal work in coded aperture imaging includes the development of masks based on Hadamard transform optics [18] and pseudorandom phase masks [19]. Modified uniformly redundant array (MURAs) [20] are generally accepted as optimal mask patterns for coded aperture imaging. These mask patterns (which we denote by $\mathbf{H}^{MURA}$) are binary, square patterns, whose *grid size matches the spatial resolution of the FPA* and whose side-length is a prime integer number of grid cells. Each mask pattern is specifically designed to have a complementary pattern $\mathbf{H}^{recon}$ such that $\mathbf{H}^{MURA} * \mathbf{H}^{recon}$ is a single peak with flat side-lobes (i.e., a Kronecker δ function).

In practice, the resolution of a detector array dictates the properties of the mask pattern and, hence, resolution at which $\mathbf{G}^{true}$ can be reconstructed. We model this effect as $\mathbf{G}^{true}$ being downsampled to the resolution of the detector array and then convolved with the mask pattern $\mathbf{H}^{MURA}$, which has the same resolution as the FPA and the downsampled $\mathbf{G}^{true}$, that is,

$$\mathbf{Y} = (\mathcal{D}(\mathbf{G}^{true})) * \mathbf{H}^{MURA} + \mathbf{N} \qquad (22.5)$$

where $*$ denotes convolution, $\mathbf{n}$ corresponds to noise associated with the physics of the sensor, and $\mathcal{D}(\mathbf{G}^{true})$ is the downsampling of the scene, which consists of partitioning $\mathbf{G}^{\text{true}}$ into m uniformly sized blocks and measuring the total intensity in each block. (Recall m is the size of the detector array and, hence, the size of $\mathbf{Y}$.) This is sometimes referred to as *integration downsampling*.

Because of the construction $\mathbf{H}^{\text{MURA}}$ and $\mathbf{H}^{\text{recon}}$, $\mathcal{D}(\mathbf{G}^{\text{true}})$ can be reconstructed using

$$\mathbf{G}^{\text{est}} = \mathbf{Y} * \mathbf{H}^{\text{recon}}$$

However, the resulting resolution is often lower than what is necessary to capture some of the desired details in the image. Clearly, the estimates from MURA reconstruction are limited by the spatial resolution of the photodetector. Thus, high-resolution reconstructions cannot generally be obtained from low-resolution MURA-coded observations. It can be shown that this mask design and reconstruction result in minimal reconstruction errors *at the FPA resolution* and *subject to the constraint that linear, convolution-based reconstruction methods would be used*.

22.3.2
Compressive-Coded Apertures

A recent study by the authors [12] addresses the accurate reconstruction of a high-resolution static image which has a sparse representation in some basis from a single low-resolution observation using *compressive*-coded aperture (CCA) imaging. In this study, we designed a coded aperture imaging mask such that the corresponding observation matrix $\mathbf{A}^{\text{CCA}}$ satisfies the RIP. Here, the measurement matrix $\mathbf{A}^{\text{CCA}}$ associated with CCAs can be modeled as

$$\mathbf{A}^{\text{CCA}} \mathbf{g}^{\text{true}} = \text{vect}(\mathcal{D}(\mathbf{G}^{\text{true}} * \mathbf{H}^{\text{CCA}})) \tag{22.6}$$

where $\mathbf{H}^{\text{CCA}}$ is the coding mask, and $\text{vect}(\cdot)$ is an operator that converts an image into a column vector. Here the coding mask, $\mathbf{H}^{\text{CCA}}$, is the size and resolution at which $\mathbf{G}^{\text{true}}$ will be reconstructed, rather than the size and resolution of the FPA as we had with $\mathbf{H}^{\text{MURA}}$. Thus in Eq. (22.6), we model the measurements as the scene being convolved with the coded mask and *then* downsampled. In contrast, MURA masks are designed for optimal performance on a downsampled version of the scene, as in Eq. (22.5).

The coded aperture masks are designed so that $\mathbf{A}^{\text{CCA}}$ satisfies the RIP of order $2k$ as described in Eq. (22.2) with high probability when $m \geq Ck^3 \log(n/k)$ for some constant C [12, 21]. Note that this is slightly weaker than what can be achieved with a randomly generated sensing matrix that satisfies Eq. (22.2) with high probability when $m \geq Ck \log(n/k)$ for some constant C. The extra factor of k^2 comes from the fact that the m projections sensed using a coded aperture framework are dependent on one another. Stronger results based on a more complex coded aperture architecture were shown recently in Ref. [22]. Specifically, if $m \geq C(k \log(n) + \log^3(k))$ for some constant C, then $\mathbf{g}^{\text{true}}$ can be accurately recovered with high probability. In general, however, to achieve a given level of accuracy, one would need more coded aperture observations than observations from, say, the Rice single-pixel camera. This drawback, however, is mitigated by the fact that coded apertures yield compact and easily implementable snapshot optical designs and relatively little measurement noise for a fixed observation time window.

The convolution of $\mathbf{H}^{\text{CCA}}$ with an image $\mathbf{G}^{\text{true}}$ as in Eq. (22.6) can be represented as the application of the Fourier transform to $\mathbf{G}^{\text{true}}$ and $\mathbf{H}^{\text{CCA}}$, followed by element-wise matrix multiplication and application of the inverse Fourier transform. In matrix notation, this series of linear operations can be expressed as

$$\text{vect}(\mathbf{H}^{\text{CCA}} * \mathbf{G}^{\text{true}}) = \mathcal{F}^{-1}\mathbf{C}_{\mathbf{H}}\mathcal{F}\mathbf{g}^{\text{true}}$$

where $\mathcal{F}$ is the two-dimensional Fourier transform matrix and $\mathbf{C}_{\mathbf{H}}$ is the diagonal matrix whose diagonal elements correspond to the transfer function $\hat{\mathbf{H}}^{\text{CCA}} \triangleq \mathcal{F}(\mathbf{H}^{\text{CCA}})$. The matrix product $\mathcal{F}^{-1}\mathbf{C}_{\mathbf{H}}\mathcal{F}$ is block circulant and each block is in turn circulant (Figure 22.3). Block-circulant matrices, whose entries are drawn from an appropriate probability distribution, are known to be CS matrices. On the basis of recent theoretical work on Toeplitz- and circulant-structured matrices for CS, the proposed masks are fast and memory efficient to generate [12, 21, 22]. In addition, the diagonalizability of block-circulant matrices with circulant blocks by the discrete Fourier transform leads to fast matrix–vector products that are necessary for efficient reconstruction algorithms. The incorporation of the integration downsampling operator $\mathcal{D}$ does not prevent the RIP from being satisfied; a key element of the proof that the RIP is satisfied is a bound on the number of rows of $\mathbf{A}^{\text{CCA}}$ that are statistically independent. Since the downsampling operator effectively sums rows of a block-circulant matrix, downsampling causes the bound on the number of dependent matrix rows to be multiplied by the downsampling factor. Enforcing symmetry on $\mathcal{F}^{-1}\mathbf{C}_{\mathbf{H}}\mathcal{F}$ (Figure 22.3) is equivalent to assuring

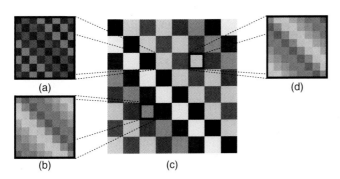

Figure 22.3 The matrix $\mathcal{F}^{-1}\mathbf{C}_{\mathbf{H}}\mathcal{F}$ in (c) is block-circulant with circulant blocks. Enforcing symmetry, that is, requiring the diagonal blocks (a) to be symmetric and opposing nondiagonal blocks (e.g., those denoted by (b) and (d)) to be transposes of each other, produces a transfer function $\hat{\mathbf{H}}^{\text{CCA}}$ that is symmetric about its center, which translates to a real-valued point-spread function, and, consequently, a physically realizable-coded aperture pattern $\mathbf{H}^{\text{CCA}}$. (Please find a color version of this figure on the color plates.)

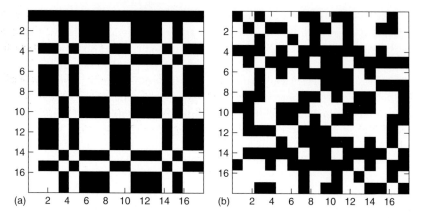

Figure 22.4 Coded aperture patterns. Here, the white blocks represent the openings in the mask pattern. (a) The 17 × 17 MURA mask pattern $\mathbf{H}^{MURA}$. The length of the side of the MURA pattern must be a prime number. (b) An example of a 16 × 16 CCA mask pattern $\mathbf{H}^{CCA}$. The pattern is not quite random – note the symmetry of the CCA pattern about the (9,9) element.

that the transfer function matrix $\hat{\mathbf{H}}^{CCA}$ is symmetric about its center, so that the resulting coding mask pattern $\mathbf{H}^{CCA} \triangleq \mathcal{F}^{-1}(\hat{\mathbf{H}}^{CCA})$ will be necessarily real [12].

Contrasting mask patterns for MURA-coded aperture imaging versus CCA imaging are displayed in Figure 22.4.

22.4
Algorithms for Restoring Compressively Sensed Images

The ℓ_1-minimization problem 22.4a is frequently posed in Lagrangian form:

$$\boldsymbol{\theta}^{est} = \arg\min_{\boldsymbol{\theta}} \frac{1}{2}\|\mathbf{y} - \mathbf{AW}\boldsymbol{\theta}\|_2^2 + \tau \|\boldsymbol{\theta}\|_1 \qquad (22.7)$$

for some regularization parameter $\tau > 0$. This formulation is known as the basis pursuit denoising problem (BPDP) [2], and it has been shown that this $\ell_2-\ell_1$ minimization problem will yield a highly accurate solution of the true underlying image $\mathbf{g}^{true}$ for an appropriate choice of the regularization parameter [3]. In this section, we describe some current methods for solving the relaxed problem 22.7 and its equivalent formulations.

22.4.1
Current Algorithms for Solving the CS Minimization Problem

To apply gradient-based methods to solve Eq. (22.7), its objective function must be made differentiable since the ℓ_1 norm is not differentiable at 0. In Ref. [23], the

BPDP is transformed to

$$\underset{u,\theta}{\text{minimize}} \quad \frac{1}{2}\|y - A\theta\|_2^2 + \tau \sum_{i=1}^{n} u_i \quad \text{subject to} \quad -u_i \leq \theta_i \leq u_i, \quad i = 1, \cdots, n$$

and an interior-point method is applied to obtain the solution. In this approach, iterates are defined in the strict interior of the feasible set, and a logarithmic barrier function is imposed for the bound constraints to prevent the iterates from nearing the boundary of the feasible set prematurely (thus, avoiding potentially very small steps along search directions).

An alternative (and often faster) approach to solving Eq. (22.7) is the gradient projection for sparse reconstruction (GPSR) algorithm [24]. In this method, the variable θ is split into its positive and negative components to make the objective function in Eq. (22.7) differentiable. However, this introduces bound constraints on the new variables. GPSR solves this bound constraint optimization problem using a gradient descent method (called *gradient projection*) that ensures that the iterates remain feasible by projecting (via simple thresholding) onto the feasible region.

The sparse reconstruction by separable approximation (SpaRSA) algorithm [25] reduces Eq. (22.7) to a series of alternating steps: (i) approximating the objective function with a regularized quadratic objective and (ii) performing regularized least squares image denoising. This framework allows users to take advantage of the plethora of fast and effective image denoising methods available. More specifically, the SpaRSA approach defines estimates $\{\theta^j\}$ of θ^{true} by solving a sequence of quadratic subproblems that approximate Eq. (22.7). If $\phi(\theta) = \frac{1}{2}\|y - AW\theta\|_2^2$ in Eq. (22.7), then these subproblems are unconstrained quadratic minimization problems of the form

$$\theta^{j+1} = \underset{\theta}{\arg\min} \ \phi(\theta^j) + (\theta - \theta^j)^t \nabla \phi(\theta^j) + \frac{\alpha_j}{2}\|\theta - \theta^j\|_2^2 + \tau\|\theta\|_1 \tag{22.8}$$

where the first three terms in the objective function correspond to the Taylor expansion of $\phi(\theta)$ at the current iterate θ^j and where the second derivative of $\phi(\theta)$ is approximated by a multiple of the identity matrix, namely, $\nabla^2 \phi(\theta^j) \approx \alpha_j I$, where α_j is defined as in Refs [24, 25]. This approach is particularly fast and effective for RIP-satisfying CS matrices since the near-isometry condition implies that $A^T A \approx \alpha I$. The subproblem 22.8 can be viewed as a denoising subproblem, which, by construction, may have an analytic solution that can easily be calculated. Therefore, the estimates $\{\theta^j\}$ can be computed quickly. We discuss this approach of solving an inverse problem by means of a sequence of denoising problems in the subsequent section where we detail how we can adapt these algorithms to the estimation of a light intensity, an inherently nonnegative quantity.

Other methods solve different formulations of the CS minimization problem. A formulation of the CS minimization problem is least absolute shrinkage and selection operator (LASSO) formulation [5] given by

$$\theta^{\text{est}} = \underset{\theta}{\arg\min} \ \frac{1}{2}\|y - AW\theta\|_2^2 \quad \text{subject to} \quad \|\theta\|_1 \leq \tau$$

In some instances, a penalty different from the ℓ_1 term is used. The Bregman iterative regularization method [26] uses a *total variation* functional $\tau \int |\nabla \mathbf{g}|$ for computing penalties. Finally, we mention (but do not detail for lack of space) two families of algorithms for finding sparse representations: iterative shrinkage/thresholding (IST) algorithms [25] and orthogonal matching pursuit (OMPs) [27]. IST methods iteratively map the objective function to a simpler optimization problem which can be solved by shrinking or thresholding small values in the current estimate of $\boldsymbol{\theta}$. OMP methods start with $\boldsymbol{\theta}^{\text{est}} = 0$ and greedily choose elements of $\boldsymbol{\theta}^{\text{est}}$ to have nonzero magnitude by iteratively processing residual errors between $\mathbf{y}$ and $\mathbf{A}\boldsymbol{\theta}^{\text{est}}$.

22.4.2
Algorithms for Nonnegativity Constrained ℓ_2–ℓ_1 CS Minimization

In optical imaging, we often estimate light intensity, which *a priori* is nonnegative. Thus, it is necessary that the reconstruction $\mathbf{g}^{\text{est}} = \mathbf{W}\boldsymbol{\theta}^{\text{est}}$ is nonnegative, which involves adding constraints to the CS optimization problem (Eq. (22.7)), that is,

$$\boldsymbol{\theta}^{\text{est}} = \arg\min_{\boldsymbol{\theta}} \frac{1}{2}\|\mathbf{y} - \mathbf{A}\mathbf{W}\boldsymbol{\theta}\|_2^2 + \tau\|\boldsymbol{\theta}\|_1 \quad \text{subject to} \quad \mathbf{W}\boldsymbol{\theta} \geq 0 \quad (22.9)$$

The addition of the nonnegativity constraint in Eq. (22.9) makes the problem more challenging than the conventional CS minimization problem and has only been addressed in CS literature recently in the context of photon-limited CS [28, 29]. Here, we discuss an approach similar to those in Refs [25, 29] to address the nonnegativity constraints in the ℓ_2–ℓ_1 CS minimization problem.

Let $\phi(\boldsymbol{\theta}) = \frac{1}{2}\|\mathbf{y} - \mathbf{A}\mathbf{W}\boldsymbol{\theta}\|_2^2$ be the quadratic term in Eq. (22.9). As in the SpaRSA algorithm, the minimization problem (Eq. 22.9) can be solved using a sequence of quadratic approximation subproblems that are easier to solve. The resulting minimization subproblem is given by

$$\boldsymbol{\theta}^{j+1} = \arg\min_{\boldsymbol{\theta}} (\boldsymbol{\theta} - \boldsymbol{\theta}^j)^t \nabla\phi(\boldsymbol{\theta}^j) + \frac{\alpha_j}{2}\|\boldsymbol{\theta} - \boldsymbol{\theta}^j\|_2^2 + \tau\|\boldsymbol{\theta}\|_1 \text{ subject to } \mathbf{W}\boldsymbol{\theta} \geq 0$$

which is similar to the SpaRSA subproblem (22.8) but with the additional nonnegativity constraint on $\mathbf{W}\boldsymbol{\theta}$. This nonnegative denoising subproblem can be written equivalently and more compactly as

$$\boldsymbol{\theta}^{j+1} = \arg\min_{\boldsymbol{\theta}} \frac{1}{2}\|\mathbf{s}^j - \boldsymbol{\theta}\|_2^2 + \frac{\tau}{\alpha_j}\|\boldsymbol{\theta}\|_1 \text{ subject to } \mathbf{W}\boldsymbol{\theta} \geq 0. \quad (22.10)$$

where $\mathbf{s}^j = \boldsymbol{\theta}^j - \frac{1}{\alpha_j}\nabla\phi(\boldsymbol{\theta}^j)$. Because the ℓ_1 norm is nondifferentiable, a change of variables must be applied to Eq. (22.10) to use gradient-based methods. By letting $\boldsymbol{\theta} = \mathbf{u} - \mathbf{v}$ where $\mathbf{u}, \mathbf{v} \geq 0$, we can write Eq. (22.10) as

$$(\mathbf{u}^{j+1}, \mathbf{v}^{j+1}) = \arg\min_{(\mathbf{u},\mathbf{v})} \frac{1}{2}\|\mathbf{s}^j - (\mathbf{u} - \mathbf{v})\|_2^2 + \frac{\tau}{\alpha_k}\mathbf{1}^t(\mathbf{u} + \mathbf{v}) \quad (22.11)$$

$$\text{subject to} \quad \mathbf{u} \geq 0, \mathbf{v} \geq 0, \mathbf{W}(\mathbf{u} - \mathbf{v}) \geq 0$$

where $\mathbb{1}$ is a vector of ones. The next iterate $\boldsymbol{\theta}^{j+1}$ is then defined as $\boldsymbol{\theta}^{j+1} = \mathbf{u}^{j+1} - \mathbf{v}^{j+1}$. Because the constraints on Eq. (22.11) are nonnegativity bounds not only on the variables $\mathbf{u}$ and $\mathbf{v}$ but also on $\mathbf{W}(\mathbf{u} - \mathbf{v})$, solving Eq. (22.11) is not straightforward. However, the *dual* formulation of this minimization problem, that is, solving for the Lagrange multipliers associated with Eq. (22.11), has simple box constraints and can be easily solved iteratively. Using convex optimization theory, we can show that the primal solution, $\boldsymbol{\theta}^{j+1}$, can be obtained from the optimal Lagrange multipliers. We refer the reader to Ref. [29] for algorithmic details. In our numerical experiments below, we solve these subproblems approximately (i.e., with a limited number of iterations) as relatively good estimates of each primal iterate $\boldsymbol{\theta}^j$ tend to be produced with only few inner iterations.

22.4.3
Model-Based Sparsity

While the majority of the CS literature has focused on the case where the scene of interest admits a sparse representation in some basis or dictionary, more recent developments have used more sophisticated models of scenes that incorporate key *structure* into the sparsity models. The basic idea has been used previously in the context of image denoising and compression. For example, it is well known that image denoising can be accomplished via wavelet coefficient-wise thresholding. However, more refined thresholding methods exploit the fact that significant wavelet coefficients tend to cluster near one another within scales and arise at similar locations between scales; this approach can yield significant improvements in accuracy [30].

CS reconstruction methods have recently been developed based upon similar principles to improve reconstruction results [31]. For instance, as we noted above, Eq. (22.10) amounts to an image denoising operation conducted during each loop of the reconstruction algorithm. Thus a variety of denoising methods can be used. Unfortunately, some of these models result in nonconvex optimization problems for which finding the globally optimal reconstruction is not computationally tractable – though locally optimal reconstructions are often very good. Nevertheless, theoretical work in Ref. [31] shows that incorporating sparsity models can reduce the number of measurements needed to achieve a desired accuracy level.

In the case where the image of interest is known to be smooth or piecewise smooth (i.e., it is compressible in a wavelet basis), we can formulate a penalty function that is a useful alternative to the ℓ_1 norm of the wavelet coefficient vector. In particular, we can build on the framework of recursive dyadic partitions (*RDP*), which are described in detail in Ref. [32]. In particular, partition-based methods calculate image estimates by determining the ideal partition of the domain of observations and by fitting a model (e.g., a constant) to each cell in the optimal partition. This gives our estimators the capability to spatially vary the resolution to automatically increase the smoothing in very regular regions of the image and to preserve detailed structure in less regular regions.

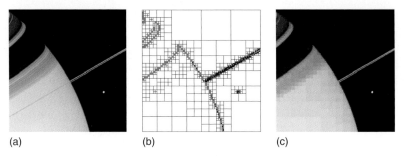

Figure 22.5 Recursive dyadic partition of an image. (a) Original image $\mathbf{G}^{\text{true}}$. (b) RDP $\mathbf{P}$, with larger partition cells corresponding to regions of more homogeneous intensity. (c) Piecewise constant approximation $\mathbf{G}(\mathbf{P})$ to original image, with constant pieces corresponding to RDP cells.

An image $\mathbf{G}^{\text{true}}$ (a), an RDP $\mathbf{P}$ (b), and an image approximation, $\mathbf{G}(\mathbf{P})$ (c), are displayed in Figure 22.5. RDP-based image estimates are computed using a very simple framework for penalized least-squares estimation, wherein the penalization is based on the complexity of the underlying partition (i.e., the number of cells in the partition). The goal here is to find the partition that minimizes the penalized squared error function. Let $\mathbf{s}^j = \mathbf{g}^j - \frac{1}{\alpha_j} \nabla \phi(\mathbf{g}^j)$ where $\phi(\mathbf{g}) = \frac{1}{2} \|\mathbf{y} - \mathbf{A}\mathbf{g}\|_2^2$, and set

$$\mathbf{P}^{j+1} = \arg\min_{\mathbf{P} \in \mathcal{P}} \|\mathbf{g}(\mathbf{P}) - \mathbf{s}^j\|_2^2 + \text{pen}(\mathbf{P}) \quad \text{subject to} \quad \mathbf{g}(\mathbf{P}) \geq 0$$

$$\mathbf{g}^{j+1} = \mathbf{g}(\mathbf{P}^{j+1})$$

where $\mathcal{P}$ is a collection of all possible RDPs of the image domain and $\mathbf{g}(\mathbf{P}) = \text{vect}(\mathbf{G}(\mathbf{P}))$. A search over partitions can be computed quickly using a dynamic program. The model coefficients for each partition cell are chosen via nonnegative least squares. Enforcing nonnegativity constraints is trivial and can be accomplished very quickly.

22.4.4
Compensating for Nonnegative Sensing Matrices

Generative models for random projection matrices used in CS often involve drawing elements independently from a zero-mean probability distribution [1, 3, 21], and likewise a zero-mean distribution is used to generate the coded aperture masks described in Section 22.3.1. However, a coded aperture mask with a zero mean is not physically realizable in optical systems. We generate our physically realizable mask by taking randomly generated, zero-mean mask pattern and shifting it so that all mask elements are in the range $[0, 1/m]$, where m is the dimension of the observation [12]. This shifting ensures that the coded aperture corresponds to a valid (i.e., nonnegative and intensity preserving) probability transition matrix which describes the distribution of photon propagation through the optical system.

This shifting, while necessary to accurately model real-world optical systems, negatively impacts the performance of the proposed $\ell_2-\ell_1$ reconstruction algorithm for the following reason. If we generate a nonrealizable zero-mean mask ($\mathbf{H}^0$) with elements in the range $[-1/2m, 1/2m]$ and simulate measurements of the form

$$\mathbf{y}^0 = \mathbf{A}^0 \mathbf{g}^{\text{true}} \triangleq \text{vect}(\mathcal{D}(\mathbf{G}^{\text{true}} * \mathbf{H}^0)) \qquad (22.12)$$

then the corresponding observation matrix $\mathbf{A}^0$ will satisfy the RIP with high probability and $\mathbf{g}^{\text{true}}$ can be accurately estimated from $\mathbf{y}^0$ using $\ell_2-\ell_1$ minimization. In contrast, if we rescale $\mathbf{H}^0$ to be in the range $[0, 1/m]$ and denote this $\mathbf{H}$, then we have a practical and realizable-coded aperture mask. However, observations of the form

$$\mathbf{y} = \mathbf{A}\mathbf{g}^{\text{true}} \triangleq \text{vect}(\mathcal{D}(\mathbf{G}^{\text{true}} * \mathbf{H}))$$

lead to an objective function whose second derivative is no longer accurately approximated by a scalar multiple of the identity matrix, thereby mitigating the effectiveness of methods that exploit this property. This problem is addressed in Ref. [12], where it is shown that the zero-mean observation $\mathbf{y}^0$ can be estimated by shifting each measurement in $\mathbf{y}$. Let $C_\mathbf{A}$ be the sum of each column of $\mathbf{A}$ and $\mu \triangleq \sum_{i=1}^{m} y_i / C_\mathbf{A}$; note $\mathbf{y}^0 \approx \mathbf{y} - (\mu/2m)\mathbb{1}_{m \times 1}$. We can then solve

$$\boldsymbol{\theta}^{\text{est}} = \arg\min_{\boldsymbol{\theta}} \frac{1}{2}\|\mathbf{y}^0 - \mathbf{A}^0 \mathbf{W}\boldsymbol{\theta}\|_2^2 + \text{pen}(\boldsymbol{\theta}) \quad \text{subject to} \quad \mathbf{W}\boldsymbol{\theta} + \mu\mathbb{1} \geq 0$$

$$(22.13)$$

where $\text{pen}(\boldsymbol{\theta})$ is either the ℓ_1 or the model-based sparsity penalty. The estimate for $\mathbf{g}^{\text{est}}$ is then given by $\mathbf{g}^{\text{est}} = \mathbf{W}\boldsymbol{\theta}^{\text{est}} + \mu\mathbb{1}$. We note that the algorithms described in Sections 22.4.2 and 22.4.3 can be modified to solve Eq. (22.13) in a straightforward manner.

22.5 Experimental Results

In this section, the utility of optical system designs based on CS theory for improving the resolution of optical systems is demonstrated via a simulation study. In particular, we consider several different optical mechanisms that could be used to image Saturn and its rings using a "ground truth" image displayed in Figure 22.5a; this image was originally obtained by the Cassini Orbiter [33] and was cropped and downsampled to size 256×256 for our numerical experiments. The pixel intensities range from 0 to 4.53. We add white Gaussian noise to the projected measurements $\mathbf{Ag}^{\text{true}}$ to model the sensor noise associated with the FPA. In the pinhole and coded aperture experiments, we acquire m measurements over a total time T s. We assume that this averaged noise has standard deviation $\sigma = 10^{-5}$. The single-pixel camera collects measurements sequentially and, consequently, it has at most T/m s to obtain each of the m measurements. Thus, the noise

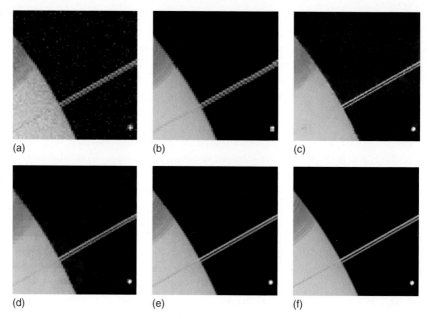

Figure 22.6 Simulation results comparing different image acquisition and reconstruction strategies. (a) Observations from a pinhole aperture, upsampled using nearest-neighbor interpolation; RMS error = 0.117. (b) Reconstruction from an MURA-coded aperture camera system, upsampled using nearest-neighbor interpolation; RMS error = 0.071. (c) Translationally invariant RDP-base reconstruction from a single-pixel camera system; RMS error = 0.064. (d) $\ell_2-\ell_1$ reconstruction from a CCA camera system; RMS error = 0.067. (e) $\ell_2-\ell_1$ (with nonnegativity constraint) reconstruction from a CCA Camera system; RMS error = 0.049. (f) Translationally invariant RDP-based reconstruction; RMS error = 0.032. Note the higher resolution in the reconstructions (d–f) from a CCA camera system and how performance can be further improved using cycle-spun RDP-based reconstruction methods.

variance associated with the single-pixel camera model must be scaled appropriately, although this disadvantage is offset by the fact that the single-pixel camera can implement arbitrary random projects and, hence, satisfy the RIP for smaller FPA dimension m.

Figure 22.6a shows the result of observing the scene with a pinhole camera and an FPA of size 128×128. This image clearly has lower resolution than the original (i.e., it is an image with 128×128 pixels, upsampled here via nearest-neighbor interpolation for visualization and computing errors). It also exhibits noise artifacts due to the relatively low-light intensity reaching the camera detectors relative to the noise level. The root-mean-squared (RMS) error of this image, $\|\mathbf{g}^{est} - \mathbf{g}^{true}\|_2 / \|\mathbf{g}^{true}\|_2$, is 0.117. Conventional coded aperture techniques, such as the MURA methods described above, can be used to improve the noise performance of a camera. This is demonstrated in Figure 22.6b, where the MURA code has a fill factor of 50% and the reconstruction has an RMS error of 0.071. However, these conventional coded aperture systems cannot resolve details smaller than the pixel size of the FPA. Note the pixelization of Saturn's moon (Mimas), rings, and edges in the MURA reconstruction.

Next, we describe the improvements obtained by taking compressive measurements. First, we simulate the image acquisition process in the single-pixel camera by taking random projections of the image and then solve the corresponding CS minimization problem. The reconstruction for the single-pixel camera is shown in Figure 22.6c. We note the improved resolution of Mimas and Saturn's rings. For brevity, we show only the reconstruction from the RDP-based method we proposed, which significantly outperformed the other methods we discussed.

We also observe that the CCA techniques described in this chapter can lead to further gains. For instance, Figure 22.6d shows the reconstructed 256×256 image computed via GPSR [24] (using the Haar wavelet basis) from CCA data collected on a 128×128 FPA with a 50% fill factor coding mask; the RMS error here is 0.067. (While a number of different convergence criteria have been successfully employed for the reconstruction algorithms, in this section we simply display the result after 100 iterations. A more detailed perspective on RMS error decay across iterations or computation time is described in Ref. [29]; the work considers a Poisson noise model but otherwise uses reconstruction methods equivalent to the ones considered here.) We note that this result has over 4200 negatively valued pixels. Additional gains are possible by incorporating positivity constraints, as shown in Figure 22.6e, where the RMS error is 0.049 and none of the pixels have negative values.

Finally, including structure in sparsity models during the reconstruction process via RDPs yields additional improvements. Using RDP-based denoising within our optimization framework yields an RMS error of 0.041 and zero pixels with negative values. However, a cycle-spun version of RDP-based denoising further improves performance, as shown in Figure 22.6f, where the RMS error is 0.032, a 73% improvement over the pinhole camera result, 55% improvement over the MURA result, and a 50% improvement over the single-pixel camera result. In addition, Figure 22.6 shows that CCAs significantly improve our ability to resolve fine features, such as the side-view of Saturn's rings, its moon, and bands in the planet's atmosphere. We note that the $\ell_2 - \ell_1$ CS minimization with nonnegativity constraints and the partition-based approaches (Figure 22.6e and f) were initialized using the GPSR result thresholded to satisfy the nonnegativity constraints.

One other advantage that CCAs have over the Rice single-pixel camera is the computational time necessary to perform reconstruction. Because the observation matrices associated with CCA involve Fourier transforms and diagonal matrices, matrix–vector products are very fast and memory efficient. In contrast, the numerical methods associated with the Rice single-pixel camera are significantly slower and more memory intensive due to the dense, unstructured projection matrix.

22.6
Noise and Quantization

While CS is particularly useful when the FPA needs to be kept compact, it should be noted that CS is more sensitive to measurement errors and noise than more direct imaging techniques. The experiments conducted in this chapter simulated very high SNR settings and showed that CS methods can help resolve high-resolution features in images. However, in low SNR settings, CS reconstructions can exhibit significant artifacts that may even cause more distortion than the low-resolution artifacts associated with conventional coded aperture techniques such as MURA.

The coded aperture work discussed earlier is focused on developing a coded aperture system with a *fill factor* of 50%; that is, 50% of the positions in the coded mask were opaque, and the remainder allowed light through to the detector. In low-light settings, this high fill factor is desirable because it allows a significant proportion of the light through to the detector and "wastes" very few photons. This approach is particularly effective when the scene is sparse in the canonical or pixel basis (e.g., faint stars against a dark sky). However, when the scene is sparse in some other basis, such as a wavelet basis, and *not* sparse in the pixel basis, a large fill factor can cause significant noise challenges. These challenges are described from a theoretical perspective in Ref. [28]. Intuitively, consider that when the fill factor is close to 50%, most of the coded aperture measurements will have the same average intensity plus a small fluctuation, and the photon noise level will scale with this average intensity. As a result, in low-light settings, the noise will overwhelm the small fluctuations that are critical to accurate reconstruction unless the scene is sparse in the pixel basis and the average intensity per pixel is low. These challenges can be mitigated slightly by using smaller fill factors, but in general limit the utility of *any* linear optical CS architecture for very low intensity images that are not sparse in the canonical basis [28].

Similar observations are made in Ref. [34], which presents a direct comparison of the noise robustness of CS in contrast to conventional imaging techniques both in terms of bounds on how reconstruction error decays with the number of measurements and in a simulation setup; the authors conclude that for most real-world images, CS yields the largest gains in high-' SNR settings. These observations are particularly relevant when considering the bit depth of FPAs, which corresponds to measurement quantization errors. Future efforts in designing optical CS systems must carefully consider the amount of noise anticipated in the measurements to find the optimal trade-off between the FPA size and image quality. (For a description of image quality assessment techniques, see Chapter 19.)

22.7
Conclusions

This chapter describes several recent efforts aimed at the exploitation of important theoretical results from CS in practical optical imaging systems. One of the main tenets of CS is that relatively few well-chosen observations can be used to form a sparse image using sophisticated image reconstruction algorithms. This suggests that it may be possible to build cameras with much smaller FPAs than are conventionally required for high-resolution imaging. However, the application of these ideas in practical settings poses several challenges.

First, directly implementing CS theory by collecting a series of independent pseudorandom projections of a scene requires either a very large physical system or observations collected sequentially over time. This latter approach is successfully used, for instance, in the Rice single-pixel camera. Alternative snapshot architectures (which capture all observations simultaneously) with a compact form factor include coded aperture techniques. These approaches impose structure upon the pseudorandom projections, most notably by limiting their independence. As a result, the number of measurements required to accurately recover an image is higher with snapshot-coded aperture systems.

A second key challenge relates to the nonnegativity of image intensities and measurements that can be collected by linear optical systems. Much of the theoretical literature on CS allows for negative measurements and does not consider nonnegativity during the reconstruction process. In this chapter, we have shown that (i) explicitly incorporating nonnegativity constraints can improve reconstruction accuracy and (ii) preprocessing observations to account for nonnegative sensing matrices improves reconstruction performance because of central assumptions underlying some fast CS algorithms. However, one important open question is whether novel approaches based on *nonlinear* optics can successfully circumvent these positivity constraints to improve performance.

Acknowledgments

This research is supported by NSF CAREER Award No. CCF-06-43947, NSF Award No. DMS-08-11062, DARPA Grant No. HR0011-07-1-003, and AFRL Grant No. FA8650-07-D-1221.

References

1. Candès, E.J. and Tao, T. (2005) Decoding by linear programming. *IEEE Trans. Inform. Theory*, **15**, 4203–4215.
2. Donoho, D.L. (2006) Compressed sensing. *IEEE Trans. Inform. Theory*, **52**, 1289–1306.
3. Haupt, J. and Nowak, R. (2006) Signal reconstruction from noisy random projections. *IEEE Trans. Inform. Theory*, **52**, 4036–4048.
4. Natarajan, B.K. (1995) Sparse approximate solutions to linear systems. *SIAM J. Comput.*, **24**, 227–234.
5. Tibshirani, R. (1996) Regression shrinkage and selection via the LASSO. *J. R. Statist. Soc. Ser. B*, **58**, 267–288.
6. Tropp, J.A. (2006) Just relax: convex programming methods for identifying sparse signals in noise. *IEEE Trans. Inform. Theory*, **52**, 1030–1051.
7. Neifeld, M. and Ke, J. (2007) Optical architectures for compressive imaging. *Appl. Opt.*, **46**, 5293–5303.
8. Duarte, M.F. *et al.* (2008) Single pixel imaging via compressive sampling. *IEEE Sig. Proc. Mag.*, **25**, 83–91.

9. Gehm, M. et al. (2007) Single-shot compressive spectral imaging with a dual-disperser architecture. *Opt. Express*, **15**, 14013–14027.
10. Maggioni, M. et al. (2006) Hyperspectral microscopic analysis of normal, benign and carcinoma microarray tissue sections, in Proceedings of SPIE, Optical Biopsy VI, San Jose, CA, vol. 6091.
11. Ashok, A., Baheti, P.K., and Neifeld, M.A. (2008) Compressive imaging system design using task-specific information. *Appl. Opt.*, **47**, 4457–4471.
12. Marcia, R.F. and Willett, R.M. (2008) Compressive Coded Aperture Superresolution Image Reconstruction. Proceedings of IEEE International Conference on Acoustics, Speech, Signal Processing, Las Vegas, NV.
13. Stern, A. and Javidi, B. (2007) Random projections imaging with extended space-bandwidth product. *IEEE/OSA J. Disp. Technol.*, **3**, 315–320.
14. Dicke, R.H. (1968) Scatter-hole cameras for X-rays and gamma-rays. *Astrophys. J.*, **153**, L101–L106.
15. Skinner, G. (1984) Imaging with coded-aperture masks. *Nucl. Instrum. Methods Phys. Res.*, **221**, 33–40.
16. Accorsi, R., Gasparini, F., and Lanza, R.C. (2001) A coded aperture for high-resolution nuclear medicine planar imaging with a conventional anger camera: experimental results. *IEEE Trans. Nucl. Sci.*, **28**, 2411–2417.
17. Levin, A. et al. (2007) Image and Depth from a Conventional Camera with a Coded Aperture. Proceedings of International Conference on Computer Graphics and Interactive Techniques, San Diego, CA.
18. Sloane, N.J.A. and Harwit, M. (1976) Masks for Hadamard transform optics, and weighing designs. *Appl. Opt.*, **15**, 107.
19. Ashok, A. and Neifeld, M.A. (2007) Pseudorandom phase masks for superresolution imaging from subpixel shifting. *Appl. Opt.*, **46**, 2256–2268.
20. Gottesman, S.R. and Fenimore, E.E. (1989) New family of binary arrays for coded aperture imaging. *Appl. Opt.*, **28**, 4344–4352.
21. Bajwa, W. et al. (2007) Toeplitz-structured Compressed Sensing Matrices. Proceedings of IEEE Statistical Signal Processing Workshop, Madison, WI.
22. Romberg, J. (2009) Compressive sensing by random convolution. *SIAM J. Imaging Sci.*, **2**, 1098–1128.
23. Kim, S.J. et al. (2007) An interior-point method for large-scale ℓ_1-regularized least squares. *IEEE J. Sel. Top. Sign. Proces.*, 606–617.
24. Figueiredo, M.A.T., Nowak, R.D., and Wright, S.J. (2007) Gradient projection for sparse reconstruction: Application to compressed sensing and other inverse problems. *IEEE J. Sel. Top. Sign. Proces.: Special Issue on Convex Optimization Methods for Signal Processing*, **1**, 586–597.
25. Wright, S., Nowak, R., and Figueiredo, M. (2009) Sparse reconstruction by separable approximation. *IEEE Trans. Signal Process.*, **57**, 2479–2493.
26. Yin, W. et al. (2008) Bregman iterative algorithms for ℓ_1-minimization with applications to compressed sensing. *SIAM J. Imag. Sci.*, **1**, 143–168.
27. Tropp, J.A. (2004) Greed is good: Algorithmic results for sparse approximation. *IEEE Trans. Inform. Theory*, **50**, 2231–2242.
28. Raginsky, M. et al. (2010) Compressed sensing performance bounds under Poisson noise. *IEEE Trans. Signal Process.*, **58**, 3990–4002.
29. Harmany, Z., Marcia, R., and Willett, R. (2009) Sparse Poisson Intensity Reconstruction Algorithms. Proceedings of IEEE Statistics of Signal Processing Workshop, Cardiff, UK.
30. Crouse, M.S., Nowak, R.D., and Baraniuk, R.G. (1998) Wavelet-based statistical signal-processing using hidden Markov-models. *IEEE Trans. Sig. Proc.*, **46**, 886–902.
31. Baraniuk, R. et al. (2008) Model-based compressive sensing. *IEEE Trans. Inform. Theory*, **56**, 1982–2001.
32. Willett, R. and Nowak, R. (2007) Multiscale Poisson intensity and density

estimation. *IEEE Trans. Inform. Theory*, **53**, 3171–3187.

33. NASA/JPL/Space Science Institute (2008) Pia10497: Saturn in recline, *http://photojournal.jpl.nasa.gov/catalog/PIA10497*.

34. Haupt, J. and Nowak, R. (2006) Compressive Sampling vs. Conventional Imaging. Proceedings of IEEE International Conference on Imaging Processing, Atlanta, GA, pp. 1269–1272.

23
Compressed Sensing: "When Sparsity Meets Sampling"

Laurent Jacques and Pierre Vandergheynst

23.1
Introduction

The 20th century had seen the development of a huge variety of sensors, which help acquire measurements that provide a faithful representation of the physical world (e.g., optical sensors, radio receivers, and seismic detector). Since the purpose of these systems was to directly acquire a meaningful "signal," a very fine sampling of this latter had to be performed. This was the context of the famous Shannon–Nyquist condition, which states that every continuous (a priori) bandlimited signal can be recovered from its discretization provided its sampling rate is at least two times higher than its cutoff frequency.

However, a recent theory named compressed sensing (or compressive sampling (CS)) [1, 2] states that this lower bound on the sampling rate can be highly reduced, as soon as, first, the sampling is generalized to any linear measurement of the signal, and, second, specific a priori hypotheses on the signal are realized. More precisely, the sensing pace is reduced to a rate equal to a few multiples of the intrinsic signal dimension rather than to the dimension of the embedding space.

Technically, this simple statement is a real revolution both in the theory of reliable signal sampling and in the physical design of sensors. It means that a given signal does not have to be acquired in its initial space as thought previously, but it can really be observed through a "distorting glass" (provided it is linear) with fewer measurements. The couple *encoder* (sensing) and *decoder* (reconstruction) are also completely asymmetric: the encoder is computationally light and linear, and hence completely independent of the acquired signal (nonadaptive), while the decoder is nonlinear and requires high CPU power.

In this chapter, we emphasize how the CS theory may be interpreted as an evolution of the Shannon–Nyquist sampling theorem. We explain that the specificity of this new theory is the generalization of the *prior* made on the signal. In other words, we associate the CS theory with the important concept of "*sparsity*," expressing the signal as the linear combination of few elements taken in a particular basis (orthogonal or redundant).

Optical and Digital Image Processing: Fundamentals and Applications, First Edition. Edited by Gabriel Cristóbal, Peter Schelkens, and Hugo Thienpont.
© 2011 Wiley-VCH Verlag GmbH & Co. KGaA. Published 2011 by Wiley-VCH Verlag GmbH & Co. KGaA.

23.1.1
Conventions

We extensively use the following mathematical notations. A discrete *signal* x denotes an N-dimensional vector $x \in \mathbb{R}^N$ of components x_j for $1 \leq j \leq N$. The support of x is supp $x = \{1 \leq j \leq N : x_j \neq 0\}$, that is, a subset of the index set $\{1, \ldots, N\}$.

When $\mathbb{R}^N$ is seen as a Hilbert space, the scalar product between two vectors in $\mathbb{R}^N$ is denoted as $\langle u, v \rangle = u^* v = \sum_{j=1}^{N} u_j v_j$, where $\langle u, u \rangle = \|u\|_2^2$ is the square of the Euclidean ℓ_2 norm of u. The ℓ_p norm of x for $p > 0$ is $\|x\|_p = (\sum_{j=1}^{N} |x_j|^p)^{1/p}$, and by extension, the ℓ_0 "quasi" norm of x is $\|x\|_0 = \#$ supp x, or the number of nonzero elements in x.

If $1 \leq K \leq N$, $x_K \in \mathbb{R}^N$ is the best K-term approximation of x in a given basis (see Sections 23.2 and 23.3). If T is a subset of $\{1, \ldots, N\}$ of size $\#T$, according to the context, x_T is either the restriction of x to T, or a thresholded copy of x to the indices in T, so that $(x_T)_j = x_j$ for $j \in T$ and $(x_T)_j = 0$ otherwise.

The identity matrix, or equivalently the canonical basis in $\mathbb{R}^N$, is written as $I \in \mathbb{R}^{N \times N}$. Given a matrix $A \in \mathbb{R}^{M \times N}$, A^* is its transposition (or adjoint), and if $M < N$, $A^\dagger$ is the pseudoinverse of A^*, with $A^\dagger = (AA^*)^{-1}A$ and $A^\dagger A^* = I$. The Fourier basis is denoted by $F \in \mathbb{R}^{N \times N}$.

For a *convex* function $f : \mathbb{R}^N \to \mathbb{R}$, arg min$_x f(x)$ returns the x that minimizes f. For constrained minimization, "s.t." is a shorthand for "*subject to*", as in arg min$_x f(x)$ s.t. $\|x\|_2 \leq 1$. The typical big-O notation $A = O(B)$ means that there exists a constant $c > 0$ such that $A \geq cB$.

23.2
In Praise of Sparsity

The concept of sparse representations is one of the central methodologies of modern signal processing and it has had tremendous impact on numerous application fields. Despite its power, this idea is genuinely simple and intuitive. Given an N-dimensional signal x, it is often easy to decompose it as a linear superposition of $K \ll N$ elementary signals, called *atoms*:

$$x = \sum_{k=1}^{K} \alpha_k \psi_k \tag{23.1}$$

The equality in Eq. (23.1) may not need to be reached, in which case a K-term approximant $\tilde{x}_K$ is found:

$$\tilde{x}_K = \sum_{k=1}^{K} \alpha_k \psi_k, \quad \text{with} \quad \|x - \tilde{x}_K\|_2 \leq \epsilon(K) \tag{23.2}$$

for some approximation error ϵ. Such an approximant is sometimes called (ϵ, K)-*sparse*. These K atoms, ψ_k, are chosen from a large collection called a *dictionary*, which can be conveniently represented by a large $N \times D$ matrix Ψ, with $D \geq N$, where each column is an atom. Strictly speaking, there is no restriction on the dictionary but usually the atoms are chosen normalized, that is, $\|\psi_k\|_2 = 1$. With these conventions, Eq. (23.1) can be written as $x = \Psi\alpha$, where $\alpha \in \mathbb{R}^D$. Note that an exact sparse representation of the form (23.1) may be a very strong requirement. In many cases, this assumption is replaced by a weaker notion of sparsity called *compressibility*. A vector α is termed compressible if its entries sorted in decreasing order of magnitude decay as a power law: $|\alpha_k| \leq c k^{-c'}$ for some constants c, $c' > 0$. Alternatively, this compressibility may be characterized by the decay of the ℓ_2-approximation error $e(K)$ obtained by the best K-term approximation $x_K = \Psi\alpha_K$ with $\|\alpha_K\|_0 = K$. This error is such that

$$e(K) = \|x - x_K\|_2 \leq \|x - x'\|_2$$

for any other $x' = \Psi\alpha'$ such that $\|\alpha'\|_0 \leq K$.

Sparse representations have been traditionally used in signal processing as a way to compress data by trying to minimize the number of atoms K in the representation. However, sparsity has recently appeared as a defining property of signals and sparse signal models are by now very common as we shall see. The success of these sparse models started out with wavelet nonlinear approximations. Indeed, many interesting signal models are sparse models involving wavelet series. For example, piecewise smooth signals or images yield wavelet decomposition coefficients that are compressible in the sense defined above: most of the information is concentrated in few big coefficients that characterize the discontinuities in the signal, or edges in the image. The main intuitive idea behind wavelet denoising, for example, is to realize that while the signal is represented by sparse wavelet coefficients, noise will induce a lot of small coefficients. They can be removed by enforcing sparsity via thresholding. There are many other signal models involving sparsity: for example, locally oscillatory signals are sparse on the MDCT[1] basis and are widely used in audio for tonal components or to model textures in images. This example also shows that complex signals cannot be well modeled by a single basis: an image contains edges, but it often contains textures as well and the latter are not represented in a sparse way by wavelets. Generating sparsity often requires the use of a collection of bases, or a dictionary.

The ultimate goal of sparse representation techniques would be to find the best, that is, the sparsest, possible representation of a signal; in other words, the goal is to solve the following problem:

$$\arg\min_u \|u\|_0 \quad \text{s.t.} \quad x = \Psi u \qquad \text{(Exact Sparse)}$$

If the dictionary is well adapted to the signal, it is very likely that a very sparse representation or approximation exists. When Ψ is an orthonormal basis, there is a unique solution to that problem: the coefficients are computed by projections on the basis $\alpha = \Psi^* x$. Unfortunately, the problem of finding a sparse expansion of a signal in a generic dictionary leads to a daunting NP-complete combinatorial

---

1) Modified Discrete Cosine Transform.

optimization problem. In Ref. [3], Chen et al. proposed to solve the following slightly different problem coined basis pursuit (BP):

$$\arg \min_{u} \|u\|_1 \quad \text{s.t.} \quad x = \Psi u \tag{BP}$$

Minimizing the ℓ_1 norm helps in finding a sparse approximation, because it prevents diffusion of the energy of the signal over a lot of coefficients. While preserving the essential property of the original problem, this subtle modification leads to a tremendous change in the very nature of the optimization challenge. Indeed, this ℓ_1 problem, called *basis pursuit* or BP, is a much simpler convex problem, which can be efficiently solved by various classical optimization techniques.

Note that the same ideas can be applied when an approximation of the signal is more suitable, that is, when solving the following convex quadratic problem known as basis pursuit denoising (BPDN):

$$\arg \min_{u} \|u\|_1 \quad \text{s.t.} \quad \|x - \Psi u\|_2 \leq \epsilon \tag{BPDN}$$

The link between the exact sparse and the BP problems is quite strong and has been thoroughly studied. When the dictionary Ψ is orthonormal, solving BP amounts to realization of a soft-thresholding of the projection coefficients $\Psi^* x$ of the original signal, which shows that minimizing the ℓ_1 norm enforces sparsity.

For a more general dictionary, we must introduce the useful concept of coherence, defined as

$$\mu(\Psi) = \sup_{i \neq j} |\langle \psi_i, \psi_j \rangle| \tag{23.3}$$

Intuitively, Eq. (23.3) shows that Ψ is not too far from being an orthonormal basis when its coherence is sufficiently small (although it may be highly overcomplete). Early results of Donoho and Huo [4], Elad and Bruckstein [5], and, later, Gribonval and Nielsen [6] have shown that if a signal has a sufficiently sparse representation, that is, if

$$\|\alpha\|_0 < \tfrac{1}{2}(1 + \mu(\Psi)^{-1})$$

then this representation is the unique solution of both the exact sparse and BP problems.

In the following sections, we show how sparse signal models are central to the idea of compressive sensing and how these induced new ways to "compressively" record images.

23.3
Sensing and Compressing in a Single Stage

23.3.1
Limits of the Shannon–Nyquist Sampling

Most of the acquisition systems built during the last 50 years have been designed under the guiding rules of the Nyquist–Shannon sampling theorem. Independent

23.3 Sensing and Compressing in a Single Stage

of their field of application, these devices implicitly relied on collection of discrete samples from the continuous reality of the signal domain. They generally obtain knowledge of this function either on specific locations or by averaging it on very localized domains (for instance, CCD[2] cameras integrate light over each pixel area).

According to the Shannon–Nyquist sampling theorem, assuming that a signal is bandlimited, or equivalently that it does not contain frequencies higher than a certain limit ν, it is indeed possible to faithfully sample the signal at a period $\Delta T = 1/2\nu$ so that a perfect interpolation procedure rebuilding the continuous signal exists. In short, no information has been lost during the sampling process since the initial continuous signal can be recovered.

As explained in Section 23.2, the concept of *sparsity* in the representation of signals in certain bases or dictionaries has provided a way to compress the acquired information.

However, *the process of sampling and then representing the signal with few coefficients in a given basis from the recorded signal samples is essentially wasteful*. Indeed, we first need to actually record and discretize the signal over N samples; second, we need to compute the coefficients in at least $O(N)$ operations (if the sparsity basis provides a fast decomposition/reconstruction algorithm, like the wavelet basis); and third, we need to retain only $K \ll N$ coefficients while discarding the others.

Is it possible to simplify this procedure? Can we avoid wasting time and computations when recording N samples and processing them, if at the end we are only retaining $K \ll N$ of them? This is what compressed sensing, also called *compressive sampling (CS)*, is all about.

As shown in the following discussion, CS answers positively to the merging of sampling and compression thanks to three main changes: (i) the signal may be assumed sparse in any kind of sparsity basis Ψ (not only in the Fourier domain as for bandlimited signal), (ii) the sampling procedure is generalized to any linear (and thus nonadaptive) measurement of the signal, and (iii), the reconstruction of the signal relies on nonlinear techniques (for instance, convex optimization).

23.3.2
New Sensing Model

Henceforth, our signal of interest is finite and we work with a vector x in the N-dimensional space $\mathbb{R}^N$. This vectorial representation can be adapted to any space, for instance, by concatenating all the columns of a $\sqrt{N} \times \sqrt{N}$ image into a vector of N components[3].

As for the example given at the end of Section 23.3.1, we assume that x has a certain structure, or a geometrical content. In other words, there exists a *sparsity basis* $\Psi \in \mathbb{R}^{N \times D}$, made of $D \geq N$ elements $\Psi_j \in \mathbb{R}^N$ with $1 \leq j \leq D$, such that x can

---

[2] Charged-coupled device.
[3] Implicitly, this "vectorization" process must be realized correspondingly on the basis elements that serve to sparsify the signal.

be represented as

$$x = \Psi\alpha = \sum_{j=1}^{D} \Psi_j \alpha_j \qquad (23.4)$$

with few nonzero or important coefficients α_j in $\alpha \in \mathbb{R}^D$. These kinds of signal transformations are implicitly used everyday, for example, when one listens to MP3 compressed songs or JPEG2000-coded images by using sparsity basis like the discrete cosine transform (DCT) or the 2D discrete wavelet transform [7].

In this framework, a signal is said to be *K-sparse* if $\|\alpha\|_0 = K$, and it is said to be *compressible* if the coefficients of α decay rapidly when sorted in decreasing order of magnitude.

For simplicity, we always assume that the sparsity basis is orthonormal ($D = N$ and $\Psi^* = \Psi^{-1}$), leading to the relation $\alpha_j = \langle \Psi_j, x \rangle$. However, the theory is validated for dictionaries and frames as explained in Ref. [8].

In compressed sensing theory, following a process that comes naturally in quantum physics, we get knowledge of x by "asking" a certain number of independent questions, or through *linear measurements*. For M measurements, the signal is thus "sampled" in M values $y_j = \langle \varphi_j, x \rangle$ ($1 \leq j \leq M$), where the vectors $\varphi_j \in \mathbb{R}^N$ form the *sensing matrix* $\Phi = (\varphi_1, \ldots, \varphi_M)^* \in \mathbb{R}^{M \times N}$. Using matrix algebra, the sensing model is thus

$$y = \Phi x = \Phi \Psi \alpha \qquad (23.5)$$

where the last equality follows from Eq. (23.4). Notice that the sparsity and the sensing basis can be merged into the global sensing basis $\Theta = \Phi\Psi$.

Equation (23.5) really models the sensing of x as if it was obtained by a physical sensor outputting y. In that model, we have no access to the components of x or those of α. What we have is only y, the knowledge of the rectangular sensing matrix Φ, and the a priori knowledge of the sparsity of x (see Figure 23.1a).

Following this framework, it is often more appropriate to consider a more realistic sensing scenario where y is corrupted by different noises:

$$y = \Theta \alpha + n \qquad (23.6)$$

This equation integrates an additional noise $n \in \mathbb{R}^M$ representing, for instance, the digitization (quantification) of Φx, for further storage and/or transmission, and the unavoidable instrumental noises (Poisson, thermal, etc). Most of the time, this noise is assumed to be identically and independently distributed over its components with a Gaussian distribution.

23.4
Reconstructing from Compressed Information: A Bet on Sparsity

Let us now study how to reconstruct the signal from its measurements. The fundamental theorem of algebra – as many equations as unknowns – teaches us

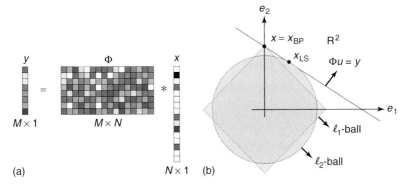

Figure 23.1 (a) Visual interpretation of the "compressed" sensing of a signal x that is sparse in the canonical basis ($\Psi = I$). (b) Explanation of the recovery of 1-sparse signals in $\mathbb{R}^2$ with BP compared to a least square (LS) solution. (Please find a color version of this figure on the color plates.)

that the recovery of α or $x = \Psi\alpha$ from y (and from the knowledge of the sensing model) is possible if $M \geq N$. This is true in all generality, whatever be the properties of x. Can we reduce M if x has a certain a priori structure? After all, if $x = \Psi\alpha$ was exactly K-sparse, with $K \ll N$, and if the support $S \subset \{1, \ldots, N\}$ of α was known, we would have

$$y = \Phi\Psi\alpha = \Theta_S \alpha_S$$

where $\Theta_S \in \mathbb{R}^{M \times K}$ represents the restrictions of the columns of Θ to those of index in S, and α_S is the vector α that is restricted to its support. Therefore, if $K \leq M$, the recovery problem is not ill posed any more.

An evident gap therefore exists between the number of measurements required for solving the known-support problem ($M \geq K$) and the one needed for the general case ($M = N$). This missing link is found by regularizing the problem, that is, by adding a prior information on the sensed x, somewhere between the full knowledge of supp x and the general "no prior information" case. In short, we must now assume the *sparsity* of x to recover it.

The signal reconstruction problem has been recast as the recovery of the sparse vector α from the observed (compressed) signal $y = \Phi x = \Theta\alpha$ given the sparsity basis $\Theta = \Phi\Psi$. According to Section 23.2, we may therefore use the ideal nonlinear recovery technique (or *decoder*)

$$\Delta_0(y) \triangleq \arg\min_{u} \|u\|_0 \text{ s.t. } y = \Theta u \qquad (23.7)$$

In spite of its combinatorial complexity, we can theoretically study the properties of this decoder Δ_0 and later observe how it can be simplified. For this purpose, we need to introduce a certain regularity on the matrix Θ. After all, if the support of α was known and equal to S, we should impose that Θ_S has rank K, that is, that the K columns of Θ_S are linearly independent. In that case, the solution is

$\boldsymbol{\alpha} = \boldsymbol{\Theta}_S^\dagger \boldsymbol{y} = (\boldsymbol{\Theta}_S^* \boldsymbol{\Theta}_S)^{-1} \boldsymbol{\Theta}_S^* \boldsymbol{y}$, with $\boldsymbol{\Theta}_S^\dagger$ being the Moore–Penrose pseudoinverse of $\boldsymbol{\Theta}_S$. The property we need now is a generalization of this concept for all the possible supports S of a given size.

Definition 1 ([9]) A matrix $\boldsymbol{\Theta} \in \mathbb{R}^{M \times N}$ satisfies the Restricted Isometry Property (or RIP(K, δ)) of order $K < M$ and isometry constant $0 \leq \delta < 1$ if, for all K-sparse signal $\boldsymbol{u} \in \mathcal{S}_K = \{\boldsymbol{v} \in \mathbb{R}^N : \|\boldsymbol{v}\|_0 = K\}$,

$$(1-\delta)\|\boldsymbol{u}\|_2^2 \leq \|\boldsymbol{\Theta}\boldsymbol{u}\|_2^2 \leq (1+\delta)\|\boldsymbol{u}\|_2^2 \tag{23.8}$$

This definition clearly amounts to impose $\boldsymbol{\Theta}$ of rank K over all the possible support $S \subset \{1, \ldots, N\}$ of size K. We later see the kinds of matrices $\boldsymbol{\Theta}$ that possess the RIP.

The RIP implies the following key result: if $\boldsymbol{y} = \boldsymbol{\Theta}\boldsymbol{\alpha}$ and $\|\boldsymbol{\alpha}\|_0 \leq K$, then

$$\boxed{\boldsymbol{\Theta} \text{ is RIP}(2K, \delta) \Rightarrow \boldsymbol{\alpha} = \Delta_0(\boldsymbol{y})}$$

The proof of this result is simple and enlightening. Denoting $\boldsymbol{\alpha}^* = \Delta_0(\boldsymbol{y})$, we must show that $\boldsymbol{x} = \boldsymbol{\Psi}\boldsymbol{\alpha} = \boldsymbol{\Psi}\boldsymbol{\alpha}^*$, or equivalently, $\boldsymbol{\alpha} = \boldsymbol{\alpha}^*$. Since $\boldsymbol{\alpha}^*$ is the solution of the minimization problem (23.7), $\|\boldsymbol{\alpha}^*\|_0 \leq \|\boldsymbol{\alpha}\|_0$ and $\|\boldsymbol{\alpha} - \boldsymbol{\alpha}^*\| \leq 2K$. Since $\boldsymbol{\Theta}$ is RIP($2K, \delta$), $(1-\delta)\|\boldsymbol{\alpha} - \boldsymbol{\alpha}^*\|_2^2 \leq \|\boldsymbol{\Theta}\boldsymbol{\alpha} - \boldsymbol{\Theta}\boldsymbol{\alpha}^*\|_2^2 = \|\boldsymbol{y} - \boldsymbol{y}\|_2^2 = 0$, proving that $\boldsymbol{\alpha} = \boldsymbol{\alpha}^*$ since $1 - \delta > 0$.

We may notice that the ideal decoder (23.7) is not guaranteed to provide a meaningful reconstruction in the cases where the signal $\boldsymbol{x}$ deviates from the exact sparsity and is just compressible, or when measurements are corrupted by a noise. These two problems are solved in the relaxed decoders that are proposed in the next section.

From the first papers about compressed sensing [10, 1], inspired by similar problems in the quest for sparse representation of signals in orthonormal or redundant bases, researcher have used a "relaxation" of Eq. (23.7). As explained in Section 23.2, the convex ℓ_1 norm can favorably replace its nonconvex ℓ_0 counterpart, and in the CS formalism, the two relaxed optimizations BP and BPDN can thus be rephrased as

$$\Delta(\boldsymbol{y}) \triangleq \arg\min_{\boldsymbol{u}} \|\boldsymbol{u}\|_1 \text{ s.t. } \boldsymbol{y} = \boldsymbol{\Theta}\boldsymbol{u} \tag{BP}$$

$$\Delta(\boldsymbol{y}, \epsilon) \triangleq \arg\min_{\boldsymbol{u}} \|\boldsymbol{u}\|_1 \text{ s.t. } \|\boldsymbol{y} - \boldsymbol{\Theta}\boldsymbol{u}\|_2 \leq \epsilon \tag{BPDN}$$

The BP decoder is suited for sparse or compressible signal reconstruction in the case of pure sensing model (23.5), while the BPDN program adds the capability to handle noisy measurements (Eq. (23.6)) with a noise power assumed bounded, that is, $\|\boldsymbol{n}\|_2 \leq \epsilon$.

In Figure 23.1b, we provide a common illustration of why in the pure sensing case the BP is an efficient way to recover sparse signals from their measurements. In this figure, signal $\boldsymbol{x}$ is assumed to be 1-sparse in the canonical basis of $\mathbb{R}^2$: $\boldsymbol{x}$ lives on one of the two axes, $\boldsymbol{e}_1$ or $\boldsymbol{e}_2$, of this space. The constraint of BP is the line, $D_y = \{\boldsymbol{u} \in \mathbb{R}^2 : \boldsymbol{\Phi}\boldsymbol{u} = \boldsymbol{y}\}$, intersecting one of the two axes, here $\boldsymbol{e}_2$, in $\boldsymbol{x}$. For a different reconstruction based on a regularization with an ℓ_2 norm, also called the least square (LS) method, the solution $\boldsymbol{x}_{\text{LS}} = \arg\min_{\boldsymbol{u}} \|\boldsymbol{u}\|_2$ s.t. $\boldsymbol{\Phi}\boldsymbol{u} = \boldsymbol{y}$ corresponds to the intersection of D_y with the smallest ℓ_2-ball, $B_2(r) = \{u \in \mathbb{R}^2 : \|u\|_2 \leq r\}$, intersecting D_y. Clearly, this point does not match the initial signal x. However, in nondegenerated situations (when D_y is not oriented at 45° with e_1 in our illustration), the solution x_{BP} of BP, which is provided by the common point between D_y and the smallest ℓ_1-ball $B_1(r) = \{u \in \mathbb{R}^2 : \|u\|_1 \leq r\}$ intersecting $D_y = D_y(0)$, is precisely the original x.

This intuitive explanation, namely, the perfect recovery of sparse signals, has been theoretically proved in Ref. [11], again with the essential RIP. More generally, considering the noisy measurement setting $y = \Theta\alpha + n$ with $\|n\|_2 \leq \epsilon$, it has been shown that

$$\Theta \text{ is RIP}(2K, \delta) \quad \text{and} \quad \delta < \sqrt{2} - 1$$
$$\Rightarrow \quad (23.9)$$
$$\|\alpha - \Delta(y, \epsilon)\|_2 \leq C\epsilon + D\frac{1}{\sqrt{K}}\|\alpha - \alpha_K\|_1$$

for C and D function of δ only. For instance, for $\delta = 0.2$, $C < 4.2$ and $D < 8.5$.

In summary, Eq. (23.9) proves the robustness of the compressed sensing setup. Perfect recovery in the pure sensing model ($\epsilon = 0$) of a K-sparse signal is still achievable but at a higher price: we must have $\delta < \sqrt{2} - 1 \simeq 0.41$ in the RIP. However, the result is fairly general since now it allows one to reconstruct the signal using noisy measurements (with a linear dependence on ϵ), and this even if x is not exactly sparse in Ψ. The deviation from the exact sparsity is indeed measured by the compressibility error $e_0(K) = \|\alpha - \alpha_K\|_1/\sqrt{K}$. For compressible signals, K is then a parameter that must be tuned to the desired accuracy of the decoder. Since Θ must be RIP, we later see that increasing K results in an increase in the number of measurements M.

23.5 Sensing Strategies Market

In Section (23.4), we detailed the required property of the sensing matrix $\Theta = \Phi\Psi$ in order to guarantee a faithful reconstruction of signal x. Now, the obvious question is, do such RIP matrices exist? The answer is "yes, there is a very high probability that they do." Indeed, deterministic construction of RIP matrices exists [12] but it suffers from very high lower bound on M with respect to the RIP(K, δ). However, as a striking result of the Concentration of Measure theory, stochastic constructions of RIP matrices exist with a precise estimation of their probability of success. Once generated, these sensing matrices have of course to be stored, or at least they must be exactly reproducible, for sensing and reconstruction purposes.

This section aims at guiding the reader through the market of available random constructions, with a clear idea of the different dimensionality conditions ensuring the RIP.

23.5.1
Random sub-Gaussian Matrices

A first random construction of sensing matrices is the one provided by the sub-Gaussian distribution class [13]. For instance, matrix $\Phi \in \mathbb{R}^{M \times N}$ can be generated as the realization of a Gaussian random variable having a variance $1/M$, identically and independently (*iid*) for each entry of the matrix:

$$\Phi_{ij} \underset{\text{iid}}{\sim} \mathcal{N}(0, 1/m)$$

Another possibility is to select one of the discrete Bernoulli distributions, for instance, $\Phi_{ij} \sim_{\text{iid}} \pm 1/\sqrt{M}$ with probability $1/2$, or $\Phi_{ij} \sim_{\text{iid}} \pm \sqrt{3/M}$ or 0 with probability $1/6$ or $2/3$, respectively. Notice that the randomness of the procedure above helps only at the creation of the matrices. These three cases, and matrices generated by other sub-Gaussian distributions, have the following two interesting properties.

First, if Φ is sub-Gaussian, then it can be shown that, for any orthonormal sparsity basis Ψ, $\Theta = \Phi\Psi$ is also sub-Gaussian. In particular, if the RIP is proved for any sub-Gaussian matrix Φ, this will also be true for any $\Theta = \Phi\Psi$. The previous stability results are thus guaranteed in this sensing strategy.

Second, in Ref. [13], it is proved that a sub-Gaussian matrix Φ satisfies the RIP(K, δ) with a controlled probability as soon as

$$M = O(\delta^{-2} K \ln N/K)$$

Even if they correspond to very general sensing strategies that are independent of the sparsity basis, there are two main problems with the random sub-Gaussian matrices above. First, their randomness makes them difficult to generate and to store even if solutions exist through the use of, for instance, pseudo-randomness. Second, their unstructured nature induces very slow matrix–vector multiplication (with complexity $O(MN)$) – an operation used implicitly and repeatedly by many reconstruction methods (like in BPDN or in greedy methods).

Fortunately, as explained later, other sensing strategies exist, even if they are often less general with respect to the class of RIP-compatible signal sparsity basis Ψ.

23.5.2
Random Fourier Ensemble

The first possibility of obtaining a "fast" sensing matrix, that is, offering both fast encoding and faster decoding techniques, occurs when one uses the Fourier transform. Here, the sensing matrix Φ is given by

$$\Phi = SF$$

where $F \in \mathbb{R}^{N \times N}$ is the (real) discrete Fourier transform on $\mathbb{R}^N$ (or on the 2D plane $\mathbb{R}^{N_1 \times N_2}$ with $N_1 N_2 = N$ for vectorized $N_1 \times N_2$ images) and the rectangular matrix $S \in \mathbb{R}^{M \times N}$ randomly picks M elements of any N-dimensional vector.

Interestingly, when Φ or Φ^* are applied to a vector, the fast Fourier transform (FFT) Cooley–Tukey algorithm can be applied, instead of the corresponding matrix multiplication. This decreases the complexity from $O(NM)$ to $O(N \ln N)$.

For a canonical sparsity basis $\Psi = I$, $\Theta = \Phi$ is RIP with overwhelming probability as soon as [10]

$$M = O(K(\ln N)^4)$$

Random Fourier ensemble matrices are, nevertheless, less general than the sub-Gaussian random constructions. Indeed, as explained in the next section, the proportionality constant implicitly involved in the last relation grows when a nontrivial sparsity basis is used.

This problem can be bypassed by altering the sparsity term of the BPDN reconstruction and replacing it by the total variation (TV) norm. This is detailed in Section 23.6

23.5.3
Random Basis Ensemble

It is possible to generalize the previous sensing basis construction to any orthonormal basis. In other words, given an orthonormal basis $U \in \mathbb{R}^{N \times N}$, we construct $\Phi = SU$ with the previous random selection matrix $S \in \mathbb{R}^{M \times N}$. In this case, the sparsity basis Ψ and the basis U from which Φ is extracted must be sufficiently incoherent. In other words, it must be difficult to express one element of Ψ as a sparse representation in U, and conversely, as measured by the quantity $\mu(U, \Psi) = \max_j |\langle U_j, \Psi_i \rangle|$.

Mathematically, the general result yields $\Theta = SU\Psi$ is RIP(K, δ) if [10]

$$M \geq C \mu(U, \Psi)^2 N K (\ln N)^4 \qquad (23.10)$$

for a certain constant $C > 0$. Beyond the combination of the Fourier basis, $U = F$, with the canonical sparsity basis, $\Psi = I$ (for which $\mu = 1/\sqrt{N}$)), a useful case is provided by the Noiselet transform [14]. This basis is maximally incoherent with the wavelet basis [7]. The coherence is, for instance, given by $\sqrt{2}$ for the Haar system, and corresponds to 2.2 and 2.9 for the Daubechies wavelets D4 and D8, respectively.

23.5.4
Random Convolution

Recently, Romberg introduced another fast sensing strategy called *random convolution* (RC) [15]. The idea of working again in the (complex) Fourier domain is simply to disturb the phase of the Fourier coefficients of a signal by multiplying them by a random complex sequence of unit amplitude before to inverse the Fourier transform; M samples of the result are subsequently randomly selected in the spatial domain. Mathematically, this amounts to considering $\Phi = S F^* \Sigma F$, with $\Sigma \in \mathbb{R}^{N \times N}$ being a complex diagonal matrix made of unit amplitude diagonal element and random phase [15].

For such a sensing procedure, it is shown that $\Theta = \Phi \Psi$ is RIP(K, δ) with high probability if

$$M \geq C K (\ln N)^5$$

for a certain constant $C > 0$. Despite a stronger requirement on the number of measurements, and contrary to the random Fourier ensemble, random convolution sensing works with any kind of sparsity basis with possibly very fast implementation. In addition, its structure seems very well suited for analog implementations in the optical domain [16], or even for CMOS imager implementing random convolution on the focal plane [17], as detailed in Section 23.7.1.

23.5.5
Other Sensing Strategies

In general, the selection of a sensing matrix for a given application depends on the following criteria: the existence of an analog model corresponding to this sensing, the availability of a fast implementation of the numerical sensing at the decoding stage, the storage of the sensing matrix in the sensors or/and in the decoder, and the coherence between this sensing and the (sparsity) class of the signal to be observed.

Nowadays, more and more different sensing strategies are developed in the field (like random modulators, Toeplitz-structured sensing matrices, and Hadamard matrices). Unfortunately, it is not possible to cover them all here. We advise the interested reader to read the comprehensive list of references given in Refs [18, 19].

23.6
Reconstruction Relatives

In this section, we briefly present some variations of the reconstruction methods commonly used in CS. These methods alter either the sparsity measure of the BP or the BPDN reconstruction, or complement their constraints by adding signal priors, or replace the often heavy optimization process by less optimal but fast iterative (greedy) algorithms.

23.6.1
Be Sparse in Gradient

As explained in Section 23.5.2, sensing based on random Fourier ensemble is perhaps fast but lacks universality as it is mainly adapted to sparse signal in the canonical basis $\Psi = I$. However, in certain applications the sensing strategy is not a choice: it can be imposed by the physics of the acquisition. For instance, in magnetic resonance imaging (MRI) [20] or in radio interferometry [21] (Section 23.7.2), the signal (an image) is acquired in the "k-space," or more specifically on a subset of the whole frequency plane in the Fourier domain.

To circumvent this problem, researchers have introduced a variation on the previous reconstruction algorithms. Rather than expressing the signal sparsity in the spatial domain using the ℓ_1-norm measure, it is indeed possible to impose the sparsity of the gradient of the image, leading to the TV quasinorm [22]. In its discrete formulation, the TV norm of a signal x representing an image is given by $\|x\|_{TV} = \sum_i |(\nabla x)_i|$, where $|(\nabla x)_i|$ is the norm of the 2D finite difference gradient of x on its ith pixel. This norm is small for "cartoon" images that are composed of smooth areas separated by (C^2 smooth) curved edges – a good approximation of many natural images showing distinct objects that are not too textured [22].

For a noisy sensing model, the TV norm leads to the new program:

$$\arg\min_u \|u\|_{TV} \text{ s.t. } \|y - \Phi u\|_2 \leq \epsilon \quad \text{(BPDN-TV)}$$

In Ref. [23], it is proved that BPDN-TV recovers with overwhelming probability the initial image for random Fourier ensemble matrices, in the absence of noise. BPDN-TV always provides an improvement in reconstruction quality compared to the results obtained with the ℓ_1 norm, even for other sensing strategies like the RCs [15, 17].

23.6.2
Add or Change Priors

In addition to the sparsity prior on the signal, other priors can enforce the stability of a reconstruction program when they are available.

For instance, if the signal is known to be positive in the spatial domain, we may alter BPDN as

$$\arg\min_u \|u\|_1 \text{ s.t. } \begin{cases} \|y - \Theta u\|_2 \leq \epsilon \\ \Psi u \geq 0 \end{cases} \quad \text{(BPDN}^+\text{)}$$

It is sometimes difficult to estimate the noise power on the measurements but easy to have a bound on the signal sparsity and to assume that there exists a τ such that $\|\alpha\|_1 \leq \tau$. In this case, the Lasso formulation [24] can be useful:

$$\arg\min_u \|y - \Theta u\|_2 \text{ s.t. } \|u\|_1 \leq \tau \quad \text{(Lasso)}$$

In certain situations, the noise level of the measurements depends on the component of the measurement vector. For random Fourier ensemble, this amounts to saying that the noise on the measurement is the result of a "colored" noise on the initial signal, that is, a nonflat spectrum noise. From the knowledge of the noise spectrum, a *whitening* of the measurement can be obtained by weighting the ℓ_2 norm of the BPDN constraint so that each weighted component of the measurement vector has a normalized standard deviation, that is,

$$\arg\min_u \|u\|_1 \text{ s.t. } \|W(y - \Theta u)\|_2 \leq \epsilon$$

$W \in \mathbb{R}^{M \times M}$ being a nonnegative diagonal matrix. The direct effect of this treatment is to give more confidence to (perhaps small) measurements with low noise level and less to those with higher standard deviation [21].

Finally, in the case where the measurement noise is non-Gaussian, the fidelity term of BPDN can be altered. For instance, if the noise comes from a uniform quantization of the measurements (the kind of digitization process that is implicitly used by any compressed sensing sensor [25, 17]) or if it follows a generalized Gaussian distribution (GGD) of shape parameter p, we can use the BP dequantization program

$$\arg\min_{u} \|u\|_1 \text{ s.t. } \|y - \Theta u\|_p \leq \epsilon, \quad \text{(with } p \geq 2\text{)} \qquad (\text{BPDQ}_p)$$

For uniformly quantized measurements, as initially suggested in Ref. [10] for $p = \infty$, it is shown in Ref. [26] that if M is higher than a bound growing with p (in oversampled situation compared to BPDN with $p = 2$), BPDQ$_p$ improves the quality of the reconstructed signal.

23.6.3
Outside Convexity

Recently, some researches have focused on nonconvex relaxation of the ideal reconstruction introduced in Section (23.4). The ℓ_0 norm of Eq. (23.7) is then replaced by the ℓ_q norm with $0 < q \leq 1$, leading to

$$\arg\min_{u} \|u\|_q \text{ s.t. } \|y - \Theta u\|_2 \leq \epsilon \qquad (\ell_p\text{-BPDN})$$

Even though the nonconvexity of this optimization program prevents us from reaching the global minimum of ℓ_q-BPDN, several authors introduced suboptimal reconstruction based on reweightings of the ℓ_1 or the ℓ_2 norm [27, 28], with reported gains in reconstruction quality compared to BPDN.

23.6.4
Be Greedy

From the advent of the sparsity concept in signal representations, and therefore from the beginning of the compressed sensing theory, researchers have tried to find fast algorithms to solve the different convex optimization programs that have been presented so far. Despite the optimality of the methods in the way they provide a global minimum, their complexity is generally high and is a function of the signal space dimension. Consequently, some suboptimal iterative methods, coined *greedy*, have been proposed.

A greedy procedure starts by setting a residual r to y and a current coefficient vector α to 0. It then checks among all the columns of $\Theta = \Phi\Psi$, or among all the subsets of columns of fixed size in this matrix, the one that is best correlated with r. This selection is included in the current coefficient vector α following various procedures, and the residual is updated by removing the influence of the selection from it. Then, the process is repeated on the new residual and the algorithm stops either naturally or after a minimal number of iterations that are needed to significantly reduce the energy of the residual.

The first greedy methods used in CS were matching pursuit (MP) and orthogonal MP. The approximation error obtained with these was unfortunately not very well controlled. However, compressive sampling MP [29], subspace pursuit [30], and few others recently filled this gap and provided stability guarantees similar to the one given in Eq. (23.9).

23.7
Some Compressive Imaging Applications

Compressed sensing is a very general concept that is reported to be adaptable to a tremendous amount of applications covering astronomy, radars, seismology, satellite telemetry, and so on. In this section, we restrict our attention to only two of them, namely, compressive imagers and radio interferometry. We refer the readers to Chapters 15 and 22 of this book, or to parse the web references [18, 19] to get a complementary overview of the huge activity presently being developed around applicative CS.

23.7.1
Compressive Imagers

The "single pixel camera" developed at Rice University (TX, USA) [25] was the first device implementing a compressed sensing imager. As explained in Chapter 22, this device uses a digital micromirror device (DMD), a key element found in a lot of digital projectors, programmed in a sequence of random configurations and focusing light on a single photodetector. As a result, the system directly acquires compressive measurements in the optical domain.

Other compressive imagers have been realized since the single pixel camera. We may cite the Georgia Tech Analog Imager of Robucci et al. [16], as well as some multispectral and coded aperture extensions, as described in more detail by Willet et al. in Chapter 22. The CMOS compressive imager presented in Figure 23.2, which was developed at the Swiss Federal Institute of Technology (EPFL), is one of them [17].

As for the Georgia Tech imager, it proceeds by embedding the compressed sensing of images in the analog domain. The selected sensing strategy relies on the RCs (Section 23.5.4) performed in the focal plane of the imager. This sensing has a very simple electronic translation. Indeed, in the spatial domain, an RC of an image $x \in \mathbb{R}^N$ is equivalent to

$$y_i = (\Phi x)_i = \sum_i a_{r(i)-j} x_j = (xa)_{r(i)} \tag{23.11}$$

where a represents the random filter and the indices $r(i)$ are uniformly selected at random from $\{1, \ldots, N\}$. For this compressive imager, the filter is not defined in the frequency domain but corresponds to a suboptimal pseudorandom Rademacher sequence, which is given by $a_i = \pm 1$, with equal probability. Practically, a one-bit flip-flop memory is integrated in each pixel of the camera. Its binary state alters the direction of the electric current that is outgoing from the photodiode electronics

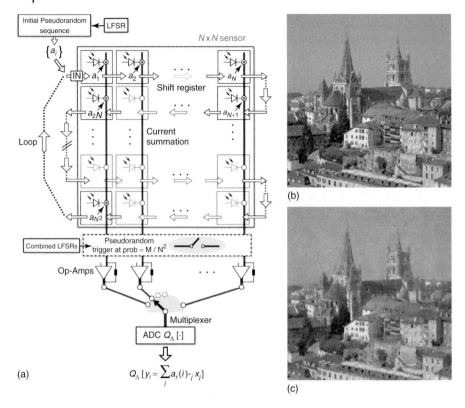

Figure 23.2 CMOS Compressive Imager. (a) The imager scheme. (b and c) Simulated example of image sensing and reconstruction: (b) original image (Lausanne Cathedral, Switzerland), (c) reconstruction from ensing of $M = \lfloor N/3 \rfloor$ 11-bits quantized measurements, corrupted by an additional Gaussian white noise (PSNR 27.3 dB).

before being gathered on column wires, thanks to the Kirchoff law. Therefore, if the N flip-flop memories are set into a random pattern driven by the filter **a**, then the sum of all the column currents provides a CS measurement of the image seen on the focal plane. The following measurements are easily obtained, since the one-bit memories are connected in a big chain, which is represented in Figure 23.2(a), creating an N shift register which is closed by connecting the last pixel memory to the first. After a few clock signals, the whole random pattern can be shifted on its chain, which yields the desired convolution operation for the following measurements. We refer the reader to Ref. [17] for more details about the electronic design of this imager.

Compared to common CMOS imagers, this compressive acquisition has some advantages: (i) data are compressed at a reasonable compression ratio, without having to integrate a digital signal processing (DSP) block in the whole scheme; (ii) no complex strategies have to be developed to numerically address the pixels/columns of the grid before data transmission; and (iii) the whole system has a low power consumption since it benefits from the parallel analog processing (as for filter shifting in the shift register). The proposed sensor is of course not designed for end-user systems (e.g., mobile phone). It, nevertheless, meets the requirements of technological niches with strong constraints (for instance, low power consumption). Indeed, the adopted CS coding involves a low computational complexity compared to systems embedding transform-based compression (like JPEG 2000).

23.7.2
Compressive Radio Interferometry

Recently, the compressed sensing paradigm has been successfully applied to the field of radio interferometry detailed in Chapter 15.

In radio interferometry, radio-telescope arrays[4] synthesize the aperture of a unique telescope, the size of which relates to the maximum distance between the two telescopes. This allows observations with otherwise inaccessible angular resolutions and sensitivities in radio astronomy. The small portion of the celestial sphere that is accessible to the instrument around the pointing direction ω tracked during observation defines the original image I_ω (or *intensity field*) to be recovered. As a matter of fact, thanks to the Van Cittert–Zernike theorem [21], the time correlation of the two electric fields E_1 and E_2 recorded by two radio telescopes that are distant by a baseline vector $\boldsymbol{b}$ corresponds to $\hat{I}_\omega(\boldsymbol{b}^\perp)$. Indeed, the evaluation of the Fourier transform of I_ω on the frequency vector $\boldsymbol{b}^\perp = \boldsymbol{b} - (\boldsymbol{\omega} \cdot \boldsymbol{b})\boldsymbol{\omega}$, which is obtained by projecting $\boldsymbol{b}$ on the plane of observation that is perpendicular to ω. Since there are $\binom{N}{2}$ different pairs in a group of N telescopes, and because of the Earth's rotation, radio interferometry finally amounts to sampling the image of interest I_ω in the Fourier domain on a list of frequencies, or *visibilities*, similar to the one represented in Figure 23.3 (bottom, left).

As explained in Chapter 15, many techniques have been designed by radio astronomers to reconstruct I_ω from the set of visibilities. One commonly used method is the CLEAN algorithm [31] and its many variants which have been generated since its first definition in 1974. Interestingly, CLEAN is actually nothing but an MP greedy procedure (Section 23.6.4) that iteratively subtracts from the visibilities the elements of the sky that are most correlated with them. However, CLEAN suffers from a lack of flexibility, and it is, for instance, not obvious to impose the positivity of the intensity field to be reconstructed. In Ref. [21], it is shown that, for random arrangements of visibilities in the frequency plane, compressed sensing reconstruction procedures like BPDN+ and BPDN-TV (Section 23.6.2) provide significative gains in the quality of the reconstructions, as illustrated in Figure 23.4 for a synthetic example.

Intrinsically, the sampling in radio interferometry is very similar to those obtained in MRI [20], where compressed sensing also provides significative improvements

---

4) As one of the Arcminute Microkelvin Imager (AMI).

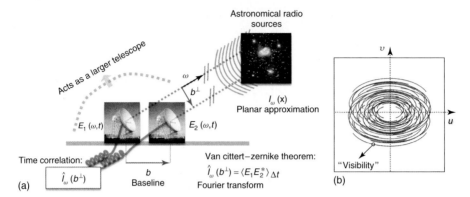

Figure 23.3 Radio Interferometry by aperture synthesis. (Left) General principles. (Right) Visibility maps in the Fourier plane. (Please find a color version of this figure on the color plates.)

over the previous reconstruction schemes. Recently, it has been shown in Ref. [32] that the modulating wave arising from the nonplanar configuration of large radio-telescope arrays on the Earth surface can advantageously be used as a "spread-spectrum" technique, further improving the image reconstruction quality.

23.8
Conclusion and the "Science 2.0" Effect

Let us conclude this chapter with an important, and somehow "sociologic," observation. The compressed sensing paradigm is born within what is often called the "Science 2.0." evolution. This term represents a real boom in the exchange of information among the members of the scientific community – a by-product of the steady development that Internet has experienced. This recent trend takes several forms. One of those is, for instance, the online publication of preprints (through central servers or directly on personal homepages) before they even have been accepted in scientific journals. More remarkable is also the writing of scientific blogs, which are regularly fed by researchers, and are often the source of fruitful online discussions with other scientists. Another aspect of this trend is the free exchange of numerical codes, guaranteeing the reproducibility of the results.

23.8.1
Information Sources

For compressed sensing, the two most visited sources of information are the compressed sensing resource webpage managed by Baraniuk and his team at the Rice University [19] and the highly active "Nuit Blanche" blog from Carron [18]. Many other blogs discuss topics related to CS, and it would be difficult

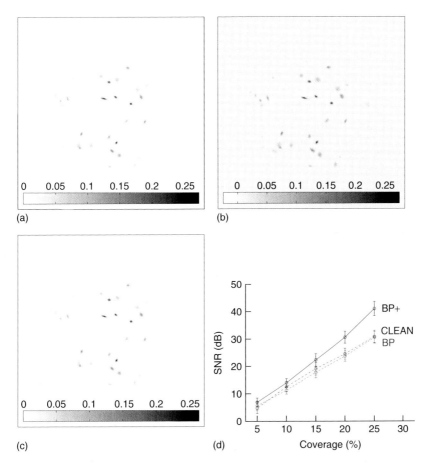

Figure 23.4 (a) Compact object intensity field I in some arbitrary intensity units. (b) and (c) CLEAN and BP$^+$ (i.e., BPDN$^+$ with $\epsilon = 0$) reconstructions for $M = N/10$ and random visibilities. (d) The graph of the mean SNR with 1σ error bars over 30 simulations for the CLEAN, BP, and BP+ reconstructions of I as a function of M (as percentage of N).

to list them all here. We can mention, for instance, the applied mathematics blog "What's new" by Terence Tao (*http://terrytao.wordpress.com*), one of the founders of CS, or the "Optical Imaging and Spectroscopy Blog" of D. Brady (*http://opticalimaging.org/OISblog*) on optical imaging and spectroscopy. Many other interesting links can be found by surfing on these blogs since most of them maintain a "blogroll" section, listing websites related to their own contents.

23.8.2
Reproducible Research

The advent of compressed sensing theory is also related to free dissemination of the numerical codes, helping the community to reproduce experiments. A lot of toolboxes have, therefore, been produced in different languages: C, C++, Matlab©, Python, and so on. One of the first tools designed to solve optimization programs such as BP, BPDN, and BPDN-TV was the ℓ_1-magic toolbox[5] in Matlab. Now, a lot of different implementations allow one to perform the same kinds of operations with more efficient methods: CVX, SPGL1, Sparco, SPOT, TwIST, YALL, and so on, to name a few. Once again, a comprehensive list is maintained in Ref. [19, 18].

Acknowledgments

LJ is funded by the Belgian National Science Foundation (F.R.S.-FNRS). LJ and PV warmly thank Pedro Correa (UCL) for his kind proofreading of the manuscript.

References

1. Donoho, D.L. (2006) Compressed sensing. *IEEE Tran. Inf. Th.*, **52** (4), 1289–1306.
2. Candès, E.J. and Romberg, J. (2006) Quantitative robust uncertainty principles and optimally sparse decompositions. *Found. Comp. Math.*, **6** (2), 227–254.
3. Chen, S.S., Donoho, D.L., and Saunders, M.A. (1998) Atomic decomposition by basis pursuit. *SIAM J. Sc. Comp.*, **20** (1), 33–61.
4. Donoho, D.L. and Huo, X. (2001) Uncertainty principles and ideal atom decomposition. *IEEE T. Inform Theory*, **47** (7), 2845–2862.
5. Elad, M. and Bruckstein, A.M. (2002) A generalized uncertainty principle and sparse representation in pairs of bases. *IEEE T. Inform. Theory.*, **48** (9), 2558–2567.
6. Gribonval, R. and Nielsen, M. (2003) Sparse representations in unions of bases. *IEEE T. Inform. Theory.*, **49** (12), 3320–3325.
7. Mallat, S. (1999) *A Wavelet Tour of Signal Processing*, 2nd edn. Academic Press.
8. Rauhut, H., Schnass, K., and Vandergheynst, P. (2008) Compressed sensing and redundant dictionaries. *IEEE Tran. Inf. Th.*, **54** (5), 2210–2219.
9. Candès, E.J. and Tao, T. (2005) Decoding by linear programming. *IEEE Trans. Inf. Th.*, **51** (12), 4203–4215.
10. Candès, E.J. and Tao, T. (2004) Near-optimal signal recovery from random projections: universal encoding strategies. *IEEE Trans. Inf. Th.*, **52**, 5406–5425.
11. Candès, E.J. (2008) The restricted isometry property and its implications for compressed sensing. *Compte Rendus Acad. Sc., Paris, Serie I*, **346**, 589–592.
12. DeVore, R.A. (2007) Deterministic constructions of compressed sensing matrices. *J. Complexity*, **23** (4–6), 918–925.
13. Baraniuk, R.G. et al. (2008) A simple proof of the restricted isometry property for random matrices. *Constructive Approximation*, **28** (3), 253–263.
14. Coifman, R., Geshwind, F., and Meyer, Y. (2001) Noiselets. *App. Comp. Harm. Anal.*, **10** (1), 27–44.
15. Romberg, J. (2007) Sensing by random convolution. 2nd IEEE International Workshop Comp. Advanced Multi-Sensor Adap. Proc., 2007, pp. 137–140.
16. Robucci, R. et al. (2008) Compressive sensing on a CMOS separable transform image sensor. In IEEE International Conference Ac. Speech Sig. Proc., 2008, pp. 5125–5128.

5) http://www.acm.caltech.edu/l1magic

17. Jacques, L. et al. (2009) CMOS compressed imaging by random convolution. In IEEE International Conference Ac. Speech Sig. Proc., 2009, pp. 1113–1116.
18. Carron, I. Nuit blanche blog. (2010) http://nuit-blanche.blogspot.com.
19. Baraniuk, R. et al. Compressive sensing resources. (2010) http://www-dsp.rice.edu/cs.
20. Lustig, M., Donoho, D.L. and Pauly, J.M. (2007) Sparse MRI: the application of compressed sensing for rapid MR imaging. *Magn. Reson. Med.*, **58** (6), 1182.
21. Wiaux, Y. et al. (2009) Compressed sensing imaging techniques for radio interferometry. *Mon. Not. R. Astron. Soc.*, **395** (3), 1733–1742.
22. Rudin, L., Osher, S., and Fatemi, E. (1992) Nonlinear total variation based noise removal. *Physica D.*, **60**, 259–268.
23. Candès, E.J., Romberg, J., and Tao, T. (2006) Robust uncertainty principles: exact signal reconstruction from highly incomplete frequency information. *IEEE Tran. Inf. Th.*, **52** (2), 489–509.
24. Tibshirani, R. (1996) Regression shrinkage and selection via the LASSO. *J. Royal Stat. Soc. Ser. B (Meth.)*, **58**, 267–288.
25. Duarte, M.F. et al. (2008) Single-pixel imaging via compressive sampling [Building simpler, smaller, and less-expensive digital cameras]. *IEEE Sig. Proc. Mag.*, **25** (2), 83–91.
26. Jacques, L., Hammond, D.K., and Fadili, M.J. (2009) Dequantizing compressed sensing: when oversampling and Non-Gaussian constraints Combine". *IEEE Trans. Trans. Inf. Th.*, **57** (1), 2011.
27. Chartrand, R. (2007) Exact reconstruction of sparse signals via nonconvex minimization. *Sig. Proc. Lett.*, **14** (10), 707–710.
28. Candès, E.J., Wakin, M., and Boyd, S. (2008) Enhancing sparsity by reweighted ℓ_1 minimization. *J. Fourier Anal. Appl.*, **14** (5), 877–905.
29. Needell, D. and Tropp, J.A. (2009) CoSaMP: iterative signal recovery from incomplete and inaccurate samples. *App. Comp. Harm. Anal.*, **26** (3), 301–321.
30. Dai, W. and Milenkovic, O. (2009) Subspace pursuit for compressive sensing signal reconstruction. *IEEE Tran. Inf. Th.*, **55** (5), 2230–2249.
31. Högbom, J.A. (1974) Aperture synthesis with a non-regular distribution of interferometer baselines. *A&AS*, **15**, 417–426.
32. Wiaux, Y. et al. (2009) Spread spectrum for imaging techniques in radio interferometry. *Mon. Not. R. Astron. Soc.*, **400**, 1029.

Further Reading

Donoho, D.L., and Elad, M. (2003) Optimally sparse representation in general (Nonorthogonal) dictionaries via ℓ_1 minimization. *Proc. Nat. Aca. Sci*, **100** (5), 2197–2202.

24
Blind Deconvolution Imaging

Filip Šroubek and Michal Šorel

24.1
Introduction

Images deteriorate during acquisition, when a signal passes through formation, transmission, and recording processes. Imaging devices have improved tremendously in the last decade, but there will always be scope for improvement. Measuring conditions are rarely ideal, which further deteriorate images irrespective of the quality of the imaging devices. In general, the degradation is a result of two phenomena. The first phenomenon is deterministic and results from defects of imaging devices (lens imperfection and CCD malfunction) and/or external effects (atmospheric turbulence, incorrect lens adjustment, and relative camera–scene motion). The second phenomenon is a random one and corresponds to noise, which accompanies any signal transmission.

Blurring, modeled as convolution, is a prominent example of the first degradation phenomenon. Removing blur from images, the so-called deconvolution, is one of the classical problems of image processing. Deconvolution is a well-studied inverse problem and a wide range of solutions has been proposed in the literature. In this chapter, we focus on blind deconvolution, which renders the problem more challenging as the convolution kernel is assumed to be unknown.

In Section 24.2, we discuss the deconvolution problem from the perspective of the calculus of variations. References to relevant earlier work are given at appropriate places. Section 24.3 covers recent advances in the single-channel case, when the solution is obtained from one degraded image. One can overcome inherent instability of this inverse problem by adding more restrictive prior knowledge. If multiple images are available, the inverse problem becomes far more stable as we show in Section 24.4. Finally, in Section 24.5, an extension to the space-variant case demonstrates applicability of multichannel (MC) deconvolution methods to images with blur varying over the field of view. Common examples are out-of-focus blur in images capturing scenes with profound depth of field, or motion blur induced by a complex relative camera-to-scene motion. Each section is enriched by several experiments on real data.

Optical and Digital Image Processing: Fundamentals and Applications, First Edition. Edited by Gabriel Cristóbal, Peter Schelkens, and Hugo Thienpont.
© 2011 Wiley-VCH Verlag GmbH & Co. KGaA. Published 2011 by Wiley-VCH Verlag GmbH & Co. KGaA.

24.2
Image Deconvolution

Image restoration refers to a methodology that removes or diminishes the effects of degradation (Chapter 26). To perform successful image restoration, it is important to correctly model the acquisition process. The following is a commonly used model. Let $u(\vec{x}) : \Omega_{\mathbb{R}^2} \to \mathbb{R}$, $\vec{x} = [x_1, x_2]$, be an original image describing a real scene, where Ω is a rectangular image support. The degradation acting on u is twofold: degradation linear operator H and additive noise n. Then, the output of acquisition is a degraded image z given by the following relation:

$$z = Hu + n \tag{24.1}$$

Since we are primarily interested in deconvolution, the deterministic degradation H will be, in our case, convolution with some point-spread function (PSF) h. We refer to PSF simply as *blur* or *blur kernel*. Let us review some examples of scenarios that lead to types of common degradation.

Blurs of cylindrical or conical shape model degradation are induced by out-of-focus lens. Degradation by a camera or an object motion results in blurs of curvy shape, where the curve reflects the motion trajectory. Effects of atmospherical turbulence are often modeled by convolution with a Gaussian blur. Figure 24.1 illustrates examples of such blurs. Noise is characterized by a probabilistic distribution. Typically, the Gaussian distribution is assumed. However, some applications require more specific ones, such as the generalized gamma distribution for radar images (speckle noise) or the Poisson distribution for tomographs and astronomical observations.

Recovering u from Eq. (24.1) is a typical example of an inverse problem. We distinguish several types of problems depending on the knowledge about the acquisition model. If H is known, we face a classical *nonblind deconvolution* problem. If H is unknown, the problem is called *blind deconvolution* and apart from recovering u we also have to estimate H. In many real scenarios, the blur varies in Ω. For example, a depth of scene influences the out-of-focus blur or lens imperfections change the orientation of motion blurs. Then, the degradation

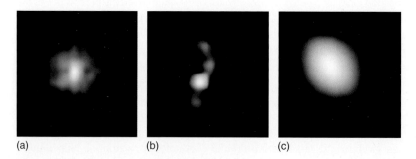

Figure 24.1 Examples of real blurs: (a) out of focus, (b) camera motion, and (c) atmospheric turbulence.

operator H is not convolution but a linear operator called *space-variant convolution* and we face a *space-variant deconvolution* problem. The complexity increases as the number of unknowns increases in the above problems.

Recovering u from z even in the nonblind case is not an easy task. A standard technique is to convert it to an energy minimization problem. The core term in the energy function is called a *data-fitting* or *fidelity term* and generally takes the following form:

$$E(u) = \int_\Omega |z - Hu|^2 d\vec{x} \tag{24.2}$$

In this case, finding the minimum of $E(u)$ is equivalent to a least-square fit. Another natural way to interpret this equation is to use probabilities. Without going into much detail, suppose that noise n is white Gaussian. Then, the minimum of $E(u)$ is the maximum likelihood realization of u for the given noise distribution. In other words, by minimizing E, we minimize the Gaussian noise variance. Different noise distributions result in different data-fitting terms, but this is beyond the scope of this chapter. The difficulty finding the minimum of Eq. (24.2) resides in the degradation operator H. Since blurring diminishes high-frequency information (image details), the spectrum of H contains zero or values close to zero. Therefore, H is generally not invertible. Solutions to such problems are numerically unstable, that is, image reconstruction becomes an ill-posed problem. To overcome this, the problem is regularized by considering a related problem that admits a unique solution.

A classical way to solve ill-posed minimization problems is to add regularization terms. Regularization conveys additional prior knowledge of the original image u to the energy function. Priors are application dependent and general rules for constructing the priors are difficult to find. Nevertheless, studying image statistics shows that the majority of natural images contain smooth regions with abrupt changes of intensity values at object boundaries that correspond to edges. An image gradient is a good tool, which can distinguish between edges and smooth regions. Therefore, regularization terms are often functions of $\nabla u = [u_{x_1}, u_{x_2}]$. From the probability point of view, regularization corresponds to an a priori distribution of u. This distribution captures local characteristics and states that similar neighboring image pixels are more likely. The L_1-norm of the image gradient, which is also called *total variation* [1], is a popular choice for the image regularization term. Then, the regularized energy is given as follows:

$$E(u) = \int_\Omega |z - Hu|^2 d\vec{x} + \lambda \int_\Omega |\nabla u| d\vec{x} \tag{24.3}$$

The parameter λ is a positive weighting constant. The first term forces the solution to be close to the observed data and the second one guarantees that the solution is sufficiently smooth in the L_1-norm sense. Noise is removed in smooth regions, while edges are not excessively penalized, since we use the L_1-norm instead of, for example, the L_2-norm.

In addition to image restoration, most of the other problems in image processing (registration, segmentation, motion detection, etc.) are formulated as minimization of energy functions, where the energy is of the form

$$E(u) = \int_\Omega f(\vec{x}, u, \nabla u) d\vec{x}$$

Note that our problem in Eq. (24.3) is of the same form. To minimize the above energy function, one can apply the calculus of variations. If the minimum of E exists, then it must satisfy an Euler–Lagrange equation $E'(u) = 0$, where

$$E'(u) = \frac{\partial f}{\partial u}(\vec{x}, u, \nabla u) - \sum_{w=1}^{2} \frac{\partial}{\partial x_w} \left(\frac{\partial f}{\partial f_{x_w}}(\vec{x}, u, \nabla u) \right) \quad (24.4)$$

and

$$\int_{\partial\Omega} \left[\sum_{w=1}^{2} \left(\frac{\partial f}{\partial f_{x_w}}(\vec{x}, u, \nabla u) n_w(\vec{x}) \right) \right] v(\vec{x}) d\vec{x} = 0 \quad (24.5)$$

Equation (24.5) is a line integral along the image support boundary $\partial\Omega$, $\vec{n} = [n_1, n_2]$ is the normal vector on $\partial\Omega$, and v is an arbitrary function, which appears on taking the variation of E. This equation gives rise to the so-called boundary conditions. To satisfy the equation, we must impose either Dirichlet or Neumann boundary conditions. The Dirichlet condition constrains intensity values of the image along its boundary, which implies that $v(\partial\Omega) = 0$. In the case of images, such constraint is not practical, since we rarely know the correct values along the image boundary. The Neumann condition constrains image derivatives in the direction of normal to be zero, that is, $\frac{\partial u}{\partial n}(\partial\Omega) = 0$. For our restoration problem in Eq. (24.3), it is easy to prove that the term in square brackets in Eq. (24.5) vanishes and the whole equation is satisfied. The Neumann condition comes very naturally in image processing. To work with the image close to its boundary, we need values outside the image support and we usually extend the support by mirroring the intensity values. Derivatives in the direction of the normal at the image boundary are then zero.

The Euler–Lagrange equation of our deconvolution problem in Eq. (24.3) is

$$H^* H u - H^* z - \lambda \mathrm{div}\left(\frac{\nabla u}{|\nabla u|} \right) = 0 \quad (24.6)$$

with the Neumann boundary condition $\frac{\partial u}{\partial n}(\partial\Omega) = 0$. The resulting equation is nonlinear owing to the divergence term. Several linearization schemes were proposed to deal with the nonlinearity, such as a fixed point iteration scheme [2], a primal-dual method [3], or a more general half-quadratic regularization scheme proposed in Ref. [4]. For example, the half-quadratic scheme introduces an auxiliary variable, which transfers Eq. (24.6) to two linear equations. Then, the steepest descent or conjugate gradient methods can be used to find the solution. Eventually, the above procedure accomplishes nonblind deconvolution formulated as a minimization of regularized energy. An overview of other classical nonblind restoration methods is given in Ref. [5]; see also Chapter 26.

Minimizing the energy using computers requires a discrete representation of Eq. (24.3). We have two choices. First, we can discretize the corresponding

Euler–Lagrange equation (24.6) using methods of finite elements or finite differences. Second, we can discretize the acquisition model in Eq. (24.1) and work in the discrete domain from the beginning. The latter procedure is more common as we often work with digital images, which are inherently discrete. A discrete version of Eq. (24.1) is given by

$$\vec{z} = \vec{H}\vec{u} + \vec{n} \tag{24.7}$$

where $\vec{z}$, $\vec{u}$, and $\vec{n}$ are vector representations of the blurred image, original image, and noise, respectively. The vector representation is obtained by stacking a discrete image columnwise. The convolution operator $\vec{H}$ is a Toeplitz-block-Toeplitz matrix performing discrete convolution with a discrete version of PSF h, which has vector representation $\vec{h}$. We can interchange the roles of $\vec{u}$ and $\vec{h}$ and write $\vec{z} = \vec{U}\vec{h} + \vec{n}$. A discrete version of Eq. (24.3) is given by

$$E(\vec{u}) = \|\vec{z} - \vec{H}\vec{u}\|^2 - \lambda \vec{u}^T \vec{L} \vec{u} \tag{24.8}$$

where $\vec{L}$ is a symmetric, positive semidefinite, block tridiagonal matrix constructed of values that depend on ∇u. The half-quadratic scheme results in an iterative algorithm, which, for the given estimation of u, recalculates $\vec{L}$ and finds a new estimation by minimizing $E(\vec{u})$ with $\vec{L}$ fixed. The derivative of $\vec{E}$ with respect to $\vec{u}$ corresponds to a discrete version of the Euler–Lagrange equation:

$$\vec{H}^T \vec{H} \vec{u} - \vec{H}^T \vec{z} - \lambda \vec{L} \vec{u} = \vec{0}$$

and this is a linear equation, which can be easily solved.

In most real applications, prior knowledge of PSFs is often difficult to obtain and it is necessary to assume the blind restoration problem. Parametric methods lie about halfway between the blind and the nonblind cases. In some applications, a parametric model of the PSF is given a priori and we search in the space of parameters and not in the full space of PSFs. An approach [6] for parametric motion and out-of-focus deblurring investigates zero patterns of the Fourier transform or of the cepstrum[1] of blurred images. More low-level parametric methods for estimating general motion blurs were proposed in Refs [7–9]. Parametric methods have two disadvantages. They are more restrictive than the fully blind ones and, surprisingly, they can be also more computationally demanding. Real PSFs always differ slightly from their parametric models and this prevents the parametric methods from finding an exact solution. Even if minimization with respect to the unknown PSF is linear, minimization with respect to one of the parameters of the PSF does not have to be linear and, thus, effective methods for solving linear problems cannot be applied.

In blind image restoration, the energy E becomes a functional of two unknowns, the original image u and the decimation H (or PSF h). If both u and h are unknown, the problem is underdetermined and some additional information or

---

[1] The cepstrum of a signal is defined as the Fourier transform of the logarithm of the squared magnitude of the Fourier transform of the signal.

different minimization strategy is necessary. In the case of a single degraded image, the most common approach is to include stronger regularization both on the image and blur and use more sophisticated estimation procedures. Section 24.3 treats such single-channel (SC) blind deconvolution methods. Another approach, extensively studied in past years, is to use multiple images capturing the same scene but blurred in a different way. Multiple images permit the estimation of the blurs without any prior knowledge of their shape, which is hardly possible in SC blind deconvolution. Section 24.4 discusses the multichannel (MC) approach. One particular MC setup attracted considerable attention only recently. Considering images with two different exposure times (long and short) results in a pair of images, in which one is sharp but underexposed and another is correctly exposed but blurred. This idea extended to the space-variant case is treated in Section 24.5.

24.3
Single-Channel Deconvolution

There has been a considerable effort in the image-processing community in the last three decades to find a reliable algorithm for single image blind deconvolution. For a long time, the problem seemed too difficult to be solved for complex blur kernels. Proposed algorithms usually worked only for special cases, such as astronomical images with uniform (black) background, and their performance depended on initial estimates of PSFs. To name a few papers from this category, consider Refs [10–12] and the survey [13]. Despite the exhaustive research, results on real-world images were rarely produced.

A partial breakthrough came only recently, when Fergus et al. [14] proposed an interesting Bayesian method with very impressive results. This triggered a furious activity of other research groups and soon several conference papers appeared on the same topic [15–18]. The key idea of the recent algorithms is to address the ill posedness of blind deconvolution by characterizing u using natural image statistics and by a better choice of minimization algorithm. We now discuss the key ideas in more detail.

SC blind deconvolution methods formulate the problem as a stochastic one. The Bayesian paradigm dictates that the inference on the original image and blur kernels should be based on the a posteriori probability distribution

$$p(\vec{u}, \vec{h}|\vec{z}) \propto p(\vec{z}|\vec{u}, \vec{h})p(\vec{u})p(\vec{h})$$

where $p(\vec{z}|\vec{u}, \vec{h})$ is the noise distribution from our observation model in Eq. (24.7) and $p(\vec{u})$, $p(\vec{h})$ are a priori distributions of the original image and blur kernel, respectively. Since the posterior distribution $p(\vec{u}, \vec{h}|\vec{z})$ cannot be found in closed form, a widely used approach is to instead find the maximum of the a posteriori distribution or equivalently to minimize $-\log(p(\vec{u}, \vec{h}|\vec{z}))$. Many distributions have exponential form and, thus, taking the logarithm of posterior is very convenient and provides an expression of the same form as the energy function in Eq. (24.8).

Then, the noise distribution $p(\vec{z}|\vec{u}, \vec{h})$ corresponds to the data-fitting term and the priors correspond to the regularization terms of the energy. Minimization is carried out exactly in the same manner as in the case of energy functions, except that this time we have to alternate between two minimization steps: with respect to u and with respect to h. This strategy is called *alternating minimization*. The maximum a posteriori (MAP) approach with alternating minimization was adopted, for example, in Ref. [17]. Levin et al. [18] pointed out that the MAP approach may not be the correct choice as she demonstrated on a scalar blind deconvolution problem, in which both u and h are scalars. The posterior distribution often has a complex shape and its maximum does not have to correspond to what we intuitively regard as a correct solution. It would be more appropriate to calculate the expected value, but since the posterior cannot be found in closed form, this can be hardly accomplished. Levin suggested the marginalization of the posterior over the image u and then maximization of $p(\vec{h}|\vec{z})$ in order to find the blur kernel. Finding the marginal posterior distribution in closed form is possible only for certain types of distributions and, thus, limits our choice of priors; otherwise, we would have to use a methodology called *Laplace distribution approximation* as, for example, in Ref. [19]. Another approach is to approximate the posterior $p(\vec{u}, \vec{h}|\vec{z})$ by a simpler distribution $q(\vec{u}, \vec{h})$, of which the expected values can be easily found. The Kullback–Leibler divergence is the criterion used to find $q(\vec{u}, \vec{h})$. The first application of this technique to blind deconvolution was proposed in Ref. [20] and the same procedure was used by Fergus et al. [14] and by Molina et al. [12].

Priors act as regularization terms and make the deconvolution problem better posed. As was discussed in Section 24.1, image priors are based on statistics of the image gradient. Let $\vec{D}_{x_1}$ denote a matrix calculating discrete derivatives in the first direction, $\vec{D}_{x_2}$ in the second direction, and $\vec{D} = [\vec{D}_{x_1}^T, \vec{D}_{x_2}^T]^T$. Traditional image priors favored smooth images and $-\log(p(\vec{u}))$ were of the type $\|\vec{C}\vec{u}\|^2$, where $\vec{C} = \vec{D}$ (the Tikhonov model) or $\vec{C} = \vec{D}^T\vec{D}$ (the simultaneous autoregression model). The Tikhonov model states that the image gradient is assumed to be normally (Gaussian) distributed. However, statistics of natural images show that this is not true and that the distribution is more heavy tailed. A natural choice was made by Rudin et al. [1] to use the total variation norm, which was also backed by a well-developed theory of functions with bounded variation. The corresponding prior is similar to the Laplace distribution: $-\log(p(\vec{u})) \propto \sum_k \sqrt{[\vec{D}_{x_1}\vec{u}.\vec{D}_{x_1}\vec{u}]_k + [\vec{D}_{x_2}\vec{u}.\vec{D}_{x_2}\vec{u}]_k}$. Generalization of total variation was done in Ref. [21], giving rise to regularization of the type $\int_\Omega \phi(\nabla u) d\vec{x}$, where $\phi(\cdot)$ is convex. State-of-the-art image priors go one step further. They calculate the image gradient distribution over a large set of natural images and then approximate it by some analytical form, such as a mixer of Gaussian distributions [14] or piecewise continuous functions [17]. In the case of blur priors, we can use the image priors mentioned above. However, it is more common to describe blur priors by an exponential distribution, that is, $-\log(p(\vec{h})) \propto \sum_k \vec{h}_k$, $\vec{h}_k \geq 0$. The exponential prior favors sparse and positive blur kernels. All the priors have one common property. They are typically i.i.d. (or close to i.i.d.), which simplifies calculations, since the covariance matrices are diagonal (or block diagonal). Jia [15] and Joshi et al. [16] tried a different approach for estimating blur kernels. First, a transparency map [15] or an edge location [16] is extracted from the blurred image and, then, the blur estimation is performed on the extracted information. This is possible since the extracted image carries all the necessary information and the advantage is that more restrictive priors (such as binary distribution) can be applied. However, the extraction step is not well defined and if it fails, the whole algorithm fails as well.

We describe Fergus's method [14] in more detail, since it seems to be currently the best SC blind deconvolution algorithm. The method assumes the acquisition model in Eq. (24.7). The basic idea is to estimate the a posteriori probability distribution of the gradient of the original image and of the blur kernel:

$$p(\vec{D}\vec{u}, \vec{h}|\vec{D}\vec{z}) = p(\vec{D}\vec{z}|\vec{D}\vec{u}, \vec{h})p(\vec{D}\vec{u})p(\vec{h}) \qquad (24.9)$$

using knowledge of independent prior distributions of the image gradient $p(\vec{D}\vec{u})$ and of the kernel $p(\vec{h})$. The likelihood $p(\vec{D}\vec{z}|\vec{D}\vec{u}, \vec{h})$ is considered Gaussian with mean $\vec{H}\vec{D}\vec{u}$ and an unknown variance. After approximating the full posterior distribution $p(\vec{D}\vec{u}, \vec{h}|\vec{D}\vec{z})$, it computes the kernel with maximal marginal probability. This computation is implemented in a multiscale mode, that is, kernel estimation is carried out by varying image resolution in a coarse-to-fine manner. Finally, the original image is restored by the classical Richardson–Lucy algorithm (Chapter 26). This final phase could obviously be replaced by an arbitrary nonblind deconvolution method.

As mentioned earlier, the approximation of the full posterior distribution by the product $q(\vec{D}\vec{u})q(\vec{h})$ is in the sense of the Kullback–Leibler divergence. The image gradient prior is considered in the form of a Gaussian mixture. In a similar way, the prior on kernel values is expressed as a mixture of exponential distributions, which reflects the fact that most kernel values for motion blur are zero. Both priors are learned from a set of natural images.

State-of-the-art SC algorithms, such as those of Fergus [14] and Shan [17], can perform remarkably well; however, they are far from foolproof. They are very sensitive to input parameters and tend to get caught in local minima, which are characterized by blur kernels that are only partially correct. The following two experiments demonstrate such vulnerability. An original "Picasso" image in Figure 24.2a was artificially blurred by a kernel in Figure 24.2b and corrupted by Gaussian noise. The resulting degraded image is shown in Figure 24.3a. The estimated image and PSF using the algorithm of Fergus and Shan are shown in Figure 24.3b and c, respectively. In this case, both algorithms estimate the PSF quite accurately, but Shan's algorithm provides a slightly more accurate estimation. The image reconstructed by Shan's method outperforms that of Fergus, but this is mainly due to weak performance of the Richardson–Lucy algorithm, which is used in Fergus's method in the nonblind deconvolution step. If we instead use the energy minimization approach in Eq. (24.8) and the PSF estimated by Fergus's method, we obtain results close to Shan's method (Figure 24.3d). The

24.3 Single-Channel Deconvolution

Figure 24.2 Single-channel blind deconvolution: (a) original "Picasso" image and (b) synthetic blur kernel.

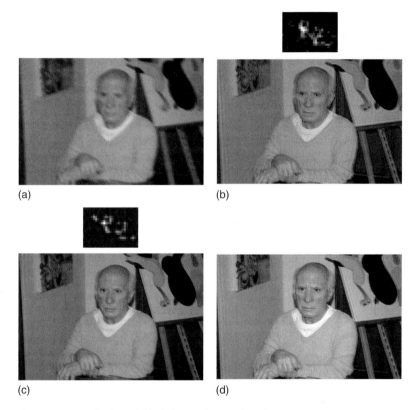

Figure 24.3 Single-channel blind deconvolution of synthetically degraded data: (a) input blurred and noisy image, (b) image and PSF reconstruction by Fergus's method, (c) image and PSF reconstruction by Shan's method, and (d) TV-based nonblind deconvolution using the PSF estimated by Fergus's method.

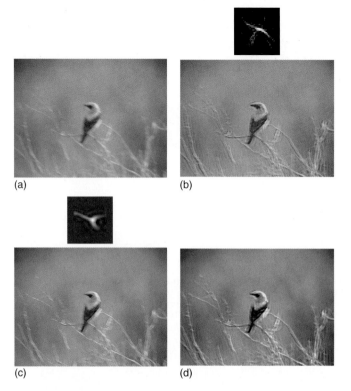

Figure 24.4 Single-channel blind deconvolution of real data: (a) input blurred and noisy image, (b) image and PSF reconstruction by Fergus's method, (c) image and PSF reconstruction by Shan's method, and (d) TV-based nonblind deconvolution using the PSF estimated by Fergus's method. (Please find a color version of this figure on the color plates.)

second experiment compares the performance of both methods on a real image as shown in Figure 24.4a. In this case, we do not know the original PSF and we can only visually assess the reconstructed images. Fergus's method seems to provide much more accurate estimation of the PSF than that provided by Shan's method as the reconstructed image shows less artifacts; compare Figure 24.4b and c. An even better result can be achieved if, again, the energy minimization approach is used with the PSF estimated by Fergus's method instead of the Richardson–Lucy algorithm (Figure 24.4d). The algorithm of Shan has many parameters that influence the result considerably and are difficult to tune. On the other hand, the algorithm of Fergus has virtually no parameters. The two experiments demonstrate that the performance of both algorithms is difficult to predict as it depends on the input data and how well the parameters were tuned.

In summary, there are methods for estimating blur kernels and subsequent restoration from a single blurred image, but the results are still far from perfect. The proposed methods exhibit some common building principles, but vary in others. There is still no thorough analysis, which would tell us which building principles are truly important and which are merely cosmetic. Is the choice of estimation methodology more important than the accurate formulation of priors? Is it the multiscale approach, which makes the difference?

24.4
Multichannel Deconvolution

In the MC scenario, we assume multiple images capturing the same scene but blurred in a different way. We can easily take such a sequence using the continuous shooting modes of present-day cameras. The important advantage over the SC case is that multiple images permit one to estimate the blurs without any prior knowledge of their shape.

Formally, the situation is described as convolution of the original image $\vec{u}$ with P convolution kernels $\vec{h}_p$

$$\vec{z}_p = \vec{H}_p \vec{u} + \vec{n}_p, \quad p = 1, \ldots, P \tag{24.10}$$

The model is equivalent to the SC case in Eq. (24.7) except that now we have more than one acquisition of the same image $\vec{u}$. Clearly, the discussion on SC deconvolution in Section 24.3 also applies for the MC case and most of the SC methods can be directly extended to the MC scenario. However, to exploit the true potential of multiple images, we have to utilize the fact that missing information about u in one degraded image can be supplemented by information in other degraded images. One of the earliest intrinsic MC blind deconvolution methods [22] was designed particularly for images blurred by atmospheric turbulence. Harikumar et al. [23] proposed an indirect algorithm, which first estimates the blur functions and then recovers the original image by standard nonblind methods. The blur functions are equal to the minimum eigenvector of a special matrix constructed by the blurred images. Necessary assumptions for perfect recovery of the blur functions are noise-free environment and channel coprimeness, that is, a scalar constant is the only common factor of the blurs. Giannakis et al. [24] developed another indirect algorithm based on Bezout's identity of coprime polynomials, which finds restoration filters and, by convolving the filters with the observed images, recovers the original image. Both algorithms are vulnerable to noise and even for a moderate noise level, restoration may break down. In the latter case, noise amplification can be attenuated to a certain extent by increasing the restoration filter order, which comes at the expense of deblurring. Pai et al. [25] suggested two MC restoration algorithms that, unlike the previous two indirect algorithms, directly estimate the original image from the null space or from the range of a special matrix. Another direct method based on the greatest common divisor was proposed by Pillai and Liang [26]. In noisy cases, the direct algorithms are more stable than the indirect ones. Interesting approaches based on the autoregressive moving average (ARMA) model are given in Ref. [27]. MC blind deconvolution based on the Bussgang algorithm was proposed in Ref. [28], which performs well on spatially uncorrelated data, such as binary text images and spiky images. Most of the algorithms lack the necessary robustness since they do not include any noise assumptions (except ARMA and Bussgang) in their derivation and do not have regularization terms. Recently, we have proposed an iterative MC algorithm [29] that performs well even on noisy images. It is based on least-squares deconvolution by anisotropic regularization of the image and between-channel regularization of the blurs.

Unfortunately, all the above-mentioned MC blind deconvolution methods contain two ultimate but unrealistic assumptions. They require exact knowledge of the PSF's support size and individual channels are supposed to be perfectly spatially aligned (registered). These strong assumptions are seldom true in practice and, in fact, they have prevented the use of MC blind deconvolution methods in real applications. Both issues were addressed in Ref. [30], where a blind deconvolution method for images, which are mutually shifted by unknown vectors has been proposed. We now describe this algorithm in more detail as it currently ranks among the best-working MC deblurring algorithms.

As in the SC situation, the algorithm can be viewed as an MAP estimator of the original image and blur kernels. It is equivalent to minimization of

$$E(\vec{u}, \vec{h}_1, \ldots, \vec{h}_P) = \frac{1}{2} \sum_{p=1}^{P} \|\vec{z}_p - \vec{H}_p \vec{u}\|^2 + \lambda_u Q(\vec{u}) + \lambda_h \sum_{i \neq j} R(\vec{h}_i, \vec{h}_j) \quad (24.11)$$

with respect to the latent image $\vec{u}$ and blur kernels $\vec{h}_1, \ldots, \vec{h}_p$. The energy function resembles the SC nonblind case in Eq. (24.8). The data-fitting term, the first term in Eq. (24.11), is a sum of measure of differences between input blurred images $\vec{z}_p$ and the original image $\vec{u}$ blurred by kernels $\vec{h}_k$. Inaccurate registration of channels can be alleviated by overestimating the size of blur kernels. The true kernel can move freely in the oversized support and, thus, compensate for possible misalignment.

The image regularization $Q(\vec{u})$ is of the form

$$Q(\vec{u}) = \vec{u}^T \vec{L} \vec{u}$$

as in Eq. (24.8), which is a result of half-quadratic linearization of $\int_\Omega \phi(\nabla u) d\vec{x}$; see the relevant discussion in Section 24.1.

If the kernel regularization term were also a smoothing term as in the case of images, it would not differ from the SC methods. Intrinsically, MC regularization couples all degraded images $\vec{z}_p$ and it is based on a simple observation. For the noiseless case, one can easily prove that quadratic terms

$$R(\vec{h}_i, \vec{h}_j) = \|\vec{z}_j * \vec{h}_i - \vec{z}_i * \vec{h}_j\|, \quad i, j = 1, \ldots, P, i \neq j \quad (24.12)$$

are zero for any blur pairs $\vec{h}_i, \vec{h}_j$, if the blurs are correctly estimated. In the case of noise, the minimum value of $R(\vec{h}_i, \vec{h}_j)$ is not zero but a value proportional to the noise variance. Taking all possible pairs of $\vec{h}_i, \vec{h}_j$ we construct a kernel regularization, the third term in Eq. (24.11), which is based solely on the observed images and does not require any prior knowledge of the blur nature. The only assumption, which is required, is a rough estimate of the maximum kernel size. Equation (24.11) is minimized by alternating minimization in the subspaces corresponding to the image and the blur kernels.

The following experiment on real data demonstrates superiority of the MC approach over the SC one. The photos in Figure 24.5a and b were taken by a handheld camera under low-light conditions. Both images are blurred but blurred

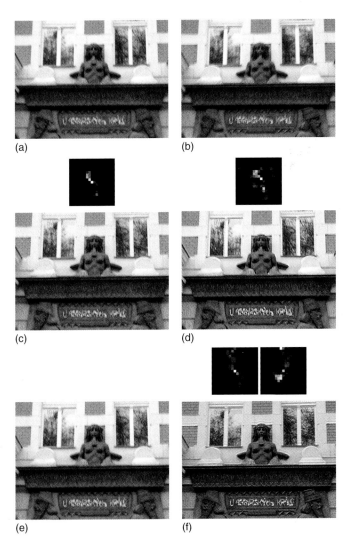

Figure 24.5 Multichannel blind deconvolution of real data: (a,b) two input images taken by a camera, (c) the image and PSF reconstruction from input image (a) using the single-channel method of Fergus, (d) the image and PSF reconstruction from the input image (a) using the single-channel method of Shan, (e) TV-based nonblind deconvolution using the PSF estimated by Fergus's method, and (f) proposed multichannel reconstruction of the image and PSFs using both input images.

in a slightly different way, because of random shaking of the cameraman's hand. Reconstruction using only the left-hand image using the SC methods of Fergus and Shan is shown in Figure 24.5c and d, respectively. An improved deconvolution result using the energy minimization approach in Eq. (24.8) with the PSF estimated by Fergus's method is shown in Figure 24.5e. The proposed MC method using both images results in an image shown in Figure 24.5f. One can see that both SC methods provide PSFs that are not complete in comparison with the PSFs estimated by the MC method. This insufficiency results in poorly reconstructed images in the case of the SC methods.

In general, the MC approach provides better results and is more stable than the SC methods. Then, the question is why not use MC methods instead of SC ones? In many situations, it is unrealistic to assume that one can capture the same scene multiple times, with only different blur kernels. The acquired images are not only misaligned but also rotated or deformed by a more complex geometric transformation. We have to first register the images and this may introduce discrepancies in the acquisition model. In addition, the blur kernel becomes space variant. As a result, the MC approach can be applied mainly to tele-lens photos if the rotational component of camera motion about the optical axis is negligible. It usually works for the central section of an arbitrary blurred image.

Nevertheless, the space-variant nature of blurs limits the applicability of both the SC and MC methods. The next section treats the extension of convolution-based algorithms to situations where the spatial variance of blur cannot be neglected.

24.5
Space-Variant Extension

We start with the corresponding degradation model, which remains linear. The operator H in Eq. (24.1) becomes

$$Hu = [u *_v h](\vec{x}) = \int_\Omega u(\vec{x} - \vec{s}) h(\vec{x} - \vec{s}; \vec{s}) d\vec{s} \qquad (24.13)$$

where h is again called the *point-spread function* (*PSF*). Note that convolution is a special case, with the function $\vec{h}$ independent of coordinates $\vec{x}$, that is, $h(\vec{x}; \vec{s}) = h(\vec{s})$. We can consider Eq. (24.13) as convolution with a kernel that changes with its position in the image. The subscript v distinguishes from ordinary convolution, denoted by an asterisk.

In the discrete case, the matrix $\vec{H}$ in Eq. (24.7) changes analogously to Eq. (24.13). Contrary to the convolution case, now each column contains a different PSF valid for the corresponding pixel.

The problem of (nonblind) restoration can be solved again in the Bayesian framework by minimization of the functionals in Eqs (24.3) and (24.8); now in the form

$$E(u) = \frac{1}{2} \|u *_v h - z\|^2 + \lambda \int_\Omega |\nabla u| \qquad (24.14)$$

Its derivative, Eq. (24.6), can be written as

$$\partial E(u) = (u *_v h - z) \circledast_v h - \lambda \operatorname{div}\left(\frac{\nabla u}{|\nabla u|}\right) \quad (24.15)$$

where $\circledast_v$ is the operator adjoint to Eq. (24.13)

$$H^* u = [u \circledast_v h](\vec{x}) = \int u(\vec{x} - \vec{s}) h(\vec{x}; -\vec{s}) d\vec{s} \quad (24.16)$$

To minimize Eq. (24.14), we can use most algorithms mentioned in the paragraph following Eq. (24.6). A variant modifying the half-quadratic iterative approach, reducing this problem to a sequence of linear subproblems, is described in Ref. [31].

An obvious problem of nonhomogeneous blur is that the PSF is now a function of four variables. Even if it is known, we must solve the problem of efficient representation. If the PSF changes in a gradual and smooth way, we can store the PSF on a discrete set of positions (a square grid) and use linear or polynomial interpolation between kernels to approximate the whole function h. Then, operations $*_v$ and $\circledast_v$ can be speeded up by Fourier transform [32] computed separately on each square corresponding to quadruples of adjacent PSFs.

If the PSF is not known, the algorithms are more complicated by the error-prone PSF estimation. We assume that the PSF changes slowly from pixel to pixel, which is often true for photos taken using handheld cameras if there are no close objects in the field of view. Then, the rotational component of camera motion is usually dominant and the blur is independent of the depth. If the PSF were not continuous, it would be necessary to detect a boundary, consider boundary effects, and so on. There are basically no satisfactory methods for this situation. As a step in this direction, we can consider methods identifying blur based on the local image statistics [33].

In principle, if the PSF changes smoothly, the above-described blind deconvolution methods could be applied locally and the results of deconvolution could be fused together. We could also use the estimated PSFs to approximate the spatially varying PSF and compute the sharp image by minimization of Eq. (24.14). The main problem of this naive procedure is that it is very slow if applied on too many positions.

One way to reduce time is to use a special setup that simplifies the PSF estimation. In Ref. [34], we tested the possibility of using two images taken with different exposure times for this purpose. Exposure time of one of the images is assumed to be so short that the image is sharp. This can be done only at the expense of noise amplification. The whole idea was explored relatively recently [35]. Most algorithms of this group first estimate the blur from the image pair and then deconvolve the blurred image.

For input, the algorithm [34] requires at least a pair of images, one of them blurred and another noisy but sharp. In the example in Figure 24.6, we can see a photo (taken at night) of a historical building taken at ISO 100 with a shutter speed 1.3 seconds. The same photo was taken at ISO 1600 with two stops underexposure to achieve a hand-holdable shutter time 1/50 seconds.

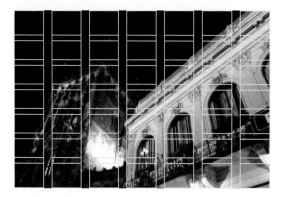

Figure 24.6 This photo was taken at night using a handheld camera at ISO 100 and a shutter speed 1.3 seconds. Another photo of the same scene was taken at ISO 1600 and two stops underexposure to achieve the hand-holdable shutter time 1/50 seconds. The algorithm working with spatially varying blur combines them to get a low-noise sharp photo. Reproduced from "Superresolution Imaging", P. Milanfar (ed.) with permission of CRC Press © 2011.

The algorithm works in three phases:

1) robust image registration;
2) estimation of convolution kernels on a grid of windows, followed by an adjustment at places where estimation failed; and
3) restoration of the sharp image by minimizing a functional such as Eq. (24.14).

Since salient points, such as edges and corners, are smeared, it is quite difficult to register precisely images blurred by motion. Consequently, it is important that this approach does not need subpixel and even pixel precision of the registration. A precision significantly better than half of the considered size of blur kernels is sufficient, because the registration error can be compensated in the next step of the algorithm by the shift of the corresponding part of the space-variant PSF. As a side effect, misalignments due to lens distortion also do not harm the algorithm.

In the second step of the algorithm, we use the fact that the blur can be locally approximated by convolution. We do not estimate the blur kernels in all pixels. Instead, we divide the image into rectangular windows (a 7×7 grid in our example in Figure 24.6) and estimate only a small set of kernels $h_{i,j}(i, j = 1..7$ in our example in Figure 24.7). The estimated kernels are assigned to centers of the windows where they were computed. In the rest of the image, the PSF h is approximated by bilinear interpolation from blur kernels in four adjacent windows.

Thus, we estimate blur kernels on a grid of windows, where the blur can be approximated by convolution in the least-squares sense as

$$h_{i,j} = \arg\min_{k} \|d_{i,j} * k - z_{i,j}\|^2 + \alpha \|\nabla k\|^2, \quad k(\vec{x}) \geq 0 \qquad (24.17)$$

where $h_{i,j}(\vec{s})$ is an estimate of $h(\vec{x}_0; \vec{s})$, $\vec{x}_0$ being the center of the current window $z_{i,j}$, $d_{i,j}$ the corresponding part of the noisy image, and k the locally valid convolution kernel.

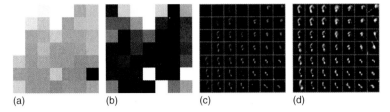

(a) (b) (c) (d)

Figure 24.7 The $49 = 7 \times 7$ kernels (c) were computed in the corresponding squares shown in white in Figure 24.6 Incorrect kernels are detected as those with energy (a) or entropy (b) above a threshold (bright squares). We replace them by the average of adjacent kernels, resulting in a new set of kernels (c, d). Details are given in Section 24.5. Reproduced from "Superresolution Imaging", P. Milanfar (ed.) with permission of CRC Press © 2011.

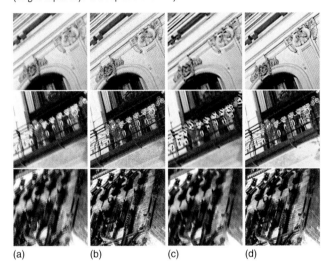

(a) (b) (c) (d)

Figure 24.8 The blurred image (a), noisy image (b), the result of deconvolution by the kernel valid in the image center (c), and our result (d). Reproduced from "Superresolution Imaging", P. Milanfar (ed.) with permission of CRC Press © 2011.

The kernel estimation procedure in Eq. (24.17) can naturally fail. In a robust system, such kernels must be identified, removed, and replaced by, for example, an average of adjacent (valid) kernels. There are basically two reasons that kernel estimation fails. Therefore, we also need two different measures to decide which kernel is wrong. To identify textureless regions, we compute entropy of the kernels and take those with entropy above some threshold. The other more serious case of failure is pixel saturation, that is, pixel values above the sensor range. This situation can be identified by computing the sum of kernel values, which should be close to 1 for valid kernels. Therefore, we simply remove kernels whose sum is too different from unity, again above some threshold. Figure 24.7c and d illustrates the importance of the kernel adjustment step of the algorithm, which handles the cases of pixel saturation or weak texture. To help reader recognize differences in quite a large photograph (1154 × 1736 pixels), we show details of the result in Figure 24.8.

In summary, the restoration of images blurred by spatially varying blur is not resolved satisfactorily in most general cases. If the PSF changes slowly, the problem can be solved but the algorithm is extremely time consuming. An especially difficult case is the situation when the PSF is not continuous, for example, it depends on the depth of scene. Such a situation was recently considered in Ref. [31], for a camera moving without rotation, but this assumption does not correspond to the real trajectory of a handheld camera.

24.6
Conclusions

Image restoration, and particularly blind image restoration, belongs to inverse problems, which has attracted considerable attention in recent decades. The problem has become tractable with an increasing computational capability of computers. A wide variety of methodologies has been proposed in literature to solve such inverse problem. One can formulate the majority of proposed approaches using the energy minimization technique with the help of calculus of variations.

Image restoration is inherently ill posed and we need to modify the problem to make it better posed. In general, there are two ways to achieve this. We must include regularization terms (priors) in the energy function, which penalize inappropriate images and blur kernels, or we must use multiple images of the same scene but each degraded in a slightly different way (different motion or out-of-focus blur).

In the single-channel scenario, it is absolutely essential to use regularization both on images and blur kernels. However, recent advances in this area show that apart from regularization, a choice of minimization strategy is also important. Techniques, such as marginalization, variational distribution approximation, or hierarchical (multiscale) minimization, tend to find the minimum more stably in comparison to classical alternating minimization over the image and blur kernel. State-of-the-art single-channel restoration methods perform well, but the results are often unpredictable and far from perfect.

The MC scenario admits much more stable performance. The advantage is that blur kernels can be estimated directly from the observed blurry images and we do not need any sophisticated PSF priors. The disadvantage is that properly registered multiple images of the same scene are necessary, which can be difficult to achieve in many circumstances.

Unfortunately, in many real situations, the blur kernel varies over the field of view and we face a very complex task of space-variant blind deconvolution. We have shown that for certain cases, we have a reliable method, which solves this problem. Using a two-channel scenario (long- and short-exposure image) and assuming slowly varying PSFs, we can estimate the original image very accurately.

Acknowledgments

Financial support for this research was provided by Czech Ministry of Education under the project 1M0572 (Research Center DAR) and the Grant Agency of the Czech Republic under the project 102/08/1593. We also wish to thank Prof. Jan Flusser for helpful comments.

References

1. Rudin, L., Osher, S., and Fatemi, E. (1992) Nonlinear total variation based noise removal algorithms. *Phys. D*, **60**, 259–268.
2. Vogel, C. and Oman, M. (1998) Fast, robust total variation-based reconstruction of noisy, blurred images. *IEEE Trans. Image Processing*, **7**, 813–824.
3. Chan, T., Golub, G., and Mulet, P. (1999) A nonlinear primal-dual method for total variation-based image restoration. *SIAM J. Sci. Comput.*, **20**, 1964–1977.
4. Aubert, G. and Kornprobst, P. (2002) *Mathematical Problems in Image Processing*, Springer Verlag, New York.
5. Banham, M. and Katsaggelos, A. (1997) Digital image restoration. *IEEE Signal Process. Mag.*, **14**, 24–41.
6. Chang, M., Tekalp, A., and Erdem, A. (1991) Blur identification using the bispectrum. *IEEE Trans. Signal Processing*, **39**, 2323–2325.
7. Shan, Q., Xiong, W., and Jia, J. (2007) Rotational Motion Deblurring of a Rigid Object from a Single Image, Proceedings of IEEE 11th International Conference on Computer Vision ICCV 2007, pp. 1–8. doi: 10.1109/ICCV.2007.4408922.
8. Stern, A. and Kopeika, N. (1997) Analytical method to calculate optical transfer functions for image motion and vibrations using moments. *J. Opt. Soc. Am. A*, **14**, 388–396.
9. Yitzhaky, Y. and Kopeika, N. (1997) Identification of blur parameters from motion blurred images. *Graph. Models Image Processing*, **59**, 310–320.
10. Ayers, G. and Dainty, J.C. (1988) Iterative blind deconvolution method and its application. *Opt. Lett.*, **13**, 547–549.
11. Chan, T. and Wong, C. (1998) Total variation blind deconvolution. *IEEE Trans. Image Processing*, **7**, 370–375.
12. Molina, R., Mateos, J., and Katsaggelos, A.K. (2006) Blind deconvolution using a variational approach to parameter, image, and blur estimation. *IEEE Trans. Image Processing*, **15**, 3715–3727. doi: 10.1109/TIP.2006.881972.
13. Campisi, P. and Egiazarian, K. (eds.) (2007) *Blind Image Deconvolution, Theory and Application*, CRC Press.
14. Fergus, R. *et al.* Removing camera shake from a single photograph, in *SIGGRAPH '06: ACM SIGGRAPH 2006 Papers*, ACM, New York, NY, pp. 787–794. doi: http://doi.acm.org/10.1145/1179352.1141956.
15. Jia, J. (2007) Single Image Motion Deblurring Using Transparency, Proceedings of IEEE Conference on Computer Vision and Pattern Recognition CVPR '07, pp. 1–8. doi: 10.1109/CVPR.2007.383029.
16. Joshi, N., Szeliski, R., and Kriegman, D.J. (2008) PSF Estimation using Sharp Edge Prediction, Proceedings of IEEE Conference on Computer Vision and Pattern Recognition CVPR 2008, pp. 1–8. doi: 10.1109/CVPR.2008.4587834.
17. Shan, Q., Jia, J., and Agarwala, A. (2008) High-quality motion deblurring from a single image, in *SIGGRAPH '08: ACM SIGGRAPH 2008 Papers*, ACM, New York, NY, pp. 1–10. doi: http://doi.acm.org/10.1145/1399504.1360672.
18. Levin, A. *et al.* (2009) *Understanding and Evaluating Blind Deconvolution Algorithms*, Proceedings of IEEE Conference on Computer Vision and Pattern Recognition CVPR '09.

19. Galatsanos, N.P., Mesarovic, V.Z., Molina, R., and Katsaggelos, A.K. et al. (2000) Hierarchical Bayesian image restoration from partially known blurs. *IEEE Trans. Image Processing*, **9**, 1784–1797. doi: 10.1109/83.869189.
20. Miskin, J. and MacKay, D.J. (2000) Ensemble learning for blind image separation and deconvolution, in *Advances in Independent Component Analysis* (ed. M. Girolani), Springer-Verlag.
21. You, Y.-L. and Kaveh, M. (1999) Blind image restoration by anisotropic regularization. *IEEE Trans. Image Processing*, **8**, 396–407.
22. Schulz, T. (1993) Multiframe blind deconvolution of astronomical images. *J. Opt. Soc. Am. A*, **10**, 1064–1073.
23. Harikumar, G. and Bresler, Y. (1999) Perfect blind restoration of images blurred by multiple filters: theory and efficient algorithms. *IEEE Trans. Image Processing*, **8**, 202–219.
24. Giannakis, G. and Heath, R. (2000) Blind identification of multichannel FIR blurs and perfect image restoration. *IEEE Trans. Image Processing*, **9**, 1877–1896.
25. Pai, H.-T. and Bovik, A. (2001) On eigenstructure-based direct multichannel blind image restoration. *IEEE Trans. Image Processing*, **10**, 1434–1446.
26. Pillai, S. and Liang, B. (1999) Blind image deconvolution using a robust GCD approach. *IEEE Trans. Image Processing*, **8**, 295–301.
27. Haindl, M. and Šimberová, S. (2002) Model-based restoration of short-exposure solar images, in *Frontiers in Artificial Intelligence and Applications*, vol. 87 (eds. L. Jain and R. Howlett), Knowledge-Based Intelligent Engineering Systems, Publisher IOS Press, Amsterdam, pp. 697–706.
28. Panci, G., Campisi, P., Colonnese, S., and Scarano, G., (2003) Multichannel blind image deconvolution using the Bussgang algorithm: spatial and multiresolution approaches. *IEEE Trans. Image Processing*, **12**, 1324–1337.
29. Šroubek, F. and Flusser, J. (2003) Multichannel blind iterative image restoration. *IEEE Trans. Image Processing*, **12**, 1094–1106.
30. Šroubek, F. and Flusser, J. (2005) Multichannel blind deconvolution of spatially misaligned images. *IEEE Trans. Image Processing*, **14**, 874–883.
31. Sorel, M. and Flusser, J. (2008) Space-variant restoration of images degraded by camera motion blur. *IEEE Trans. Image Processing*, **17**, 105–116. doi: 10.1109/TIP.2007.912928.
32. Nagy, J.G. and O'Leary, D.P. (1998) Restoring images degraded by spatially variant blur. *SIAM J. Sci. Comput.*, **19**, 1063–1082.
33. Levin, A. (2006) Blind Motion Deblurring using Image Statistics, Proceedings of NIPS, MIT Press, Cambridge, MA, USA (2007) pp. 841–848.
34. Šorel, M. and Šroubek, F. (2009) *Space-variant Deblurring using One Blurred and one Underexposed Image*, Proceedings of IEEE International Conference on Image Processing, IEEE Press, Piscataway, NJ, USA (2009).
35. Tico, M., Trimeche, M., and Vehvilainen, M. (2006) *Motion Blur Identification Based on Differently Exposed Images*, Proceedings of IEEE International Conference on Image Processing, pp. 2021–2024. doi: 10.1109/ICIP.2006.312843.

25
Optics and Deconvolution: Wavefront Sensing

Justo Arines and Salvador Bará

25.1
Introduction

Scientists have developed different techniques to achieve high-resolution imaging (see Chapters 14, 24, and 26). All of them can be broadly classified into three main groups: postdetection, predetection, and hybrid techniques. The postdetection group gathers the procedures aimed to computationally restore the frequency content of the degraded images based on the knowledge of the degradation function: it includes both linear and nonlinear restoration algorithms, from the simple inverse filter to the sophisticated blind deconvolution (BD) procedures based on neural networks, iterative algorithms (see Chapter 24) or genetic algorithms, providing an estimation of the degradation function and of the undistorted geometrical image [1–5]. Deconvolution from wavefront sensing (DWFS) and myopic deconvolution (MD) also belong to this group. The predetection approach, in turn, is based on the improvement or compensation of the degradation function before recording the image, using optical techniques like adaptive optics (AOs) or optical filters [6, 7]. Finally, the hybrid approaches are those which combine the main features of the pre- and postdetection techniques. They are based on correcting, at least partially, the degradation function before recording the image, and on using the information on the residual degradation function to perform an improvement of the recorded image by applying some of the available postdetection algorithms.

Previous chapters of this book have established the basic theory of these high-resolution imaging techniques. Chapter 9 describes the main elements of the adaptive optical systems presented in Chapters 14 and 15. Chapters 24 and 26 present several postdetection image restoration algorithms. In this chapter, we focus on the DWFS approach. First, in Section 25.2, we give a general overview of the technique. Section 25.3 describes its beginnings, development, and present state. Section 25.4 is devoted to the analysis of some key issues that are directly related to the restoration process: the estimation of the wavefront aberration and the computation of the degraded point spread function (PSF) and optical transfer function of the system (OTF). We end this section with an analysis of the effective resolution attainable in the presence of wavefront aberrations and noise

Optical and Digital Image Processing: Fundamentals and Applications, First Edition. Edited by Gabriel Cristóbal, Peter Schelkens, and Hugo Thienpont.
© 2011 Wiley-VCH Verlag GmbH & Co. KGaA. Published 2011 by Wiley-VCH Verlag GmbH & Co. KGaA.

in the image detector. Section 25.5 describes some examples of application of DWFS in different fields of research. And finally, in Section 25.6, the DWFS approach is compared with other alternative techniques for obtaining high-resolution images.

25.2
Deconvolution from Wavefront Sensing (DWFS)

DWFS is a high-resolution imaging technique that has been developed in the framework of linear shift-invariant systems under the assumption that the image degradation depends on the optical aberrations and noise of the imaging system. Under these conditions, the recorded image $i(x',y')$ can be expressed as the convolution of the geometrical image of the object $o_g(x',y')$ with the optical system PSF $H(x',y')$, plus an additive noise term $n(x',y')$:

$$i(x',y') = \iint_\infty H(x'-\xi, y'-\eta) o_g(x',y') d\xi d\eta + n(x',y') \tag{25.1}$$

By Fourier transforming Eq. (25.1), we obtain the spectrum $I(f_x, f_y)$ of the degraded image as a filtered and noise-corrupted version of the spectrum $O_g(f_x, f_y)$ of the undistorted geometrical image of the object.

$$I(f_x, f_y) = \text{OTF}(f_x, f_y) O_g(f_x, f_y) + N(f_x, f_y) \tag{25.2}$$

where $\text{OTF}(f_x, f_y)$ is the OTF of the imaging system (i.e., the Fourier transform of the incoherent PSF) and $N(f_x, f_y)$ is the Fourier transform of the noise, with f_x, f_y being the spatial frequencies. Thus, if the system OTF can be determined by some means, the spectrum of the geometrical undegraded image can, in principle, be estimated from the spectrum of the recorded image and the noise can be determined by a suitable inversion of Eq. (25.2). These initial basic assumptions are common to most deconvolution approaches. What makes DWFS unique is that the degraded PSF used for the construction of the restoration filter is estimated from the measurement of the system's wave aberrations made with a wavefront sensor, simultaneously with the acquisition of the degraded images. This feature makes DWFS appropriate for environments or optical systems where the optical aberrations vary with time, for example, turbulent media like the atmosphere or the aberrated human eye [8–11].

A general DWFS system consists of two channels: one for image acquisition (image channel) and the other for measuring the wave aberration at the exit pupil of the imaging system (sensor channel). Data gathering in both channels is synchronized in order to record the degraded image and the wavefront measurement at the same time (Figure 25.1). The aberration estimation provided by the wavefront sensor is then used to compute the degraded PSF or OTF, and, subsequently, to build the restoration filter.

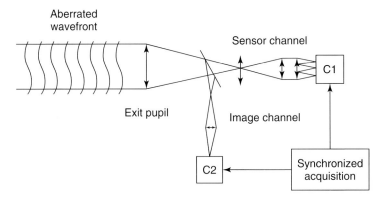

Figure 25.1 Basic scheme of a DWFS imaging system.

25.3
Past and Present

Two precursors of DWFS were developed by Harris and McGlamery [5]. They presented the first results about image restoration of long- and short-exposure turbulence-degraded images. In his study, Harris used a mathematical model for the PSF, adjusted to the strength of the turbulence, while McGlamery employed the recorded image of a point source in order to obtain information on the degradation function. This technique was named *speckle holography*. Other studies preceding DWFS were the those of A. Labeyrie and by K. T. Knox and B. J. Thompson, which established the technique of speckle interferometry. The development of speckle holography was continued, among others, by C. Y. C. Liu, A. W. Lohmann, and Gerd P. Weigelt. The readers are referred to Ref. [5] for more information.

It was in 1985 that Fontanella finally described the technique of DWFS [8]. In that work, he proposed the use of a wavefront sensor to estimate the degradation function and build the restoration filter. Posterior studies performed by Fried, Primot *et al.*, Rousset *et al.*, Primot, Gonglewski *et al.*, and Michau *et al.* contributed to the initial development of DWFS as a potential alternative to AO [5, 12, 13].

The success of DWFS in astronomical imaging encouraged its application to other areas of research, in particular to the high-resolution imaging of the eye fundus. The images of the living human retina provided by clinical ophthalmoscopic instruments (flood illumination cameras, scanning laser ophthalmoscopes, OCTs, etc.) become noticeably blurred by the physiological aberrations of the human eye. Although the relative magnitude of these aberrations and their spatial and temporal statistics are different from those induced by atmospheric turbulence, the basic problem to solve is very similar. This contributed to the rapid expansion of DWFS to the eye fundus imaging field [10, 11, 14, 15]. In the last years, the development of more accurate and reliable AO systems has relegated DWFS to the second place, with AO being the first-choice solution for high-resolution imaging through time-varying aberrated media. However, DWFS is nowadays a mature technique that has found many applications in different areas of research, and whose simplicity, robustness, and affordability warrant its present and future application for high-resolution imaging under random time-varying blurs, at a fraction of the cost of more sophisticated AO systems.

25.4
The Restoration Process

DWFS is based on the construction of the restoration filter from the measurement of the wavefront aberrations in the exit pupil of the imaging system at the instant that the image is recorded. The main steps of this procedure are outlined in Figure 25.2.

In this section, we analyze several issues related to this procedure, in particular, the estimation of the wave aberration with gradient-based wavefront sensors,

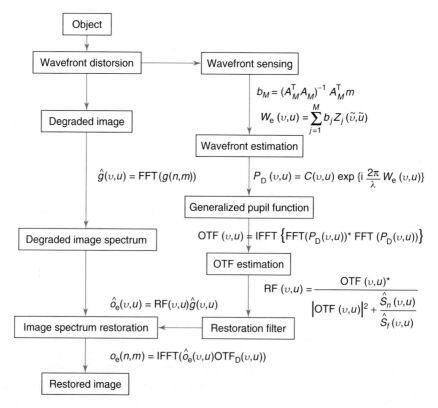

Figure 25.2 Block diagram of a basic DWFS reconstruction procedure.

some factors affecting the precision and bias in the PSF and OTF estimation, the significance of the restoration filter, and the effective spatial resolution of the restored image.

25.4.1
Estimating the Wave Aberration with Gradient-Based Wavefront Sensors

Wavefront sensors are devices that provide information about the amplitude and phase of an optical field based on measurements of irradiance. Depending on their configuration, they give information about the modulus-2π phase itself (using self-referenced interferometers, e.g., the point diffraction interferometer) [16], or about the local wavefront slopes (lateral- and radial-shear interferometers, Hartmann or Hartmann–Shack (HS) setups) [16], or about the local wavefront Laplacians (curvature sensors) [16]. Among them, the HS wavefront sensor has found widespread use.

The HS sensor is a conceptually simple system that consists of an array of wavefront sampling elements (microlenses, in most cases) and an irradiance detector, placed at a known distance from the sampling array (not necessary at the microlens focal plane), which records the corresponding spots [16]. The displacements of the spot centroids with respect to the ones obtained for a given reference wavefront are directly related to the difference between the irradiance-weighted, spatially averaged, wavefront slopes of both wavefronts [17, 18]. This information is then used as an input to the wavefront estimation algorithm. Figure 25.3 shows the basic scheme of an HS wavefront sensor.

The literature presents different measurement models that increase in complexity and accuracy with the number of years. Most of them were obtained in the framework of the parabolic approximation of the scalar wave equation for nonabsorbing homogeneous media, considering also point like-sources and no cross talk between sampling elements. With these assumptions, it is easy to obtain the relationship between the local wavefront gradient and the centroid displacement [18]:

$$m_\alpha^s = \frac{\int_{A_s} I_s(x, y) \partial_\alpha W(x, y) dx dy}{\int_{A_s} I_s(x, y) dx dy} = \frac{X_c}{d} \text{ with } \begin{cases} \alpha = x \text{ if } s \leq N_s \\ \alpha = y \text{ if } s > N_s \end{cases} \quad (25.3)$$

where m_α^s is the averaged wavefront slope in the α-direction at the sth microlens, x and y are the pupil plane coordinates, $I_s(x,y)$ is the irradiance distribution at the

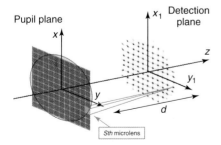

Figure 25.3 Basic scheme of a Hartmann–Shack wavefront sensor.

sth sampling element, of area A_s, ∂_α is the gradient operator along the α-direction, $W(x,y)$ is the wavefront's optical path length, d is the distance between the sampling array and the detector plane, and X_c is the α-direction coordinate of the microlens spot centroid.

Notice that d can be different from the microlens focal length. In some cases, it may be advisable to locate the detector away from the microlens focal plane, in order, for example, to improve the signal-to-noise ratio (SNR) of the spot centroids. As demonstrated in the optical version of the Eherenfest's theorem, within the parabolic approximation in nonabsorbing homogeneous media, the wavefront centroid propagates along a straight line, so that the centroid slope, as measured from the microlens center, is constant regardless of the location of the detection plane.

Equation (25.3) takes the effects of the local irradiance distribution over each sampling element into account. Applying this model correctly requires the measurement of $I_s(x,y)$, which, in principle, offers no difficulty if a separate channel for pupil imaging is available. However, in many practical cases one can get a reasonably good approximation to the sensor measurements by considering $I_s(x,y)$ as a constant. The expression (25.3) then reduces to the one presented by Primot et al. [3], which is a commonly used measurement model.

If, in addition, the wavefront behaves as locally plane within the sth microlens area (a sufficient, but not necessary, condition), expression (25.3) reduces significantly. In this case, the centroid displacement can be taken as directly related to the wavefront slope at the center of the sampling element:

$$m_\alpha^s = \partial_\alpha W(x,y)\big|_{(0,0)} = \frac{X_c}{d} \tag{25.4}$$

More complex measurement models include the effects of thresholding the detected irradiance distribution, using extended sources, or considering the spot diagram from an interferometric point of view [19].

There are various sources of error affecting the accuracy and precision of the estimated wavefront, and hence the deconvolution process. They can be classified into two main groups: measurement errors and estimation errors.

Typical measurement systematic errors are associated with the determination of the following parameters: the distance between the sampling and detection planes; the effective pixel size at the detection plane when using relay optics to image the focal spots onto the detector; and the measurement of the normalized coordinates of the sampling elements within the unit radius pupil. These parameters are linearly related to the measured wavefront gradient, so their effects can be minimized with simple calibration procedures. The most common calibration procedure consists of measuring a set of known parabolic wavefronts induced by different lenses of known focal length or Badal systems (optical systems consisting of two lenses, where the amount of defocus of the light beam after the second lens can be controlled by the separation between the two lenses). The use of phase plates inducing high-order aberrations has also been proposed, showing their relevance in detecting systematic errors in laser ray tracing sensor devices. Measurement random errors depend mainly on the detector (read noise, pixelization, etc.) and on the illumination source (photon noise, speckle noise, etc.) [20].

The performance of the HS sensors depends crucially on the capability of determining the centroids of the focal irradiance distributions. Researchers have proposed different alternatives to overcome or at least to reduce the impact of these error sources. Image processing algorithms, Gaussian fitting of the irradiance distribution, neural networks, or iterative Fourier algorithms are some of the proposals [20–22].

Estimation errors are those induced in the data reduction process that is used to obtain the wavefront or its modal coefficients from the measurements. They depend on the propagation of the measurement error and on the modeling of the wavefront and the sensor's measurements [6, 17, 23]. Wavefront estimation from wavefront slope sensors can be done using several algorithms, which are conventionally classified into two main groups: zonal and modal. While zonal algorithms search for the local reconstruction of the wavefront using the information of its gradient in the surrounding sampling points, modal algorithms look for the coefficients of the modal expansion of the wavefront into a set of orthogonal polynomials [16]. Among the different linear and nonlinear estimation algorithms used in wavefront sensing, those based on the least squares and minimum variance (MV) criteria are the most relevant ones, especially in DWFS [16, 23, 24].

Let us now recall the basic modal least squares quadratic (LSQ) estimation approach. We start by considering that the actual (W) and estimated (W_e) wavefronts can be expanded into a linear combination of Zernike polynomials:

$$W(x,y) = \sum_{i=1}^{Q} a_i Z_i(\tilde{x}, \tilde{y}) \quad W_e(x,y) = \sum_{i=1}^{M} b_i Z_i(\tilde{x}, \tilde{y}) \tag{25.5}$$

where $W(x,y)$ and $W_e(x,y)$ are the actual and estimated phases, a_i and b_i are the corresponding actual and estimated Zernike coefficients (see Chapter 14), and $\tilde{x}$ and $\tilde{y}$ are the spatial coordinates normalized to the pupil radius. The LSQ criterion searches for a set of b_i (unknowns) that minimize the squared differences between the actual sensor measurements m_α^s and those computed from the estimated wavefront $m_{e\alpha}^s$.

$$\min \left\{ \sum_{s=1}^{2N_S} (m_\alpha^s - m_{e\alpha}^s)^2 \right\} \tag{25.6}$$

where $m_{e\alpha}^s$ is defined as

$$m_{e\alpha}^s = \frac{\int_{A_S} I_s(x,y) \partial_\alpha \left(\sum_{i=1}^{M} b_i Z_i(\tilde{x}, \tilde{y}) \right) dxdy}{\hat{I}_s} + v_\alpha^s \tag{25.7}$$

where $\hat{I}_s = \int_{\Omega_s} I_s(x,y) dx dy$, and v_α^s is the measurement noise of the sth microlens along the α-direction. Hence, Eq. (25.7) becomes

$$\min \left\{ \sum_{s=1}^{2N_s} \left(m_\alpha^s - \frac{1}{\hat{I}_s} \int_{\Omega_s} I_s(x,y) \partial_\alpha \sum_{i=1}^{M} b_i Z_i(\tilde{x},\tilde{y}) dx dy + v_\alpha^s \right)^2 \right\} \quad (25.8)$$

Minimizing Eq. (25.8) with respect to b_i we get the LSQ solution in matrix form, which, for zero-mean uncorrelated noise of identical variance for all measurements, takes the simple form:

$$b_M = \left(A_M^T A_M \right)^{-1} A_M^T m = Rm \quad (25.9)$$

where m is the measurement column vector and A_M is a matrix of dimensions $2N_s \times M$ defined as

$$A_M =$$

$$\begin{pmatrix} \frac{1}{\hat{I}_1} \int_{\Omega_1} I_1(x,y) \partial_\alpha Z_1(\tilde{x},\tilde{y}) dx dy & \frac{1}{\hat{I}_1} \int_{\Omega_1} I_1(x,y) \partial_\alpha Z_2(\tilde{x},\tilde{y}) dx dy & \cdots & \frac{1}{\hat{I}_1} \int_{\Omega_1} I_1(x,y) \partial_\alpha Z_M(\tilde{x},\tilde{y}) dx dy \\ \frac{1}{\hat{I}_2} \int_{\Omega_2} I_2(x,y) \partial_\alpha Z_1(\tilde{x},\tilde{y}) dx dy & \cdot & \cdots & \cdot \\ \cdot & \cdot & \cdots & \cdot \\ \frac{1}{\hat{I}_{2N_s}} \int_{\Omega_{2N_s}} I_{2N_s}(x,y) \partial_\alpha Z_1(\tilde{x},\tilde{y}) dx dy & \cdot & \cdots & \frac{1}{\hat{I}_{2N_s}} \int_{\Omega_{2N_s}} I_{2N_s}(x,y) \partial_\alpha Z_M(\tilde{x},\tilde{y}) dx dy \end{pmatrix}_{2N_s \times M}$$

(25.10)

R is the reconstruction matrix, which, in this case, is equal to the pseudoinverse of A_M. On the other hand, the vector m can be also expressed in matrix form as $m = Aa + v$, where A is a matrix with the exact measurement model and v is the actual measurement noise vector. Since the measurement noise averages to zero, the expected value and variance of b_M are given by Eqs. (25.11) and (25.12):

$$\langle b_M \rangle = \left\langle \left(A_M^T A_M \right)^{-1} A_M^T (Aa + v) \right\rangle = RA \langle a \rangle \quad (25.11)$$

$$\left\langle (b_M - a)(b_M - a)^T \right\rangle$$

$$= \underbrace{R \left[AC_a A^T + AC_{av} + C_{av}^T A^T + C_v \right] R^T - R \left[AC_a + C_{av}^T \right] - \left[C_a A^T + C_{av} \right] R^T + C_a}_{1}$$

$$= \underbrace{(I - RA) C_a (I - RA)^T + RC_n R^T}_{2} \quad (25.12)$$

where $C \langle vv^T \rangle$ are the second-moment matrices of the actual modal coefficients and of the measurement noise, respectively; $C_{av} = \langle av^T \rangle$ is the cross-products matrix of actual coefficients and measurement noise; and I is the identity matrix. In Eq. (25.12), to arrive at 2 from 1, we have made the assumption that a and v are uncorrelated and that either a or v (or both) are zero-mean variables.

Modal coupling is an issue to take into account in any wavefront estimation procedure. The noiseless measurements can be written as $m = Aa$, where A is a

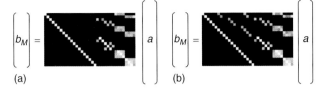

Figure 25.4 First 35 columns of the coupling matrix for a square grid of 69-square microlenses assuming that only the first $M = 20$ Zernike modes are included in the estimator R: (a) with A_M constructed using the correct model of measurements (wavefront slopes spatially averaged over each subpupil) and (b) with an incorrect measurement model (wavefront slopes evaluated at the center of each subpupil). The values are shown in a logarithmic gray scale comprising four decades (white = 1, black < 0.0001). See Ref. [25] for more details.

matrix of dimensions $2N_s \times Q$ (with Q possibly being infinite), the elements of which depend on the exact relationship between the spot-centroid displacements and the actual wavefront modal coefficients (vector a). Equation (25.9) can be rewritten as

$$b_M = \left(A_M^T A_M\right)^{-1} A_M^T A a = C a \qquad (25.13)$$

where C is the coupling matrix linking the estimated modal coefficients to the actual ones. Notice that if $A_M = A$, then we would get an unbiased perfect estimation of the unknown wavefront. However, if either the dimensions of A_M and A, or the value of their elements, or both, are different, then b_M turns out to be a biased estimation of a. This effect is graphically shown in Figure 25.4, where we show the values of the first 20×35 elements of the matrix C, calculated for an HS sensor with 69 square microlenses arranged in a square lattice, assuming that $M = 20$ modes are estimated. Figure 25.4a has been obtained with A_M, constructed using the correct model of measurements (wavefront slopes spatially averaged over each subpupil), while Figure 25.4b corresponds to the use of an incorrect measurement model (wavefront slopes evaluated at the center of each subpupil).

Note that the first $M \times M$ square box in Figure 25.4a is the identity (there is no cross-coupling, nor wrong self-modal coupling) but there are nonzero elements in the columns of index greater than M. These nonzero elements give rise to aliasing of high-order modes to low-order ones. By contrast, in Figure 25.4b we observe the presence of several nondiagonal elements in the first $M \times M$ box, inducing cross-coupling between modes with index smaller than M. The magnitude of the nondiagonal elements depends on the difference between the real measurements and the model used for their description [25]. Most of these elements are generally of small magnitude (note that in Figure 25.4 the values are shown in a logarithmic scale).

Recent studies have also analyzed the influence of the geometry of the sampling grid on the performance of the LSQ estimator [23, 25, 26]. In particular, Soloviev et al. arrived at the interesting conclusion that the rank of A_M can be increased by using random sampling patterns.

The LSQ estimator is widely used in practice. No knowledge of the statistics of the wavefronts is required and its implementation is straightforward. However, if the statistics of the unknown variables and the measurement noise are known, a better choice is the MV estimator. Different studies have shown the superiority of this criterion with respect to LSQ, most of them related being to atmospheric turbulence [5, 23, 27].

We showed in the previous paragraphs as to how to estimate the modal coefficients from the HS measurements. In the next section, we show the procedure for computing the estimated PSF and OTF of the aberrated optical system.

25.4.2
PSF and OTF: Estimation Process and Bias

The computation of the OTF is an important step in the construction of the restoration filter used in DWFS. It is defined as the autocorrelation of the generalized pupil function normalized to its maximum value [5]:

$$\mathrm{OTF}(f_x, f_y) = \frac{\int_{-\infty}^{\infty} P(x, y) P^*(x - \lambda df_x, y - \lambda df_y) dx dy}{\int_{-\infty}^{\infty} |P(x, y)|^2 dx dy} \quad (25.14)$$

where x and y are pupil coordinates and the generalized pupil function has the following form:

$$P(x, y) = C(x, y) \exp\left\{ i \sum_{j=1}^{N} a_j Z_j(\tilde{x}, \tilde{y}) \right\} \quad (25.15)$$

The estimated modal coefficients of the incident wavefront should be substituted in Eq. (25.15) for the unknown ones, in order to get the estimated version of the OTF.

Deconvolution algorithms are developed to deal with digital images. So, it is necessary to compute the discrete representation of $P(x,y)$, the OTF, and the PSF functions in order to get a meaningful reconstruction. Therefore, it is important to find the relation that links the continuous and discrete magnitudes. This expression can be easily obtained by direct comparison of the kernels of the continuous and discrete Fourier transforms:

$$\left. \begin{array}{l} \tilde{g}(k) = \sum_{n=0}^{N-1} g(n) \exp\left\{-i2\pi \dfrac{kn}{N}\right\} \\[6pt] \tilde{g}(x_1) = \int_{\Pi_o} g(x_0) \exp\left\{-i2\pi \dfrac{x_1 x_0}{\lambda d}\right\} dx_0 \end{array} \right\} \xrightarrow{\text{Exponential kernel comparison}} \dfrac{kn}{N} = \dfrac{x_0 x_1}{\lambda d}$$

(25.16)

where n and x_0 are the discrete and continuous position variables at the pupil plane, k and x_1 correspond to the image plane, N is the number of pixels of the discretized function, Π_o is the definition domain of the continuous function, and λ and d are the wavelength and the distance between the exit pupil and the image plane, respectively. By defining $x_1 = km_1$ and $x_0 = nm_0$ where m_1 and m_0 are the pixel sizes in the image and pupil planes, respectively, we get the desired equation:

$$m_0 = \frac{\lambda d}{N m_1} \quad (25.17)$$

From a practical point of view, Eq. (25.17) allows us to calculate the pupil radius in pixel units as a function of some physical parameters (wavelength, image distance, and image pixel size) and discrete parameters (dimensions of the extended pupil and image matrices). The computation of the OTF and the PSF can be easily performed by using the discrete Fourier transform of the discrete pupil function, $P_D(v, u)$:

$$\text{OTF}(v, u) = \text{FFT}^{-1}\left\{\text{FFT}(P_D(v, u))^* \cdot \text{FFT}(P_D(v, u))\right\} \quad (25.18)$$

$$\text{PSF}(n, m) = \text{FFT}^{-1}(P_D(v, u))^* \cdot \text{FFT}^{-1}(P_D(v, u)) \quad (25.19)$$

The estimated OTF is computed from the estimated wavefront, and so, its resemblance to the actual OTF depends on the performance of the wavefront sensing procedure. The induced correlations and the limited number of wavefront modal coefficients induce a biased reconstruction of the OTF and thus cause the modification of its statistical properties [5]. These errors limit the performance of DWFS algorithms, because of an over- or underestimation of certain spatial frequencies of the object [20].

25.4.3
Significance of the Restoration Filter

In DWFS, as in other image restoration techniques, we try to obtain the best estimation of the geometrical image by using some knowledge of the degradation function (it may be recalled that, in DWFS, this information is gathered from the wavefront sensor). Among all the restoration filters used in DWFS, the three more relevant ones are the Wiener filter (WF); the regression filter (RF); and the vector Wiener filter (VWF) [3, 5, 9, 20], all of them obtained from the minimization of specific criteria developed in the framework of linear least squares estimation theory. Table 25.1 shows the functional form of these restoration filters and the related minimization criteria. In this table, $o(n,m)$ is the object of interest; $o_e(n,m)$ is the estimated object; $h(n,m)$ is the PSF; $g(n,m)$ is the image; $g_e(n,m)$ is the estimated image; $\hat{o}(v, u)$, $\hat{o}_e(v, u)$, $\hat{g}(v, u)$, $\hat{g}_e(v, u)$, and $\hat{h}(v, u)$ are the Fourier transforms of $o(n,m)$, $o_e(n,m)$, $g(n,m)$, $g_e(n,m)$, and $h(n,m)$ respectively; $\hat{S}_n(v, u)$ and $\hat{S}_f(v, u)$ are the power spectral densities of the noise and the geometrical image, respectively. P is the number of averaged samples. The quotient $\hat{S}_n(v, u)/\hat{S}_f(v, u)$ is usually reduced to a constant parameter that can be estimated from prior information of the spectrum of the noise and the geometrical image, or can be selected by trial and error. This factor is important for controlling the amplification of noise, but an erroneous high value can also excessively smooth the restored image by limiting its high-frequency content.

Table 25.1 Minimization criteria and restoration filters.

	Minimization criteria	Restoration filter
Wiener filter	$\min\left\{\left\langle \|o(n,m) - o_e(n,m)\|^2 \right\rangle\right\}$ $o_e(n,m) = h(n,m) * g(n,m)$	$\hat{o}_e(v,u) = \dfrac{\hat{h}(v,u)^* \hat{g}(v,u)}{\|\hat{h}(v,u)\|^2 + \dfrac{S_n(v,u)}{S_f(v,u)}}$
Regressive filter	$\min\left\{\sum_{j=1}^{P} \|\hat{g}(v,u) - \hat{g}_e(v,u)\|^2\right\}$ $\hat{g}_e(v,u) = \hat{h}(v,u)\hat{o}_e(v,u)$	$\hat{o}_e(v,u) = \dfrac{\sum_{j=1}^{P} \hat{h}_j(v,u)^* \hat{g}_j(v,u)}{\sum_{j=i}^{P} \|\hat{h}_j(v,u)\|^2}$
Vector Wiener filter	$\min\left\{\left\langle \|o(n,m) - o_e(n,m)\|^2 \right\rangle\right\}$ $o_e(n,m) = \dfrac{1}{P}\sum_{j=1}^{P} h_j(n,m) * g_j(n,m)$	$\hat{g}_e(v,u) = \dfrac{\sum_{j=1}^{P} \hat{h}_j(v,u)^* \hat{g}_j(v,u)}{\sum_{j=i}^{P} \|\hat{h}_j(v,u)\|^2 + \dfrac{S_n(v,u)}{S_f(v,u)}}$

From Table 25.1, it is clear that the minimization of different criteria provides different restoration filters, each of them optimized for a specific situation. Therefore, the selection of the restoration filter should be made in terms of the circumstances in which the degraded images were obtained.

Different works have demonstrated the relevance of the filter selection in the field of DWFS [28]. These works show the superiority of the VWF when operating with several images obtained in the presence of random time-varying blur, for example, when observing through a turbulent atmosphere or the human eye.

25.4.4
Resolution of the Restored Image: Effective Cutoff Frequency

The SNR of the degraded images is a relevant factor in the DWFS process, limiting (in a certain fashion) the resolution of the restored image. In this section, we formalize this issue by presenting a simple criterion that links the image SNR, the system OTF, and the maximum effective cutoff frequency of the restored image, in the framework of linear shift-invariant systems and additive zero-mean Gaussian random noise.

Under these assumptions, the SNR of the degraded image spectrum $SNR_s(v,u)$ can be expressed as

$$SNR_S(v,u) = \frac{|\hat{g}(v,u)|}{|\hat{g}(0,0)|} MTF(v,u) SNR_i \qquad (25.20)$$

where $\hat{g}(v,u)$ is the spectrum of the geometrical image, $MTF(v,u)$ is the modulus of the OTF, and SNR_i is the mean SNR of the degraded image. This expression shows that $SNR_s(v,u)$ depends on the object, the imaging system, and the image detector.

We can define the SNR of the of image observation–detection system as $\text{SNR}_{\text{obs-det}} = \text{MTF}(v, u)\text{SNR}_i$. We can now set up, following the scheme of Idell [29], a threshold value for the $\text{SNR}_{\text{obj-det}}$, below which the system does not provide useful information on the object spatial frequencies. This threshold can be established as

$$\text{SNR}_{\text{obs-det}} \geq 1 \Rightarrow \text{MTF}(v, u) \geq \frac{1}{\text{SNR}_i} \qquad (25.21)$$

With this criterion, the *effective cutoff frequency* (f_c) is defined as the spatial frequency from which the value of the $\text{MTF}(v, u)$ is smaller than the inverse of the SNR_i. Thus, the maximum frequency that can be efficiently detected depends on both the optical system and the image detector. An additional conclusion is that in order to increase f_c one must search for procedures that give an improvement in the OTF and/or an increase in the SNR_i.

Without modifying the observation system, an increase in the SNR_i can be obtained by averaging several frames. This procedure was used, for example, by Catlin and Dainty for obtaining high-resolution images of the foveal cones in human eyes [10]. On the other hand, the improvement in the OTF can be achieved by reducing the magnitude of the wavefront aberration. This idea is the basis of the compensated DWFS [5, 11].

Figure 25.5 shows an example of application of this criterion. We analyze the case of a DWFS system designed for human eye fundus imaging. In the figure,

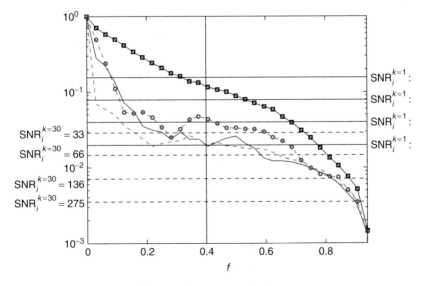

Figure 25.5 Example of application of the criterion for the determination of the system-effective cutoff frequency. See Ref. [14] for more details. (Please find a color version of this figure on the color plates.)

the MTF of the observation system, including the ocular wavefront aberration of two subjects (SB-blue, SM-red), is shown and the MTF of the observation system with the eye aberrations is partially compensated by customized phase plates (SB, blue squares; SM, red circles). The horizontal lines represent different SNR_i levels: solid lines correspond to the labels at the right-hand side and represent the $SNR_i^{k=1}$ without frame averaging; the dashed lines correspond to those on the left-hand side, obtained after averaging 30 frames. Static wavefront aberrations, corresponding to the measured average value for each subject, were assumed in the calculation. The vertical line represents the normalized spatial frequency of the human cone mosaic. Notice that for the case $SNR_i, {}^{k=1} = 25$ and subject SB (blue solid line), the normalized spatial frequency of the cone mosaic is beyond the f_c, which prevents its observation.

Thus, in order to observe this anatomic structure one needs to increase the f_c of the observation system. One option is to improve the system MTF by compensating the wavefront aberration. By doing this, we got an increase in the f_c (for subject SB, single frame) from 0.55 to 0.78 for $SNR_i^{k=1} = 50$, and from 0.17 to 0.67 for $SNR_i^{k=1} = 25$. The improvement obtained for subject SM was less significant, owing to the smaller aberration compensation achieved in this case.

The other option is to increase of the f_c by raising the SNR_i through, for example, frame averaging. See, for instance, how in the case of SB (blue solid line) and $SNR_i^{k=1} = 12$, the f_c is 0.12, but after averaging 30 frames the $SNR_i^{k=30}$ rises to 66 and f_c becomes 0.6. So, an improvement in the effective resolution is also achievable in this way.

In conclusion, we have shown that frame averaging is a good strategy for increasing the effective spatial resolution of the system. Its implementation, however, can be somewhat difficult in systems seeking near-real-time imaging applications. On the other hand, compensating the wavefront aberrations with static elements manufactured using currently available technology provides a similar range of improvement, but requiring only a single-frame approach. This makes achieving near-real-time imaging potentially easier. When real-time imaging is not a goal, the combination of aberration compensation and frame averaging allows for obtainment of increased benefits in terms of restored image quality.

As suggested previously, the system's effective cutoff frequency should be higher than the characteristic spatial frequency of the structure of interest; otherwise, its observation would be very difficult. Besides, an increase in f_c also contributes to the reduction in the amplification of the noise spectrum induced by the restoration filter, and so it helps improve the quality of the restored image.

25.4.5
Implementation of the Deconvolution from Wavefront Sensing Technique

In the previous sections, we have presented the main concepts involved in DWFS. Now, we summarize the common procedure used to achieve the DWFS-restored image. In Section 25.4, we presented (Figure 25.2) the block diagram of the algorithm. We refer to this figure in the following paragraphs.

The first step of the DWFS is the computation of the estimated discrete generalized pupil function ($P_D(v, u)$) from the measurements of the wavefront sensor. Then, we compute the estimated system OTF by the autocorrelation of $P_D(v, u)$. This OTF is then used to build the restoration filter (see the right branch of the block diagram of Figure 25.2).

Once the restoration filter has been computed, the following step is its element-wise multiplication with the discrete Fourier transform of the degraded image, obtaining the restored image spectrum in this way. The procedure ends by computing the inverse Fourier transform of the restored spectrum.

Additional steps are normally used in the DWFS process, but they are not essential and so they are usually not commented on in detail. Perhaps one of the most frequent choices is the selection of a specific part of the degraded image by its element-wise multiplication with a binary mask. This kind of mask is easy to build but induces a hard-border diffraction-like effect that manifests itself in the appearance of high-frequency oscillating components in the restored image. In order to avoid this unwanted effect, an apodizing mask with soft borders is used, for example, the hypergaussian mask (Eq. (25.22)).

$$\text{mask}(x, y) = \exp\left\{-\left(\frac{\sqrt{(x-x_o)^2 + (y-y_o)^2}}{\sigma_x^2 + \sigma_y^2}\right)^{14}\right\} \qquad (25.22)$$

where x_o and y_o are the coordinates of the center of the hypergaussian, and σ_x^2 and σ_y^2 are related to its horizontal and vertical widths, respectively.

As a final comment, we want to point out that some researchers have proposed the apodization of the restored spectrum using the diffraction-limited OTF (OTF_{Diff}). This apodization function is useful to limit the erroneous amplification of the spatial frequencies of the restored image [5].

25.5
Examples of Application

In this section, we present some examples of application of DWFS in two different fields (astronomical imaging and eye fundus imaging), in order to show the performance of the technique and the improvement level that may be expected.

25.5.1
Astronomical Imaging

Among the first practical applications of DWFS, we highlight the first experimental results presented by Primot et al. and the first astronomical imaging performed by Gonglewski et al. [3]; Primot et al. [5] analyzed the restoration of a complex object

 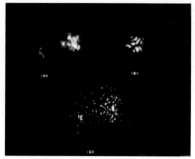

Figure 25.6 The first astronomical images obtained with DWFS. The observed stars are α Aurigae and α Geminorum. Please find a color version of this figure on the color plates. Courtesy of Dr. J. Gongleswki [9]. Taken from Gongleswki J.D., Volez D.G., Fender J.S., Dayton D.C., Spielbusch B.K., Pierson R.E., "First astronomical application of postdetection turbulence compensation: images of α Aurigae, ν Ursae Majoris, and α Germinorum using self-referenced speckle holography," *Appl. Opt.*, **29** (31), pp. 4527–4529, 1990. © 1990 OSA.

degraded by a turbulent simulation cell. They achieved a moderate improvement in the resolution through DWFS. The results were conditioned by the loss of information in certain frequencies where the OTF approached zero. However, they showed how this information could be recovered by averaging different images, exploiting the fact that for atmospheric turbulences the zero crossing of the OTF does not occur always at the same frequencies.

Meanwhile, Gonglewski *et al.* restored the images of several stars degraded by atmospheric turbulence. Figure 25.6 shows the degraded image, the short-exposure atmospheric estimated PSF, and restored image for the observation of α Aurigae and α Geminorum. Similar improvements were later presented by other researchers [13, 30].

Despite the encouraging beginnings of DWFS, its use for astronomical purposes has nowadays succumbed to the charm of AO. DWFS is mainly employed as an additional tool to improve AO images, correcting the residual wavefront errors caused by the delays in the compensation loop and the high-order aberration terms that the AO systems do not compensate for.

25.5.2
Eye Fundus Imaging

By the beginning of the 2000, different scientists had started to study the possibility of using DWFS for obtaining high-resolution images of the eye [10, 11, 14]. The first experimental results were limited by the reduced SNR of the retinal images. This question was partially overcome (as can be seen in Figure 25.7) by averaging several frames, increasing the value of f_c above that of the characteristic frequency of the cone mosaic, and thus enabling its observation.

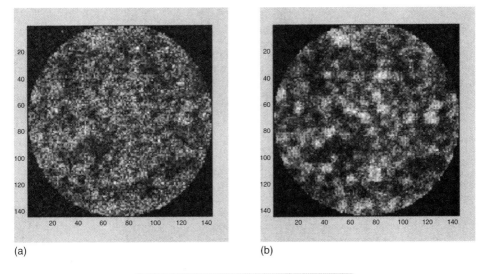

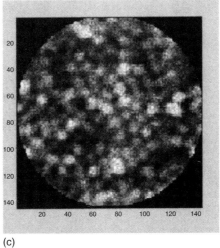

Figure 25.7 Cone mosaic observed with DWFS. Courtesy of Dr. Catlin and Dainty. Taken from Catlin D., Dainty C., "High resolution imaging of the human retina with a Fourier deconvolution technique," *J. Opt. Soc. Am. A* **19** (8), pp. 1515–1523, 2002. [10] © 2002 OSA.

Soon afterward, we proposed the adaptation of the compensated DWFS to the observation of the eye fundus [11]. We suggested the use of customized phase plates manufactured to correct the main part of the subjects' ocular aberrations. With these elements, it is possible to increase the SNR of the wavefront sensing and imaging channels, and thus to obtain high-resolution images of the eye fundus

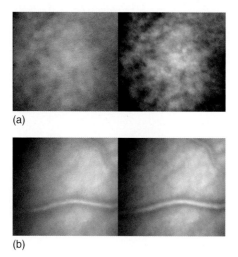

Figure 25.8 (a) Human eye fundus and (b) retinal vessels. See Ref. [20] for more details.

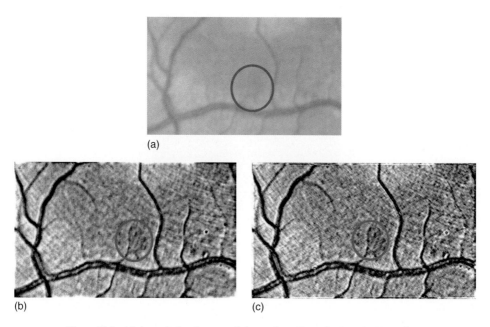

Figure 25.9 High-resolution images of the eye fundus obtained with BD combined with compensated DWFS: (a) the original image; (b) image restored from blind deconvolution; (c) image restored from the combination of interactive wavefront-based and blind deconvolution. Courtesy of S.Yang et al. Taken from Yang S., Erry G., Nemeth S., Mitra S., Soliz P., "Image restoration with high resolution adaptive optical imaging System," *Proceedings of the 17th IEEE Symposium on Computer-Based Medical Systems (CBMS'04)*, pp. 479–484, 2004. [15] © 2004 IEEE.

with just one frame. These results are one step forward in obtaining near-real-time DWFS-restored images (Figure 25.8).

As already stated, DWFS is currently combined with other techniques such as AO and/or BD [15, 31]. In particular, in Ref. [15], the authors present a technique that applies BD to AO-compensated DWFS images; Figure 25.9 shows the improvement achieved with their procedure.

25.6
Conclusions

Previous chapters of this book have presented the AO and BD techniques. In this chapter, we presented the DWFS approach, and some examples of its application. All these techniques present advantages and disadvantages when compared among themselves, which should be taken into account when deciding which of them is the most appropriate for use in a specific application.

AO provides real-time high-resolution imaging at the cost of increased system complexity. Even if a certain amount of wavefront errors are not completely compensated for in actual AO systems, because of several error sources thoroughly addressed in the literature, raw AO images present significantly higher signal-to-noise ratios (SNR_i) than the ones that are not compensated. This allows for obtainment of further improvements in image resolution by using AO images as high-quality inputs to DWFS or BD algorithms. The demonstrated successes of this technique in astronomy and ophthalmology make it the first choice for obtaining high-resolution images [5].

BD is an elegant, efficient, and complete postprocessing technique. It requires high computational cost [2], which makes its real-time application difficult. The quality of the restored image depends highly on the skills of the user, who must choose the appropriate constraints and algorithms in order to get a satisfactory result. However, this technique can handle a wide range of image degradation functions (including scattering), and does not need the inclusion of any additional optical element in the imaging system, so that its maintenance and complexity are not increased.

DWFS, in turn, is a relatively simple technique, which uses conventional wavefront sensing measurements to build the restoration filter. The cost and complexity of the system are higher than those of BD but substantially lower than those of AO. The quality of the restored image depends highly on the user's choice of the restoration filter and on the SNR of the imaging and wavefront sensing channels. As with any other technique relying on wavefront sensing measurements (AO included), a certain amount of bias is unavoidable because of the step of wavefront reconstruction from a limited set of data, which gives rise to differences in the statistical properties between the actual wavefronts and the estimated ones [5]. DWFS may suffer, much like other computational image restoration procedures, from the undesirable erroneous amplification of certain object frequencies, its reduction being one of the main issues that is still under study.

The combinations of these techniques (AO, BD, and DWFS) have also been exploited, providing significant benefits. For example, DWFS of AO images may help reduce the residual wavefront error due to delay in the compensation loop, and allow for the correction of high-order modes that the active element of the AO system cannot compensate for. Besides, DWFS algorithms benefit from the increased SNR of AO-compensated images, which significantly improves the quality of the restoration. AO has also been used in combination with BD algorithms, achieving relevant improvements [31]. DWFS has been combined with BD in what was called *myopic deconvolution*. This technique introduces the DWFS algorithm in the BD process, increasing the rate of convergence and robustness of the solution [4].

Acknowledgments

We wish to thank Dr. Gonglewski, Dr. Catlin, Dr. Yang, and specially Dr. Primot for providing some of the images included in this chapter. We want to acknowledge financial support from the *Isidro Parga Pondal* Program 2009 (Xunta de Galicia, Spain). This work has been supported by the Spanish MICINN, grant FIS2008-03884.

References

1. Meinel, E.S. (1986) Origins of linear and nonlinear recursive restoration algorithms. *J. Opt. Soc. Am. A*, **3** (6), 787–799.
2. Pantin, E., Starck, J.L., and Murtagh, F. (2007) in *Blind Image Deconvolution: Theory and Applications* (eds. K. Egiazarian and P. Campisi), CRC Press, pp. 277–317.
3. Primot, J., Rousset, G., and Fontanella, J.C. (1990) Deconvolution from wavefront sensing: a new technique for compensating turbulence-degraded images. *J. Opt. Soc. Am. A*, **7** (9), 1598–1608.
4. Conan, J.M. *et al.* (1998) Myopic deconvolution of adaptive optics images by use of object and point-spread function power spectra. *Appl. Opt.*, **37** (21), 4614–4622.
5. Roggemann, M.C. and Welsh, B. (1996) *Imaging through Turbulence*, CRC Press, Boca Raton.
6. Tyson, R.K. (1991) *Principles of Adaptive Optics*, Academic Press, San Diego.
7. Rousset, G. *et al.* (1990) First diffraction-limited astronomical images with adaptive optics. *Astron. Astrophys.*, **230**, 29–32.
8. Fontanella, J.C. (1985) Analyse de surface d'onde, déconvolution et optique active. *J. Opt. (Paris)*, **16**, 257–268.
9. Gongleswki, J.D. *et al.* (1990) First astronomical application of postdetection turbulence compensation: images of α Aurigae, ν Ursae Majoris, and α Germinorum using self-referenced speckle holography. *Appl. Opt.*, **29** (31), 4527–4529.
10. Catlin, D. and Dainty, C. (2002) High resolution imaging of the human retina with a Fourier deconvolution technique. *J. Opt. Soc. Am. A*, **19** (8), 1515–1523.
11. Arines, J. and Bará, S. (2003) Hybrid technique for high resolution imaging of the eye fundus. *Opt. Express*, **11**, 761–766.
12. Primot, J. (1989) Application des techniques d'analyse de surface d'onde a la restauration d'images degradees par la

turbulence atmospherique, Ph.D. Thesis, Universitè de Paris-Sud, Orsay, France.

13. Michau, V. et al. (1991) High-resolution astronomical observations using deconvolution from wavefront sensing. *Proc. SPIE*, **1487**, 64–71.

14. Iglesias, I. and Artal, P. (2000) High resolution images obtained by deconvolution from wave-front sensing. *Opt. Lett.*, **25**, 1804–1806.

15. Yang, S. et al. (2004) Image restoration with high resolution adaptive optical imaging System. Proceedings of the 17th IEEE Symposium on Computer-Based Medical Systems (CBMS'04), pp. 479–484.

16. Malacara, D. (2007) *Optical Shop Testing*, 3rd edn, John Wiley & Sons, Inc., New York.

17. Bará, S. (2003) Measuring eye aberrations with Hartmann–Shack wave-front sensors: Should the irradiance distribution across the eye pupil be taken into account? *J. Opt. Soc. Am. A*, **20** (12), 2237–2245.

18. Arines, J. and Ares, J. (2004) Significance of thresholding processing in centroid based gradient wavefront sensors: effective modulation of the wavefront derivative. *Opt. Commun.*, **237**, 257–266.

19. Primot, J. (2003) Theoretical description of Shack–Hartmann wave-front sensor. *Opt. Commun.*, **222**, 81–92.

20. Arines, J. (2006) Imagen de alta resolución del fondo de ojo por deconvolución tras compensación parcial, Ph. D. Thesis, Universidade de Santiago de Compostela, Galicia, Spain.

21. Montera, D.A. et al. (1996) Processing wave-front-sensor slope measurements using artificial neural networks. *Appl. Opt.*, **35** (21), 4238–4251.

22. Talmi, A. and Ribak, E.N. (2004) Direct demodulation of Hartmann–Shack patterns. *J. Opt. Soc. Am. A*, **21** (4), 632–639.

23. Liebelt, P.B. (1967) *An Introduction to Optimal Estimation*, Addison-Wesley, Washington, DC.

24. Voitsekhovich, V.V. et al. (1998) Minimum-variance phase reconstruction from Hartmann sensors with circular subpupils. *Opt. Commun.*, **148**, 225–229.

25. Díaz-Santana, L., Walker, G., and Bará, S. (2005) Sampling geometries for ocular aberrometry: a model for evaluation of performance. *Opt. Express*, **13**, 8801–8818.

26. Soloviev, O. and Vdovin, G. (2005) Hartmann–Shack test with random masks for modal wavefront reconstruction. *Opt. Express*, **13**, 9570–9584.

27. Maethner, S.R. (1997) Deconvolution from wavefront sensing using optimal wavefront estimators, Ph.D. Thesis, Air Force Institute of technology (Wright–Patterson Air Force Bace), Ohio.

28. Conan, J.M., Michau, V., and Rousset, G. (1996) Signal-to-noise ratio and bias of various deconvolution from wavefront sensing estimators. *Proc. SPIE*, **2828**, 332–339.

29. Idell, P.S. and Webster, A. (1992) Resolution limits for coherent optical imaging: signal-to-noise analysis in the spatial-frequency domain. *J. Opt. Soc. Am. A*, **9** (1), 43–56.

30. Dayton, D., Gonglewski, J., and Rogers, S. (1997) Experimental measurements of estimator bias and the signal-to-noise ratio for deconvolution from wave-front sensing. *Appl. Opt.*, **36** (17), 3895–3903.

31. Christou, J.C., Roorda, A., and Williams, D.R. (2004) Deconvolution of adaptive optics retinal images. *J. Opt. Soc. Am. A*, **21** (8), 1393–1401.

Further Reading

Rodríguez, P. et al. (2006) A calibration set for ocular aberrometers. manufacture, testing and applications. *J. Ref. Surg.*, **22** (3), 275–284.

Roggeman, M.C., Welsh, B.M., and Devey, J. (1994) Biased estimators and object-spectrum estimation in the method of deconvolution from wave-front sensing. *Appl. Opt.*, **33** (24), 5754–5763.

26
Image Restoration and Applications in Biomedical Processing

Filip Rooms, Bart Goossens, Aleksandra Pižurica, and Wilfried Philips

26.1
Introduction

In research, continuously better research equipment is being designed, for example, telescopes and microscopes with increased resolution every new generation. These devices expand the sensory skills of humans enormously. They also capture implicit information that is not directly accessible for interpretation, but still present in the measurements in a hidden form. Researchers always try to explore the limits of their equipment, and also desire to exploit this hidden information and visualize it to facilitate interpretation. Taking into account the physical processes in the equipment when gathering the information, we can model which information is "lost" and which information is still present in a "hidden" form. Then, we can design solutions to recover this hidden information.

Here, we focus on optical systems such as microscopes connected to a detector. The optical system has its limitations: it images an ideal mathematical point, not as a point but as a smeared-out spot, because of the wave nature of light (see Chapters 1, 12, and 13 for more details). This image of an ideal point is the so-called point spread function (PSF) of the imaging system. This causes image blurring, even with a well-aligned system: all optical systems have a finite resolution, even when optimally tuned; of course, with misalignment of the optical components, the situation degrades even further.

The optical system thus produces a blurred image of the object, which is then recorded by some detector system, like a charge-coupled device (CCD). The reasons that noise occurs in recorded images can be classified into three categories:

- *Photon or shot noise:* Owing to the particle nature of light, image detectors are, in Fact, particle-counting devices. The photons that hit the detector are converted into charge carriers, such as electrons. However, this detection/conversion process is not deterministic because not every photon is actually detected in the detector.
- *Sensor noise:* This is caused by a variety of noise processes in the detector, such as dark current (charges that are generated in the absence of a signal), and charges

---

Optical and Digital Image Processing: Fundamentals and Applications, First Edition. Edited by Gabriel Cristóbal, Peter Schelkens, and Hugo Thienpont.
© 2011 Wiley-VCH Verlag GmbH & Co. KGaA. Published 2011 by Wiley-VCH Verlag GmbH & Co. KGaA.

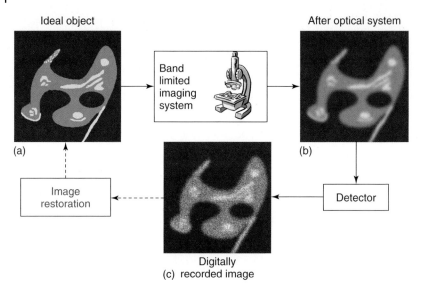

Figure 26.1 Block diagram of the flow of information during the imaging process and the corresponding degradation. (a) A synthetic cell image; (b) the image, blurred because of bandlimited optics; and (c) the image as actually captured.

that are captured in a wrong detection unit, spatial sensitivity variations between the detector pixels.
- *Read-out or amplifier noise:* Since the signal from the charge carriers has to be amplified and converted into an intensity number (through an analog-to-digital converter (ADC)), additional noise occurs in the final image.

We illustrate the flow of information during the imaging process in Figure 26.1; an ideal object (in fact, the image under ideal circumstances) is shown in Figure 26.1a. The ideal object is blurred because of the bandlimited properties of the optical system, which is modeled by convolution with the PSF. After the recording process, we finally obtain a degraded image, blurred and noisy.

Summarizing, the final recorded image has in fact been blurred and corrupted by noise, which results in the loss of actual information. However, it is not because some feature is invisible that the feature is lost [1]:

- Quite often, details are hidden in the noise or masked by other features.
- Artifacts may confuse the viewer.
- Information may be present in implicit form, that is, it can only be retrieved when one imposes prior knowledge (i.e., knowledge that one already can assume about an image before observation, e.g., the fact that all pixel intensities must be positive).

So, the goal of image restoration is to recover the "hidden" information in observed images as well as we can by approximating the inversion of the degradation process (because no exact solution exists, as explained further). This will facilitate interpretation by human observers, and further automatic postprocessing, such as segmentation and object recognition.

While the term *image restoration* is widely used in connection with the removal of various types of degradations, sometimes just to enhance the visual perception of the observed images, we address image restoration in the classical sense, which means estimating the underlying degradation-free image from the observations affected by noise and blur. This estimation process makes use of the degradation model of the image formation (for noise and blurring), which is often referred to as *data distribution*. Next to this degradation model, the restoration process often makes use of a prior model, which encodes our prior knowledge about the image to be estimated, for example, some of its statistical properties. The use of these models, and of clearly defined criteria for optimally estimating true values of the underlying degradation-free data, is what makes image-restoration techniques essentially different from image-enhancement techniques (which only aim at improving the visual perception of the image).

In Ref. [2], a two-step approach was proposed: first, a Fourier-based constrained deblurring (cf Wiener deconvolution) is performed, after which a wavelet-based shrinkage is applied.

In Ref. [3], a prefilter combined with a multiscale Kalman filter has been applied to the deconvolution problem.

In Ref. [4], a regularized image-restoration method is presented using a local Gaussian scale mixture(GSM) model both for the image and the degradation using an efficient noniterative Bayes least-square estimator. However, an important drawback in their method was the limitation to certain types of blur kernels. In Ref. [5], this drawback was solved by an improved combination of a global method followed by a local denoising of the prefiltered image.

A class of techniques combines the classical Richardson–Lucy (RL) deconvolution algorithm (explained in Section 26.2.3.1 in more detail) with wavelet-based noise suppression. The oldest reference we found of such a combination was for astronomical image restoration [6]. In Ref. [7], an iterative technique is described, where RL iterations alternated by wavelet-based hard thresholding, using the Stein unbiased risk estimate (SURE) threshold from Donoho. A very similar algorithm was used to restore confocal microscopy images [7].

In Ref. [8], a more theoretical foundation for these hybrid algorithms was established: their hybrid iterative method was derived from the Bayesian maximum a posteriori (MAP) framework that resulted in a method in which the deconvolution method is a Landweber-like deconvolution algorithm, and the wavelet regularization is based on the L1 penalty term to induce sparseness of the wavelet coefficients. Their algorithm, known in literature as IST or iterative soft thresholding, is essentially the same iterative as the one developed independently in Ref. [9]. Figueiredo *et al.* [8, 9] was the inspiration for many authors to apply soft thresholding and deblurring alternating in an iterative procedure. Now, the current trend is to enforce constraints using split-Bregman methods to impose smoothness.

This chapter is organized as follows: first, a selection of classical image-restoration techniques is discussed. Next, we present our own combined restoration method called $SPERRIL$ (steerable pyramid-based estimation and regularized Richardson–Lucy restoration) (which was published in Ref. [10]). Then, we present an evaluation of the restoration performance of SPERRIL on synthetic and real biological data. Finally, we present our conclusions.

26.2
Classical Restoration Techniques

In this section, we discuss a number of image-restoration algorithms that are based on the minimization of a functional with a data-fitting term and a simple smoothness term; hence, we do not discuss more complex prior models like the ones based on Markov random fields or multiresolution representations. We start from the inverse filter, and discuss the most common problem in image restoration, that is, ill-posedness, or instability, of the solution with respect to small changes in the input. This section is divided in two subsections. Section (26.2.1) approaches image restoration in a rather intuitive way, whereas Section 26.2.2 considers the restoration problem starting from Bayes' rule.

A value that is frequently used to evaluate the performance of restoration algorithms, is the peak signal-to-noise ratio (PSNR). A more recent image quality measure based on the human visual system is structural similarity (SSIM). More information about these measures can be found in Chapter 19.

26.2.1
Inverse Filter and Wiener Filter

26.2.1.1 Inverse Filter

The most basic approach toward image restoration is a naive inversion of the blurring process. It is easy to explain in the Fourier domain, because convolution in the spatial domain is transformed into multiplication in the Fourier domain. The image formation equation in the spatial domain is given by[1]

$$g(x, y) = (f * h)(x, y) + n(x, y)$$

where $g(x, y)$ is again the degraded image, $f(x, y)$ is the ideal unknown image, $h(x, y)$ is the PSF, and $n(x, y)$ is the noise in the image. Transformation to the Fourier domain yields

$$G(u, v) = F(u, v)H(u, v) + N(u, v) \qquad (26.1)$$

with $G(u, v)$, $F(u, v)$, $H(u, v)$, and $N(u, v)$ the Fourier transforms of $g(x, y), f(x, y)$, $h(x, y)$, and $n(x, y)$, respectively, and u and v coordinates in the Fourier space. So, a naive estimate for the ideal (Fourier transform of the) image $\hat{F}(u, v)$ would be

$$\hat{F}(u, v) = F(u, v) + \frac{N(u, v)}{H(u, v)} \qquad (26.2)$$

---

1) For the sake of compactness of notations, we consider 2D images here.

26.2 Classical Restoration Techniques

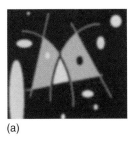
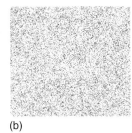
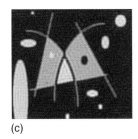

(a) (b) (c)

Figure 26.2 Some examples of inverse filtering. (a) Blurred with Gaussian noise ($\sigma = 5$). (b) Inverse filter of (a), no frequency truncation. (c) Inverse filter of (a), optimal frequency truncation.

From this expression, it is clear that this estimate does not contain useful information at frequencies where $H(u, v)$ is very small (or even zero), because $G(u, v)$ and $N(u, v)$ are then divided by these small values, which are in the order of magnitude of $0 \leq H(u, v) \ll 1$, that is, the corresponding frequency components are amplified strongly. Figure 26.2a shows a blurred synthetic image with Gaussian noise. Figure 26.2b shows the result of the inverse filter. It is clear that the inverse filter is highly unstable in the presence of noise.

In scientific terms, this inversion is called an *ill-posed* or *ill-conditioned* problem. Therefore, it is highly desirable to stabilize the methods to solve the restoration problem by imposing prior knowledge. This is done by defining certain criteria that an image must satisfy to be a likely solution of the restoration problem. Images whose intensities fluctuate between black and white from pixel to pixel (e.g., Figure 26.2c) are less likely than images where larger connected areas have similar intensities. The corresponding criterion is an example of a smoothness criterion, which is used under different forms based on prior knowledge.

The simplest stabilization procedure for the inverse filter is only to allow the spectral components of the signal above the noise level to be amplified, and to set the other spectral components to zero. An example of an image restored with this technique is given in Figure 26.2c.

26.2.1.2 Wiener Filter

The Wiener filter has a long history that goes back to the Wiener–Hopf equations derived by Norbert Wiener and Eberhard Hopf in 1930. The Wiener filter is in fact a solution of these equations.

In essence, the Wiener filter behaves in a manner similar to the truncated inverse filter, except that an optimal trade-off is made in the area where the noise power becomes of the same order of magnitude as the signal power. In fact, the truncated inverse filter amplifies the amplitude of the spectral components of the signal whose energy dominates that of the noise; spectral components with an energy smaller than the noise energy are set to zero (which is, in fact, a kind of global hard thresholding in Fourier space).

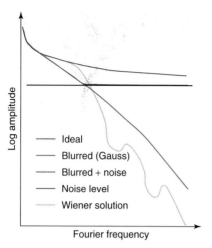

Figure 26.3 Spectra of ideal image, degraded image, and Wiener restored image. (Please find a color version of this figure on the color plates.)

So the Wiener filter applies a least-square rescaling of the spectral components. The effect is that it amplifies the amplitude of the spectral components of the signal whose energy dominates the noise energy, and shrinks the spectral components where the noise starts to dominate the signal (see Figure 26.3).

Let us recall the degradation equation:

$$g(x, y) = (f * h)(x, y) + n(x, y)$$

or in the Fourier domain:

$$G(u, v) = F(u, v)H(u, v) + N(u, v) \qquad (26.3)$$

The Wiener filter searches for a least squares solution for the restoration, that is, it computes an optimal filter $w(x, y)$ for which the solution

$$\hat{f}(x, y) = (g * w)(x, y)$$

minimizes the mean square error (MSE):

$$\text{MSE} = E\left(\left|\hat{f}(x, y) - f(x, y)\right|^2\right)$$

where $\hat{f}(x, y)$ is the estimated solution and $f(x, y)$ is the unknown ideal image. Since the same information is present in the spatial domain as well as the Fourier domain, it is equivalent to minimizing

$$\text{MSE} = E\left(\left|\hat{F}(u, v) - F(u, v)\right|^2\right)$$

$$\Rightarrow \hat{F}(u, v) = \frac{H^*(u, v)}{|H(u, v)|^2 + S_{nn}(u, v)/S_{ff}(u, v)} \cdot G(u, v)$$

with $S_{\text{ff}}(u, v)$ the power spectrum of the ideal signal (which is unknown, and has to be estimated) and $S_{\text{nn}}(u, v)$ the power spectrum of the noise. This is the well-known formula for the Wiener filter in the Fourier domain. From this expression, it is clear that spectral components with $S_{\text{nn}}(u, v) \ll S_{\text{ff}}(u, v)$ are amplified because

$$\frac{S_{\text{nn}}(u, v)}{S_{\text{ff}}(u, v)} \ll |H(u, v)|^2$$

On the other hand, when $S_{\text{nn}}(u, v) \approx S_{\text{ff}}(u, v)$, all the spectral components (signal and noise) are attenuated to reduce the noise energy in the restored image (at the expense of the suppression of the signal components).

26.2.2
Bayesian Restoration

Most image restoration algorithms aim at finding the most probable solution f, given a degraded observation g. In this case, f is the unknown ideal image as if recorded in ideal circumstances (no blur, noise, etc.), while g is the degraded image we actually can observe. To reconstruct f from g, we need a prior distribution $P(f)$, which is used to incorporate our knowledge about the expected structure of the image.

We also need a conditional distribution $P(g|f)$, that is, a degradation model that relates the observed, degraded image to an ideal object as input (with a good model for the optics and the detector, it is fairly easy to calculate how an ideal object is imaged). On the other hand, $P(f)$ is more difficult to model or estimate, since it reflects what we expect a plausible image to look like.

Knowing $P(g|f)$ and the prior probability $P(f)$, it is possible to infer the posterior probability $P(f|g)$ by using Bayes' rule:

$$P(f|g) = \frac{P(g|f)P(f)}{P(g)}$$

Bayesian methods for image restoration can use this information in several ways. Methods like *minimum mean square error* (MMSE) minimize the quadratic error between the estimated and the true solutions, while the MAP looks for a solution that maximizes the posterior probability. Here, we focus on the latter class of methods since MAP usually results in less-complex estimators.

In the case of MAP methods, we wish to maximize $P(f|g)$, that is, find the most probable likely image f, given the observation g. In practice, however, it is often easier to minimize minus the logarithm of the following expression:

$$-\log\left(P\left(f|g\right)\right) = -\log\left(P\left(g|f\right)\right) - \log\left(P\left(f\right)\right) + \log\left(P\left(g\right)\right)$$

where the term $\log\left(P\left(g\right)\right)$ is irrelevant in the minimization. It is common to call $L\left(f\right) = \log\left(P\left(g|f\right)\right)$ the log-likelihood and $R\left(f\right) = \log\left(P\left(f\right)\right)$ the regularization term, which carries the prior information that we impose on the solution.

Maximizing only the log-likelihood (i.e., minimizing minus the log-likelihood) is called *maximum likelihood estimation* (ML); this is often used when no prior information is available. Maximizing the full expression (or minimizing minus the expression) is called *maximum a posteriori* (*MAP*) estimation. Methods that use an entropy prior model for the image are referred to as *maximum entropy methods* (*MEM*). More information and references can be found in Ref. [11].

26.2.2.1 Gaussian Noise Model

The most common algorithms are derived from this expression for the cases of Gaussian or Poisson noise, combined with a variety of regularization functions. In this subsection, we start with Gaussian noise, which is the simplest case.

Tikhonov–Miller Restoration versus Wiener Filter The likelihood functional in case of Gaussian noise is

$$P(g|f) \sim \exp\left(-\frac{1}{2}\|g - (h*f)\|^2\right)$$

in which the square error between the observed image g and the estimated restored image reblurred with the blur kernel is to be minimized. The corresponding negative log-likelihood function $-\log P(f|g)$ is then proportional to

$$-\log P(g|f) \sim \frac{1}{2}\|g - (h*f)\|^2 \tag{26.4}$$

In many image restoration methods (as in Tikhonov–Miller (TM) restoration), a smoothness constraint is used as a regularization functional in order to reduce oscillations in the image due to noise amplification. In fact, regularization imposes prior knowledge on the solution, that is, we want to impose which kind of solutions are more likely. For example, we expect solutions with smooth areas to be more likely than solutions with random oscillations in the image. Therefore, we impose a limit on the high-frequency content of the image by adding a fraction of high-frequency energy as a penalty in the total functional to minimize, that is, $\log P(f)$ is

$$\log(P(f)) \sim \|(c*f)(x,y)\|^2$$

where f is the estimated solution and c is a high-pass filter (e.g., the Laplacian filter kernel is a popular choice for c) that is to be convolved with the solution f. The TM functional to be minimized is, therefore,

$$\Phi(\hat{f}) = \|g(x,y) - (h*\hat{f})(x,y)\|^2 + \lambda\|(c*\hat{f})(x,y)\|^2 \tag{26.5}$$

where $\|x\|^2$ is the Euclidean norm of x. The constant λ is the so-called regularization parameter, which determines the relative importance of the regularization to the importance of the data (the larger λ, the more importance is given to the regularization; the smaller λ, the more importance is given to the observed data). Again, this expression is minimized in the Fourier domain, which results in

$$\hat{F}(u,v) = \frac{H^*(u,v)}{|H(u,v)|^2 + \lambda|C(u,v)|^2}G(u,v) \tag{26.6}$$

which is equivalent to the solution of the Wiener filter when λ and C are chosen such that

$$\lambda \left| C(u, v) \right|^2 = \frac{S_{nn}(u, v)}{S_{ff}(u, v)}$$

Iterative Constrained Tikhonov–Miller The Wiener filter and the TM restoration algorithm are examples of linear algorithms that have two significant disadvantages:

1) Only those spectral components of the signal that have not been removed completely by the degradation process can be recovered.
2) Linear models cannot restrict the set within the subspace of possible solutions. An image, for example, can only have positive gray values. The algorithms we discussed previously do not guarantee positivity, even though negative gray values have no physical meaning. It would be advantageous to restrict the domain possible solutions even further to only those solutions that have a physical meaning (e.g., images that contain negative values, or those that are too "rough," like the one in Figure 26.2b should not be accepted as real physical solutions).

In this paragraph, we discuss a variant of the TM restoration algorithm that imposes nonnegativity – the *iterative constraint Tikhonov–Miller* (ICTM) algorithm. The practical implementation of the ICTM minimizes Eq. (26.4) in an iterative way using the well-known conjugate gradient technique. The nonlinearity here is imposed as a hard constraint: after every iteration, the negative values of the current estimated solution are set to zero. Applying this constraint is a highly nonlinear operation, and introduces the extra prior knowledge that an image contains only positive intensities.

In the absence of this positivity constraint operator, the value $\beta^{(k)}$ can be calculated analytically [12].

However, in the presence of the positivity constraint operator, this expression cannot be calculated analytically anymore. Nevertheless, an approximate expression can be derived.

Choice of the Regularization Parameter What we have not discussed yet, is how to choose a proper value for the regularization parameter λ. For the sake of completeness, we briefly mention some popular choices for λ [12]:

1) *Methods that require the knowledge of the noise variance:*
   - SNR method: This method sets λ equal to the inverse of the SNR (signal-to-noise ratio), that is, the ratio of the energy of the noise (i.e., $\sum_{x, y} n^2(x, y)$) and the energy of the signal (i.e., $\sum_{x, y} f^2(x, y)$). Both the numerator and the denominator are not known, and need to be estimated. Using this regularization parameter has a clear intuitive interpretation – the higher this value, the more we need to suppress the noise and more weight is given to the regularization than to the data fit.

- **Method of constrained least squares (CLS):** This method determines the value λ_{CLS} for which the total squared error between the observed image and the blurred restored image is minimal. The restored image is estimated by the TM algorithm for various values of λ_{CLS}, and the minimum is found by using a one-dimensional line-search method.

2) Methods not requiring knowledge of the noise variance:
   - **Method of generalized cross-validation:** This method determines the value λ_{GCV} from the so-called *leave-one-out* principle: for every pixel the TM restoration result is calculated using all pixels, except the one under consideration and the left-out pixel is predicted from this reduced, restored image. λ_{GCV} is chosen as the minimizer of the cross-validated prediction error.
   - **Maximum likelihood method:** This method calculates the value λ_{ML} on the basis of the stochastic assumption that both the noise and the regularization function $\sqrt{\lambda_{ML}}(c*f)(x, y)$ have a Gaussian distribution with variance σ_n^2 (where $\sqrt{\lambda_{ML}}$ scales the variance of $(c*f)(x, y)$ to σ_n^2). The derived likelihood function is then minimized.

26.2.3
Poisson Noise Model

26.2.3.1 Richardson–Lucy Restoration

The RL algorithm is an iterative algorithm with a relatively simple iteration formula to restore a blurred image in the presence of Poisson noise. It was introduced independently by Lucy [13] and Richardson [14], and is related to the expectation maximization (EM) algorithm [15]. EM was first put in an iterative algorithm and applied in image reconstruction in Ref. [16], which lead to an algorithm equivalent to the RL algorithm. Background information and references can be found Ref. in [11]. This algorithm is common practice in confocal image restoration [12, 17, 18].

To obtain the RL algorithm, the log-likelihood function is minimized in the presence of Poisson noise. The probability of getting an observed image intensity distribution $g(x)$, when the original fluorescence distribution in the object is given by $f(x)$, is given by [12, 18, 19]

$$P(g|f) = \prod_x \left(\frac{[(h*f)(x)]^{g(x)} e^{-(h*f)(x)}}{g(x)!} \right)$$

which is, in fact, the mathematical formulation for stating that the probability for observing an image g given an ideal "image" f is given by the probability that the first observed pixel intensity is $g(x_1)$ in case the "true" intensity was $f(x_1)$, multiplied by the probability that a second observed pixel intensity is $g(x_2)$ in case the "true" intensity there was $f(x_2)$, and so on for all pixel intensity probabilities.

The negative log-likelihood function $L(f)$ of getting the observed g when $(h*f)$ is expected, is then given by

$$\log L(f) \approx \sum_x g(x) \log\left(\frac{(h*f)(x)}{g(x)}\right) + ((h*f)(x) - g(x)) \quad (26.7)$$

where, in the second formula, the approximation formula of Stirling for factorials was used.

Minimizing this log-likelihood with respect to f leads to the Richardson–Lucy algorithm[2] with the following iterative formula:

$$f^{(k+1)}(x) = f^{(k)}(x)\left(h^* * \frac{g}{h*f^{(k)}}\right)(x) \qquad (26.8)$$

where $f^{(k+1)}(x)$ is our new estimate for iteration $(k+1)$, $*$, denotes the convolution operator (with h^* means with a point-mirrored version of h), and all multiplications and divisions are point by point.

Csiszár then examined a number of discrepancy measures between two positive functions $f_1(x)$ and $f_2(x)$, like the MSE and the Kullback–Leibler distance (sometimes also referred to as *cross entropy*). He considered the possibility that $f_1(x)$ and $f_2(x)$ are not probability distributions, that is, not normalized to one, but to arbitrary constants. Therefore, Csiszár extended the Kullback–Leibler distance measure and thus obtained the I-divergence [20]:

$$I(f_1(x)|f_2(x)) = \sum_x f_1(x) \log\left(\frac{f_1(x)}{f_2(x)}\right) - \sum_x (f_1(x) - f_2(x))$$

Csiszár concluded that for functions that have both positive and negative values, the MSE is the only discrepancy measure consistent with the axioms mentioned above, while for functions that have only nonnegative values, the I-divergence is the only consistent measure. In Ref. [20], it is proven that minimizing $I\left(g(x)|\left(h*\hat{f}\right)(x)\right)$ is equivalent to minimizing the negative log-likelihood function for Poisson statistics, which leads to the RL algorithm. So, in a way, the RL algorithm is also related to maximum entropy restoration.

The RL algorithm has the following basic properties:

- Every estimate $f^{(k)}(x)$ is nonnegative;
- the total intensity of $h*f^{(k+1)}(x)$ corresponds to the total number of counts in the observed image $g(x)$;
- the log-likelihood function $L(f)$ is nondecreasing as the iterations increase;
- $f^{(k)}$ makes $L(f)$ converge to its maximum.

Note that the standard RL algorithm does *not* include regularization (except for the fact that it keeps total mean intensity constant and it keeps zero values at zero, so noise amplification is somehow limited). In Section 26.2.3.2, we discuss some regularization schemes for the standard RL algorithm.

26.2.3.2 Classical Regularization of Richardson–Lucy

An extended discussion is given for the minimization of likelihood functions both in the case of Poisson data as well as Gaussian data, regularized with a set of different regularization priors in Ref. [18].

---

2) Note that the last term does not depend on f, and can therefore be dropped in the minimization.

Here, we limit ourselves to a few common schemes of regularized RL, based on Refs [12, 18, 19]. All the schemes we discuss now were developed for confocal microscope image restoration.

- **RL with slight postblur** The simplest method to regularize the RL algorithm is to apply a blurring after every iteration with a small Gaussian kernel ($\sigma = 0.5$). This imposes some smoothness to the solution by blurring singularities after every iteration.
- **RL with Tikhonov–Miller (TM) regularization** [19]: The TM regularization functional is given by

$$R_{TM}(f) = \sum_{x} |\nabla f(\mathbf{x})|^2 \tag{26.9}$$

As mentioned before, this regularization functional imposes a degree of smoothness on the solution by limiting the high-frequency content of the image. With this regularization functional, the iterative scheme becomes

$$\hat{f}^{(k+1)}(x, y) = \left(\frac{g(x, y)}{\left(h * \hat{f}^{(k)}\right)(x, y)} * h^*(x, y) \right) \frac{\hat{f}^{(k)}(x, y)}{1 + 2\lambda \Delta \hat{f}^{(k)}(x, y)}$$

with Δ the Laplacian operator,

$$\Delta f(x, y) = \frac{\partial^2 f(x, y)}{\partial x^2} + \frac{\partial^2 f(x, y)}{\partial y^2}$$

- **RL Conchello** [12] Conchello proposes a regularization functional that penalizes only the excessively bright spots, but does not impose smoothness on the restored image. Equation (26.7) gives the log-likelihood function for Poisson noise. Pure RL deconvolution minimizes this function. In the Conchello algorithm, the following regularization functional is added to the functional to be minimized:

$$R(f) = \sum_{x} |f(\mathbf{x})|^2$$

This results in a regularized iterative scheme of the form

$$\hat{f}^{(k+1)}_{Conchello}(x, y)) = \frac{-1 + \sqrt{1 + 2\lambda \hat{f}^{(k+1)}_{unregularized}(x, y)}}{\lambda}$$

where $\hat{f}^{(k+1)}_{unregularized}(x, y)$ is the result applying a single unregularized RL iteration step to the previous regularized estimation $\hat{f}^{(k)}_{Conchello}(x, y)$. Conchello describes also an estimation for an appropriate value for λ in this scheme [21]: $1/\lambda$ is chosen to be equal to the largest pixel value of the estimated specimen function. For images with bright pixels in isolated areas comparable in size to the diffraction limited spot, $1/\lambda$ can be approximated by the maximum value in the image divided by the maximum value of the PSF. For images with bright pixels in linelike areas, $1/\lambda$ can be approximated by the maximum value in the image divided by the maximum value of the line spread function (i.e., the PSF integrated orthogonally to the direction of the line structure).

- RL with total variation (TV) regularization [19]: The total variation (TV) regularization functional is given by

$$R_{TV}(f) = \sum_x |\nabla f(\mathbf{x})|$$

Using the L_1 norm instead of the L_2 norm as in Eq. (26.9) allows edge-preserving smoothing. It can be shown that the smoothing process introduced by this regularization functional acts only in the direction tangential to the edge gradients, and not in the direction orthogonal to these gradients [19]. In fact, this regularization method is closely related to anisotropic diffusion [22], where smoothing is not allowed across large image gradients. TV is very efficient for objects with smooth regions and no texture, since it smoothes homogeneous regions and restores sharp edges. However, fine texture is destroyed. Minimizing the functional composed of the likelihood functional of Eq. (26.7) with the total variation functional results in the next explicit regularized iterative scheme:

$$\hat{f}_{TV}^{(k+1)}(x) = \left[\frac{g(x)}{(h * \hat{f}^{(k)})(x)} * h^*(x) \right] \frac{\hat{f}^{(k)}(x)}{1 - \lambda \nabla \cdot \left(\frac{\nabla \hat{f}^{(k)}(x)}{|\hat{f}^{(k)}(x)|} \right)} \qquad (26.10)$$

The usage of $|\nabla f(\mathbf{x})|$ imposes edge-preserving smoothing of the solution.

In the next section, we discuss a new method to regularize the RL algorithm, based on multiresolution. In general, multiresolution-based methods allow us to make a better distinction between signal and noise than classical restoration methods.

26.3
SPERRIL: Estimation and Restoration of Confocal Images

26.3.1
Origin and Related Methods

Here, we describe a new method for joint restoration and estimation of the degradation of confocal microscope images. The observed images are degraded due to two sources: first, we have blurring due to the bandlimited nature of the optical system (modeled by the PSF); second, Poisson noise contaminates the observations due to the discrete nature of the photon detection process. The proposed method iterates between noise reduction and deblurring.

This method is automatic, and no prior information on the image is required. We have named our method SPERRIL (steerable pyramid-based estimation and regularized Richardson–Lucy restoration). The performance of SPERRIL is comparable to or even better than that of of the existing techniques that use both objective measures and visual Observation. [10].

Hybrid restoration methods (deblurring in the spatial or the Fourier domain, regularization in the wavelet domain) have already been in existence for quite some time [2, 7, 8]. Here, we tried to integrate the existing RL algorithm for deblurring with the existing wavelet shrinkage algorithm of Şendur for regularization [23] (he only used the algorithm for denoising). Furthermore, the Anscombe transform has already been used to deal with Poisson data [24], and has been applied in the field of astronomy, but rarely in microscopy (we only found one instance, i.e., Ref. [25]).

This section is organized as follows: first, the outline of our algorithm is presented. In particular, we discuss the noise reduction, the deblurring step, and a stopping criterion. Next, experimental results are shown and discussed. Finally, a conclusion is given.

26.3.2
Outline of the Algorithm

First, the image is prefiltered to reduce Poisson noise. This prefilter is equivalent to the "denoise" steps further in the algorithm, except that the "denoise" is applied to the image transformed by the Anscombe transform to convert Poisson noise into Gaussian noise [24] instead of to the normal image. The "denoise" steps in the algorithm are performed by computing a steerable pyramid decomposition of the image and applying noise reduction to the sub-bands of the decomposed image. After denoising, the filtered sub-bands of the steerable pyramid are recombined, which produces a partially denoised, but not deblurred image. Finally, RL deblurring is applied. These steps are iterated alternatively until convergence is reached.

26.3.2.1 Noise Reduction

The wavelet-based denoising method in Ref. [26] has already been applied in microscopy by Boutet de Monvel [7] and Stollberg [27]. However, computation of the SURE threshold involves sorting the coefficients in the different sub-bands, the number of operations for which is on the order of $N \log N$ per sub-band, with N the number of coefficients in a sub-band; this is quite time consuming.

This kind of noise reduction is suboptimal for two reasons:

1) It assumes Gaussian noise. However, in confocal fluorescence imaging, the major source of errors is Poisson noise [12]. Unlike Gaussian noise, Poisson noise is intensity dependent, which makes separating image from noise very difficult.
2) It only exploits the fact that useful coefficients should be large, and does not exploit information contained in spatial correlation between wavelet coefficients, and information contained in correlation across scales.

We approach these problems in the following way. To deal with Poisson noise, we used the Anscombe transform [24]: $t(I(x)) = 2\sqrt{I(x) + \frac{3}{8}}$, with $I(x)$ the raw image intensities. The Anscombe transform has the property to convert the Poisson data approximately into data with a Gaussian distribution with unit standard deviation. So, this transformation allows us to use well-studied methods for additive Gaussian noise on data with signal-dependent Poisson noise, and has been applied in the field of astronomy, but rarely in microscopy [25]. We have chosen to apply the simple yet powerful bivariate shrinkage from Ref. [23], which is based on a bivariate distribution modeling relations between parent and child wavelet coefficients:

$$\hat{w}_1(x) = \frac{\left(\sqrt{v_1^2(x) + v_2^2(x)} - \frac{\sqrt{3}\sigma_n^2}{\sigma(x)}\right)_+}{\sqrt{v_1^2(x) + v_2^2(x)}} \cdot v_1(x) \tag{26.11}$$

with $(a)_+ = \max(a, 0)$. A denoised coefficient $\hat{w}_1(x)$ is calculated from the corresponding noisy coefficient $v_1(x)$ and its parent[3] coefficient $v_2(x)$. σ_n^2 denotes the noise variance, and $\sigma(x)$ denotes the marginal standard deviation for the coefficient at location x (in practice, $\sigma(x)$ is estimated as the standard deviation of the wavelet coefficients in a local window around the coefficient at position x). This algorithm is simple to implement, has low computational cost, and yet provides a powerful noise reduction, since

- it adapts locally to the presence of edges (due to the presence of $\sigma(x)$);
- the factor $(v_1^2(x) + v_2^2(x))$ captures the relation between coefficients across scales. It tends to keep wavelet coefficients unchanged when both the current coefficient $v_1(x)$ and its parent $v_2(x)$ are large, which means that they probably originate from a significant image feature.

The first time, the denoising step is combined with the Anscombe transform as a prefiltering step to reduce Poisson noise. Later, the denoising step is applied as a regularization step (without the Anscombe transform) after each deblurring step.

26.3.2.2 Deblurring Step
For the deblurring step, we used the RL algorithm (which is discussed in Section 26.2.3.1). This algorithm is already common practice in confocal image restoration [12, 18].

Since the standard RL algorithm is obtained by maximizing just the log-likelood, no explicit regularization is applied. We already discussed some classical regularization schemes in Section 26.2.3.2. We now replaced the classical regularization by the wavelet-based denoising from Section 26.3.2.1, which imposes smoothness of the result by imposing sparseness in the wavelet domain.

26.3.2.3 SPERRIL as RL with a Prior?
As far as we know, the first paper that applied this hybrid *ad hoc* combination of an existing deconvolution method with a wavelet-based regularization was Ref. [2]. Later, it was refined and founded more theoretically in Ref. [28]. Shortly after SPERRIL was born [29], papers appeared with attempts to find a more theoretical foundation for these hybrid algorithms. One of these papers was Ref. [8], where a method was derived from the Bayesian MAP framework, that resulted in a method that is very similar to ours (except that the deconvolution method is a Landweber-like algorithm, and the wavelet regularization is based on a very simple l_1 penalty term to induce sparseness of the wavelet coefficients).

---

3) That is, the coefficient at the same spatial location and same orientation band, but in the next coarser scale.

The RL algorithm is a method that maximizes the likelihood of recovering data that are corrupted by a Poisson process, as expressed in Eq. 26.7. Şendur's method [23], on the other hand, maximizes the probability that clean wavelet coefficients follow a bivariate parent–child conditional probability function. Therefore, this algorithm alternates maximizing the Poisson probability and the prior probability that assumes a bivariate model for a "clean" solution.

26.3.3 Experimental Results

26.3.3.1 Colocalization Analysis: What and Why?

To test the biological validity of any restoration algorithm is always challenging, because there is no so-called ground truth about the cell features of dimensions near the resolution limit of the microscopes. So, we have chosen to compare different image restoration algorithms as a preprocessing step before colocalization analysis, and evaluate the restoration results by using the results of the colocalization analysis. In the following two paragraphs, the principle of colocalization is discussed. They are based on Ref. [30].

Specimens can be labeled with one or more fluorochromes (or fluorescent contrast agents). Fluorochromes with a single duo of excitation/emission wavelengths are often used to highlight structures in the cell-like microtubules (tubelike structures used for transport inside the cell, serves also like a kind of skeleton of the cell), mitochondria (for energy production in the cell), or nuclei (where the genetic information is stored). This is fine when a researcher is only interested in a specific structure within the field of view. However, often two or more labels are used to study many different structures at the same time, but keeping them separated (in different wavelength bands). The specimen is imaged at each fluorescent wavelength independently and then combined with the others.

Colocalization describes the presence of two or more types of molecules at the same physical location. The reason may be that the fluorochromes are attached to the same receptor, or are attached to molecules that are interacting with each other. In image processing, this translates to the fact that the intensities emitted by the fluorochromes contribute to the same pixel in the image.

In Ref. [31], the topic of deconvolution as a preprocessing step to colocalization is discussed, and significant improvements in the analysis results were obtained when deconvolution was applied. Separate objects seemed to have a certain overlap due to image blurring and the presence of noise in the image. The better the restoration technique that is applied, the more this false overlap is reduced, thus improving the accuracy of the colocalization analysis.

26.3.3.2 Experimental Setup

Here, we evaluate the performance of our algorithm on the colocalization analysis of two fluorescently tagged molecules in human A431 skin cancer cells [32].

The first fluorescently labeled molecule is a transmembrane receptor protein. This molecule spans the cell membrane and can bind with molecules outside the cell to pass on information to the cell. The erbB family of receptors include the epidermal growth factor (EGF) receptor erbB1 and erbB3 that, when activated, cause skin cells to grow. The erbB1 receptors were labeled with green fluorescent protein (GFP), causing formation of erbB1-eGFP; erbB3 was labeled with mCitrine, which is an improved yellow fluorescent protein mutant.

The second molecule is the so-called ligand, which is a molecule like a hormone that passes certain information to the cell by binding to the transmembrane proteins. The ligand used here is EGF. When the EGF receptor binds with this ligand, the cell gets the signal to start growing. The ligand EGF was labeled with fluorescent quantum dots (QD), thus resulting in EGF-QD.

One way to test if the receptor and the ligand interact, is to fluorescently label both molecules as described above and monitor the colocalization (or overlap) of these two fluorescent signals in the image from the receptor and the ligand molecules [32].

The test set of images used in our analysis were two sets of four single confocal sections of living cells. The first set of four images were from cells expressing erbB1-eGFP to which EGF-QDs have been added. After activation by EGF-QDs, the erbB1 internalizes, that is, it is transported from the membrane to the inside of the cell. Since the QD-EGF remains attached to the erbB1-eGFP after internalization, the colocalization should be high.

The second set were A431 cells expressing erbB3-mCitrine. In this case, EGF-QD does not bind directly to erbB3, but still binds to the native, unlabeled erbB1 present in the cell. Upon activation of erbB1 by EGF-QD, no complexes with the labeled erbB3 are formed; therefore in these images, the colocalization should be low.

Our aim is to determine colocalization in these two cases experimentally, and to judge whether the results agree with what is expected. We will demonstrate a better agreement after restoration.

Following Ref. [31], we applied and compared the result of colocalization analysis on the raw image to the result of colocalization analysis applied after three different

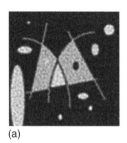
(a)
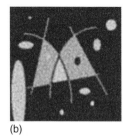
(b)
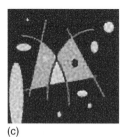
(c)

Figure 26.4 Comparison of unregularized versus regularized RL deconvolution. (a) RL, Gaussian blur as regularization; (b) RL, Conchello regularization; and (c) RL, with TV regularization.

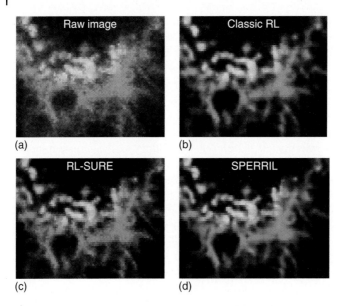

Figure 26.5 Detailed views of the different restoration results for a cell from the test set. (a) The raw image; (b) the results after classical RL; (c) the result after RL-SURE; and (d) the result of SPERRIL. Note that (b) and (c) still contain noise, while (d) maintains similar sharpness and the noise is better suppressed. (Please find a color version of this figure on the color plates.)

image restoration methods (classical RL, RL-SURE, and SPERRIL) combined each time with standard background correction. Each time, the different color channels were processed independently for the sake of simplicity, and the PSF was estimated separately for the different color channels. In Figure 26.6, we show the results of our colocalization analysis. For the different methods, we calculate the overlap coefficient R_{overlap}, which has been defined by Manders et al. [30]:

$$R_{\text{overlap}} = \frac{\sum_{x,y} \left(I_1(x,y) I_2(x,y) \right)}{\sqrt{\left(\sum_{x,y} (I_1(x,y))^2 \right) \left(\sum_{x,y} (I_2(x,y))^2 \right)}} \tag{26.12}$$

where $I_1(x, y)$ and $I_2(x, y)$ are the intensities in the red and green channels, respectively.

The R_{overlap} values for the raw data (plotted in a graph in Figure 26.6) show only a small difference between the two cell types. Restoration with the classical RL already improves the result of the colocalization analysis, in that the difference is clearer. However, when applying SPERRIL restoration prior to analysis, the largest difference between cell types is seen; this is consistent with what is expected from the underlying biochemical process in the cells [32]. In Figure 26.5, a detail of a cell from our test set is shown. Again, we can conclude that for classical RL, the regularization is rather poor. The results of RL-SURE also do not remove all the noise in the bright areas, while the result of SPERRIL provides a better suppression of the noise.

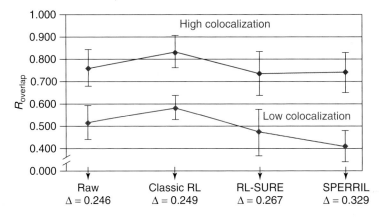

Figure 26.6 Colocalization results visualized in a graph. For each class of cells (low/high colocalization) and for each restoration method, the sample mean is shown as a dot and the sample variance is plotted as an error bar.

26.4
Conclusions

In this chapter, we discussed some classical image restoration techniques and illustrated that these techniques can improve image quality significantly. We then discussed how the addition of multiresolution priors to classical image restoration techniques further improve the quality. This was explained with the SPERRIL restoration algorithm. We illustrated that in a real biological experiment, it is possible to improve the analysis results in a colocalization experiment where interactions between different proteins were studied. In our case, we were able to distinguish the different classes of images (low colocalization versus high colocalization) better than when applying classical restoration. Since the underlying biological process was known, this confirms that better restoration facilitates colocalization analysis. In a later stage, this can be applied to facilitate colocalization analysis when the underlying biological process is not known.

Acknowledgment

The experimental work in Section 26.3.3.2 was carried out in collaboration with Dr Diane S. Lidke[4], who kindly provided the images and feedback about the image processing results.

4) At the time of this research, she worked at the Department of Molecular Biology of the Max Planck Institute for Biophysical Chemistry (Göttingen, Germany); she now works at the Health Sciences Center of the University of New Mexico, USA.

References

1. Van der Voort, H.T.M. Image restoration: getting it right. Online presentation at http://www.svi.nl/imagerestoration/.
2. Neelamani, R., Choi, H., and Baraniuk, R.G. (1999) Wavelet-based deconvolution for ill-conditioned systems, in *Proceedings of IEEE Conference on Acoustics, Speech, and Signal Processing (ICASSP)*, vol. 6, IEEE, pp. 3241–3244.
3. Banham, M.R. and Katsaggelos, A.K. (1996) Spatially adaptive wavelet-based multiscale image restoration. *IEEE Trans. Image Processing*, **54**, 619–634.
4. Portilla, J. and Simoncelli, E.P. (2003) Image restoration using Gaussian scale mixtures in the wavelet domain, in *Proceedings of the 9th IEEE Int'l Conference on Image Processing (ICIP)*, vol. II, IEEE, pp. 965–968.
5. Guerrero-Colon, J.A., Mancera, L., and Portilla, J. (2008) Image restoration using space-variant Gaussian scale mixtures in overcomplete pyramids. *IEEE Trans. Image Processing*, **171**, 27–41.
6. Starck, J.-L. and Murtagh, F. (1994) Image restoration with noise suppression using the wavelet transform. *Astron. Astrophys.*, **2881**, 342–348.
7. Boutet de Monvel, J., Le Calvez, S., and Ulfendahl, M. (2001) Image restoration for confocal microscopy: improving the limits of deconvolution, with application to the visualization of the mammalian hearing organ. *Biophys. J.*, **805**, 2455–2470.
8. Figueiredo, M.A.T. and Nowak, R.D. (2003) An EM algorithm for wavelet-based image restoration. *IEEE Trans. Image Processing*, **128**, 906–916.
9. Daubechies, I., Defrise, M., and De Mol, C. (2004) An iterative thresholding algorithm for linear inverse problems with a sparsity constraint. *Commun. Pure Appl. Math.*, **5711**, 1413–1457.
10. Rooms, F., Philips, W., and Lidke, D.S. (2005) Simultaneous estimation and restoration of confocal images and performance evaluation by colocalization analysis. *J. Microsc.*, **2181**, 22–36.
11. Molina, R., Núñez, J., Cortijo, F.J., and Mateos, J. (2001) Image restoration in astronomy, a Bayesian perspective. *IEEE Signal Process. Mag.*, **182**, 11–29.
12. van Kempen, G.M.P. (1999) Image restoration in Fluorescence Microscopy. PhD thesis, Delft University.
13. Lucy, L.B. (1974) An iterative technique for the rectification of observed distributions. *Astron. J.*, **796**, 745–754.
14. Richardson, W.H. (1972) Bayesian-based iterative method of image restoration. *J. Opt. Soc. Am.*, **621**, 55–59.
15. Dempster, A.O., Laird, N.M., and Rubin, D.B. (1977) Maximum likelihood from incomplete data via the EM algorithm. *J. R. Stat. Soc. B*, **391**, 1–38.
16. Shepp, L.A. and Vardi, Y. (1982) Maximum likelihood reconstruction for emission tomography. *IEEE Trans. Med. Imaging*, **12**, 113–122.
17. van der Voort, H.T.M. and Strasters, K.C. (1995) Restoration of confocal images for quantitative image analysis. *J. Microsc.*, **178**, 165–181.
18. Verveer, P.J. (1998) Computational and optical methods for improving resolution and signal quality in fluorescence microscopy, PhD thesis, Delft University.
19. Dey, N., Blanc-Féraud, L., Zimmer, C., Roux, P., Kam, Z., Olivo-Marin, J.C., and Zerubia, J. (2004) 3D microscopy deconvolution using Richardson-Lucy algorithm with total variation regularization. http://www-sop.inria.fr/ariana/BIBLIO-ENG/Author/DEY-N.html.
20. Snyder, D., Schultz, T.J., and O'Sullivan, J.A. (1992) Deblurring subject to non-negativity constraints. *IEEE Trans. Signal Processing*, **405**, 1143–1150.
21. Conchello, J.-A. and McNally, J.G. (1996) Fast regularization technique for expectation maximization algorithm for optical sectioning microscopy, in *Proceedings of SPIE Vol. 2655, Three-Dimensional Microscopy: Image Acquisition and Processing III*, SPIE, pp. 199–208.
22. Rudin, L.I., Osher, S., and Fatami, E. (1992) Nonlinear total variation based noise removal algorithms. *Phys. D*, **60**, 259–268.

23. Şendur, L. and Selesnick, I.W. (2002) Bivariate shrinkage functions for wavelet-based denoising exploiting interscale dependency. *IEEE Trans. Signal Processing*, **50**11, 2744–2756.
24. Starck, J.-L., Murtagh, F., and Bijaoui, A. (2000) *Image Processing and Data Analysis, The Multiscale Approach*, Cambridge University Press.
25. Homem, M.R.P., Mascarenhas, N.D.A., Costa, L.F., and Preza, C. (2002) Biological image restoration in optical-sectioning microscopy using prototype image constraints. *Real-Time Imaging*, **8** (6), 475–490.
26. Donoho, D.L. and Johnstone, I.M. (1994) Ideal spatial adaptation by wavelet shrinkage. *Biometrika*, **81**3, 425–455.
27. Stollberg, H., Boutet de Monvel, J., Holmberg, A., and Hertz, H.M. (2003) Wavelet-based image restoration for compact X-ray microscopy. *J. Microsc.*, **211**2, 154–160.
28. Neelamani, R., Choi, H., and Baraniuk, R. (2004) Forward: Fourier-wavelet regularized deconvolution for ill-conditioned systems. *IEEE Trans. Signal Processing*, **52**2, 418–433.
29. Rooms, F., Philips, W., and Van Oostveldt, P. (2003) Integrated approach for estimation and restoration of photon-limited images based on steerable pyramids, in *Proceedings of the 4th EURASIP Conference focused on Video / Image Processing and Multimedia Communications, Zagreb, Croatia* (M. Grcic and S. Grcic), EURASIP, pp. 131–136.
30. Manders, E.M., Verbeek, F.J., and Aten, J.A. (1993) Measurement of co-localization of objects in dual colour confocal images. *J. Microsc.*, **169**, 375–382.
31. Landmann, L. (2002) Deconvolution improves colocalization analysis of multiple fluorochromes in 3D confocal data sets more than filtering techniques. *J. Microsc.*, **208**2, 134–147.
32. Lidke, D.S., Nagy, P., Heintzmann, R., Arndt-Jovin, D.J., Post, J.N., Grecco, H.E., Jares-Erijman, E.A., and Jovin, T.M. (2004) Quantum dot ligands provides new insights into erbB/HER receptor-mediated signal transduction. *Nature Biotechnol.*, **22**2, 198–203.

27
Optical and Geometrical Super-Resolution

Javier Garcia Monreal

27.1
Introduction

Every imaging system presents a limited capability in resolution, which can be expressed as a function of the minimal distance at which two infinitely small spatial features can be positioned in proximity to each other while remaining separable in the image provided by the system [1]. But an imaging system is a medium that connects the optical input signal with the electronic output signal provided by the detector. T, the term for *"imaging system,"* must be divided into three parts; each one of them defines a different resolution limit.

First, one can find the medium in which the optical signal propagates from the input plane through the optical system toward a detector. Here, the angular span of diffracted beams is linearly proportional to the optical wavelength and inversely proportional to the size of the feature that generates the diffraction. Thus, only the angles arriving within the diameter of the imaging lens are imaged at the detection plane and the object's spectrum is trimmed by the limited aperture of the imaging system. The achievable resolution becomes restricted by diffraction and it is called *diffractive optical resolution* [2].

After this, the input signal is captured by a digital device (typically a CCD camera). Once again, the spatial information regarding the optical signal is distorted by the detector array geometry. This limitation (named as *geometrical resolution*) can be divided into two types of constraints. The first type is related to the number of the sampling points and the distance between two such adjacent points. The denser the two-dimensional (2D) spatial sampling grid, the better the quality of the sampled image. The second one is related to the spatial responsivity of each sampling pixel. Since each pixel integrates the light that impinges on its area, the 2D sampling array is not an ideal sampling array. This local spatial integration done at each pixel results in low-pass filtered image [3].

After this process, the optical signal is converted into electronic signal by the digital conversion of the electrons collected in the electronic capacitor of every sampling pixel. In this third group, the quality of the detector comes into play and features such as sensitivity, dynamic range, and different types of noise again

Optical and Digital Image Processing: Fundamentals and Applications, First Edition. Edited by Gabriel Cristóbal, Peter Schelkens, and Hugo Thienpont.
© 2011 Wiley-VCH Verlag GmbH & Co. KGaA. Published 2011 by Wiley-VCH Verlag GmbH & Co. KGaA.

damage the final output signal [3]. The resolution defined in this stage is called *noise equivalent resolution*.

In any case, the resolution of the final electronic signal is affected by the three previously defined factors. And resolution improvements of the final electronic readout signal can come from a combination of improvements achieved in each one of the three stages. The whole process of improvement is the real meaning of the term super-resolution. In this chapter, we review the state of the art in optical and geometrical super-resolution, with emphasis on the first works concerning such approaches. Thus, Section 1.3 analyzes an optical system from an information theory point of view and provides the theoretical tools to improve the resolution in an optical imaging system. After this, Section 1.4 presents an overview of different approaches that are capable of overcoming the *diffractive optical resolution* in conventional (nonholographic), holographic, and fluorescent imaging. In a similar way, Section 1.5 includes a collection of methods that are aimed at achieving *geometrical super-resolution*. Finally, Section 1.6 presents a basic bibliography regarding optical super-resolution.

27.2
Fundamental Limits to Resolution Improvement

Many researchers had used information theory to establish a relation between resolution and the number of degrees of freedom of an optical system. In 1966, Lukosz proposed an invariance theorem to explain the concepts underlying all super-resolution approaches [4]. This theorem states that, for an optical system, it is not the spatial bandwidth but the number of degrees of freedom (N_F) of the system that is fixed. Such an N_F value is, in fact, the number of points necessary to completely specify the system in the absence of noise and is given by

$$N_F = 2\left(1 + L_x B_x\right)\left(1 + L_y B_y\right)\left(1 + TB_T\right) \qquad (27.1)$$

where B_x and B_y are the spatial bandwidths, L_x and L_y are the dimensions of the field of view in the (x,y) directions, respectively, T is the observation time, and B_T is the temporal bandwidth of the optical system. Factor 2 is due to the two orthogonal independent polarization states.

Using this invariance theorem, that is, $N_F =$ constant, Lukosz theorized that any parameter in the system could be extended above the classical limit if any other factor represented in Eq. (27.1) is proportionally reduced, provided that some a priori information concerning the object is known. However, this theorem is not complete because it does not consider noise in the optical system. The signal-to-noise ratio (SNR), as well as three spatial dimensions, two independent polarizations states and the temporal dimension, may be included:

$$N = (1 + 2L_x B_x)(1 + 2L_y B_y)(1 + 2L_z B_z)(1 + 2TB_T)\log(1 + \text{SNR}) \qquad (27.2)$$

where L_x, L_y, B_x, B_y, B_T, and T are as defined in Eq. (27.1), L_z is the depth of field, and B_z is the spatial bandwidth in the z direction. Once again, factor 2 in each term of Eq. (27.2) refers to the two independent polarization states of each independent dimension.

Then, the invariance theorem of information theory states that it is not the spatial bandwidth but the information capacity of an imaging system that is constant [5]. Thus, provided that the input object belongs to a restricted class, it is in principle possible to extend the spatial bandwidth (or any other desired parameter) by encoding–decoding additional spatial-frequency information onto the independent (unused) parameter(s) of the imaging system. And this is the fundamental underlying all super-resolution schemes.

But as mentioned previously, resolution improvement requires a priori knowledge about the input object. One can classify the objects using a priori information into different types allowing different super-resolution strategies. Thus, one finds angular multiplexing for nonextended objects [4], time multiplexing for temporally restricted objects [4, 6], spectral encoding for wavelength-restricted objects [7], spatial multiplexing with one-dimensional (1D) objects [4], polarization coding with polarization restricted objects [8], and gray-level coding for objects with restricted intensity dynamic range [9].

Aside from the purely optical aspects of the imaging system (light propagation, lenses, object optical characteristics, etc.), the sensor itself exhibits nonideal properties (mainly pixelation noise and sensitivity issues) that greatly influence the recorded image. In principle, a detector with large pixels is preferred, from the light collection capability point of view, but this results in a low-resolution image. This limitation is referred to as *geometrical resolution*. Several directions have been devised to overcome the geometrical resolution limit. They mostly deal with introducing subpixel information obtained by means of relative shift between the image and the sensor [10].

27.3
Diffractive Optical Super-Resolution

Resolving power limited by diffraction in imaging systems dates back to 1873 when German physicist Ernst Abbe published that the resolution of an optical imaging system is limited by the wave nature of light, through the wavelength and the numerical aperture (NA) of lenses [2].

27.3.1
Optical System Limitations and Super-Resolution Strategy

The Rayleigh resolution criterion establishes that two closed points having the same relative intensity are just resolved when the first minimum of the diffraction pattern in the image of one point source coincides with the maximum of the other. In such cases, the distance between the central maximum and the first minimum in the intensity point spread function (PSF) of the system is called as Rayleigh resolution distance (RRD). In paraxial approximation, the RRD is $\delta X_{dif} = 1.22 \lambda F/D$, with

λ being the illumination wavelength, and F and D being the focal distance and the diameter of the imaging system, or equivalently $\delta X_{\text{dif}} \cong 0.6\lambda/\text{NA}$, with NA being defined as $\text{NA} = n\sin\theta$, where n is the refractive index of the surrounding medium and θ is the half angle of the illumination cone of light collected by the lens. Thus, a direct way to increase the image resolution consists in decreasing the illumination wavelength or increasing the NA of the imaging lens or both.

In general, only a portion of the diffracted components that are generated when an object is illuminated can pass through an optical system because of the limited size of the input pupil. Figure 27.1 depicts this situation where, aside from the undiffracted beam, only two additional diffracted beams build the final image up. The cutoff frequency gives the truncation in the spatial-frequency content of the input object and is defined as

$$f_c^{\text{coh}} = \frac{\text{NA}}{n\lambda} \quad \text{and} \quad f_c^{\text{incoh}} = \frac{2\text{NA}}{n\lambda} \tag{27.3}$$

Once again, improvement in the cutoff frequency (extending the transfer function in the frequency domain) can be achieved by an increase in the NA of the optical system or by a decrease in the illumination wavelength or both. But this procedure is not always possible and new techniques that are able to improve the resolution without changes in the physical properties of the optical system are desired. The super-resolution techniques define a synthetic numerical aperture (SNA) higher than the conventional NA, which implies a reduction in the RRD and, therefore, an improvement in the resolution of the imaging system. Figure 27.2 depicts the SNA generation integrating a high number of diffracted beams.

The key to achieving generation of an SNA is the introduction of additional optical elements (or masks) or the use of a special illumination procedure in the imaging system. Some examples of such masks are diffraction gratings (physical or projected) or prisms, while tilted beams are the most appealing illumination procedure. The selection of the aided optical element is related to a certain a priori knowledge about the input object, that is, with the invariance theorem of information theory. In any case, either the masks used in the encoding should have a critical size below the diffraction limit of the imaging system or the illumination angle of the tilted beam must be higher than the NA of the imaging lens. Then,

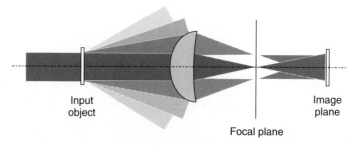

Figure 27.1 Image formation in the scope of the Abbe's theory: the object's spectrum is trimmed by the system's aperture when on-axis illumination is used.

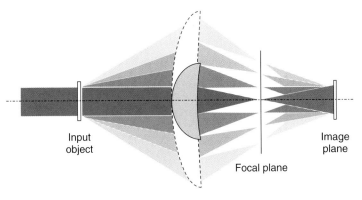

Figure 27.2 Generation of the SNA (dashed big white lens) in comparison with the conventional aperture (solid small gray lens).

the object's high spectral content is diverted toward the system-limited aperture, allowing its transmission through it. Then, the additional information that now is passing through the aperture, must be recovered and relocated to its original position in the object's spectrum, by using the proper decoding algorithm.

In general, the super-resolving effect is achieved in at least two basic steps. First, in an *encoding stage* the higher spatial-frequency content is redirected through the imaging lens aperture by means of an encoding mask. Then, in a *decoding stage*, recovery and correct replacement of the additional frequency content transmitted during the encoding stage is performed. A third stage, the *digital postprocessing stage*, may be needed to reconstruct the final super-resolved imaging.

27.3.2
Nonholographic Approaches

Because of historical and chronological reasons, in this section we first report on classical methods to improve the resolution of imaging systems without considering holographic tools.

27.3.2.1 Time Multiplexing

When the object's amplitude distribution does not vary with time or it has a slow variation, we say that the object is temporally restricted or time independent. Then, super-resolution can be achieved by time multiplexing. In essence, part of the spatial-frequency information of the object is encoded into the temporal degree of freedom because the information of the object is constant during the observation time.

Back in 1952 [11], Françon proposed a time-multiplexing approach based on the synchronized movement of two pinholes: one over the object and another at the image plane. By spatial scanning the object field of view, the method provides 2D imaging by point-by-point imaging of the input object: the pinhole position and the

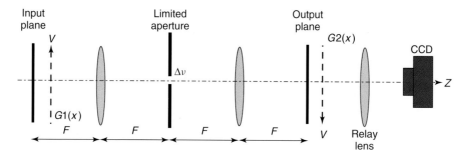

Figure 27.3 Optical setup for obtaining expanded aperture with two moving gratings [7].

intensity in each measurement determine the spatial location and the transmittance of each point at the reconstructed image. However, this technique provides low image intensity and low scanning velocity because of the high integration times during imaging.

Lukosz [4] replaced the scanning pinholes by synchronized moving gratings, alleviating the low light efficiency problem, and the experimental setup proposed by Lukosz is sketched in Figure 27.3. The input object is illuminated with a monochromatic plane wave, and a Ronchi grating (encoding grating) is placed after the input object. Then, different frequency slots of the object's spectrum can pass through the limited aperture of the imaging system. The Ronchi grating diffracts the different frequency slots of the object's spectrum by means of its different diffraction orders so that they are diverted to the optical axis and can pass throughout the system's aperture.

To attain super-resolution, a correct replacement of each frequency band is needed. Therefore, two conditions must be fulfilled. First, another Ronchi grating must be considered in order to shift each frequency slot back to its original position in the object's spectrum. After this second Ronchi grating, each transmitted frequency band propagates with the proper diffraction angle as it originated at the image plane. Besides the super-resolution effect, the second grating is also responsible for replicas of each frequency band. So, a second condition must be imposed to avoid distortion: both gratings must be moved in a synchronous way with constant velocity and time integration, in an imaging condition with matching magnification. The approach has some disadvantages, such as the complex experimental setup due to the synchronized movement of both gratings and a high-resolution relay optics to create a final super-resolved image with a CCD camera because the decoding grating is placed after the output plane.

To alleviate these drawbacks, many modifications of the Lukosz's setup have been proposed over the years [3]. A significant simplification was reported by Shemer et al. [6]. The problem of synchronization between both gratings is eliminated by using a *digital grating* introduced by computer and not optically. Figure 27.4 shows the experimental results achieved through the approach presented by Shemer et al. when considering a test object composed of gratings having different periods, for both coherent and incoherent illuminations [6].

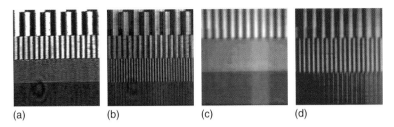

Figure 27.4 Experimental results for a grating test: (a,c) the low-resolution images and (b, d) the super-resolved images for coherent and incoherent cases, respectively [11].

Another way to achieve time multiplexing super-resolution is by using speckle patterns instead of grating patterns [12]. Basically, the object is illuminated with high-resolution speckle patterns (encoding mask) that are laterally shifted in time sequence. The decoding pattern is the same speckle pattern but corrected with the magnification of the imaging system. The decoding is performed for each image illuminated with the set of shifted encoding patterns. Finally, the reconstructed image is obtained by averaging the set of obtained images. Assuming $g(x)$, $s(x)$, and $s'(x)$ as the input object, the speckle pattern on the sample plane, and the decoding mask functions, respectively, the imaging system produces a low-pass filtering that can be modeled by the impulse response $h(x)$ as

$$o_\xi(x) = [g(x)s(x-\xi)] * h(x) = \int [g(x')s(x'-\xi)] h(x-x')dx' \quad (27.4)$$

with ξ being the displacement of the speckle pattern at a given instant, $o_\xi(\xi)$ being the captured image at a given instant, which is characterized by the displacement ξ, and $*$ denoting convolution. This single image is multiplied by the decoding pattern that is displaced by the same amount as the encoding pattern and added for all the displacements. The final reconstructed image is

$$o(x) = \int o_\xi(x)s'(x-\xi)d\xi = \int\int \{[g(x')s(x'-\xi)] h(x-x')\} s'(x-\xi)dx'd\xi \quad (27.5)$$

Changing the integration order and defining $\gamma(x-x')$ as

$$\gamma(x-x') = \int s(x'-\xi)s'(x-\xi)d\xi = \int s(v)s'(v+(x-x'))dv \quad (27.6)$$

Eq. (27.5) can be rewritten as

$$o(x) = \int g(x')h(x-x')\gamma(x-x')dx' = \int g(x')h'(x-x')dx' = g(x)*h'(x) \quad (27.7)$$

that is, the final reconstructed image is the convolution of the input object's amplitude distribution with the impulse response associated with the full process: $h'(x) = h(x)\gamma(x)$, where $\gamma(x)$ is the correlation between the encoding and

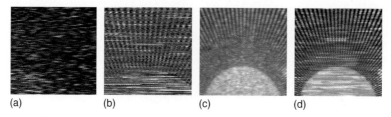

Figure 27.5 (a, c) Low-resolution and (b, d) super-resolved images under coherent and incoherent illuminations, respectively, when using the super-resolution by speckle pattern projection [13].

decoding masks. Since the encoding pattern is a random distribution, the autocorrelation is a function that is highly concentrated at the origin. Furthermore, if the autocorrelation peak is small as compared to the lens impulse response, the autocorrelation can be approximated by a delta function centered at the origin: $h'(x) \approx h(x)\delta(x) = h(0)\delta(x)$. Thus, under these assumptions, a high-resolution image is reconstructed according to Eq. (27.7). A similar analysis can be done for incoherent illumination [12]. Figure 27.5 depicts the experimental results achieved using this approach [13], for both coherent (cases (a) and (b)) and incoherent (cases (c) and (d)) illuminations.

27.3.2.2 Angular Multiplexing

Bachl and Lukosz [14] presented a super-resolving optical system that is capable of overcoming the spatial resolution limit imposed by diffraction by reducing the object field of view. Two static gratings are inserted into conjugate planes at the object and image space of the experimental setup. The encoding mask (at the object space) allows the transmission of additional diffracted object waves through the limited system aperture and generates replications of the object while each replica contains different spectral information. The role of the decoding mask (at the image space) is to redefine the propagation direction of the new diffracted components as they were generated in the input object. The object field of view should be limited around the object region of interest since no movement is required for the gratings and to avoid image distortion from the ghost images produced in the encoding–decoding process. Figure 27.6 shows experimental results obtained using the Bachl and Lukosz approach. When the object is spatially limited (case (a)), the ghost images do not overlap and the approach provides super-resolution (central white dashed square in case (c)) when compared with the low-resolution image (case (b)).

However, the condition for obtaining super-resolution implies that the two gratings are not positioned in the volume that is between the object and the image plane. Some modifications to the original idea proposed by Bachl and Lukosz considering different configurations of static gratings were also proposed when considering three gratings allocated within the space between the object and the

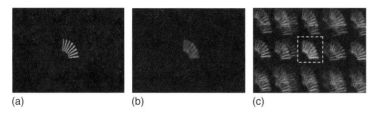

Figure 27.6 Experimental results according to the Bachl and Lukosz approach: (a) object with limited field of view, (b) low-resolution image provided by the imaging system, and (c) super-resolved image (central dashed white square) and ghost images that do not overlap with the central image because of reduction in the field of view. Images are from Ref. [15].

image plane [15–17]. But in all the cases, super-resolved imaging was attained by severe limitation over the object field of view.

The fact that one needs to sacrifice the field of view is also somewhat problematic. One solution to the restriction on the field of view was recently reported and in this white light illumination is used for averaging the ghost images (undesired replicas) that are obtained outside the region of interest since the positions of those images are wavelength sensitive [16]. However, the increase in resolution is obtained at the price of reducing the dynamic range since the reconstructed high-resolution image appears on top of an averaged background level.

27.3.2.3 Multiplexing in Other Degrees of Freedom

Aside from time and angular degrees of freedom, there are other domains that can be exploited to obtain super-resolution imaging by multiplexing the object's spatial-frequency information. In the following discussion, we summarize the most important achievements.

Wavelength multiplexing allows super-resolved imaging when the object has only gray-level information, that is, it is a monochromatic object. A typical approach involves illumination of an object through a grating that causes wavelength dispersion around the first order of diffraction. This way the various spatial features of the object are illuminated and encoded by different wavelengths. Then, a lens collects the light (performs spatial integration) and directs it into a single-mode fiber (capable of transmitting only a single degree of freedom). On the other side of the fiber, the spatial information of the object is decoded using the same dispersion grating. A 2D diffractive optical element instead of a grating can be used to allow dispersion of 2D wavelengths [7].

Another type of multiplexing is *spatial multiplexing* that can be applied to objects that have a restricted shape. This is the case of 1D objects or objects having a cigar shape in the Fourier domain. Super-resolution of 1D objects can also be based on spatial multiplexing by using spectral dilations [17]. Zalevsky *et al.* reported a method based on the various dilations in the period of a special mask for obtaining the required encoding and decoding of information. Here, the axis that carries no information is used for the different dilations. This encoding–decoding process performed by spectral dilation ensures that the unwanted spectral information does not overlap with the desired spectral bands, and thus it may be filtered out.

The gray levels or the *dynamic range* domain is another example of a domain that may be used in order to encode, transmit, and decode the spatial information of the imaged object [9]. Assume that the input object is restricted in dynamic range (a priori information). By attaching a gray-level coding mask to the input object, each pixel of the input object is encoded with a different transmission value while the ratio between each one of those values is 2^M, M being the a priori known and limited number of bits spanning the dynamic range of the imaged object. Obviously, the imager should have sufficient number of dynamic range bits. It should be at least $M \times K^2$ where K is the super-resolution improvement factor in every spatial dimension.

And finally, another way to achieve super-resolution is by exploiting the *degree of coherence* of the illumination light. In spatial coherence coding, the mutual intensity function of the illumination beam is used to code spatial information in a way analogous to time multiplexing but with multiplexing time slots that are given by the coherence time of the illumination beam. On the other hand, instead of transversally generating different coherence distributions, the longitudinal axis is used to code the transversal spatial information in the temporal coherence coding by using incoherent light interferometry. After this, interferometric image plane recording with postprocessing digital stage allows the needed decoding to achieve super-resolution effect [18].

27.3.3
Holographic Approaches

Holography is a wavefront recording process that is produced when a given beam interferes with another beam that is coherent to it. Typically, one beam (the object beam) contains information about the input object while the second one (the reference beam) is used to encode the diffracted wavefront from the input object because the relative phase between both beams varies along the recording medium in steps of 2π with a period of $d = \lambda/\sin\theta$, where λ is the illumination wavelength and θ is the angle between the two interferometric beams. The relative phase of the object and reference beams is encoded in a sinusoidal fringe pattern. Thus, with the proper decoding, it is possible to obtain any of the two interferometric beams. In particular, the object's complex wavefront can be recovered by means of optical or digital reconstruction process. In this section, we review the main methods that are applied to super-resolution imaging when considering holographic tools.

27.3.3.1 Holographic Wavefront Coding
Holography started as early as when Gabor proposed a new method to achieve imaging in electron microscope working without lenses [19]. In its basic architecture, Gabor's setup proposed an in-line configuration where two waves interfere at the output plane: the imaging wave caused by diffraction at the sample's plane and the reference wave incoming from the nondiffracted light passing through the

sample. However, this procedure is restricted to weak diffractive samples. Only under this assumption, the light diffracted by the sample can be considered as a perturbation of the reference beam and the underlying Gabor's principle becomes true.

This dichotomy can be easily removed by inserting an external reference beam at the recording plane. Thus, regardless of the sample being considered or not considered as a weak diffractive one, interference permits the holographic recording. In this case, the sample information is placed in one interferometric beam and the reference beam in a different one and both are brought together to produce an interference pattern. One can find different schemes to reinsert the reference beam based on an off-line holographic architecture. This procedure avoids the distortion caused by overlapping, in the direction of observation, of the three holographic terms incoming from the in-line scheme.

Also, the problem concerning Gabor's concept can also be solved by inserting the reference beam in on-axis mode and applying phase-shifting technique [20] over one of the interferometric beams (typically the reference one). By means of the proper algorithm, it is possible to remove the twin image and the zero order of the holographic recording. However, phase-shifting method implies the use of digital recording devices (typically CCD or CMOS sensors) to evaluate the different holograms that are recorded in time sequence when the reference beam is varied in optical path length. Nowadays, digital holography combines the classical optical holographic implementation with digital recording and reconstruction resulting from the development of modern solid-state image sensors with numerical processing capabilities provided by computers, and a lot of new configurations with enhanced capabilities are constantly appearing in the literature [13, 21].

27.3.3.2 Multiplexed Holograms

The basic idea behind super-resolution is that of obtaining a hologram in which a coherent superposition of a set of sequential images is recorded during the exposure time interval. If each hologram is recorded using different tilted illuminations onto the object and complementary off-axis reference waves at the hologram plane, a super-resolved image can be obtained in the hologram reconstruction [22]. A main difference should be noticed between holographic approaches and those proposed by Lukosz and their derivatives. In holographic approaches, tilted beam illumination is the key to downshift the object's high-order spectral content and not diffraction gratings. An intuitive explanation is obtained by observing the equivalence between the different diffraction orders generated by a grating and a set of tilted beams (on-axis and off-axis tilted beams) incoming from a set of point sources and having the same propagation angles as those diffracted by the grating. Figure 27.7 is representative of this analogy. Thus, the *encoding stage* can be defined as a set of high-order frequency bands that are directed toward the aperture of the imaging system either by the grating action or as a function of the number of illumination beams (on-axis and off-axis) impinging the input object, where, for a given wavelength, the shift in the object's spectrum is a function of the period of the grating or the oblique illumination angle, respectively.

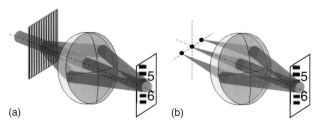

Figure 27.7 Analogy between a 1D diffraction grating (a) and a 1D set of tilted beams (b).

Notice that, although in a strict way the equivalence between diffraction grating and tilted beams should consider a set of coherence point sources, the possibility that such sources will be incoherent to each other offers new capabilities to holographic approaches in super-resolution imaging. Super-resolution can be performed using spatial multiplexing, that is, all the sources lighted on at the same time [23], or using time multiplexing, that is, by sequential activation of the different sources [24].

27.3.3.3 Digital Holography

In the previous section, we have presented the first experimental approaches to time multiplexing super-resolution in holography. In fact, Sato et al.'s work [22] used digital devices instead of holographic mediums as recording mediums and postprocessing stages to assemble the final super-resolved image. Nowadays, digital holography is a well-known and widely used technique that allows fast, nondestructive, full-field, high-resolution quantitative complex amplitude measurements of imaged objects. However, both the finite number and the size of the CCD pixels limit the resolution of the digital holographic approach. Image-resolution improvements in digital holography have been experimentally achieved by generating a synthetic aperture (SA) from the combination of different holograms recorded at different camera positions in order to construct a larger digital hologram or by digital holographic implementation of the Lukosz approach [25]. A diffraction grating redirects high-order frequency bands of the object's spectrum toward the imaging system. Then, each additional band is recovered by spatial filtering at the spatial-frequency domain by reinserting a reference beam at the recording plane using a Fourier holographic configuration. And finally, an SA is generated by digital postprocessing that yields a super-resolved imaging.

On the other hand, microscopy in combination with digital holography provides high-resolution quantitative phase imaging of complex specimens. Although low NA microscope lenses define lower resolutions than high NA ones, they have different properties that make them useful for many microscopy applications. Some of these advantages are long working distance, large field of view, large depth of focus, and low cost. Several super-resolved approaches have been developed for different fields of application such as interferometric imaging, interferometric lithography, and holographic microscopy.

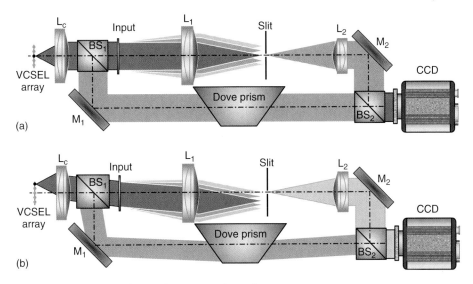

Figure 27.8 Experimental setup for super-resolution by tilted illumination in interferometric imaging. (a) on axis illumination. (b) off axis illumination.

The underlying principle of a super-resolution approach based on holographic tools and tilted illumination is presented in Figure 27.8 [23]. Essentially, an interferometric architecture is assembled in which the imaging system is placed in one of the branches while the other provides the reference beam for the holographic recording. By oblique illumination over the input plane, the object's spectrum is downshifted and high-frequency components that are not transmitted through the limited system aperture (Figure 27.8a) are now transmitted through it (Figure 27.8b). Tilted illumination can be provided sequentially [24] or in a single illumination shot [23]. After this, the way to upshift the high-order frequency bandpasses back to their original spatial-frequency position is by using interferometric image plane recording. Signal processing is then used to reconstruct the complete super-resolved image that can be understood as a synthetic enlargement of the system's aperture produced by incoherent addition of the individual recorded intensities.

Tilted beam illumination can be provided by an array of vertical cavity surface emitting laser (VCSEL) sources with two important advantages. First, the VCSEL sources are individually coherent but incoherent to each other. So, it is possible to light all the VCSEL sources at once, enabling super-resolution in a single shot of the illumination array. And second, the VCSELs may be temporally modulated at modulation rates of up to several gigahertz. Since the SA is the convolution of the VCSEL line array and the coherent transfer function (CTF) of the system, by temporally varying the relative amplitudes of each source in the line array, any synthetic transfer function may be realized at will. In addition, VCSEL sources exhibit high energy efficiency and high optical intensity.

From a theoretical point of view, we can say that the complex amplitude distribution arriving at the CCD from the imaging branch and due to a single m-VCSEL source is

$$U_m(x) = \left[f(-x)e^{-j2\pi m\Delta v x}\right] * \operatorname{sinc}(x\Delta v) \qquad (27.8)$$

where $f(-x)$ is representative of the input object amplitude distribution, $\exp(-j2\pi m\Delta v x)$ is the tilted illumination beam incoming from the off-axis position of the m-VCSEL source (notice that when $[m = 0]$, we obtain on-axis illumination), Δv is the spacing produced in the Fourier domain due to the VCSELs separation in the linear array source, $\operatorname{sinc}(x\Delta v)$ is the Fourier transformation of the CTF of the imaging system (rectangular pupil of size of width Δv), and x and v are the spatial and spatial-frequency coordinates, respectively.

After this, the CCD performs intensity recording of Eq. (27.8) plus a reference beam incoming from the reference branch. Attention needs to be paid only to the term of the hologram containing the multiplication of the amplitude distribution arriving from the imaging branch and the one arriving from the reference branch:

$$I_m(x) = |U_m(x) + R(x)|^2 = \cdots + \left[f(-x)e^{-j2\pi m\Delta v x}\right] * \operatorname{sinc}(x\Delta v)\, e^{j2\pi(m\Delta v + Q)x} + \cdots \qquad (27.9)$$

$\exp(j2\pi \Delta v + Q)$ is the reference beam and Q is the bias carrier frequency of the off-axis holographic recording (for instance, by tilting the mirror M1 in Figure 27.8). This bias carrier frequency is needed to avoid overlapping between the different orders in the Fourier transformation of the recorded hologram. If we examine the Fourier transform of Eq. (27.9), we obtain that the object's spectrum $\tilde{f}(v)$ is shifted by the action of the m-VCSEL source, multiplied by the CTF of the imaging system, and convolved with a delta function incoming from the reference beam:

$$\tilde{T}_3(v) = \left[\tilde{f}(v - m\Delta v) \times \operatorname{rect}\left(\frac{v}{\Delta v}\right)\right] * \delta(v + m\Delta v + Q) \qquad (27.10)$$

Notice that different contiguous frequency slots are transmitted through the system aperture because of the off-axis illumination provided that the source spacing equals the CTF width. On the other hand, the delta function contains shifts related to the m-VCSEL tilt ($m\Delta v$ spatial frequency) and the tilt in the mirror M1 (Q).

By considering that the full VCSEL array is activated and owing to the mutual incoherence of the holograms, the final output becomes the addition of all terms in Eq. (27.9) for all the VCSEL sources that can be written (after some manipulation) as

$$\tilde{T}_{3\Sigma}(v) = \sum_{m=-\infty}^{\infty}\left[\tilde{f}(v) \times \operatorname{rect}\left(\frac{v + m\Delta v}{\Delta v}\right)\right] * \delta(v + Q)$$

$$= \left\{\tilde{f}(v) \times \operatorname{SA}(v)\right\} * \delta(v + Q) \qquad (27.11)$$

where $\operatorname{SA}(v)$ is the generated synthetic aperture of the system which is actually the convolution of the VCSEL array and the CTF of the imaging system:

$$\operatorname{SA}(v) = \operatorname{rect}\left(\frac{v}{\Delta v}\right) * \sum_{m=-\infty}^{\infty}[\delta(v + m\Delta v)] \qquad (27.12)$$

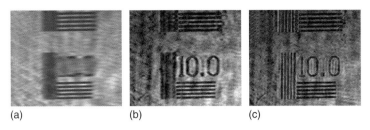

Figure 27.9 Experimental results of an NBS resolution test target: (a) low-resolution image, (b) super-resolved image with three VCSEL sources, and (c) super-resolved image with five VCSEL sources.

Finally, the last delta function in Eq. (27.11) is representative of a shift in the object's spectrum. This shift can be removed digitally and the resulting distribution centered digitally. Thus, a new Fourier transform recovers $f(-x)$, so that the super-resolved input object can be completely reconstructed. Figure 27.9 depicts the experimental results that can be achieved when using the approach presented by Micó et al. [23]. A National Bureau Standard (NBS) test target is used to show the resolution improvement when considering single centered VCSEL (on-axis) illumination (Figure 27.9a), three VCSEL sources (Figure 27.9b), and five VCSEL sources (Figure 27.9c) simultaneously.

In general, the whole process that allows super-resolution imaging by tilted beam illumination is summarized in the schematic chart depicted in Figure 27.10.

Different angular directions of the object's spectrum are time multiplexed by a set of tilted illumination beams. Each tilted beam allows the recording of an image plane hologram corresponding to a different bandpass image of the input object. The spatial-frequency content of each bandpass image is recovered by Fourier transforming the recorded hologram and the SA is assembled by proper reallocation of each recovered elementary aperture.

27.3.3.4 Axial Super-Resolution

Moreover, as a nonnegligible effect, the definition of a synthetic enlargement in the system's aperture also affects the axial resolution of the imaging lens. As the axial extent of the focal spot is proportional to NA^{-2}, a number of different methods have been developed to increase the limited axial resolution provided by microscope lenses. Also, the generation of an SA yields an improvement in axial resolution since the axial resolution can be expressed as a function of the NA of the imaging lens. Thus, if the NA of the imaging lens is synthetically increased, the axial resolution of the system will also be increased as a quadratic function of the SNA. Figure 27.11 depicts experimental results concerning the approach reported by Micó et al. [26], showing high optical sectioning of the sample under test. Axial super-resolution obtained by SA generation using tilted illumination has also been demonstrated [27]. In both cases, a standard configuration of the digital holographic microscope operating in transmission mode and considering time-multiplexing tilted beam illumination is used.

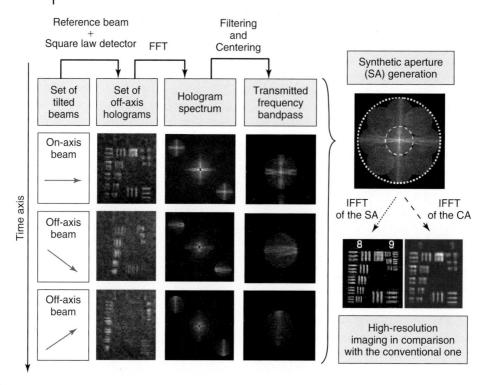

Figure 27.10 Schematic chart corresponding to any super-resolution approach based on time multiplexing and tilted beam illumination.

27.4
Geometrical Super-Resolution

One of the basic resolution limitations is related to the geometry of the sensing array. This limitation is affected by both the number and the pitch of the sampling pixels, that is, the density of the sampling grid as well as the spatial responsivity of each pixel. The spatial responsivity determines the PSF while the pitch is related to the Nyquist sampling condition. Thus, the ideal sampling case is obtained by considering large number of pixels with small pitch while each pixel is a delta function. This way owing to the small pitch, the Nyquist frequency can be high enough to recover all the spectral content from the image, and if the sampling pixels are delta functions, this is an ideal sampling and no low-pass effects on the spatial responsivity of the sampling points are generated. However, owing to practical reasons of energy efficiency, this ideal case is not the geometrical structure of commonly available sensors. Thus, when one refers to geometrical super-resolution, he/she refers to approaches aiming to overcome the two above-mentioned factors.

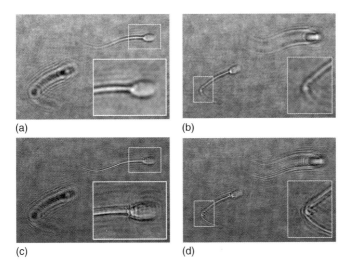

Figure 27.11 Experimental results of 3D super-resolution imaging for a swine sperm sample.

In general, geometrical super-resolution can be tackled by digital processing methods, by image sequence approaches, and by techniques involving physical components. In this section, an introduction to geometrical super-resolution based on physical components is provided, while digital processing methods and image sequence approaches are presented in detail in Chapter 28. The limitation related to the shape of each sampling pixel can be resolved by applying various approaches. One way of observing the problem of resolution enhancement is by looking at each pixel in the sampling array as a subpixel matrix functioning as an averaging operator, so that for each shift of the camera, some new subpixels are added and some are removed from the averaging. This is equivalent to the convolution of a superresolved image and a function having the spatial shape of a pixel. In this way, it is possible to solve the problem of image enhancement by inverse filtering, dividing the image in the frequency domain, by using the Fourier transform of the pixel's shape. The solution to this problem of inverse filtering is not always possible due to zeros existing in this function. Addition of special spatial mask can modify the Fourier transform of the pixel's shape and allow the realization of the inverse filtering [28]. Instead of physically attaching the mask to the sensor, one may use projection to create the desired distribution on top of the object [29].

Another related physical technique involves the position of the physical mask, which is not near the sensor but rather in the intermediate image plane. Owing to the spatial blurring generated by the spatial responsivity of the pixels of the sensor, one actually has more variables to recover (the high-resolution information) than the number of equations (the sampled readout coming from the detector) and thus the insertion of the mask inserts a priori knowledge that increases the number of equations, allowing us to eventually perform improved reconstruction of the high-resolution image out of its blurred version by applying matrix inversion procedure [30].

After performing a microscanning operation, that is, increasing the number of spatial sampling points (or in other words oversampling), the reduction in the geometrical resolution relates only to the blurring generated due to the local averaging that is performed by each one of the pixels in the detection array. This averaging is basically a spatial blurring which in general can be mathematically treated using a simple matrix representation:

$$A \cdot x = b \qquad (27.13)$$

where the 2D object is expressed by the 1D vector x (the 2D information is folded into a much longer 1D vector having the number of terms that is equal to the overall number of terms in the 2D image), the blurring or the reduction in resolution is expressed by a matrix A, and the vector b is related to the imaged information obtained at the detector array (once again the 2D information is folded into a longer 1D vector).

In order to reconstruct the high-resolution information that is related to the original object, we need to perform a matrix inversion operation. In the general case, the blurring matrix A is not a square matrix and thus its inversion is not feasible. The addition of a high-resolution, random a priori known mask that is positioned in the intermediate image plane adds more equations, and thus the matrix A gains more rows corresponding to this additional information and becomes a square matrix. Accordingly, the output readout vector b also gains more terms and its length now becomes equal to that of the original vector x containing the high-resolution information about the object.

The random mask that we add is a binary mask, and therefore it blocks the transmission of light in certain given spatial positions and transmits the light as is in the remaining locations. Usually the percentage of the blocked spatial positions is about 50% of the overall mask area. The additional equations that are added and which are expressed as the additional rows in A or the additional terms in b are related to the fact that some of the spatial values in the input object x are blocked while others are not according to the a priori known structure of the added random mask.

Although the set of equations represented by Eq. (27.13) contains a mathematically invertible matrix, the extraction of the original high-resolution vector x is performed by applying a pseudoinverse based upon Tikhonov regularization process:

$$x = \left(A^T A + \alpha^2 I\right)^{-1} A^T b \qquad (27.14)$$

where α is a constant and I is the unit matrix. In Figure 27.12, we present a numerical example for this geometrical superresolution approach and it is further detailed in Ref. [30].

In Figure 27.12a, one may see the original high-resolution object. In Figure 27.12b, we present the high-resolution image obtained in the intermediate image plane (before having the blurring caused because of the spatial averaging

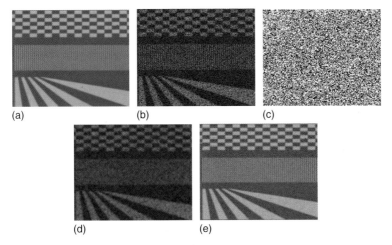

Figure 27.12 Simulations demonstrating the approach of Ref. [30]: (a) the high-resolution object; (b) this high-resolution object as seen in the intermediate image plane, after being multiplexed by the high-resolution encoding mask; (c) an example of the random encoding mask having 50% of energy blocking; (d) the low-resolution image as obtained at the detection array because of its spatial blurring; (e) the obtained super-resolved reconstruction.

that is performed during the capture of the image) after being multiplied by the random encoding mask that was positioned in this plane. An example of such a mask is seen in Figure 27.12c. After being sampled and spatially averaged by the detection array, one obtains the image as depicted in Figure 27.12d. After we apply the pseudoinversion algorithm that takes the enlarged A matrix (consisting of the additional rows due to the added encoding mask) and the extended vector b into account, one obtains the result seen in Figure 27.12e. One may clearly see the improvement in resolution that is obtained in comparison to the reduced-resolution image of Figure 27.12d.

Note that although the matrix inversion operation of $(A^T A + \alpha^2 I)^{-1} A^T$ involves some computational complexity, it may be computed off line since it is not object dependent and thus it is performed only once.

References

1. Goodman, J.W. (1996) *Introduction to Fourier Optics*, 2nd edn, McGraw-Hill, New York.
2. Abbe, E. (1873) Beitrage zur theorie des mikroskops und der mikroskopischen wahrnehmung. *Arch. Mikrosk. Anat.*, **9**, 413–468.
3. Zalevsky, Z. and Mendlovic, D. (2002) *Optical Super Resolution*, Springer.
4. Lukosz, W. (1967) Optical systems with resolving powers exceeding the classical limits II. *J. Opt. Soc. Am.*, **57**, 932–941.
5. Cox, I.J. and Sheppard, J.R. (1986) Information capacity and resolution in an optical system. *J. Opt. Soc. Am.*, **A3**, 1152–1158.
6. Shemer, A. *et al.* (1999) Superresolving optical system with time multiplexing

and computer decoding. *Appl. Opt.*, **38**, 7245–7251.

7. Mendlovic, D. *et al.* (1997) Wavelength multiplexing system for a single mode image transmission. *Appl. Opt.*, **36**, 8474–8480.

8. Zlotnik, A., Zalevsky, Z., and Marom, E. (2005) Superresolution with nonorthogonal polarization coding. *Appl. Opt.*, **44**, 3705–3715.

9. Zalevsky, Z., García-Martínez, P., and García, J. (2006) Superresolution using gray level coding. *Opt. Express*, **14**, 5178–5182.

10. Ben-Ezra, M. and Nayar, S.K. (2004) Motion-based motion deblurring. *IEEE Trans. Pattern Anal. Mach. Intell.*, **26**, 689–698.

11. Françon, M. (1952) Amélioration the résolution d'optique, *Nuovo Cimento*, (Suppl. 9), 283–290.

12. García, J., Zalevsky, Z., and Fixler, D. (2005) Synthetic aperture superresolution by speckle pattern projection. *Opt. Express*, **13**, 6073–6078.

13. Kreis, T. (2005) *Handbook of Holographic Interferometry: Optical and Digital Methods*, Wiley-VCH Verlag GmbH & Co. KGaA, Weinheim.

14. Bachl, A. and Lukosz, W. (1967) Experiments on superresolution imaging of a reduced object field. *J. Opt. Soc. Am.*, **57**, 163–169.

15. Sabo, E., Zalevsky, Z., Mendlovic, D., Konforti, N., and Kiryuschev, I. (2001) Superresolution optical system using three fixed generalized gratings: experimental results. *J. Opt. Soc. Am.*, **A18**, 514–520.

16. García, J. *et al.* (2008) Full field of view superresolution imaging based on two static gratings and white light illumination. *Appl. Opt.*, **47**, 3080–3087.

17. Zalevsky, Z. *et al.* (2004) Super resolving optical system based on spectral dilation. *Opt. Commun.*, **241**, 43–50.

18. Micó, V. *et al.* (2007) Spatial information transmission using axial temporal coherence coding. *Opt. Lett.*, **32**, 736–738.

19. Gabor, D. (1948) A new microscopic principle. *Nature*, **161**, 777–778.

20. Yamaguchi, I. and Zhang, T. (1997) Phase-shifting digital holography. *Opt. Lett.*, **22**, 1268–1270.

21. Schnars, U. and Jueptner, W.P. (2005) *Digital Holography*, Springer.

22. Sato, T., Ueda, M., and Yamagishi, G. (1974) Superresolution microscope using electrical superposition of holograms. *Appl. Opt.*, **13**, 406–408.

23. Micó, V. *et al.* (2004) Single-step superresolution by interferometric imaging. *Opt. Express*, **12**, 2589–2596.

24. Micó, V. *et al.* (2006) Superresolved imaging in digital holography by superposition of tilted wavefronts. *Appl. Opt.*, **45**, 822–828.

25. Granero, L. *et al.* (2009) Superresolution imaging method using phase-shifting digital lensless Fourier holography. *Opt. Express.*, **17**, 15008–15022.

26. Micó, V. *et al.* (2008) Superresolution digital holographic microscopy for three-dimensional samples. *Opt. Express*, **16**, 19261–19270.

27. Micó, V., Zalevsky, Z., and García, J. (2008) Axial superresolution by synthetic aperture generation. *J. Opt. A: Pure Appl. Opt.*, **10**, 125001–125008.

28. Zalevsky, Z., Shamir, N., and Mendlovic, D. (2004) Geometrical super-resolution in infra-red sensor: experimental verification. *Opt. Eng.*, **43**, 1401–1406.

29. Fixler, D. *et al.* (2007) Pattern projection for subpixel resolved imaging in microscopy. *Micron*, **38**, 115–120.

30. Borkowski, A., Zalevsky, Z., and Javidi, B. (2009) Geometrical superresolved imaging using nonperiodic spatial masking. *J. Opt. Soc. Am.*, **A26**, 589–601.

28
Super-Resolution Image Reconstruction considering Inaccurate Subpixel Motion Information
Jongseong Choi and Moon Gi Kang

28.1
Introduction

Since the development of charge-coupled device (CCD) imaging sensor three decades ago, imaging systems employing digital imaging sensors and digital image processing techniques have been widely used for various purposes. In almost all digital imaging applications, high-resolution images are desired and have often been required. High resolution means that the pixel density within a given image is high, and therefore these images offer more critical visual information. Although digital imaging sensors have been continuously improved to overcome their spatial resolution limitations, the current resolution level does not yet satisfy the demand. Consumers demand inexpensive high-resolution digital imaging systems, and scientists often require very high resolution levels with no visible artifacts. Thus, finding a way to increase current resolution levels is very important.

The easiest way of increasing spatial resolution is to increase the number of pixels in an imaging sensor by using device physics and circuit technology. In order to integrate more pixels in imaging sensors, the pixel size has to be decreased or the chip size has to be increased. As the pixel size decreases, however, shot noise that severely degrades the image quality is generated because the amount of light available also decreases. The increase in capacitance as the chip size enlarges is an obstacle that makes it difficult to speed up charge transfer rate. The high cost for high precision optics and imaging sensors is also an important concern in many commercial applications that require high-resolution imaging. Therefore, the number of pixels integrated in imaging sensors is limited and a new approach to increasing spatial resolution is necessary to overcome this limitation.

One alternative approach is to use signal-processing techniques to reconstruct high-resolution images from observed multiple low-resolution images. Recently, a resolution enhancement approach called super-resolution image reconstruction has been one of the most active research areas in this field [1]. The major advantage of super-resolution image reconstruction is that it generally costs less, because existing low-resolution imaging systems can be utilized. To increase the level of spatial resolution in super-resolution image reconstruction, multiple low-resolution

Optical and Digital Image Processing: Fundamentals and Applications, First Edition. Edited by Gabriel Cristóbal, Peter Schelkens, and Hugo Thienpont.
© 2011 Wiley-VCH Verlag GmbH & Co. KGaA. Published 2011 by Wiley-VCH Verlag GmbH & Co. KGaA.

images have to be subsampled as well as shifted with different subpixel motion information. Accurate subpixel registration between low-resolution images is very important for the reconstruction of high-resolution images. However, subpixel motion estimation is very complicated and inaccurate subpixel motion information can cause distortion in reconstructed high-resolution images. To reduce this distortion, inaccurate subpixel motion information has to be considered. In this chapter, the fundamentals of super-resolution algorithm are described and an approach that considers inaccurate subpixel registration is presented. The recent development and the applications of super-resolution algorithms are also demonstrated with the experimental results.

28.2
Fundamentals of Super-Resolution Image Reconstruction

28.2.1
Basic Concept of Super-Resolution

The spatial resolution that represents the number of pixels per unit area in a given image is the principal factor in determining the quality of that image. However, since an imaging sensor is composed of a number of photodiodes, the spatial resolution is limited. Super-resolution image reconstruction produces one or a set of high-resolution images using multiple low-resolution images that are captured from the same scene.

In super-resolution image reconstruction, typically, low-resolution images contain different information at the same scene. In other words, low-resolution images can be shifted with subpixel precision and then subsampled. In general, if the low-resolution images are shifted by integer precision, then each image contains exactly the same information, and thus there is no new information that can be used to reconstruct a high-resolution image. If the low-resolution images are shifted with subpixel precision among themselves and then aliased, however, then each image contains new information that can be exploited to obtain a high-resolution image. To obtain different viewpoints of the same scene, relative scene motions must exist from frame to frame via multiple scenes or video sequences. Multiple scenes can be obtained from one camera with several captures or from multiple cameras located in different positions. These scene motions can occur because of the controlled motions in imaging systems, for example, images acquired from orbiting satellites. The same is true of uncontrolled motions, for example, the movement of local objects or vibrating imaging systems. If these scene motions are known or can be estimated with subpixel accuracy, super-resolution image reconstruction is possible (as illustrated in Figure 28.1).

Also, the image interpolation technique is related to super-resolution reconstruction, because it has been used to increase the size of a single image [2]. Although this field has been widely studied, the quality of an image that is magnified from a single aliased low-resolution image is inherently limited even if the ideal sinc basis

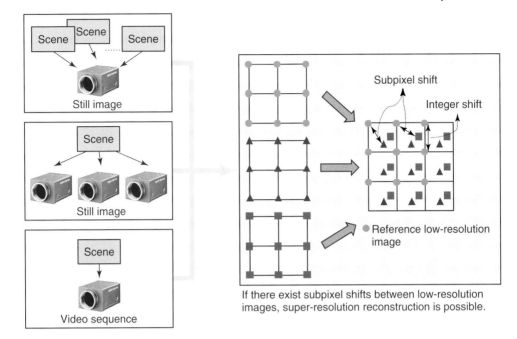

Figure 28.1 Basic premise of super-resolution image reconstruction.

function is employed. That is, image interpolation cannot recover high-frequency components that get lost or degraded during the image acquisition process. To achieve further improvements in this field, the next step requires the utilization of multiple datasets that contain additional data of the same scene. The fusion of information from multiple observations of the same scene allows for super-resolution reconstruction of that scene.

In the acquisition process of digital images, there is a natural loss of spatial resolution, caused by optical distortions, motion blur due to limited shutter speed, and noise that occurs within the sensor or during transmission. Therefore, super-resolution image reconstruction covers image restoration techniques that produce high-quality images from noisy, blurred images, as well as reconstruct high-resolution images from undersampled low-resolution images. In this respect, another related problem is image restoration, which is a well-established area in image processing applications (see also Chapter 26) [3]. Image restoration only recovers degraded (e.g., blurred, noisy) images. It does not change the size of those images. In fact, restoration and super-resolution reconstruction are closely related theoretically, and super-resolution reconstruction can be considered a second-generation development of image restoration.

Most super-resolution image reconstruction methods consist of three stages: registration, interpolation, and restoration. These stages can be implemented separately or simultaneously according to the reconstruction methods adopted.

The estimation of motion information is referred to as registration, which has been extensively studied in various fields of image processing. In this stage, the relative shifts in the low-resolution images compared to the reference low-resolution image are estimated in subpixel precision. Obviously, accurate subpixel motion estimation is a very important factor in the success of the super-resolution image reconstruction algorithm. Since the shifts between the low-resolution images are arbitrary, the registered high-resolution images do not always match up to a uniformly spaced high-resolution grid. Thus, nonuniform interpolation is necessary to obtain a uniformly spaced high-resolution image from a nonuniformly spaced composite of low-resolution images. Finally, image restoration can be applied to the upsampled image to remove blurring and noise factors.

28.2.2
Observation Model

The first step when comprehensively analyzing super-resolution image reconstruction methods is to formulate an observation model that relates the original high-resolution image to the observed low-resolution images. Various observation models have been proposed in the literature, and they can be broadly divided into models for still images and models for video sequences. In order to present a basic concept of super-resolution reconstruction techniques, the observation model for still images is employed, since it is rather straightforward to extend the still image model to the video sequence model.

Let **x** denote the ideal undegraded image that resulted from the sampling of a continuous scene, which was assumed to be bandlimited, at or above the Nyquist rate. This is represented in lexicographic notation by the vector $\mathbf{x} = [x_1, x_2, \ldots, x_N]^t$ of size $N(= N_1 \times N_2)$. Thus, each observed low-resolution image has a size of $M(= M_1 \times M_2)$. The downsampling factors for the horizontal and vertical directions are represented by $L_1 = N_1/M_1$ and $L_2 = N_2/M_2$, respectively. Let the number of observed low-resolution images be p. The kth low-resolution image can be denoted in lexicographic notation as $\mathbf{y}_k = [y_{k,1}, y_{k,2}, \ldots, y_{k,M}]^t$, for $k = 1, \ldots, p$. The observed low-resolution images, which are degraded by motion, blur, downsampling, and noise factors, are acquired from the high-resolution image **x**. Then, the observation model can be written as [4, 5]

$$y_{k,m} = \sum_{r=1}^{N} w_{k,m,r}(\mathbf{s}k) x_r + n_{k,m} \tag{28.1}$$

where $m = 1, 2, \ldots, M$, $w_{k,m,r}(\mathbf{s}k)$ represents the contribution (including motion, blurring, and downsampling) of the rth high-resolution pixel x_r to the mth low-resolution observed pixel of the kth frame. The vector $\mathbf{s}_k$ denotes the motion parameters of the kth observation with respect to the desired high-resolution grid and $n_{k,m}$ is the additive zero-mean Gaussian noise. The relationship between the high-resolution image and its low-resolution observations is rewritten in

28.2 Fundamentals of Super-Resolution Image Reconstruction

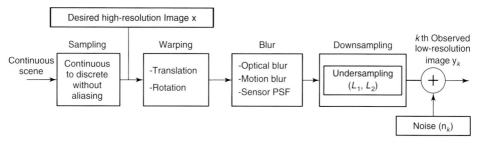

Figure 28.2 Block diagram of the observation model.

matrix–vector notation as

$$y_k = Wk(s_k)x + n_k$$
$$= DB_k M_k(s_k)x + nk, \quad \text{for } k = 1, 2, \ldots, p \quad (28.2)$$

where the matrix $\mathbf{W}_k$ and the vector $\mathbf{n}k$ are formed by the coefficients $w_{k,m,r}(s_k)$ and $n_{k,m}$ represented in Eq. (28.1), respectively. By assuming that $\mathbf{B}_k$ and $\mathbf{M}_k$ are linear, and there is no quantization error, the matrix $\mathbf{W}k$ can be decomposed into a combination of $\mathbf{D}$, $\mathbf{B}_k$, and $\mathbf{M}_k(s_k)$, which represent the downsampling, blurring, and motion operations, respectively. A block diagram for the observation model is illustrated in Figure 28.2.

28.2.3
Super-Resolution as an Inverse Problem

Super-resolution image reconstruction is a typical example of an inverse problem, because high-resolution images are estimated using observed low-resolution images. However, most cases of super-resolution image reconstruction are ill-posed problems because of an insufficient number of low-resolution images and ill-conditioned blur operators. Procedures adopted to stabilize the inversion of those ill-posed problems are called *regularization*. In this section, the constrained least squares (CLS) and Bayesian approaches are introduced.

28.2.3.1 Constrained Least Squares Approach

The observation model in Eq. (28.2) can be completely specified with estimates of the registration parameters. The CLS approach solves the inverse problem presented in Eq. (28.2) by using prior information about the solution that can be used to make the problem well posed. The CLS approach can be formulated by choosing an $\mathbf{x}$ to minimize the functional

$$F(\alpha_k, \mathbf{x}) = \left[\sum_{k=1}^{p} \alpha_k ||\mathbf{y}_k - \mathbf{W}_k \mathbf{x}||^2 + ||\mathbf{C}\mathbf{x}||^2 \right] \quad (28.3)$$

where α_k denotes the regularization parameter and the operator $\mathbf{C}$ is generally a high-pass filter. In Eq. (28.3), a priori knowledge concerning a desirable solution is represented by a smoothness constraint which suggests that most images are naturally smooth with limited high-frequency activity, and therefore it is appropriate to minimize the amount of high-pass energy in the restored image. The choice of the regularization parameter is important since it controls the balance between fidelity to the data, expressed by the term $\sum_{k=1}^{p} ||\mathbf{y}_k - \mathbf{W}_k\mathbf{x}||^2$, and smoothness of the solution, expressed by $||\mathbf{Cx}||^2$. Smaller values of α_k generally lead to smoother solutions. This is useful when the problem is underdetermined, that is, only a small number of low-resolution images are available or the fidelity of the observed data is low due to registration error and noise factors.

However, if a large number of low-resolution images are available and the amount of noise is small, large α_k values lead to a good solution. The cost functional in Eq. (28.3) is convex and differentiable with the use of a quadratic regularization term. Therefore, a unique image $\hat{x}$ can be estimated to minimize the cost functional in Eq. (28.3). One of the most basic deterministic iterative techniques considers solving the following equation:

$$\left[\sum_{k=1}^{p} \alpha_k \mathbf{W}_k^t \mathbf{W}_k + \mathbf{C}^t\mathbf{C}\right]\hat{x} = \sum_{k=1}^{p} \alpha_k \mathbf{W}k^t \mathbf{y}_k \qquad (28.4)$$

and this leads to the following iteration for $\hat{x}$:

$$\hat{x}^{n+1} = \hat{x}^n + \beta\left[\sum_{k=1}^{p} \alpha k \mathbf{W}_k^t(\mathbf{y}_k - \mathbf{W}_k\hat{x}^n) - \mathbf{C}^t\mathbf{C}\hat{x}^n\right] \qquad (28.5)$$

where β represents the convergence parameter, and $\mathbf{W}_k^t$ contains an upsampling (with interpolation) operator and a type of blur and warping operator.

A multichannel-regularized super-resolution approach has been gradually developed. Kang formulated the generalized multichannel deconvolution method including the multichannel-regularized super-resolution approach [6]. Also, a super-resolution reconstruction method obtained by minimizing a regularized cost functional was proposed by Hardie et al. [7]. These researchers defined an observation model that incorporated knowledge about the optical system and the detector array. They used an iterative gradient-based registration algorithm, and considered both gradient descent and conjugate-gradient optimization procedures to minimize the cost functional. Bose et al. [8] pointed out the important role of the regularization parameter, and proposed CLS super-resolution reconstruction to generate the optimum value of the regularization parameter, using the L-curve method [9].

28.2.3.2 Bayesian Approach

Bayesian estimation methods are used when the a posteriori probability density function (PDF) of the original image can be established. The maximum a posteriori (MAP) estimator of $\mathbf{x}$ maximizes the a posteriori PDF $P(\mathbf{x}|\mathbf{y}_k)$ with respect to $\mathbf{x}$:

$$\mathbf{x} = \arg\max P(\mathbf{x}|\mathbf{y}_1, \mathbf{y}_2, \ldots, \mathbf{y}_p) \qquad (28.6)$$

By taking the logarithmic function and applying Bayes' theorem to the conditional probability, the maximum a posteriori optimization problem can be expressed as

$$\mathbf{x} = \arg\max\{\log P(\mathbf{y}_1, \mathbf{y}_2, \ldots, \mathbf{y}_p|\mathbf{x}) + \log P(\mathbf{x})\} \qquad (28.7)$$

Here, both the a priori image model $P(\mathbf{x})$ and the conditional density $P(\mathbf{y}_1, \mathbf{y}_2, \ldots, \mathbf{y}_p|\mathbf{x})$ are defined by a priori knowledge concerning the high-resolution image $\mathbf{x}$ and the statistical information of the noise factor. Bayesian estimation distinguishes between possible solutions by utilizing the a priori image model, and the use of an edge-preserving image prior model is a major advantage of the Bayesian framework. In general, the image is assumed to be globally smooth, and is incorporated into the estimation problem through a Gaussian prior. In this case, the regularization term $\log P(\mathbf{x})$ (a potential function in the (Markov random field) (MRF) approach) takes the form of a quadratic, and it encounters serious difficulties in dealing with edges, since there is no precise control over the smoothness constraint. In order to overcome this problem, the MRF priors, which provide a powerful method for image prior modeling, are often used. Using the MRF priors, $P(\mathbf{x})$ is described by a Gibbs prior that represents piecewise smooth data, with the probability density defined as

$$P(\mathbf{x}=x) = \frac{1}{Z}\exp\{-U(\mathbf{x})\} = \frac{1}{Z}\exp\left\{-\sum_{c\in S}\varphi_c(\mathbf{x})\right\} \tag{28.8}$$

where Z is simply a normalizing constant, $U(\mathbf{x})$ represents an energy function, and $\varphi_c(\mathbf{x})$ represents a potential function that depends only on the values of pixels that are located within clique c. By defining $\varphi_c(\mathbf{x})$ as a function of the derivative of the image (the clique of a specified neighborhood system may determine the order of the difference), $U(\mathbf{x})$ is used to measure the cost due to irregularities of the solution. With the Gaussian prior, the potential function takes the quadratic form $\varphi_c(\mathbf{x}) = (D^{(n)}\mathbf{x})^2$ where $D^{(n)}$ is an nth order difference. Though the quadratic potential function made the algorithm linear, it penalizes the high-frequency components severely. In other words, the more irregular $D^{(n)}\mathbf{x}$ is, the larger $|D^{(n+1)}\mathbf{x}|$ is, and this results in larger potential which is contributed to $U(\mathbf{x})$. As a result, the solution became oversmoothed. However, if a potential function is modeled such that it penalizes the large difference in $\mathbf{x}$ to a lesser degree, an edge-preserving high-resolution image can be obtained.

In general, if the error between frames is assumed to be independent, and the noise factor is assumed to have independent identically distributed (i.i.d) zero-mean Gaussian distribution, the optimization problem can be expressed more compactly as

$$\hat{\mathbf{x}} = \arg\min\left[\sum_{k=1}^{p}\alpha_k||\mathbf{y}_k - \mathbf{H}_k\hat{\mathbf{x}}||^2 + \sum_{c\in S}\varphi_c(\mathbf{x})\right] \tag{28.9}$$

where α_k represents the regularization parameter. Finally, the estimate defined in Eq. (28.3) is equal to a MAP estimate using the Gaussian prior in Eq. (28.9).

A (maximum likelihood) (ML) estimation has also been applied to the super-resolution reconstruction. The ML estimation is a special case of MAP estimation with no prior term. If the magnification ratio is very low (i.e., if the pixels are overdetermined), then the effect of a priori knowledge is insignificant, and the MAP estimate becomes the same as the ML estimate. However, if the

Figure 28.3 Partially magnified "shop" images of the MAP super-resolution reconstruction results (a) with small α_k, and (b) with large α_k.

magnification ratio is high (i.e., if the pixels are underdetermined), then the effect of a priori knowledge gives a reasonable estimate of the high-resolution image. Owing to the ill-posed nature of super-resolution inverse problems, MAP estimation is usually used in preference to ML.

The simulation results of the MAP method are shown in Figure 28.3. In these simulations, the original 256×256 "shop" image was shifted with one of the subpixel shifts $\{(0,0),(0,0.5),(0.5,0),(0.5,0.5)\}$ and decimated by a factor of 2 in both the horizontal and vertical directions. In this experiment, only sensor blur was considered and a 20 dB Gaussian noise was added to these low-resolution images. MAP super-resolution results using a Gaussian prior with small and large regularization parameters appear in Figure 28.3a and b, respectively. In fact, these estimates can be considered as a part of CLS reconstruction.

Tom and Katsaggelos [10] proposed the ML super-resolution image estimation problem to estimate the subpixel shifts, the noise variances of each image, and the high-resolution image simultaneously. The proposed ML estimation problem was solved by the (expectation-maximization) (EM) algorithm. Super-resolution reconstruction from a low-resolution video sequence using the MAP technique was proposed by Schultz and Stevenson [11]. These researchers proposed a discontinuity-preserving MAP reconstruction method using the Huber–Markov Gibbs prior model, which resulted in a constrained optimization problem with a unique minimum value. Here, they used the modified hierarchical (block matching algorithm) (BMA) to estimate the subpixel displacement vectors. They also considered independent object motion and inaccurate motion estimates, which were modeled by Gaussian noise factors. A MAP framework for the joint estimation of image registration parameters and the high-resolution image was presented by Hardie et al. [12]. The registration parameters, horizontal and vertical shifts, in this case, were iteratively updated along with the high-resolution image in a cyclic optimization procedure.

Robustness and flexibility in modeling noise characteristics and a priori knowledge about the solution are some major advantages of the stochastic super-resolution approach. Assuming that the noise process is white Gaussian, a MAP estimation with convex *priors* can be used to ensure the uniqueness of the solution. Therefore, efficient gradient descent methods can also be used to estimate the high-resolution image. It is also possible to estimate the motion information and high-resolution images simultaneously.

28.2.4
The Frequency Domain Interpretation

The frequency domain approach is helpful to comprehend the effect of the aliasing that exists in low-resolution images to reconstruct high-resolution images [13]. The frequency domain approach is based on the following three principles: (i) the shifting property of the Fourier transform, (ii) the aliasing relationship between the (continuous Fourier transform) (CFT) of an original high-resolution image and the (discrete Fourier transform) (DFT) of observed low-resolution images, (iii) and the assumption that an original high-resolution image is bandlimited. On the basis of these properties, the system equation relating the aliased DFT coefficients of the observed low-resolution images to a sample of the CFT of an unknown image can be formulated. For example, we assume that there are two 1D low-resolution signals that are sampled below the Nyquist sampling rate. From the above three principles, the aliased low-resolution signals can be decomposed into the unaliased high-resolution signal (as shown in Figure 28.4).

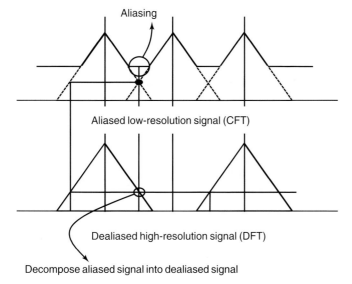

Figure 28.4 Aliasing relationship between low-resolution images and high-resolution images.

Let $X(w_1, w_2)$ denote the CFT of a continuous high-resolution image $x(t_1, t_2)$. The global translations (the only motions considered in the frequency domain approach) yielded the kth shifted image of $x_k(t_1, t_2) = x(t_1 + \delta_{k1}, t_2 + \delta_{k2})$, where δ_{k1} and δ_{k2} are arbitrary but known values, and $k = 1, 2, \ldots, p$. By shifting property of the CFT, the CFT of the shifted image, $X_k(w_1, w_2)$ can be written as

$$X_k(w_1, w_2) = \exp[i2\pi(\delta_{k1} w_1 + \delta_{k2} w_2)] X(w_1, w_2) \tag{28.10}$$

The shifted image $x_k(t_1, t_2)$ is sampled with the sampling periods T_1 and T_2 to generate the observed low-resolution image $y_k[n_1, n_2]$. From the aliasing relationship and the assumption of bandlimitedness of the continuous high-resolution image $X(w_1, w_2)$, the relationship between the CFT of the high-resolution image and the DFT of the kth observed low-resolution image can be written as

$$Y_k[\Omega_1, \Omega_2] = \frac{1}{T_1 T_2} \sum_{n_1 = -L_1}^{L_1 - 1} \sum_{n_2 = -L_2}^{L_2 - 1} X_k \left(\frac{2\pi}{T_1} \left(\frac{\Omega_1}{N_1} + n_1 \right), \frac{2\pi}{T_2} \left(\frac{\Omega_2}{N_2} + n_2 \right) \right) \tag{28.11}$$

By using lexicographic ordering for the indices n_1, n_2 on the right-hand side and k on the left-hand side, a matrix vector form is obtained as

$$\mathbf{y} = \Phi \mathbf{x} \tag{28.12}$$

where $\mathbf{y}$ represents a $p \times 1$ column vector with the kth element of the DFT coefficients of $y_k[n_1, n_2]$, $\mathbf{x}$ represents a $4L_1 L_2 \times 1$ column vector with the samples of the unknown CFT of $x(t_1, t_2)$, and Φ represents a $p \times 4L_1 L_2$ matrix that relates the DFT of the observed low-resolution images to samples of the continuous high-resolution image.

Therefore, the reconstruction of a desired high-resolution image requires us to determine Φ and then solve this inverse problem. The frequency domain approach can be extended for blurred and noisy images using a weighted least squares formulation. A (discrete cosine transform) (DCT)-based method was proposed by Rhee and Kang [14]. These researchers reduced memory requirements and computational costs by using DCT instead of DFT. They also applied multichannel-adaptive regularization parameters to overcome ill-posedness such as underdetermined cases or insufficient motion information cases.

A major advantage of the frequency domain approach is the simplicity of the theory. That is, the relationship between low-resolution images and high-resolution images can be clearly demonstrated in the frequency domain. The frequency method is also convenient for parallel implementation that is capable of reducing hardware complexity. However, the observation model is restricted to only global translational motion and linear spatially invariant blurring. Owing to the lack of data correlation in the frequency domain, it was also difficult to apply the spatial domain a priori knowledge for regularization.

28.3
Super-Resolution Image Reconstruction considering Inaccurate Subpixel Motion Estimation

Registration is very important for the success of super-resolution image reconstruction. Therefore, accurate registration methods, based on robust motion models (including multiple object motion, occlusions, and transparency), are needed. However, when the performance of the registration algorithms cannot be ensured (for example, in certain environments), the errors caused by inaccurate registration should be considered in the reconstruction procedure. Although most super-resolution algorithms implicitly model registration errors as an additive Gaussian noise, more sophisticated models for these errors are needed.

Ng et al. [15] considered the error generated by inaccurate registration in the system matrix $\mathbf{W}_k$, and proposed the total least squares method to minimize this error. This method was useful for improving the accuracy of the solution when errors existed not only in the recording process but also in the measurement matrix. Ng and Bose analyzed displacement errors on the convergence rate of the iterations used to solve the transform-based preconditioned system [16]. Another approach to minimize the effect of the registration error is based on channel-adaptive regularization [17, 18]. The basic concept of this approach is that low-resolution images with a large amount of registration error should contribute less to the estimate of the high-resolution image than reliable low-resolution images. The misregistration error is generally modeled as Gaussian noise which has a different variance according to the registration axes, and channel-adaptive regularization is performed with a directional smoothing constraint. Kang et al. [17] extended these works on the basis of a set-theoretic approach. These researchers proposed the regularization functional that was performed on data consistency terms.

28.3.1
Analysis of the Misregistration Error

In practical super-resolution image reconstruction, accurate estimation of $\mathbf{M}_k(\mathbf{s}_k)$ in the observation model (Eq. (28.2)) is very difficult. On the basis of empirical evidence, the misregistration error has previously been modeled as a zero-mean additive Gaussian process, resulting in reasonable high-resolution estimates. This error can be numerically analyzed and the analysis results can be used later to obtain improved high-resolution estimates.

The misregistration error is analyzed in a continuous domain, since it is generally difficult to represent subpixel shifts in discrete domains. Let a continuous image of size $N(= N_1 \times N_2)$ be denoted by $\tilde{x}(h, v)$ and its shifted version by $\tilde{x}(h - \delta_h, v - \delta_v)$. Here, h and v represent the horizontal and vertical directions, respectively, and δ_h and δ_v represent the horizontal and vertical motion estimation errors, respectively. $\tilde{x}(h, v)$ represents the registered image to a reference image and $\tilde{x}(h - \delta_h, v - \delta_v)$ represents the shifted or perturbed image due to the registration error. In other words, δ_h and δ_v mean a global translational registration error. With this model, $|\delta_h| \leq 0.5$ and $|\delta_v| \leq 0.5$, which are utilized later in the analysis. Thus, the noise factor caused by the misregistration error (NME) can be represented by $n(h, v, \delta_h, \delta_v) = \tilde{x}(h, v) - \tilde{x}(h - \delta_h, v - \delta_v)$. The mean of $n(h, v, \delta_h, \delta_v)$ is equal to

$$\overline{n(h, v, \delta_h, \delta_v)} = \frac{1}{N} \int_0^{N_1} \int_0^{N_2} \left(\tilde{x}(h, v) - \tilde{x}(h - \delta_h, v - \delta_v) \right) dh\,dv \qquad (28.13)$$

In order to obtain a computable approximation of Eq. (28.13), with the Taylor series expansion of $\tilde{x}(h - \delta_h, v - \delta_v)$, the equation can be approximated as

$$\overline{n(h, v, \delta_h, \delta_v)} \approx \frac{1}{N} \int_0^{N_1} \int_0^{N_2} \left(\delta_h \tilde{x}_h(h, v) + \delta_v \tilde{x}_v(h, v) \right) dh\,dv$$

$$= \delta_h \overline{\tilde{x}_h(h, v)} + \delta_v \overline{\tilde{x}_v(h, v)} \qquad (28.14)$$

where $\overline{\tilde{x}_h(h, v)}$ and $\overline{\tilde{x}_v(h, v)}$ represent the mean values of the first differentials in the horizontal and vertical directions, respectively. The approximate equality in Eq. (28.14) is achieved by ignoring the higher order terms, that is, above the second order of the Taylor series expansion. If the image contains enough intensity variability and if it is of adequate size, $\overline{\tilde{x}_h(h, v)} = \overline{\tilde{x}_v(h, v)} \approx 0$ is generally a reasonable assumption. Then, Eq. (28.14) can be written as

$$\overline{n(h, v, \delta_h, \delta_v)} \approx 0 \qquad (28.15)$$

On the basis of $\overline{n(h, v, \delta_h, \delta_v)}$ in Eq. (28.15), the variance of $n(h, v, \delta_h, \delta_v)$ can now be evaluated as

$$\text{var}\left(n(h, v, \delta_h, \delta_v)\right) \approx \delta_h^2 \overline{\tilde{x}_h^2(h, v)} + \delta_v^2 \overline{\tilde{x}_v^2(h, v)} \qquad (28.16)$$

where $\overline{\tilde{x}_h^2(h, v)}$ and $\overline{\tilde{x}_v^2(h, v)}$ represent the mean values of the squares of the first differentials in the horizontal and vertical directions, respectively. Utilizing the "resolution chart" image shown in Figure 28.5, an example of the distribution of the horizontal misregistration error for three different values of δ_h is shown in Figure 28.6a. The distribution can be modeled by a zero-mean generalized Gaussian or a Laplacian distribution. The relationship between the standard deviation of the NME and the horizontal motion estimation error δ_h is shown in Figure 28.6b. This distribution can be modeled by a zero-mean generalized Gaussian or Laplacian distribution. As expected, according to Eq. (28.16), this relationship is linear.

28.3.2
Multichannel-Regularized Super-Resolution Image Reconstruction Algorithm

In general, super-resolution image reconstruction is an ill-posed problem because of an insufficient number of low-resolution images, ill-conditioned blur operators, and inaccurate motion information. Various regularized multichannel techniques have been proposed to address these problems. The CLS approach has been widely used in this task. In multichannel CLS using Eq. (28.2), the functional to be minimized can be defined by

$$F(\alpha, \mathbf{x}) = \sum_{k=1}^{p} \alpha_k \|\mathbf{y}_k - \mathbf{W}_k \mathbf{x}\|^2 + \|\mathbf{C}\mathbf{x}\|^2 \qquad (28.17)$$

28.3 Super-Resolution Image Reconstruction considering Inaccurate Subpixel Motion Estimation

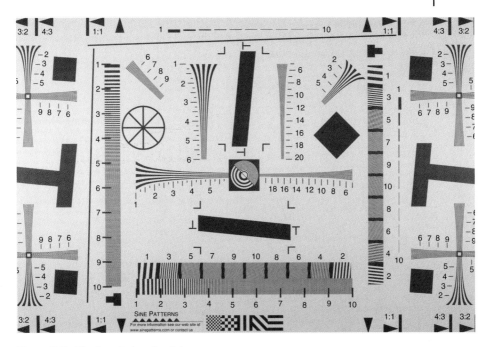

Figure 28.5 The "resolution chart" image.

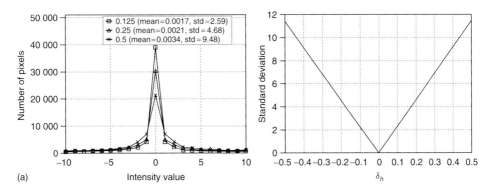

Figure 28.6 Analysis of the misregistration error for the "resolution chart" image: (a) histograms of misregistration error for $\delta_h \in \{0.125, 0.25, 0.5\}$, (b) standard deviation of the NME versus δ_h.

where α_k denotes the regularization parameter and $\mathbf{C}$ is a high-pass operator. The choice of the regularization parameter is important since it controls the balance between fidelity to the data, expressed by the term $\|\mathbf{y}k - \mathbf{W}k\mathbf{x}\|^2$, and the smoothness of the solution, expressed by the term $\|\mathbf{C}\mathbf{x}\|^2$ (see also Chapter 24). Since the motion error is typically different for each low-resolution image (also referred to as a *channel*) and the observation noise is typically different per channel, a different regularization parameter should be used for each channel.

Thus, Eq. (28.17) is generalized as

$$F(\alpha k(\mathbf{x}), \mathbf{x}) = \sum_{k=1}^{p} \alpha k(\mathbf{x}) \|\mathbf{y}_k - \mathbf{W}_k \mathbf{x}\|^2 + \gamma \|\mathbf{C}\mathbf{x}\|^2 \tag{28.18}$$

where $\alpha k(\mathbf{x})$ represents the regularization functional for the kth channel and γ represents a normalization parameter. Originally, the balance between fidelity and smoothness is fully controlled by $\alpha k(\mathbf{x})$. However, since the range of $\alpha_k(\mathbf{x})$ is limited, γ is necessary to control the balance between fidelity and smoothness totally.

With set-theoretic regularization approaches, the regularization functional $\alpha k(\mathbf{x})$ in each channel is generally proportional to $\|\mathbf{C}\mathbf{x}\|^2$ and inversely proportional to $\|\mathbf{y}_k - \mathbf{W} k\mathbf{x}\|^2$. In other words,

$$\alpha_k(\mathbf{x}) = \frac{\|\mathbf{C}\mathbf{x}\|^2}{\|\mathbf{y}_k - \mathbf{W}_k \mathbf{x}\|^2 + \delta k} \tag{28.19}$$

where δ_k represents a parameter the prevents the denominator from becoming zero. This regularization functional is reasonable because it controls the trade-off between fidelity to the data $\|\mathbf{y}_k - \mathbf{W}_k \mathbf{x}\|^2$ and the smoothness of the solution $\|\mathbf{C}\mathbf{x}\|^2$. However, if one channel has no motion estimation error, the residual term $\|\mathbf{y}_k - \mathbf{W}_k \mathbf{x}\|^2$ decreases rapidly as the iteration progresses since one channel must be chosen as the reference frame for motion estimation, and at least one channel shows no motion estimation error in most cases. Often, because of the rapid decrease in the residual term $\|\mathbf{y}_k - \mathbf{W}_k \mathbf{x}\|^2$, $\alpha k(\mathbf{x})$ becomes very large; thus this channel dominates the quality of the high-resolution estimate by reducing the contribution of the remaining channels. In order to address this problem and also to take the cross-channel correlations into account, two regularization functionals were used. However, this approach did not result in a functional with a global minimum, which made the final estimate depend on the initial condition. For example, if the initial condition is an image with constant values (i.e., a white or black image), then $\|\mathbf{C}\mathbf{x}\|^2$ is zero, and therefore $\alpha k(\mathbf{x}) = 0$. The regularization functional can usually be determined by considering the misregistration error in the channel and its relationship to the rest of remaining channels. Let the motion estimation error in the kth channel be denoted by e_k. The regularization functional should be inversely proportional to e_k. The use of e_k^2 instead of e_k is more appropriate, since $\alpha k(\mathbf{x})$ should always be positive. Therefore, the inverse proportionality between $\alpha k(\mathbf{x})$ and e_k^2 is set as

$$\alpha_k(\mathbf{x}) \propto \exp(-e_k^2) \tag{28.20}$$

As shown in Section 28.3.2, the variance of the NME caused by e_k is proportional to the product of e_k^2 and the high-frequency energy of the image. Equivalently, e_k^2 is proportional to the variance of the NME and inversely proportional to the high-frequency energy of the image. Therefore, Eq. (28.20) can be rewritten as

$$\alpha_k(\mathbf{x}) \propto \exp\left(-\frac{\|\mathbf{y}_k - \mathbf{W}_k \mathbf{x}\|^2}{\|\mathbf{C}\mathbf{x}\|^2}\right) \tag{28.21}$$

28.3 Super-Resolution Image Reconstruction considering Inaccurate Subpixel Motion Estimation

Since $\|\mathbf{C}x\|^2$ does not depend on the channel with index k, it can be changed to a parameter P_G, which is chosen appropriately so that the functional to be minimized is always convex and has a global minimum.

Therefore, the final form of the regularization functional $\alpha k(\mathbf{x})$ becomes

$$\alpha_k(\mathbf{x}) = \exp\left(-\frac{\|\mathbf{y}_k - \mathbf{W}_k\mathbf{x}\|^2}{P_G}\right) \tag{28.22}$$

Guaranteeing a global minimum requires the convexity of the proposed functional $F(\mathbf{x})$ to be minimized in Eq. (28.18). In general, a functional $f(\mathbf{x})$ is said to be convex if

$$f(a\mathbf{x}_1 + (1-a)\mathbf{x}_2) \leq af(\mathbf{x}_1) + (1-a)f(\mathbf{x}_2) \tag{28.23}$$

for any $\mathbf{x}_1, \mathbf{x}_2 \subset R^n$ and $0 \leq a \leq 1$. In the following, without lack of generality, the number of channels can be assumed as equal to one, or $p = 1$. The proposed minimization functional is considerable to be convex if the condition:

$$P_G \geq 3 \times \max(\|\mathbf{y} - \mathbf{W}x_1\|^2, \|\mathbf{y} - \mathbf{W}x_2\|^2) \tag{28.24}$$

is satisfied, where $\max(\|\mathbf{y} - \mathbf{W}x_1\|^2, \|\mathbf{y} - \mathbf{W}x_2\|^2)$ represents the maximum value of $\|\mathbf{y} - \mathbf{W}x_1\|^2$ and $\|\mathbf{y} - \mathbf{W}x_2\|^2$. Since the single-channel ($p = 1$) minimization functional is convex under the condition in Eq. (28.24), the sum is also convex. To illustrate the convexity of the proposed minimization functional with a simple example, a scalar image (which has the same value in a whole image) instead of a vector image (which has a different value for every pixel) is used when the condition $P_G \geq (3 \times \max(\|\mathbf{y}_1 - \mathbf{W}_1\mathbf{x}\|^2, \ldots, \|\mathbf{y}_p - \mathbf{W}p\mathbf{x}\|^2))$ is satisfied. Figure 28.7 shows the convex form of the proposed minimization functional $F(\alpha(\mathbf{x}), \mathbf{x})$ as a function of a scalar $\mathbf{x}$. A gradient descent method can be used for minimizing the

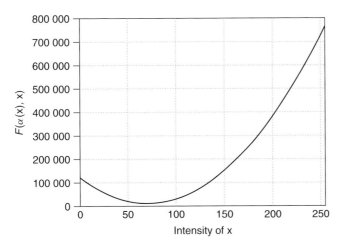

Figure 28.7 Shape of the minimization functional $F(\alpha(\mathbf{x}), \mathbf{x}))$ as a function of x (scalar case).

functional in Eq. (28.18), and owing to this convexity and the sufficient condition of convergence of the gradient descent method, a global minimum is obtained at convergence.

28.3.3
Experimental Results

In this section, performance of super-resolution image reconstruction considering inaccurate motion information is explained using a combination of synthetically generated and nonsynthetic image sequences. In all the experiments, the gradient descent optimization was used for reconstructing high-resolution images. P_G was chosen according to the sufficient condition for a global minimum to be equal to $(3 \times \max(\|\mathbf{y}_1 - \mathbf{W}_1\mathbf{x}\|^2, \ldots, \|\mathbf{y}p - \mathbf{W}_p\mathbf{x}\|^2))$. The criterion $\|\mathbf{x}^{n+1} - \mathbf{x}^n\|^2/\|\mathbf{x}^n\|^2 \leq 10^{-6}$ was used for terminating the iteration. In order to quantitatively measure the improvement in the estimated images, the (peak signal-to-noise ratio) (PSNR) and the (modified peak signal-to-noise ratio) (MPSNR) were used. The MPSNR is defined as the PSNR in the edge regions, since these mainly determined the quality of the high-resolution reconstructed images. The edge regions were detected by thresholding the magnitude of the gradient of the original image. Two other reconstruction algorithms were used for comparison with the proposed algorithm: CM1, which is the CLS-based super-resolution reconstruction algorithm described in Ref. [7], and CM2 is the reconstruction algorithm described in Ref. [17] with an irrational regularization functional.

In the first experiment, four low-resolution images were synthetically generated from the high-resolution "resolution chart" image. The spatial resolution of this high-resolution image was 512×512 pixels, and it was translated with one of the subpixel shifts $\{(0, 0), (0, 0.5), (0.5, 0), (0.5, 0.5)\}$, blurred, and decimated by a factor of 2 in both the horizontal and vertical directions.

The (point spread function) (PSF) of the imaging sensor was determined from the following equation:

$$B[h - \Delta_h, v - \Delta_v] = \begin{cases} \frac{1}{P_n}\left(1 - \frac{\sqrt{\Delta_h^2 + \Delta_v^2}}{ms}\right) & \text{if } \sqrt{\Delta_v^2 + \Delta_h^2} \leq ms, \\ 0 & \text{otherwise} \end{cases} \quad (28.25)$$

where $[h, v]$ represents a central point of the PSF kernel. Δ_h and Δ_v represent shifts in the horizontal and vertical directions, respectively, and P_n and ms are a normalizing parameter and the mask size of the blurring function, respectively. Although the motion information of the simulated low-resolution images is known exactly, it is assumed that the motion estimation is inaccurate, as demonstrated in the four cases shown in Table 28.1. The partially magnified reconstructed high-resolution images are shown in Figure 28.8. These results were obtained using the parameter set listed as "Case2" in Table 28.1. Figure 28.8a shows the enlarged image of one of the low-resolution observations obtained by bicubic interpolation. Figure 28.8b and c shows the high-resolution images reconstructed

28.3 Super-Resolution Image Reconstruction considering Inaccurate Subpixel Motion Estimation

Table 28.1 Accurate and Inaccurate Subpixel Motion Information $(\delta_{h,k}, \delta_{v,k})$, for $k = 1, 2, 3, 4$.

Motion	Real	Case1	Case2	Case3	Case4
$(\delta_{h,1}, \delta_{v,1})$	(0.0,0.0)	(0.0,0.0)	(0.0,0.0)	(0.2,0.2)	(0.0,0.0)
$(\delta_{h,2}, \delta_{v,2})$	(0.0,0.5)	(0.1,0.2)	(0.1,0.4)	(0.0,0.2)	(−0.1,0.2)
$(\delta_{h,3}, \delta_{v,3})$	(0.5,0.0)	(0.4,0.1)	(0.4,0.2)	(0.5,0.0)	(0.3,−0.1)
$(\delta_{h,4}, \delta_{v,4})$	(0.5,0.5)	(0.5,0.5)	(0.3,0.3)	(0.4,0.4)	(0.4,0.3)

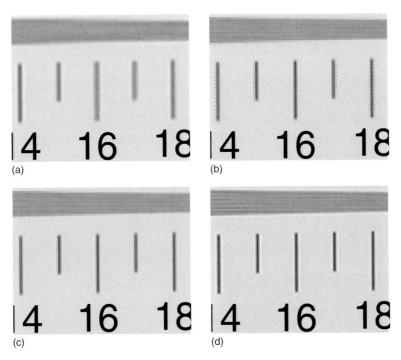

Figure 28.8 Partially magnified images of the reconstructed "resolution chart" by (a) bicubic interpolation, (b) CM1, (c) CM2, and (d) PM.

by the CM1 and CM2 algorithms, respectively. Finally, the result obtained with the proposed algorithm is shown in Figure 28.8d. The bicubic interpolated image is the poorest among these results, since only one low-resolution image is considered in the reconstruction process. Compared to this method, the CM1 result improved. However, since CM1 does not generally consider inaccurate motion information in each channel, the image in Figure 28.8b shows visual artifacts like white or black dots near the edges. These visual artifacts were reduced by increasing the value of the regularization parameter. However, this caused oversmoothing as well as a loss of important high-frequency components. To obtain a better result with $CM1$, many trial and error tests or additional information about the motion is needed. However, the $CM2$ and PM results provided better solutions in that the high frequency of the reconstructed image was conserved while suppressing the noise caused by the inaccurate motion information. The PSNR and MPSNR values of the reconstructed high-resolution image for the four cases in Table 28.1 are presented in Table 28.2. This table shows that the algorithm that considered inaccurate motion information outperformed the conventional methods. The proposed algorithm improved the PSNR values by 2.0 dB over CM1 and 1.0 dB over CM2, and improved the MPSNR values by 3.0 dB over CM1 and 2.0 dB over CM2. Another experiment using images captured by real imaging systems was performed. The images shifted with subpixel accuracy at the same scene can be captured using two different methods: by using multiple cameras (as shown in Figure 28.9a) and by obtaining the sequence of images from one camera with several captures. In this

Table 28.2 The PSNR and MPSNR of the Conventional Reconstruction Algorithms and the Proposed Algorithm for the Subpixel Motion in Table 28.1.

	PSNR(dB)				MPSNR(dB)			
	Case1	Case2	Case3	Case4	Case1	Case2	Case3	Case4
Bicubic	24.271	24.271	24.271	24.271	17.210	17.210	17.210	17.210
CM1	28.542	27.846	27.646	27.321	20.321	19.865	19.321	19.987
CM2	29.865	29.145	29.084	28.854	21.654	21.312	20.541	20.135
PM	30.746	30.612	29.854	29.479	22.326	22.154	21.896	21.541

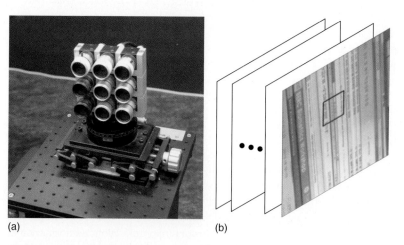

(a)   (b)

Figure 28.9 A multicamera system: (a) with nine 640 × 480 resolution cameras and (b) an example of low-resolution images of common region.

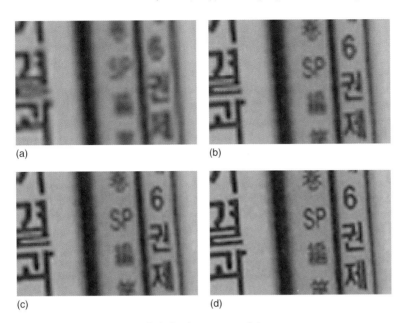

Figure 28.10 Partially magnified "books" images of the reconstructed images by (a) bicubic interpolation, (b) CM1, (c) CM2, and (d) PM.

experiment, nine real low-resolution images, shown in Figure 28.9b, captured with multicamera system were used to perform nonsynthetic resolution enhancement, and the region bordered by the bold square in Figure 28.9b is magnified in the result images. The size of all nine low-resolution images was 640 × 480 pixels. The integer-level motions were estimated with the hierarchical block-matching and the low-resolution observations were registered and cropped to 256 × 256 pixels, since all observations must have the common region without integer-level shift. The relative subpixel motions between the low-resolution observations were estimated with the gradient-based method at the subpixel level [7]. It is assumed that the blur is given by Eq. (28.25). The results of bicubic interpolation, CM1, CM2, and PM are shown in Figure 28.10. As in the first experiment, the bicubic interpolated image proved to be the poorest one. The CM1 result in Figure 28.10b contained visual artifacts like white or black dots near the edges. However, reconstruction when considering the misregistration error produced improved image details while suppressing the noise factors caused by the inaccurate motion information.

28.4
Development and Applications of Super-Resolution Image Reconstruction

Super-resolution image reconstruction has proven to be useful in many practical cases where multiple frames of the same scene can be obtained, including medical imaging, satellite imaging, and video applications.

One application has been to reconstruct a high-quality digital image from low-resolution images obtained with an inexpensive low-resolution camera/camcorder for printing or frame-freeze purposes.

Synthetic zooming of regions of interest is another important application in surveillance, forensic, scientific, medical, and satellite imaging. For surveillance or forensic purposes, it is often necessary to magnify objects in scenes such as the face of a criminal or the license plate of a car. The super-resolution technique is also useful in medical imaging since the acquisition of multiple images is possible when the spatial resolution is limited. In satellite imaging applications, such as remote sensing, several images of the same area are usually provided, and the super-resolution technique can then be used to improve the spatial resolution of the target. Super-resolution image reconstruction originally referred to the methods of overcoming the limited spatial resolution of low-resolution imaging systems. But the meaning of the term has become wider, and it now refers to techniques that are used to overcome not only limited spatial resolution but also all the physical limitations of imaging sensors in the signal-processing-based approach. For example, in order to reconstruct color information as well as enhance the spatial resolution of a single color imaging sensor, color interpolation schemes based on super-resolution image reconstruction have been suggested. In many image acquisition processes, observations suffer from limitations of both spatial resolution and dynamic range. For this reason, signal-processing techniques to simultaneously enhance the spatial resolution and dynamic range have been developed. In this section, recent applications of super-resolution image reconstruction such as super-resolution for color imaging systems and video systems are introduced.

28.4.1
Super-Resolution for Color Imaging Systems

When an image is acquired by a color imaging system, it is degraded by spatial downsampling, blurring, and additive noise factors. In addition to these effects, most commercial digital cameras use a (color filter array) (CFA) to sample one of the color components at each pixel location. This process downsamples the spectral component of a color image and the additional color restoration process is then used to obtain the color image. Therefore, the super-resolution of color imaging systems is aimed at restoring the spatial and the spectral resolution of color images. For color imaging systems, the image formation process in Eq. (28.2) can be modified as

$$\mathbf{y}_k^i = \mathbf{W}_k^i(\mathbf{s}_k)\mathbf{x}^i + \mathbf{n}_k^i$$
$$= \mathbf{D}^i \mathbf{B}_k \mathbf{M}_k(\mathbf{s}_k)\mathbf{x}^i + \mathbf{n}_k^i, \qquad (28.26)$$

for $k = 1, 2, \ldots, p$ and $i = R, G, B$. The vectors $\mathbf{x}^i$ and $\mathbf{y}_k^i$ represent the ith band of the high-resolution color frame and the kth low-resolution frame, respectively. By considering the red, green, and blue channels of the color image, the high-resolution and kth low-resolution frames in Eq. (28.26) were represented as $\mathbf{x} = \{\mathbf{x}^R, \mathbf{x}^G, \mathbf{x}^B\}^t$

and $\mathbf{y}_k = \{\mathbf{y}_k^R, \mathbf{y}_k^G, \mathbf{y}_k^B\}^t$, respectively. The downsampling operator $\mathbf{D}^i$ included both the color filtering and CFA downsampling operations. Following the formulation model of Eq. (28.26), the problem was to estimate the high-resolution image $\mathbf{x}$ from the observed data $\mathbf{y}_k$. Vega et al. proposed a single-frame super-resolution technique that considered the blur and only Bayer CFA sampling without the spatial downsampling process [19]. By considering four downsampled images that were obtained from a single-frame image, these researchers formulated a demosaicing problem as a reconstruction of a high-resolution color image from the given four single-channel low-resolution images.

The high-resolution color image was reconstructed by utilizing the Bayesian paradigm. In general, there are two ways to reconstruct high-resolution color images from multiple low-resolution frames. One is to construct low-resolution color images by using demosaicing methods and then enlarge the spatial resolution by using multichannel super-resolution techniques. The other is to reconstruct high-resolution color images within a single process. Farsiu et al. [20] proposed a "multiframe demosaicing and super-resolution method" as a single process for high-resolution color images. In this process, the multichannel super-resolution technique was applied to the demosaicing problem. On the basis of the MAP estimator, Farsiu et al. obtained the solution using the following equation:

$$\hat{\mathbf{x}} = \arg \min_{\mathbf{x}}[\rho(\mathbf{y}_1, \ldots, \mathbf{y}_p, \mathbf{W}_1(\mathbf{s}_1)\mathbf{x}, \ldots, \mathbf{W}_p(\mathbf{s}_p)\mathbf{x}) + \alpha \Upsilon(\mathbf{x})] \quad (28.27)$$

where $\rho(\cdot)$ is the data fidelity term used to enforce the similarities between the observed low-resolution images and the high-resolution estimate, and Υ is the regularization cost function that imposes a penalty on the unknown $\mathbf{x}$. The regularization parameter α weights the first term against the second term. By considering the heavy tailed distribution of the noise in the presence of the motion error, the researchers mentioned above modeled the noise distribution as Laplacian. Therefore, the data fidelity term $J_0(\mathbf{x})$ was alternatively used based on the L_1 norm:

$$J_0(\mathbf{x}) = \rho(\mathbf{y}_1, \ldots, \mathbf{y}_p, \mathbf{W}_1(\mathbf{s}_1)\mathbf{x}, \ldots, \mathbf{W}_p(\mathbf{s}_p)\mathbf{x})$$
$$= \sum_{i=R,G,B} \sum_{k=1}^{p} \|\mathbf{W}_k^i(\mathbf{s}_k)\mathbf{x}^i - \mathbf{y}_k^i\|_1 \quad (28.28)$$

For the regularization cost function, they introduced a multiple cost function with constraints on the luminance and chrominance values of the high-resolution image. When the high-resolution color image $\mathbf{x}$ was decomposed into the luminance component $\mathbf{x}^l$ and the chrominance components $\mathbf{x}^{c1}$ and $\mathbf{x}^{c2}$, each component was constrained according to the characteristics of human eyes. Since the (human visual system) (HVS) is more sensitive to details in the luminance component of an image, the luminance component often contains sharp edges. However, the chrominance components are generally penalized as smooth since the HVS is less sensitive to the resolution of these bands. By applying a bilateral filter [21] to the constraints of the luminance components, the similarity between the pixels of both the intensity and the spatial location is exploited. The spatial luminance penalty term $J_1(\mathbf{x})$ is given by

$$J_1(\mathbf{x}) = \sum_{l=-P}^{P} \sum_{m=-P}^{P} \alpha^{|m|+|l|} \|\mathbf{x}^l - \mathbf{s}_h(l)\mathbf{s}_v(m)\mathbf{x}^l\|_1 \qquad (28.29)$$

where $0 < \alpha < 1$, parameter P defines the size of the bilateral kernel, and $\mathbf{s}_h(l)$ and $\mathbf{s}_v(m)$ represent the shifting operators corresponding to shifting the image by l and m pixels in the horizontal and vertical directions, respectively. Also, the spatial chrominance penalty term $J_2(\mathbf{x})$ is

$$J_2(\mathbf{x}) = \|\mathbf{C}\mathbf{x}^{c1}\|_2^2 + \|\mathbf{C}\mathbf{x}^{c2}\|_2^2 \qquad (28.30)$$

where $\mathbf{C}$ represents a smoothness constraint on the chrominance channel. In addition to these constraints, Farsiu et al. considered the correlation between the red, green, and blue channels of high-resolution images. In other words, they penalized the mismatch between the locations or orientations of the edges across the color bands. To reduce these mismatches, they minimized the outer product of the vectors at each pixel location which consisted of red, green, and blue intensities. The intercolor dependencies penalty term $J_3(\mathbf{x})$ is

$$J_3(\mathbf{x}) = \sum_{l=-1}^{1} \sum_{m=-1}^{1} \|\mathbf{x} \otimes \mathbf{s}_h(l)\mathbf{s}_v(m)\mathbf{x}\|_2^2 \qquad (28.31)$$

where $\otimes$ represents the pixelwise vector outer product. By using the intercolor dependencies penalty term, the color artifacts that can be produced in the demosaicing process can be reduced.

As a result, the overall cost function is the summation of the cost functions described above:

$$\hat{\mathbf{x}} = \arg\min_{\mathbf{x}}[J_0(\mathbf{x}) + \alpha_1 J_1(\mathbf{x}) + \alpha_2 J_2(\mathbf{x}) + \alpha_3 J_3(\mathbf{x})] \qquad (28.32)$$

The regularization parameters α_1, α_2, and α_3 controlled the weights between the fidelity of the data, the constraints on the sharpness of the edges of the luminance, smoothness of the chrominance, and the high correlation between the color channels. The resulting images are shown in Figure 28.11. In Figure 28.11a, the Bayer CFA-sampled image is demosaiced using Kimmel's [22] method and upsampled by an independent process. The reconstructed image shows the high spatial resolution and reduced color artifacts, as shown in Figure 28.11b.

28.4.2
Simultaneous Enhancement of Spatial Resolution and Dynamic Range

The limited dynamic range of imaging sensors is another obstacle which generally results in the loss of information in captured images. The light intensity range of natural scenes is very broad and the HVS can adapt to an extremely high range of light intensity levels, from the scotopic threshold to the glare limit. Digital imaging sensors, however, have a narrow limited dynamic range compared to the HVS. When capturing a scene with a very broad intensity range of light that

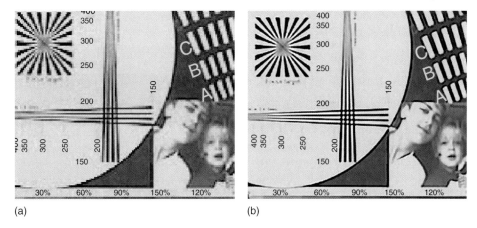

Figure 28.11 Comparisons between (a) an interpolated color image and (b) a super-resolved color image.

exceeds the dynamic range of an imaging sensor, it is inevitable to lose detail in the dark areas, or the bright areas, or both. Therefore, several signal-processing techniques to increase the dynamic range of images using multiple frames with different exposures of a scene have been proposed. Most previous approaches using multiple frames have several limitations, such as the assumption that the image capture devices have a linear response and there is no saturation at the highest exposure. Debevec and Malik proposed an algorithm that considered the non-linearity of the response function and saturation of the pixel value for the highest exposure [23].

Since the time imaging sensors were invented, the enhancement of spatial resolution and dynamic range have been researched independently. However, in many practical situations, since observations suffer from limitations of both spatial resolution and dynamic range, it is necessary to simultaneously enhance the spatial resolution and dynamic range at once. Recently, approaches that simultaneously enhance the spatial resolution and the dynamic range have been proposed [24, 25]. On the basis of the spatially varying pixel exposures, an algorithm that enhances the resolution of multisampled images was proposed by Narasimhan and Nayar [24]. Multisampled images mean that a scene has been sampled in multiple dimensions such as space, time, spectrum, and brightness. Gunturk and Gevrekci proposed the MAP approach to enhance the spatial resolution and the dynamic range with prior knowledge of an imaging system's response function [26]. Choi *et al.* developed a technique for concurrent estimation of high-resolution images with a high dynamic range and the imaging system's response function [25].

In order to model the image degradation process, reciprocity and Hurter–Driffield curve can be applied to the image observation model of Eq. (28.2). Reciprocity means that the exposure can be defined as the product of the luminance r at the surface of an imaging sensor and the exposure time Δt. The Hurter–Driffield curve is a graph of the optical density of the imaging sensor versus the logarithm of the exposure $r\Delta t$. With these physical characteristics, the observation model Eq. (28.2) was rewritten as

$$\mathbf{y}_k = f(\Delta t_k \mathbf{W}_k \mathbf{x} + \mathbf{n}_k^a) + \mathbf{n}_k^q \quad \text{for } k=1,\ldots,p \tag{28.33}$$

where $\mathbf{n}_k^a$ is a lexicographically ordered noise vector, and $\mathbf{n}_k^q$ represents a lexicographically ordered quantization error vector.

The nonlinear function f includes the characteristics of image capture devices and the quantization process during image acquisition.

To reconstruct the high-resolution and high dynamic range image based on the image formation model, the inverse function of f was required. In general, even if it is a reasonable assumption that the response function of an imaging system is monotonically increasing, the inverse function f^{-1} does not exist because f represents a many-to-one mapping due to the quantization process. Therefore, the reverse mapping function g was defined as the function that maps the pixel intensity to the mid-level of the corresponding quantization interval, and it was used instead of f^{-1}. With the definition of the reverse mapping function g and Taylor series expansion, Eq. (28.33) was rewritten as

$$g(\mathbf{y}_k) = g(f(\Delta t_k \mathbf{W}_k \mathbf{x} + \mathbf{n}_k^a) + \mathbf{n}_k^q)$$
$$\simeq \Delta t_k \mathbf{W}_k \mathbf{x} + \mathbf{n}_k^t \tag{28.34}$$

where $\mathbf{n}_k^t = \mathbf{n}_k^a + g'(\mathbf{y}_k) \cdot \mathbf{n}_k^q$ represents the total noise term including the additive noise factors and quantization errors. Since the reverse mapping function g was unknown, as well as the high-resolution and high dynamic range image $\mathbf{x}$, $\mathbf{x}$ and g had to be simultaneously reconstructed. In Ref. [25], g was approximated using polynomial fitting, and Gauss–Seidel relaxation was used to determine the solution for both $\mathbf{x}$ and g.

The resulting images and the partially magnified results are shown in Figure 28.12. Figure 28.12a shows one frame of the low-resolution and low dynamic range images which were interpolated to the size of the reconstructed image. In Figure 28.12b, the information from both the dark and bright regions is preserved in comparison with Figure 28.12a. Figure 28.12c shows one nearest-neighbor-interpolated frame of low-resolution images. Figure 28.12d shows the result of successive application of the dynamic range enhancement and spatial resolution enhancement. Figure 28.12e depicts the simultaneously reconstructed high-resolution and high dynamic range image, obtained without prior knowledge about the imaging system.

These results show that super-resolution image reconstruction is very effective for the reconstruction of high-resolution and high dynamic range images.

28.4.3
Super-Resolution for Video Systems

There are many different video signals in use that are not of the same resolution, and the typical resolution of display devices is often much higher than that of the

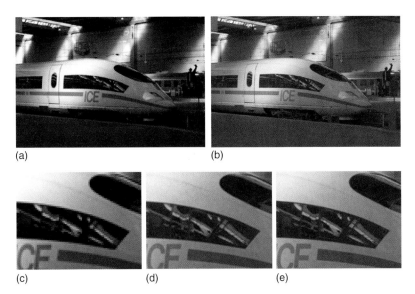

Figure 28.12 The enhancement of spatial resolution and dynamic range: (a) one low-resolution and low dynamic range image, (b) the reconstructed high-resolution and high dynamic range image, (c) the partially magnified frame of low-resolution images, (d) the partially magnified result of successive enhancement, and (e) the partially magnified result of simultaneous enhancement.

video signals. Therefore, a video scaler is often used to convert video signals from one resolution to another: usually "upscaling" or "upconverting" a video signal to display the signals on a high-resolution display device. Techniques for video scalers can be roughly classified into two groups according to the type of information used. The first group of techniques uses information only from the current frame, for example, the cubic-spline interpolation method. The second group consists of super-resolution-based scalers with multiple neighboring frames. These methods are more advanced, and they require higher hardware complexity. They produce high-quality images even in motion areas, provided that the estimated motion information is reliable.

In general, super-resolution-based techniques often suppose that low-resolution images contain only global motions such as translation and rotation. However, real video sequences have a lot of local motions that cause occlusions, and it is not always easy to estimate the local motions accurately. Various motion estimation schemes have been investigated to treat these local motions including pel-recursive, object-based approaches and BMAs. Among them, BMAs have become widely accepted in real-time video applications because of their simplicity in terms of efficient hardware implementation. However, the fundamental problem of BMAs is that a block at object boundaries generally has an unreliable motion vector since the block may contain multiple objects with different motions, which leads to visual artifacts. Since human eyes are very sensitive to temporal irregularities, it is very important to suppress artifacts by designing additional constraint cost functions or sophisticated arbitration methods.

Real-time processing is another task that should be achieved to apply super-resolution techniques to video processing. Modern super-resolution algorithms require huge computational complexity and many of them use iterative approaches to obtain optimal solutions. These shortcomings restrict the algorithms' field of practical application, that is, they are unsuitable for real-time or near-real-time resolution enhancement of video sequences. Among the various super-resolution approaches, the nonuniform interpolation-based approach, which is the simplest method for super-resolution that is described in Refs [27, 28], is often used to achieve a low-cost and real-time solution for the implementation of super-resolution algorithms.

As mentioned above, conventional super-resolution algorithms often require accurate motion information to obtain high-resolution images. The need for precise subpixel motion estimates has limited its applicability to video sequences, and most methods perform poorly in the presence of complex motions, which are commonly found in image sequences. In order to overcome these difficulties, a super-resolution approach based on multidimensional kernel regression has been developed [29]. The objective of this work is to achieve super-resolution for general sequences, while avoiding explicit (accurate subpixel) motion estimation, which is quite different from existing super-resolution methods that require highly accurate subpixel motion estimation. The new method also requires a motion estimation process, but this motion estimation can be quite rough (e.g., integer-pixel accuracy). This rough motion estimate is used to compensate for the large motion, leaving behind a residual of small motions, which can be implicitly captured within the 3D orientation kernel. A detailed explanation of the fundamental framework of (kernel regression) (KR) is represented in Ref. [29]. The KR framework defines its data model as

$$v_i = z(\mathbf{C}_i) + \varepsilon_i, \quad i = 1, \ldots, P, \quad \mathbf{C}_i = [h_i, v_i, t_i]^t \qquad (28.35)$$

where v_i is a noisy input at spatio-temporal index $\mathbf{C}_i$, $z(\cdot)$ is the regression function to be estimated, ε_is are the noise values, and P is the number of samples in the 3D neighborhood of interest, which are around the motion trajectory of a position $\mathbf{C}$. KR provides a rich mechanism for computing pointwise estimates of the function with minimal assumptions about global signal or noise models. If $\mathbf{C}$ is near the sample at $\mathbf{C}_i$, we have the N-term Taylor series of $z(\mathbf{C}_i)$ as in

$$\begin{aligned} z(\mathbf{C}_i) &\approx z(\mathbf{C}) + \{\nabla z(\mathbf{C})\}^t (\mathbf{C}_i - \mathbf{C}) + \frac{1}{2}(\mathbf{C}_i - \mathbf{C})^t \{Hz(\mathbf{C})\}(\mathbf{C}_i - \mathbf{C}) + \cdots \\ &= \beta_0 + \beta_1^t(\mathbf{C}_i - \mathbf{C}) + \beta_2^t \mathrm{vech}\{(\mathbf{C}_i - \mathbf{C})^t(\mathbf{C}_i - \mathbf{x})\} + \cdots \end{aligned} \qquad (28.36)$$

where ∇ and $\mathcal{H}$ are the gradient (3×1) and Hessian (3×3) operators, respectively, and vech($\cdot$) is the half-vectorization operator that lexicographically orders the lower triangular portion of a symmetric matrix into a column-stacked vector.

For a sequence with only small motions (e.g., subpixel motions), a weighted least-square formulation of the fitting problem is as follows:

$$\min_{\{\beta_n\}_{n=0}^{N}} \sum_{i=0}^{P} [v_i - \beta_0 - \beta_1^t(\mathbf{C}_i - \mathbf{C}) - \beta_2^t \text{vech}\{(\mathbf{C}_i - \mathbf{C})^t(\mathbf{C}_i - \mathbf{C})\} - \cdots]^2 K_{\mathbf{H}_i}(\mathbf{C}_i - \mathbf{C}) \quad (28.37)$$

with

$$K_{\mathbf{H}_i}(\mathbf{C}_i - \mathbf{C}) = \frac{1}{\det(\mathbf{H}_i)} K\left(\mathbf{H}_i^{-1}(\mathbf{C}_i - \mathbf{C})\right) \quad (28.38)$$

where N represents the regression order, $K(\cdot)$ represents the kernel function, and $\mathbf{H}_i$ represents the smoothing matrix. In Ref. [29], the steering kernel was adopted for kernel regression.

For the small motion case, the KR with the steering kernel achieved the effects of resolution enhancement by using the information of neighboring frames. However, in the presence of large displacements, similar pixels, though close in the time dimension, were found far away in space because of the effect of aliasing. Therefore, it was necessary to neutralize the motion of the given video data v_i by using rough motion estimation to produce a new sequence of data $v(\mathbf{C}_i)$ as follows:

$$\widetilde{\mathbf{C}}_i = \mathbf{C}_i + \begin{bmatrix} \mathbf{M}_i \\ 0 \end{bmatrix}(t_i - t) \quad (28.39)$$

where $\mathbf{M}_i$ is a motion vector at $\mathbf{C}_i$. Then, Eq. (28.37) was rewritten as

$$\min_{\{\beta_n\}_{n=0}^{N}} \sum_{i=0}^{P} [v(\widetilde{\mathbf{C}}) - \beta_0 - \beta_1^t(\widetilde{\mathbf{C}}_i - \mathbf{C}) - \beta_2^t \text{vech}\{(\widetilde{\mathbf{C}}_i - \mathbf{C})^t(\widetilde{\mathbf{C}}_i - \mathbf{C})\} - \cdots]^2 K_{\widetilde{\mathbf{H}}_i}(\widetilde{\mathbf{C}}_i - \mathbf{C}) \quad (28.40)$$

where $\widetilde{\mathbf{H}}_i$ was computed from the motion-compensated sequence $v(\mathbf{C}_i)$. To summarize, 3D KR with motion compensation can be regarded as a two-tiered approach to handle a wide variety of transitions in video, where rough motions are first neutralized and then the smaller motions are implicitly captured by the 3D kernel.

Figure 28.13 shows an example of the super-resolution algorithm applied to the video sequence. Figure 28.13a represents the original frame. The image has increased the spatial resolution by a factor of 4 in Figure 28.13b and c. Figure 28.13b was obtained by performing the cubic-spline interpolation method that uses only a current frame, and the result in Figure 28.13c was reconstructed by nonuniform interpolation-deblurring method and three frames (previous, current, and next frames) were then used to obtain the final super-resolution results. The super-resolution method improved the spatial resolution in comparison with the bilinear or cubic-spline interpolation method and provided a high-quality image in most parts of the picture.

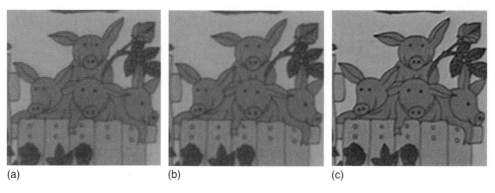

Figure 28.13 Comparisons between (a) input image, (b) cubic-spline interpolation, and (c) nonuniform-deblur scheme.

28.5
Conclusions

Recently, super-resolution image reconstruction has been in the limelight, because it can overcome the physical limitations of imaging systems and improve the performance of most digital image processing applications. Super-resolution image reconstruction approaches increase spatial resolution using multiple low-resolution images that are subsampled as well as shifted with different subpixel motion. Accurate subpixel registration between low-resolution images is very important for the reconstruction of high-resolution images, because inaccurate subpixel motion information results in distortion in the reconstructed high-resolution image. To reduce this distortion, the inaccurate subpixel motion information has to be considered in the process of the super-resolution image reconstruction. In this chapter, the fundamentals of super-resolution image reconstruction and an approach considering inaccurate subpixel registration were described. The effect of inaccurate subpixel motion information was modeled as the error added, respectively, to each low-resolution image, and the high-resolution image was reconstructed based on multichannel regularization to minimize the energy of the error.

Acknowledgments

This work was supported by Mid-career Researcher Program through NRF grant funded by the MEST (No. 2010-0000345) and supported in part by the IT R&D program of MKE/MCST/IITA (KI001820, Development of Computational Photography Technologies for Image and Video Contents).

References

1. Park, S.C., Park, M.K., and Kang, M.G. (2003) Superresolution image reconstruction - a technical overview. *IEEE Signal Process. Mag.*, **20** 21–36.
2. Unser, M., Aldroubi, A., and Eden, M. (1995) Enlargement or reduction of digital images with minimum loss of information, *IEEE Trans. Image Process.*, **4**, 247–258.
3. Katsaggelos, A.K. (ed.) (1991) *Digital Image Restoration*, Springer Series in Information Sciences, vol. 23, Springer Verlag, Heidelberg, Germany.
4. Elad, M. and Feuer, A. (1997) Restoration of a single superresolution image from several blurred, noisy, and undersampled measured images, *IEEE Trans. Image Process.*, **6**, 1646–1658.
5. Nguyen, N., Milanfar, P., and Golub, G. (2001) Efficient generalized cross-validation with applications to parametric image restoration and resolution enhancement, *IEEE Trans. Image Process.*, **10**, 1299–1308.
6. Kang, M.G. (1998) Generalized multichannel image deconvolution approach and its applications, *SPIE Opt. Eng.*, **37**, 2953–2964.
7. Hardie, R.C. et al. (1998) High-resolution image reconstruction from a sequence of rotated and translated frames and its application to an infrared imaging system, *SPIE Opt. Eng.*, **37**, 247–260.
8. Bose, N.K., Lertrattanapanich, S., and Koo, J. (2001) Advances in superresolution using L-curve, *Proc. IEEE Int. Symp. Circ. Syst.*, **2**, 433–436.
9. Hansen, P.C. and O'Leary, D.P. (1993) The use of the L-curve in the regularization of discrete ill-posed problems, *SIAM J. Sci. Comput.*, **14**, 1487–1503.
10. Tom, B.C. and Katsaggelos, A.K. (1995) Reconstruction of a high-resolution image by simultaneous registration, restoration, and interpolation of low-resolution images, *Proc. Int. Conf. Image Process.*, **2**, 539–542.
11. Schulz, R.R. and Stevenson, R.L. (1996) Extraction of high-resolution frames from video sequences, *IEEE Trans. Image Process.*, **5**, 996–1011.
12. Hardie, R.C. et al. (1997) Joint map registration and high-resolution image estimation using a sequence of undersampled images, *IEEE Trans. Image Process.*, **6**, 1621–1633.
13. Kim, S.P., Bose, N.K., and Valenzuela, H.M. (1990) Recursive reconstruction of high-resolution image from noisy undersampled multiframes, *IEEE Trans. Acoust.*, **38**, 1013–1027.
14. Rhee, S.H. and Kang, M.G. (1999) Discrete cosine transform based regularized high-resolution image reconstruction algorithm, *SPIE Opt. Eng.*, **38**, 1348–1356.
15. Ng, M.K., Koo, J., and Bose, N.K. (2002) Constrained total least squares computations for high-resolution image reconstruction with multisensors, *Int. J. Imaging Syst. Technol.*, **12**, 35–42.
16. Ng, M.K. and Bose, N.K. (2002) Analysis of displacement errors in high-resolution image reconstruction with multisensors, *IEEE Trans. Circuits Syst. I Fundam. Theory Appl.*, **49**, 806–813.
17. Lee, E.S. and Kang, M.G. (2003) Regularized adaptive high-resolution image reconstruction considering inaccurate subpixel registration, *IEEE Trans. Image Process.*, **12**, 826–837.
18. Park, M.K., Kang, M.G., and Katsaggelos, A.K. (2007) Regularized high-resolution image reconstruction considering inaccurate motion information, *SPIE Opt. Eng.*, **46**, 117004.
19. Vega, M., Molina, R., and Katsaggelos, A.K. (2006) A Bayesian super-resolution approach to demosaicing of blurred images, *EURASIP J. Appl. Signal Process.*, **2006** 1–12.
20. Farsiu, S., Elad, M., and Milanfar, P. (2006) Multiframe demosaicing and super-resolution of color images, *IEEE Trans. Image Process.*, **15**, 141–159.
21. Paris, S. et al. (2009) Bilateral filtering: theory and applications, *Found. Trends Comput. Graph. Vis.*, **4**, 1–73.
22. Kimmel, R. (1999) Demosaicing: image reconstruction from color CCD samples, *IEEE Trans. Image Process.*, **8**, 1221–1228.

23. Debevec, P.E. and Malik, J. (1997) Recovering high dynamic range radiance maps from photographs. SIGGRAPH 97 Conference Proceedings, Los Angeles, CA, pp. 369–378.
24. Narasimhan, S.G. and Nayar, S.K. (2005) Enhancing resolution along multiple imaging dimensions using assorted pixels, *IEEE Trans. Pattern Anal. Mach. Intell.*, **27**, 518–530.
25. Choi, J., Park, M.K., and Kang, M.G. (2009) High dynamic range image reconstruction with spatial resolution enhancement, *The Oxford Journals - The Computer Journal*, **52**, 114–125.
26. Gunturk, B.K. and Gevrekci, M. (2006) High-resolution image reconstruction from multiple differently exposed images, *IEEE Signal Process. Lett.*, **13**, 197–200.
27. Shah, N.R. and Zakhor, A. (1999) Resolution enhancement of color video sequences, *IEEE Trans. Image Process.*, **8**, 879–885.
28. Alam, M.S. *et al.* (2000) Infrared image registration and high-resolution reconstruction using multiple translationally shifted aliased video frames, *IEEE Trans. Instrum. Meas.*, **49**, 915–923.
29. Takeda, H. *et al.* (2009) Super-resolution without explicit subpixel motion estimation, *IEEE Trans. Image Process.*, **18**, 1958–1975.

29
Image Analysis: Intermediate-Level Vision

Jan Cornelis, Aneta Markova, and Rudi Deklerck

29.1
Introduction

Segmentation is the process by which an original image I is partitioned into N regions R_i, $1 \leq i \leq N$, which are homogeneous/similar in one or more image data attributes/features (e.g., brightness, gray value, texture, color, or any local image feature), so that

$$\bigcup_{i=1}^{N} R_i = I$$
$$R_i(i = 1, 2, \ldots, N) \text{ is a 4 or 8 connected set}^{1)}$$
$$R_i \cap R_j = \phi, 1 \leq i, j \leq N, i \neq j \quad (29.1)$$
$$H(R_i) = \text{TRUE}, 1 \leq i \leq N$$
$$H(R_i \cup R_j) = \text{FALSE}, 1 \leq i, j \leq N, i \neq j$$

Most of the *classical textbooks* on digital image processing [1–6] adopt a restrictive definition of segmentation in which contextual information about the scene or the task is ignored (i.e., the operator H is purely data dependent and expresses a homogeneity condition derived from image features, such as gray value, color, texture, shape, etc.). Segmentation is treated independently of object recognition. There is no theory of segmentation, no gold standard. Instead, a collection of popular *ad hoc* generic methods exists. These are primarily based on two properties of the considered image features: local discontinuity and similarity.

As *a complement to segmentation*, textbooks contain chapters on image feature extraction and shape description (i.e., parameters that can be used as input to the segmentation task itself and subsequently for labeling of pixels or segments).

---

1) In 4-connectedness, pixel $p(i, j)$ has four neighbors, namely those positioned at $(i-1, j)$, $(i+1, j)$, $(i, j-1)$, and $(i, j+1)$. In 8-connectedness, pixel $p(i, j)$ has eight neighbors: we add the pixels in the diagonal positions $(i-1, j-1)$, $(i-1, j+1)$, $(i+1, j-1)$, and $(i+1, j+1)$ to the 4-connected neighbors.

Optical and Digital Image Processing: Fundamentals and Applications, First Edition. Edited by Gabriel Cristóbal, Peter Schelkens, and Hugo Thienpont.
© 2011 Wiley-VCH Verlag GmbH & Co. KGaA. Published 2011 by Wiley-VCH Verlag GmbH & Co. KGaA.

The *goals of segmentation* might be very diverse, but the definition given above remains valid, whether it is for region of interest extraction, edge detection, region segmentation, precise object delineation, or texture segmentation.

Extracting and describing semantic content of a scene (e.g., the physical objects in a scene) from images requires contextual and scene modeling [7]. It is outdated to describe a scene understanding system as a purely sequential process [7], involving imaging (physical image formation), digitalization (time and spatial sampling, quantization of samples), segmentation into regions, characterization by regional descriptors, extraction of relational descriptions, and object formation, so that finally facts about the physical world can be extracted from this pipeline of operations. In Ref. [8], a comprehensive overview is given on concepts and methods to extract 3D scene information from 2D images as well as image understanding schemes going beyond simple data-related shape representation and description. As such, the book is a reference work on *intermediate-level vision*.

In this chapter, we describe data-driven operators for segmentation that are compatible with the view that a mixture of data-driven bottom up and model-driven top down processings are taking place either under the control of a hierarchical scheme or a nonhierarchical data/hypothesis-dependent supervision [8]. In *intermediate-level vision*, segmentation then becomes a mixed delineation, estimation, and recognition problem as schematically illustrated in Figure 29.1. Even if scene modeling is involved, the segmentation definition given in Eq. (29.1) remains valid, but the operator *H* can become very complex and difficult to describe explicitly (e.g., it can be the result of an iterative and/or learning process). Most approaches start from a data-driven preprocessing to set up an appropriate data structure (often including a presegmentation step) and refine the segments afterward by including context models (Figure 29.1).

For keeping things simple, we deliberately emphasize 2D gray-value images although the concepts are often applicable on video- and vector-valued (e.g., color) images too. Most of our own work has been developed in a multiscale framework, because whenever recognition is involved, the scale interval of interest should be specified. In this chapter, we stripped most references to multiscale image

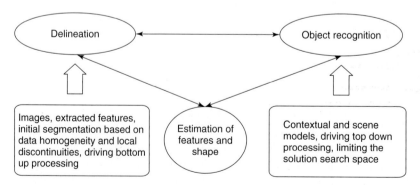

Figure 29.1 Segmentation in an intermediate-level vision context.

Table 29.1 Choice made in this chapter for categorizing the presented segmentation concepts (*Type* in column 1 refers to the main driving forces in the segmentation algorithms: in rows 1 and 2 these driving forces are the image properties of pixels/regions and edges, in row 3 they are the image and scene models based on a mixture of local and global features). Several other types of categorizations exist.

Segmentation type	Instantiation example	Typical scene modeling
Pixel and region based	*Supervised classification*: Maximizing the A posteriori Probability (MAP); *Unsupervised clustering*: K-means, Expectation-Maximization (EM)	selection of neighborhood, features, local model, and its parameters
Edge based	*Deformable models*: active contours	control by external forces and a priori shape
Model based	*Markov random fields*: graph matching and neighborhood modeling	a mixture of local and global features

representations and readers are referred to Chapter 8 for an extensive overview. We also rely on background knowledge concerning elementary image-processing operators described in Chapter 4. Table 29.1 summarizes the content of this chapter.

29.2
Pixel- and Region-Based Segmentation

Undoubtedly, *thresholding* is the most popular form of pixel-based segmentation. In fact, segmentation is then treated as solving a classification problem: pixels (and/or image objects containing grouped pixels) are treated as independent image objects to be classified. *The classification is based on the homogeneity (similarity) of image features within the region, which should be distinct from other regions.* What happens is that a single threshold or multiple gray-value thresholds are applied on an image histogram and every pixel is attributed to a class depending on the result of the comparison of the gray values with the thresholds. Comparison does not need to be limited to gray-value histograms; it can also be applied to one- or more dimensional histograms of image values (intensity, color, multispectral component), or any local image features derived from the spatial neighborhood of the pixel. In images with high contrast or images from calibrated acquisition, this might work well but, in general, the regions resulting from threshold operations are fragmented and lack connectivity, since the image histogram is a representation that destroys any local and global image structure [3]. Methods to improve the connectivity of the regions are discussed in Section 29.2.3.

With respect to control the segmentation process, the threshold values might be derived interactively by the user or by minimizing a chosen objective function, derived from a parameterized histogram model, which is assumed to be valid and whose parameters have to be estimated in a learning phase (supervised approach), as illustrated in Section 29.2.1. Nonsupervised approaches also exist (Section 29.2.2), but here the selection of a distance measure and the number of classes or an appropriate stop criterion for the amount of classes emerging from an iterative process has to be decided upon. If the local/global histogram contains overlapping peaks, selecting threshold values is a difficult task. Depending on the goals of the segmentation, a selection of local features has to be proposed or selected on the basis of maximal discriminative power.

Several classifiers exist—for example, Bayesian classifiers, neural networks, fuzzy clustering [9] – and all of them have been used for image segmentation by applying them on individual pixels. The approaches usually suffer from the limited capabilities to take contextual (scene) information into account during the classification process.

29.2.1
Supervised Approaches

Supervised approaches are based on a design/learning phase in which the class membership probabilities, and the histogram models as well as their parameters that characterize the distribution of the measured features are estimated, on the basis of a sample of labeled images (i.e., images for which the pixel class memberships are known).

29.2.1.1 Classification Based on MAP (Maximizing the A posteriori Probability)

A first approach to the segmentation and labeling of images considers pixels or other image objects, characterized by a feature vector $\mathbf{x} = (x_1, x_2, \ldots, x_N)^\tau$, which have to be assigned to a class that is uniquely identified by labels l_i, $L = \{l_1, l_2, \ldots, l_K\}$. The classes may correspond to meaningful regions in the image (e.g., in biomedical image analysis, regions defined by different tissue types). When one is dealing with a color image, the most obvious choice for the pixel features is the color components. More than three features may be proposed, even for monochrome images, when the local image properties in the neighborhood of the pixel to be classified are calculated. Statistical classification criteria are often derived from Bayes' theorem. If the set of classes is exhaustive and the classes are mutually exclusive, Bayes' theorem can be formulated as follows [10]:

$$P(l_i|\mathbf{x}) = \lim_{\bigcap_{n=1}^{N} dx_n \to 0} P\left(l_i \bigg| \bigcap_{n=1}^{N} \left(x_n - \frac{dx_n}{2} \leq x_n \leq x_n + \frac{dx_n}{2}\right)\right)$$

$$= \lim_{\bigcap_{n=1}^{N} dx_n \to 0} \frac{P(l_i) P\left(\bigcap_{n=1}^{N}\left(x_n - \frac{dx_n}{2} \leq x_n \leq x_n + \frac{dx_n}{2}\right)|l_i\right)}{\sum_{k=1}^{K} P(l_k) P\left(\bigcap_{n=1}^{N}\left(x_n - \frac{dx_n}{2} \leq x_n \leq x_n + \frac{dx_n}{2}\right)|l_k\right)}$$

$$= \lim_{\bigcap_{n=1}^{N} dx_n \to 0} \frac{P(l_i) p(\mathbf{x}|l_i) dx_1 \ldots dx_N}{\sum_{k=1}^{K} P(l_k) p(\mathbf{x}|l_k) dx_1 \ldots dx_N}$$

$$= \frac{P(l_i) p(\mathbf{x}|l_i)}{\sum_{k=1}^{K} P(l_k) p(\mathbf{x}|l_k)} = \frac{P(l_i) p(\mathbf{x}|l_i)}{p(\mathbf{x})} \qquad (29.2)$$

where $P(l_i)$ is the prior probability of class l_i, $p(\mathbf{x}|l_i)$ is the probability density of $\mathbf{x}$ for class l_i and $\bigcap_{n=1}^{N}$ has the meaning of an AND function over all components of $\mathbf{x}$. Assigning a feature vector to its most probable class l_i is equivalent to assigning the feature vector to the class corresponding to the largest value of the numerator:

$$d_{l_i}(\mathbf{x}) = P(l_i) p(\mathbf{x}|l_i) \qquad (29.3)$$

since the denominator is common to all classes. Choosing a model for the pixel classifier implies the choice of a model for the probability distribution of the features.

Both quadratic and linear discriminant analysis are based on the normal distribution model [10], that is,

$$p(\mathbf{x}|l_i) = \frac{1}{\sqrt{2\pi}^N \sqrt{|\Sigma_{l_i}|}} \exp\left(-\frac{1}{2}(\mathbf{x} - \mu_{l_i})^\tau \Sigma_{l_i}^{-1} (\mathbf{x} - \mu_{l_i})\right) \qquad (29.4)$$

with μ_{l_i} being the mean vector for the class with label l_i and with Σ_{l_i} being the covariance matrix for the class with label l_i. After replacing $p(\mathbf{x}|l_i)$ in Eq. (29.3) by the right-hand side of Eq. (29.4) and after simplification, one obtains quadratic discriminant functions $d'_{l_i}, i = \{1, 2, \ldots, K\}$:

$$d'_{l_i}(\mathbf{x}) = -\frac{1}{2} \log |\Sigma_{l_i}| - \frac{1}{2}(\mathbf{x} - \mu_{l_i})^\tau \Sigma_{l_i}^{-1}(\mathbf{x} - \mu_{l_i}) + \log P(l_i) \qquad (29.5)$$

A further simplification can be realized when equality of the class covariance matrices is assumed, yielding linear discriminant functions:

$$d''_{l_i}(\mathbf{x}) = (\mu_{l_i}^\tau \Sigma^{-1})\mathbf{x} - \frac{1}{2}\mu_{l_i}^\tau \Sigma^{-1} \mu_{l_i} + \log P(l_i) \qquad (29.6)$$

with Σ being the global covariance matrix. Σ_{l_i} or Σ, μ_{l_i} and $P(l_i)$ have to be estimated in the learning phase.

29.2.2
Unsupervised Approaches

In contrast to the supervised approach, unsupervised pixel analysis does not rely on prior knowledge about class membership. In unsupervised pixel analysis, most often, two basic processing steps can be identified.

The *first step* consists of a clustering – a vector quantization of the feature vector distribution V derived from the finite set of pixels composing the image. Clustering within the feature vector distribution may be considered as a mapping of the distribution V on a set W of N-dimensional class prototype (feature) vectors ξ_{l_i}, $i = \{1, 2, \ldots, K\}$:

$$V(p(\mathbf{x})) \to W = \{\xi_{l_1}, \xi_{l_2}, \ldots, \xi_{l_K}\} \tag{29.7}$$

This transformation introduces an index mapping function $F(\mathbf{x})$ that maps a given feature vector $\mathbf{x}$ to the label l_i of the corresponding class prototype vector ξ_{l_i}:

$$F: \quad V \to \{l_1, l_2, \ldots, l_K\} : F(\mathbf{x}) = l_i \tag{29.8}$$

The label l_i can be regarded as the index of the centroid of a subset of V for which vector ξ_{l_i} is the prototype vector. An approximation for such mapping can be found by minimizing the expectation value of the squared error between the feature vectors of the distribution V and the corresponding class prototype vector $\xi_{F(\mathbf{x})}$:

$$W_{\min} = \arg\min_W \int_V |\mathbf{x} - \xi_{F(\mathbf{x})}|^2 p(\mathbf{x}) dx_1 \ldots dx_N \tag{29.9}$$

During the *second step* of unsupervised pixel analysis, a labeled image is obtained by replacing each pixel in the image by a unique label derived from the prototype vector that fulfills the best matching principle for the feature vector of the pixel.

29.2.2.1 K-means Clustering

This is a vector quantization algorithm that leads to a set of prototype vectors that represents a global minimum of the objective function Eq. (29.9). During K-means clustering, each prototype vector ξ_{l_i} is updated after a sequential scan of all the feature vectors $\mathbf{x}$ of the image, by replacing each prototype vector with the mean vector of those feature vectors for which the prototype vectors fulfilled the best matching principle, that is,

$$\xi_{l_i}^{(t+1)} = \frac{\sum_{m=1}^{M} \left(\delta_{l_i, F^{(t)}(\mathbf{x}_m)} \cdot \mathbf{x}_m \right)}{\sum_{m=1}^{M} \delta_{l_i, F^{(t)}(\mathbf{x}_m)}} \tag{29.10}$$

where t denotes the iteration step, $\mathbf{x}_m$ is a feature vector corresponding to a pixel m in the image, and δ represents the Kronecker delta operator. It can be shown that in the case of the choice of a Euclidean distance measure as a matching criterion between a given prototype vector ξ_{l_i} and the feature vectors $\mathbf{x}$, that is,

$$D(\mathbf{x}, \xi_{l_i}) = \sqrt{\sum_{n=1}^{N} (x_n - \xi_{l_i, n})^2} \tag{29.11}$$

the iterative repetition of the adaptation procedure Eq. (29.10) results in the minimization of the squared distance error Eq. (29.9) between the whole distribution and the prototype vector set.

A labeled image is obtained by replacing each pixel in the image by a unique label derived from the prototype vector for which the Euclidian distance measure Eq. (29.11) is minimal.

29.2.2.2 Expectation-Maximization (EM)

This approach finds clusters by determining a mixture of Gaussians ($\sum_{k=1}^{K} \beta_{l_k} G(\mu_{l_k}, \Sigma_{l_k})$ with β_{l_k}, μ_{l_k} and Σ_{l_k}, the prior probability, mean vector, and covariance matrix of the class l_k, respectively) that minimizes an objective function expressing how well that mixture of Gaussians fits an image histogram [11, 12]. The algorithm converges to a locally optimal solution by iteratively updating the maximum likelihood estimates, that is, iteratively updating the values for the prior probabilities, mean vectors, and covariance matrices in two steps:

Step 1: Computing the expectation of the likelihood according to Bayes' theorem Eq. (29.2):

$$\omega_{ml_i}^{(t+1)} = P(l_i|\mathbf{x}_m, (\mu_{l_i}, \Sigma_{l_i})) = \frac{P(l_i)p(\mathbf{x}_m|(\mu_{l_i}, \Sigma_{l_i}))}{\sum_{k=1}^{K} P(l_k)p(\mathbf{x}_m|(\mu_{l_k}, \Sigma_{l_k}))} \qquad (29.12)$$

where $P(l_i)$ corresponds to β_{l_i}.

Step 2: Calculating the maximum likelihood estimates of the parameters by maximizing the likelihood expectation:

$$\beta_{l_i}^{(t+1)} = \frac{1}{M} \sum_{m=1}^{M} \omega_{ml_i}^{(t)}$$

$$\mu_{l_i}^{(t+1)} = \left(\frac{1}{\sum_{m=1}^{M} \omega_{ml_i}^{(t)}}\right) \sum_{m=1}^{M} \omega_{ml_i}^{(t)} \mathbf{x}_m \qquad (29.13)$$

$$\Sigma_{l_i}^{(t+1)} = \left(\frac{1}{\sum_{m=1}^{M} \omega_{ml_i}^{(t)}}\right) \sum_{m=1}^{M} \omega_{ml_i}^{(t)} (\mathbf{x}_m - \mu_{l_i}^{t+1})^{\tau} (\mathbf{x}_m - \mu_{l_i}^{t+1})$$

where M is the number of pixels in the image and K is the number of Gaussians, each corresponding to a class l_i. ω_{ml_i} is the membership weight expressing the degree to which a certain feature vector $\mathbf{x}_m$, corresponding to the pixel m, belongs to class l_i.

A labeled image is obtained by replacing each pixel m in the image – after a predefined number of iterations or when all parameters remain within a certain tolerance from their previous values – by the unique label l_i for which the membership weight ω_{ml_i} is the highest. In general, however, only a random fraction of the pixels will be used during the iterative part of the algorithm for computational reasons, while all pixels need to get a label during the labeling process.

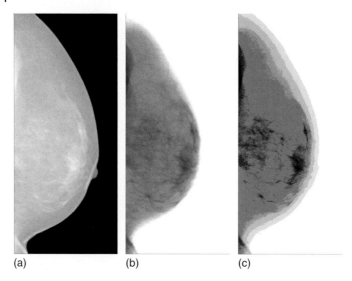

(a)   (b)   (c)

Figure 29.2 (a) Original image meant for visualization; (b) inverted original image meant for processing; (c) the image in (b) segmented with the EM algorithm using eight clusters.

Therefore, to speed up the labeling process, ω_{ml_i} can be replaced by a mapping function $F(\mathbf{x}_m)$ based on the quadratic discriminant functions Eq. (29.5). When the feature vector $\mathbf{x}_m$ is one or two dimensional, a serious speedup can also be obtained by expressing $\mathbf{x}_m$ in Eq. (29.13) for bins in the histogram rather than for individual pixels and by precalculating a 1D or 2D mapping table for the labeling step.

The parameters can be initialized by (i) randomly selecting the Gaussians' parameters; (ii) using the output of the K-means algorithm; or (iii) by profiling the histogram in the 1D case and estimating its maxima and the extrema in its first derivative. If the initial parameters of the approximation model are chosen well, fast convergence and better precision can be achieved. When using random initialization, it is worthwhile for K-means and EM to run the algorithm a number of times and choose the most frequently occurring label as the final one. An example, illustrating the segmentation of a mammogram by the EM algorithm is provided in Figure 29.2, where the segments are displayed with their average gray value.

29.2.3
Improving the Connectivity of the Classification Results

Several methods, usually based on domain heuristics, exist to counteract a deficient spatial connectivity within the pixel class that received the same class label. Two of

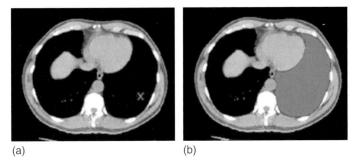

Figure 29.3 (a) A 2D lung CT scan with a seed point annotated by a cross and (b) the result of a region growing from the seed point (homogeneously gray segmentation).

them are briefly presented here: (i) seeded region growing [3], and (ii) mathematical morphology [13].

29.2.3.1 Seeded Region Growing

If one is interested only in one or a few objects of the image dataset, a better time performance and a guaranteed connectivity can be obtained by combining seeded region growing, together with a pixel classification method [14]. In its simplest form, the method starts with one pixel (the seed), characterized by a specific class label, and the region grows until no more candidate connected pixels having the same label are left to join the region, as shown in Figure 29.3. Hence, instead of scanning all pixels in the image, only neighboring pixels are iteratively grouped as long as they are classified to the same label. This approach is especially advantageous for small objects, since only a limited amount of pixels have to be visited and a full connected component labeling is avoided.

In an interactive system, the user has the possibility to select training samples for the design of discriminant functions, in an optimal way, directly inside the object that has to be extracted.

Also for 3D datasets, the method is extremely powerful (e.g., 3D medical images), since it is able to follow branches and arcs; that is, regions appearing visually separated in a 2D image because of the out-of-plane connections, will be automatically added to the seeded region in 3D.

29.2.3.2 Mathematical Morphology

Once a segmentation has been performed by a pixel classification algorithm, morphological operators [3, 13] can be extremely useful for further refinement: for example, after conversion of the segmentation results into a binary image, very small segments containing only a single or a few pixels can be removed. This approach is also used to disconnect loosely coupled segments from each other, for instance, the separation of the brain from the skull in magnetic resonance images. The closing (dilation followed by erosion) operation is also helpful to fill small gaps or to smooth the borders of the segments.

29.3
Edge-Based Segmentation

In case the homogeneity (similarity) of the image features does not allow to separate the different regions, one has to rely on local properties characterizing steep transitions (local discontinuities) in image features that most probably delineate objects and on the shape geometry. The classification is based on the local discontinuities of image components, which indicate edges that are part of region contours. In general, elementary edge detectors produce fragmented contours with a lack of connectivity, and hence linking and grouping algorithms [3, 6] or more complex schemes involving a priori scene information have to be used to obtain meaningful closed contours.

$$\text{Roberts mask}$$
$$\begin{pmatrix} -1 & 0 \\ 0 & 1 \end{pmatrix} \quad \text{and} \quad \begin{pmatrix} 0 & -1 \\ 1 & 0 \end{pmatrix} \tag{29.14}$$

$$\text{Prewitt mask}$$
$$\begin{pmatrix} -1 & -1 & -1 \\ 0 & 0 & 0 \\ 1 & 1 & 1 \end{pmatrix} \quad \text{and} \quad \begin{pmatrix} -1 & 0 & 1 \\ -1 & 0 & 1 \\ -1 & 0 & 1 \end{pmatrix} \tag{29.15}$$

$$\text{Sobel mask}$$
$$\begin{pmatrix} -1 & -2 & -1 \\ 0 & 0 & 0 \\ 1 & 2 & 1 \end{pmatrix} \quad \text{and} \quad \begin{pmatrix} -1 & 0 & 1 \\ -2 & 0 & 2 \\ -1 & 0 & 1 \end{pmatrix} \tag{29.16}$$

$$\text{Diagonal mask}$$
$$\begin{pmatrix} -1 & -1 & 0 \\ -1 & 0 & 1 \\ 0 & 1 & 1 \end{pmatrix} \quad \text{and} \quad \begin{pmatrix} 0 & 1 & 1 \\ -1 & 0 & 1 \\ -1 & -1 & 0 \end{pmatrix} \tag{29.17}$$

If the border is characterized by a large step in intensities, the magnitude of a gradient operator will yield high values for the border pixels, which can be selected via *thresholding*. The approximations of the gradient components via finite differences give rise to spatial edge filtering masks (such as the Roberts cross gradient Eq. (29.14), Prewitt Eq. (29.15) and Sobel Eq. (29.16) masks, which are very popular linear filtering operators in the image-processing community [3]). The magnitude of the gradient can be approximated via the sum of the absolute values or via the maximum of the absolute values of the gradient components, each of them obtained via convolution with their respective edge filtering mask (see Chapter 4 for an explanation of convolution). This is computationally less expensive than computing the square root of the squares of the gradient components. The continuous gradient magnitude is rotation invariant. For a gradient magnitude derived from discrete filtering masks, this is no longer true, but lower rotation variance can be achieved by considering extra filtering masks (e.g., Prewitt diagonal masks Eq. (29.17) in addition to the basic horizontal and vertical Prewitt masks Eq. (29.15)) or by extending the size of the basic masks (e.g., from 3×3 to 5×5).

The result of edge detection is either a binary image indicating where the edges are, or a list of the edges, possibly with attached descriptors such as direction and contrast strength. To form closed boundaries, linking or grouping of the edge elements that correspond to a single boundary may be required.

Also, second-order derivatives can be used, since their zero-crossings are characteristic for edge transitions. The continuous Laplacian, which is also rotation invariant, can be approximated via the convolution with a discrete filtering mask. Several near rotation invariant masks exist for the Laplacian as well.

In order to avoid the noise amplification by applying difference operators (which are equivalent to high-pass or band-pass filters in the frequency domain), the difference operators are often combined with weighted averaging masks (which are equivalent to low-pass filters). The Laplacian mask is often used in combination with a Gaussian mask. Since both operators are linear they can easily be merged into a single filter operation, known as the Laplacian of Gaussian (LoG), also called *Mexican hat filter* because of its particular shape in the spatial domain [15].

The edges obtained by *zero-crossing detection* have the advantage to directly form closed contours. However, a major problem is that this method does not take into account the strength of the edge so that the zero crossings produced by slight variations in the intensity will be detected as well. These kind of unwanted edges are commonly called *spurious edges*. They can be eliminated by choosing a larger σ for the Gaussian filter, which means, in fact, that the Mexican hat filter operates in a larger neighborhood. However, as σ grows, edges will become less well localized (see Figure 29.4).

Another problem for the *zero-crossing detection* of the LoG, is that it might detect false edges for certain edge topographies, also known as *phantom edges*. There are two reasons for this:

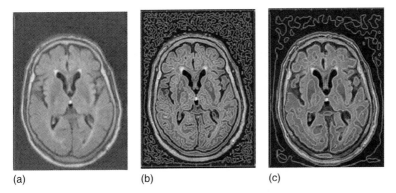

Figure 29.4 (a) Original image; (b) and (c) LoG: Edges obtained after zero-crossing detection for increasing values of sigma, superimposed over the original image.

1) Zero-crossings correspond to extremal values, hence *minima* in the gradient magnitude will be detected as well.
2) Owing to limited numerical precision, extra zero crossings may be introduced.

An improvement to the LoG filter was made by Haralick [16], by locating the edges at the zero-crossings of the second-order derivative in the direction of the gradient.

29.3.1
Optimal Edge Detection and Scale-Space Approach

Canny [17] was the first to analyze the optimal filter for edge detection by variational calculus. He tried to find the optimal filter for finding the edge location of a step edge in the presence of additive Gaussian noise. Given a certain filter, Canny succeeded in computing the signal to noise ratio (SNR) and a localization parameter (LOC). Wider filters give a better SNR, but a worse localization and vice versa. Canny succeeded in obtaining a scale or filter width invariant parameter by multiplying the SNR and LOC related terms. Yet the optimization of this parameter on its own does not take multiple response errors into account. A third criterion was then added to ensure that the detector has only one response to a single edge. According to these criteria, Canny showed that in 1D, the optimal filter for a step edge is a linear combination of four exponentials that can be quite well approximated by the first derivative of a Gaussian. The criteria proposed by Canny should, however, not be considered as the ultimate or unique ones.

As stated before, edges are becoming less well localized as the smoothing factor gets larger. This can, however, be overcome when the trajectory of the edges is tracked over different resolutions and scales (i.e., different values of σ). This kind of approach is known as the *scale-space* approach [18], which has good mathematical foundations (see Chapter 8).

29.4
Deformable Models

In 2D, deformable models are flexible contours with elasticity properties, often called *active contours* or *snakes* [19], whereas in 3D they are represented as elastic surfaces or membranes, commonly called *active surfaces* or *balloons* [20]. The theory of deformable models is based on continuum mechanics and elasticity theory. The active contours and surfaces are deformed by virtual forces, which are derived from:

– image features such as lines and edges, which are considered as strong attractors;
– differences between images such as highly dissimilar regions exerting strong forces; and
– user-defined attractors, repulsers, or springs.

Owing to these virtual forces, the model will deform until some equilibrium state is obtained, that is, the inner elastic forces of the model will be in balance with

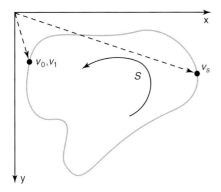

Figure 29.5 Deformable contour parameterized with the normalized perimeter variable s.

the external image-related forces. Extra goals of the internal forces are to retain the model as one entity during deformation and favor a priori shape properties (e.g., convexity, semicircularity, etc.). The external force vector field has to be defined in such a way that the snake is attracted toward features such as lines and edges in the image and reaches an energy minimum (i.e., becomes stable) on top of the features, hence localizing them accurately. Some preprocessing may be needed to eliminate the noise and enhance the edges in the image; an adequate edge detector has to be chosen to obtain a suitable force field. Tuning of the elasticity/rigidity parameters might be necessary to obtain realistic boundaries, which are not sensitive to jags in the edges that are due to artifacts or noise. Active contours or surfaces are appropriate to achieve subpixel accuracy. The mathematical formulation of the continuous case is given in Section 29.4.1, while the discrete case is explained on the website of this book.

29.4.1
Mathematical Formulation (Continuous Case)

A deformable contour $\mathbf{v}(s) = (x(s), y(s))$ is defined from a domain $s \in [0, 1]$ into the image plane $\mathfrak{R}^2$, with $x(s)$, $y(s)$ being the coordinates of the contour, see Figure 29.5.

Next, consider a deformation energy $E_{int}()$ related to the internal forces and a potential $P() = E_{ext}(x, y)$ related to the external forces acting upon the contour, derived from the image and/or some artificially defined energy sources introduced interactively by the user. In this energy formulation, the snake has to minimize the total energy:

$$E(\mathbf{v}) = E_{int}(\mathbf{v}) + E_{ext}(\mathbf{v}) \tag{29.18}$$

For a snake, we can write the internal deformation energy as follows:

$$E_{int}(\mathbf{v}) = \int_0^1 \left(w_1(s) \frac{|\mathbf{v}_s|^2}{2} + w_2(s) \frac{|\mathbf{v}_{ss}|^2}{2} \right) ds$$

where $v_s(s) = (x_s(s))^2 + (y_s(s))^2$, $x_s = \dfrac{dx}{ds}$ and $y_s = \dfrac{dy}{ds}$ (29.19)

and $v_{ss}(s) = (x_{ss}(s))^2 + (y_{ss}(s))^2$, $x_{ss} = \dfrac{d^2x}{ds^2}$ and $y_{ss} = \dfrac{d^2y}{ds^2}$

The parameter $w_1(s)$ corresponds to the tension of the deformable contour, whereas $w_2(s)$ is related to the rigidity of the contour. The external energy can be defined as

$$E_{ext}(\mathbf{v}) = \int_0^1 E_{ext}(\mathbf{v}(s))\, ds = \int_0^1 P(\mathbf{v}(s))\, ds = \int_0^1 P(x(s), y(s))\, ds \quad (29.20)$$

For a gray-level image $I(x, y)$, viewed as a function of continuous position variables (x, y), typical external energies designed to lead an active contour toward step edges are

$$E^{(1)}_{ext}(x, y) = -|\nabla I(x, y)|$$
$$E^{(2)}_{ext}(x, y) = -|\nabla(G_\sigma(x, y) * I(x, y))| \quad (29.21)$$

which correspond to the magnitudes of the image gradient. In $E^{(2)}_{ext}(x, y)$, the image is convolved with a Gaussian G_σ characterized by the standard deviation σ in case the image is too noisy, in order to make the derivatives or the gradient well behaving. The total energy of the snake can then be written as

$$E_{tot} = E_{int} + P$$
$$= \int_0^1 \left(w_1(s)\dfrac{|\mathbf{v}_s(s)|^2}{2} + w_2(s)\dfrac{|\mathbf{v}_{ss}(s)|^2}{2} + P(\mathbf{v}(s)) \right) ds \quad (29.22)$$

Applying the Euler–Lagrange differential equations leads to the following static equations:

$$\dfrac{\partial P}{\partial x} - \dfrac{d}{ds}(w_1(s)x_s(s)) + \dfrac{d^2}{ds^2}(w_2(s)x_{ss}(s)) = 0$$
$$\dfrac{\partial P}{\partial y} - \dfrac{d}{ds}(w_1(s)y_s(s)) + \dfrac{d^2}{ds^2}(w_2(s)y_{ss}(s)) = 0 \quad (29.23)$$

In case w_1 and w_2 are both constants, the equations will reduce to

$$\left. \begin{array}{l} \dfrac{\partial P}{\partial x} = w_1 x_{ss} - w_2 x_{ssss} \\ \dfrac{\partial P}{\partial y} = w_1 y_{ss} - w_2 y_{ssss} \end{array} \right\} \quad \nabla P = w_1 \mathbf{v}_{ss} - w_2 \mathbf{v}_{ssss} \quad (29.24)$$

These equations have the form $\mathbf{F}_{int} + \mathbf{F}_{ext} = 0$ with $\mathbf{F}_{ext} = -\nabla E_{ext} = -\nabla P$. The internal force $\mathbf{F}_{int}$ discourages stretching and bending while the external force $\mathbf{F}_{ext}$ pulls the snake toward the desired image edges. For an arbitrary initial contour chosen some distance away from the edges, the static equation Eq. (29.23) will not be satisfied, meaning that the internal forces will not be in equilibrium with the external forces. Hence, the deformable contour cannot keep its position and will move to a new position. Therefore, the equations should be made dynamic, that is, time dependent. This can be done by treating (x, y) as a function of time t and s, (i.e., $x(s, t)$, $y(s, t)$) and by setting the partial derivative of x, y with respect to t equal to the left-hand side of Eq. (29.23) as follows:

$$-\gamma x_t(s, t) = (w_1 x_s)_s - (w_2 x_{ss})_{ss} - \dfrac{\partial P}{\partial x}$$
$$-\gamma y_t(s, t) = (w_1 y_s)_s - (w_2 y_{ss})_{ss} - \dfrac{\partial P}{\partial y} \quad (29.25)$$

The parameter γ controls the speed with which the snake will move into a new position. When the solution $x(s, t)$, $y(s, t)$ stabilizes, the term $x_t(s, t)$, $y_t(s, t)$ vanishes and we obtain the solution for Eq. (29.23).

In order to better understand Eq. (29.25), one can compare them to the mass-spring equation:

$$m\frac{d^2x}{dt^2} + r\frac{dx}{dt} + kx = G_{\text{ext}} \qquad (29.26)$$

with m being a point mass, r the mechanical resistance, k the spring constant (Hooke's constant), and G_{ext} the external force.

Consider a point mass hanging down in a nonequilibrium starting position $x(t=0) = x_0$ at one end of a spring, whose other end is attached to the ceiling. The gravity force G_{ext} will start stretching the spring, and depending on the values of m, r, k, the system may arrive in the equilibrium situation ($kx(t \to \infty) = G_{\text{ext}}$) via a damped oscillation or without oscillations. If we do not consider second-order time derivatives, this means that we assume the snake to be massless. As seen in the dynamic snake formula of Eq. (29.25), γ plays the role of r. The second-order time derivative is absent (i.e., $m = 0$), which causes equilibrium to be reached without oscillations (at least for the continuous case). G_{ext} is related to $\frac{\partial P}{\partial x}$ and the rigidity term $(w_2 x_{ss})_{ss}$ is absent in the mass-spring equation Eq. (29.26), which is in fact completely linearized in the spatial term (kx), being related to $(w_1 x_s)_s$.[2]

29.4.2
Applications of Active Contours

For volumetric medical images, the 2D image in the slice above or below will not be too different from the current one. In such a case, user interventions can be limited to defining one snake contour in the current slice. Once this snake contour has converged to the real border of an object in the image, it can be reused as an initial contour in the slice above or below. By repeatedly using the converged contours as initial contours in adjacent slices, the whole volume can be segmented. For a dynamic volumetric set (i.e., volumetric images acquired at different time instants) the user interaction can also be kept minimal. If slices in similar positions are quite similar to each other at neighboring time instants, the converged contours of one time instant can be used as initial templates at the neighboring time instants. This is, for instance, the case for a dynamic sequence of volumetric MR cardiac images,

---

[2] In continuum mechanics $\ddot{\epsilon} = \frac{1}{2}((\nabla u_X))^\tau + (\nabla u_X)$ is the infinitesimal strain tensor, and $u(X, t) = x(X, t) - X$ is the displacement vector for position X toward a new position $x(X, t)$. The general equation for Hooke's law is $\ddot{\sigma} = \ddot{C} \otimes \ddot{\epsilon}$ with $\ddot{\sigma}$ being the Cauchy stress tensor and $\ddot{C}$ being the fourth-order tensor of material stiffness. Hence (x_s, y_s) in Eq. (29.25) is related to $u(X, t)$ because it expresses a first-order derivative, and (x_{ss}, y_{ss}) to $\ddot{\epsilon}$ because it expresses a second-order derivative.

where snakes can be used to segment the left ventricular volume for the different time bins [21].

29.4.3
The Behavior of Snakes

Figure 29.6 illustrates the dynamic evolution of the active contour. In general, one has to perform some experiments and observe how the contour behaves for different values of the parameters, to make a good choice for the w_1 and w_2 parameters.

If w_2 is chosen too high, it will be difficult to bend the contour, and it will not follow indentations. For very high values of w_2, the contour will take a circular shape. If w_1 is chosen too high, it will be difficult to stretch the contour, and it will also not be suited to follow indentations. For very high values of w_1 the contour will become very small, as stretching requires a considerable force (see Figure 29.7).

A shortcoming of the presented snake approach is that it cannot follow sharp indentations (see Figure 29.8) or be attracted from a reasonable distance toward the desired border (see Figure 29.9). A solution to these problems is the gradient vector flow snake [22]. Here the idea is to extend the range of the forces, by diffusing them from within their limited region around the actual border. The diffusion process will also create vectors that point into boundary concavities (see Figure 29.10). The classical external force field needs to be replaced by the Gradient Vector Flow (GVF) field: $\mathbf{F}_{ext}^{(g)} = \mathbf{g}(x, y)$. The GVF field $\mathbf{g}(x, y) = (u(x, y), v(x, y))$ can be computed from the classical gradient-based forces $-\nabla P = -(P_x, P_y)$ by two diffusion equations for the x and y components (i.e., in fact, the u and v components) as follows:

$$\mu \nabla^2 u - (u - P_x)(P_x^2 + P_y^2) = \gamma u_t$$
$$\mu \nabla^2 v - (v - P_y)(P_x^2 + P_y^2) = \gamma v_t$$
(29.27)

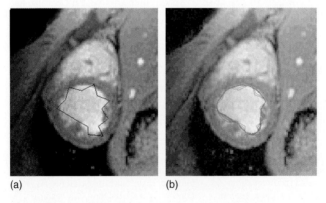

(a) (b)

Figure 29.6 (a) Initial snake contour interactively defined on a cardiac MR image and (b) visualization of the evolution of the snake contour.

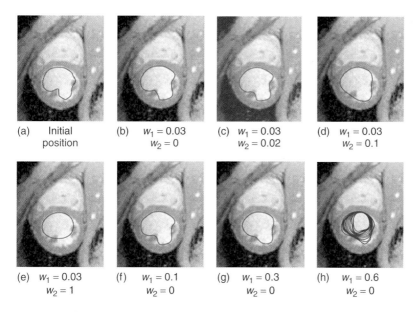

Figure 29.7 The influence of the rigidity and elasticity parameters on the behavior of the snake (w_1 represents the elasticity parameter and w_2 the rigidity parameter).

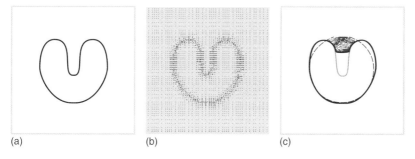

Figure 29.8 (a) Artificial edge image $E(x, y)$; (b) force field $-\nabla P = -\nabla |G_\sigma (x, y) * E(x, y)|$; and (c) the structure of the force field hampers the snake from following the indentation.

The stationary solution of these partial differential equations will correspond to the minimum of the energy functional:

$$E_{tot} = \frac{1}{2} \int \int \mu \left(u_x^2 + u_y^2 + v_x^2 + v_y^2 \right) + |\nabla P|^2 \cdot |\mathbf{g} - \nabla P|^2 \, dx \, dy \quad (29.28)$$

From this energy functional, it is clear that it will be minimized for strong edges by keeping $|\mathbf{g} - \nabla P|^2$ small; that is, by keeping $\mathbf{g} \approx \nabla P$ in the neighborhood of strong edges.

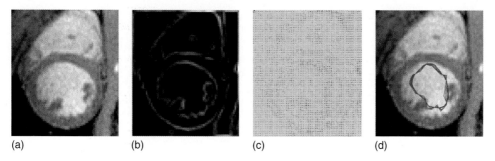

Figure 29.9 (a) Original image $I(x, y)$; (b) gradient magnitude $P = |\nabla(G_\sigma(x, y) * I(x, y))|$; (c) force field $-\nabla P$; and (d) the initial contour cannot be attracted toward the edge by the limited force field.

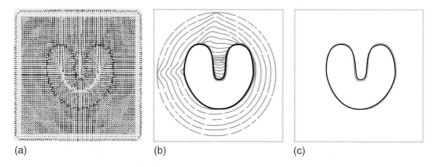

Figure 29.10 (a) Extended force field $F_{ext}^{(g)} = \mathbf{g}(x, y)$ for the artificial object of Figure 29.8(a); (b) the structure of the force field leads the snake into the indentation; and (c) final result.

When GVF snakes are used, the diffusion has to be applied on a binary image; otherwise, forces resulting from nonboundary pixels may disturb the diffused forces. Owing to some noise and inhomogeneities in the intensities, an edge detector (e.g., Canny) may find edges inside the blood volume of the left ventricle. If those edges lie within the diffusion distance (determined empirically by the number of iterations of the GVF algorithm), their diffused forces will interfere with each other and as a result, snake contours lying in between these borders will not be attracted (Figure 29.11 d).

A method (nonlinear in nature) that can reduce noise and at the same time preserve the edges is the anisotropic diffusion approach of [23]. We do not discuss the details of this algorithm in this chapter. The approach using anisotropic diffusion leads to satisfactory results [21] (see Figure 29.12).

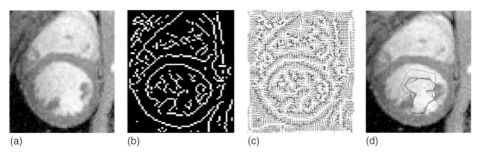

Figure 29.11 (a) Original image; (b) edge image obtained with Canny edge detection; (c) GVF force field $F_{ext}^{(g)} = \mathbf{g}(x, y)$; and (d) interference with weak edges.

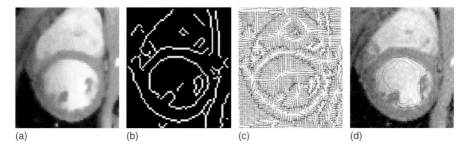

Figure 29.12 (a) Denoised image $I_{aniso}(x, y)$ obtained via anisotropic diffusion of the original image in Figure 29.11; (b) edge image obtained with Canny edge detection; (c) GVF force field $F_{ext}^{(g)} = \mathbf{g}(x, y)$; and (d) correct attraction of an initial contour lying far away from the target border.

29.5
Model-Based Segmentation

In this section, we briefly describe the outlines of a fairly general approach, that our research group has exploited in several applications [24–29][3]. The image analysis problem is modeled as a statistical inference problem (infer labels l_i from the observed data **x**). The approach consists of two main steps: the generation of an appropriate problem-specific data structure extracted from the original image(s), and a statistical labeling process based on Bayesian decision theory that uses Markov random fields (MRFs) as a convenient tool to express probabilities describing the prior knowledge about the observation model and the contextual scene information.

3) Examples can be found on the website of the book.

In this section, we only discuss the principles and do not provide any solution recipes.

29.5.1
Statistical Labeling

Let $\mathbf{S} = \{s_1, \ldots, s_M\}$ be a set of sites (i.e., image pixels, line segments, regions, complete images, etc.) and $\mathbf{L} = \{l_1, \ldots, l_K\}$ be the set of labels. The labeling process consists of assigning a label from $\mathbf{L}$ to each site from $\mathbf{S}$. Let $\mathbf{R} = \{R_1 \ldots, R_M\}$ be a family of random variables (i.e., a random field) defined on $\mathbf{S}$, in which each R_i can take a value in the set $\mathbf{L}$. A realization of $\mathbf{R}$ is called a *labeling* (a configuration of $\mathbf{R}$) and is denoted by $\mathbf{r} = \{r_1, \ldots, r_M\}$ with r_i belonging to $\mathbf{L}$ [30]. The set of sites $\mathbf{S}$ is characterized by a set of observations $\mathbf{x} = \{\mathbf{x}_1, \ldots, \mathbf{x}_M\}$ with $\mathbf{x}_i$ being the observation or feature vector corresponding to the site s_i.

29.5.2
Bayesian Decision Theory

Bayes' theorem makes use of observed data $\mathbf{x}$ and knowledge expressed via probability functions $p()$ and probabilities $P()$ (see Section 29.2.1):

$$P(\mathbf{R} = \mathbf{r}|\mathbf{X} = \mathbf{x}) = \frac{P(\mathbf{R} = \mathbf{r})p(\mathbf{X} = \mathbf{x}|\mathbf{R} = \mathbf{r})}{p(\mathbf{X} = \mathbf{x})} \tag{29.29}$$

$$\text{or in short notation} \quad P(\mathbf{r}|\mathbf{x}) = \frac{P(\mathbf{r})p(\mathbf{x}|\mathbf{r})}{p(\mathbf{x})} \tag{29.30}$$

Knowledge is contained in the observation model described by the conditional likelihood function $p(\mathbf{x}|\mathbf{r})$ and in the prior information about $\mathbf{r}$ carried by $P(\mathbf{r})$. The denominator of Bayes' theorem is a marginal probability function of $P(\mathbf{r}, \mathbf{x})$ and can be treated as a normalizing constant. Note that if the random field variables R_m are independent of each other and their ordering is also unimportant (i.e., all variables follow the same statistical distribution) this approach falls back to the one outlined earlier in Section 29.2.1.

29.5.3
Graphs and Markov Random Fields Defined on a Graph

The theoretical approach considering the labels from all sites as being dependent on each other is untreatable in practice. Therefore, a strong simplification needs to be introduced: that is, only labels of neighboring sites will be considered as being dependent of each other. In this case, $\mathbf{R}$ will be assumed to be an MRF. We now describe the general concepts of MRF as a tool for expressing priors for the labeling of images and/or sites, organized as a graph. Let $\mathbf{G} = \{\mathbf{S}, \mathbf{A}\}$ be a graph, where $\mathbf{S} = \{s_1, s_2, \ldots, s_M\}$ is the set of nodes and $\mathbf{A}$ is the set of arcs containing them.

We define a neighborhood system on $\mathbf{G}$, denoted by

$$\mathbf{N} = \mathbf{N}(s_1), \mathbf{N}(s_2), \ldots, \mathbf{N}(s_M) \tag{29.31}$$

where $\mathbf{N}(s_i)$ is the set of all nodes that are neighbors of s_i, such that (i) $s_i \in \mathbf{N}(s_i)$ and (ii) if $s_j \in \mathbf{N}(s_i)$ then $s_i \in \mathbf{N}(s_j)$.

The criterion that is used to decide whether a site is a neighbor of another one can be based on any distance measure.

R is an MRF on **G**, with respect to the neighborhood system **N** if and only if

$$P(r_i) > 0, \forall r_i \in L$$
$$P(r_i|r_j) = P(r_j|r_i), \forall j \neq i, \forall j : s_j \in \mathbf{N}(s_i)$$

where $P(r_i)$ and $P(r_i|r_j)$ are the prior and conditional probabilities, respectively.

29.5.4
Cliques

The neighborhood system **N** is used to define cliques, which are sets of sites, where each site is neighbor of all the others. A clique c, associated with the graph **G**, is a subset of **S** such that it contains either a single node, or several nodes, that are all neighbors of each other. If we denote the collection of all the cliques of **G**, with respect to the neighborhood system **N**, as **C(G, N)**, then the general form $P(\mathbf{r}|\mathbf{x})$ of a realization of **R** can be expressed by means of a Gibbs distribution.

29.5.5
Models for the Priors

Theory supports the equivalence of MRFs and Gibbs distribution [30]. By associating an energy function $U(\mathbf{x}|\mathbf{r})$ to $p(\mathbf{x}|\mathbf{r})$ and $U(\mathbf{r})$ to $P(\mathbf{r})$ respectively, we can express the posterior probability as

$$P(\mathbf{r}|\mathbf{x}) = \frac{1}{Z} \exp(-U(\mathbf{r}|\mathbf{x})) \qquad (29.32)$$

where $U(\mathbf{r}|\mathbf{x}) = U(\mathbf{x}|\mathbf{r}) + U(\mathbf{r})$

where $U(\mathbf{r}) = \sum_c V_c(\mathbf{r})$ and $U(\mathbf{x}|\mathbf{r}) = \sum_c V_c(\mathbf{x}|\mathbf{r})$

$U(\mathbf{r}|\mathbf{x})$ is called the *Gibbs energy* and $V_c(\mathbf{r})$ and $V_c(\mathbf{x}|\mathbf{r})$ the *clique potentials* defined on the corresponding cliques $c \in \mathbf{C(G, N)}$. Z is a normalizing constant. Given a configuration, we estimate the Gibbs energy as a sum of clique potentials $V_c(\mathbf{r})$ and $V_c(\mathbf{x}|\mathbf{r})$ established for each clique separately.

For a discrete labeling problem $V_c(\mathbf{r})$ and $V_c(\mathbf{x}|\mathbf{r})$ can be expressed by a number of parameters, as illustrated in Ref. [27] and mentioned in the supplementary section included on the website of the book.

29.5.6
Labeling in a Bayesian Framework based on Markov Random Field Modeling

In case of a *labeling problem*, when we have prior information and knowledge about the distribution of our data, the most optimal labeling of the graph **G** can be obtained on the basis of the selection of the appropriate *loss function* [31]. The MAP-MRF framework is a popular choice in which the most optimal labeling, given the observation field $\mathbf{x}$ is $\hat{\mathbf{r}} = \arg\min(U(\mathbf{r}|\mathbf{x}))$.

29.6
Conclusions

Three main classes of segmentation types controlled by the knowledge obtained by observation from the scene models have been presented. We have emphasized concepts and outlined recipes without going into algorithmic or computational details. Evaluation of segmentation and labeling results strongly depends on the particular target one wants to achieve. Literature is available both for domain-specific evaluation [32] and for more generic evaluation [33, 34]. No gold standard exists.

References

1. Pratt, W.K. (2007) *Digital Image Processing*, 4th edn, John Wiley & Sons, Inc., Rochester, USA.
2. Jahne, B. (1997) *Digital Image Processing*, 6th edn, Springer, Berlin/Heidelberg, Germany.
3. Gonzalez, R.C. and Woods, R.E. (2007) *Digital Image Processing*, 3rd edn, Prentice Hall Press, Upper Saddle River, NJ, USA.
4. Petrou, M. and Bosdogianni, P. (1999) *Image Processing: The Fundamentals*, 1st edn, John Wiley & Sons, Inc., Chichester, Sussex, GB.
5. Castleman, K.R. (1996) *Digital Image Processing*, Prentice Hall Press, Upper Saddle River, NJ, USA.
6. Rosenfeld, A. and Kak, A. (1982) *Digital Picture Processing*, 2nd edn, Academic Press, Inc., London, UK.
7. Hall, E.L. (1979) *Computer Image Processing and Recognition*, Academic Press, Inc., New York, USA.
8. Sonka, M., Hlavac, V., and Boyle, R. (2007) *Image Processing, Analysis and Machine Vision*, Thomson-Engineering, London, UK.
9. Mitchell, T.M. (1997) *Machine Learning*, 1st edn, McGraw-Hill Education, New York, USA.
10. Duda, R.O. and Hart, P.E. (1973) *Pattern Classification and Scene Analysis*, John Wiley & Sons, Inc., Chichester, Sussex, UK.
11. Hartley, H.O. (1958) Maximum likelihood estimation from incomplete data. *Biometrics*, **14**, 174–194.
12. Dempster, A.P., Laird, N.M., and Rubin, D.B. (1977) Maximum likelihood from incomplete data via the EM algorithm. *J. R. Stat. Soc. Series B Methodol.*, **39**, 1–38.
13. Cornelis, J. et al. (2000) Statistical models for multidisciplinary applications of image segmentation and labelling. Proceedings of the 16th World Computer Conference 2000 - 5th International Conference on Signal Processing, Beijing, China, vol. III/III, pp. 2103–2110.
14. Krishnamachari, S. and Chellapa, R. (1996) Delineating buildings by grouping lines with MRFs. *IEEE Trans. Image Process.*, **5**, 164–168.
15. Marr, D. and Hildreth, E. (1980) Theory of edge detection. *Proc. R. Soc. London, Ser. B, (Biological Sciences)*, **207**, 187–217.
16. Haralick, R. (1984) Digital step edges from zero crossing of second directional derivatives. *IEEE Trans. Pattern Anal. Mach. Intell.*, **6**, 58–68.
17. Canny, J. (1986) A computational approach to edge detection. *IEEE Trans. Pattern Anal. Mach. Intell.*, **8**, 679–697.

18. Lindeberg, T. (1990) Scale-space for discrete signals. *IEEE Trans. Pattern Anal. Mach. Intell.*, **12**, 234–254.
19. Kass, M., Witkin, A., and Terzopoulos, D. (1988) Snakes: active contour models. *Int. J. Comput. Vis.*, **1**, 321–331.
20. Singh, A., Terzopoulos, D., and Goldgof, D.B. (1998) *Deformable Models in Medical Image Analysis*, IEEE Computer Society Press, Los Alamitos, CA, USA.
21. Gavrilescu, A. et al. (2001) Segmentation of the left ventricle of the heart in volumetric cardiac MR images: Evaluation study of classical and GVF-snakes. Proceedings of the 8th International Workshop on Systems, Signals and Image Processing, Bucharest, Romania, pp. 73–76.
22. Xu, C. and Prince, J.L. (1998) Snakes, shapes, and gradient vector flow. *IEEE Trans. Image Process.*, **7**, 359–369.
23. Perona, P. and Malik, J. (1990) Scale-space and edge detection using anisotropic diffusion. *IEEE Trans. Pattern Anal. Mach. Intell.*, **12**, 629–639.
24. Katartzis, A., Vanhamel, I., and Sahli, H. (2005) A hierarchical Markovian model for multiscale region-based classification of vector valued images. *IEEE Trans. Geosci. Remote Sens.*, **43**, 548–558.
25. Katartzis, A. and Sahli, H. (2008) A stochastic framework for the identification of building rooftops using a single remote sensing image. *IEEE Trans. Geosci. Remote Sens.*, **46**, 259–271.
26. Katartzis, A. et al. (2001) A model-based approach to the automatic extraction of linear features from airborne images. *IEEE Trans. Geosci. Remote Sens.*, **39**, 2073–2079.
27. Deklerck, R., Suliga, M., and Nyssen, E. (2004) An intelligent tool to aid in understanding cortical brain anatomy visualized by means of flat maps. Proceedings of the 3rd IASTED International Conference on Web-Based Education, Innsbruck, Austria, pp. 474–479.
28. Katartzis, A. et al. (2002) Model-based technique for the measurement of skin thickness in mammography. *Med. Biol. Eng. Comput.*, **40**, 153–162.
29. Katartzis, A. et al. (2005) A MRF-based approach for the measurement of skin thickness in mammography, *Medical Image Analysis Methods*, CRC Taylor and Francis Group, Boca Raton, Florida, USA, pp. 315–340.
30. Li, S.Z. (1995) *Markov Random Field Modeling in Computer Vision*, Springer-Verlag, Berlin/Heidelberg, Germany.
31. Figueiredo, M.T. (2002) *Bayesian Methods and Markov Random Fields*. Department of Electrical and Computer Engineering, Instituto Superior Tecnico, Lisboa, Portugal, http://www.lx.it.pt/~mtf/ FigueiredoCVPR.pdf.
32. Chalana, V. and Kim, Y. (1997) A methodology for evaluation of boundary detection algorithms on medical images. *IEEE Trans. Med. Imaging*, **16**, 642–652.
33. Bagon, S., Boiman, O., and Irani, M. (2008) What is a good image segment? A unified approach to segment extraction. Proceedings of the 10th European Conference on Computer Vision, Marseille, France, vol. 5305, pp. 30–44.
34. Zhang, Y. (1997) Evaluation and comparison of different segmentation algorithms. *Pattern Recognit. Lett.*, **18**, 963–974.

30
Hybrid Digital–Optical Correlator for ATR

Tien-Hsin Chao and Thomas Lu

30.1
Introduction

Human beings are capable of recognizing objects and understanding images easily. However, a digital computer has a hard time in recognizing even a simple object in an unknown environment. Optical Fourier domain correlation architectures have the inherent advantage of parallel processing of images at the speed of light [1–3]. However, in the past, the optical system was perceived as bulky and fragile, and not practical for real-world applications. The advancement in the laser and electro-optic spatial light modulators (EO-SLMs) have made it possible to construct a miniaturized optical correlator (OC) with low power consumption and high-speed information processing [4–5]. Multiple composite filters can be synthesized to recognize broad variations in object classes, viewing angles, scale changes, and background clutters [6–11]. Digital neural networks (NNs) are used as a postprocessor to assist the OC to identify the objects and to reject false alarms [12]. A multistage automated target recognition (ATR) system has been designed to perform computer vision tasks with adequate proficiency in mimicking human vision. The system can detect, identify, and track multiple targets of interest in high-resolution images. The approach is capable of handling a large number of object variations and filter sets. Real-world experimental and simulation results are presented and the performance is analyzed. The tests results show that the system was successful in substantially reducing the false alarm rate while reliably recognizing targets.

To date, one of the most successful system architectures developed for optical processing is the OC [13]. OC has been developed for various pattern recognition applications. Owing to the inherent advantages of the vast parallelism and shift invariance, the data throughput rate of the state-of-the-art OC is 2–3 orders of magnitude higher than that of the digital counterpart. The schematic diagram of a typical updatable OC is shown in Figure 30.1. The laser beam emanating from a point laser source is used as the light source. Two spatial light modulators (SLMs) [14], one placed in the input plane and the other in the Fourier plane, are used for incoherent-to-coherent data conversion for the input data and the correlation filter

Optical and Digital Image Processing: Fundamentals and Applications, First Edition. Edited by Gabriel Cristóbal, Peter Schelkens, and Hugo Thienpont.
© 2011 Wiley-VCH Verlag GmbH & Co. KGaA. Published 2011 by Wiley-VCH Verlag GmbH & Co. KGaA.

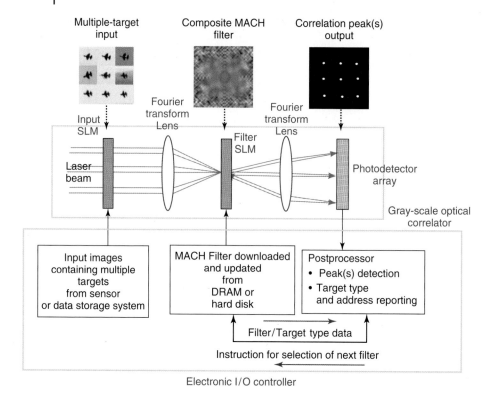

Figure 30.1 System functional block diagram of a spatial-light-modulator-based updatable optical correlator.

data encoding, respectively. A pair of thin diffraction-limited lenses is used for Fourier and inverse Fourier transforms. A two-dimensional photodetector array is inserted into the output plane for the detection of correlation data. During operation, the input data from a 2D sensor are fed into the input SLM. A set of precomputed correlation filter data is loaded into the filter SLM. The correlation output data are detected by the photodetector array. A digital controller is used to synchronize the input and filter downloading and the correlation output detection. Postprocessing, including peak detection and peak to side lobe ratio (PSR) computation, is also performed to identify the correlation peak signals. The location and target type data are also coregistered by the postprocessor. A detailed functionality description of the correlator is shown in Figure 30.2.

Since the correlator takes advantage of the optical Fourier transform that takes place at the speed of light, the correlation system's data throughput rate is limited only by the space–bandwidth product and the update speed of the input and the filter SLMs as well as the output photodetector array.

Initially, owing to the limitations of the modulation capabilities of the SLM used for the correlation filter encoding, the correlator filter had restricted the filter encoding to binary phase-only filters (BPOFs) and phase-only filters (POFs) [8].

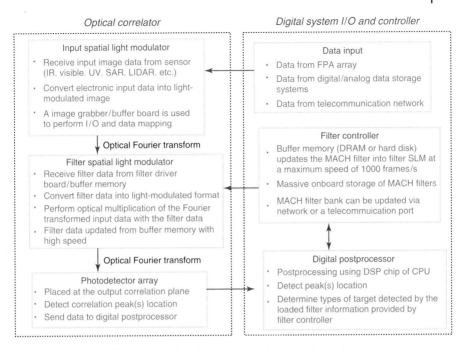

Figure 30.2 Detailed functionality of each component of an optical correlator.

Both BPOFs and POFs construct edge-enhanced filters for high-pass matched spatial filtering. Although both BPOFs and POFs provide high discrimination, they lack distortion tolerance among the same class of input images under recognition. Moreover, the BPOF and POF algorithms limited the development of optimum composite filter algorithms for distortion-invariant correlation. With the recent advancement in SLM technology, bipolar-amplitude modulation is realizable using a ferroelectric liquid crystal (FLC) SLM developed by the industry. Recently, researchers at the Jet Propulsion Laboratory (JPL) have taken advantage of this SLM to develop a bipolar-amplitude filter (BAF) technique for OC applications [15].

A gray-scale optical correlator (GOC) has been developed to replace the binary-phase only optical correlator (BPOC) [16–17]. In this correlator, a gray-scale input SLM is used to replace the previously used binary SLM. This has eliminated the need for a binarization preprocessing step. An FLC SLM (e.g., made by Boulder Nonlinear Systems) that is capable of encoding real-valued data is also used for the correlation filter implementation. This gray-scale optical correlator has enabled a direct implementation of a gray-scale-based correlator composite filter algorithm.

30.1.1
Gray-Scale Optical Correlator System's Space–Bandwidth Product Matching

The number of pixels in an SLM is described as the space–bandwidth product (SBP). SBP of the input SLM and that of the filter SLM should be well matched.

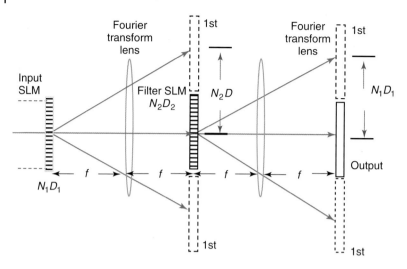

Figure 30.3 Matching the SBP of the pair of SLMs used in an optical correlator.

As illustrated in Figure 30.3, the formula is as follows:

$$\lambda f \frac{1}{D_2} = N_1 D_1 \tag{30.1}$$

where λ is the wavelength of the laser source, f is the focal length of the Fourier transform lens (assuming it is the same for both transform lenses), and D_1 is the pixel pitch of the input SLM. In order to avoid the aliasing problem, as shown in Eq. (30.1), λ and f are chosen such that the first-order diffraction of the input SLM, at the backfocal plane of the Fourier transform lens, is displaced by a distance equal to the aperture size, $N_2 D_2$, of the filter SLM. This can be expressed as

$$\lambda f \frac{1}{D_1} = N_2 D_2 \tag{30.2}$$

where D_2 is the pixel pitch of the filter SLM.
Therefore,

$$N_1 = N_2 \tag{30.3}$$

Thus, in order to optimize the spatial frequency matching between the input and the filter SLMs, and to avoid aliasing in the output plane (i.e., nonoverlapping of the correlation planes associated with zero and first orders of diffraction), the number of pixels (SBP) of the input and that of the filter SLM must be equal. Otherwise, the SLM that has less number of pixels will be the bottleneck of the correlator system. It is, therefore, very important to pick a pair of matching SLMs for optimum correlation performance.

30.1.2
Input SLM Selection

The primary system issues involved in the design of a high-resolution GOC (e.g., 512 × 512 pixel) include the following: gray-scale resolution, filter modulation method, and frame rate.

The first step is the selection of the SLM. Candidate SLMs for building a compact, 512 × 512 GOC, based on a recent survey, are listed in Table 30.1.

As shown in Table 30.1, both the transmission SLMs (Kopin) and the reflective SLMs (BNS) are qualified for building the 512 × 512 GOC. Two possible generic system architectures, as shown in Figure 30.4a,b, are available. Figure 30.4a shows the architecture utilizing a transmission input SLM coupled with a reflective filter SLM. Figure 30.4b shows the architecture utilizing reflective SLMs for both input and a filter encoding.

The selection of the GOC architecture will be determined by the requirements with respect to the SLM gray-scale resolution, frame rate (video rate or higher), and SLM pitch size (and therefore the ultimate system size). As shown in Table 30.1,

Table 30.1 Candidate SLMs for building a 512 × 512 GOC.

SLM vendor	Function	Speed (Hz)	SBP	Gray-scale resolution (bit)	Modulation	Pixel pitch (μm)
Kopin	Input SLM	60	640 × 480	8	Positive-real	24
	Input SLM[a]	60	1280 × 1024	8	Positive-real	15
BNS	Input/filter SLM	1000	512 × 512	8	Positive-real or bipolar-amplitude	7

[a] A laser source emanating at 632.8 nm wavelength is used for this estimation.

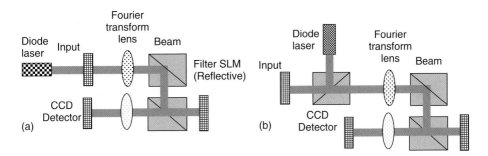

Figure 30.4 Two key architectures for 512 × 512 gray-scale optical correlator: (a) transmissive input SLM and reflective filter SLM and (b) reflective input SLM and reflective filter SLM.

Kopin SLM possesses high gray-scale resolution (up to 1280 × 1024) and modest frame rate (60 Hz). The minimum pixel pitch is currently 15 µm. Since a Kopin SLM exhibits only positive-real modulation of input data, it is primarily used as the input SLM. The lack of bipolar-amplitude modulation capability makes it less applicable for filter SLM applications. The BNS SLM, on the other hand, possesses performance characteristics of high-speed (1000 frames/s), positive-real or bipolar-amplitude modulation capability. It is usable for both input and filter SLMs. The current minimum pixel pitch of Kopin is also 15 µm. However, a new model with a 7 µm pixel pitch has recently become available.

As we have discussed in a previous section, in a typical $4f$ OC system, The SBP of the input SLM should be the same as that of the filter SLM. Otherwise, the SLM with the lower SBP will be the throughput bottleneck. This can be expressed as

$$N_1 = N_2 \tag{30.4}$$

where N_1 and N_2 are the SBPs of the input and filter SLMs, respectively.

In order to avoid aliasing effect, the system parameters of the OC would be selected as follows:

$$\lambda f = \frac{1}{D_2} = N_1 D_1 \tag{30.5}$$

where λ is the wavelength of the laser source, f is the focal length of the Fourier transform lens (assuming it is the same for both transform lenses), and D_1 and D_2 are the pixel pitch of the input SLM and filter SLM, respectively. The focal length f of the OC can be expressed as

$$f = \frac{N_1 D_1 D_2}{\lambda} \tag{30.6}$$

Thus, the focal length of the Fourier transform lens pair used in the OC will be proportional to the product of the pixel pitch of the input and the filter SLM. It is essential to shorten the pixel pitch to miniaturize the system length of a $4f$ OC.

The overall lengths of an OC versus the SLM pixel size are shown in Table 30.2. The most compact OC can be built using a pair of BNS SLMs with 512 × 512

Table 30.2 Length of optical correlator (OC) versus SLM pixel size.

Input SLM	Filter SLM	Focal length f (mm)	Overall length ($4f$) (mm)
Kopin 640 × 480 (24 µm)	BNS 512 × 512 (15 µm)	267.1	1068.4
Kopin 640 × 480 (24 µm)	BNS 512 × 512 (7 µm)	124.7	498.8
Kopin 1280 × 1024 (15 µm)	BNS 512 × 512 (15 µm)	167	668
Kopin 1280 × 1024 (15 µm)	BNS 512 × 512 (7 µm)	77.9	311.6
BNS 512 × 512 (15 µm)	BNS 512 × 512 (15 µm)	167	668
BNS 512 × 512 (7 µm)	BNS 512 × 512 (7 µm)	36.4	145.6

resolution, and system length of 145.6 mm. By folding the optical path twice, the dimension of the OC can be miniaturized to be less than 40 mm² ($\sim 2'' \times 2''$).

30.2
Miniaturized Gray-Scale Optical Correlator

Recently, a compact portable 512 × 512 GOC [17] was developed by integrating a pair of 512 × 512 ferroelectric liquid crystal spatial light modulators (FLCSLMs), a red diode laser, Fourier optics, and a CMOS photodetector array. The system is designed to operate at the maximum speed of 1000 frames/s. An field programmable gate array (FPGA) card was custom programmed to perform peak detection and PSR postprocessing to accommodate the system throughput rate. Custom mechanical mounting brackets were designed to miniaturize the optics head of the GOC to a volume of $6'' \times 3.5'' \times 2''$. The hardware and software for device drivers are installed in a customized PC. The GOC system's portability has been demonstrated by shipping it to various locations for target recognition testing.

30.2.1
512 × 512 GOC System Architecture

The architecture of the optical system is shown in Figure 30.5. A 30 MW laser diode emitting at 635 nm is used as the light source. Lens #1 and lens #2 are selected as the laser beam collimators. Lens #1 is mechanically mounted with adjustment capability in the z direction. This enables the minor tuning of collimated beam divergence and, therefore, the scale of the Fourier spectrum. The collimated beam passes through a mirror, a polarizing beam splitter cube, and waveplate #1, and impinges upon the 512 × 512 input FLCSLM. The reflected readout data passes through waveplate #1, Fourier transform lens #3, a second beam cube, and waveplate #2, and impinge upon the 512 × 512 filter FLCSLM. The readout beam from the filter SLM passes through the beam cube, a mirror, lens #4, a second mirror, and lens #5, and impinges upon the CMOS photodetector array. Lens #4 and lens #5 jointly perform scaled inverse Fourier transform to accomplish the optical correlation process. The scaling factor is designed to match the pixel difference between the input SLM and the CMOS detector (the pixel pitches of the input SLM and the CMOS detector are 5 and 12 µm, respectively).

Several auto-alignment mechanisms have been added to the system architecture to enable fine alignment. First, the collimator lens #1 and the Fourier transform lens #3 can be linearly translated in the z direction. This ensures that the Fourier transformed imager scale can be matched to the input plane of the filter SLM. One of the inverse Fourier transform lenses (lens #5) can also be translated in the z direction to permit sharp correlation plane (i.e., the input plane of the CMOS photodetector array) imaging as well as precise size matching. Furthermore, a piezo controller is mounted with the input SLM to provide auto-alignment with

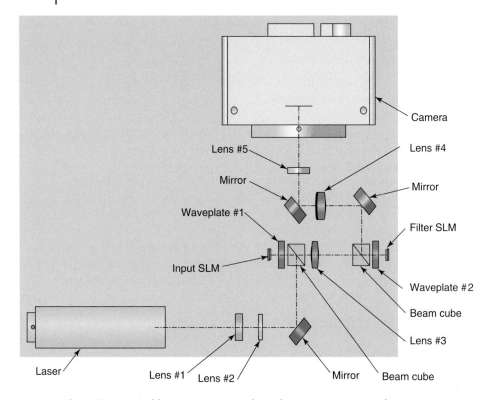

Figure 30.5 Optical layout: top view; mechanical components are not shown.

tip and tilt adjustability. A set of micrometers are also added to the filter SLM to provide manual 3D translation and rotation adjustment capability.

The system's mechanical design has been customized to develop a rugged portable GOC system. The overall packaged volume of the GOC hardware (not including the driver boards) is 6″ × 3.5″ × 2″. For more detailed illustrations, two photos of the 512 × 512 GOC system are shown in Figure 30.6a (side view) and Figure 30.6b (top view), respectively.

30.2.2
Graphic User Interface of the GOC System

A graphic user interface (GUI) has been developed as an interface between the GOC hardware and the controlling software to enable user-friendly operations. When this program is opened, the dialog shown in Figure 30.7 is displayed. This program is meant to demonstrate two modes of operation. The first mode, referred to as *Alignment Mode*, allows the user to hold the input and filter SLMs on an individual image and to view the correlation. The second mode, referred to as *Run Mode*, allows the user to sequence through a list of input and filter image combinations. With this program, the user has the ability to drive two SLMs, grab frames from the

Figure 30.6 (a) Photos of the 512 × 512 GOC system: (a) side view, the volume is 6″ × 3.5″ × 2″; (b) top view.

camera, analyze the correlation images, perform system synchronization, start and stop the program from sequencing, and control the frame rate, settling periods, and laser power. In addition, the GUI allows the user to easily generate an entire list of images to sequence through, each image with its own defined set of parameters used to drive the laser. This software was developed using kernel-mode interrupt drivers for the fastest possible system operation.

30.2.3
Gray-Scale Optical Correlator Testing

We have tested the GOC performance using a set of aircraft boneyard images. As shown in Figure 30.7, samples from one class of aircraft were selected and trained

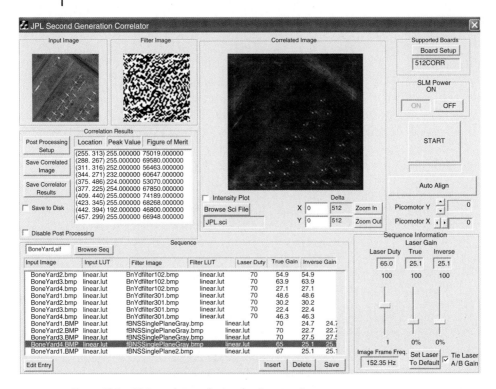

Figure 30.7 GOC emulator software showing experimental results of multiple target recognition of a type of aircraft from a boneyard image.

using the optimum-trade-off maximum average correlation height (OT-MACH) correlation filter algorithm [18]. The resultant optically implemented correlation filters were loaded in the filter SLM of the GOC. The correlation output results were postprocessed using the aforementioned PSR approach. The selected correlation peak signals are displayed and marked in square boxes.

The results are accurate with the selection of the PSR and peak value. The complete target recognition receiver operating characteristic (ROC) curves can be obtained by varying the PSR and peak values.

30.2.4
Summary

A rugged portable 512 × 512 GOC system for the automatic target recognition applications has been described in this section. The system employs two state-of-the-art 512 × 512 FLC SLMs and a CMOS photodetector array to handle the high-speed input and output data flow. Engineering designs in both optics and packaging mechanics have been customized to result in a compact and rugged system. The portable GOC has been successfully used as an ATR accelerating processor.

30.3 Optimization of OT-MACH Filter

Object recognition attempts to match the human ability to recognize targets in noisy environments. A good system must recognize real targets while minimizing false detections [19]. General filters are crucial in these systems for condensing a large image to smaller regions of interest (ROIs), where more specific and computationally intense techniques can then identify targets with better accuracy. The avenues of target recognition using gray-scale optical correlation (GOC) have been researched. One established GOC filter is the OT-MACH filter [18]. This optical Fourier filter allows nearly instantaneous calculation of the likely target locations in an image. The filter generalizes well and has the advantageous features of shift invariance, high speed, and parallelism. Computer simulations allow the most flexibility for developing the filter and testing its performance. The simulation uses the Fourier transforms of training images to construct a filter based on the measures of average correlation height (ACH), average similarity measure (ASM), average correlation energy (ACE), and output noise variance (ONV). The filter depends on the characteristic measures that are mentioned above, which are connected through the energy function:

$$E(\mathbf{h}) = \alpha(\text{ONV}) + \beta(\text{ACE}) + \gamma(\text{ASM}) - \delta(\text{ACH}) \tag{30.7}$$

The filter equation simplifies to one based on the α, β, and γ terms since the γ term is mostly insignificant. Thus, by choosing different values for these coefficients the filter can have different properties, allowing the OT-MACH filter to be flexible enough to emphasize different features for different types of target images.

While there is an optimal value for α, β, and γ for a given training set, it is unfeasible to solve this analytically. Presently, a manual solution is used to estimate the necessary changes in α, β, or γ to achieve improved detection rates. This approach finds solutions but it is time consuming and requires a user to repeat an iterative process.

Selection of α, β, and γ can be automated to speed up and improve the performance of the OT-MACH [19]. The goal of the automated program is to approach the optimal trade-off values of α, β, and γ in as few iterations as possible. The program approaches the optimum through an adaptive step gradient descent algorithm. The optimization is measured by the performance metrics of Correlation Peak Height (PK) and PSR, which are associated with the filter's target detection in true positive rate (TPR) and false positive reduction [20, 21].

30.3.1 Optimization Approach

The OT-MACH composite filter is the first step to identify potential targets in the images. The performance of the OT-MACH filter is sensitive to the values of the parameters α, β, and γ. It is critical to construct an effective filter that is broad enough to catch every target but is narrow enough to reject as many nontargets as possible. The first step taken to automate the filter generation process was to select the training images and properly size them to perform a correlation with a test image. Target images are cut out from the real images to make training images. Sometimes, abstract models of the targets may be used as training images since the models may be better representations of the general features of the real targets.

From these training images, half are used to generate the filter using initial seed values for α, β, and γ. An incremental constant is added to these values to measure the effect on the performance metrics in order to use the adaptive step gradient descent algorithm. The derivative is assumed to be linear since the change in α, β, and γ is small. The performance measures are calculated by testing the newly generated filters on the other half of the trained images, the ones not used to build the filter. The filter is correlated with the test images and the performance score is counted from the known target position from the test images. In order to reduce the performance score to a single metric, the correlation peak and PSR measures are combined in an equation favoring optimization of the peak or PSR value. The adaptive gradient descent algorithm maintains a user input minimum PSR threshold while the peak score is optimized. The increased peak score benefits the TPR at the cost of an increase in false positives. A higher PSR score makes the filter capable of discerning between true positives and false positives better. In some alternate cases, the filter can be optimized via PSR values, while retaining a minimum peak value.

The performance measures of correlation peak and PSR are used to calculate a first-order derivative that is used in the gradient descent algorithm to calculate new values for α, β, and γ. The algorithm repeats this process again using the same training images, except that this time the new values for α, β, and γ rather than the seed values are used. This loops through a set number of iterations as the filter is optimized by gradient descent [19]. A flowchart of the filter generation and optimization process is shown in Figure 30.8.

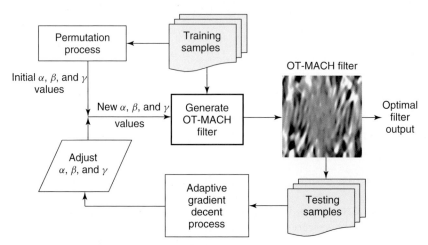

Figure 30.8 The OT-MACH automated filter generator and α, β, and γ optimizer.

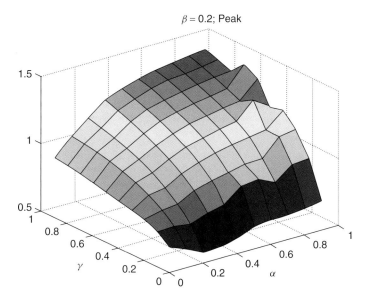

Figure 30.9 3D plot of peak value versus α versus γ with constant β. (Please find a color version of this figure on the color plates.)

Sometimes the OT-MACH filter optimizer converged to a local maximum, leaving room for error; thus, a more detailed examination was needed to find the best filter. To see the relationship between the OT-MACH parameters α, β, and γ and the correlation peak height and PSR values, a permutation method was run that tested a broad set of parameter values and graphed the results for easy recognition. The method holds one parameter constant while it varies the other two and graphs the corresponding output, as shown in Figures 30.9–30.11. The peak location indicates the global optimal values of the OT-MACH parameters α, β, and γ, as measured by the highest PSR value.

Once a performance method was chosen, the starting values of the parameters were obtained by approximating the plots in the Figures 30.9–30.11. The final values for the filters were obtained by using the filter optimization program. After the performance characteristics were set, the local region of optimization was chosen.

Filter optimization starts with creating a set of filters for a range of permutations of α, β, γ between 0 and 1 and measuring the performance of each filter on a set of test images containing the target. Performance was measured according to the weighted sum:

$$\text{Performance} = a(\text{peak}) + b(\text{PSR}) \tag{30.8}$$

where a and b are weights for the peak and PSR. Peak is the height of the correlation peak due to the target, and PSR is the peak-to-side lobe ratio, which is defined by

$$\text{PSR} = (\text{peak} - \text{mean})/\sigma \tag{30.9}$$

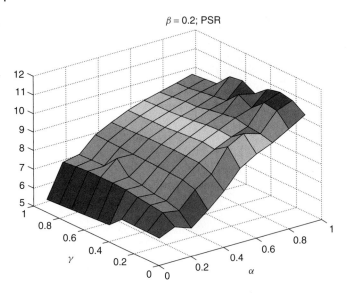

Figure 30.10 3D plot of PSR value versus α versus γ with constant β. (Please find a color version of this figure on the color plates.)

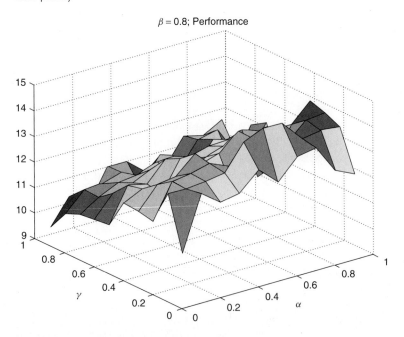

Figure 30.11 Sample PSR output from permutation procedure for filter optimization. Filter performance as a function of parameters α and γ, for $\beta = 0.8$. (Please find a color version of this figure on the color plates.)

In the equation above, peak is the correlation peak height, mean is the ACH of the region immediately around the peak, and σ is the standard deviation of the same region. It has been found in previous research that PSR is a better predictor of filter performance than peak, so the PSR value is given a higher weight. In many cases, for example, the parameters $[a, b] = [0.1, 1]$. Figures 30.9–30.11 show the performance of peak and PSR values with changes in the parameters α, β, and γ. In Figures 30.9 and 30.10, the peak and PSR values change smoothly. However, in Figure 30.11, we observe that the PSR value has many local peaks and valleys. In order to find the global optimal value, we need to use a permutation process before using gradient descent method. This permutation process produces a coarse estimation of the filter performance as a function of the parameters α, β, and γ. The parameter values with the highest performance are then used as a first guess for gradient descent.

The gradient descent algorithm attempts to maximize the performance of the filter by iteratively moving α, β, and γ in the direction of greatest performance increase. From an initial starting point, gradient at that point is approximated by measuring the performance of nearby points. The values of the filter parameters are adjusted in the direction of the gradient and the same process is repeated. This process undergoes a preset number of repetitions, converging to a maximum performance value. Since the gradient descent methods used can often converge to a local, rather than a global, maximum performance, it is very important to start with a "best guess" generated by the permutation process outlined above.

30.4
Second Stage: Neural Network for Target Verification

Once the OT-MACH filter has detected the likely ROIs that contain the target, the ROIs can go through more rigorous image processing and feature extraction and can be sent to a NN for verification. The objective was to design an efficient computer vision system that can learn to detect multiple targets in large images with unknown backgrounds. Since the target size is small relative to the image size in many problems, there are many regions of the image which could potentially contain a target. A cursory analysis of every region can be computationally efficient, but may yield too many false positives. On the other hand, a detailed analysis of every region can yield better results, but may be computationally inefficient. The multistage ATR system was designed to achieve an optimal balance between accuracy and computational efficiency by incorporating both models [22, 23]. NN has various advantages that are ideal for pattern recognition; some advantages of a biological NN that the artificial ones mimic are parallelism, learning ability, generalization ability, adaptability, fault tolerance, and low energy consumption [24, 25]. The NN must be trained on a specific training target set in order to have high accuracy. The combination of GOC and NNs provides speed and accuracy improvements to current ATR systems, allowing benefits for various applications.

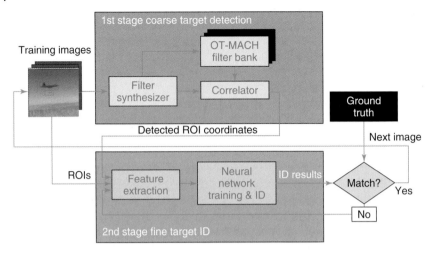

Figure 30.12 A multistage ATR system architecture.

As illustrated in Figure 30.12, the detection stage first identifies potential ROIs where the target may be present by performing a fast Fourier domain OT-MACH filter-based correlation. Since the threshold for this stage is usually set low with the goal of detecting all true positives, a number of false positives are also detected. The verification stage then transforms the ROIs into feature space and eliminates false positives using artificial NN classifiers. This section focuses on the design test of the feature extraction algorithms, the verification stage, and their performance in eliminating false positives.

30.4.1
Feature Extraction Methods

The ROIs identified by the detection stage represent small windows of the image which might contain a target. The dimensionality of these regions is the area of the window in pixel space. Even with small windows, this dimensionality is too large for inputting into a NN. Much of the pixel data from each window also contains redundancies which lack substantial information about the target itself. Thus, feature extraction techniques must be applied to transform the pixel data of each region into a smaller set of points that retains information about the target. Various feature extraction techniques have been described in the literature and the choice of technique to use is an important aspect in achieving optimal performance. Horizontal and vertical binning is a relatively simple feature vector that works relatively well to provide features for targets with low background noise.

30.4.1.1 Horizontal and Vertical Binning
Horizontal and vertical binning can be used as the feature extraction method for simple targets. As shown in Figure 30.13, the image windows representing the ROIs in Figure 30.13a were binned into horizontal and vertical columns.

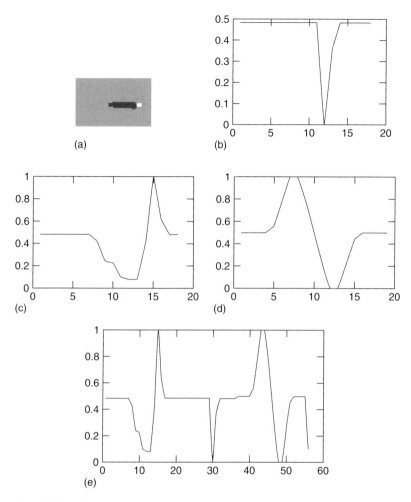

Figure 30.13 (a) A typical target model and its features; (b) vertical binning; (c) horizontal binning; (d) vertical binning of edge-enhanced image; and (e) combined feature for neural network training.

The intensities along these bins were summed to obtain points of the feature vector. Feature vectors of true positives were characterized by two well-defined dips representing areas of low intensity at the center of the horizontal and vertical bin profiles corresponding to the target shadow, as illustrated in Figure 30.13b,c. The edge feature is also binned, as shown in Figure 30.13d. All three features are combined into a single vector for NN identification, as shown in Figure 30.13e.

In this example, the first step was to test various window sizes for the best fit for the targets. After trying various window sizes, the one that was chosen was 56 × 24 pixels. This window size was big enough to capture most of the targets while minimizing the background size. When binned, this gives a vector of length 80.

To reduce the size of the vector, the image is resized to a smaller window before binning so that it contains all the features of the original image without as many pixels. This size of window was chosen to fit the entire target into it; the size is large compared to the target, but some targets may have different shapes and sizes. Since the targets would have various shapes, sizes, and contrasts, more than one ROI was binned for the feature vector to try to acquire as many features of the targets as possible. Many different filters were then tested to find the one that created the most unique feature vector for the targets. In this example, two different edge filters were tested: the Prewitt and Sobel filters. These edge filters generated too much noise in the image when applied to the ROI; so, a blur filter was applied before the edge filter, but the resulting image was still too noisy to give a unique feature vector of the target. The next filter tested was a custom match filter. The general features of the target in this example in Figure 30.13a were taken to produce the match filter with a light head and a long dark tail that contrasted with the background. A custom edge filter was also tested at this time. Of the custom filters, the edge filter had the best correlation with the targets. Various blur filters and unsharp filters were also tested, and the final feature vector is explained below.

The original ROI was first filtered with a blur filter and binned to make the first section of the feature vector. A custom edge filter was created and applied to the blurred image to append as the second part of the feature vector. The last section of the feature vector was the wavelet transform of the window. A sample feature vector is described below. Statistical measures were also taken for the sonar feature windows. The feature vector of each window was processed, and ranges outside the acceptable level were set to zero.

30.4.1.2 Principal Component Analysis

Principal component analysis (PCA) is a form of dimension reduction that attempts to preserve the variance of the original training data. Given a sphered training data matrix in the form:

$$X^T = \begin{bmatrix} x^1 \\ \vdots \\ x^n \end{bmatrix} \tag{30.10}$$

where x^i represents the ith training vector, the covariance matrix C is calculated as follows:

$$C = \frac{1}{n-1} \sum_{i=1}^{n} (x^i - \mu)(x^i - \mu)^T \tag{30.11}$$

where μ represents the column means of X^T. The diagonal eigenvalue matrix D and the corresponding eigenvector matrix V are then computed:

$$C = VDV^{-1} \tag{30.12}$$

Intuitively, the columns of V represent directions in the feature space in which the training data vary. The degree to which the training data vary in that direction

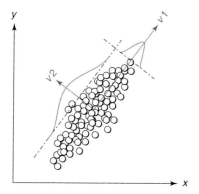

Figure 30.14 PCA performed on a two-dimensional dataset.

is determined by each vector's corresponding eigenvalue, found in D. In PCA, a reduced matrix V_k is formed by selecting columns of V that have large eigenvalues. Data are then projected onto this reduced eigenbasis using the transform:

$$Y = V_k^T X \tag{30.13}$$

This effectively reduces the dimension of the data while maintaining much of the original variance, information, and data structure.

Figure 30.14 illustrates the PCA process. It attempts to find the largest common vector **v1**, then find the next common vector **v2** that is orthogonal to **v1**, and so on. With only a small number of PCA vectors, one can reconstruct the original image very well, as shown in Figure 30.15. Figure 30.15a is the original image, and Figure 30.15b is the image reconstructed from only 18 PCA vectors. There is very little difference between the original targets and the targets reconstructed using only 18 out of 1100 eigenvectors. This represents a reduction of 98% of information.

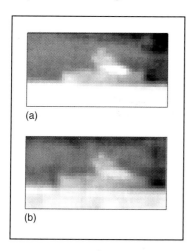

Figure 30.15 (a) Original ROI and (b) ROI reconstructed after PCA with 18 components.

30.4.2
Neural Network Identification

An NN has the characteristics of emulating the behavior of a biological nervous system and draws upon the analogies of adaptive biological learning [12]. Because of their adaptive pattern classification paradigm, NNs possess the significant advantages of a complex learning capability plus a very high fault tolerance. They are not only highly resistant to aberrations in the input data but are also excellent at solving problems too complex for conventional technologies such as rule-based and determinative algorithm methods. A unique contribution of NNs is that they overcome many of the obstacles experienced with conventional algorithms. Figure 30.16 shows the NN ATR process.

30.4.2.1 Neural Network Algorithm

An artificial NN can be used to classify feature vectors as true or false positives [25]. The standard feed-forward back-propagation variant was chosen for this task. Such NNs have been successfully employed in image retrieval and object tracking systems. This computational model operates under the supervised learning paradigm. A set of input–output pairs was classified by hand to serve as the training set. Two separate NNs were created and trained to classify feature vectors of the datasets as true or false positives. Cross-validation was used to improve generalization. The trained NNs substantially reduced the number of false positives per image from the detection stage results in both datasets.

The variant used in an NN system is the classic feed-forward back-propagation model. It is composed of separate layers of connected units called *neurons*. Every neuron of one layer is connected to every neuron of the next layer and each connection has an assigned weight value w. The output of a neuron y is calculated by a weighted sum of the inputs, x, into that neuron [26]:

$$y_j = f(X) = K(\text{Sum}(w_i x_i)) \qquad (30.14)$$

where K is a transfer function that maps in the range (0, 1).

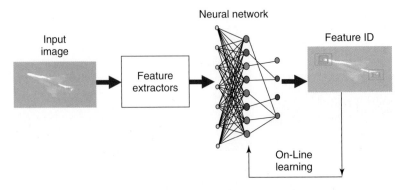

Figure 30.16 Functional block diagram of a neural network ATR process.

The feature vector serves as the initial input into the neurons of the first layer. The output of neurons of one layer then feeds into the neurons of the next layer until the output layer returns a value in the interval [0, 1]. This architecture is known as *feed-forward*.

Since the connection weights are initially randomized, an untrained NN models a completely random function. In order to classify data, it must learn from a training set that contains pairs of input, a feature vector, and an output, either 0 or 1. We arbitrarily choose to denote false positives by 0 and true positives by 1. The back-propagation algorithm performs gradient descent on the NN to shift the weights in a direction that minimizes the error between the network and training set outputs using a performance metric such as mean square error.

The training and test dataset consisted of images, each containing targets and nontargets. For example, sonar images of mines were characterized primarily by the presence of their shadows. Owing to the lack of many distinguishing features as well as the varying shapes and sizes of the shadows, the detection stage correlation filter detects many "mine-like" objects on this dataset and threshold was lowered to insure the initial detection of the mines. This resulted in hundreds of ROIs per image, of which the vast majority were false positives. A training set was designated by randomly selecting a subset of images from the initial training set. This provided an initial set of true positive and many false positive ROIs. The false positive regions were culled by only taking false positives from selected few images to prevent redundant vectors from being overexpressed. The true positive regions were offset by a number of pixels in all four directions to improve generalization in the cases where regions were off center and this was repeated until a roughly equal number of true and false positive vectors were obtained.

The last part of the process is training the NN. A program was created to visualize the groups of positive and negative feature vectors to aid in training. After analysis of both groups, a set of positive and negative feature vectors was chosen for training. The training sets were split into three fractions: 70% was trained on using back-propagation, 15% was used for validation, and the remaining 15% was used for independent verification of the networks ability to generalize. The feature vector data were then input to a feed-forward back-propagation NN and simulations were run.

30.5
Experimental Demonstration of ATR Process

30.5.1
Vehicle Identification

The following is a simulated vehicle identification problem. The aerial imaging camera looks for vehicles in a low contrast image. The vehicle has 3D models. The filter needs to be tolerant to the rotation and aspect changes. Using the above-mentioned OT-MACH filter algorithm, one can generate distortion-tolerant composite filters. FFT-based filters are shift invariant, but they are still sensitive to scale, rotation, perspective, and lighting conditions. In addition, targets may also be partially occluded or camouflaged, or may be decoys. In order to achieve the highest level of assurance that the targets are identified with minimal errors while the number of filters is reasonably small, all desired variations and distortions in targets need to be carefully included systematically when generating filters. A filter bank, in which every filter covers a small range of variations, is generated for the targets, and a set of filters covers the full expected range of those variations.

A software tool is used to streamline the filter synthesis/testing procedure effectively while maintaining filter performance. In order to cover all potential target perspectives with up to six degrees of freedom, a large bank of composite filters has to be generated. The filter training images can be obtained either from real-world examples or from a 3D synthesizer. Figure 30.17 shows a subset of the example training images with changing rotation angles. The corresponding OT-MACH filters were generated using the optimal filter process, as shown in Figure 30.18.

It may take well over 1000 filters to cover a range of aspect and scale changes in 3D space of an object. Filter bank management software is needed to keep track of the filters and update the filter bank after new filters are generated.

A set of training object images was extracted from the 3D models. The sampling interval is 5° angular and 7% scale. For a given input image, on the basis of its camera information, a set of 24 filters is generated to cover 15° roll and pitch angle variation and 360° yaw angle. The objects are embedded in arbitrary background

Figure 30.17 A set of training images for composite filter synthesis.

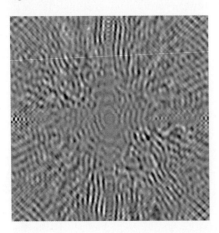

Figure 30.18 An example of the OT-MACH filter.

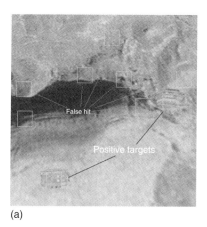 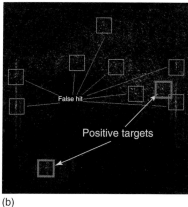

Figure 30.19 (a) An input image with objects embedded and (b) the correlation plane shows the detection of the object and the false alarm of the clutters. (Please find a color version of this figure on the color plates.)

images. The intensity of the image areas varies dramatically to simulate the climate variation and urban changes. Figure 30.19a shows an image with two embedded objects. Figure 30.19b shows the correlation plane. The peaks are located in the object coordinates, indicating the detection of the objects. However, some clutters also cause false peaks in the correlation plane. A PSR method can increase the reliability of the correlation detection. However, owing to the complexity of the background clutters, further postprocessing is needed to eliminate the clutters.

A multilayer feed-forward NN is used to train with the object features. Figure 30.20 shows an example of the gradient features extracted from the targets and nontargets. The gradient features capture the edge information of a generally rectangular object. The relative locations, magnitudes, and the signs of the edges are shown as the peaks of the feature vectors. The clutters show random patterns in the gradient feature vectors. The object features are used to train the NN. The clutter features are used as counterexamples of the NN training. The NN is then used to examine the content of each peak location in the correlation plane. In the preliminary experiments, over 90% of the false alarms are eliminated by the NN. From the initial experimental results, we can see that the NN, as a postprocessor, is capable of reducing false alarms significantly from the correlation results, increasing the sensitivity and reliability of the OC.

30.5.2
Sonar Mine Identification

In another example, a sonar image set was used to test the NN classification. For the final results, we used a frequency-relative operating characteristic (FROC) curve to test the performance of the system, plotting the average false positives per image

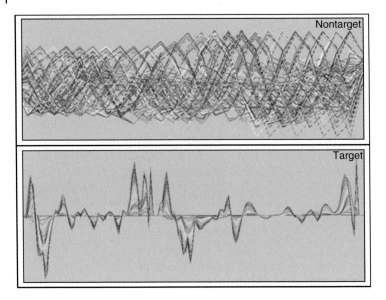

Figure 30.20 Examples of the target and nontarget gradient features for the neural network. (Please find a color version of this figure on the color plates.)

against the TPR. The threshold of the NN output was varied between 0 and 1 while measuring the TPR versus false alarms.

The NN is repeatedly trained with missing targets and false positives after each successive recognition operation. The NN gets better after more training routines, as shown in Figure 30.21. The points of interest on FROC curve are at 90% detected rate with 2.5 false positives per image and 96.5% detected with 10 false positives per image. The FROC curve in Figure 30.22 shows the performance of various NN training stages. The performance increases as training is carried out and better training sets are used.

Different NN training algorithms were tested to find the best performance. The FROC curve in Figure 30.22 shows the performance of the best four algorithms. The algorithms tested were cyclical order weight/bias, gradient descent, resilient back-propagation, and the Levenberg–Marquardt training algorithm [25]. As seen from the plot, the Levenberg–Marquardt algorithm has the best average performance.

30.6
Conclusions

The combination of the OT-MACH filter and the NN is effective in ATR systems with appropriate optimizations. The optimization of the OT-MACH filter and the trained NNs substantially reduced the number of false positives per image from

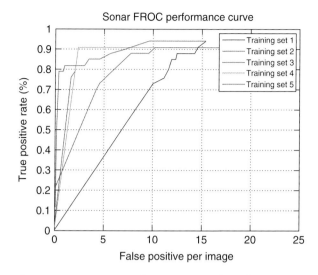

Figure 30.21 FROC curve of sonar targets. (Please find a color version of this figure on the color plates.)

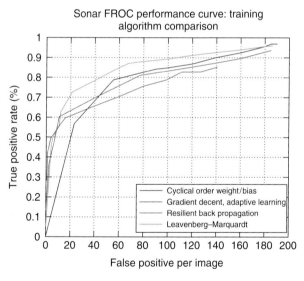

Figure 30.22 Comparing the performance of different NN learning algorithms. (Please find a color version of this figure on the color plates.)

the detection stage results in both datasets. The false positive rate per image was significantly reduced by using this strategy. To improve accuracy, targets with a wide variety of shapes require a more sophisticated approach to feature extraction techniques. The binning of filtered images did not give a unique enough feature vector for the NN to completely discriminate targets from background features and the large amounts of noise in the sonar images. We also showed that the process is generalizable to work with targets of varied structure.

Future work could improve the accuracy of both the detection stage and the verification stage. The reduction in false positives in the detection stage will lead to improved accuracy in both the verification stage and overall performance. Another area for improvement is the reduction in computational time in the algorithm. The work was carried out by a simulation that was much slower than the system implemented in hardware, but making improvements in the algorithm would benefit the overall timing of the system. The major areas for improvement are the Fourier transform and the feature extraction phase, which are computationally expensive. The algorithm is set up for a very specific type and size of image. This could be made more adaptable for all types of images. A process that could be more easily generalized to multiple images and targets would be very beneficial. OC implementation could speed up the ATR process by 2–3 orders of magnitude.

Acknowledgments

This research was carried out at the Jet Propulsion Laboratory, California Institute of Technology, under a contract with the National Aeronautics and Space Administration (NASA). The authors would like to acknowledge Dr. Jay Hanan, Dr. H. Zhou, Mr. C Hughlett, Mr. H. Braun, Mr. W. Edens, Mr. D. Yee, Mr. W. Greene, and Mr. O. Johnson for useful help and discussions.

References

1. VanderLugt, A. (1964) Signal detection by complex spatial filtering. *IEEE Trans. Inf. Theory*, **IT-10**, 139–145.
2. Weaver, C.S. and Goodman, J.W. (1966) A technique for optically convolving two functions. *Appl. Opt.*, **5**, 1248–1249.
3. Horner, J. L. (1987) *Optical Signal Processing*, Academic Press, New York.
4. Vijaya Kumar, B.V.K. (1992) Tutorial survey of correlation filters for optical pattern recognition. *Appl. Opt.*, **31**, 4773–4801.
5. Yu, F.T.S. and Gregory, D.A. (1996) Optical pattern recognition: architectures and techniques *Proc. IEEE*, **84** (5), 733–752.
6. Casasent, D. and Psaltis, D. (1976) Scale invariant optical correlation using Mellin transforms. *Opt. Commun.*, **17**, 59–63.
7. Hsu, Y.N. and Arsenault, H.H. (1982) Optical pattern recognition using circular harmonic expansion. *Appl. Opt.*, **21**, 4016–4019.
8. Casasent, D. and Chao, T.-H. (eds.) (1991–2008) *Optical Pattern Recognition I-XVIII*, Proceedings of the SPIE, SPIE.
9. Towghi, N., Javidi, B., and Li, J. (1998) Generalized optimum receiver for pattern recognition with multiplicative, additive, and nonoverlapping noise. *J. Opt. Soc. Am. A*, **15**, 1557–1565.
10. Haider, M.R., Islam, M.N., and Alam, M.S. (2006) Enhanced class associative generalized fringe-adjusted joint transform correlation for multiple target detection. *Opt. Eng.*, **45** (4), 048201.
11. Poon, T.C. and Kum, T. (1999) Optical image recognition of three-dimensional objects. *Appl. Opt.*, **38** (2), 370–381.
12. Jain, A.K., Mao, J., and Mohiuddin, K.M. (1996) Artificial neural networks: a tutorial. *IEEE*, 0018-9162/96 March 29 (03), 31–44.

13. Chao, T.H., Reyes, G., and Park, Y.T. (1998) Grayscale optical correlator, *Optical Pattern Recognition IX*, Proceedings of SPIE, SPIE, Vol. 3386, 60–64.
14. Serati, S.A., Sharp, G.D., and Serati, R.A. (1995) 128 × 128 analog liquid crystal spatial light modulator *Proc. SPIE*, **2490**, 378–387.
15. Chao, T.H., Reyes, G., and Zhou, H. (1999) Automatic target recognition field demonstration using a grayscale optical correlator. *Proc. SPIE*, **3715**, 399–406.
16. Chao, T.H., Zhou, H., and Reyes, G. (2001) Grayscale optical correlator for real-time onboard automatic target recognition *Proc. SPIE*, **4387**, 10–15.
17. Chao, T.H. and Zhou, H.R. (2000) 512 × 512 high-speed grayscale optical correlator. *Proc. SPIE*, **4043**, 90–93.
18. Zhou, H. and Chao, T.-H. (1999) MACH filter synthesizing for detecting targets in cluttered environment for gray-scale optical correlator. *SPIE*, **0229**, 399–406.
19. Johnson, O.C., Edens, W., Lu, T., and Chao, T.-H. (2009) Optimization of OT-MACH filter generation for target recognition. SPIE defense, Security, and Sensing, Optical Pattern Recognition XX, Vol. 7340-6.
20. Hanan, J.C., Chao, T.H., Assad, C., Lu, T., Zhou, H., and Hughlett, C. (2005) Closed-loop automatic target recognition and monitoring system. *Proc. SPIE*, **5816**, 244–251.
21. Lu, T., Hughlett, C., Zhou, H., Chao, T.H., and Hanan, J.C. (2005) Neural network postprocessing of grayscale optical correlator, *Optical Information Processing III*, Proceedings of the SPIE, SPIE, Vol. 5908, 291–300.
22. Ye, D., Edens, W., Lu, T., and Chao, T.-H. (2009) Neural network target identification system for false alarm reduction. SPIE Defense, Security, and Sensing, Optical Pattern Recognition XX, Vol. 7340-17.
23. Lu, T. and Mintzer, D. (1998) Hybrid neural networks for nonlinear pattern recognition, in *Optical Pattern Recognition* (eds. F.T.S. Yu and S. Jutamulia), Cambridge University Press, 40–63.
24. Chao, T.-H. and Lu, T. (2010) Accelerated sensor data processing using a multichannel GOC/NN processor. *Proc. SPIE*, **7696C**.
25. Lin, T.H., Lu, T., Braun, H., Edens, W., Zhang, Y., Chao, T.-H., Assad, C., and Huntsberger, T. (2010) Optimization of a multi-stage ATR system for small target identification. *Proc. SPIE*, **7696C**, 76961Y–76961Y-10.
26. Greene, W.N., Zhang, Y., Lu, T., and Chao, T.-H. (2010) Feature extraction and selection strategies for automated target recognition, SPIE symposium on defense, security & sensing conference, independent component analyses, wavelets, neural networks, biosystems, and nanoengineering. *Proc. SPIE*, **7703**, 77030B–77030B-11.

31
Theory and Application of Multispectral Fluorescence Tomography

Rosy Favicchio, Giannis Zacharakis, Anikitos Garofalakis, and Jorge Ripoll

31.1
Introduction

Optical imaging technologies are being increasingly employed in various biomedical and biological applications where the localization and quantification of fluorescence signal is required. In this respect, optical tomographic methodologies are an optimal approach owing to the volumetric and quantitative capabilities they offer. The low cost and the nonionizing nature of the radiation represent further incentives toward this research field and optical technologies, in many cases, are thus being considered very attractive alternatives to established techniques such as X-ray CT, MRI, and PET. The development of optical tomographic approaches has been driven by technological advances in light-capturing techniques (more sensitive CCD cameras) and advances in theoretical optics for the description of the propagation of light in tissue. Particular attention must be directed toward understanding the physics involved in the propagation of light inside biological media, which is governed by absorption and scattering and is highly diffusive. Furthermore, nonspecific tissue fluorescence can be a significant impairing factor, masking the signal from deep-seated fluorescence targets. It is therefore of high importance to design experimental arrangements that can take advantage of specific spectral features enabling the separation of different fluorescence activities as well as three-dimensional rendering. Spectral unmixing is a method that can be used to simultaneously detect emission from fluorescence targets with overlapping spectra. In this chapter, investigations *in vitro* as well as *in vivo* are used to demonstrate the ability of such a combined methodology in imaging of biological applications. In the future, multimodal technological advancements will undoubtedly improve sensitivity, resolution, and applicability of optical methods in both biological and translational research.

Fluorescence molecular tomography (FMT) incorporates the principles of diffuse optical tomography (DOT) with the use of fluorescent probes or fluorescing proteins as a source of contrast [1–4]. In DOT, the inherent optical properties of the subject, such as the scattering and absorption coefficients and the changes they experience during the investigated biological processes, are the source of contrast. Over the

Optical and Digital Image Processing: Fundamentals and Applications, First Edition. Edited by Gabriel Cristóbal, Peter Schelkens, and Hugo Thienpont.
© 2011 Wiley-VCH Verlag GmbH & Co. KGaA. Published 2011 by Wiley-VCH Verlag GmbH & Co. KGaA.

last decade, the field of biomedical optics has witnessed the use of diffuse light to quantify oxy- and deoxy-hemoglobin [5, 6], to characterize constituents in tissue [7, 8], to detect and characterize breast cancer [9–11], and to be used as a diagnostic tool for cortical studies [5], for arthritic joint imaging [12], and for functional activity in the brain [13, 14] and muscle [15]. Optical tomography of tissues could further become a valuable tool for drug development studies [3, 16].

In FMT, the sample that carries the fluorescent probe is exposed to light from different source positions and the emitted light is captured by detectors arranged in a spatially defined order (typically a CCD), either in reflectance or transmittance geometry. The accumulated information is then mathematically processed, resulting in a reconstructed tomographic image. In many of the current biological and biomedical studies, more than one fluorescing protein or fluorescent probe needs to be imaged simultaneously and more importantly distinguished and followed independently. To accomplish this, multispectral optical imaging needs to be applied. Multispectral imaging is a relatively new approach in the biomedical field, which combines the advantages of two already established optical modalities: optical tomography and optical spectroscopy [17, 18]. The combined use of these two technologies provides the capability to image accurately, quantitatively, and independently, the processes that have overlapping emissions and dependent temporal evolutions.

In this chapter, we discuss the principles of fluorescence tomography combined with a multispectral approach for the detection of multiple fluorescing targets. The spectral unmixing method is described and, finally, some applications are discussed for both *in vitro* and *in vivo* studies.

31.2
Fluorescence Molecular Tomography (FMT)

FMT is an important tool for imaging the spatial distribution of the concentration of fluorescent probes inside small animal models [2, 4]. FMT can resolve molecular signatures in deep tissues by combining modeling of light propagation in tissues by means of the diffusion approximation, with the use of fluorescent probes or proteins. For FMT image acquisition, the sample that carries the fluorescent probe is exposed to light at multiple source positions and the emitted light is captured by photon detectors, such as a CCD camera. Even though the spatial resolution of FMT is limited to approximately 1 mm, the fluorescent probes used offer information of interactions occurring at the molecular level [2, 3].

31.2.1
FMT Principle

The general scheme for a noncontact fluorescence imaging setup [19, 20] is shown in Figure 31.1. A laser beam of wavelength λ_a illuminates a subject that contains a spatially distributed concentration of specific fluorophores. The light propagates

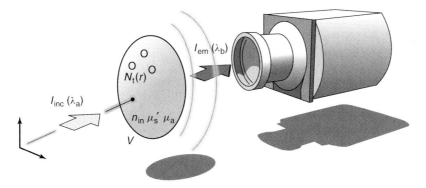

Figure 31.1 Schematic representation of a fluorescence tomography arrangement. I_{inc} is the excitation source illuminating the subject of volume V at different positions and producing the emitted light I_{em} due to the fluorescent molecules with concentration $N_t(\mathbf{r})$ that is then captured by the CCD detector.

inside the medium and excites the fluorophores, which, in turn, emit light at a different wavelength λ_b. The emitted light is captured by an objective lens, which is focused on the sample and recorded by a CCD camera. Different wavelengths are distinguished by appropriate bandpass filters placed in front of the camera lens.

The tomography setup is composed of three main components: the source, the subject, and the detector. The most common illumination source is a laser beam I_{inc} in position $\mathbf{r}$, pointing at a direction $\mathbf{s}$, with a wavelength λ_a. The medium of volume V is a highly scattering medium, such as tissue, and contains fluorophores with concentration $N_t(\mathbf{r})$. The optical properties of the medium are given by the absorption coefficient $\mu_a(\mathbf{r})$, the reduced scattering coefficient $\mu'_s(\mathbf{r})$, and the average index of refraction n_{in}. These optical properties, in addition to the fluorescent concentration, are assumed constant during an experiment. Finally, the term I_{em} expresses the light that is emitted from the fluorophores and leaves the surface of the sample.

31.2.2
Theoretical Background

The propagation of light in biological tissue is typically modeled by the radiative transfer equation (RTE) [21] for the specific intensity I:

$$\frac{n}{c}\frac{\partial I(\mathbf{r},\mathbf{s})}{\partial t} = -\mathbf{s}\nabla I(\mathbf{r},\mathbf{s}) - \mu_t I(\mathbf{r},\mathbf{s}) + \frac{\mu_t}{4\pi}\int_{4\pi} p(\mathbf{s},\mathbf{s}')I(\mathbf{r},\mathbf{s}')d\Omega' + \varepsilon(\mathbf{r},\mathbf{s}) \quad (31.1)$$

where n is the refractive index of the medium and c the speed of light in vacuum, μ_t is the transport coefficient, $\varepsilon(\mathbf{r},\mathbf{s})$ is the power radiated by the medium per unit volume and per unit solid angle in direction $\hat{\mathbf{s}}$, $p(\mathbf{s},\mathbf{s}')$ is the scattering function that relates the incoming angle $\mathbf{s}'$ with the outgoing direction $\mathbf{s}$, and $d\Omega'$ is the differential solid angle. In Eq. (31.1) $I(\mathbf{r},\mathbf{s})$ is the specific intensity, which is defined as the average power flux at the position **r**, which flows in the direction **ŝ** and has units of power per area and stereoradian. The specific intensity in more physical terms represents the magnitude of the pointing vector averaged over all directions within a differential volume.

The RTE is a conservation equation, where in Eq. (31.1) the first term in the right-hand side describes the change in I in the direction **s**, the second term shows the intensity loss in direction **s** due to the absorption and the scattering, the third term represents gain owing to scattering from other directions into the **s** direction, and finally the last term represents the gain due to a source in **r**.

The equation for flux conservation can be obtained by integrating Eq. (31.1) over the entire solid angle 4π, resulting in

$$\frac{1}{c}\frac{\partial U(\mathbf{r})}{\partial t} + \nabla J(\mathbf{r}) + \mu_a U(\mathbf{r}) = E(\mathbf{r}) \tag{31.2}$$

where $U(\mathbf{r}) = \int_{4\pi} I(\mathbf{r},\mathbf{s})d\Omega$ is the average intensity, $\mathbf{J}(\mathbf{r}) = \int_{4\pi} I(\mathbf{r},\mathbf{s})\mathbf{s}d\Omega$ is the total flux intensity vector, and $E(\mathbf{r}) = \int_{4\pi} \varepsilon(\mathbf{r},\mathbf{s})d\Omega$ is the source term that represents the power generated per unit volume.

31.2.2.1 Optical Parameters

There are several important parameters that describe the propagation of light in turbid media, the main ones being the scattering coefficient, the absorption coefficient, the index of refraction, and the anisotropy factor. All other parameters can be obtained from these [21]. The transport coefficient or total attenuation coefficient μ_t describes light attenuation due to either scattering or absorption per unit length of its travel through the medium and is defined as

$$\mu_t = \mu_a + \mu_s \tag{31.3}$$

where μ_a is the absorption coefficient and μ_s is the scattering coefficient. The way the particles are present in the medium and scatter light is described by the scattering phase function $p(\mathbf{s},\mathbf{s}')$, which represents the average light scattered into direction **s** when light is incident from direction **s'**. When strong anisotropy in scattering exists, the effective scattering coefficient is reduced since most of the light is scattered in the forward direction. In this case, we define the reduced scattering coefficient μ_s', which takes into account the anisotropy of the scattered light as

$$\mu_s' = \mu_s(1-g) \tag{31.4}$$

where g is the anisotropy factor and is defined as the average of the cosine of the scattered angle:

$$g = \langle \cos\theta \rangle = \frac{\int_{4\pi} p(\hat{\mathbf{s}},\hat{\mathbf{s}}')\hat{\mathbf{s}} \cdot \hat{\mathbf{s}}' d\Omega'}{\int_{4\pi} p(\hat{\mathbf{s}},\hat{\mathbf{s}}')d\Omega'} \tag{31.5}$$

which represents the average anisotropy of the scattered light on interaction with the particles and takes values from -1 to 1 depending on the direction of scattering (complete backscattering or forward scattering). The value $g = 0$

represents isotropic scattering. In tissue optics, the Henyey–Greenstein function [22] for g is the most commonly used, yielding values in the range of 0.8–0.9:

$$p(\mathbf{s}, \mathbf{s}') = p(\mathbf{s} \cdot \mathbf{s}') = \frac{1}{4\pi} \frac{\mu_s}{\mu_t} \frac{(1 - g^2)}{(1 + g^2 - 2g\mathbf{s} \cdot \mathbf{s}')^{3/2}} \qquad (31.6)$$

In terms of the described parameters, we can introduce the definition of the transport mean free path length l_{tr} as

$$l_{tr} = \frac{1}{\mu_s'} = \frac{l_{sc}}{1 - g} \qquad (31.7)$$

which represents the average distance that a photon travels before its direction is completely randomized by a series of scattering events.

31.2.2.2 The Diffusion Equation

The use of the RTE for tissue measurements imposes several practical limitations due to its integrodifferential nature. Approximations have been developed to convert the transport equation to more manageable but functional forms [21, 23]. If we assume a highly scattering medium where the direction of each individual photon is randomized after a few scattering events, the specific intensity under these conditions can be described by a first-order expansion series on s as

$$I(\mathbf{r}, \mathbf{s}) \approx \frac{1}{4\pi} U(\mathbf{r}) + \frac{3}{4\pi} \mathbf{J}(\mathbf{r}) \cdot \mathbf{s} \qquad (31.8)$$

in which case, the average intensity U obeys the diffusion equation. Introducing Eq. (31.8) into the RTE, we obtain the diffusion equation:

$$\frac{1}{c} \frac{\partial U(\mathbf{r})}{\partial t} - D\nabla^2 U(\mathbf{r}) + \mu_a U(\mathbf{r}) = E(\mathbf{r}) \qquad (31.9)$$

where the diffusion coefficient D is defined as [21]

$$D = \frac{1}{3\mu_s'} = \frac{1}{3\mu_s(1 - g)} = \frac{l_{tr}}{3} \qquad (31.10)$$

Note that the definition of the diffusion coefficient Eq. (31.7) is correct only for low values of the absorption coefficient. In the presence of high absorption, this expression needs to be reformulated [24].

31.2.2.3 Solutions of the Diffusion Equation for Infinite Homogeneous Media

Although FMT deals with objects that have certain dimensions, it is useful to start the analysis for the case of an infinite medium and later on to add the contribution of the boundaries of the object. In the simple schematic arrangement of Figure 31.1, we can identify two sources of light. The first source is the laser emitting at a specific wavelength λ_a for the illumination of the medium. The second source arises from the fluorescence molecules, after being excited by the laser light, and they emit light at a different wavelength λ_b. Taking these two events/processes into consideration, we must first derive an expression for the propagation of both the excitation light and the corresponding fluorescence term.

31.2.2.4 The Excitation Source Term

Let us consider a continuous wave (cw) laser source, which can irradiate the sample as shown in Figure 31.1. For a laser spot impinging on the medium, the source term in Eq. (31.9) can be approximated to a point source located one transport mean free path inside the medium, at r_s:

$$E(\mathbf{r}, t) = S_0 \delta(\mathbf{r} - \mathbf{r}_s) \tag{31.11}$$

By substituting Eq. (31.11) into the diffusion equation Eq. (31.9), assuming a cw source, and rearranging terms, we obtain Helmholtz's equation [25]:

$$\nabla^2 U(\mathbf{r}) + \kappa_0^2 U(\mathbf{r}) = -\frac{S_0 \delta(\mathbf{r} - \mathbf{r}_s)}{D} \tag{31.12}$$

where κ_0 is the wave number defined as

$$\kappa_0 = \sqrt{-\frac{\mu_a}{D}} \tag{31.13}$$

Therefore, assuming that the incident intensity can be approximated to a point source, the distribution of the average intensity U in an infinite homogeneous medium is given by

$$U(\mathbf{r}) = \frac{S_0}{4\pi D} \frac{\exp(i\kappa_0 |\mathbf{r} - \mathbf{r}_s|)}{|\mathbf{r} - \mathbf{r}_s|} \tag{31.14}$$

which can be easily generalized to a distribution of sources $S(\mathbf{r})$ by integrating over the volume occupied by this distribution.

31.2.2.5 The Fluorescence Source Term

To complete all the elements in the scheme described in Figure 31.1, the next step is to derive the expression for the emitted fluorescent light. The incident light with wavelength λ_a defined in Eq. (31.14) will diffuse inside the medium and will excite the fluorescent particles. This will result in the emission of light with a different wavelength λ_b, which will then diffuse within the medium. In this case, we have to solve the diffusion equation Eq. (31.10) by applying an appropriate expression for the fluorescence source term.

By describing the fluorescence process as a two-level system (Figure 31.2) and by assuming that the surrounding medium has no gain, the number of the excited molecules per unit volume can be calculated by the following expression:

$$\frac{\partial N_e(\mathbf{r}, t)}{\partial t} = -\Gamma N_e(\mathbf{r}, t) + \sigma^{fluo} U^{exc}(\mathbf{r}, t)[N_t(\mathbf{r}, t) - 2N_e(\mathbf{r}, t)] \tag{31.15}$$

where $N_e(\mathbf{r}, t)$ is the number of fluorescence molecules per unit volume in the excited state for time t and at the position $\mathbf{r}$. $N_t(\mathbf{r}, t)$ is the total number of the fluorescent molecules per unit volume, σ^{fluo} is the absorption cross section of a molecule, $U^{exc}(\mathbf{r}, t)$ is the excitation average intensity (equivalent to Eq. (31.14)), and Γ is the total radiative decay rate from the excited state into the ground state.

In the case of a cw excitation source, a steady state is reached, which means $\partial N_e/\partial t = 0$, and Eq. (31.15) can give the solution for the density of the fluorescence

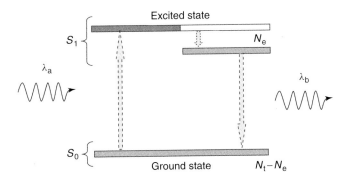

Figure 31.2 Schematic representation of a two-level system illustrating the processes of the absorption of a photon with wavelength λ_a and emission of a fluorescence photon with wavelength λ_b.

molecules per unit volume in the excited state:

$$N_e(\mathbf{r}, \mathbf{r}_s) = \frac{\sigma^{\text{fluo}} N_t(\mathbf{r}, \mathbf{r}_s)}{\Gamma + 2\sigma^{\text{fluo}} U^{\text{exc}}(\mathbf{r}, \mathbf{r}_s)} U^{\text{exc}}(\mathbf{r}, \mathbf{r}_s) \tag{31.16}$$

Typical values of the absorption cross section are on the order of 10^{-6} cm^2, and typical fluorescence lifetimes are on the order of nanoseconds, yielding $\Gamma \sim 10^9$. For intensities on the order of ~ 1 mW cm^{-2}, which is the case in our experiments, $\Gamma \gg 2\sigma^{\text{fluo}} U^{\text{exc}}(\mathbf{r}, \mathbf{r}_s)$ and Eq. (31.16) can be simplified to

$$N_e(r, r_s) = \frac{\sigma^{\text{fluo}} N_t(\mathbf{r}, \mathbf{r}_s)}{\Gamma} U^{\text{exc}}(\mathbf{r}, \mathbf{r}_s) \tag{31.17}$$

The source term of the fluorescent light that represents the fluorescent light emitted per unit volume and per second will therefore be

$$S^{\text{fluo}}(\mathbf{r}, \mathbf{r}_s) = \Gamma \gamma N_e(\mathbf{r}, \mathbf{r}_s) = \gamma \sigma^{\text{fluo}} N_t(\mathbf{r}, \mathbf{r}_s) U^{\text{exc}}(\mathbf{r}, \mathbf{r}_s) \tag{31.18}$$

where γ is the fluorescence quantum yield.

The fluorescence radiation is assumed to be well separated in energy from that of incident light, so that we can safely ignore the possibility of a re-excitation of the fluorophores by the emitted light. The fluorescent light distribution is also described by the diffusion equation for the wavelength λ_b. Assuming a weakly absorbing spatial distribution of fluorophores, we can use the Born approximation, which is described in the following section. In this case, the coupled diffusion equations that result from excitation and emission can be solved independently, and the contribution of the fluorescence light can be determined as the sum over the contribution from all fluorophores as

$$U^{\text{fluo}}(\mathbf{r}) = \frac{\sigma^{\text{fluo}} \gamma}{4\pi D_{\text{fluo}}} \int_V N_t(\mathbf{r}') U^{\text{inc}}(\mathbf{r}', \mathbf{r}_s) \frac{\exp\left(i\kappa_{\text{fluo}} |\mathbf{r} - \mathbf{r}'|\right)}{|\mathbf{r} - \mathbf{r}'|} d^3 \mathbf{r}' \tag{31.19}$$

where $\kappa_{\text{fluo}} = \sqrt{-\mu_a^{\text{fluo}}/D_{\text{fluo}}}$, with μ_a^{fluo} and D_{fluo} being the absorption and diffusion coefficients corresponding to the wavelength of emission, and where

$g_{\text{fluo}}(\mathbf{r}, \mathbf{r}') = \exp(i\kappa_{\text{fluo}} |\mathbf{r} - \mathbf{r}'|) / |\mathbf{r} - \mathbf{r}'|$ is the Green function for the emission wavelength. Substituting the expression for the excitation source term from Eq. (31.14) into Eq. (31.19), we finally obtain

$$U^{\text{fluo}}(\mathbf{r}_s, \mathbf{r}) = \frac{\sigma^{\text{fluo}} \gamma S_0}{16\pi^2 D_{\text{exc}} D_{\text{fluo}}}$$

$$\times \int_V \frac{\exp(i\kappa_{\text{exc}} |\mathbf{r}' - \mathbf{r}_s|)}{|\mathbf{r}' - \mathbf{r}_s|} N_t(\mathbf{r}') \frac{\exp(i\kappa_{\text{fluo}} |\mathbf{r}' - \mathbf{r}|)}{|\mathbf{r}' - \mathbf{r}|} d^3\mathbf{r}' \quad (31.20)$$

Finally, we can write Eq. (31.20) in a more general way in terms of the Green function of a general geometry as

$$U^{\text{fluo}}(\mathbf{r}_s, \mathbf{r}) = \frac{\sigma^{\text{fluo}} \gamma S_0}{16\pi^2 D_{\text{exc}} D_{\text{fluo}}} \int_V G_{\text{exc}}(\mathbf{r}_s, \mathbf{r}') N_t(\mathbf{r}') G_{\text{fluo}}(\mathbf{r}', \mathbf{r}) d^3\mathbf{r}' \quad (31.21)$$

31.2.2.6 The Born Approximation for the Excitation Term

In the previous subsection, to derive the excitation light propagation and the consequent excitation of the fluorophores, we assumed a weakly absorbing distribution of fluorophores. The intensity that excites the fluorophores can be expressed as the sum of two terms – the scattered intensity $U^{\text{sc}}(\mathbf{r}, \mathbf{r}_s)$ reaching the fluorophores after interacting with the other absorbers and the incident intensity, which is the intensity that would be present in point $\mathbf{r}$, in the absence of the fluorophore. We may then express the total excitation as

$$U^{\text{exc}}(\mathbf{r}, \mathbf{r}_s) = U^{\text{inc}}(\mathbf{r}, \mathbf{r}_s) + U^{\text{sc}}(\mathbf{r}, \mathbf{r}_s) \quad (31.22)$$

In the Born approximation, we do not consider the nonlinear effect that the presence of the other fluorophores causes on the excitation intensity [25]. We therefore assume that the effect of the absorption by the fluorophore is negligible and does not affect the intensity of the propagating excitation light. In other words, we may assume that the average intensity at the location of the fluorophores is the same as in the absence of the fluorophores:

$$U^{\text{exc}}(\mathbf{r}, \mathbf{r}_s) \approx U^{\text{inc}}(\mathbf{r}, \mathbf{r}_s) \quad (31.23)$$

31.2.2.7 Boundary Conditions

To predict the measurement collected by a setup equivalent to Figure 31.1, we also have to take into consideration the geometry of the sample and to correctly account for the interaction of the diffusive photons with the interface. The expression of the flux at the surface of the medium is found through the boundary conditions. In the diffusion approximation, the boundary condition considers all the flux traversing the interface to be outward toward the nondiffusive medium:

$$J_{\text{in}}(\mathbf{r})|_S = 0 \quad (31.24)$$

$$J_{\text{out}}(\mathbf{r})|_S = J_n(\mathbf{r})|_S \quad (31.25)$$

where J_n is the total flux traversing the interface. Using Fick's law for the flux **J** [21, 25], we obtain

$$\mathbf{J}(\mathbf{r}) = -D\nabla U(\mathbf{r}) \tag{31.26}$$

After accounting for the differences in index of refraction and using Eqs (31.24) and (31.25) [26], we find the following Robin boundary condition:

$$U(\mathbf{r}) = C_{nd}\mathbf{J}(\mathbf{r}) \cdot \mathbf{n} = C_{nd} J_n(\mathbf{r}) \tag{31.27}$$

where **n** is the surface normal pointing outward from the diffuse medium, and C_{nd} is a coefficient related to the refractive index mismatch between the two media [26]. Typical values of biological tissue are $C_{nd} = 5$. One way of using this boundary condition is assuming that the average intensity is zero at an extrapolated distance $z = C_{nd}D$ [23], which in the index-matched case gives the well-known expression $z = 2D = 2/3\, l_{sc}$.

31.2.2.8 Inverse Problem

So far, we have dealt with the excitation and emission photon propagation inside a diffuse medium and their interaction with the boundaries. We have derived the corresponding solutions to the diffusion equation and we have also presented the expressions of light flux at the interface, which determine the light distribution inside the subject's volume and surface. Once we have described the light distribution at the surface, we can use the noncontact or free-space propagation equations [20] to relate these expressions with the signal on the CCD. This derivation is cumbersome and therefore it will not be shown here, and the reader is referred to Refs [19, 20] for further reading. Overall, we have set and presented the parameters of the forward problem. To proceed to the reconstruction of the spatial distribution of the fluorescent sources, we have to solve the inverse problem. There are several approaches to solve this problem, and in this work we present the iterative approach (see Ref. [27] for a review on inverse approaches in diffuse imaging). To that end, we discretize the volume of interest into small elementary volumes, called *voxels*. We then derive the linear equations presented in the previous section that describe the contribution of each voxel separately on the measured data, these contributions being the unknowns (more specifically, the fluorophore concentration N_t is the unknown). The area of detection of our imaging system (CCD) will also be separated into small elementary detectors (typically binning the pixels into detectors of area sizes of 0.1 cm × 0.1 cm) in which case the measured signal will be determined as the sum of the contribution of the voxels from the entire subject on each binned area of the CCD for each source position.

31.2.2.9 The Normalized Born Approximation

In this section, we present an approximate method that reconstructs accurately the distribution of fluorophores inside a homogeneous medium. The Born approximation [25] has been proposed for reconstructing fluorescent, scattering, or absorption heterogeneities in diffuse media. In the case of fluorescence reconstruction, the normalization of this expression has been found to yield the best results, which has been termed the *normalized Born approximation* [28]. The main idea of this approximation is using the excitation measurement in order to account for the intrinsic properties of the subject, which will also be present in the fluorescence measurement. The contribution due to the spatial distribution of the fluorophores is determined by dividing the fluorescent signal measured (at wavelength λ_b) by the incident excitation measurement (at wavelength λ_a) as is demonstrated here. By normalizing, the experimental measurement is autocalibrated and now accounts for deviations from the theory used, since the light traveled by the excitation and by the emission will traverse tissues with relatively similar optical properties. The expression for the normalized quantity, U^{nB}, becomes [28]

$$U^{nB}(\mathbf{r}_s, \mathbf{r}_d) = \frac{U^{fluo}(\mathbf{r}_s, \mathbf{r}_d)}{U^{exc}(\mathbf{r}_s, \mathbf{r}_d)} = \frac{q^{fluo}\sigma^{fluo}\gamma}{4\pi q^{exc}D_{fluo}} \\ \times \int_V \frac{G_{exc}(\mathbf{r}_s, \mathbf{r}')N_t(\mathbf{r}')G_{fluo}(\mathbf{r}', \mathbf{r}_d)d^3\mathbf{r}'}{G_{exc}(\mathbf{r}_s, \mathbf{r}_d)} \quad (31.28)$$

where q^{exc} is a factor accounting for the detector quantum efficiency at the excitation wavelength and the attenuation caused by the bandpass filter used to collect the excitation light and q^{fluo} is a factor accounting for the detector quantum efficiency at the emission wavelength, including the attenuation caused by the bandpass filter used to collect the emission light. In Eq. (31.28), we recall that D_{fluo} is the diffusion coefficient at the fluorescence wavelength, $N_t(\mathbf{r}')$ is the fluorophore concentration, and G_{fluo} is the Green's function solution to the diffusion equation and describes the propagation of the emitted light from the fluorophore to the detector. It should be emphasized that the intensity of the excitation source has been canceled out after normalization. Another important advantage is that some of the position-dependent terms, such as the filter's spatial dependence, are canceled out.

To solve the inverse problem and obtain the values of $N_t(\mathbf{r}')$ in Eq. (31.28), we apply the Born approximation by expressing the integral over V as a sum over N_v voxels of volume ΔV:

$$U^{nB}(\mathbf{r}_s, \mathbf{r}_d) = \frac{q^{fluo}\sigma^{fluo}\gamma}{4\pi q^{exc}D_{fluo}} \sum_{i=1}^{N_v} \frac{G_{exc}(\mathbf{r}_s, \mathbf{r}_i)N_t(\mathbf{r}_i)G_{fluo}(\mathbf{r}_i, \mathbf{r}_d)\Delta V}{G_{exc}(\mathbf{r}_s, \mathbf{r}_d)} \quad (31.29)$$

which when written in matrix form can be expressed as

$$[U^{nB}]_{s \times d} = [W]_{s \times d}^{N_v}[N]_{N_v} \quad (31.30)$$

where we have grouped all known factors into what is termed the *weight* or *sensitivity matrix* W. Note that the Born approximation is considered while discretizing the volume integral and is not yet present in Eq. (31.28), which simply represents a normalized quantity.

In the analysis presented above, we explained how the problem of extracting the fluorescence distribution can be transformed into a matrix-inversion problem. Typical matrix-inversion problems present in FMT and DOT are ill posed and can be approached with different methods [27], the most common being singular value decomposition (SVD) or iterative approaches such as the algebraic reconstruction technique (ART) [29], which is the method employed in this chapter.

We have now described how light excitation and emission may be modeled through the diffusion approximation, and how a linear set of equations may be built and solved by numerically inverting Eq. (31.30). In the next section, we describe how the experimental measurements are taken.

31.2.3
Experimental Setup

In biomedical imaging, the use of more than one contrast agent is very common, especially when monitoring cell number changes associated with biological processes, such as distinct specificities during an immune response [30]. In those studies, the ability to excite the probes in optimum conditions can be crucial, especially as the research on new fluorescing proteins and smart probes is constantly providing novel, improved constructs that do not necessarily share the same spectral characteristics. Consequently, it has become essential to develop a system that is able to cover a wide range of excitation wavelengths. To account for and incorporate these features, we developed the instrument depicted schematically in Figure 31.3, which consists of an arrangement of different laser sources covering the entire visible and near-infrared part of the spectrum that can be very easily interchanged by means of flip mirrors. The laser sources used in the current setup are an Argon-ion CW multiline laser (LaserPhysics, Reliant 1000m, USA) emitting

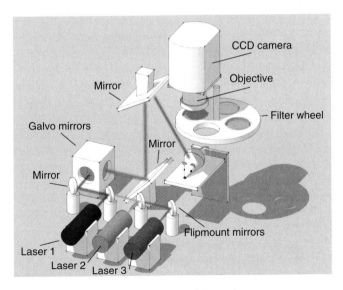

Figure 31.3 Schematic representation of the multiwavelength excitation experimental setup that allows the use of many laser sources that can be alternated by means of flip mirrors.

at 458, 488, and 514 nm, a 592 nm solid-state laser (Soliton, Germany), a 635 nm solid-state laser (Coherent, USA), and a 786 nm laser (Hamamatsu, PLP-10, Japan). The power reaching the subject can be adjusted to the desired levels by a set of neutral density filters and measured in real time by a power meter.

The laser light is directed to the imaging chamber and into the laser scanning device (Scancube 7, Scanlab, Germany), which incorporates a system of mirrors mounted on miniaturized galvanometer motors. The scanning head is controlled by software that has been developed in-house; the laser beam can thus scan an $X-Y$ plane normal to the optical path on the subject with the additional aid of large rectangular mirrors (first surface mirrors, 4–6 Wave, 73 mm 116 mm, Edmund Optics). The first mirror is mounted permanently on the ceiling of the imaging chamber (not shown in Figure 31.3 for simplicity) box, while the other lies on the optical table and can be moved along a rail between two fixed positions, altering the geometry of the experiment between reflection and transmission. For measurements in the reflection geometry, the bottom mirror is moved forward so that the laser beam is directed to the top mirror and then to the sample from the same side of the camera. For measurements in transmission geometry, the bottom mirror is moved to the back position, so that the laser light illuminates the sample from below.

The specimen is placed on a transparent glass plate with an antireflection (AR) coating (Glassplate, high AR coated 96–99% (400–700 nm)). Depending on the desired resolution of each measurement, the distance of the sample from the camera can be modified by placing the glass plate at different fixed positions along the Z-axis. Images are captured by a thermoelectrically cooled 16 bit CCD camera with a chip size of 1024×1024 pixels (Andor Corp., DV434, Belfast, Northern Ireland), which is mounted on the upper plate of the imaging box. The CCD camera is equipped with a SIGMA 50 mm $f/2.8$ objective (Sigma Corporation, Tokyo, Japan), which is focused on the specimen's surface. A filter wheel mounted in front of the CCD camera lens with different bandpass interference filters enables the choice of the desired spectral range of detection.

31.3
Spectral Tomography

In biological research, it is common to simultaneously image different fluorescent probes. Although fluorescent proteins are available in a broad palette of colors, the spectral overlap of most fluorophores is significant and thus distinction between two reporters, if at all possible, is complex and inaccurate. In fact, cyan/green and far-red proteins are the only colors that can be accurately separated by the sole use of filters. This problem is especially evident if the two fluorophores are present at significantly different concentrations and, even at microscopy level, emission of green and yellow dyes can be confused in much the same way that intense emission from a yellow fluorescent protein (YFP) can be detected by a red filter. Quantitative analysis of such data requires the use of a multispectral technique that takes into account the relative contribution of each fluorophore within the detection window used for the other.

Such spectral unmixing techniques have been developed mainly for confocal microscopy. For small animal imaging applications, these methodologies are still novel and need to be implemented in the tomographic approach. This section describes the experiments performed to test whether the development of a linear unmixing algorithm can distinguish between overlapping fluorophores, thus providing improved contrast, sensitivity, and higher quantification accuracy.

31.3.1
Spectral Deconvolution

When performing *in vivo* studies using biological models, it is very important to be able to detect and distinguish the fluorescence activity of different fluorochromes and/or fluorescing proteins simultaneously. This can be accomplished by realizing a tomographic system and method capable of producing multispectral three-dimensional images. The process of detecting, imaging, and reconstructing multiple fluorescing targets requires the unmixing of the fluorescence signals of each of the different targets based on their specific spectral signatures. Such methods are very common in fluorescence microscopy where they are applied pixel by pixel in the so-called spectral cube to produce two-dimensional images [17]. In the following, we describe how these methods can be applied in a tomographic approach.

The decomposition of the signals of different fluorochromes is based on a linear unmixing algorithm, which is applied to the reconstructed data obtained at each spectral region [4]. The algorithm takes into consideration the relative strengths of the fluorophores within the spectral window used for the detection. These spectral strengths are calculated after measuring the fluorescence emission of the different fluorophores with a commercial fluorimeter (Varian, Cary Eclipse) and integrating the signal under the part of the curve that corresponds to the spectral band allowed by each filter. The reconstructed signal at each spectral region is written as a linear combination of all unmixed signals multiplied with the corresponding spectral strengths. In the case of m fluorophores and n filters, the calculation of the unknown concentrations is based on solving a set of linear equations, which can be represented by using matrices as

$$\begin{vmatrix} I_1 \\ I_2 \\ \vdots \\ I_n \end{vmatrix} = \begin{vmatrix} s_{G_1} & s_{R_1} & \cdots & s_{M_1} \\ s_{G_2} & s_{R_2} & \cdots & s_{M_2} \\ \vdots & & & \vdots \\ s_{G_n} & s_{R_n} & \cdots & s_{M_n} \end{vmatrix} * \begin{vmatrix} C_G \\ C_R \\ \vdots \\ C_M \end{vmatrix} \quad (31.31)$$

were $I_1, I_2 \ldots I_n$ are the reconstructed three-dimensional images obtained from FMT, $s_{G_1}, s_{G_2} \ldots s_{G_n}, s_{R_1}, s_{R_2} \ldots s_{R_n}$ and $s_{M_1}, s_{M_2} \ldots s_{M_n}$ are the strengths of the fluorochromes in the spectral regions measured as described above, and C_G,

$C_R, \ldots, C_M$ are the unknown unmixed images. The uncoupled reconstructed data can be then calculated by solving the linear $n \times m$ system, basically calculating the unmixed values for every voxel in the mesh.

In the case of two fluorescing molecules, the uncoupled images can be calculated by solving the following linear 2×2 system:

$$\begin{vmatrix} I_1 \\ I_2 \end{vmatrix} = \begin{vmatrix} s_{G_1} & s_{R_1} \\ s_{G_2} & s_{R_2} \end{vmatrix} \cdot \begin{vmatrix} C_G \\ C_R \end{vmatrix} \Rightarrow \begin{vmatrix} C_G \\ C_R \end{vmatrix} = \begin{vmatrix} s_{G_1} & s_{R_1} \\ s_{G_2} & s_{R_2} \end{vmatrix}^{-1} \cdot \begin{vmatrix} I_1 \\ I_2 \end{vmatrix} \qquad (31.32)$$

For illustration purposes, let us consider the fluorescent dyes, CFSE and ATTO590, the spectral profiles of which are shown in Figure 31.4. The aim of our experiment is to separate CFSE from ATTO590. Spectral deconvolution of CFSE and ATTO590 was performed acquiring FMT data using 540 ± 20 and 615 ± 45 nm filters. As mentioned above, the unmixing system of equations can be solved using the relative fluorophore contributions to each filter given in Table 31.1.

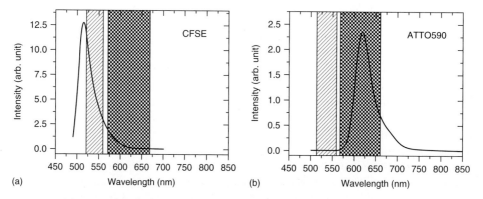

Figure 31.4 CFSE (a) and ATTO590 (b) spectra with the spectral regions corresponding to the filter bands used for the recording of the tomography data. Integration of the intensities within these spectral regions provides the weighting factors for the unmixing algorithm.

Table 31.1 The relative spectral strengths in arbitrary units for CFSE and ATTO590 calculated from spectra obtained with a commercial fluorimeter and integrating over the bandpass regions of the detection filters.

	540 ± 20 nm	615 ± 50 nm
CFSE	0.2058	0.0761
ATTO590	0.000418	0.8798

31.4
Multitarget Detection and Separation

Multispectral imaging can provide useful insight into the choice of filters, fluorophores, and the most appropriate excitation and emission wavelengths to use, depending on the targeted tissue. Using spectral unmixing, multiple color detection is also possible by choosing a combination of partially overlapping fluorophores. In this section, multiwavelength information is added to the FMT reconstruction algorithms and used during the data analysis stage to recover the concentrations of two distinct fluorophores. Results are presented from *in vitro* and *in vivo* experiments in which the volumetric spectral unmixing algorithm was used to separate signals from different probes and the unmixed data were then used for the 3D rendering of fluorescence tomography data.

31.4.1
Multicolor Phantom

31.4.1.1 *In vitro* Fluorophore Unmixing

To distinguish between two partially overlapping fluorophores, the unmixing algorithm was first applied to a phantom model with high autofluorescence contribution. CFSE and ATTO590 were the two dyes used in this experiment. Their emission spectra (recorded with a commercial fluorimeter at 514 nm excitation) are shown in Figure 31.4, together with the regions corresponding to the filters used for FMT acquisition. Although the maximum peaks of the two fluorophores are separated, the tails of both fluorophores are detectable in both of the emission filters. Depending on their relative intensities, each fluorophore will thus contribute to a fraction of the total detected signal for each filter; this implies that it is not possible to quantitatively separate the two fluorophores using filters alone.

In the following experiments, the total signal originating from the green (CFSE) and the red (ATTO590) dye was measured in two separate spectral regions. The unmixing algorithm, described in the previous section, was then applied to the three-dimensional data to discriminate between the two fluorophores. The aim of this experiment was for the unmixed data to correctly reconstruct the concentration of the CFSE tube in the green channel and the concentration of the ATTO590 in the red channel.

31.4.1.2 Methodology

Two capillary tubes filled with equal concentrations (50 µM) of CFSE and ATTO590 were placed at 3 mm depth in a gel phantom (2% agarose, 1% Intralipid-20%, and 0.02% India ink). The tubes were located approximately 2 cm apart and a 5 × 13 source grid was used for tomographic data acquisition. Spectral information was recorded using (i) a 488 nm excitation light and emission gathered using a 540/40 and a 615/90 nm filter and (ii) using 514 nm excitation light and 540/40 and 615/90 nm emission filters. Three-dimensional reconstructions for each of the filter combinations were obtained using a 16 × 40 detector grid, covering a field

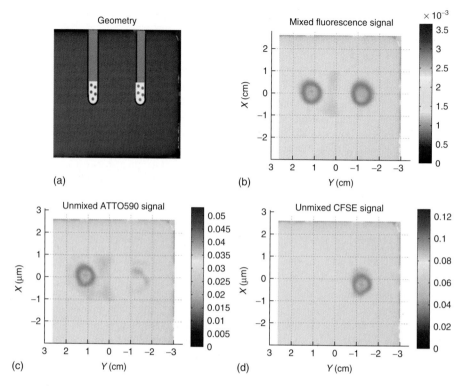

Figure 31.5 *In vitro* unmixing of CFSE and ATTO590 in a phantom model. (a) Geometry image of the phantom and position of the capillary sample tubes. Equal concentrations of the two fluorophores were imaged in two spectral channels (green and red). (b) The mixed signal in the red channel detects two tubes of similar intensities. (c, d) After spectral unmixing, the red and the green components were separated: the unmixed signals for CFSE and ATTO590 could then be individually rendered. The CFSE unmixed signal was approximately twofold stronger than the ATTO590, as would be expected from their spectral profiles. (Please find a color version of this figure on the color plates.)

of view of 1.6 × 4 cm² and 10 ART iterations with positive restriction. Spectral unmixing was then carried out on the reconstructed data, and the unmixed data were then displayed separately for the green and red channels.

The results are presented in Figure 31.5, where coronal slices of the reconstructions of the original mixed signals (excitation at 514 nm, emission at 615 nm) and the unmixed CFSE and ATTO590 signals are shown. In the mixed fluorescence data, two tubes are visible and it is not possible to distinguish which of the two contains the CFSE and which the ATTO590. The spectral unmixing algorithm delivers clear separation between the two fluorophores and, for each channel, only one tube is visible. The unmixed data therefore allow discriminating the CFSE tube in the green channel from the ATTO590 tube in the red channel. As expected, using 514 nm as the excitation source, the ATTO590 signal is weaker than that displayed by CFSE; however, this difference in relative intensity is only detectable in the unmixed data (see the difference in scale), whereas it would go unnoticed in the mixed image. This type of component separation relies on specific spectral properties displayed by tissue and fluorophores and can therefore be used to improve contrast and remove autofluorescence from *in vivo* data.

31.4.2
In vivo Study

31.4.2.1 *In vivo* Fluorophore Unmixing
In the previous section, we showed results from homogeneous phantoms. The effect that heterogeneous properties of tissue have on fluorescence detection will be tested in this section, by applying the unmixing FMT approach *in vivo*. To this end, tubes containing the same concentrations of CFSE and ATTO590 (as above) were placed subcutaneously in a mouse and then imaged with FMT. The aim of this experiment was to spectrally decompose the two fluorophores, as shown for the *in vitro* experiments, and test the effectiveness of the algorithm for *in vivo* applications.

31.4.2.2 Methodology
Two capillary tubes, measuring 0.5 cm length, were filled with equal concentrations (50 µM) of CFSE and ATTO590 and placed subcutaneously in a C57bl mouse, previously shaved. Two skin flaps were opened at the height of the mammary glands and the tubes inserted approximately 2 cm apart (Figure 31.6a). Care was taken so that the glass capillaries were completely covered by skin. Data were acquired using an 8×7 source grid. Spectral information was acquired using (i) a 488 nm excitation light and emission gathered using 540/40 and 615/90 nm filters and (ii) 514 nm excitation light and 540/40 and 615/90 nm emission filters. Three-dimensional reconstructions for each of the filter combinations were obtained using a 25×15 detector grid, covering a field of view of 1.5×2.5 cm^2 and 10 ART iterations with positive restriction. Spectral unmixing was carried out on the FMT reconstructions and the unmixed data were then resolved individually for the green and red channels, as explained in the previous section.

The three-dimensional reconstruction from the mixed data in the red channel (Figure 31.6b–d) shows how both fluorophores are detected and reconstructed. As seen in this figure, the linear spectral unmixing algorithm performs well in a mouse model. The presence of tissue with highly diverse spectral characteristics represents a technical obstacle for light penetration. Tissue absorption properties, for example, vary significantly between the green and the red spectra and results in the green fluorophore being detected at lower intensity levels than the red fluorophore. Red fluorophores are overall expected to perform better for *in vivo* applications as light travels further in the far-red, resulting in enhanced sensitivity and reduced depth limitations. Basic analysis on the mixed data would incorrectly conclude that ATTO590 and CFSE were present at different concentrations. This difference had not been observed in the *in vitro* experiments and is thus undoubtedly due to

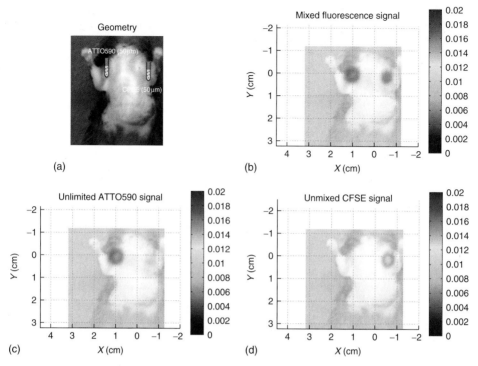

Figure 31.6 *In vivo* unmixing of CFSE and ATTO590 capillary tubes placed subcutaneously in a C57bl mouse and imaged with the FMT. (a) Geometry image of the mouse showing the position in which the capillary sample tubes were inserted. Equal concentrations (50 μM) of the two fluorophores were imaged in two spectral regions (green and red). (b) The mixed signal in the red channel correctly detects two tubes and the CFSE tube emission appears weaker than the ATTO590. (c,d) After spectral unmixing, the red and the green components were separated: the unmixed signals for both dyes were then rendered. The CFSE unmixed signal was approximately twofold stronger than the ATTO590, consistent with what was shown for the *in vitro* data. (Please find a color version of this figure on the color plates.)

the spectral properties of tissue. Data unmixing using spectral strengths and filter weighting allows the unmixed images to correctly recover the original fluorophore intensities and in the resulting "corrected" images the relative intensity between the green and red fluorophores is equal to the one measured in the *in vitro* sample tested previously, presenting further evidence of the robustness of the algorithm itself.

31.5 Conclusions

The combination of tomographic and spectroscopic methodologies can be very important in improving and expanding the capabilities of fluorescence tomography as a stand-alone technique when it comes to its applicability to *in vivo* biological studies. On-going research in the field will further improve a technology that is still in its initial stages. The spectral information needed for the implementation of the relative strengths of different spectral contributions in the wavelength-dependent weight matrix can be obtained by spectrally resolved tomographic measurements of the fluorophore emission, initially in phantoms and then inside the animals. Using the same principles and procedures, the contribution of skin autofluorescence, a significant limitation when measuring in reflection geometry, could also be subtracted. Autofluorescence imposes one of the main hindering factors for *in vivo* optical imaging and the ability to remove its contribution could be extremely important, especially in cases where the true fluorescing signal is weak or masked by signal from surrounding tissue and organs. Finally, the synthesis of new, stable, and bright far-red or near-infrared fluorescent proteins will play a significant role in the improvement of sensitivity, accuracy, and fidelity of *in vivo* optical imaging.

Spectral-FMT provides the means for detecting, imaging, distinguishing, and following multiple fluorescence signals, even when they are spectrally overlapping. A great number of biological processes can be quantified and three-dimensional spatial information can be extracted *in vivo*. The efficiency of *in vivo* imaging can thus be greatly improved since animals can be imaged in more than one time point and longitudinal studies can be significantly simplified, decreasing at the same time the total number of sacrificed animals. Studies involving immunologic responses where different cell populations need to be followed and quantified, cancer development and metastasis, drug delivery and therapy, as well as gene expression pathways, will benefit from the aforementioned methodologies.

References

1. Cherry, S.R. (2004) In vivo molecular and genomic imaging: new challenges for imaging physics. *Phys. Med. Biol.*, **49**, R13–R48.
2. Ntziachristos, V., Tung, C., Bremer, C., and Weissleder, R. (2002) Fluorescence-mediated molecular tomography resolves protease activity in vivo. *Nat. Med.*, **8**, 757–760.
3. Ntziachristos, V., Schellenberger, E.A., Ripoll, J., Yessayan, D., Graves, E.E., Bogdanov, A., Josephson, L., and Weissleder, R. (2004) Visualization of antitumor treatment by means of fluorescence molecular tomography with an Annexin V-Cy5.5 conjugate. *Proc. Natl. Acad. Sci. U.S.A*, **101**, 12294–12299.
4. Zacharakis, G., Kambara, H., Shih, H., Ripoll, J., Grimm, J., Saeki, Y., Weissleder, R., and Ntziachristos, V. (2005) Volumetric tomography of fluorescent proteins through small animals in vivo. *Proc. Natl. Acad. Sci. U.S.A*, **102**, 18252–18257.
5. Boas, D., Brooks, D.H., Miller, E.L., DiMarzio, C.A., Kilmer, M., Gaudette, R.J., and Zhang, Q. (2001) Imaging the body with diffuse optical tomography. *IEEE Signal Process. Mag.*, **18**, 57–75.
6. Culver, J.P., Durduran, T., Furuya, D., Cheung, C., Greenberg, J.H., and Yodh, A.G. (2003) Diffuse optical tomography of cerebral blood flow, oxygenation, and metabolism in rat during focal ischemia. *J. Cereb. Blood Flow Metab.*, **23**, 911–924.
7. Srinivasan, S., Pogue, B.W., Jiang, S., Dehghani, H., Kogel, C., Soho, S., Gibson, J.J., Tosteson, T.D., Poplack,

S.P., and Paulsen, K.D. (2003) Interpreting hemoglobin and water concentration, oxygen saturation, and scattering measured in vivo by near-infrared breast tomography. *Proc. Natl. Acad. Sci. U.S.A*, **100**, 12349–12354.

8. Durduran, T., Choe, R., Culver, J.P., Zubkov, L., Holboke, M.J., Giammarco, J., Chance, B., and Yodh, A.G. (2002) Bulk optical properties of healthy female breast tissue. *Phys. Med. Biol.*, **47**, 2847–2861.

9. Hawrysz, D.J., and Sevick-Muraca, E.M. (2000) Developments toward diagnostic breast cancer imaging using near-infrared optical measurements and fluorescent contrast agents. *Neoplasia*, **2**, 388–417.

10. Ntziachristos, V., Yodh, A.G., Schnall, M., and Chance, B. (2000) Concurrent MRI and diffuse optical tomography of breast after indocyanine green enhancement. *Proc. Natl. Acad. Sci. U.S.A.*, **97**, 2767–2772.

11. Pogue, B.W., Poplack, S.P., McBride, T.O., Wells, W.A., Osterman, K.S., Osterberg, U.L., and Paulsen, K.D. (2001) Quantitative hemoglobin tomography with diffuse near-infrared spectroscopy: pilot results in the breast. *Radiology*, **218**, 261–266.

12. Hielscher, A.H., Klose, A.D., Scheel, A.K., Moa-Anderson, B., Backhaus, M., Netz, U., and Beuthan, J. (2004) Sagittal laser optical tomography for imaging of rheumatoid finger joints. *Phys. Med. Biol.*, **49**, 1147–1163.

13. Benaron, D.A., Hintz, S.R., Villringer, A., Boas, D., Kleinschmidt, A., Frahm, J., Hirth, C., Obrig, H., van Houten, J.C., Kermit, E.L., Cheong, W.F., and Stevenson, D.K. (2000) Noninvasive functional imaging of human brain using light. *J. Cereb. Blood Flow Metab.*, **20**, 469–477.

14. Culver, J.P., Siegel, A.M., Stott, J.J., and Boas, D.A. (2003) Volumetric diffuse optical tomography of brain activity. *Opt. Lett.*, **28**, 2061–2063.

15. Hillman, E.M.C., Hebden, J.C., Schweiger, M., Dehghani, H., Schmidt, F.E.W., Delpy, D.T., and Arridge, S.R. (2001) Time resolved optical tomography of the human forearm. *Phys. Med. Biol.*, **46**, 1117–1130.

16. Rudin, M. and Weissleder, R. (2003) Molecular imaging in drug discovery and development. *Nat. Rev. Drug. Discov.*, **2**, 123–131.

17. Zimmermann, T., Rietdorf, J., and Pepperkok, R. (2003) Spectral imaging and its applications in live cell microscopy. *FEBS Lett.*, **546**, 87–92.

18. Srinivasan, S., Pogue, B.W., Jiang, S., Dehghani, H., and Paulsen, K.D. (2005) Spectrally constrained chromophore and scattering near-infrared tomography provides quantitative and robust reconstruction. *Appl. Opt.*, **44**, 1858–1869.

19. Ripoll, J. and Ntziachristos, V. (2004) Imaging scattering media from a distance: theory and applications of non-contact optical tomography *Mod. Phys. Lett. B*, **18**, 1403–1431.

20. Ripoll, J., Schulz, R.B., and Ntziachristos, V. (2003) Free-space propagation of diffuse light: theory and experiments. *Phys. Rev. Lett.*, **91**, 103901.

21. Ishimaru, A. (1978) *Wave Propagation and Scattering in Random Media*, vol. 1, Academic Press.

22. Henyey, L.G. and Greenstein, J.L. (1941) Diffuse radiation in the galaxy. *Astro. Phys. J.*, **93**, 70–83.

23. Patterson, M.S., Chance, B., and Wilson, B.C. (1989) Time-resolved reflectance and transmittance for the noninvasive measurement of tissue optical properties. *Appl. Opt.*, **28**, 2331–2336.

24. Aronson, R. and Corngold, N. (1999) Photon diffusion coefficient in an absorbing medium. *J. Opt. Soc. Am. A*, **16**, 1066–1071.

25. Born, M. and Wolf, E. (1999) *Principles of Optics*, 7th edn, Cambridge University Press, Cambridge.

26. Aronson, R. (1995) Boundary conditions for diffusion of light. *J. Opt. Soc. Am. A*, **12**, 2532–2539.

27. Arridge, S.R. (1999) Optical tomography in medical imaging. *Inverse Probl.*, **15**, R41–R93.

28. Ntziachristos, V. and Weissleder, R. (2001) Experimental three-dimensional

fluorescence reconstruction of diffuse media using a normalized Born approximation. *Opt. Lett.*, **26**, 893–895.
29. Kak, C. and Slaney, M. (1988) *Principles of Computerized Tomographic Imaging*, IEEE, New York.
30. Martin, M., Aguirre, J., Sarasa-Renedo, A., Tsoukatou, D., Garofalakis, A., Meyer, H., Mamalaki, C., Ripoll, J., and Planas, A.M. (2008) Imaging changes in lymphoid organs in vivo after brain ischemia with three-dimensional fluorescence molecular tomography in transgenic mice expressing green fluorescent protein in T lymphocytes. *Mol. Imaging*, **7**, 157–167.

32
Biomedical Imaging Based on Vibrational Spectroscopy

Christoph Krafft, Benjamin Dietzek, and Jürgen Popp

32.1
Introduction

Histology is the study of the microscopic anatomy of cells and tissues of plants and animals. In the context of disease diagnosis, histology is called *histopathology*. Histologic or histopathologic diagnosis is generally performed by examining a thin slice (a section) of tissue under a light microscope or an electron microscope. The ability to visualize or differentially identify microscopic structures is frequently enhanced through the use of histological stains. Trained medical doctors, pathologists, are the personnel who perform histopathological examination. New research techniques, such as immunohistochemistry and molecular biology, have expanded the horizons of histological and pathological inspection. This chapter describes the use of vibrational microspectroscopic imaging as an innovative emerging tool in biomedicine.

Numerous developments in lasers, spectrometers, detectors, and image-processing algorithms in the past decade have enabled significant progress of vibrational spectroscopy in biological and life sciences. Raman- and infrared (IR)-based spectroscopies are rich in fine spectral features and enable studies on the cellular and molecular level from solid and liquid specimens without the use of external agents such as dyes, stains, or radioactive labels. Their principle is that they probe inherent molecular vibrations that correlate with the composition and structure of the samples. Diseases and other pathological anomalies lead to chemical and structural changes, which also change the vibrational spectra and can be used as sensitive, phenotypic markers of the disease. Their advantages are diffraction-limited spatial resolution in the micrometer and even submicrometer range, nondestructive sampling, use of nonionizing radiation, and rapid data collection for high time resolution of dynamic processes. A disadvantage of all optical techniques is their limited penetration in tissue, which depends on its absorption and scattering properties. Therefore, biomedical *in vivo* and real-time applications require an exposed field of view or fiber optic probes for endoscopy.

IR microspectroscopy, Raman microspectroscopy, and coherent anti-Stokes–Raman scattering (CARS) microscopy are so far the most commonly

Optical and Digital Image Processing: Fundamentals and Applications, First Edition. Edited by Gabriel Cristóbal, Peter Schelkens, and Hugo Thienpont.
© 2011 Wiley-VCH Verlag GmbH & Co. KGaA. Published 2011 by Wiley-VCH Verlag GmbH & Co. KGaA.

applied vibrational spectroscopic methods to assess tissues. The methodological section 32.3 describes the fundamentals of these techniques and the principles for recording images. The resulting hyperspectral images are measurements that include information not only of high quality but also in high quantity. Preprocessing, unsupervised and supervised approaches are summarized for image processing to obtain maximum information content and to visualize the results. Finally, recent applications of vibrational-spectroscopy-based imaging in biomedicine are presented.

32.2
Vibrational Spectroscopy and Imaging

The energy transitions of IR, Raman, and CARS spectroscopy are depicted in term schemes (Figure 32.1). Molecular vibrations can be probed by absorption of mid-infrared (MIR) radiation, which is the principle of IR spectroscopy, and inelastic scattering of monochromatic light, which is the principle of Raman spectroscopy. Depending on the symmetry of molecular bonds and the involved atoms, vibrations are IR- or Raman-active or both. Consequently, IR and Raman spectra are not identical and complement each other to some extent. To overcome the disadvantage of low Raman signal intensities from most biomolecules, enhancement effects such as resonance Raman spectroscopy and surface-enhanced Raman spectroscopy were utilized. The first technique probes vibrations of chromophores, the second technique depends on short-range interactions with metal surfaces in the nanometer range. Both the techniques require careful selection of the excitation wavelength and careful preparation, respectively. For brevity, they are not considered here. CARS is a nonlinear variant of Raman spectroscopy. A coherent anti-Stokes-scattered photon is emitted as a result of a four-photon mixing process involving two pump photons and one probe photon. The process is described in more detail in Section 32.2.3. The CARS technique has several advantages, making it an interesting candidate for chemically sensitive microscopy. Besides signal enhancement by more than 4 orders of magnitude, CARS provides a highly directional signal in the absence of an autofluorescence background, as the CARS signal is generated on the shortwave side of the excitation-laser pulses. Vibrational spectroscopic images represent a particular type of measurement that contains both spatial and spectral – and, thus, chemical – information about a sample. The sample is compartmented into small surface or volume areas, referred to as *pixels* or *voxels*, respectively. Acquisition of vibrational spectroscopic images is introduced next.

32.2.1
Infrared Spectroscopy

Whereas most Raman spectrometers operate in a dispersive mode, MIR spectrometers use the interferometric Fourier transform (FT) principle, which has the multiplex, throughput, and wavenumber accuracy advantages. The basic setup consists of a broadband radiation source, an interferometer, a sample chamber, which can also be a microscope, and a fast detector. The detection range depends on the radiation source, the transmission of the beam splitter inside the interferometer, the sample substrate, and the detector.

Many applications in biomedicine require the acquisition of images. The spectroscopic data can be combined with the lateral information in FTIR spectrometers with single-channel detectors by restricting radiation at the sample plane with an aperture and scanning this aperture over the area of interest with an automated translation stage. According to Abbe's law, $d = 0.612\lambda/\text{NA}$, the resolution d in spectroscopic imaging is limited by diffraction (wavelength λ and numerical aperture NA of the microscope objective). To optimize the sensitivity using microscope apertures near the diffraction limit, high brilliant IR radiation from synchrotron sources is used instead of IR radiation from thermal illumination sources in standard FTIR spectrometers. As typical Cassegrain IR objectives have NA between 0.4 and 0.6, the diffraction limit is in the order of the wavelength of MIR radiation of 2.5–25 μm.

FTIR spectrometers with multichannel detectors, termed *focal plane array* (FPA) detectors, offer another way to collect FTIR images. The entire field of view is illuminated and imaged on such an FPA detector, which segments radiation at the detection plane. Without apertures and without moving the samples, the lateral information is collected in parallel, whereas the spectral information is obtained serially, operating the interferometer in a special collection mode. As depicted in Figure 32.2a, an image is recorded for each position of the moving mirror. After data acquisition is completed, the interferogram for each pixel is Fourier

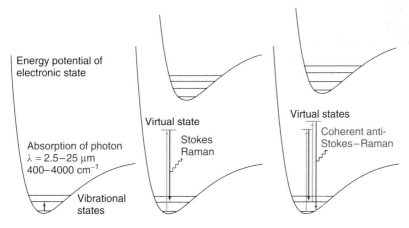

Figure 32.1 Energy transitions of IR, Raman, and CARS spectroscopy. IR spectroscopy is a one-photon process in which vibrations are probed by absorption of a photon. Spontaneous Raman spectroscopy is a two-photon process in which the first photon excites the molecule to a virtual state and upon emission of a second photon the molecule returns to an excited vibrational state. CARS microscopy is a four-photon process, involving two pump photons and one probe photon before emission of an anti-Stokes photon. (Please find a color version of this figure on the color plates.)

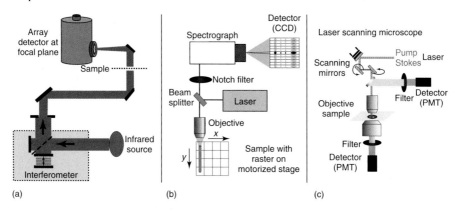

Figure 32.2 Principle of Fourier-transform infrared imaging using focal plane array detectors (a). Point-mapping and line-mapping modes in Raman imaging (b). In point-mapping, the laser is focused to a spot (blue) and spectra are registered sequentially in x and y dimensions, whereas in line-mapping mode the laser generates a line (gray) and the spectral information is registered sequentially in x direction and parallel in y direction using both dimensions of the CCD detector. Diagram of a CARS microscope operating in epi and forward detection (c). The short-pulse laser system was omitted for simplicity. (Please find a color version of this figure on the color plates.)

transformed, giving a full IR spectrum. The main advantage of FTIR spectrometers with multichannel detection is that an entire image can be acquired in a time comparable with that of conventionally acquiring a single spectrum. When a 128 × 128 FPA is used, each image contains 16 384 individual spectra and a single image can be acquired in less than a minute.

Owing to the high water content of cells, tissues, and body fluids and the strong absorption of MIR radiation by water, the penetration depth is limited to a few micrometers. Although cells are sufficiently thin and fluids can be filled in cuvettes with 5–10 μm path length, tissue samples for most MIR spectroscopic studies in transmission mode are cut into 5–20 μm thick sections, transferred onto MIR transparent substrates such as calcium fluoride or barium fluoride, and subsequently dried. Highly reflective, metal-coated glass slides constitute a less expensive alternative class of substrates from which IR spectra can be collected in the so-called reflection absorption mode [1]. Attenuated total reflection (ATR) offers another way in IR spectroscopy to record spectra from nontransparent biomedical samples [2]. At the interface between an ATR crystal of high refractive index and the sample of lower refractive index, an evanescent wave penetrates a few micrometers into the sample. ATR crystals are made of MIR transparent materials such as zinc selenide, germanium, and diamond.

32.2.2
Raman Spectroscopy

The basic setup of a dispersive Raman spectrometer consists of a laser as an intense and monochromatic light source, a device that separates the elastically (Rayleigh) scattered light of the sample from the inelastically (Raman) scattered light, a spectrograph, and a detector. Multichannel detectors such as charge-coupled device (CCD) cameras enable registration of the whole Raman spectrum simultaneously within a fraction of seconds. Owing to the problem of intense autofluorescence that often masks the Raman signals, Raman spectra of unprocessed biological material are usually excited with near-IR lasers. As most tissues and body fluids show minimum absorption in the wavelength interval from 700 to 900 nm, the excited autofluorescence is at a minimum, and the penetration of the exciting radiation and the scattered radiation is at a maximum. Fluorescence in Raman spectra from biological samples can also be avoided by laser excitation in the deep UV because a fluorescence-free window exists with excitation below 270 nm. However, deep UV excitation also harbors the risk of inducing photodegradation damage. The inherent weak Raman intensities of biomolecules require the use of high-throughput optics and sensitivity-optimized detectors. Because the wavelength is shorter and the NA of microscope objectives is larger, the diffraction limit of Raman microscopy is below 1 μm and higher lateral resolution can be achieved than in FTIR microscopy.

Most Raman images are collected in the serial point-mapping mode (Figure 32.2b). Here, the laser is focused onto the sample, the scattered light is registered, and subsequently the focus or the sample is moved to the next position. In the case of laser line illumination of the sample (Figure 32.2b), the spatial data can be registered on the detector on a line parallel to the entrance slit of the spectrometer and the spectral information is dispersed perpendicularly. The second spatial dimension of an image is recorded by scanning in the direction perpendicular to that line. This so-called line-mapping registration mode is faster because only one dimension instead of two dimensions in the point-mapping mode has to be scanned. The parallel registration approach, called *direct* or *wide-field Raman imaging*, employs intense, global sample illumination. The inelastically scattered light from the sample is projected onto a two-dimensional CCD detector. Most wide-field Raman imaging spectrometers use filters to select the wavelength such as dielectric, acousto-optic tunable, and liquid-crystal tunable filters.

32.2.3
Coherent Anti-Stokes–Raman Scattering Microscopy

The description of the CARS process starts with the expression of the intensity of a Raman line observed at the frequency ω_R upon scattering at the frequency ω_0:

$$I_{\text{Stokes}} = \text{const} \cdot I_0 \cdot (\omega_0 - \omega_R)^4 \cdot |\alpha|^2 \tag{32.1}$$

Equation (32.1) can be derived classically considering the sample as an ensemble of harmonic oscillators. However, to obtain a quantitatively correct value for the Raman intensity, the molecular polarizability α has to be calculated quantum mechanically using second-order perturbation theory.

In CARS, three pulses interact with the sample, where the interaction of the first two pulses creates a macroscopic molecular polarization pattern in the sample, which is proportional to

$$q = \frac{1}{2m} \cdot (\partial_q \alpha)_0 \cdot E_L^0 E_S^0 \cdot \frac{1}{\omega_R^2 - (\omega_L - \omega_S)^2 + i\Gamma(\omega_L - \omega_S)} \cdot e^{i[(\omega_L - \omega_S)t - (k_L - k_S)z]}$$

(32.2)

where q denotes the amplitude of the coherently driven vibration. This molecular vibration is driven by the simultaneous interaction of two different laser fields, that is, the pump (E_L^0) and the Stokes laser (E_S^0), which are detuned so that their energetic difference matches a Raman-active vibration. The resulting amplitude resonates, if the difference between ω_L and ω_S, that is, the frequencies of the pump and the Stokes laser, matches the frequency of a Raman-active vibration ω_R.

The third pulse interacts with the vibrationally excited sample creating a third-order nonlinear polarizability, which can be expressed as

$$P_{NL} = \varepsilon_0 \cdot (\partial_q \alpha)_0 N q E_L$$

(32.3)

Using Eq. (32.3) together with the explicit time dependence of the third electric field to derive an analytic expression of the nonlinear polarization, which constitutes the source for the CARS signal, one arrives at

$$P_{NL} = \varepsilon_0 \cdot \frac{N}{2m} \cdot (\partial_q \alpha)_0^2 \cdot (E_L^0)^2 E_S^0 \cdot \frac{1}{\omega_R^2 - (\omega_L - \omega_S)^2 + i\Gamma(\omega_L - \omega_S)}$$
$$\cdot e^{i[(2\cdot\omega_L - \omega_S)t - (2\cdot k_L - k_S)z]}$$

(32.4)

It can be seen from the exponential term that this nonlinear polarization, which forms the molecular source term for the CARS signal, oscillates at the CARS frequency $\omega_{CARS} = 2\omega_L - \omega_S$ and is emitted in the phase-matched direction $k_{CARS} = 2 \cdot k_L - k_S$, which is fulfilled in gaseous media, in which dispersion does not play a significant role by collinear excitation geometry. However, doing CARS spectroscopy (or CARS microspectroscopy) in dispersive media, that is, liquids or solids, particular phase-matching geometries are applied. The CARS-signal generation in the tight-focusing geometry of CARS microscopy using collinear pump and Stokes beams was discovered by Xie and coworkers [3]. Here, the tight and steep focus produced when focusing light with a high-numerical aperture objective causes a cone of pump and Stokes light to propagate within the sample. In this cone, phase-matching conditions for various pairs of k-vectors are fulfilled, giving rise to a strong CARS signal detected in the forward scattered direction. Owing to its coherent nature, the CARS process gives rise to a stronger and directed signal as compared to the Raman-scattered light. When coupled to laser scanning microscopes (Figure 32.2c), CARS enables to collect images at video-time rates. Therefore, this variant of nonlinear chemically sensitive microscopy is a very promising candidate for real-time Raman studies of cells and tissues.

32.3
Analysis of Vibrational Spectroscopic Images

The analysis of vibrational spectroscopic images is a typical hyperspectral problem because of the extended spectral dimension, which depends on the spectral range and the spectral resolution, and the high number of spectra, which depends on the sample area and the spatial resolution. Both the spectral and the lateral dimensions can become large, yielding large amounts of data. The basic idea of hyperspectral imaging is to derive one single spectral parameter for each spectrum of a given dataset. This parameter is then color scaled and plotted as a function of the spatial coordinates. Hyperspectral images are often displayed as data cubes where two dimensions are the pixel coordinates and the third dimension is the spectral one. When considering the diversity of existing and in-progress methodologies for image analysis, it would be far too ambitious to attempt to explain all of them in a single book chapter. Instead, a general description of some widely applied tools is given here. For more details, the reader is referred to a textbook covering this field [4]. A selection of tools is available in a commercial software package called *Cytospec* (*www.cytospec.com*) and in an open-source software package called *hyperSpec* for programming language R (*http://hyperspec.r-forge.r-project.org*).

32.3.1
Preprocessing

The quality of the raw measurements is often affected by instrumental variations that hamper the analysis. Typical problems encountered in image measurements are the high noise level, the presence of intense and irregularly shaped baseline contributions, or the existence of anomalous pixel spectra that may influence the results of the image analyzed. For this reason, an effective preprocessing system can significantly improve the results obtained from any image data analysis tool.

32.3.1.1 Quality Test
The initial preprocessing step is a quality test. Spectra with negative test results are excluded from further evaluations. A common criterion is the threshold methodology to detect unexpectedly high or low results that are well below or above a representative reference value. Median values for spectral reading or median spectra taken from regions of interest in the image can be used as references for this detection. High intensities can cause nonlinear detector response or even detector saturation. Low intensities are consistent with holes or fissures in samples. Negative test results also point to outliers and diverse instrumental artifacts. Prominent outliers in Raman spectra originate from cosmic spikes. As the spike signals are usually narrow, they can be corrected by interpolated values obtained by taking the readings of normal neighboring spectral channels. Prominent outliers in FTIR images can be due to dead pixels of the FPA detector. Algorithms have been developed to interpolate the response of dead pixels from normal neighboring pixels (Chapter 6).

32.3.1.2 Denoising

Many methods have been designed to denoise spectroscopic datasets, all of which can be applied to hyperspectral images. Some of these rely on smoothing procedures whether by averaging or fitting neighboring spectral channels to a polynomial function, such as the Savitzky–Golay approach, or on mathematical signal filtering such as Fourier-transform-based filters or wavelet-based filters (Chapter 7). When applying any of these methodologies, it is vital that none of the relevant features of the dataset is removed together with the noise. Another method based on principal component analysis (PCA) is described in Section 32.3.2.2.

32.3.1.3 Background and Baseline Correction

Subtraction of background signal, for example, from the substrate or the optical components in the light path, is another commonly used preprocessing. Spectral contributions also originate from atmospheric gas in the light path such as water vapor and carbon dioxide in IR spectra or oxygen and nitrogen in Raman spectra. As the gas signals always occur at the same spectral region, they can be corrected by reference spectra.

Baseline correction works better if it is performed after image denoising and background subtraction. In this way, the baseline is better defined and can, as a consequence, be better subtracted. A typical baseline correction can be based on linear or polynomial functions. Calculation of first or second derivatives also helps to eliminate instrumental variations such as constant and linear spectral contributions that are unrelated to the chemical composition of the image. However, differentiation also diminishes the signal-to-noise ratio, which can be compensated by smoothing of spectra. A popular method of calculating derivatives is by applying the Savitzky–Golay algorithm. In this method, nth order derivatives are obtained, while data are smoothed at the same time to reduce the noise.

32.3.1.4 Normalization

Normalization compensates for intensity fluctuations due to different sample thicknesses or focusing onto the sample. It is mandatory to do this before comparing the shapes of the pixel spectra independently from their global intensity. Minimum–maximum normalization indicates that all spectra of the source data are scaled between 0 and 1, that is, the maximal signal of the spectrum in the selected spectral region is equal to one and the minimum zero. Vector normalization is carried out in three steps. First, the average value of the intensities is calculated for the spectral region indicated. Then, this value is subtracted from the spectrum such that the new average value is equal to 0. Finally, the spectra are scaled such that the sum-squared deviation over the indicated wavelengths is equal to 1. Vector normalization is often performed for derivative spectra.

An extended multiplicative signal correction (EMSC) can also be applied for normalization [5]. In the basic form of EMSC, every spectrum $z(\nu)$ is a function of the wavenumber ν, which can be expressed as a linear combination of a baseline shift a, a multiplicative effect b times a reference spectrum $m(\nu)$, and linear and quadratic wavenumber-dependent effects $d\nu$ and $e\nu^2$, respectively:

$$z(v) = a + bm(v) + dv + ev^2 + \varepsilon(v)$$

The term $\varepsilon(v)$ contains the unmodeled residues. The reference spectrum $m(v)$ is calculated by taking the sample mean of the dataset. In the simple form of multiplicative signal correction (MSC), only the baseline shift and the multiplicative effect are considered. The parameters a and b can be estimated from ordinary least squares, and finally the spectra can be corrected according to

$$z_{\text{corr}}(v) = (z - a)/b$$

32.3.1.5 Image Compression

Very often, the large size of raw images results in unreasonable computation times. Depending on the size of the image and the computation tool of the data analysis selected, the original image should be compressed, which can be achieved by cutting the spectral or the spatial dimension. The simplest compression is to aggregate (the term *bin* is also used) spectral or spatial points, which decrease the spectral or lateral resolution, respectively. Aggregation (or binning) means adding the readings on an interval of neighboring spectral or spatial channels to form a single value. For example, if an aggregation factor of 4 is chosen, four frequency intervals or data pixels are merged into one. Aggregation also reduces the noise in datasets. The direction (spectral/spatial) of binning will depend on the resolution of the image and on the type of information preferred. Although the binned image will enclose the information in all spectral channels or pixels, it will suffer a loss of spatial and/or spectral resolution. Another way of accelerating the computing time is based on the selection of a reduced number of pixels or spectral channels. This is the case if feature selection is performed in the spectral direction or if the analysis is performed on a representative pixel subset/region of interest.

32.3.2
Exploratory Image Analysis

32.3.2.1 Classical Image Representations

The classical approach is the so-called chemical mapping technique (also referred to *functional group mapping*). Here, the IR absorbance, Raman intensity, half-width, or frequency of a given vibrational band is color encoded and plotted against the spatial coordinates. These univariate imaging techniques permit visualization of the spatial distribution of functional groups or specific chemical substances and are extensively applied in biomedical science. The risk of this approach is that images from natural or complex samples may contain unknown components, so that assuming the presence of selective spectral channels without any solid evidence could provide incorrect distributions images. Another classical representation of images is the global intensity plot, which is obtained by adding up the reading of all spectral channels. In this way, each pixel is represented by a single number related to the total spectral intensity measured. These plots allow for the observation of certain spatial features in the image, but do not provide any information about the chemical information of the image. At most, they provide a first approximation of the physical characteristics of the image.

As an example of a classical image representation, Figure 32.3 plots the integrated intensities from 2800 to 3000 cm^{-1} of a Raman image from a liver tissue section. A photomicrograph is included for comparison in this figure. Two Raman spectra are displayed that represent the average of the white regions and the colored regions. The white regions correlate with cracks and the corresponding spectrum contains spectral contributions of the substrate calcium fluoride at 322 cm^{-1}, atmospheric nitrogen and oxygen at 2330 and 1555 cm^{-1}, respectively, and broad bands between 1100 and 1500 cm^{-1} of the glass elements in the light path. Subtraction of this spectrum from each spectrum in the Raman image compensates for these constant contributions. The average spectrum of liver tissue can be divided into a high wavenumber portion above 2800 cm^{-1} and a low wavenumber portion below 1800 cm^{-1}. The high wavenumber portion is mainly assigned to stretch vibrations of CH_2, CH_3, NH, and OH moieties of biomolecules. Although their intensities are more intense, they are less informative than the low wavenumber bands. Owing to the high number of bands of deformation and stretch vibrations of CH, CC, CN, CO, OH, NH, and so on, the low wavenumber region is called *fingerprint region*. Therefore, the forthcoming examples focus on the analysis of the spectral fingerprint. The molecular contrast in the Raman image shown in Figure 32.3 is low. Red color indicates higher intensities at the top and blue color lower intensities at the bottom. The circular feature (see arrow) correlates with a morphological feature in the sample. However, the signal intensities in the selected spectral range are not sufficient to assign the feature.

32.3.2.2 Principal Component Analysis

In contrast to classical approaches, PCA belongs to the multivariate image-analysis tools. They enable to utilize more information contained in the vibrational spectroscopic image in the spatial and spectral dimensions and provide ways to visualize images. PCA is a chemometric method that decomposes a data table into a bilinear model of latent variables, the so-called principal components. From a geometrical point of view, PCA projects data to a new coordinate system. The new coordinate axes are PCs or loadings, which are uncorrelated or orthogonal to each other. The values for each PC are called *scores* in the new coordinate system. PCs are ordered so that PC1 exhibits the greatest amount of variation, PC2 the second greatest amount of variation, and so on. In this way, PCA allows to describe as much as possible of the variance in the dataset by the first significant PCs, while all subsequent PCs are so low as to be virtually negligible. For spectroscopic data, the latter PCs are dominated by noise. PCA can be used for feature reduction and spectral noise reduction. Although an image can have thousands of spectral channels, the relevant information in all these channels is contained in a very small number of PCs and the dataset can be reproduced by taking only the relevant components (e.g., reassembling of spectra on the basis of low order PCs). In this way, higher order PCs that are supposed to contain mainly noise are omitted.

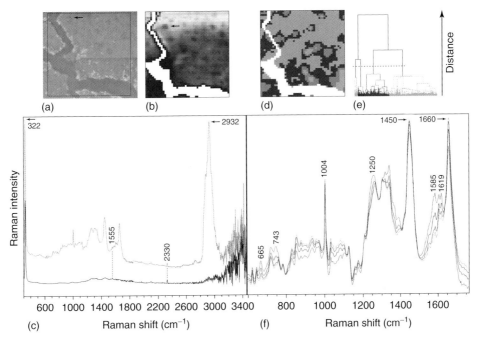

Figure 32.3 Photomicrograph of a liver tissue section on calcium fluoride substrate, with a box indicating the sampling area (a). Intensity plot of the band near 2932 cm^{-1} of the 49 × 49 Raman image, with a step size of 2.5 μm (b). Unmanipulated Raman spectrum (gray trace) and background spectrum representing the crack (black trace) (c). Color-coded class membership of K-means clustering (d) and dendrogram of hierarchical clustering (e) from the liver Raman image. Four-cluster-averaged Raman spectra are overlaid for comparison (f). (Please find a color version of this figure on the color plates.)

PCA is demonstrated in the Raman image of the liver tissue section, which is shown in Figure 32.4. The score plots and the loadings of PC1–PC4 are shown in Figure 32.4 for the spectral range 595–1765 cm^{-1}. PC1 accounts for the largest variations, while PC2 accounts for the second largest variation, and so on. The positive and negative contributions in loadings are correlated with vibrations of chemical constituents. Negative features in the loading of PC1 at 665, 743, 1250, 1585, and 1619 cm^{-1} are associated with spectral contributions of hemoglobin. Therefore, the blue spots in the score plot of PC1 can tentatively be assigned to red blood cells. Negative features in the loading of PC2 at 728, 783, 1100, 1336, 1483, 1577, and 1679 cm^{-1} are associated with spectral contributions of nucleic acids. Therefore, the blue spots in the score plot of PC2 can tentatively be assigned to cell nuclei. The spectral information is mixed in PC3 and PC4 and the variances of constituents overlap making assignments more difficult. A comparison of loadings PC1–PC4 confirms that the signal-to-noise ratio decreases with increasing PCs.

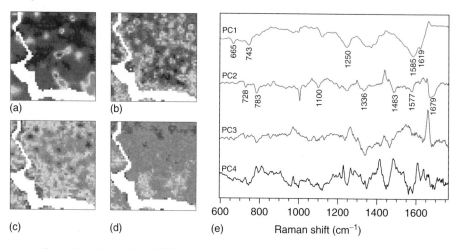

Figure 32.4 Illustration of PCA and the relation between scores t, loadings p, average x̄, and data matrix X of N samples with k variables (a). Three-dimensional space with a straight line fitted to the points as an example of a one-component PC model. The PC score of an object t_i is its orthogonal projection on the PC line (b). Score plots of PC1 (a), PC2 (b), PC3 (c), PC4 (d), and loadings (e) of a PCA from the liver Raman image in Figure 32.3. Labeled negative values of PC1 are correlated with spectral contributions of heme, negative values of PC2 with nucleic acids. (Please find a color version of this figure on the color plates.)

32.3.3
Image Segmentation: Cluster Analysis

One of the goals of image analysis is to identify groups of similar pixels in an image that correspond to similar composition and chemical or biological properties. This operation is referred to as *image segmentation*. It is used to separate zones in biomedical samples or to obtain an insight into the variation of chemical composition along the surface scanned. Clustering multispectral images is dealt with in Ref. [6]. In general, clustering is the partitioning of a dataset into subsets (the clusters) so that the differences between the data within each cluster are minimized and the differences between clusters are maximized according to some defined distance measure. The Euclidean distance can be used to assess the proximity of data:

$$d(x_i, x_j) = \sqrt{\sum_{l=1}^{d} (x_{il} - x_{jl})^2}$$

where x_i and x_j are two pixels and d is the total number of variables (e.g., wavenumbers). Distances can also be defined differently from a mathematical point of view.

In K-means clustering, the distances are calculated from a pixel to the centroid and the dataset is grouped into a preselected number (K) of clusters. The initial centroids are chosen randomly. Then, each spectrum is assigned to the cluster whose centroid is nearest. New centroids are computed, being the average of all spectra in the cluster. The two previous steps are repeated until the solution converges. The results of the *K*-means cluster analysis are the centers of each cluster and the cluster membership map. *K*-means clustering belongs to the partitional methods. Their advantages include that they are fast and applicable to large datasets. Disadvantages include that the selection of the number of clusters is not easy and, often, different results can be obtained depending on the starting points (centroids) of the clustering process.

The results of *K*-means clustering on the Raman image of the liver tissue section are included in Figure 32.3. The Raman image is segmented into four clusters. The centroid spectra reveal a gradual increase in the spectral contributions of red blood cells from the orange, vine, blue, to the green cluster as evident from the bands at 665, 743, 1250, 1585, and 1619 cm^{-1}. Spectral contributions of proteins show an inverse tendency, which means that bands at 1004 and 1660 cm^{-1} decrease in intensity.

Hierarchical cluster analysis (HCA) is an agglomerative clustering method. It calculates the symmetric distance matrix (size $n \times n$) between all considered spectra (number n) as a measure of their pairwise similarity. The algorithm then searches for the minimum distance, collects the two most similar spectra into a first cluster, and recalculates spectral distances between all remaining spectra and the first cluster. In the next step, the algorithm performs a new search for the most similar objects, which now can be spectra or clusters. This iterative process is repeated $n - 1$ times until all spectra have been merged into one cluster. The result is displayed in a treelike, two-dimensional dendrogram, in which one axis refers to the reduction of clusters with increasing number of iterations and the other axis to the respective spectral distances (Figure 32.3). With this method, it is possible to examine the clustering arrangements with different number of groups and to select a scheme that may be more easily interpretable. The dotted line in the dendrogram indicates the distance for the segmentation into four clusters. The drawback is the intensive computation required because new distances are calculated between pixels and clusters at each step. In contrast to partitional clustering, agglomerative clustering provides very robust clustering schemes.

32.3.4
Supervised Image Segmentation: Linear Discriminant Analysis

Supervised image segmentation methods operate by defining the different pixel classes beforehand. These approaches use reference knowledge, that is, class membership of training data. Each class is described by a model and these models are used to assign unknown data to the predefined classes. In the context of IR and Raman spectroscopy, several algorithms have been applied so far such as linear discriminant analysis (LDA), partial least squares (PLS), soft independent modeling of class analogies, artificial neural networks, and support vector machines. As they are summarized in a recent review [7], only LDA is described as an example next.

LDA and PLS use a common transformation principle, which is based on calculation of the Mahalanobis distance. The Mahalanobis distance measures distances in units of standard deviation. Therefore, it can be used to determine the distance between a point and a group or a class of points. The calculation projects the dataset into a new coordinate system so that the new point cloud has unit covariance, that is, the directions are uncorrelated and have variance of 1. This projection is achieved by a linear combination of the original coordinate axis (e.g., the intensity or absorbance at wavenumbers where the spectrum is measured). In this new coordinate system, the Euclidean distance is the Mahalanobis distance.

A number of objects belong exactly to one out of k similar classes and the class membership is known for each object. Each object is defined by characteristic parameters. LDA uses this information to calculate $(k-1)$ linear discriminant functions that optimally discriminate k classes and subsequently to assign unknown objects to classes. The discriminant functions describe a separation hyperplane. The normal vector of this separation plane is the direction that maximizes the ratio of the difference between classes (interclass variance) to the differences within the classes (intraclass variance). This direction of the vector is simply the direction that connects the class means if the intraclass variance is 1 in all directions. In the present vibrational spectroscopic context, the objects are IR or Raman spectra, the classes are sample types, and the characteristic parameters or features are band intensities or intensities ratios. In the case of imaging datasets, the resulting class memberships can be represented by colors, and color-coded images are assembled for visualization.

32.4
Challenges for Image Analysis in CARS Microscopy

The use of CARS for imaging imposes two challenges on image analysis, which are subsequently discussed. As they are connected with the coherent nonlinear image-generation process, they are also applicable to related nonlinear optical contrast techniques, for example, second- and third-harmonic generation microscopy.

As Enejder and coworkers have laid out [8], the central information in linear microscopy originates from the knowledge about the point-spread function (PSF). The PSF is convoluted with the light-matter interaction voxels, in which the signal is generated in a linear interaction process. The convolution of the voxel-related signal with the PSF yields the contributions to the image and with the accurate knowledge of the PSF the true contour of the spatial structure of the sample can be reconstructed. However, CARS microscopy generates image contrast owing to a coherent nonlinear interaction of three laser fields with the sample. Hence, the individual oscillators in the sample cannot be considered as independent any more. Thus, the emitted light, that is, the signal used for image generation, follows a fundamentally different generation processes compared with the signal in linear microscopy. Consequently, it is not a simple sum of contributions from individual oscillators and the signal intensity per voxel can be written as

$$S(\mathbf{r}) = \left|\chi^{(3)}(\mathbf{r})\right|^2 G(\mathbf{r}) P_P^2 P_S$$

where S(**r**) denotes the signal intensity arising from the interaction of the pump- and Stokes pulses with intensities P_P and P_S, respectively. $\chi^{(3)}$ refers to the third-order nonlinear optical susceptibility, which includes the sum of vibrationally resonant contributions $\chi_R^{(3)}$ and the so-called nonresonant background $\chi_{NR}^{(3)}$. The latter originates from the presence of any polarizable electron system in the sample. G(**r**) contains the influence of linear effects such as phase-matching effects, absorption, and refraction and is the source of the topological contrast, which is superimposed to the chemical contrast provided by the CARS process.

The fact that the CARS signal is the coherent sum of $\chi_R^{(3)}$ and $\chi_{NR}^{(3)}$ constitutes the driving force for image analysis in CARS microscopy up to now. The approach to image analysis has been focused on two different, yet related aspects: (i) the adaptation of methods for particle identification, which are well known from linear microscopy, for nonlinear microscopy and (ii) numerical reduction or suppression of the nonresonant background to extract pure chemical information not contaminated by contributions from morphology.

32.4.1
Particle Identification in Nonlinear Microscopy

In conventional linear microscopy, a variety of image-analysis tools have been established, which are based on the global intensity histogram of the images, a kernel convolution operation, the calculation of an intensity gradient, or the minimization of a cost function. Enejder and coworkers evaluated the use of these image-analysis methods for CARS microscopy [8]. Global threshold classifies individual pixels either as background or as signal based on the brightness value of the pixel. Cut-off criteria need to be defined. Local threshold, on the other hand, is based on relating pixel intensity to the mean of its immediate surrounding to separate background from signal. A sliding window is used and scanned over the image to perform the numerical operation. However, the size of this window and its relation to the image and object sizes determines with the performance of the algorithm. Watershed segmentation refers to the numerical realization of flooding the intensity landscape of the image with water. The islands, which remain *dry land* in this operation, are considered objects; the flooded pixels, on the other hand, are the background. Finally, level-set algorithm determines image borders by minimizing a cost function, which establishes the boarder length and position. For using the level-set segmentation, the weighting factors of the aforementioned contributions to the cost function need to be chosen prior to the run of the algorithm. This gives an object identification, as the choice of parameters affects the identification scoring of the algorithm.

Enejders and coworkers focused on the applicability of the various image correction methods for object identification and size determination [8]. Using model samples, it was found that both global threshold and the full-width-half maximum of an intensity distribution, a metric conventionally used in linear microscopy, are insufficient to determine particle sizes in nonlinear microscopy. In contrast, watershed and level-set algorithms were found to yield good results for well-defined simply structured objects but failed on being applied to samples more complex than polystyrene beads. Local threshold leads to the best identification rates and estimates for particle sizes and performs in the most robust manner.

32.4.2
Numerical Reduction or Suppression of Nonresonant Background

The presence of nonresonant nondispersive contributions in CARS imaging is conventionally considered to be a negative factor, reducing the chemical contrast. Hence, experimental methods suppressing the nonresonant background have been proposed [9]. Unfortunately, some of those methods reduce the amplitude of the signal or result in a significant complication of the setup. Recently, an image-analysis approach using the nonresonant nondispersive part of the signal was proposed [10]. Experimentally this nonresonant contribution is accessible by detuning the frequency difference $\Delta\omega = \omega_P - \omega_S$ from a Raman resonance. By recording a nonresonant image in addition to a CARS image, the spatial distribution of the signal allows for specification of $\chi_{NR}^{(3)}(r)$-related contrast. In the absence of any electronic two-photon resonances, $\chi_{NR}^{(3)}$ is a real quantity and the dispersion of the topology factor $G(\mathbf{r})$ can be ignored using this procedure as Raman vibrational resonances in contrast to electronic transitions are spectrally quite narrow.

Hence, the subtraction of the nonresonant image $S^{NR}(\mathbf{r})$ from the CARS image, that is, calculation of $S^R(\mathbf{r}) - S^{NR}(\mathbf{r})$, results in a corrected image, which, after normalization to the laser power, can be expressed as

$$M(\mathbf{r}, \omega_R) = \left[\left| \chi_R^{(3)}(\mathbf{r}, \omega_R) \right|^2 + 2\chi_{NR}^{(3)}(\mathbf{r}) \, \mathrm{Re} \, \chi_R^{(3)}(\mathbf{r}, \omega_R) \right] G(\mathbf{r})$$

If the frequency difference between the pump and the Stokes laser is chosen to match a Raman transition, $\chi_{NR}^{(3)}$ becomes negligible, that is,

$$M(\mathbf{r}, \omega_R) \approx \mathrm{Im}^2 \, \chi_R^{(3)}(\mathbf{r}, \omega_R) \, G(\mathbf{r}) \tag{32.5}$$

Thus, Eq. (32.5) provides a method to calculate images that are not contaminated by nonresonant contrast. However, this particular routine for image processing yields images in which topological information encoded in $G(\mathbf{r})$ is convoluted with chemical contrast ($\chi_R^{(3)}(\mathbf{r}, \omega_R)$).

For purification of CARS images from topological contrast, which arises from linear interactions of the sample with either of the laser beams, that is, distortion of beam geometry, phase-matching effects, and refraction, it is instructive to normalize the corrected image $M(\mathbf{r})$ by the nonresonant intensity distribution. Thus, we arrive at

$$\frac{S^R(\mathbf{r}) - S^{NR}(\mathbf{r})}{S^{NR}(\mathbf{r})} \approx \frac{\mathrm{Im}^2 \, \chi_R^{(3)}(\mathbf{r}, \omega_R)}{\left(\chi_{NR}^{(3)}(\mathbf{r}) \right)^2} \tag{32.6}$$

32.4 Challenges for Image Analysis in CARS Microscopy

Image processing is of fundamental importance, as any CARS image provides the desired chemical sensitivity, which, however, is convoluted with nonresonant image contributions. Furthermore, the morphological signal contains information about linear effects, which can significantly change the signal strength and result in the appearance of topological contrast. A correction method based on using nonresonant images provides the possibility to distinguish linear and nonlinear contrasts within the images.

This is illustrated in Figure 32.5 on sections of a plant leaf. Images of a cut of a *Paniflora spec.* leaf were recorded at 2918 cm^{-1} (Figure 32.5a), revealing the spatial distribution of C–H oscillators. These are correlated to the presence of waxes and lipids. For comparison, a nonresonant image was recorded at 3088 cm^{-1} (Figure 32.5b). The comparison of both images reveals areas of resonant signal due to high concentrations of CH bonds. The bright zigzag line, for example, corresponds to cuticular waxes, while the presence of a liposome is revealed by the bright spot in the left bottom corner of the image. Figure 32.5c displays the image after nonresonant contrast was eliminated. However, it still shows the convolution of the resonant signal and the topological factor $G(\mathbf{r})$. The second image-processing approach displayed in Figure 32.5d disentangles chemical from topological contrast as suggested in Eq. (32.6).

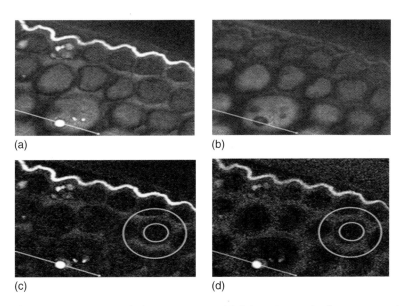

Figure 32.5 Correction of CARS images using the analytical expressions discussed in the text. (a) CARS images of the section prepared from a fresh leaf (*Paniflora spec.*) recorded at a Raman shift of 2918 cm^{-1}. (b) Nonresonant image of the same sample at a Raman shift of 3088 cm^{-1}. (c) Subtraction result of nonresonant signal distribution (b) from resonant one (a). (d) Corrected CARS image by means of Eq. (32.6). White bordered empty circles indicate a single cell within the sample (inner circle) and the corresponding cell wall (outer circle).

32.4.3
Outlook – Merging CARS Imaging with Chemometrics

Up to now, we focused on the discussion of image analysis of single-color CARS images, that is, a narrowband pump pulse and a narrowband probe pulse are used to generate the CARS signal at a well-defined spectral position. However, recent technological advantages have shown the possibility to obtain CARS-microspectroscopic images by recording a CARS spectrum over a broad wavenumber range of $\sim$4000 cm^{-1} at each pixel of the image. This approach called *multiplex CARS* leads to a huge demand for computational power for handling extensive data files. However, it allows for numerical separation of the vibrationally resonant and nonresonant contributions of the signal by either invoking Kramers–Kronig relation or by means of maximum-entropy methods. Multiplex CARS opens fundamentally novel possibilities for CARS microscopy, as it allows one to obtain the full vibrational information at each sample point. Nonetheless, it should be pointed out that up-to-date multiplex CARS imaging has barely been put forward outside a physics laboratory. Proof-of-principle experiments have been published, showing the capability of multiplex CARS imaging. To exploit the full potential of multiplex CARS for biomedical imaging, it will be necessary to adopt the chemometric image-analysis approaches for Raman and IR microspectral imaging, discussed above, for multiplex CARS microscopy. We anticipate that such an approach will be pursued by many research groups in the near future and will significantly enhance the applicability of multiplex CARS to relevant questions in medicine and the life sciences.

32.5
Biomedical Applications of Vibrational Spectroscopic Imaging: Tissue Diagnostics

This section describes an example of how FTIR, Raman, and CARS imaging have been applied to characterize colon tissue (Figure 32.6). More applications including papers dealing with vibrational spectroscopic imaging of single cells have been summarized elsewhere [9]. The photomicrograph indicates a transition from the muscular layer (top) to mucosa (bottom) with some droplets resolved in the upper portion. Similar to Figure 32.3, the FTIR and Raman images were segmented by cluster analysis. The muscular layer (blue/cyan) is distinguished from mucosa (yellow/brown) and droplets (black). A representative Raman spectrum of the droplets (black trace) is compared with the Raman spectrum of unsaturated fatty acids (magenta trance). The agreement of both spectra is consistent with high fatty acid content in these droplets. Typical Raman bands at 1060 and 1128 cm^{-1} are assigned to C–C vibrations, bands at 1294 and 1439 cm^{-1} to CH$_2$ deformation vibrations, and bands at 2851, 2881, and 2929 cm^{-1} to CH$_2$ stretch vibrations. As the stretch vibrations have high Raman intensities in fatty acids, the CARS image was collected near 2880 cm^{-1}. The droplets were well resolved as red spots, indicating maximum CARS signal. The IR difference spectrum between the black and the blue clusters confirms the high fatty acids content in the droplets.

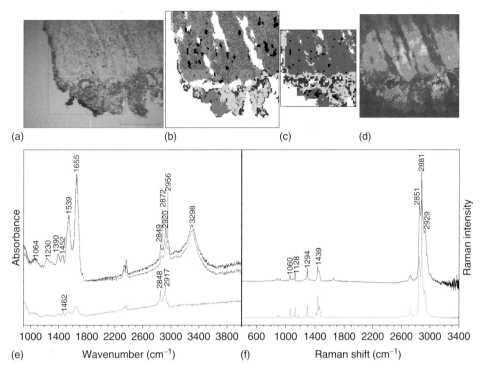

Figure 32.6 Photomicrograph (a), FTIR image (b), Raman image (c), and CARS image (d) of an unstained colon tissue section on calcium fluoride slide showing the transition from muscle (top) to mucosa (bottom). Droplets are resolved as black clusters in Raman and FTIR images and as red spots in the CARS image. IR difference spectrum in (e) (green = black minus blue cluster) and Raman spectrum in (f) show high fatty acid content in droplets. Raman spectrum of unsaturated fatty acids is included as reference (magenta trace). (Please find a color version of this figure on the color plates.)

Difference bands at 1462 and 2848–2917 cm^{-1} are assigned to deformation and stretch vibrations of CH_2, respectively. The IR spectra are dominated by spectral contributions of proteins near 1539 cm^{-1} (amide II), 1655 cm^{-1} (amide I), and 3298 cm^{-1} (amide A) due to vibrations of the peptide backbone. Further bands at 1064, 1230, 1390, and 1452 cm^{-1} contain overlapping spectral contributions of nucleic acids, phospholipids, proteins, and carbohydrates. More detailed band assignments are given elsewhere [11]. The observation that spectral contributions of proteins dominate in the IR spectra of droplets points to spectral dilution effects. The CARS image indicates that most droplets are smaller than 10 μm. As the lateral resolution in FTIR imaging is larger than the size of the droplets, each IR spectrum also "sees" the surrounding material, which is, in this case, muscular layer containing proteins. This resolution effect is also consistent with the extended size of the black clusters in the FTIR image compared with the Raman and CARS image. It is evident from the close agreement between the Raman spectrum of the droplets and the reference spectrum that the laser can be focused on single droplets and the

Raman spectra obtained are virtually free from spectral contribution of proteins. Besides the different lateral resolution, FTIR, Raman, and CARS imaging differ in acquisition times. Whereas CARS images at a single Stokes shift can be acquired at video-time frame rates (~seconds and less), FTIR images require exposure times in the range of minutes. As acquisition of each Raman spectrum from tissue takes typically seconds, total exposure times for a Raman image of thousands of individual spectra might take up to several hours. A more detailed comparison of Raman and CARS microscopy of colon tissue has previously been published [12].

32.6
Conclusions

Biomedical applications of vibrational microspectroscopic imaging are still in their infancy and will continue to grow with the development of new instruments and innovative image-processing approaches. The progress in this field is described in some recent textbooks [13–17] and a review [9], which are recommended for further reading. Improvements are expected in the sensitivity of instruments, which will enable collection of nonresonant, spontaneous Raman images in minutes or even seconds. This can be realized by line-mapping acquisition routines using scanning mirror or cylindrical lenses and including intensified CCD detection. Innovative short-pulse lasers and CCD detectors are to be developed to utilize signal enhancement due to nonlinear inelastic scattering processes such as CARS. Of particular interest is multiplex CARS microscopy probing an extended spectral interval. These instruments will allow to study dynamic cellular processes in real time with sub-second time resolution. These developments will also give larger data volumes that require more sophisticated image-processing strategies including algorithms and more powerful computing technology. In summary, biomedical imaging based on vibrational spectroscopies is a highly multidisciplinary task. The current contribution has introduced the reader to the state of the art and pointed to future directions.

Acknowledgments

Financial support of the European Union via the Europäischer Fonds für Regionale Entwicklung (EFRE) and the "Thüringer Ministerium für Bildung, Wissenschaft und Kultur (TMBWK)" (Projects: B714-07037, B578-06001, 14.90 HWP) is highly acknowledged. Furthermore, B.D. thanks the Fonds der Chemischen Industrie for financial support.

References

1. Diem, M. et al. (2004) A decade of vibrational micro-spectroscopy of human cells and tissue (1994–2004). *Analyst*, **129**, 880–885.

2. Kazarian, S.G. and Chan, K.L. (2006) Applications of ATR-FTIR spectroscopic imaging to biomedical samples. *Biochim. Biophys. Acta*, **1758**, 858–867.

3. Zumbusch, A. et al. (1999) Three-dimensional vibrational imaging by coherent anti-stokes Raman scattering. *Phys. Rev. Lett.*, **82**, 4142–4145.
4. Grahn, H.F. and Geladi, P. (2007) *Techniques and Applications of Hyperspectral Image Analysis*, John Wiley & Sons, Ltd, Chichester.
5. Kohler, A. et al. (2005) Extended multiplicative signal correction as a tool for separation and characterization of physical and chemical information in Fourier transform infrared microscopy images of cryo-sections of beef loin. *Appl. Spectrosc.*, **59**, 707–716.
6. Tran, T.N. et al. (2005) Clustering multispectral images: a tutorial. *Chemom. Intell. Lab. Syst.*, **77**, 3–17.
7. Krafft, C. et al. (2008) Disease recognition by infrared and Raman spectroscopy. *J. Biophoton.*, **2**, 13–28.
8. Hagmar, J. et al. (2008) Image analysis in nonlinear microscopy. *J. Opt. Soc. Am. A Opt. Image Sci. Vis.*, **25**, 2195–2206.
9. Krafft, C. et al. (2009) Raman and CARS microspectroscopy of cells and tissues. *Analyst*, **134**, 1046–1057.
10. Akimov, D. et al. (2009) Different contrast information obtained from CARS and non-resonant FWM images. *J. Raman Spectrosc.*, **40**, 941–947.
11. Naumann, D. (2000) in *Encyclopedia of Analytical Chemistry* (ed. R.A. Meyers), John Wiley & Sons, Ltd, Chichester, pp. 102–131.
12. Krafft, C. et al. (2009) A comparative Raman and CARS imaging study of colon tissue. *J. Biophoton.*, **2**, 303–312.
13. Diem, M. et al. (2008) *Vibrational Spectroscopy for Medical Diagnosis*, John Wiley & Sons, Ltd, Chichester.
14. Lasch, P. and Kneipp, J. (2008) *Biomedical Vibrational Spectroscopy*, John Wiley & Sons, Inc., Hoboken.
15. Salzer, R. and Siesler, H.W. (2009) *Infrared and Raman Spectroscopic Imaging*, Wiley-VCH, Weinheim.
16. Bhargava, R. and Levin, I.W. (2005) *Spectrochimical Analysis Using Infrared Multichannel Detector*, Blackwell Publishing, Oxford.
17. Srinivasan, G. (2009) *Vibrational Spectroscopic Imaging for Biomedical Applications*, McGraw-Hill.

33
Optical Data Encryption
Maria Sagrario Millán García-Varela and Elisabet Pérez-Cabré

33.1
Introduction

Information security is an integral part of our lives, which affects the simplest ordinary transactions among individuals through to the most complex operations concerning all kinds of institutions (industrial companies, health-care organizations, research institutes, and governments). For example, a secure piece of information is often required for personal identification in a variety of formats, such as a passport, credit card, driving license, and many other identity (ID) cards. Another important aspect of security involves information gathering, storage, transmission, recovery, processing, and analysis. In this scenario, there is a need to protect valuable data against deception (forgery, counterfeit) or attack (tampering, intrusion) and, to this end, the encryption and decryption of data (image, message, signature, etc.) play a vital role. In the last two decades, there has been an increasing demand and an intense research in the field. A significant progress in optoelectronic devices has made optical technologies attractive for security. Optical-processing systems have been proposed for a number of security applications, including encryption–decryption, anticounterfeiting, and authentication. They have been widely studied as evidenced by the extensive literature published on the issue [1–7]. In addition, the most commercial implementations have been protected under patents.

Optical security technology is based on complex information processes in which the signals are, first, hidden from human perception or easy conversion into visible signals (to keep them secret); second, extremely difficult to reproduce with the same properties (to avoid counterfeiting); and, third, automatically, in real-time, robustly, and often remotely readable by compact processors that validate authorized signatures. Optical security techniques involve tasks such as encoding, encryption, recognition, secure identification and (or) verification, biometric imaging, and optical keys. The security strength of optical cryptography stems from the ability of optics to process the information in a hyperspace of states, where variables such as amplitude, phase, polarization, wavelength, spatial position, and fractional

---

Optical and Digital Image Processing: Fundamentals and Applications, First Edition. Edited by Gabriel Cristóbal, Peter Schelkens, and Hugo Thienpont.
© 2011 Wiley-VCH Verlag GmbH & Co. KGaA. Published 2011 by Wiley-VCH Verlag GmbH & Co. KGaA.

spatial frequency domain can all be used to serve as keys for data encryption or to hide the signal. Moreover, optical processing has the valuable property of inherent parallelism, which allows for fast encryption and decryption of large volumes of complex data, that is, amplitude and phase. High-resolution optical materials can store encoded data in small volumes (Chapter 10).

This chapter is organized into two parts. The first part reviews the main optical techniques involved in encryption–decryption systems, paying particular attention to the widely known double-random phase encryption (*DRPE*) technique and its variants. In the second part, some applications based on specific combinations of the optical techniques described in the first part are discussed. Although optical image processing is mainly formulated as a two-dimensional analysis, all the mathematics is developed using one-dimensional notation for simplicity. Here, it is not the aim to include a comprehensive review of all the aspects treated in the research on optical data encryption. For a complementary insight, the interested reader is referred to additional content accessible through *ftp.wiley-vch.de* as well as to the pioneering works, sound reviews, and other papers detailed in the list of references.

33.2
Optical Techniques in Encryption Algorithms

33.2.1
Random Phase Mask (RPM) and Phase Encoding

A random noise pattern, which is obtained with statistically independent realizations, maximizes entropy and is the most difficult pattern to regenerate. Random noise has been extensively used in security applications with a high degree of success. More specifically, random noise is used to generate a random phase mask (RPM) that is jointly encoded with the primary image. One of the first proposals concerning this technique was introduced by Javidi and Horner in 1994 [1]. The basic idea was to permanently and irretrievably bond an RPM to a primary amplitude pattern, which was a representative identification feature for a given person. Let $\phi_m(x) = \exp[i2\pi m(x)]$ denote the RPM generated by a random sequence $m(x)$ uniformly distributed in [0, 1] and let $f(x)$ denote the primary image. The resultant distribution $f(x)\phi_m(x)$, where x represents the coordinate in the spatial domain, is used in an ID card, similarly to how a picture of a person or a fingerprint is displayed on conventional ID cards. Any attempt to remove the RPM from the primary image would destroy the mask and would likely damage the primary image too. A phase mask is invisible under conventional illumination and cannot be recorded by a CCD camera. Thus, this procedure permits to secure data without modifying the appearance of the ID card. The verification procedure can be carried out using optical correlation in a joint transform correlator (JTC) [8] (Figure 33.1). The reference pattern introduced in the processor corresponds either to the RPM $\phi_m(x)$, if one wants to validate the authenticity of the card, or to the primary image $f(x)$, if one wants to identify the image on the card.

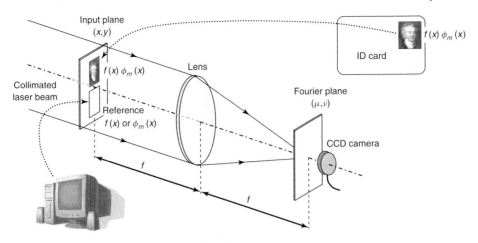

Figure 33.1 Random phase mask bounded to the picture of an ID card. A joint transform correlator is the optical processor used to validate the authenticity of the ID card.

Javidi and Sergent [9] investigated the performance of the all-phase-encoding scheme, which consists of phase encoding the primary image $f(x)$ prior to multiplying it by the RPM, $\phi_m(x)$. This configuration permits to completely hide the information of both the primary image and the RPM on a card or an object. This information cannot be noticed by human inspection or common recording by intensity-sensitive cameras. Functions $m(x)$ and $f(x)$, both normalized in [0,1], are introduced in the final phase-encoded distribution with different weights. The phase distribution used to validate the information is $r(x) = \exp\{i\pi[f(x) + 2am(x)]\}$. The weighting parameter a is introduced to control the influence of the RPM in the final verification result. The all-phase-encoded distribution $r(x)$ is included in an ID card and used as the reference function in a nonlinear JTC [10].

33.2.2
Double-Random Phase Encryption (DRPE)

DRPE was introduced by Réfrégier and Javidi in their pioneering paper [2]. The algorithm is briefly described here. The primary image $f(x)$ is an intensity representation and, consequently, it can be assumed to be positive. Let $\psi(x)$ denote the encrypted function, and $n(x)$ and $m(x)$ denote two independent white sequences uniformly distributed in [0,1]. To encode $f(x)$ into a white stationary sequence, two RPMs are used, $\phi_n(x) = \exp[i2\pi n(x)]$ and $\phi_m(x) = \exp[i2\pi m(x)]$. The second RPM, $\phi_m(u)$, is the Fourier transform of function $h(x)$, that is,

$$\text{FT}\{h(x)\} = \hat{h}(u) = \phi_m(u) = \exp[i2\pi m(u)] \qquad (33.1)$$

Hereafter, FT$\{\cdot\}$ and a hat ($\hat{\cdot}$) denote Fourier transform and u the coordinate in the spatial frequency domain. The encryption process consists of multiplying the primary image by the first RPM $\phi_n(x)$. The result is then convolved by the function

$h(x)$. The encrypted function is complex, with amplitude and phase, and is given by the following expression:

$$\psi(x) = \{f(x)\phi_n(x)\} * h(x) = \{f(x)\phi_n(x)\} * \mathrm{FT}^{-1}\{\phi_m(u)\} \qquad (33.2)$$

where the symbol (*) denotes convolution. The encrypted function of Eq. (33.2) has a noiselike appearance that does not reveal the content of the primary image. Regarding the amplitude-coded primary image $f(x)$, Eq. (33.2) is a linear operation.

In the decryption process, $\psi(x)$ is Fourier transformed, multiplied by the complex conjugate of the second RPM $\phi_m^*(u)$ that acts as a key, and then inverse Fourier transformed. As a result, the output is

$$\mathrm{FT}^{-1}\{\mathrm{FT}[\psi(x)]\phi_m^*(u)\} = \mathrm{FT}^{-1}\{\mathrm{FT}[f(x)\phi_n(x)]\phi_m(u)\phi_m^*(u)\} = f(x)\phi_n(x) \qquad (33.3)$$

whose absolute value turns out the decrypted image $f(x)$. The whole encryption–decryption method can be implemented either digitally or optically. The optical hardware can be a classical 4f-processor [8] (Figure 33.2). In the encryption process, the 4f-processor has the first RPM stuck to the primary image in the input plane and the second RPM in its Fourier plane. In the output plane, the encrypted function is recorded, in amplitude and phase, using holographic techniques. In the decryption process, the 4f-processor has the encrypted function in the input plane, and the key, that is, the complex conjugate of the second RPM, in its Fourier plane. In the output plane, the decrypted primary image is recovered by using an intensity-sensitive device such as a CCD camera.

The DRPE proposed by Réfrégier and Javidi [2] is the basis of a number of variations that intend to increase security by introducing additional keys that make the unauthorized deciphering even more difficult than with the original proposal.

Optical information can be hidden either in the complex-amplitude form or in the phase-only form or in the amplitude-only form. If the encrypted data $\psi(x)$ are complex (amplitude and phase) functions, such as those described in the method originally proposed by Réfrégier and Javidi [2], then there are some practical constraints to encode them. However, if the encrypted data can be either phase or amplitude only, then the recording and storage is easier.

The phase is often chosen to encode, convey, and retrieve information for many reasons such as higher efficiency, invisibility to the naked eye, and more security than with the amplitude. Towghi et al. [11] modified the linear encoding technique of DRPE [2] by introducing a nonlinear (full-phase) encoding, for which a phase-only version of the primary image $f(x)$ is encoded. Thus, the fully phase encrypted image is given by the following equation:

$$\psi_P(x) = \{\exp[i\pi f(x)]\phi_n(x)\} * h(x) = \{\exp[i\pi f(x)]\phi_n(x)\} * \mathrm{FT}^{-1}\{\phi_m(u)\} \qquad (33.4)$$

and it can be generated either optically or electronically in a way similar to that described for Eq. (33.2). The same optical setup is used for decryption (Figure 33.2), but in this case, the complex conjugate of both RPMs $\phi_n^*(x) = \exp[-i2\pi n(x)]$ and $\phi_m^*(u) = \exp[-i2\pi m(u)]$, referred to as keys, are necessary for decryption. The Fourier phase key $\phi_m^*(u)$ is placed in the Fourier plane, whereas the output phase

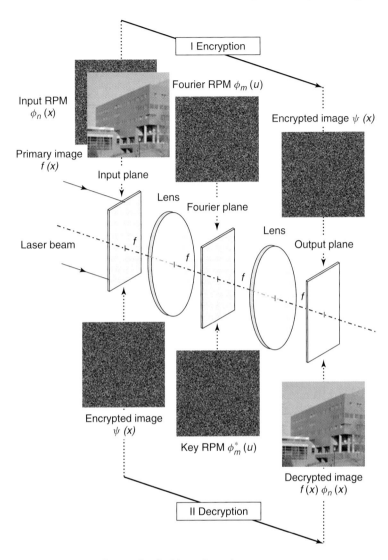

Figure 33.2 Optical setup for double-random phase encryption (top) and decryption (bottom).

key $\phi_n^*(x)$ is placed at the output plane of the optical processor. Just over there, the phase-only version of the primary image $\exp[i\pi f(x)]$ is recovered in the spatial domain. The primary image $f(x)$ can be visualized as an intensity distribution by extracting the phase of $\exp[i\pi f(x)]$ and dividing by the normalization constant π.

Both encoding methods, linear (with amplitude-coded primary image) [2] and nonlinear (with phase-coded primary image) [11], have been analyzed in the presence of perturbation of the encrypted image and the results show their beneficial properties regarding noise robustness [11, 12]. In the presence of additive noise and using the mean-squared-error metric, the results obtained in Ref. [11] show that fully phase performs better than amplitude-based encoding.

Since the optical system to carry out the encryption–decryption process is a holographic system, it requires a strict optical alignment, which is difficult to attain in practice. To alleviate this constraint, a method for optical DRPE using a JTC architecture [8] was proposed in Ref. [13]. In the encryption process (Figure 33.3a), the input RPM $\phi_n(x)$ is placed on the image $f(x)$ (Figure 33.3b) and then, the product $[\phi_n(x)f(x)]$ and $h(x)$ (whose Fourier transform is $\phi_m(u)$, Eq. (33.1)) are placed side by side on the input plane at coordinates $x = a$ and $x = b$, respectively. The joint power spectrum $\hat{e}(u)$, also named encrypted power spectrum, is given by

$$\hat{e}(u) = \left|\text{FT}\left[\phi_n(x-a)f(x-a) + h(x-b)\right]\right|^2 = \left|\hat{\phi}_n(u) * \hat{f}(u)\right|^2 + 1$$
$$+ \left[\hat{\phi}_n(u) * \hat{f}(u)\right]^* \phi_m(u)\exp\left[-i2\pi(b-a)u\right]$$
$$+ \left[\hat{\phi}_n(u) * \hat{f}(u)\right] \phi_m^*(u)\exp\left[-i2\pi(a-b)u\right] \quad (33.5)$$

and is recorded as the encrypted data. The Fourier transform of an RPM is represented by $\hat{\phi}_i(u) = \text{FT}\{\phi_i(x)\}, i = n, m$ in Eq. (33.5). The inverse Fourier transform of $\hat{e}(u)$ is

$$e(x) = \left[\phi_n(x)f(x)\right] \star \left[\phi_n(x)f(x)\right] + \delta(x) + h(x) \star \left[\phi_n(x)f(x)\right] * \delta(x-b+a)$$
$$+ \left[\phi_n(x)f(x)\right] \star h(x) * \delta(x-a+b) \quad (33.6)$$

where the autocorrelation (AC) of $[\phi_n(x)f(x)]$ and the cross correlations of $[\phi_n(x)f(x)]$ and $h(x)$, with noiselike appearances, are obtained at coordinates $x = 0, x = a - b$, and $x = b - a$, respectively. The symbol ($\star$) indicates correlation in Eq. (33.6). The primary image $f(x)$ cannot be recovered from these distributions without the knowledge of either $h(x)$ or $\phi_n(x)$. In the decryption process (Figure 33.3a), the keycode $h(x)$ is placed at coordinate $x = b$ and, consequently, in the Fourier plane, the encrypted power spectrum $\hat{e}(u)$ is illuminated by $\phi_m(u)\exp[i2\pi bu]$ as shown in Figure 33.3e. By Fourier transforming the product $\hat{d}(u) = \hat{e}(u)\phi_m(u)\exp[i2\pi bu]$, in the output plane, we obtain

$$d(x) = h(x) * \left[\phi_n(x)f(x)\right] \star \left[\phi_n(x)f(x)\right] * \delta(x-b) + h(x) * \delta(x-b)$$
$$+ h(x) * h(x) \star \left[\phi_n(x)f(x)\right] * \delta(x-2b+a) + \left[\phi_n(x)f(x)\right] * \delta(x-a) \quad (33.7)$$

The fourth term on the right-hand side of this equation produces the primary image, given that $f(x)$ is positive and the RPM $\phi_n(x)$ is removed by an intensity-sensitive device such as a CCD camera. The primary image is obtained at the coordinate $x = a$ (Figure 33.3c), whereas the undesired terms appear spatially separated from it, at coordinates $x = b$ and $x = 2b - a$. The decrypted image obtained with incorrect keys is shown in Figure 33.3d. There are some advantages of using the JTC architecture. Note that the decryption process uses the same Fourier RPM $\phi_m(u)$ as in the encryption process. The original DRPE, when it was applied in a 4f optical processor [2], required the complex conjugate Fourier phase key $\phi_m^\star(u)$ to

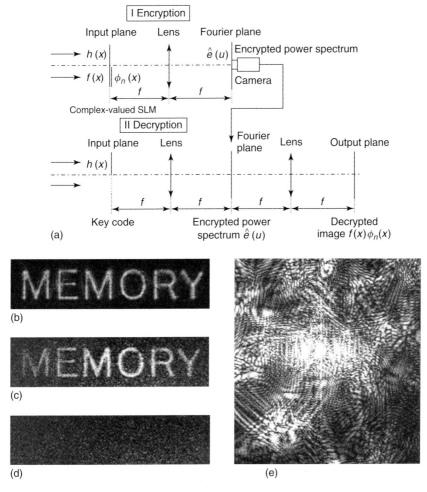

Figure 33.3 (a) JTC architecture encryption and decryption schemes; (b) primary image used in the optical experiment; (c) decrypted image with correct key codes; (d) decrypted image with incorrect key codes; (e) encrypted power spectra. ((b)–(e) from Ref. [13].)

decrypt the image. With the JTC system, there is no need to produce such an exact complex conjugate key. Moreover, because of the shift-invariance property of the JTC architecture, if $h(x)$ is shifted to a coordinate $x = c$, the decrypted image can still be obtained at coordinate $x = a - b + c$. In the optical experiments described in Ref. [13], the encrypted power spectrum is recorded into a photorefractive crystal (LiNbO$_3$, doped with iron ions). The change in the refractive index of the crystal is proportional to the modulation depth. This property makes the crystal operate as a volume hologram (we discuss it in Section 33.2.5). In the decryption stage, when the appropriate key is used, the fourth term of Eq. (33.7) fulfills the Bragg condition.

33.2.3
Resistance of DRPE against Attacks

The amplitude encoding for DRPE (Eq. (33.2)), as proposed by Réfrégier and Javidi [2], is a linear process and this linearity leads to a security flaw in the encryption scheme that facilitates some kinds of attacks. DRPE is much more secure when employed in optical systems because it frequently involves some nonlinear effects. Moreover, it involves some additional experimental parameters (optical storage materials, positions, wavelength, and polarization of the light beam) that need to be known precisely to retrieve the primary image.

The DRPE can be seen as a cryptographic algorithm, which can be alternatively implemented digitally. From this point of view, its security has been tested through the results to various attacks from cryptoanalysis. The scheme is found to be resistant against brute force attacks but vulnerable to various plaintext attacks [14]. In cryptoanalysis, it is assumed that attackers already know the encryption algorithm and other resources. For example, in known plaintext attacks [14–16], the attacker uses a number of known – but not chosen – image pairs consisting of a primary image (plaintext) and the corresponding encrypted image (ciphertext) to obtain the decryption key. One of the most dangerous attacks requires only two known plaintext images with their corresponding ciphered images [14].

In chosen-plaintext attacks, it is assumed that an opponent is able to choose some images to encrypt and has access to either the encrypting or the decrypting parts of the system. We briefly illustrate these types of attacks as first reported by Carnicer *et al.* [17]. They show two different chosen-plaintext attacks. In the first one, the opponent has unauthorized access to the encrypting part of the system and obtains the ciphered image corresponding to a centered Dirac delta function as plain image ($f(x) = \delta(x)$). In other words, the primary image is completely black except for a single centered pixel. In such a case, the resulting ciphered image is precisely the Fourier transform of the second key, $\psi(x) = \text{FT}\{\phi_m(u)\} = \text{FT}\{\exp[i2\pi m(u)]\}$. From this particular encrypted function, the second key, necessary for decryption, can be straightforwardly extracted by inverse Fourier transformation. It is remarkable that this type of attack would succeed with just a single unauthorized access to the encryption system. The second and more sophisticated chosen-plaintext attack is described in Ref. [17] for the case when only the decryption part of the system is accessible to the opponent. In such a case, the attacker uses a set of ciphertexts defined by the expression $\psi_b(x) = (1/2)[\exp(i2\pi f_{x1}x) + \exp(i2\pi f_{x2}x)]$, whose Fourier transform $\hat{\psi}_b(u)$ consists of two Dirac deltas centered at frequencies f_{x1} and f_{x2}. After the key (see decryption setup in Figure 33.2), we have $\hat{\psi}_b(u)\phi_m^*(u)$ and, therefore, at the output

$$h_b(x) = \text{FT}\left\{\hat{\psi}_b(u)\phi_m^*(u)\right\}$$
$$= \frac{1}{2}\left\{\exp[-i2\pi m(f_{x1})]\exp[-i2\pi f_{x1}x] + \exp[-i2\pi m(f_{x2})]\exp[-i2\pi f_{x2}x]\right\}$$
(33.8)

In the optical decryption system, the captured intensity would have a cosine-type distribution, such as the following one:

$$|h_b(x)|^2 = \frac{1}{2} + \frac{1}{2}\cos\left[2\pi\left(f_{x1} - f_{x2}\right)x + m\left(f_{x2}\right) - m\left(f_{x1}\right)\right] \tag{33.9}$$

Since the frequencies f_{x1} and f_{x2} are known, by adjusting the distribution $|h_b(x)|^2$ to a cosine profile, the value of the phase difference $m(f_{x2}) - m(f_{x1})$ can be computed. A repeated application of this attack, changing the value of the frequencies f_{x1} and f_{x2} in the expression of $\psi_b(x)$, enables the attacker to obtain the phase difference between any two pixels of the phase key. Keeping one of the frequencies as a reference, for instance, taking $f_{x1} = 0$, one can compute the phase difference between any value of the key $\phi_m^*(u)$ and the value of the key at the origin $\phi_m^*(0)$. Setting an arbitrary value for $\phi_m^*(0)$, which can be done with no loss of generality, the decryption key can be completely recovered. Although the second type of chosen-plaintext attack concerning the ciphertext allows unlimited access to the decryption part of the system, full key recovery requires as many applications of the algorithm as the number of pixels of the key. However, in the case of a DRPE system, the full key is not necessary. The key recovering algorithm can be applied to only the inner pixels of a square centered at the origin and the effect is equivalent to applying a low-pass filter to the primary image. Figure 33.4 shows the decryption of the original image of Lena (primary image or plaintext) in four cases (Figure 33.4a–d) of full and partially recovered phase key. Even if only the central pixels (a square of 16 × 16 pixels in the key of 128 × 128 pixel size) are known, the decrypted image can be recognized (Figure 33.4d).

The attacks proposed in Refs [15–18] assume that the primary image is amplitude coded. Several attacks reported in Ref. [14] are applicable to phase-encoded images as well, so they can be used against the fully phase encryption [11]. In general, all the attacks assume an ideal mathematical version of the DRPE technique, where all data are available in digital form. Some of the attacks require a large number of images, which make them impractical. In real optical implementation, the physical complexity of the system contributes with additional challenges to the opponent. Large keys (of at least 1000 × 1000 pixels) have been recommended [14], and the variants of DRPE that use Fresnel domain keys (Section 33.2.7) [19] or fractional

  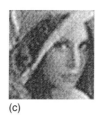 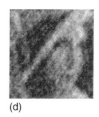

(a)   (b)   (c)   (d)

Figure 33.4 The image of Lena reconstructed from partially recovered keys: (a) 128 × 128 pixels, (b) 64 × 64 pixels, (c) 32 × 32 pixels, and (d) 16 × 16 pixels. (From Ref. [17].)

Fourier transform (Section 33.2.8) [20] are more secure and, therefore, preferred to the original DRPE. It is very important that multiple uses of the same keys for different primary images should be avoided.

33.2.4
Encryption Algorithms Based on Real (Phase-Only and Amplitude-Only) Functions

One of the earliest encryption–decryption methods that used either phase-only or amplitude-only functions was introduced by Wang et al. [21]. Their algorithm is not based on DRPE. They used an RPM $\phi(x)$ as the input of a system that produced an intensity function $|f(x)|^2$ corresponding to the image $f(x)$ for verification. If both the input and output images are known to the authorized person, it is possible to find the phase solution function $\exp[i\zeta(u)]$ that satisfies the following equation:

$$\left|\mathrm{FT}^{-1}\left\{\mathrm{FT}\left[\phi(x)\right]\exp\left[i\zeta(u)\right]\right\}\right| = |f(x)| \tag{33.10}$$

The phase solution function $\exp[i\zeta(u)]$ can be calculated using an iterative algorithm via Fourier transformation for phase retrieval [21]. Once the phase solution function $\exp[i\zeta(u)]$ is calculated for a given pair of functions $\phi(x)$ and $|f(x)|$, it is saved and uploaded in the database of the central computer. The decoding and verification process represented by Eq. (33.10) can be carried out using the classical 4f architecture (Figure 33.5). The RPM $\phi(x)$ is first Fourier transformed by the first lens of the processor and multiplied by the phase function $\exp[i\zeta(u)]$, which acts as a filter that has been previously downloaded from the central computer and displayed on an SLM at the Fourier plane. This product is Fourier transformed again by the second lens of the processor to produce the expected image $f(x)$ whose intensity $|f(x)|^2$ is recorded by a CCD camera at the output plane. The system is double secured. Two keys are required to

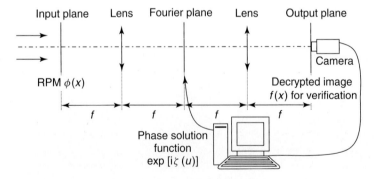

Figure 33.5 Optical-processing system for verification. It uses phase-only and amplitude-only functions and is based on the computation of the phase solution function via an iterative algorithm for phase retrieval.

decipher the encrypted image because the input RPM $\phi(x)$ contains no information of the identification image $|f(x)|$ or the phase-only filter (POF) $\exp[i\zeta(u)]$. If the input RPM is a counterfeit, the output intensity distribution captured by the CCD camera is a noise-type image. The optical processor can then be used to authenticate the image produced by both the input RPM and the POF. Some computer simulations reported in Ref. [21] show the feasibility of this method.

The projection onto constraint set (POCS) technique [22] and the far-field assumption [8] can be used to design the phase-only mask.

33.2.5
Holographic Memory

The idea of combining a holographic optical memory and DRPE for secure data storage was introduced by Javidi et al. [23]. Multiple images can be recorded on the same memory with different RPMs that are independent of each other. Therefore, a specific data image can be decrypted from the memory with the corresponding key, whereas other data remain encrypted. As noticed by the authors, the decoding process requires very good alignment. In practice, an electrically addressed SLM needs to be used to display the key RPM in the Fourier plane. Then, the alignment requirements of the system are alleviated. Because of the finite size of the optical elements and the limited number of pixels of the RPMs, the encryption optical system is bandwidth limited and the codes have a limited space–bandwidth product. These characteristics directly affect the quality of the experimental recovery of the decrypted image. A remarkable improvement is achieved if the decryption is done by the generation of a conjugate of the encrypted function through optical phase conjugation in a photorefractive crystal [24]. Photorefractive materials are attractive because of their large storage capacity and high-speed readout of two-dimensional data sheets. Phase conjugation leads to aberration correction of the optical system, resulting in near-diffraction-limited imaging. In addition to this, the key RPM that is used during encryption can be used for decrypting the data. Consequently, there is no need to produce a conjugate of the key, as it was required in the original version of DRPE. In the optical implementation, the conjugate of the encrypted image used for decryption is generated through four-wave mixing in a photorefractive crystal (Figure 33.6). The beam carrying the encrypted image (I) interferes with the reference plane wave (R) so that the encrypted information is fringe modulated. The transmitted part of beam R is reflected by the mirror placed behind the crystal, producing the plane wave counterpropagating to the reference beam (R^*). This beam R^* interacts inside the photorefractive crystal to produce a fourth beam (I^*) with a complex amplitude proportional to the conjugate of the image-bearing beam. The phase-conjugate beam is counterpropagating to the image beam, I, and remains the phase conjugate of the image everywhere. Therefore, the phase introduced during encryption is canceled if the same RPM is used at the same position in the Fourier plane. The intensity-sensitive detector (camera 1) registers the decrypted image.

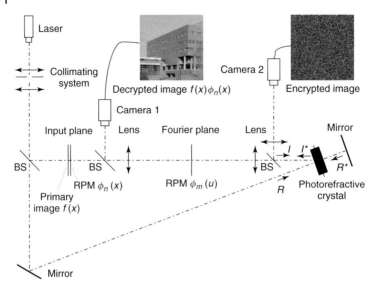

Figure 33.6 Experimental setup of an encryption–decryption system that uses a photorefractive crystal for a secure holographic memory. Four-wave mixing technique: I, the beam bearing the encrypted image; R, the reference plane wave; R^*, the retroreflected plane wave; and I^*, the diffracted phase-conjugate beam used for decryption. BS, beam splitter. Camera 1 registers the decrypted image. Camera 2 allows the visualization of the encrypted function and its use in the system is optional.

33.2.6
Wavelength Multiplexing and Color Image Encryption

When the optical implementation is considered, the system is very sensitive to the wavelength used for decryption. Successful decryption requires a readout beam with the same wavelength as that used in the recording of the encryption function (hologram), in addition to the correct key (RPM). Only noise will be obtained if the recording and the readout beams are of different wavelengths. Consequently, the wavelength can be used as an additional code that performs well as a key for encryption and decryption [25]. Moreover, it is possible to utilize the wavelength-multiplexing technique to encrypt multiple primary images. In such a case, each decrypted image would be obtained by decoding a single encrypted image with its corresponding wavelength [26].

In other lines of work, it will be interesting to consider color images as primary images for encryption–decryption. Some recent examples can be found in Ref. [27]. The color primary image is classified into three channels, namely, red, green, and blue, and the information conveyed by each channel is independently encrypted by wavelength multiplexing. If all the keys are correct in all the channels, the three components are recovered and, by merging them, the full color-decrypted image can be obtained. When RPMs and system parameters are wrong during decoding, the final result will not give correct and complete information about the color.

33.2.7
Fresnel Domain

The two RPMs can be introduced in the Fresnel domain [7] instead of placing them in the input and Fourier planes as it was originally done in the DRPE [2]. In addition to the RPMs, their locations in the optical setup can be considered as additional keys (Figure 33.7). These new parameters make the encryption procedure more secure, even though they also add more severe practical constraints to optically decrypt the hidden information.

Figure 33.7 shows the optical setup with all the planes involved in this technique and their corresponding spatial coordinates. Considering that a plane wavefront of wavelength λ impinges the primary image $f(x)$ located at the input plane, and calculating Fresnel propagation [8] to the plane where the first RPM is placed, the complex field distribution can be expressed as

$$u_1(x_1) = \frac{\exp\left(i\frac{2\pi z_1}{\lambda}\right)}{i\lambda z_1} \int f(x) \exp\left[i\frac{\pi}{\lambda z_1}(x_1 - x)^2\right] dx \tag{33.11}$$

For simplicity, Eq. (33.11) can be rewritten

$$u_1(x_1) = \text{FrT}_\lambda \{f(x); z_1\} \tag{33.12}$$

where $\text{FrT}_\lambda\{\cdot\}$ stands for the Fresnel transform with respect to λ. This distribution is multiplied by the RPM $\phi_n(x_1) = \exp[i2\pi n(x_1)]$ and Fresnel transformed to the lens plane

$$u_L(x_L) = \text{FrT}_\lambda \{u_1(x_1) \phi_n(x_1); f - z_1\} \tag{33.13}$$

When the lens is in paraxial approximation, it introduces a quadratic phase factor $L(x_L) = \exp\left\{-i\frac{\pi}{\lambda f}x_L^2\right\}$ by which the distribution of Eq. (33.13) is multiplied. Propagation to the plane where the second RPM is located provides

$$u_2(x_2) = \text{FrT}_\lambda \{u_L(x_L) L(x_L); f - z_2\} \tag{33.14}$$

and, after multiplying for the second mask $\phi_m(x_2) = \exp[i2\pi m(x_2)]$, it is propagated again toward the back focal plane of the lens (Fourier plane in Figure 33.7).

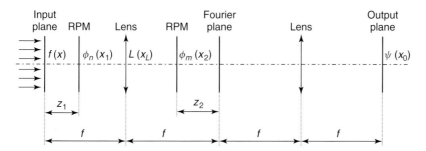

Figure 33.7 Setup for the double-random phase encryption in the Fresnel domain.

A second lens performs a Fourier transform that provides the final encrypted distribution:

$$\psi(x_0) = \text{FT}^{-1}\{\text{FrT}_\lambda\{u_2(x_2)\phi_m(x_2); z_2\}\} \tag{33.15}$$

The encryption in the Fresnel domain has been experimentally proved by holographically registering three different primary images in an optical memory (photorefractive crystal) using angular multiplexing [19]. Each single image is retrieved in the correct angular direction only when the appropriate RPMs are introduced in the proper location of the setup.

A much simpler optical implementation for encryption in Fresnel domain is presented in Ref. [28]. The authors use a lensless optical system to encrypt the information (Figure 33.8). They take into account free-space light propagation and introduce an RPM, $\phi_n(x) = \exp[i2\pi n(x)]$, in the input plane where the primary image, $f(x)$, is located. Under Fresnel approximation, the complex amplitude $u(x')$

$$u(x') = \text{FrT}_\lambda\{f(x)\phi_n(x); z_1\} \tag{33.16}$$

is obtained in the so-called transform plane at a distance z_1 from the input plane.

In the transform plane, the distribution $u(x')$ is multiplied by a second RPM, $\phi_m(x') = \exp[i2\pi m(x')]$, and then, it is Fresnel transformed to the output plane at a distance z_2

$$\psi(x_0) = \text{FrT}_\lambda\{u(x')\phi_m(x'); z_2\} \tag{33.17}$$

The function $\psi(x_0)$ of Eq. (33.17) corresponds to the encrypted information.

For a given primary image and RPMs, a different group of parameters (z_1, z_2, λ) can result in a different distribution for the encrypted image $\psi(x_0)$. Therefore, the positions of the significant planes and the illuminating wavelength are used as additional keys, together with the RPMs, to encrypt and retrieve the primary image.

The optical setup for decryption is the same as that for encryption (Figure 33.8) but in the reverse direction. Since the inverse Fresnel transform cannot be optically obtained, the complex conjugation of $\psi(x_0)$ should be used for decryption. If the encrypted data are stored in a photorefractive crystal, $\psi^*(x_0)$ can be easily generated through phase conjugation [24] as discussed in Section 33.2.5. Then, $\psi^*(x_0)$ is Fresnel transformed to the transform plane

$$v(x') = \text{FrT}_\lambda\{\psi^*(x_0); z_2\} \tag{33.18}$$

multiplied by the RPM $\phi_m(x')$, and the product is Fresnel transformed to the system output (Figure 33.8)

$$g(x) = \text{FrT}_\lambda\{v(x')\phi_m(x'); z_1\} \tag{33.19}$$

Since the primary image $f(x)$ is a positive function, the intensity distribution $|g(x)|^2$ captured by a CCD camera yields the information about the hidden primary image $f(x)$ when the correct RPMs are used in the appropriate location.

Note that in the decryption step, the same RPM $\phi_m(x') = \exp[i2\pi m(x')]$, but not its complex conjugated $\phi_m^*(x') = \exp[-i2\pi m(x')]$, is needed. As discussed in

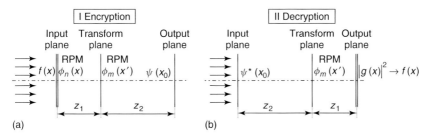

Figure 33.8 (a) Encryption and (b) decryption lensless systems to obtain DRPE technique in the Fresnel domain.

Sections 33.2.2 and 33.2.5, this property makes the deciphering easier by avoiding the fabrication of a new phase mask.

Lensless DRPE in the Fresnel domain [28] has been shown to be vulnerable to the chosen-plaintext attack discussed in Ref. [18], where it is described how an opponent can access both random phase keys in the input and Fresnel domains provided that the wavelength and the propagation distances are known.

Encryption in the Fresnel domain has also been used to cipher multiple images in a single encrypted distribution [29]. A detailed study of the space–bandwidth product of the encrypted signal is given in Ref. [30], permitting to deduce the most efficient way of using the bulk optical system and recording method available. They show how to deduce the maximum spatial extent and spatial frequency extent of the primary image such that the encrypted image can be fully represented after recording. Otherwise, the primary image cannot be recovered. Following their study, many important parameters (minimum size of the lens apertures, need for magnification at critical stages in the system, and sampling condition of the images) can be identified and determined.

Following the line of the work first proposed by Wang et al. [21] (Section 33.2.4 and Figure 33.5), other contributions exist in the literature that extend their encryption method to the Fresnel domain [31].

33.2.8
Fractional Fourier Transforms

An encryption system was proposed by Unnikrishnan et al. [20] that modifies the classical architecture of DRPE so that the input, the encryption, and the output planes are in fractional Fourier domains (Chapter 12). The distributions in these planes are related by an FRFT with three parameters: the input and output scale factors and the order of the FRFT. The two RPMs as well as the three parameters of the FRFT constitute the key to the encrypted image. The optical implementation of the encryption system consists of two lenses separated by some distances d_1, d_2, d_3, and d_4 (Figure 33.9a). The input and the encryption planes are, in general, two planes located asymmetrically with respect to the first lens. Similarly, the encryption plane and the output plane are asymmetrically placed with respect to

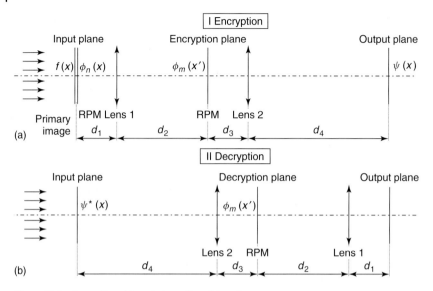

Figure 33.9 Encryption (a) and decryption (b) systems based on fractional Fourier transform.

the second lens. In the input plane, the first RPM $\phi_n(x)$ is placed against the primary image $f(x)$. In the first fractional domain of order α, the distribution is encoded by the second RPM $\phi_m(x')$, placed in the encryption plane. A second FRFT of order β yields the encrypted function $\psi(x)$ in the output plane. This encrypted function, which is a wide-sense stationary noise, is recorded using holography so that the information of both the amplitude and the phase can be retrieved later. The scale and fractional order parameters of the two FRFTs can be related to the focal lengths of the lenses and the distances $d_1,\ldots,d_4$ using simple expressions [20]. Decryption is simply the inverse of the encryption process and can be implemented using the optical system of Figure 33.9b that is the reverse of the system shown in Figure 33.9a. The complex conjugate of the encrypted function $\psi^*(x)$ is placed in the input plane of the decryption system and the second RPM $\phi_m(x')$ is placed in the fractional domain of order β. The decrypted signal can be obtained using an intensity detector in the output plane of the decryption system. To decrypt the image correctly, it is essential to know the FRFT parameters and the two RPMs used for encryption. This method was experimentally demonstrated using an optical setup with a photorefractive crystal as the holographic medium to record the encrypted image [20].

The recently proposed programmable optical fractional Fourier processor [32] is a flexible system that is able to perform FRFT with no additional scaling and phase factors depending on the fractional orders, which can be tuned continuously. This system shows promising properties in its application to optical encryption.

33.3
Applications to Security Systems

In the field of securing data, digital techniques are widely applied and there are several cryptographic protocols. For example, the public-key cryptosystem and the exclusive-OR (XOR) encryption are worth mentioning. However, optical methods have several advantages in information security applications. Using optical techniques, a large amount of data can be stored or retrieved in parallel and at high speed. Additionally, they take advantage of high-resolution optical materials, which are capable of storing a considerable amount of data in small areas.

Some optical security systems are designed to encrypt and decrypt a signal (primary image, plaintext, text, etc.) so that the decrypted signal replicates the primary image (information recovery). Many examples of such systems have been presented in former sections. In some other cases, the system is intended to perform a more complex operation of authentication. The system is intended to compare the decrypted signal with a reference signal and recognize their similarity. Pattern recognition algorithms based on matched filtering and optical correlation techniques have been successfully applied to these systems for encryption and validation purposes. Some encryption–authentication systems of this type are detailed below.

Regarding the primary images or signals used for the identification of a person, biometric images such as fingerprints, face, hand, iris, and retina are increasingly considered in authentication mechanisms because biometrics is based on something intrinsic to a person (something the person *is*) in contrast to other schemes based on either something a person *knows* (e.g., a password) or *has* (e.g., a metal key, an ID card, etc.) [33]. Security can be enhanced by combining different authenticators (multifactor authentication). In such a case, a Boolean AND operation has to be performed for each factor's authentication results, so all these must be affirmative before final authentication is satisfied [33].

Optical ID tags [34] have been shown as a good application of the optical encryption techniques for real-time remote authentication tasks. They enable surveillance or tracking of moving objects and they can be applied in a number of situations in transportation or industrial environments. A distortion-invariant ID tag permits the verification and authentication of the information provided the ID tag can be read even if it appears rotated or scaled. As an additional degree of security, the ID tag can be designed so that it is only captured in the NIR (near infrared region) spectrum. In this way, data are no longer visible to the naked eye and it is possible to obtain the correct information only by using an adequate sensor.

33.3.1
Optical Techniques and DRPE in Digital Cryptography

In the field of electronic ciphering, there are many encryption algorithms [35], some of which are widely used for securing data, such as the XOR encryption or the public-key cryptography.

Recently, a proposal for combining the DRPE technique with public-key cryptography has been published in the literature [36]. In the designed technique, the DRPE is used to scramble the information so that it becomes confusing for unauthorized users, and allows a certain degree of resistance to occlusion of the encrypted information. Two phase masks and a random 8-bit gray-scaled input image, which allow deciphering the secret information by using a classical 4f setup, are computed by a phase-retrieval algorithm. The random 8-bit gray-scaled image is split into three bit-planes. Each bit-plane is further encrypted by the Rivest–Shamir–Sdleman (RSA) public-key cryptography [36], so that only the owner of the private key of this encryption method will decipher the information. The authors propose hiding of the encrypted signal and the two phase masks used for decryption in a host image similarly to watermarking techniques. Thus, the combination of optical techniques along with the public-key cryptosystem ensures security and permits simultaneous transmission of the encrypted data along with the phase masks acting as keys for decryption.

33.3.2
Multifactor Identification and Verification of Biometrics

In this section, we describe a multifactor encryption/authentication technique that reinforces optical security by allowing the simultaneous AND verification of more than one primary image [37]. The method is designed to obtain four-factor authentication. The authenticators are as follows: two different primary images containing signatures or biometric information and two different RPMs that act as the key codes. The multifactor encryption method involves DRPE, fully phase-based encryption, and a combined nonlinear JTC and a classical 4f-correlator for simultaneous recognition and authentication of multiple images.

There is no a priori constraint on the type of primary images to encode. As an example, we consider a combination of stable and accurate biometric signals such as retina images (Figure 33.10). Two reference images, double-phase encoded and ciphered in the encrypted function, are compared with the actual input images extracted in situ from the person whose authentication is required. The two key RPMs are known to the authentication processor. Let $r(x)$ and $s(x)$ be the reference primary images (Figure 33.10), and let $b(x)$ and $n(x)$ be two independent random white sequences used to mask and encrypt the information in function $\psi(x)$. All the four, $r(x)$, $s(x)$, $b(x)$, and $n(x)$, are normalized positive functions distributed in [0,1]. These images can be phase encoded to yield $t_r(x)$, $t_s(x)$, $t_{2b}(x)$, and $t_{2n}(x)$ that are generically defined by $t_f(x) = \exp\{i\pi f(x)\}$. The complex-amplitude encrypted function $\psi(x)$ (Figure 33.11a) containing the multifactor authenticators is given by the following equation:

$$\psi(x) = t_{r+2b}(x) * t_s(x) * \mathrm{FT}^{-1}[t_{2n}(u)] \tag{33.20}$$

where $t_{r+2b}(x) = t_r(x)t_{2b}(x) = \exp\left[i\pi r(x)\right]\exp[i2\pi b(x)]$.

Let $p(x)$ and $q(x)$ denote the positive and normalized input images that are to be compared with the reference images $r(x)$ and $s(x)$ (Figure 33.10), respectively.

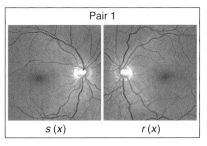
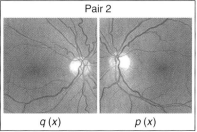

Figure 33.10 Retina images: Pair 1 corresponds to the eyes of an authorized person and Pair 2 corresponds to the eyes of a nonauthorized person. In all the cases, image size is 188 × 188 pixels.

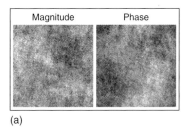
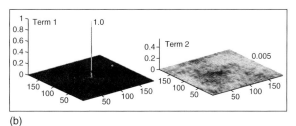

Figure 33.11 (a) Magnitude and phase distributions of the encrypted image $\psi(x)$. Its noiselike appearance does not reveal the content of any primary image. (b) Output intensity distributions. (Please find a color version of this figure on the color plates.)

A possible realization of the optical processor combines a nonlinear JTC and a classical 4f-correlator (Figure 33.12). In the first step, the encrypted function $\psi(x - a)$ and one phase-encoded input image, for instance, $t_p(x + a) = \exp[i\pi p(x + a)]$, are displayed side by side at a distance $2a$ apart on the input plane of the nonlinear JTC illuminated by coherent light. In the first approach, let us assume that $t_{2b}(x)$ and $t_{2n}(u)$ are key phase codes (RPMs) known to the processor. The RPM $t_{2b}(x + a) = \exp[i2\pi b(x + a)]$ is placed against the screen where the input $t_p(x + a)$ is displayed. Consequently, the amplitude distribution in the input plane is $\psi(x - a) + t_{p+2b}(x + a)$, where $t_{p+2b}(x) = t_p(x)t_{2b}(x)$. A CCD sensor placed in the Fourier plane of the JTC captures the intensity distribution $I(u)$ of the joint power spectrum:

$$I(u) = \left|\mathrm{FT}[\psi(x - a) + t_{p+2b}(x + a)]\right|^2 \qquad (33.21)$$

The expansion of Eq. (33.21) gives the classical four terms, two of which are interesting because they convey the cross-correlation signals that lead to spatially

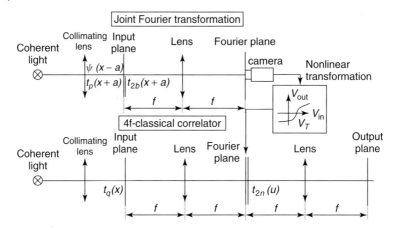

Figure 33.12 Optical processor for multifactor authentication.

separated distributions in the output plane. These two terms are

Term 1: $\text{FT}^*[\psi(x)]\text{FT}\left[t_{p+2b}(x)\right]\exp\{i2au\} = \hat{t}^*_{r+2b}(u)\hat{t}^*_s(u)\hat{t}^*_{2n}(u)\hat{t}_{p+2b}(u)\exp\{i2au\}$

Term 2: $\text{FT}[\psi(x)]\text{FT}^*\left[t_{p+2b}(x)\right]\exp\{-i2au\} = \hat{t}_{r+2b}(u)\hat{t}_s(u)t_{2n}(u)\hat{t}^*_{p+2b}(u)\exp\{-i2au\}$

(33.22)

Terms 1 and 2 of Eq. (33.22) can be modified based on a number of nonlinear techniques [10, 38] that are useful to adjust the discrimination capability of a recognition system and to improve noise resistance, among other properties. Let us consider, for instance, the phase extraction pattern recognition also called pure phase correlation (PPC) [38]. The resultant nonlinearly modified joint power spectrum is displayed on the Fourier plane of a 4f-classical correlator (Figure 33.12). Here, a transparency with the RPM $t_{2n}(u)$ is placed against the screen. The second input image $q(x)$ is phase encoded and displayed on the input plane of the 4f-correlator. Behind the Fourier plane, terms 1 and 2 of Eq. (33.22) are, respectively, converted into

Term 1: $\left[\hat{t}_q(u)\dfrac{\hat{t}^*_s(u)}{|\hat{t}_s(u)|}\right]\left[\dfrac{\hat{t}^*_{r+2b}(u)\hat{t}_{p+2b}(u)}{\left|\hat{t}_{r+2b}(u)\hat{t}_{p+2b}(u)\right|}\right]\left[t^*_{2n}(u)t_{2n}(u)\right]\exp\{i2au\}$

Term 2: $\left[\hat{t}_q(u)\dfrac{\hat{t}_s(u)}{|\hat{t}_s(u)|}\right]\left[\dfrac{\hat{t}_{r+2b}(u)\hat{t}^*_{p+2b}(u)}{\left|\hat{t}_{r+2b}(u)\hat{t}_{p+2b}(u)\right|}\right]\left[t_{4n}(u)\right]\exp\{-i2au\}$ (33.23)

If the information contained in the encrypted function corresponds to a positive validation, then the AND condition $r(x) = p(x)$ and $s(x) = q(x)$ is fulfilled. In such a case, and provided the system is free of noise and distortions, term 1 of Eq. (33.23) simplifies into $|\hat{t}_s(u)|\exp\{i2au\}$, which represents a wavefront with all its curvature canceled [8] that focuses on a sharp multifactor AC peak centered at $x = -2a$ of the output plane (Figure 33.12). From Eq. (33.23), the output intensity distribution corresponding to term 1 (Figure 33.11b) is the cross-correlation of AC signals

Table 33.1 Maximum value of the output intensity (term 1).

q	Person category	p	Key b	Key n	Normalized peak
s	Authorized	r	True	True	1.0
$q \neq s$	Unauthorized	$p \neq r$	True	True	0.0138
s	Authorized	r	False	True	0.0050
s	Authorized	r	True	False	0.0051
s	Authorized	r	False	False	0.0051

given by

$$\left| AC_{POF}[t_s(x)] \star AC^*_{PPC}[t_{r+2b}(x)] \star AC^*_{CMF}[\hat{t}_{2n}(x)] * \delta(x+2a) \right|^2 \quad (33.24)$$

where subindexes classical matched filter (CMF), POF, and PPC indicate the type of filter involved in the AC signal. Since AC peaks are usually sharp and narrow, particularly those for POF and PPC, we expect that the cross-correlation of such AC signals will be sharper and narrower. Consequently, the information contained in term 1 allows reinforced security verification by simultaneous multifactor authentication.

On the other hand, when the condition $r(x) = p(x)$ and $s(x) = q(x)$ is fulfilled, term 2 of Eq. (33.23) becomes $\left(\hat{t}_s^2(u) / |\hat{t}_s(u)| \right) t_{4n}(u) \exp\{-i2au\}$, which does not yield an interesting result for recognition purposes (Figure 33.11b). If $p(x) \neq r(x)$ or $q(x) \neq s(x)$, term 1 contains a cross-correlation signal that is, in general, broader and less intense than the multifactor AC peak of Eq. (33.24). Furthermore, the key codes known to the processor, that is, the RPMs represented by $t_{2b}(x)$ and $t_{2n}(u)$, play an important role in optical security as additional authenticators with the properties described in previous sections (Table 33.1).

Some other techniques can be combined with this, including Fresnel domain encoding (Section 33.2.7), to increase the degrees of freedom of the system or the number of keys.

33.3.3
ID Tags for Remote Verification

Optical ID tags [34] are attached to objects under surveillance. They enable tracking of moving objects and they can be applied in a number of cases such as for surveillance of vehicles in transportation, control of restricted areas for homeland security, item tracking on conveyor belts or other industrial environments, inventory control, and so on.

An optical ID tag is introduced as an optical code containing a signature (i.e., a characteristic image or other relevant information of the object), which permits its real-time remote detection and identification. To increase security and avoid counterfeiting, the signature is introduced in the optical code as an encrypted function following the DRPE technique. The encrypted information included in the ID tag is verified by comparing the decoded signal with a reference by optical correlation.

The combination of distortion-invariant ID tags with multifactor encryption (Section 33.3.2) has been proposed recently as a satisfactory compact technique for encryption-verification [7, 39]. This highly reliable security system combines the advantages of the following four elements: multifactor encryption, distortion-invariant ID tag, NIR readout, and optical processor. More specifically, the multifactor encryption permits the simultaneous verification of up to four factors, the distortion-invariant ID tag is used for remote identification, NIR writing and readout of the ID tag signal allows invisible transmission, and an optical processor, based on joint transform pattern recognition by optical correlation, carries out the automatic verification of the information.

This security system is designed to tackle situations such as the one illustrated in Figure 33.13. Let us consider the surveillance and tracking of classified parcels that have been confiscated and which are located on a conveyor belt for inspection. The parcels may have different sizes and orientations when passing through the machine. This situation may require the control of a number of elements: the person who is responsible for delivering the parcels (retina image, $s(x)$), their origin ($n(x)$), and their content ($r(x)$). One of the factors is an RPM that can be used as a random code ($b(x)$) to control the access of one person to a given area or for a given period of time (Figure 33.13).

The information corresponding to the set of chosen factors or primary images (Figure 33.13) is previously ciphered following the scheme of the multifactor encryption technique [37] described in Section 33.3.2. Distortion invariance is achieved by both multiplexing the information included in the ID tag and taking advantage of the ID tag topology, without unnecessarily increasing the system complexity [40, 41]. Synthesis of the optical ID tag is depicted in Figure 33.14. Let us consider the encrypted function $\psi(x)$ obtained from Eq. (33.20) in array

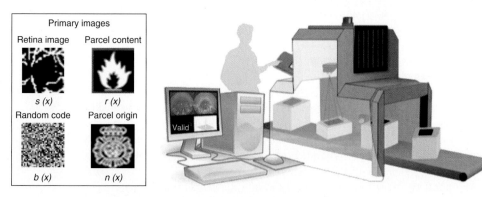

Figure 33.13 Classified parcel surveillance by using multifactor authentication and NIR distortion-invariant optical ID tags. (Left) A set of chosen primary images.

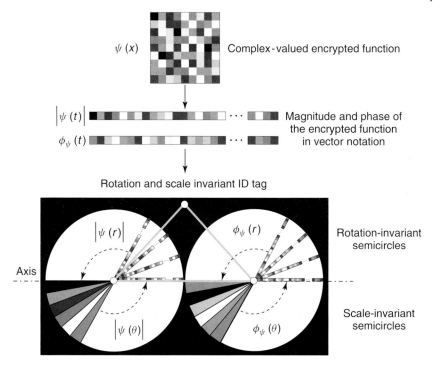

Figure 33.14 Synthesis of a rotation and scale-invariant ID tag from the encrypted function $\psi(x)$.

notation $\psi(t) = |\psi(t)| \exp\{i\phi_\psi(t)\}$ where $t = 1, 2,\ldots, N$ and N is the total number of pixels of the ciphered function. We build two real valued vectors that are to be encoded gray scale: the magnitude vector $|\psi(t)|$ and the phase vector $\phi_\psi(t)$. It is convenient to print the phase content of $\psi(x)$ in gray-scale variations rather than in phase to avoid sabotage by, for instance, sticking an adhesive transparent tape on the tag. The information included in the ID tag is distributed in two optical circles following the structure of a wedge-ring detector [41]. One of them corresponds to the magnitude of the encrypted function (the circle on the left-hand side). The other contains its phase distribution (the circle on the right-hand side). One half of each circle (upper semicircles in Figure 33.14) includes either the magnitude or the phase distribution written in a radial direction and repeated angularly so that rotation invariance can be achieved. The other semicircle of both circles (bottom semicircles) contains either the magnitude or the phase written circularly and repeated in concentric rings. Therefore, the information of a given pixel of the encrypted function will correspond to an angular sector in the optical code. Thus, the readout of the ciphered information will be tolerant to variations in scale up to a certain range. For encrypted functions with a large number of pixels, such as in Figure 33.15, information of the scale-invariant ID tag have to be distributed by using different concentric semicircles to ensure a minimum number of pixels for each sector to recover the information properly. Consequently, the tolerance to

Figure 33.15 Optical distortion-invariant ID tag (rotation angle 7°) experimentally captured by using an NIR XEVA camera. Incandescent illumination was used to capture the image.

scale variations will be affected in accordance with the number of concentric circles used in the ID tag. A mathematical description of the distortion-invariant ID tag can be found in Ref. [41].

As an additional degree of security, the resistance of the system to counterfeiting can be increased by gathering the data of the ID tag from the NIR region of the spectrum. The NIR ID tag is built by printing the ID tag gray-level distribution with a common laser printer on a black cardboard. In the visible spectrum, the entire tag looks like a black patch and, therefore, the whole information is completely hidden to either the naked eye or common cameras operating in the visible region of the spectrum. Consequently, it is not possible to know either the kind of information included in the ID tag or the exact position of this ID tag over the item under surveillance. Only NIR InGaAs cameras or conventional monochrome CCD cameras without the IR cut-off filter are able to detect the appropriate information to decode and verify the multifactor information. Figure 33.15 shows the distortion-invariant ID tag as it is experimentally captured by a camera sensitive to the NIR spectrum. To show the robustness of the system against rotation, the ID tag was rotated 7° from the horizontal position. The scale and rotation-invariant regions are clearly detected.

In the situation depicted in Figure 33.13a as an example, some tolerance to scratches or other sources of noise can be expected to a certain extent. The redundant information multiplexed in the tag allows signal verification despite the scratches owing to, for instance, common friction. Moreover, previous works have shown that information redundancy in the design of the ID tag allows a certain tolerance to the presence of additive noise [41, 42] in the capturing process. On the other hand, a system for an automatic destruction of the ID tag should be applied to prevent unauthorized manipulation or tampering of tags and parcels. For example, a reservoir of black ink (black in terms of NIR illumination) under the ID tag could act as a simple autodestructive mechanism. If the tag was cut, the ink would spread throughout it. The tag cannot be then properly read, and the system would raise an alarm.

Since the NIR receiver resolution is not generally known a priori, the image of the triangular pattern consisting of three white spots (Figure 33.14) can be used as a reference to know whether the receiver has enough resolution to read the encrypted information. Since the triangular pattern can provide information about scale and rotation, one might feel that there is no need to codify the encrypted distribution in the distortion-invariant ID tag. However, we must take into account that if the encrypted function, written in a matrix array, is affected by rotation and/or scale variations, then it needs to be sampled again and rearranged into the matrix form before decryption. This operation entails interpolations that can produce errors such as aliasing. For this reason, we consider that the distortion-invariant ID tag, provided it is correctly built, allows more accurate readouts of the encrypted information under rotation and/or scale variations.

The triangular pattern permits to allocate the two circles (Figure 33.14). The base of the isosceles triangle determines the border line between the two sectors of each circle. The vertex of the triangle distinguishes between the semiplanes where rotation invariance (upper semiplane in Figure 33.14) and scale invariance (bottom semiplane in Figure 33.14) are achieved. The ciphered information in vector notation $\psi(t)$ can be decoded by reading out the information of the optical code from either the rotation-invariant or the scale-invariant areas. From the semicircles corresponding to the rotation-invariant areas, the magnitude and the phase could be read out by using a linear array detector placed in any radial direction of the imaged semicircles [41]. Not only is a single pixel value read along a unique radius, but also a set of different radial codes are read and their median value is computed to increase their robustness against noise. From the semicircles corresponding to the scale-invariant areas, the magnitude and phase in vector notation are recovered by reading out the pixels of the ID tag in semicircular rings. To minimize errors in the reading process, the median value of a set of pixels located in neighbor concentric rings is considered. The retrieved pixel information should be written back into matrix notation prior to verifying the multifactor authentication. Following this procedure, the encrypted distribution can be recovered when the ID tag is captured either in its original orientation and size or in rotated or scaled formats.

The four scrambled factors (primary images $s(x)$, $r(x)$, $b(x)$, and $n(x)$) (Figure 33.13b) are decrypted and introduced as a reference for the validation of the set of input images $q(x)$, $p(x)$, $d(x)$, and $m(x)$. In this work, we have used the nonlinear technique known as phase extraction [10, 38]. Figure 33.16 shows the output planes of the processor for different situations that correspond to the most relevant identification results obtained in the experiment. The maximum correlation intensity value of the output planes has been normalized to the case for which a satisfactory verification is achieved. For comparison, the output correlation signal of the processor is depicted for both an ideally captured ID tag and, on its side, for the experimentally captured ID tag (Figure 33.16). A high and sharp multifactor AC peak is essential for the method to be reliable. This situation occurs when the system is free of noise and distortions.

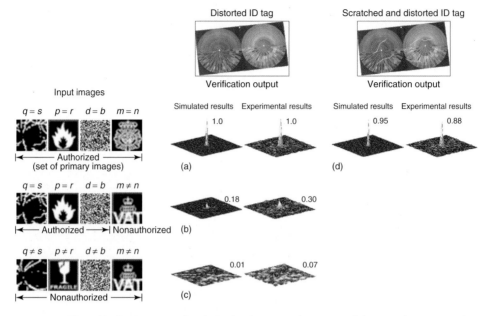

Figure 33.16 Experimental and simulated results for the verification system by using distorted NIR multifactor ID tags: (a) positive validation when the four identifying factors coincide with the information included in the ID tag; negative results obtained when one (b) or more factors (c) do not coincide with the set of primary images; and (d) positive validation with a partially scratched ID tag. In all cases, verification outputs are normalized to the positive validation (a).

The first analyzed case (Figure 33.16a) corresponds to a positive validation, for which the four input images coincide with the primary images included in the ID tag. As a result, an intense and sharp intensity peak projects over a low background on the output plane of the optical processor. This peak stands for the correct identification of an authorized person and parcel.

When the parcel origin does not match the one included in the ID tag (Figure 33.16b), the output plane obtained shows an insignificant intensity peak. If another signal among the factors (biometric, parcel content, or key code) or even the whole set of input images does not correspond to the set introduced into the ID tag, the resulting output planes are very similar to the one plotted in Figure 33.16c. A low energy background is produced at the system output. In all these cases, an appropriate threshold value would indicate that at least one factor corresponds to an unauthorized signal.

Finally, if the ID tag is slightly scratched due to friction (Figure 33.16d), a positive verification result is still obtained when the whole set of input images coincides with the authorized factors included in the ID tag, as proved by the sharp and intense peak.

In all the analyzed examples, there is a satisfactory agreement between the experimental and the predicted verification results.

33.4
Conclusions

This work has overviewed the potential of optical techniques in data encryption and security applications. A series of encryption methods with origin in the DRPE have been presented and analyzed. Their optical implementation has also been detailed. Optical data encryption has many attractive advantages and is a powerful technique for providing multidimensional keys to the task of securing information. Certainly, when using optical encryption, many degrees of freedom are available to manipulate the physical parameters of the optical waves that convey a piece of information. Consequently, a high level of security can be achieved in the information processed by both the encryption and the decryption systems. The encrypted data can be stored in optical material or in digital format. The algorithms described can be optically or electronically implemented. However, the state-of-the-art optical cryptography also has some drawbacks: its computational flexibility is far from that of electronics and it is relatively difficult to transmit ciphered information or keys through communication links. In addition, stringent alignment is required when using a purely optical implementation of cryptographic algorithms. Certain vulnerability has also been pointed out in the case of the theoretical expression of DRPE. The progress in optical devices and photonic technologies contributes to overcome some of these difficulties.

Some applications that combine several of the techniques described above have been presented. They mainly deal with optical–digital cryptosystems, multifactor authentication, and the use of secure ID tags. Multiple signatures (biometrics included), matched filtering, optical correlation, DRPE, and infrared imaging are combined not only to secure information strongly but also to authenticate the encrypted data in comparison with a stored reference. In this process, the ID tag plays an important role. It has been specifically designed to convey the encrypted data and to allow remote readout with in-plane distortion invariance.

Acknowledgments

This work has been funded by the Spanish Ministerio de Ciencia e Innovación y Fondos FEDER (projects DPI2006-05479, PET2008-0156, DPI2009-08879).

References

1. Javidi, B. and Horner, J.L. (1994) Optical pattern recognition for validation and security verification. *Opt. Eng.*, **33**, 1752–1756.
2. Réfrégier, P. and Javidi, B. (1995) Optical image encryption based on input plane and Fourier plane random encoding. *Opt. Lett.*, **20**, 767–769.
3. (1996) Special issue on optical security. *Opt. Eng.*, **35** (9).
4. Javidi, B. (1997) Securing information with optical technologies. *Phys. Today*, **50** (03). 27–32.

5. (1999) Special issue on optical security. *Opt. Eng.*, **38** (1).
6. Javidi, B. (ed.) (2001) *Smart Imaging Systems*, SPIE Press, Bellingham.
7. Matoba, O. et al. (2009) Optical techniques for information security. *Proc. IEEE J.*, **97**, 1128–1148.
8. Goodman, J.W. (1996) *Introduction to Fourier Optics*, 2nd edn, McGraw-Hill, New York.
9. Javidi, B. and Sergent, A. (1997) Fully phase encoded key and biometrics for security verification. *Opt. Eng.*, **36**, 935–941.
10. Javidi, B. (1994) Nonlinear joint transform correlators, in *Real-Time Optical Information Processing* (eds. B. Javidi and J.L. Horner), Academic Press, San Diego, 115–183.
11. Towghi, N. et al. (1999) Fully-phase encryption image processor. *J. Opt. Soc. Am. A*, **16**, 1915–1927.
12. Goudail, F. et al. (1998) Influence of a perturbation in a double phase-encoding system. *J. Opt. Soc. Am. A*, **15**, 2629–2638.
13. Nomura, N. and Javidi, B. (2000) Optical encryption using a joint transform correlator architecture. *Opt. Eng.*, **39**, 2031–2035.
14. Frauel, Y. et al. (2007) Resistance of the double random phase encryption against various attacks. *Opt. Express*, **15**, 10253–10265.
15. Peng, X. et al. (2006) Known-plaintext attack on optical encryption based on double random phase keys. *Opt. Lett.*, **31**, 1044–1046.
16. Gopinathan, U. et al. (2006) A known-plaintext heuristic attack on the Fourier plane encryption algorithm. *Opt. Express*, **14**, 3181–3186.
17. Carnicer, A. et al. (2005) Vulnerability to chosen-ciphertext attacks of optical encryption schemes based on double random phase keys. *Opt. Lett.*, **30**, 1644–1646.
18. Peng, X. et al. (2006) Chosen-plaintext attack on lensless double random phase encoding in the Fresnel domain. *Opt. Lett.*, **31**, 3261–3263.
19. Matoba, O. and Javidi, B. (1999) Encrypted optical memory system using three-dimensional keys in the Fresnel domain. *Opt. Lett.*, **24**, 762–764.
20. Unnikrishnan, G. et al. (2000) Optical encryption by double-random phase encoding in the fractional Fourier domain. *Opt. Lett.*, **25**, 887–889.
21. Wang, R.K. et al. (1996) Random phase encoding for optical security. *Opt. Eng.*, **35**, 2464–2469.
22. Rosen, J. (1993) Learning in correlators based on projections onto constraint sets. *Opt. Lett.*, **14**, 1183–1185.
23. Javidi, B. et al. (1997) Encrypted optical memory using double-random phase encoding. *Appl. Opt.*, **36**, 1054–1058.
24. Unnikrishnan, G. et al. (1998) Optical encryption system that uses phase conjugation in a photorefractive crystal. *Appl. Opt.*, **37**, 8181–8186.
25. Matoba, O. and Javidi, B. (1999) Encrypted optical storage with wavelength-key and random phase codes. *Appl. Opt.*, **38**, 6785–6790.
26. Situ, G. and Zhang, J. (2005) Multiple-image encryption by wavelength multiplexing. *Opt. Lett.*, **30**, 1306–1308.
27. Amaya, D. et al. (2008) Digital colour encryption using a multi-wavelength approach and a joint transform correlator. *J. Opt. A: Pure Appl. Opt.*, **10**, 104031–104035.
28. Situ, G. and Zhang, J. (2004) Double random-phase encoding in the Fresnel domain. *Opt. Lett.*, **29**, 1584–1586.
29. Situ, G. and Zhang, J. (2006) Position multiplexing for multiple-image encryption. *J. Opt. A: Pure Appl. Opt.*, **8**, 391–397.
30. Hennelly, B.M. and Sheridan, J.T. (2005) Optical encryption and the space bandwidth product. *Opt. Commun.*, **247**, 291–305.
31. Meng, X.F. et al. (2007) Hierarchical image encryption based on cascaded iterative phase retrieval algorithm in the Fresnel domain. *J. Opt. A: Pure Appl. Opt.*, **9**, 1070–1075.
32. Rodrigo, J.A. et al. (2009) Programmable two-dimensional optical fractional Fourier processor. *Opt. Express*, **17**, 4976–4983.
33. O'Gorman, L. (2003) Comparing passwords, tokens, and biometrics for

user authentication. *Proc. IEEE J.*, **91**, 2021–2040.
34. Javidi, B. (2003) Real-time remote identification and verification of objects using optical ID tags. *Opt. Eng.*, **42**, 2346–2348.
35. Schneier, B. (1996) *Applied Cryptography – Protocols, Algorithms and Source Code in C*, 2nd edn, John Wiley & Sons, Inc., New York.
36. Sheng, Y. *et al.* (2009) Information hiding based on double random-phase encoding and public-key cryptography. *Opt. Express*, **17**, 3270–3284.
37. Millán, M.S. *et al.* (2006) Multifactor authentication reinforces optical security. *Opt. Lett.*, **31**, 721–723.
38. Pérez, E. *et al.* (1997) Dual nonlinear correlator based on computer controlled joint transform processor: digital analysis and optical results. *J. Mod. Opt.*, **44**, 1535–1552.
39. Pérez-Cabré, E. *et al.* (2007) Near infrared multifactor identification tags. *Opt. Express*, **15**, 15615–15627.
40. Pérez-Cabré, E. and Javidi, B. (2005) Scale and rotation invariant optical ID tags for automatic vehicle identification and authentication. *IEEE Trans. Veh. Technol.*, **54**, 1295–1303.
41. Pérez-Cabré, E. *et al.* (2007) Design of distortion-invariant optical ID tags for remote identification and verification of objects, in *Physics of the Automatic Target Recognition* (eds. F. Sadjadi and B. Javidi), Springer-Verlag, New York, 207–226.
42. Pérez-Cabré, E. *et al.* (2005) Remote optical ID tag recognition and verification using fully spatial phase multiplexing. *Proc. SPIE*, **5986**, 598602-1–598602-13.

34
Quantum Encryption

Bing Qi, Li Qian, and Hoi-Kwong Lo

34.1
Introduction

The evolution of cryptography has been propelled by the endless war between code-makers and code-breakers, among whom are some of the brightest minds in human history. The holy grail of cryptography is to develop an absolutely secure coding scheme that is secure against eavesdroppers (Eves) with unlimited computational power. Surprisingly, this goal was achieved, at least in principle, when Gilbert Vernam invented the one-time pad (OTP) encryption in 1917 [1], which was later proved by Shannon to be absolutely secure [2]. The encoding key used in OTP has to be completely random and secure – known only to the sender and the receiver (known as *Alice and Bob*, respectively). Therefore, Alice and Bob need to preestablish the secure key they will use before encoding/decoding their messages.

Although OTP is unbreakable in principle, there is a problem in applying this scheme in practice: once the legitimate users have used up the preestablished secure key, the secure communication will be interrupted until they can acquire a new key. This is the well-known key distribution problem, which typically involves two unachievable tasks in classical physics: truly random number generation and unconditionally secure key distribution through an insecure channel.

First, the deterministic nature of classical physics, which is implied by Albert Einstein's famous quotation "God doesn't play dice," rules out the existence of truly random numbers in chaotic, but classical, processes. In contrast, as we show in Section 34.3, truly random numbers can be generated from elementary quantum processes.

Secondly, in a world where information is encoded classically, there is no secure scheme to distribute a key through an insecure channel. The fundamental reason is that in classical physics, information can be duplicated. Alice and Bob cannot prove that a key established through an insecure channel has not been copied by a malevolent Eve. The only conceivable but cumbersome way to perform key distribution is by sending trusted couriers. Owing to this key distribution problem, the OTP has been adopted only when extremely high security is required.

Optical and Digital Image Processing: Fundamentals and Applications, First Edition. Edited by Gabriel Cristóbal, Peter Schelkens, and Hugo Thienpont.
© 2011 Wiley-VCH Verlag GmbH & Co. KGaA. Published 2011 by Wiley-VCH Verlag GmbH & Co. KGaA.

To solve the key distribution problem, public key cryptographic protocols, including the famous RSA scheme [3], have been invented. Unfortunately, their security rests upon unproven mathematical assumptions. For example, the security of RSA is based on the assumption that there is no efficient way to find the prime factors of a large integer. However, this assumption has not been proved despite the tremendous efforts from mathematicians. Moreover, an efficient factoring algorithm running on a quantum computer exists [4]. This suggests that as soon as the first large-scale quantum computer switches on, most of today's cryptographic systems could collapse overnight.

It is interesting to learn that one decade before people realized that a quantum computer could be used to break public key cryptography, they had already found a solution against this quantum attack – quantum key distribution (QKD). On the basis of the fundamental principles in quantum physics, QKD provides an unconditionally secure way to distribute secure key through insecure channels. The secure key generated by QKD could be further applied in the OTP scheme or other encryption algorithms to enhance information security.

In this chapter, we introduce the fundamental principles behind various QKD protocols and present the state-of-the-art QKD technologies. Bearing in mind that the majority of our readers are from engineering and experimental optics, we focus more on experimental implementations of various QKD protocols rather than on security analysis. Special attention is given to security studies of real-life QKD systems, where unconditional security proofs of theoretical protocols cannot be applied directly owing to imperfections. For a comprehensive review on QKD, see Ref. [5].

34.2
The Principle of Quantum Cryptography

34.2.1
Quantum No-Cloning Theorem

The foundation of QKD is the quantum no-cloning theorem [6, 7], which states that an arbitrary quantum state cannot be duplicated perfectly. This theorem is a direct result of the linearity of quantum physics. Quantum no-cloning theorem is closely related to another important theorem in quantum mechanics, which is stated as follows: if a measurement allows one to gain information about the state of a quantum system, then, in general, the state of this quantum system will be disturbed, unless we know in advance that the possible states of the original quantum system are orthogonal to each other.

Instead of providing a mathematical proof of quantum no-cloning theorem, we simply discuss two examples to show how it works. In the first case, we are given a photon whose polarization is either vertical or horizontal. To determine its polarization state, we can send it through a properly placed polarization beam splitter, followed by two single-photon detectors (SPDs). If the detector at the reflection path clicks, we know the input photon is vertically polarized; otherwise, it is horizontally polarized. Once we know the polarization state of the input photon, we can prepare arbitrary number of photons in the same polarization state. Equivalently, we have achieved perfect cloning of the polarization state of the input photon. In the second case, we are given a photon whose polarization is randomly chosen from a set of {horizontal, vertical, $45°, 135°$}. Since the four polarization states given above are linearly dependent, it is impossible to determine its polarization state from any experiment. For example, if we use the same polarization beam splitter mentioned above, a $45°$ polarized photon will have a 50/50 chance to be either reflected or transmitted; therefore, it cannot be determined with certainty.

One common question is why an optical amplifier cannot be used to copy photons. As pointed out by Wootters, the impossibility of making perfect copies of photon through stimulated emission process originates from the unavoidable spontaneous emission: while the stimulated photon is a perfect copy of the incoming one, the spontaneous emitted photon has a random polarization state [6].

At first sight, the impossibility of making perfect copies of unknown quantum states seems to be a shortcoming. Surprisingly, it can also be an advantage. It turned out that by using this impossibility smartly, unconditionally secure key distribution could be achieved: any attempts by the Eve to learn the information encoded quantum mechanically will disturb the quantum state and expose her existence. In this section, we present a few of the most common QKD protocols.

34.2.2
The BB84 Quantum Key Distribution Protocol

The first attempt at using quantum mechanics to achieve missions impossible in classical information started in the early 1970s. Stephen Wiesner, a graduate student at Columbia University, proposed two communication modalities not allowed by classical physics: quantum multiplexing channel and counterfeit-free banknote. Unfortunately, his paper was rejected and could not be published until a decade later [8]. In 1980s, Charles H. Bennett and Gilles Brassard extended Wiesner's idea and applied it to solve the key distribution problem in classical cryptography. In 1984, the well-known Bennett–Brassard-1984 (BB84) QKD protocol was published [9].

The basic idea of the BB84 protocol is surprisingly simple. Let us assume that both Alice and Bob stay in highly secured laboratories that are connected by an insecure quantum channel, for example, an optical fiber. Here, we use the term *quantum channel* to emphasize the fact that information through this channel is encoded on the quantum state of photons. Eve is allowed to fully control the quantum channel, but she is not allowed to sneak into Alice's or Bob's laboratory to steal information directly. Furthermore, Alice and Bob share an authenticated public channel (a channel that Eve can listen in but in which she cannot interfere) and they can generate random numbers locally. The question is how they can establish a long string of shared random numbers as a secure key.

Being aware of quantum no-cloning theorem, Alice encodes her random bits on the polarization state of single photons and sends them to Bob through the quantum channel. As the first try, Alice uses a horizontally polarized photon to encode bit "0" and a vertically polarized photon to encode bit "1." Bob decodes Alice's random bit by performing polarization measurement. However, this scheme turns out to be insecure: when Alice's photon travels through the channel, Eve can intercept it and measure its polarization state. After that, Eve can prepare a new photon according to her measurement result and send it to Bob. By the end of the day, Eve has a perfect copy of whatever Bob has. No secure key can be generated. Then, Alice realizes what the problem is: the quantum no-cloning theorem cannot be applied to a set of orthogonal states. After some thinking, she introduces a new concept – "basis," which represents how the random bits are encoded. In basis one (rectilinear basis), she uses horizontal polarization to represent bit "0" and vertical polarization to represent bit "1." In basis two (diagonal basis), she uses $45°$ polarization to represent bit "0" and $135°$ polarization to represent bit "1." For each transmission, Alice randomly chooses to use either rectilinear or diagonal basis to encode her random number. Now, the polarization of each photon is randomly chosen from a set of {horizontal, vertical, $45°$, $135°$} and it is impossible for Eve to determine its polarization state. If Eve uses a polarization beam splitter to project the input photon into either horizontal or vertical polarization state (we call it a *measurement in rectilinear basis*), then she will destroy information encoded in diagonal basis, since a $45°$ or $135°$ polarized photon has the same chance to be projected into either horizontal or vertical polarization state. Eve is puzzled, and so is Bob, because neither of them knows which basis Alice will choose in advance.

Without the knowledge of Alice's basis selection, Bob randomly chooses either rectilinear or diagonal basis to measure each incoming photon. If Alice and Bob happen to use the same basis, they can generate correlated random bits. Otherwise, their bit values are uncorrelated. After Bob has measured all the photons, he compares his measurement bases with that of Alice through the authenticated public channel. They only keep random bits generated with matched bases, which are named as *sifted keys*. Without environmental noises, system imperfections, and Eve's disturbance, their sifted keys are identical and can be used as a secure key.

The existence of an authenticated channel between Alice and Bob, which gives Bob an advantage over Eve, is essential to the security of QKD. The authentication can be assured if Alice and Bob share a short secure key in advance. In this sense, a QKD protocol is actually a quantum key expansion protocol: it takes a short shared key and expands it into an information-theoretically secure long shared key.

Let us see what happens if Eve launches a simple "intercept and resend" attack: for each photon from Alice, Eve performs a measurement in a randomly chosen basis and resends a new photon to Bob according her measurement result. Let us focus on those cases when Alice and Bob happen to use the same bases since they will throw away other cases anyway. If Eve happens to use the correct basis, then both she and Bob will decode Alice's bit value correctly. On the other hand, if Eve uses the wrong basis, then both she and Bob will have random measurement results. This suggests that if Alice and Bob compare a subset of the sifted key, they will see a significant amount of errors. It can be shown that the intercept and resend attack will introduce a 25% quantum bit error rate (QBER). This example illustrates the basic principle behind QKD: *Eve can only gain information at the cost of introducing errors, which will expose her existence.*

In practice, errors can be originated from either the intrinsic noises of the QKD system or Eve's attack, so Alice's sifted key is typically different from Bob's, and Eve may have partial information about it. To establish the final secure key, error correction and privacy amplification algorithms can be applied [10]. Alice and Bob estimate the information gained by Eve during the quantum transmission stage from the observed QBER and some parameters of the QKD system. If Eve's information is too much, no secure key can be generated and they have to restart the whole QKD protocol. On the other hand, if Eve's information is below certain threshold, Alice and Bob can perform privacy amplification algorithm to generate a shortened final key on which Eve's information is exponentially small.

Typically, a QKD protocol can be separated into a quantum transmission stage and a classical postprocessing stage. The former includes quantum state preparation, transmission, and detection, while the latter includes bases comparison, error correction, and privacy amplification. The procedures of the BB84 protocol are summarized in Table 34.1.

We remark that any communication channel (such as telecom fiber) has loss and the efficiencies of practical SPDs are typically low. This suggests that a significant portion of photons sent by Alice will not be registered by Bob. Since Alice's photons carry only random bits but not useful message, this loss-problem can be easily resolved by employing a postselection process: Bob publicly announces which photons he registers and Alice only keeps the corresponding data.

The basic idea of the BB84 QKD protocol is beautiful and its security can be intuitively understood from the quantum no-cloning theorem. On the other hand, to apply QKD in practice, Alice and Bob need to find the upper bound of Eve's information quantitatively, given the observed QBER and other system parameters. This is the primary goal of various QKD security proofs and it has been extremely difficult to find this bound. One major challenge comes from the fact that Eve could launch attacks way beyond today's technologies and our imagination. Nevertheless,

Table 34.1 Summary of the BB84 QKD protocol.

Alice's random bits	1	0	0	1	0	1	1	1	0	1
Alice's encoding bases	×	×	+	×	+	+	+	×	+	+
Alice's photon polarization	135°	45°	H	135°	H	V	V	135°	H	V
Bob's measurement bases	+	×	+	×	×	+	×	+	+	+
Bob's measurement result	H	45°	H	*	135°	V	45°	*	H	V
Bob's raw data	0	0	0	–	1	1	0	–	0	1
Sifted key from matched bases	–	0	0	–	–	1	–	–	0	1

+, rectilinear basis; ×, diagonal basis; H, horizontal polarization; V, vertical polarization; *, photon lost.

more than a decade after its invention, QKD was proved to be unconditionally secure [11–12]. This is one of the most significant achievements in quantum information.

The BB84 protocol had been ignored by almost the whole scientific community until early 1990s when its founders decided to build a prototype. In October 1989, the first QKD experiment was demonstrated over a distance of 32.5 cm in air [13]. Since then, significant progresses have been achieved and even commercial QKD systems are available. We present more details about the recent progresses in Section 34.3.

34.2.3
Entanglement-Based Quantum Key Distribution Protocol

Ever since Einstein, Podolsky, and Rosen published the well-known "EPR" paper in 1935 [14], entanglement has been one of the most puzzling yet attractive features in quantum mechanics. To illustrate how entanglement can be used to achieve secure key distribution, we first review some properties of a polarization-entangled photon pair.

An arbitrary polarization state of a single photon can be described by a superposition of two basis states

$$|\psi\rangle_s = \alpha |\updownarrow\rangle + \beta |\leftrightarrow\rangle \qquad (34.1)$$

where $|\updownarrow\rangle$ and $|\leftrightarrow\rangle$ represent the vertical and horizontal polarization states, which constitute a set of orthogonal bases. α and β are complex numbers satisfying the normalization condition $\alpha\alpha^* + \beta\beta^* = 1$.

In Eq. (34.1), we have assumed that the polarization of photon is described by a pure state – there is a well-defined phase relation between the two basis components. In contrast, a nonpolarized photon is in a mixed state, which can only be interpreted as a statistical mixture of basis states.

Similarly, the most general polarization state (pure state) of a photon pair can be described by a superposition of four basis states:

$$|\psi\rangle_{\text{pair}} = \alpha_1 |\updownarrow\rangle_1 |\updownarrow\rangle_2 + \alpha_2 |\updownarrow\rangle_1 |\leftrightarrow\rangle_2 + \alpha_3 |\leftrightarrow\rangle_1 |\updownarrow\rangle_2 + \alpha_4 |\leftrightarrow\rangle_1 |\leftrightarrow\rangle_2 \qquad (34.2)$$

Here $|\updownarrow\rangle_1 |\updownarrow\rangle_2$ represents a basis state that both photons are in vertical polarization state. The other three terms in Eq. (34.2) can be understood in a similar way. In the special case when $\alpha_1 = \alpha_4 = 1/\sqrt{2}$ and $\alpha_2 = \alpha_3 = 0$, we have one type of polarization-entangled EPR photon pair:

$$|\Phi\rangle_{\text{pair}} = 1/\sqrt{2} \, (|\updownarrow\rangle_1 |\updownarrow\rangle_2 + |\leftrightarrow\rangle_1 |\leftrightarrow\rangle_2) \qquad (34.3)$$

One special feature of the above state is that it cannot be described by a tensor product $|\Phi\rangle_{\text{pair}} \neq |\psi\rangle_1 \otimes |\psi\rangle_2$, where $|\psi\rangle_1$ and $|\psi\rangle_2$ are arbitrary single-photon polarization states. In other words, the two photons are "entangled" with each other. Entangled photons can present nonlocal correlation, which does not exist in classical physics.

Suppose we send one photon of an EPR pair to Alice and the other one to Bob. If Alice measures her photon in the rectilinear basis, she will detect either a

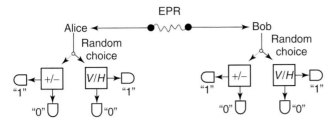

Figure 34.1 Entanglement-based QKD. Each of Alice and Bob measures half of an EPR pair in a randomly chosen basis. V/H, rectilinear basis; +/−, diagonal basis.

vertical or a horizontal polarized photon with the same probability. Depending on Alice's measurement result, Bob's photon will be projected to the corresponding polarization state. If Bob subsequently measures his photon in the same basis, his measurement result will be perfectly correlated to Alice's result. On the other hand, if Bob measures in diagonal basis, no correlation exists. The above arguments are also applicable if Bob performs his measurement first.

The question is, what happens if both Alice and Bob measure in the diagonal basis? Surprisingly, they still get perfect correlation. This can be shown by rewriting Eq. (34.3) into the following form:

$$|\Phi\rangle_{\text{pair}} = 1/\sqrt{2}(|+\rangle_1|+\rangle_2 + |-\rangle_1|-\rangle_2) \tag{34.4}$$

where $|+\rangle = 1/\sqrt{2}(|\updownarrow\rangle + |\leftrightarrow\rangle)$ represents the $45°$ polarization state and $|-\rangle = 1/\sqrt{2}(|\updownarrow\rangle - |\leftrightarrow\rangle)$ represents the $135°$ polarization state.

The discussion above suggests that Alice and Bob can implement BB84-type QKD based on an EPR source, as shown in Figure 34.1. The EPR source can be placed between Alice and Bob. One photon of each EPR pair is sent to Alice and the other one to Bob. For each incoming photon, Alice and Bob randomly and independently choose their measurement bases to be either rectilinear or diagonal. After they have measured all the photon pairs, Alice and Bob will compare their measurement bases and only keep random bits generated with matched bases. Similar to the BB84 QKD based on a single-photon source, they can further perform error correction and privacy amplification to generate the final secure key. Note that, before Alice and Bob perform the measurement, the polarization of each photon is undetermined. More precisely, each photon of the EPR pair is in a maximally mixed state (fully nonpolarized). Eve cannot gain any information from the photon when it transmits from the EPR source to the user, because there in no information encoded. The random bits are generated during the measurement processes.

The above-mentioned QKD protocol based on EPR pairs is quite similar to the original BB84 QKD. However, there is much more physical insight in the first entanglement-based QKD protocol proposed by Ekert in 1991 [15], especially the strong connection between entanglement and the security of QKD. In his original proposal, Ekert suggested that Alice and Bob can verify entanglement by testing a certain type of Bell's inequalities [16]. As long as they can verify the existence of entanglement, it is possible to generate a secure key. Without discussing more details about Bell's inequality, we simply remark that Alice and Bob can perform Bell's inequalities test without knowing how the measurement results are acquired. This has inspired the so-called self-testing QKD, which is discussed in Section 34.4.

Entanglement-based QKD could yield better security under certain circumstances. Furthermore, the long-distance QKD scheme based on quantum repeaters, can only be implemented with entanglement-based QKD protocols. However, the technical challenges in implementing entangled QKD are much bigger than that of the BB84 QKD.

34.2.4
Continuous Variable Quantum Key Distribution Protocol

In the BB84 QKD protocol, Alice's random bits are encoded in a two-dimensional space such as the polarization state of a single photon. More recently, QKD protocols working with continuous variables have been proposed. Among them, the Gaussian-modulated coherent state (GMCS) QKD protocol has drawn special attention [17].

The basic scheme of the GMCS QKD protocol is shown in Figure 34.2. Alice modulates both the amplitude quadrature and phase quadrature of a coherent state with Gaussian-distributed random numbers. In classical electromagnetics, these two quadratures correspond to the in-phase and out-of-phase components of electric field, which can be conveniently modulated with optical phase and amplitude modulators. Alice sends the modulated coherent state together with a strong local oscillator (a strong laser pulse that serves as a phase reference) to Bob. Bob randomly measures one of the two quadratures with a phase modulator and a homodyne detector. After performing his measurements, Bob informs Alice which

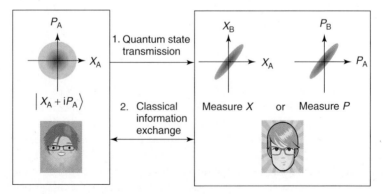

Figure 34.2 The Gaussian-modulated coherent state (GMCS) QKD. X, amplitude quadrature and P, phase quadrature.

quadrature he actually measures for each pulse and Alice drops the irrelevant data. At this stage, they share a set of correlated Gaussian variables that are called the *raw key*. Given the variances of the measurement noises below certain thresholds, they can further work out perfectly correlated secure key by performing reconciliation and privacy amplification.

The security of the GMCS QKD can be comprehended from the uncertainty principle. In quantum optics, the amplitude quadrature and phase quadrature of a coherent state form a pair of conjugate variables, which cannot be simultaneously determined with arbitrarily high accuracies due to Heisenberg uncertainty principle. From the observed variance in one quadrature, Alice and Bob can upper bound Eve's information about the other quadrature. This provides a way to verify the security of the generated key. Recently, an unconditional security proof of the GMCS QKD has appeared [18].

34.3
State-of-the-Art Quantum Key Distribution Technologies

A QKD system is built upon cutting edge technologies, such as SPD, truly random number generator, EPR photon source, phase-stabilized interferometer, and so on. In pursuit of higher secure key rate and longer key distribution distance, many conventional techniques have been significantly improved and new tools have been developed. In this section, we present state-of-the-art QKD technologies.

34.3.1
Sources for Quantum Key Distribution

34.3.1.1 Single-Photon Source
Although the term *single photon* has been used extensively in quantum optics and quantum communication, devices generating individual photons on demand are still not ready for practical applications despite the significant progress achieved in the last decade.

A perfect single-photon source is sometimes called a *photon gun*: each time the user fires, a light pulse containing exactly one photon will be emitted. The single-photon state is a special "photon-number" state – a state contains a precise number of photons. The conventional laser source, regardless of how stable it is, cannot produce photon-number states. This can be intuitively understood as follows: the gain medium of a conventional laser contains a large number of individual atoms or molecules and each of them has a certain probability of emitting a photon during a fixed time period. Thus, a statistical fluctuation of the total photon number is expected. To generate a single-photon state, it is more desirable that a microscopic source containing a single emitting object should be employed: the single emitter can be pumped from its ground state to an excited state during an excitation cycle; after that, a single photon could be generated through spontaneous emission. So far, various single emitters have been investigated for single-photon generation, including atoms or ions in gas phase, organic molecules, color centers in crystals, semiconductor nanocrystals, quantum dots, and so on. For a comprehensive review, see Ref. [19].

34.3.1.2 EPR Photon Pair

Entangled photon pairs can be generated through nonlinear optical processes, such as spontaneous parametric down-conversion (SPDC) [20]. In this process, a pump photon spontaneously decays into a pair of daughter photons in a nonlinear medium. The conservation of energy and momentum implies that the generated daughter photons are entangled in spectral and spatial domains. Like other phase-sensitive nonlinear processes, SPDC requires phase matching to be efficient. Phase matching means that a proper phase relationship between pump light and down-converted light should be maintained throughout the nonlinear medium, so the probability amplitudes of the SPDC process at different locations will add up coherently. The phase-matching requirement and the birefringence of nonlinear crystal suggest that it is possible to selectively generate photon pairs with a certain polarization state. Entangled photon pairs can also be generated in other nonlinear processes, such as four-wave mixing in optical fiber. We will not discuss more details but simply remark that nonlinear optics has been playing an important role in both classical optical communication and quantum communication. The knowledge we have accumulated in classical domain could be the most valuable resource as we venture into the quantum domain.

34.3.1.3 Attenuated Laser Source

In the original BB84 QKD protocol, Alice encodes her random bits on single-photon states. However, an efficient single-photon source with high quality is not available in practice. Instead, heavily attenuated laser sources have been employed in most practical QKD systems.

The output of an attenuated laser source can be modeled as a coherent state, which is a superposition of different photon-number states. In practice, we can decrease the average photon number of a laser pulse by sending it through an optical attenuator. However, no matter how weak the laser pulse is, there is always a nonzero probability that it contains more than one photon. This opens a door for the so-called photon-number splitting (PNS) attack [21]. In the PNS attack, Eve performs a special "quantum nondemolition" (QND) measurement to determine the photon number of Alice's laser pulse without destroying it or disturbing the encoded quantum information. If the laser pulse contains one photon, Eve simply blocks it and Bob will not receive anything. On the other hand, if the laser pulse contains more than one photon, Eve splits out one photon and sends the rest to Bob through a lossless channel. Eve stores the intercepted photons in quantum memories until Bob announces his measurement bases. Then, she measures her photons in the same bases as Bob. In the end, Eve has an exact copy of whatever Bob has. No secure key can be generated.

In a commonly used security proof for the BB84 protocol implemented in a practical QKD system [22], one crucial argument is as follows: Among all of Bob's detection events, only those originated from single-photon pulses contribute to the secure key. In other words, all the multiphoton pulses in BB84 are insecure. The multiphoton probability of a coherent state can be effectively suppressed by applying heavy attenuation. However, this also significantly increases the proportion of vacuum state (laser pulses contain no photon), which, in turn, lowers the efficiency of the QKD system. As a result, the final secure key rate scales quadratically with the transmittance of the quantum channel. This is much worse than a QKD system employing a single-photon source, where a linear relation between the secure key rate and the channel transmittance is expected.

One breakthrough came in 2003 when the "decoy" idea was proposed [23]. The principle behind the decoy idea is that the PNS attack can be detected by testing the quantum channel during QKD process. Alice and Bob can conduct QKD with laser pulses having different average photon numbers (which are named as either *signal state* or *decoy states*) and evaluate their transmittances and QBERs separately. A PNS attack by Eve will inevitably result in different transmittances for signal state and decoy states and, thus, can be detected. It has been shown that the secure key rate of the decoy-state BB84 QKD scales linearly with the transmittance of the quantum channel [24], which is comparable to the secure key rate of the BB84 QKD with a perfect single-photon source. Today, the decoy-state idea has been widely adopted in BB84 QKD systems implemented with weak coherent state sources.

34.3.2
Quantum State Detection

34.3.2.1 Single-Photon Detector

Conventional photon counters, such as photomultiplier tubes (PMTs) and avalanche photodiodes (APDs), have been widely adopted in QKD systems as SPDs. More recently, superconducting single-photon detectors have been invented and these have demonstrated superior performances.

The most important parameters of an SPD include detection efficiency, dark count probability, dead time, and time jitter. Detection efficiency is the probability that an SPD clicks upon receiving an incoming photon. Dark counts are detection events registered by an SPD while no actual photon hits it (i.e., "false alarm"). Dead time is the recovery time of an SPD after it registers a photon. During this period, the SPD will not respond to input photons. Time jitter is the random fluctuation of an SPD's output electrical pulse in time domain. The performances of two commonly encountered SPDs are summarized below.

APD-SPD APDs operated in Geiger mode are commonly used as SPDs. In Geiger mode, the applied bias voltage on an APD exceeds its breakdown voltage and a macroscopic avalanche current can be triggered by a single absorbed photon.

In free-space QKD systems, which are typically operated at the wavelength range of 700–800 nm, silicon APDs are the natural choice and high-performance commercial products are available. The overall efficiency of a silicon APD-SPD can be above 60% with a dark count rate less than 100 counts per second. The time jitter can be as small as 50 ps and the typical dead time is in the range of tens of nanoseconds.

In QKD systems operated at telecom wavelengths, InGaAs/InP APDs have been routinely applied. When compared with silicon APDs, InGaAs/InP APDs operated in Geiger mode have particularly high after pulsing probabilities: a small fraction of charge carriers generated in previous avalanche process can be trapped in the APD and trigger additional "false" counts later on. To reduce the after pulsing probability, the gated Geiger mode has been introduced: most of the time, the bias voltage on an APD is below its breakdown voltage; thus, the APD will not respond to incoming photons. Only when laser pulses are expected, the bias voltage is raised above the breakdown voltage for a short time window to activate the SPD. After the SPD registers a photon, the bias voltage will be kept below the breakdown voltage for a long time (typically, a few microseconds) to release the trapped carriers completely. This is equivalent to introduce an external "dead time," which significantly limits its operating rate. Typically, an InGaAs/InP SPD has an efficiency of 10%, a dark count rate of 10^{-6} per nanosecond, and a maximum count rate less than 1 MHz.

Superconducting SPD SPDs exploiting superconductivities, such as superconducting nanowire single-photon detector (SNSPD) and transition edge sensors (TESs), have been successfully developed. Both of them have wide spectral response range, covering the visible and telecom wavelengths. The TES has both a high intrinsic efficiency and an extremely low dark count rate. However, owing to its long relaxation time, it cannot be operated above a few megehartz. SNSPD, on the other hand, has shown a very small relaxation time and has been employed in QKD systems operated at 10 GHz repetition rate. Currently, these detectors have to be operated at temperatures below a few kelvin, which may not be convenient for practical applications.

34.3.2.2 Optical Homodyne Detector

In principle, the homodyne detectors used in the GMCS QKD are no different from the one used in classical coherent communication system. In either case, the sender encodes information on either the in-phase or quadrature component of electric field (in terms of quantum optics, the information is encoded on either the amplitude quadrature or phase quadrature of a coherent state), while the receiver decodes the information by mixing the signal with a local oscillator. However, the performance of a GMCS QKD is much more sensitive to the electrical noise of the homodyne detector.

When compared with an SPD, a homodyne detector normally has a higher efficiency and a larger bandwidth. This is due to the fact that a strong local oscillator is used to produce interference with the weak quantum signal during homodyne detection. The resulting interference signals are strong enough to be detected with high speed photodiodes.

34.3.3
Quantum Random Number Generator

Quantum random number generator (QRNG) is both a critical device in most QKD systems and a stand-alone product by itself offered by quantum physics. Conventional pseudorandom number generators based on algorithms or physical random number generators based on chaotic behaviors of complex systems cannot generate truly random numbers: given the seeds of a pseudorandom number generator or the initial condition of a classical chaotic system, their future outputs are completely predictable. In contrast, the probabilistic nature of quantum mechanics suggests that truly random numbers could be generated from elementary quantum processes.

Most of today's QRNGs are implemented at single-photon level. The basic principle can be easily understood from Figure 34.3: Single-photon pulses are sent to a symmetric beam splitter and two SPDs are used to detect transmitted and reflected photon, respectively. Depending on which SPD clicks, the registered event is assigned as either bit "1" or bit "0."

Another promising quantum random number generation scheme is based on measuring random field fluctuation of vacuum, or literally speaking, to generate random numbers from "nothing." This scheme can be implemented by sending a strong laser pulse through a symmetric beam splitter and detecting the differential signal of the two output beams with a balanced receiver. To acquire high-quality random numbers, the electrical noise of the balanced receiver has to be orders of magnitude lower than the shot noise.

34.3.4
Quantum Key Distribution Demonstrations

34.3.4.1 QKD Experiments through Telecom Fiber
The availability of a worldwide fiber network suggests that single-mode fiber (SMF) could be the best choice as the quantum channel for practical QKD systems.

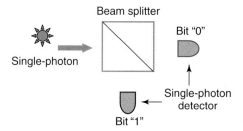

Figure 34.3 The basic principle of a QRNG based on sending a single photon through a beam splitter and observing which path it emerges from.

There are two basic requirements on the quantum channel: low loss and weak decoherence. The first requirement is obvious. The secure key can only be generated from photons detected by Bob. Any channel loss will lower the efficiency of the QKD system. The second requirement, weak decoherence, means that the disturbance of the quantum state by the channel should be as small as possible.

While the loss of standard SMF is reasonably low, the decoherence introduced by a long fiber link is sensitive to both the encoding method and the environmental noise. Owing to the residual birefringence in optical fiber, the polarization of a photon will change as it propagates through a long fiber. This is one of the reasons why most of fiber-based QKD systems adopt phase-coding scheme: Alice encodes her random bit on the relative phase between two laser pulses and sends both of them to Bob through a quantum channel while Bob decodes the phase information by performing interference experiment. It can be shown that, in principle, a phase-coding BB84 QKD is equivalent to a polarization-coding BB84 QKD.

Recently, phase-coding BB84 QKD systems incorporating decoy states have been implemented at gigahertz operating rate [25]. The achieved secure key rate is about 1 Mbit/s over a 20 km fiber link and 10 kbit/s over a 100 km fiber link. Several companies have brought QKD into the commercial market, including id Quantique in Switzerland, MagiQ in the United States, and SmartQuantum in France.

34.3.4.2 QKD Experiments through Free Space

Limited by the intrinsic loss of optical fiber, it is unlikely that a fiber-based QKD system can go beyond a few hundreds of kilometers. Quantum repeaters could come to the rescue. However, they are still far from being practical. One major motivation behind free-space QKD is that a global quantum communication network could be built upon ground-to-satellite and satellite-to-satellite QKD links.

There are several advantages of transmitting photons in free space. First of all, the atmosphere has a high transmission window around 770 nm, which coincides with the spectral range of silicon APDs. Currently, the performance of silicon APDs is superior to that of InGaAs APDs, which operate at telecom wavelength (Section 34.3.2). Secondly, the atmosphere shows very weak dispersion and is essentially nonbirefringence at the above wavelength range, which suggests that the relatively simple polarization-coding scheme can be applied. On the other hand, challenges in free-space QKD are also enormous. As a laser beam propagates in free space, the beam size will increase with distance due to diffraction. To effectively collect the photon at the receiver's side, cumbersome telescopic system is required. Moreover, the atmosphere is not a static medium. Atmospheric turbulence will cause random beam wandering and sophisticated beam tracing system is required to achieve stable key transmission. To date, both polarization-coding decoy-state BB84 QKD and entanglement-based QKD have been demonstrated over a 144 km free-space link [26].

34.4
Security of Practical Quantum Key Distribution Systems

It is important to realize that, in various security proofs, what have been proved to be unconditionally secure are mathematical models of QKD systems built upon certain assumptions. On the other hand, practical QKD systems are full of imperfections. While people can keep improving their mathematical models to get better descriptions of the actual QKD systems, a fully quantum mechanical description of a macroscopic device is not practical. In this sense, "absolutely secure" can only exist in the Hilbert space.

34.4.1
Quantum Hacking and Countermeasures

Quantum cryptography is a competitive game between the legitimate users and Eve. If Eve always follows the rules made by Alice and Bob, she has no hope of winning. On the other hand, if Eve could think differently, the game could become more interesting. Here, we present one specific example, time-shift attack [27], to highlight how Eve could break a real-life QKD system by exploring imperfections ignored by Alice and Bob.

In a typical BB84 QKD setup, two separate SPDs are employed to detect bit "0" and bit "1," respectively. To reduce dark count rate, an InGaAs/InP APD-SPD is typically operated at gated Geiger mode: It is only activated for a narrow time window when Alice's laser pulse is expected. Therefore, the detection efficiency of each SPD is time dependent. In practice, owing to various imperfections, the activation windows of the two SPDs may not fully overlap. Figure 34.4 shows an extreme case: At the expected signal arrival time T_0, the detection efficiencies of the two SPDs are equal. However, if the signal has been chosen to arrive at t_0 or t_1, the efficiencies of the two SPDs are substantially different.

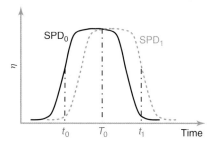

Figure 34.4 The efficiency mismatch of two single-photon detectors in time domain. η, detection efficiency. Note that at the expected signal arrival time T_0, the detection efficiencies of SPD_0 and SPD_1 are equal. However, if the signal has been chosen to arrive at t_0 or t_1, the efficiencies of the two SPDs are substantially different.

The basic idea of the time-shift attack is quite simple: Eve randomly shifts the time delay of each laser pulse (but does not perform any measurement) to make sure that it arrives at Bob's SPDs at either time t_0 or t_1. If Bob registers a photon (recall that Bob will announce whether he has a detection or not for each incoming pulse), Eve will know that it is more likely from the SPD with a high efficiency and consequently she can gain partial information of Bob's bit value. In the case of a complete efficiency mismatch (which means only one of the two SPDs has a nonzero efficiency at certain time), Eve will gain as much information as Bob does; thus, no secure key can be generated. Time-shift attack is simple to implement because it does not require any measurement. Furthermore, since Eve does not attempt to measure the quantum state, in principle, she will not introduce errors and cannot be detected.

Once a security loophole has been discovered, it is often not too difficult to develop countermeasures. There are several simple ways of resisting the time-shift attack. For example, in the phase-coding BB84 QKD system, Bob could use a four-setting setup for his phase modulator, that is, the phase modulator is used not only for basis selection but also for randomly choosing which of the two SPDs is used to detect the bit "0" for each incoming pulse. Thus, Eve cannot learn the bit value from the knowledge of "which detector fires." However, the difficulty remains with discovering the existence of an attack in the first place. It is, thus, very important to conduct extensive research in any implementation details of a QKD system to see if they are innocent or fatal for security. This highlights the fundamental difference between a QKD protocol and a real-life QKD system: While the security of the former can be proved rigorously based on quantum mechanics, the security of the latter also relies on how well we understand the QKD system.

34.4.2
Self-Testing Quantum Key Distribution

So far, we have assumed that we can trust the whole chain of QKD development, from components suppliers, system manufacturer to deliverer. This might be a reasonable assumption, but in the mean time it also gives rise to a concern: Is it possible that a certificated QKD manufacture leaves a trapdoor in the QKD system on purpose so that they can take this advantage to access private information? Historically, it was rumored that during the development of Data Encryption Standard (DES), National Security Agency (NSA) asked IBM to adopt a short 56 bits key for easy breaking. Now, armed with quantum mechanics, is it possible for Alice and Bob to generate secure keys from a "black-box" QKD system that could be made by Eve? This is the motivation behind self-testing QKD [28], which is also inspired by the strong connection between the security of QKD and entanglement in quantum physics.

As we have discussed in Section 34.2, as long as Alice and Bob can verify the existence of entanglement, it is possible to generate a secure key. In the mean time, entanglement can be verified through the violation of certain Bell-type inequalities. The key point is as follows: Alice and Bob can perform Bell inequalities test without knowing how the device actually works. Imagining the QKD system as a pair of black boxes, to perform a Bell test, each user inputs a random number for basis selection to his (her) black box, which in turn outputs an experimental result. By repeating this process many times, the two users can gain enough data to test a Bell inequality. If Alice and Bob's measurement results have been predetermined before they make the basis selection (e.g., Eve could prestore random bits in the two black boxes), then they cannot violate a Bell inequality. On the other hand, if the measurement results do violate a Bell inequality, they can only be generated during the measurement processes and Eve cannot have full knowledge about them. Thus, secure key generation is possible.

Intuitively the above argument works. However, to go from intuition to a rigorous security proof is a very hard question. It is still an open question whether it is possible to unconditionally generate secure key from "untrusted" QKD systems.

We conclude this section by remarking that not only QKD but also any classical cryptographic system suffers from various human mistakes and imperfections of actual implementations in practice. The investigation of security loopholes and the corresponding countermeasures in practical QKD systems plays a complementary role to security proofs.

34.5 Conclusions

Twenty-five years have passed since the publication of the celebrated "BB84" paper [10]. Thanks to the lasting efforts of physicists and engineers, QKD has evolved from a toy model operated in a laboratory into commercial products. Today, the highest secure key rate achieved is on the order of megabits per second, which is high enough for using OTP in certain applications. At the same time, QKD over a distance of a few hundreds of kilometers has been demonstrated through both optical fiber links and free-space links. Furthermore, QKD over arbitrarily long distances could be implemented by employing quantum repeaters.

It is very difficult to predict the future of a rather unconventional technology such as QKD. In the long term, academic researchers may be more interested in fundamental issues, such as a better understanding of the security of various QKD protocols, the implementation of long-distance QKD with quantum repeaters, and so on. On the other hand, in the near future, one of the most urgent tasks is to integrate QKD into classical cryptographic systems and to demonstrate its usefulness in real life. Efforts by European Telecommunications Standards Institute (ETSI) to provide standardization for QKD systems are currently underway. The Swiss government even adopted QKD in their 2007 election.

Currently, there are still heated arguments about whether QKD is useful in practice. The central argument is not about its security or efficiency; it is more about whether QKD is solving a real problem. Quoted from Bruce Schneier, a well-known security expert, "Security is a chain; it is as strong as the weakest link. Mathematical cryptography, as bad as it sometimes is, is the strongest link in most security chains." It might be true that under attacks of a classical Eve (who has no access to a quantum computer), the security of public key distribution protocol is strong. However, cryptography is a technology that is evolving at a fast pace. Historically, whenever a new attack appeared, the security of the old cryptographic system had to be reevaluated and, quite often, people had to develop a new coding scheme to reassure the security. Shor's factoring algorithm has shown that the security of many public cryptographic systems is compromised once a quantum computer is available. This is just one example that demonstrates the impact of quantum mechanics on classical cryptography. Considering the fundamental difference between quantum physics and classical physics, the security of each link in the whole chain of cryptographic system has to be carefully examined to see if it still survives in the quantum era.

In a larger scope, the concept of "quantum Internet" has been proposed [29]. Quantum computers, various quantum sources (such as EPR source), and quantum memories could be connected by the quantum Internet, which extends its reach with the help of quantum repeaters. Although its full power is still to be seen, global QKD can be easily implemented through such a quantum network.

Quantum physics, one of the weirdest and most successful scientific theories in human history, has changed both our view of nature and our everyday life fundamentally. The foundation stones of our information era, including transistor, integrated circuits, and lasers, and so on, are direct applications of quantum mechanics on a macroscopic scale. With the rise of quantum computation and quantum information, we expect the second quantum revolution.

Acknowledgments

Support of the funding agencies CFI, CIPI, the CRC program, CIFAR, MITACS, NSERC, OIT, and QuantumWorks is gratefully acknowledged. The authors' research on QKD was supported in part by the National Science Foundation under Grant No. PHY05-51164.

References

1. Vernam, G.S. (1926) Cipher printing telegraph systems for secret wire and radio telegraphic communications. *J. Am. Inst. Elecr. Eng.*, **45**, 109–115.
2. Shannon, C.E. (1949) Communication theory of secrecy systems. *Bell Syst. Tech. J.*, **28**, 656–715.
3. Rivest, R.L., Shamir, A., and Adleman, L.M. (1978) A method for obtaining digital signatures and public-key cryptosystems. *Commun. ACM*, **21**, 120–126.
4. Shor, P.W. (1994) in Algorithms for quantum computation: discrete logarithms and factoring, *Proceedings of the 35th Symposium on Foundations of Computer Science* (ed. S. Goldwasser), IEEE Computer Society, Los Alamitos, CA, pp. 124–134.
5. Gisin, N., Ribordy, G., Tittel, W., and Zbinden, H. (2002) Quantum cryptography. *Rev. Mod. Phys.*, **74**, 145–195.
6. Wootters, W.K. and Zurek, W.H. (1982) A single quantum cannot be cloned. *Nature*, **299**, 802–803.

7. Dieks, D. (1982) Communication by EPR devices. *Phys. Lett.*, **92A**, 271–272.
8. Wiesner, S. (1983) Conjugate coding. *SIGACT News*, **15**, 78–88.
9. Bennett, C.H. and Brassard, G. (1984) Quantum cryptography: public key distribution and coin tossing, *Proceedings of the IEEE International Conference on Computers, Systems and Signal Processing, Bangalore, India*, IEEE, New York, pp. 175–179.
10. Bennett, C.H. et al. (1995) Generalized privacy amplification. *IEEE Trans. Inf. Theory*, **41**, 1915–1923.
11. Mayers, D. (1996) in *Advances in Cryptology – Proceedings of Crypto'96*, Lecture Notes in Computer Science, Vol. 1109 (ed. N. Koblitz), Springer, New York, pp. 343–357.
12. Lo, H.-K. and Chau, H.F. (1999) Unconditional security of quantum key distribution over arbitrary long distances. *Science*, **283**, 2050–2056.
13. Bennett, C.H. et al. (1992) Experimental quantum cryptography. *J. Cryptology*, **5**, 3–28.
14. Einstein, A., Podolsky, B., and Rosen, N. (1935) Can quantum mechanical description of physical reality be considered complete? *Phys. Rev.*, **47**, 777–780.
15. Ekert, A.K. (1991) Quantum cryptography based on Bell's theorem. *Phys. Rev. Lett.*, **67**, 661–663.
16. Bell, J.S. (1964) On the Einstein Podolsky Rosen paradox. *Physics (Long Island City, NY)*, **1**, 195–200.
17. Grosshans, F. et al. (2003) Quantum key distribution using Gaussian-modulated coherent states. *Nature*, **421**, 238–241.
18. Renner, R. and Cirac, J.I. (2009) de Finetti representation theorem for infinite-dimensional quantum systems and applications to quantum cryptography. *Phys. Rev. Lett.*, **102**, 110504.
19. Lounis, B. and Orrit, M. (2005) Single-photon sources. *Rep. Prog. Phys.*, **68**, 1129–1179.
20. Kwiat, P.G. et al. (1999) Ultrabright source of polarization-entangled photons. *Phys. Rev. A*, **60**, R773–R776.
21. Lütkenhaus, N. (2000) Security against individual attacks for realistic quantum key distribution. *Phys. Rev. A*, **61**, 052304.
22. Gottesman, D. et al. (2004) Security of quantum key distribution with imperfect devices. *Quant. Inf. Comput.*, **4**, 325–360.
23. Hwang, W.-Y. (2003) Quantum key distribution with high loss: toward global secure communication. *Phys. Rev. Lett.*, **91**, 057901.
24. Lo, H.-K., Ma, X., and Chen, K. (2005) Decoy state quantum key distribution. *Phys. Rev. Lett.*, **94**, 230504.
25. Yuan, Z.L. et al. (2009) Practical gigahertz quantum key distribution based on avalanche photodiodes. *New J. Phys.*, **11**, 045019.
26. Schmitt-Manderbach, T. et al. (2007) Experimental demonstration of free-space decoy-state quantum key distribution over 144 km. *Phys. Rev. Lett.*, **98**, 010504.
27. Qi, B. et al. (2007) Time-shift attack in practical quantum cryptosystems. *Quant. Inf. Comput.*, **7**, 73–82.
28. Mayers, D. and Yao, A. (1998) Quantum cryptography with imperfect apparatus. Proceedings of 39th IEEE Symposium on Foundations of Computer Science, pp. 503–509 (arXiv:quant-ph/9809039v1).
29. Kimble, H.J. (2008) The quantum internet. *Nature*, **453**, 1023–1030.

35
Phase-Space Tomography of Optical Beams

Tatiana Alieva, Alejandro Cámara, José A. Rodrigo, and María L. Calvo

35.1
Introduction

Many applications of optical signals in science and technology require adequate characterization of the optical field, which in turn includes the information about the spectral content of light, its polarization, intensity distribution, coherence properties, and so on. Here, we consider only a small part of this important task that is related to the determination of the spatial coherence status of light. This problem is treated in the framework of the paraxial scalar monochromatic field approximation. In this case, coherent or partially coherent beam is described by its amplitude and phase or mutual intensity, relatively, at the plane transversal to the propagation direction. The phase recovery is of interest, for example, in the field of adaptive optics and metrology [1, 2], whereas the determination of the coherence features of light is important in microscopy, analysis of highly scattering medium, and so on [3, 4]. Thus, for beam characterization we need to know more than just its intensity distribution that can be measured in the experiments. There exist various interferometric [5, 6] and iterative [7–9] techniques for phase retrieval and light coherence status estimation. The practical realization of most of them, especially the ones related to the analysis of partially coherent light, is rather cumbersome and time consuming. Here, we consider a technique for the determination of the coherence status of the paraxial field which includes the application of the formalism of the Wigner distribution (WD) that characterizes coherent as well as partially coherent fields.

The method of reconstruction of the optical field WD, called *phase-space tomography*, was established more than 10 years ago [10, 11]. It is based on the rotation of the WD (see Chapter 5) during its propagation through the optical system performing the fractional Fourier transform (FT) [12] (see Chapter 12). The WD projections coincide with the intensity distribution at the output plane of the system. The application of the inverse Radon transform to the set of these projections allows the reconstruction of the WD and, therefore, of the phase or mutual intensity of the beam. The main obstacle in the wide implementation of this method has been a lack of flexible optical setup for the measurements of relatively large number of projections. In the case of two-dimensional optical beams that are separable in Cartesian coordinates, one can use the optical setup consisting of four glass cylindrical lenses. The appropriate lens rotations permit one to acquire the required number of projections. This setup can also be used in graduate student laboratories for the phase-space tomography experiments. The practical work proposed in this chapter underscores the fundamental concepts of spatial coherence theory and phase retrieval.

The chapter begins with a review of the phase-space tomography method. Then, we establish the technique for the reconstruction of the WD of the optical beams that are separable in Cartesian coordinates. A short review of the direct and inverse Radon transforms is given. In the following section, the reconstruction of the WD of a Gaussian beam is considered as an example. Further, an optical setup that is suitable for the measurement of the corresponding WD projections is described and the experimental results of the WD recovery are demonstrated. Finally, in the Appendix we propose a practical work on the phase-space tomography, which is designed for postgraduate students. The code of the Matlab program for the reconstruction of the WD from the projection measurements is provided in Wiley-vch web page: ftp.wiley-vch.de.

35.2
Fundamentals of Phase-Space Tomography

A coherent paraxial monochromatic optical signal is characterized by the complex field amplitude, which is defined as

$$f(\mathbf{x}) = a(\mathbf{x}) \exp\left[i\varphi(\mathbf{x})\right] \tag{35.1}$$

where $\mathbf{x} = [x, y]^t$ is a two-dimensional position vector at the plane transversal to the propagation direction. The intensity distribution $|f(\mathbf{x})|^2 = a^2(\mathbf{x})$ and therefore the amplitude $a(\mathbf{x}) \geq 0$ can be relatively easily measured using, for example, a CCD camera. Nevertheless, the determination of the phase distribution $\varphi(\mathbf{x})$ requires some special techniques because of the high temporal frequency of the optical signals.

The partially coherent field is characterized by mutual intensity, also called *two-point correlation function*,

$$\Gamma(\mathbf{x}_1, \mathbf{x}_2) = \langle f(\mathbf{x}_1) f^*(\mathbf{x}_2) \rangle \tag{35.2}$$

where the brackets $\langle \cdot \rangle$ here and in other instances indicate the ensemble averaging, which for the case of ergodic process coincides with temporal averaging (see Chapter 12 for details), whereas * stands for complex conjugation. For coherent fields, the mutual intensity is reduced to the product of the complex field amplitudes: $\Gamma(\mathbf{x}_1, \mathbf{x}_2) = f(\mathbf{x}_1) f^*(\mathbf{x}_2)$.

Both cases can be treated together by applying the formalism of the WD (see Chapter 12). The WD, $W_f(\mathbf{x}, \mathbf{p})$, is the spatial FT of the mutual intensity Γ with respect to the coordinate difference, $x' = x_1 - x_2$:

$$W(x, p) = \int \Gamma(x + \tfrac{1}{2}x', x - \tfrac{1}{2}x') \exp(-i2\pi p^t x') \, dx' \qquad (35.3)$$

where $p = [p_x, p_y]^t$ denotes spacial frequencies. The vectors x and p have dimensions of length and inverse of length, respectively. Note that the WD of two-dimensional field is a function of four Cartesian coordinates x, y, p_x, and p_y, which form the so-called phase space.

The integration of the WD over p leads to the WD projection:

$$I(x) = \Gamma(x, x) = \int W(x, p) \, dp \qquad (35.4)$$

which is associated with the intensity distribution in a transversal plane and therefore can be easily measured.

Another important property of the WD is its rotation under the FT [12]. It is the basis of the Fourier optics that the complex field amplitude $f_o(x_o)$ at the back focal plane of the thin convergent spherical lens, with focal distance f, is a FT of the complex field amplitude at the front focal plane $f_i(x_i)$ except for a constant factor $-i$ and a scaling that depends on the wavelength λ and f:

$$f_o(x_o) = \mathcal{R}^{FT}[f_i(x_i)](x_o) = \frac{1}{i\lambda f} \int f_i(x_i) \exp\left(-i2\pi \frac{x_i x_o}{\lambda f}\right) dx_i \qquad (35.5)$$

For the general case of the partially coherent fields, the following expression holds for the mutual intensities at the front and back focal planes of the lens [6]:

$$\Gamma_{f_o}(x_{o1}, x_{o2}) = \frac{1}{(\lambda f)^2} \int \Gamma_{f_i}(x_{i1}, x_{i2}) \exp\left[-i2\pi \frac{(x_{i1} x_{o1} - x_{i2} x_{o2})}{\lambda f}\right] dx_{i1} dx_{i2} \qquad (35.6)$$

The relation between the field WDs at the focal plane for both cases is the following:

$$W_{f_o}(x, p) = W_{f_i}(-sp, s^{-1}x) \qquad (35.7)$$

where $s = \lambda f$. Thus, if we introduce the normalized dimensionless coordinates $x_n = x s^{-1/2}$ and $p_n = s^{1/2} p$, we observe that W_{f_o} corresponds to the W_{f_i}, rotated at an angle $\pi/2$ at planes $x_n - p_{xn}$ and $y_n - p_{yn}$ of phase space. Correspondingly, the integration of the $W_{f_o}(x, p)$ over p leads to the intensity distribution:

$$I_{f_o}(x) = \int W_{f_o}(x, p) \, dp = \int W_{f_i}(-sp, s^{-1}x) \, dp = \left\langle \left| \mathcal{R}^{FT}[f_i(x_i)](x) \right|^2 \right\rangle \qquad (35.8)$$

which corresponds to the Fourier power spectrum of the field f_i apart from some scaling factor s. Therefore, using a simple optical system we can obtain another projection of the WD of the analyzed field.

We remind the reader that any two-dimensional regular function $g(q, p)$ can be reconstructed from the set of its projections, known as *Radon transform* of $g(q, p)$, in different directions associated with angles γ which cover π-interval. The algorithm of the function reconstruction from the projections is called *tomographic reconstruction* and it is discussed in Section 35.4.

To apply this method for the reconstruction of the WD of optical field, we need an optical system that is able to rotate the WD for different angles at certain planes of phase space. Moreover, since the WD of a two-dimensional field is a function of four variables, a pair of independent rotations at two phase-space planes $x_n - p_{xn}$ and $y_n - p_{yn}$ is required. These rotations can be produced by applying the optical system performing the separable fractional FT (see Chapter 12).

The 2D fractional FT [13], which can be implemented optically, is defined by

$$F^{\gamma_x,\gamma_y}(\mathbf{x}_o) = \int f_i(\mathbf{x}_i)\, K^{\gamma_x,\gamma_y}(\mathbf{x}_i, \mathbf{x}_o)\, d\mathbf{x}_i \qquad (35.9)$$

The kernel $K^{\gamma_x,\gamma_y}(\mathbf{x}_i, \mathbf{x}_o)$ of the fractional Fourier transformation is separable for orthogonal coordinates: $K^{\gamma_x,\gamma_y}(\mathbf{x}_i, \mathbf{x}_o) = K^{\gamma_x}(x_i, x_o)\, K^{\gamma_y}(y_i, y_o)$, where

$$K^{\gamma_q}(q_i, q_o) = (is \sin \gamma_q)^{-1/2} \exp\left[i\pi \frac{(q_o^2 + q_i^2)\cos\gamma_q - 2q_i q_o}{s \sin \gamma_q}\right] \qquad (35.10)$$

q is a placeholder for x and y, whereas the parameter s has a dimension of length squared whose value depends on the optical system. For $\gamma_q = \pi n$, the kernel Eq. (35.10) reduces to $-\delta(q_o + q_i)$ or $\delta(q_o - q_i)$ for odd or even n, respectively. It is easy to see that the kernel $K^{\gamma_q}(q_i, q_o)$ corresponds to the common direct (inverse) FT for $\gamma_q = \pm\pi/2$ apart from a constant phase and scaling factor s. Note that $K^{-\gamma_q}(q_i, q_o) = [K^{\gamma_q}(q_i, q_o)]^*$. If $\gamma_x = -\gamma_y$ is set in Eq. (35.9), the antisymmetric fractional FT is obtained.

Since the fractional FT produces the rotation of the WD:

$$\begin{array}{ccc} \Gamma_f(\mathbf{x}_1, \mathbf{x}_2) & \Rightarrow & W_f(\mathbf{x}, \mathbf{p}) \\ \Downarrow & & \Downarrow \\ \Gamma_{F^{\gamma_x,\gamma_y}}(\mathbf{x}_1, \mathbf{x}_2) & \Rightarrow & W_{F^{\gamma_x,\gamma_y}}(\mathbf{x}, \mathbf{p}) = W_f(\mathbf{x}', \mathbf{p}') \end{array} \qquad (35.11)$$

where

$$\Gamma_{F^{\gamma_x,\gamma_y}}(\mathbf{x}_{o1}, \mathbf{x}_{o2}) = \iint \Gamma_f(\mathbf{x}_{i1}, \mathbf{x}_{i2}) K^{\gamma_x,\gamma_y}(\mathbf{x}_{i1}, \mathbf{x}_{o1}) \\ \times K^{-\gamma_x,-\gamma_y}(\mathbf{x}_{i2}, \mathbf{x}_{o2})\, d\mathbf{x}_{i1}\, d\mathbf{x}_{i2}$$

and the coordinates of vectors $\mathbf{x}' = [x', y']^t$ and $\mathbf{p}' = [p'_x, p'_y]^t$ are given by

$$q' = q \cos \gamma_q - s p_q \sin \gamma_q$$
$$p'_q = s^{-1} q \sin \gamma_q + p_q \cos \gamma_q$$

the squared modulus of the fractional FT $\left\langle \left|F^{\gamma_x,\gamma_y}(\mathbf{x})\right|^2 \right\rangle$ (known as *fractional power spectrum*) corresponds to the projection of the WD of the input signal $f(\cdot)$ at the plane associated with the angles γ_x and γ_y.

$$S^{\gamma_x,\gamma_y}(\mathbf{x}) = \Gamma_{F^{\gamma_x,\gamma_y}}(\mathbf{x}, \mathbf{x}) = \left\langle \left|F^{\gamma_x,\gamma_y}(\mathbf{x})\right|^2 \right\rangle = \int W_f(\mathbf{x}', \mathbf{p}')\, d\mathbf{p} \qquad (35.12)$$

The set of these projections for angles from π-interval $\gamma_q \in [\pi/2, 3\pi/2]$, also called the *Radon–Wigner transform*, can be easily measured by using the flexible fractional FT processor proposed in Ref. [14]. Then, $W_f(\mathbf{x}, \mathbf{p})$ of the input signal can be recovered by applying the inverse Radon transform or other algorithms used in the tomography. From the inverse FT of the WD, the mutual intensity can be obtained:

$$\Gamma_f(\mathbf{x}_1, \mathbf{x}_2) = \int W_f\left[(\mathbf{x}_1 + \mathbf{x}_2)/2, \mathbf{p}\right] \exp\left[i2\pi \mathbf{p}(\mathbf{x}_1 - \mathbf{x}_2)\right] d\mathbf{p} \qquad (35.13)$$

If the mutual intensity can be presented as a product $\Gamma_f(x_1, x_2) = f(x_1) f^*(x_2)$, then the field $f(x)$ is coherent and its phase can be defined up to constant factor as

$$\varphi(x) = \arg\left[\frac{\Gamma_f(x, x_0)}{f^*(x_0)}\right] \tag{35.14}$$

where x_0 is fixed. If it is known a priori that the optical field is spatially coherent, then the phase retrieval can be achieved using other methods such as Gerchberg–Saxton algorithm which are also based on the acquisition of the WD projections. The number of the required projections depends on the complexity of the signal and the noise level (see Chapter 12).

In spite of the complexity it is possible to reconstruct the WD of any two-dimensional signal; its analysis and representation are difficult. Moreover, the experimental setup for the acquisition of the required WD projections demands the application of spatial light modulators (SLMs) for digital lens implementation, which might not be affordable for a graduate student laboratory. Therefore, in the next section we propose a simplified method for WD reconstruction of optical fields that are separable in the Cartesian coordinates, which can be implemented by using four cylindrical glass lenses [15].

35.3
Phase-Space Tomography of Beams Separable in Cartesian Coordinates

If the mutual intensity of the input field is separable with respect to Cartesian coordinates x and y – which leads to $\Gamma_f(x_1, x_2) = \Gamma_{f_x}(x_1, x_2)\Gamma_{f_y}(y_1, y_2)$, where $\Gamma_{f_q}(q_1, q_2) = \langle f(q_1)f^*(q_2)\rangle$ – then the fractional FT spectra are also separable:

$$S_f^{\gamma_x,\gamma_y}(x) = \Gamma_{F^{\gamma_x,\gamma_y}}(x, x) = \iint \Gamma_{f_x}(x_1, x_2) K^{\gamma_x}(x_1, x) K^{-\gamma_x}(x_2, x) dx_1 dx_2$$
$$\times \iint \Gamma_{f_y}(y_1, y_2) K^{\gamma_y}(y_1, y) K^{-\gamma_y}(y_2, y) dy_1 dy_2$$
$$= S_{f_x}^{\gamma_x}(x) S_{f_y}^{\gamma_y}(y)$$

as derived from Eqs (35.12) and (35.14). Moreover, the WD of a separable field is also separable: $W_f(x, p) = W_{f_x}(x, p_x) W_{f_y}(y, p_y)$.

The Parseval theorem for the one-dimensional fractional FT [13] takes the following form:

$$\int S_{f_q}^{\alpha}(q) dq = \int S_{f_q}^{\beta}(q) dq = A_q \tag{35.15}$$

Since A_q is a constant, which indicates the energy conservation law, we derive that

$$\int S_f^{\gamma_x,\gamma_y}(x) dy = A_y S_{f_x}^{\gamma_x}(x) \tag{35.16}$$

holds for any value of angle γ_y. Correspondingly, we obtain

$$\iint W_{f_x}(x, p_x) W_{f_y}(y, p_y) dy dp_y = A_y W_{f_x}(x, p_x) \tag{35.17}$$

and therefore the problem of the WD reconstruction is reduced to the one-dimensional case. Thus, from the knowledge of the one-dimensional Radon–Wigner transform:

$$S_{f_q}^{\gamma_q}(q) = \int W_{f_q}(q\cos\gamma_q - sp_q\sin\gamma_q, s^{-1}q\sin\gamma_q + p_q\cos\gamma_q)dp_q \qquad (35.18)$$

the $W_{f_q}(q, p_q)$ can be found by applying the inverse Radon transform. A consecutive reconstruction of $W_{f_x}(x, p_x)$ and $W_{f_y}(y, p_y)$ allows the recovery of the entire WD.

Note that the Parseval relation (Eq. (35.18)) can also be used to test the field separability. Indeed, only for fields separable in the Cartesian coordinates, the following holds:

$$\int S_f^{\gamma_x,\gamma_y}(x)dy \int S_f^{\gamma_x,\gamma_y}(x)dx = S_f^{\gamma_x,\gamma_y}(x) \int S_f^{\gamma_x,\gamma_y}(x)dxdy \qquad (35.19)$$

As has been mentioned above, for the practical realization of this phase-space tomographic method an optical setup performing the fractional FT for tunable angles is needed. For a field separable in the Cartesian coordinates, the required experimental WD projections can be measured by using the optical setup associated with the antisymmetric fractional FT ($\gamma_x = -\gamma_y = \gamma$). Such an experimental setup, as discussed later, consists of four glass cylindrical lenses and allows relatively easy modification of the transformation angle γ.

35.4
Radon Transform

In 1917 mathematician Johann Radon proposed a method for the recovery of a function from its projections, called *tomographic reconstruction*. The representation of a function as a set of its projections was later denoted in his honor as *Radon transform*. The tomographic methods are now widely used in the different areas of science, technology, and medicine, such as radiology, nuclear medicine, quantum mechanics, and signal analysis (see Ref. [16]). As we have previously mentioned, it is also used for the characterization of optical fields.

Let us consider the Radon transform of a two-dimensional function denoted as $g(\mathbf{q}) = g(q, p)$. In order to display this function, we need a three-dimensional space with Cartesian coordinates (q, p, g). In Figure 35.1 the corresponding values of $g(\mathbf{q})$ are represented using different gray levels. The integral of $g(\mathbf{q})$ along a line $\mathbf{nq} = x$, which is characterized by its normal $\mathbf{n} = (\cos\gamma, \sin\gamma)$, where γ is the angle formed by $\mathbf{n}$ with the axis q and x is the distance from the coordinate origin, is given by

$$S_g^\gamma(x) = \int g(\mathbf{q})\delta(x - \mathbf{nq})d\mathbf{q} = \int g(x\cos\gamma - \xi\sin\gamma, x\sin\gamma + \xi\cos\gamma)d\xi \qquad (35.20)$$

Here, we have used the parametric definition of a line

$$q(\xi) = x\cos\gamma - \xi\sin\gamma$$
$$p(\xi) = x\sin\gamma + \xi\cos\gamma$$

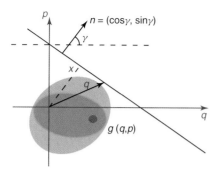

Figure 35.1 Illustration of the projection operation.

For fixed **n**, and therefore γ, we obtain the one-dimensional sinogram function $S_g^\gamma(x)$ of the variable x which is a projection of $g(\mathbf{q})$ in the direction **n**. A set of these projections for different angles γ forms a Radon transform. From Eq. (35.24), it follows that $S_g^\gamma(x) = S_g^{\gamma+\pi}(-x)$. Consequently, the Radon transform is completely defined by the projections for the angles γ that cover an interval of π radians.

Note that the projections $S_g^\gamma(x)$ of a single-point signal, $g(\mathbf{q}) = \delta(\mathbf{q} - \mathbf{q}_0)$, represented in the Cartesian coordinates (x, γ), result in a cosine function:

$$S_\delta^\gamma(x) = \int \delta(\mathbf{q} - \mathbf{q}_0)\delta(x - \mathbf{nq})d\mathbf{q} = \delta(x - \mathbf{nq}_0) = \delta\left[x - q_0 \cos(\gamma_0 - \gamma)\right] \quad (35.21)$$

where $\mathbf{q}_0 = (q_0 \cos \gamma_0, p_0 \sin \gamma_0)$ (see Figure 35.2). Since the Radon transform is linear, by defining an arbitrary function $g(\mathbf{q})$ as

$$g(\mathbf{q}) = \int g(\mathbf{q}_0)\delta(\mathbf{q} - \mathbf{q}_0)d\mathbf{q}_0 \quad (35.22)$$

we obtain that $S_g^\gamma(x)$ is a superposition of cosine functions with different amplitudes q_0, phases γ_0, and intensity levels $g(\mathbf{q}_0)$, often called *sinogram*:

$$S_g^\gamma(x) = \int g(\mathbf{q}_0)\delta\left[x - q_0 \cos(\gamma_0 - \gamma)\right] d\mathbf{q}_0 \quad (35.23)$$

Note that the set of the WD projections presented in a similar way is referred to as Radon–Wigner transform of a signal.

The Radon transform is closely related to the FT. Indeed, the unidimensional FT of $S_g^\gamma(x)$, with respect to the variable x, corresponds to the two-dimensional FT of

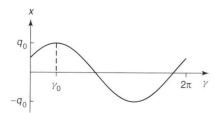

Figure 35.2 Sinogram of a single point.

$g(\boldsymbol{q})$ where the vector $n\upsilon$ appears in the place of spatial frequencies:

$$\mathcal{R}^{FT}[S_g^\gamma(x)](\upsilon) = \int S_g^\gamma(x) \exp(-i2\pi\upsilon x)\,dx$$

$$= \int d\boldsymbol{q}\,g(\boldsymbol{q}) \int dx \exp(-i2\pi\upsilon x)\,\delta(x - \boldsymbol{nq})$$

$$= \int d\boldsymbol{q}\,g(\boldsymbol{q}) \exp(-i2\pi\boldsymbol{qn}\upsilon)$$

It is easy to see that for fixed $\boldsymbol{n}$ the points $\boldsymbol{n}\upsilon = $ constant on the Fourier plane describe a line that passes through the origin of coordinates. Because of this, Eq. (35.28) is known as *central slice theorem*.

Since there exists an inverse FT, it follows that the Radon transform is also invertible. Writing $\mathcal{R}^{FT}[S_g^\gamma(x)](\upsilon)$ in the polar coordinates, $\tilde{S}_g^\gamma(\upsilon)$, and then applying the inverse two-dimensional FT, we can recover the function $g(\boldsymbol{q})$ as

$$g(\boldsymbol{q}) = \int_0^\infty \upsilon\,d\upsilon \int_0^{2\pi} d\gamma \exp\left[i2\pi\upsilon\left(q\cos\gamma + p\sin\gamma\right)\right] \tilde{S}_g^\gamma(\upsilon) \tag{35.24}$$

Using the symmetry of the Radon transform, $S_g^\gamma(x) = S_g^{\gamma+\pi}(-x)$, the latter equation can be written as

$$g(\boldsymbol{q}) = \int_0^\pi d\gamma \int_{-\infty}^\infty d\upsilon\,|\upsilon| \exp(i2\pi\boldsymbol{qn}\upsilon)\,\tilde{S}_g^\gamma(\upsilon) \tag{35.25}$$

We recall that $\tilde{S}_g^\gamma(\upsilon)$ is the FT of $S_g^\gamma(x)$, and therefore

$$S_g^\gamma(x) = \int_{-\infty}^\infty d\upsilon \exp(i2\pi\boldsymbol{qn}\upsilon)\,\tilde{S}_g^\gamma(\upsilon) \tag{35.26}$$

while

$$P_g^\gamma(x) = \int_{-\infty}^\infty d\upsilon\,|\upsilon| \exp(i2\pi\boldsymbol{qn}\upsilon)\,\tilde{S}_g^\gamma(\upsilon) \tag{35.27}$$

Comparing these integrals, we conclude that $P_g^\gamma(x)$ is obtained from $S_g^\gamma(x)$ by filtering in the Fourier domain (multiplying $\tilde{S}_g^\gamma(\upsilon)$ by $|\upsilon|$). Then, the reconstruction formula can be written in a short form, known as *filtered backprojection*:

$$g(\boldsymbol{q}) = \int_0^\pi d\gamma\,P_g^\gamma(\boldsymbol{qn}) \tag{35.28}$$

This approach for the recovery of WD from the measured intensity distributions, associated with its projections, is utilized in the following sections.

35.5
Example: Tomographic Reconstruction of the WD of Gaussian Beams

In order to demonstrate how the tomographic reconstruction of the WD, discussed in Section 35.3, including the inverse Radon transform, is performed in detail, let us consider the separable Gaussian beam:

$$g(\boldsymbol{x}) = \frac{\sqrt{2}}{\sqrt{w_x w_y}} \exp\left[-\pi\left(\frac{x^2}{w_x^2} + \frac{y^2}{w_y^2}\right)\right] \tag{35.29}$$

whose WD is also a Gaussian function,

$$W_g(\mathbf{x}, \mathbf{p}) = W_{g_x}(x, p_x) W_{g_y}(y, p_y)$$
$$= 4\exp\left[-2\pi\left(\frac{x^2}{w_x^2} + w_x^2 p_x^2\right)\right]\exp\left[-2\pi\left(\frac{y^2}{w_y^2} + w_y^2 p_y^2\right)\right]$$

Then, the projections of the WD, which correspond to the data measured at the output of the antisymmetric fractional FT system at an angle γ characterized by the parameter s (see Eq. (35.14)), are given by

$$S_g^{\gamma,-\gamma}(\mathbf{x}) = \iint_{\mathbb{R}^2} d\mathbf{p}\, W_g(\mathbf{x}', \mathbf{p}')$$
$$= \frac{2}{w_x' w_y'}\exp\left\{-2\pi\left[\frac{x^2}{(w_x')^2} + \frac{y^2}{(w_y')^2}\right]\right\}$$

where $\mathbf{x}' = (x', y')^t$, $\mathbf{p}' = (p_x', p_y')^t$ with

$$q' = q\cos\gamma_q - sp_q\sin\gamma_q$$
$$p_q' = s^{-1}q\sin\gamma_q + p_q\cos\gamma_q$$

and w_q' depends on the angle γ by the relation

$$w_q' = w_q\sqrt{\cos^2\gamma + \frac{s^2}{w_q^4}\sin^2\gamma} \qquad (35.30)$$

To recover the WD analytically, we impose the condition $w_x^2 = w_y^2 = s$, and therefore,

$$S_g^{\gamma,-\gamma}(\mathbf{x}) = \frac{2}{s}\exp\left[-2\pi\frac{(x^2+y^2)}{s}\right] \qquad (35.31)$$

The reconstruction process starts from the integration of the WD projections over the x and y coordinates, leading to the projections of the WDs for the 1D functions $g_x(x)$ and $g_y(y)$. As both coordinates are equivalent, apart from a sign in the transformation angle, we focus on the reconstruction of $g_x(x)$. The integration over y yields

$$S_{g_x}^\gamma(x) = \int_\mathbb{R} dy\, S_g^{\gamma,-\gamma}(\mathbf{x}) = \frac{\sqrt{2}}{\sqrt{s}}\exp\left(-2\pi\frac{x^2}{s}\right) \qquad (35.32)$$

The next step is to perform the inverse Radon transform according to Eq. (35.29):

$$W_{g_x}(x, p_x) = \frac{1}{\sqrt{s}}\int_0^\infty dv\, v\, \tilde{S}_{g_x}^\gamma(v)$$
$$\times \int_0^{2\pi} d\gamma \exp\left[i2\pi\frac{v(x\cos\gamma + sp_x\sin\gamma)}{s}\right] \qquad (35.33)$$

where

$$\tilde{S}_{g_x}^\gamma(v) = \frac{1}{\sqrt{s}}\int_\mathbb{R} dx\, S_{g_x}^\gamma(x)\exp\left(-i2\pi\frac{xv}{s}\right) = \exp\left(-\frac{\pi v^2}{2s}\right) \qquad (35.34)$$

is the FT of the WD projection for angle γ. Note that we have used the 1D optical FT (see Eq. (35.9) for $\gamma = \pi/2$). Also, we stress that $\tilde{S}_g^\gamma(x)$ does not depend on γ. Consequently, using the following particularized version of Eqs (3.937-1) and (8.406-3) from Ref. [17], we get

$$\int_0^{2\pi} d\alpha \exp\left[i\left(x\cos\alpha + y\sin\alpha\right)\right] = 2\pi J_0\left(\sqrt{x^2 + y^2}\right) \tag{35.35}$$

The expression for the reconstructed WD can be simplified to

$$W_{gx}(x, p_x) = \frac{2\pi}{\sqrt{s}} \int_0^\infty dv\, v\, \tilde{S}_{gx}^\gamma(v) J_0\left(2\pi \frac{v}{s}\sqrt{x^2 + s^2 p_x^2}\right) \tag{35.36}$$

Now, taking into account Eq. (6.631-4) from Ref. [17]:

$$\int_0^\infty dx\, x e^{-\alpha x^2} J_0(\beta x) = \frac{1}{2\alpha} \exp\left(-\frac{\beta^2}{4\alpha}\right) \tag{35.37}$$

one finds the WD of the 1D Gaussian signal:

$$W_{gx}(x, p_x) = 2\exp\left[-2\pi\left(\frac{x^2}{s} + s p_x^2\right)\right] \tag{35.38}$$

Repeating the algorithm for the y coordinate by taking into account the change of the sign of γ, we obtain the WD of the entire 2D Gaussian beam: $W_g(x, p_x, y, p_y) = W_{gx}(x, p_x) W_{gy}(y, p_y)$.

35.6
Experimental Setup for the Measurements of the WD Projections

As discussed in Chapter 12, the antisymmetric fractional FT can be performed by an optical system consisting of three generalized lenses, where the corresponding distance z between them is fixed. Each generalizad lens L_j (with $j = 1, 2$) is an assembled set of two identical thin cylindrical lenses with fixed focal length $f_j = z/j$ (see Figure 35.3). Meanwhile, the last generalized lens, L_3, and the first one, L_1, are identical. However, since the WD is reconstructed from measurements of the output intensity distribution, the last generalized lens L_3 is not required. Therefore, the experimental setup consists of two generalized lenses (four cylindrical lenses) as displayed in Figure 35.4.

To change the transformation angle γ, these cylindrical lenses have to be properly rotated at angles $\phi_1 = \varphi_j + \pi/4$ and $\phi_2 = -\phi_1$ according to the following equations:

$$\sin 2\varphi_1 = \frac{\lambda z}{s} \cot(\gamma/2)$$
$$\sin 2\varphi_2 = \frac{s}{2\lambda z} \sin\gamma \tag{35.39}$$

where s is the normalization factor and λ is the wavelength. In our case, we chose $s = 2\lambda z$ with $z = 50$ cm and $\lambda = 532$ nm (Nd:YAG laser with 50 mW output power). From Eq. (35.47) we derive that the angle interval $\gamma \in [\pi/2, 3\pi/2]$ needed for the WD reconstruction is covered.

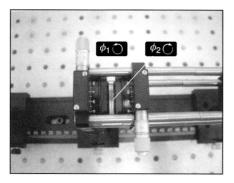

Figure 35.3 Generalized lens constructed by using two identical plano-convex cylindrical lenses. Lens rotation is performed in scissor fashion with angles $\phi_1 = \varphi_j + \pi/4$ and $\phi_2 = -\phi_1$ according to Eq. (35.47).

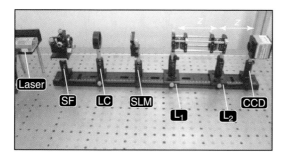

Figure 35.4 Experimental setup. The input signal is generated by modulating a collimated laser beam with wavelength $\lambda = 532$ nm. In particular, we use a spatial filter (SF), collimation lens (LC), and a transmissive SLM with pixel size of 19 μm. Generalized lenses L_1 and L_2 and the camera CCD (pixel size of 4.6 μm) are set according to $z = 50$ cm.

We underline that each generalized lens consists of two identical plano-convex cylindrical lenses. Notice that the glass–air interface between the cylindrical lenses results in reflection losses that can be minimized by using an appropriate refractive index-matching fluid. Moreover, in order to obtain an accurate assembly, the cylindrical lenses were set in a cage system with rotation mount (see Figure 35.3). This system can also be automatically controlled by using a driving rotation mount if needed.

Different input signals $f(x_i, y_i)$ can be studied. Here, we consider HG beams that can be generated by means of a laser cavity or holograms. Specifically, we use an SLM where the input signal is encoded as a phase-only hologram, which offers a higher optical throughput than an amplitude-only hologram. In this case, the amplitude and phase distributions of the input signal, $f(x_i, y_i)$, are addressed onto the phase-only hologram $H(x_i, y_i)$ as follows:

$$H(x_i, y_i) = \exp\left\{i\left|f(x_i, y_i)\right|\left(\arg[f(x_i, y_i)] + 2\pi x_i/\Lambda\right)\right\} \quad (35.40)$$

where the phase and amplitude ranges are $[-\pi, \pi]$ and $0 < |f(x_i, y_i)| < 1$, respectively [18]. The additional linear phase factor $2\pi x_i/\Lambda$ yields a phase grating consisting of fringes parallel to the y axis. It is used to redirect the input signal into the first diffraction order and, therefore, avoid the overlapping with the zero-order beam. We underline that the grating period has to be set according to the Nyquist–Shannon sampling theorem: at least a sampling of two pixels per fringe is needed. This approach has been demonstrated, for example, in Ref. [19] for the generation of various Gaussian beams. Notice that here we consider linear polarized light.

35.7
Reconstruction of WD: Numerical and Experimental Results

To demonstrate the feasibility of the phase-space tomography method, the WD reconstruction explained in Section 35.3, it is applied to a known field that is separable in Cartesian coordinates: $f(x, y) = f_x(x)f_y(y)$.

It is advantageous to consider an Hermite–Gaussian (HG) beam, whose complex amplitude in the plane of the beam waist is defined as

$$\mathrm{HG}_{m_x,m_y}(\mathbf{x}; \mathbf{w}) = \sqrt{2}\frac{H_{m_x}\left(\sqrt{2\pi}\frac{x}{w_x}\right)H_{m_y}\left(\sqrt{2\pi}\frac{y}{w_y}\right)}{\sqrt{2^{m_x}m_x!w_x}\sqrt{2^{m_y}m_y!w_y}}\exp\left[-\pi\left(\frac{x^2}{w_x^2}+\frac{y^2}{w_y^2}\right)\right] \quad (35.41)$$

where $H_m(\cdot)$ denotes the Hermite polynomial of order m and $\mathbf{w} = [w_x, w_y]^t$ are the beam waists in the x and y directions, respectively. It is indeed separable, $\mathrm{HG}_{m_x,m_y}(\mathbf{x}; \mathbf{w}) = \mathrm{HG}_{m_x}(x; w_x)\mathrm{HG}_{m_y}(y; w_y)$, with

$$\mathrm{HG}_{m_q}(q; w_q) = 2^{1/4}\frac{H_{m_q}\left(\sqrt{2\pi}\frac{q}{w_q}\right)}{\sqrt{2^{m_q}m_q!w_q}}\exp\left(-\pi\frac{q^2}{w_q^2}\right) \quad (35.42)$$

where q is again a placeholder for x and y. We use the $\mathrm{HG}_{2,1}(\mathbf{x}; \mathbf{w})$ field with sizes $w_x = 1.25\sqrt{s}$, $w_y = 0.75\sqrt{s}$, and $s = 0.266$ mm^2. The intensity and the phase of this mode are displayed in Figure 35.5. Eventually, we make use of the two one-dimensional HG modes, $\mathrm{HG}_2(x; w_x)$ and $\mathrm{HG}_1(y; w_y)$, in which the two-dimensional HG mode factorizes.

It is important to notice that the HG field is shape invariant during fractional FT, apart from a quadratic phase and a possible scale [13]. Since we measure only the intensity distributions, the phase factor is not important. Therefore, the intensity at the output of the fractional FT system is given by

$$S^{\gamma_q}_{\mathrm{HG}_{m_q}}(q) = \left|\mathrm{HG}_{m_q}(q; w'_q)\right|^2 \quad (35.43)$$

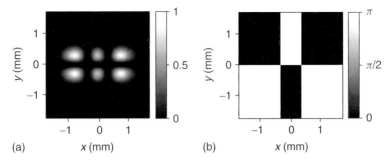

Figure 35.5 (a) Intensity and (b) phase distributions of a $HG_{2,1}(x;w)$ mode with beam waists $w_x = 1.25\sqrt{s}$ and $w_y = 0.75\sqrt{s}$, where $s = 0.266$ mm^2.

where w'_q is of the same size as that found for the fundamental Gaussian mode (see Eq. (35.38)), whose expression is

$$w'_q = w_q \sqrt{\cos^2 \gamma_q + \frac{s^2}{w_q^4} \sin^2 \gamma_q} \qquad (35.44)$$

where s is the coordinate normalization factor of the fractional FT system (see Section 35.6), w_q is the beam waist of the input HG mode, and γ_q is the fractional FT angular parameter. Therefore, it is clear from Eq. (35.52) that if the beam waist of the input mode is $w_q = \sqrt{s}$, then the intensity at the input and the output planes of the fractional FT system is the same for any angle γ_q. Here, we have chosen $w_x = 1.25\sqrt{s}$ and $w_y = 0.75\sqrt{s}$.

Besides this, the analytic expression for WD of an HG mode is also known [20]:

$$W_{HG_{mq}}(q, p_q) = (-1)^{mq} 4 \exp\left[-2\pi \left(\frac{q^2}{w_q^2} + w_q^2 p_q^2\right)\right] L_{mq}\left[4\pi \left(\frac{q^2}{w_q^2} + w_q^2 p_q^2\right)\right] \qquad (35.45)$$

where $L_{mq}(\cdot)$ is the Laguerre polynomial of order m, and q and p_q are the coordinates of the phase space. The WD of our one-dimensional HG modes, $HG_2(x; w_x)$ and $HG_1(y; w_y)$, are displayed in Figure 35.6. Notice that as has been mentioned in Chapter 12, the WD takes both positive and negative values.

First, let us estimate a minimum number of projections needed for a successful reconstruction of the WD. The numerical simulations of the fractional FT spectra of the considered mode and the further WD recovery from 80, 20, 10, and 5 projections by using the inverse Radon transform indicate that 20 projections are sufficient for a qualitative analysis of the WD (see Figure 35.7). Therefore, for the experimental reconstruction of the WD we have limited the number of projections to $N = 20$, sweeping a π-interval from $\pi/2$ to $3\pi/2$. After the numerical integration of each of the 20 measured projections along the y and x coordinates, the Radon transforms for the two WDs of the unidimensional signals, $HG_2(x; w_x)$ and $HG_1(y; w_y)$, respectively, are obtained. Both are represented in Figure 35.8

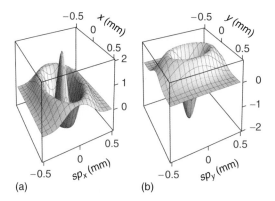

Figure 35.6 WD of (a) $HG_2(x; w_x)$ with $w_x = 1.25\sqrt{s}$ and (b) $HG_1(y; w_y)$ with $w_y = 0.75\sqrt{s}$, where $s = 0.266$ mm^2.

a and b. They are compared with the simulated Radon–Wigner transforms in Figure 35.8 c and d. Note that, since for the input field $w_x > \sqrt{s}$, the size w'_x is firstly enlarged and afterward gets smaller according to Eq. (35.52). The opposite is observed for $w_y < \sqrt{s}$.

Applying the digital inverse Radon transform to the experimental sinograms, the WDs for both parts of the HG field are obtained. The simulated and the experimental WDs recovered from $N = 20$ projections are compared with the ones calculated from Eq. (35.53) (see Figure 35.9). A good agreement is observed between them.

Certainly the number of projections needed for the WD reconstruction depends on the signal complexity, noise level, and the characteristics of the experimental setup. Nevertheless, even for a relatively small number of them, the main features of the reconstructed WD of the HG beam are preserved as can be concluded from its comparison with the analytically calculated WD.

35.8
Practical Work for Postgraduate Students

The practical work on the phase-space tomographic reconstruction of the WD of optical field underscores the different aspects of classical optics, signal processing, and programming. It treats such fundamental concepts as coherence, phase space, angular spectrum, stable modes, and digital holography. Note that similar formalism is also used in quantum optics, and therefore the proposed task can be considered as a bridge between classical and quantum approaches.

It is assumed that the optical setup for the antisymmetric fractional FT power spectral measurements has been built by a supervisor, since this task involves a certain expertise in system alignment. The separable input field considered for

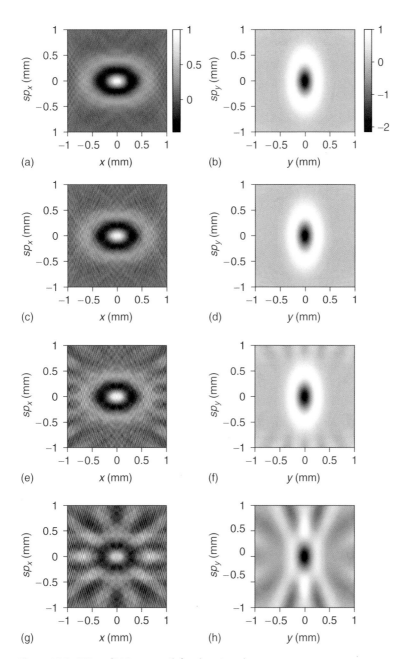

Figure 35.7 WDs of $HG_2(x;w_x)$ (left column) and $HG_1(y;w_y)$ (right column) obtained by numerical simulation from (a,b) $N = 80$, (c,d) $N = 20$, (e,f) $N = 10$, and (g,h) $N = 5$ projections.

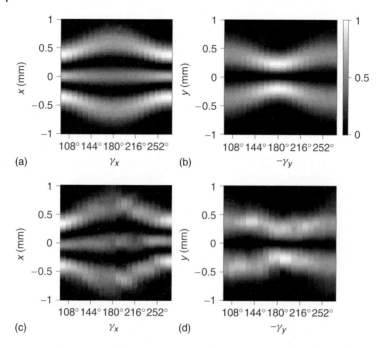

Figure 35.8 Simulated (top) and experimental (bottom) Radon–Wigner transforms for 20 projections of the (a,c) $HG_2(x;w_x)$ and (b,d) $HG_1(y;w_y)$ modes.

analysis (an HG mode in our case) can be implemented by using a computer generated hologram (CGH). In Section 35.6, we have proposed to use a phase-only hologram addressed onto an SLM which can be designed by the students as an additional practical work. A CGH can also be created by using an amplitude-only hologram as described in Ref. [21]. A commercial hologram or a diffractive optical element (DOE) is a good alternative as well.

The experimental and numerical work can be achieved by a group of students. Two students can make the experimental measurements together at one setup. Further, one of them can perform the reconstruction of a part of the WD for coordinate x while the other can do it for variable y.

The procedure of the WD reconstruction of separable fields is as follows:

1) *Measurement of the antisymmetric fractional FT spectra, $S_f^{\gamma,-\gamma}(x)$, for angles $\gamma \in [\pi/2, 3\pi/2]$*. The students rotate the cylindrical lenses following a table similar to Table 35.1, the values of which have been calculated using Eq. (35.47), in order to register the required intensity distributions corresponding to the WD projections by means of a CCD camera. The table corresponds to the rotation angles of the cylindrical lenses, associated with the transformation parameter γ of the fractional FT for $s = 2\lambda z$. Note that for further application

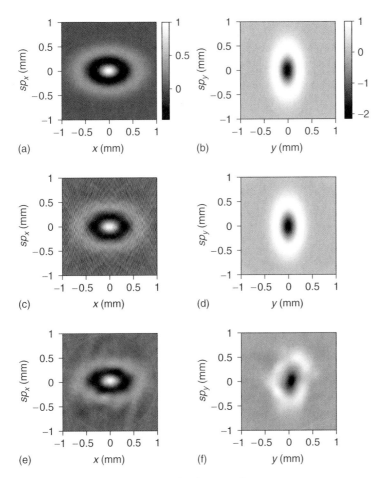

Figure 35.9 WDs of $HG_2(x;w_x)$ (left column) and $HG_1(y;w_y)$ (right column) obtained (a,b) theoretically, (c,d) by tomographic reconstruction using 20 projections, and (e,f) by tomographic reconstruction using 20 experimental projections.

of the program calculating the inverse Radon transform (which can be downloaded from ftp.wiley-vch.de), the WD projections have to be measured at the equidistant angle interval $\Delta\gamma$. The acquisition of 20 projections is enough to obtain reasonably good results.

2) *Integration of the acquired spectra over the y or x coordinate to obtain $A_y S_{f_x}^{\gamma}(x)$ or $A_x S_{f_y}^{-\gamma}(y)$ for angles $\gamma \in [\pi/2, 3\pi/2]$. The results of this integration are presented in the form of Radon–Wigner maps (see, for example, Figure 35.8).*

3) *Validation of the hypothesis about the field separability.* It consists of the proof of Eq. (35.22) for $\gamma_x = -\gamma_y = \gamma$.

Table 35.1 List of the cylindrical lens rotation angles $\phi_1 = \varphi_j + \pi/4$ for each generalized lens L_j corresponding to the antisymmetric fractional FT at angle γ. Notice that $\phi_2 = -\phi_1$ holds in each case. These values of angle γ (given in degrees) were used for the numerical and experimental reconstruction of the WD presented in Section 35.7.

Angle γ	ϕ_1 (Lens L_1)	ϕ_1 (Lens L_2)
90	60.00	90.00
99	57.60	94.50
108	55.70	99.00
117	53.90	103.50
126	52.40	108.00
135	51.00	112.50
144	49.70	117.00
153	48.40	121.50
162	47.30	126.00
171	46.10	130.50
180	45.00	135.00
189	43.90	139.50
198	42.70	144.00
207	41.60	148.50
216	40.30	153.00
225	39.00	157.50
234	37.60	162.00
243	36.10	166.50
252	34.30	171.00
261	32.40	175.50
270	30.00	180.00

4) Application of the inverse Radon transform for the reconstruction of the two parts of the WD, $W_{f_x}(x, p_x)$ and $W_{f_y}(y, p_y)$, from $S_{f_x}^{\gamma}(x)$ and $S_{f_y}^{-\gamma}(y)$ respectively. A qualitative comparison between experimental results and the numerically simulated ones (program for numerical simulations is available at ftp.wiley-vch.de) is the final task of the work.

There are two additional tasks, not considered here, which might also be included in this practical work. The first task is the phase retrieval for a coherent field, achieved by the application of Eqs (35.15) and (35.16). The second one concerns the application of the antisymmetric fractional FT setup as an optical mode converter (see Chapter 12). In such cases, the input signal is a HG mode rotated at 45°, obtained by the corresponding rotation of the hologram. Thus, by repeating the scrolling of the angle γ for the same interval, $[\pi/2, 3\pi/2]$, other paraxial stable modes different from the HG one are obtained as output signals [14]. For example, in the case of $\gamma = 3\pi/4$, such an HG input mode is transformed into the well-known helicoidal Laguerre–Gaussian mode.

35.9
Conclusions

The tomographic method for the reconstruction of the WD of paraxial beams is a useful tool for the characterization of both the fully and partially spatially coherent fields. This technique can be used for the analysis of random media by measuring the changes in the coherence properties of the beam during its propagation through it. The algorithm for the reconstruction of the WD, which is demonstrated in this chapter for the case of separable beams, can be used in the postgraduate student laboratories. The description of the possible practical work is given in the Section 35.8.

Acknowledgments

The financial support of the Spanish Ministry of Science and Innovation under projects TEC2005–02180 and TEC2008–04105 and of the Santander–Complutense project PR–34/07–15914 are acknowledged. José A. Rodrigo gratefully acknowledges a 'Juan de la Cierva' grant.

References

1. Brady, G.R. and Fienup, J.R. (2004) Improved optical metrology using phase retrieval, in *Frontiers in Optics*, Optical Society of America, p. OTuB3.
2. Restaino, S.R. (1992) Wave-front sensing and image deconvolution of solar data. *Appl. Opt.*, **31**, 7442–7449.
3. Yura, H.T., Thrane, L. and Andersen, P.E. (2000) Closed-form solution for the Wigner phase-space distribution function for diffuse reflection and small-angle scattering in a random medium. *J. Opt. Soc. Am. A*, **17**, 2464–2474.
4. Baleine, E. and Dogariu, A. (2004) Variable coherence tomography. *Opt. Lett.*, **29**, 1233–1235.
5. Mandel, L. and Wolf, E. (1995) *Optical Coherence and Quantum Optics*, Cambridge University Press, Cambridge.
6. Goodman, J.W. (2000) *Statistical Optics*, Wiley-Interscience, New York.
7. Gerchberg, R.W. and Saxton, W.O. (1972) A practical algorithm for the determination of the phase from image and diffraction plane pictures. *Optik*, **35**, 237.
8. Fienup, J.R. (1978) Reconstruction of an object from the modulus of its Fourier transform. *Opt. Lett.*, **3**, 27–29.
9. Zalevsky, Z., Mendlovic, D., and Dorsch, R.G. (1996) Gerchberg-Saxton algorithm applied in the fractional Fourier or the Fresnel domain. *Opt. Lett.*, **21**, 842–844.
10. Raymer, M.G., Beck, M., and McAlister, D.F. (1994) Complex wave-field reconstruction using phase-space tomography. *Phys. Rev. Lett.*, **72**, 1137–1140.
11. McAlister, D.F. et al. (1995) Optical phase retrieval by phase-space tomography and fractional-order Fourier transforms. *Opt. Lett.*, **20**, 1181–1183.
12. Lohmann, A.W. (1993) Image rotation, Wigner rotation, and the fractional Fourier transform. *J. Opt. Soc. Am. A*, **10**, 2181–2186.
13. Ozaktas, H.M., Zalevsky, Z., and Kutay, M.A. (2001) *The Fractional Fourier Transform with Applications in Optics and Signal Processing*, John Wiley & Sons, Ltd, New York.
14. Rodrigo, J.A., Alieva, T., and Calvo, M.L. (2009) Programmable two-dimensional optical fractional Fourier processor. *Opt. Express*, **17**, 4976–4983.

15. Cámara, A. *et al.* (2009) Phase space tomography reconstruction of the Wigner distribution for optical beams separable in Cartesian coordinates. *J. Opt. Soc. Am. A*, **26**, 1301–1306.
16. Barrett, H.H. and Myers, K.J. (2004) *Foundations of Image Science*, Wiley-Interscience, New Jersey.
17. Gradshteyn, I.S. and Ryzhik, I.M. (1980) *Table of Integrals, Series, and Products*, Academic Press.
18. Davis, J.A. *et al.* (1999) Encoding amplitude information onto phase-only filters. *Appl. Opt.*, **38**, 5004–5013.
19. Bentley, J.B. *et al.* (2006) Generation of helical Ince-Gaussian beams with a liquid-crystal display. *Opt. Lett.*, **31**, 649–651.
20. Simon, R. and Agarwal, G.S. (2000) Wigner representation of Laguerre–Gaussian beams. *Opt. Lett.*, **25**, 1313–1315.
21. Cámara, A. and Alieva, T. (2008) *Practical Work: Digital Holography and Beam Analysis. Proceedings of the Topical Meeting on Optoinformatics*, St. Petersburg, Russia.

36
Human Face Recognition and Image Statistics using Matlab
Matthias S. Keil

36.1
Introduction

Human face recognition still performs in a superior way when compared to artificial face-recognition systems. Without any noticeable effort, we recognize faces at various distances or under difficult lighting conditions. The likely reason for this superiority is that human face recognition reveals special properties that are not yet incorporated in artificial systems.

Imagine, for example, that you are scanning the faces of a crowd because you are looking for a friend. Depending on the viewing distance, the faces of the people you are looking at project images of different sizes on your retina. For reliable and successful face recognition (you eventually want to recognize your friend as soon as you spot his or her face), your brain needs to select the proper scale from the retinal image [1].

Psychophysicists usually refer to (spatial frequency) *channels*, in order to describe the perceptual correlate of scale-sensitive (or spatial-frequency-sensitive) mechanisms in the visual brain (e.g., simple and complex cells, [2]). Because it is a priori uncertain at which distance from the observer an object will appear, an analysis by means of several spatial frequency channels can render object recognition independent of scale and viewing distance, respectively. To this end, the visual system needs to select the most informative scale for an object, which is tantamount to an observer establishing an object-based description of spatial frequency content [1]. Object-based spatial frequencies are expressed in *"cycles per object,"* which is in contrast to viewing-distance-dependent units such as *"cycles per pixel," "cycles per image,"* or *"cycles per degree."*

Face recognition is a prime example for studying object-based spatial frequencies. A considerable number of psychophysical studies arrived at the consensus that the brain prefers a narrow band of spatial frequencies (about two octaves bandwidth) from 8 to 16 cycles per face for identifying individuals [3–10]. This *critical band* of spatial frequencies is invariant with turning a face upside down (face inversion, [11]) and is not influenced by image background [12].

Optical and Digital Image Processing: Fundamentals and Applications, First Edition. Edited by Gabriel Cristóbal, Peter Schelkens, and Hugo Thienpont.
© 2011 Wiley-VCH Verlag GmbH & Co. KGaA. Published 2011 by Wiley-VCH Verlag GmbH & Co. KGaA.

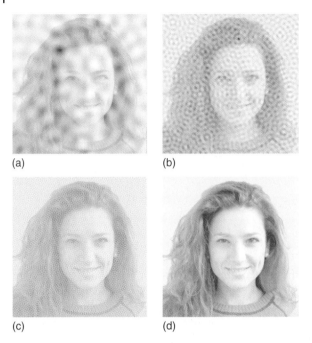

Figure 36.1 Noise with different spatial frequencies interacts with face perception. In each of the four images, noise with a different spatial frequency was superimposed on a face (original image is shown in (d)). If the frequency of the noise matches the critical spatial frequency band (which is preferred by the brain for face recognition), then the "visibility" of the internal face features is degraded, which impairs the identification process (b). Conversely, if the noise lies above (c) or below (a) the critical band, the face is easier to recognize.

In a typical psychophysical experiment, subjects are instructed to learn faces from photographs of a number of persons. Usually, subjects are shown only *internal face features* (the eyes, nose, and mouth), in order to exclude additional recognition cues such as hair, ears, and shoulders (*external face features*). Then, a test face is chosen from the previously learned set and presented on a monitor, but with a selected spectral sub-band of the face being degraded (Figure 36.1). Subjects then need to indicate the identity of the test face. Most studies agree in that face-recognition performance drops upon degrading the *critical range* from 8 to 16 cycles per face.

Face recognition does not exclusively depend on the critical frequencies, as faces are still somewhat recognizable without them [7, 9]. In the latter case, the brain must thus rely on higher or lower spatial frequencies. This indicates that observers have a certain degree of flexibility in selecting informative frequency channels depending on the cognitive task at hand [1]. This, however, comes at a cost, because recognition time increases when the critical band is not available [9].

Notice that the identification of the girl in Figure 36.1 could be facilitated by her eye-catching hairstyle. When internal face features are too unreliable (as in real life at big viewing distances or when the person is not familiar), then the brain seems to rely especially on the external face features [13–15].

In the following, we present some of the techniques and mathematical tools that permit to derive the spatial frequency preference just described by analyzing nothing but face images – we do not carry out any psychophysical or neurophysiological experiments. This means that we connect basic signal processing with human cognitive processing. It is certainly worthwhile to consider biological results for designing artificial face-recognition systems, as biology may give the necessary inspiration for the engineer to bring artificial systems closer to human performance.

36.2
Neural Information-Processing and Image Statistics

Accumulated evidence supports the hypothesis that the front-end stages for the processing of sensory information adapted – on phylogenetic or shorter timescales – to the signal statistics of the specific environment inhabited by an organism [16–21]. Optimal adaptation is reflected in terms of efficiency of the code that is used to transfer and process sensory signals ("economical encoding," [22, 23]). Efficiency usually is related to a specific goal. These goals include, for example, the minimization of energy expenditure [24–26], to achieve a sparse activation of neurons in response to the input signal [27–32] (but see Ref. [20]), to minimize wiring lengths [33], or to reduce spatial and/or temporal redundancies in the input signal [17, 22, 34, 35], while at the same time encoding the maximum amount of information about the input signal in cell responses [22, 23, 36–38].

In order to produce an efficient code (or an efficient representation of a natural image in terms of neuronal activity), two information-processing strategies were proposed [39] : *spatial decorrelation* and *response equalization* (described below in the corresponding sections). Both strategies refer to how visual neurons exploit the inherent statistical properties of natural images.

The possible values that pixels of a natural image can assume are not arbitrary. The physical properties of the visual world introduce relationships between pixels. The purpose of natural image statistics is to find diagnostic statistical properties, which distinguish natural images from completely random ones. Perhaps the most striking finding is that the spectra of natural images reveal a characteristic k^{-2} fall off of power with spatial frequency k [27, 40, 41] (Figure 36.2). Before discussing decorrelation and response equalization, respectively, we take a look at the origin of the k^{-2} characteristic (or equivalently k^{-1} in amplitude spectra, which are just the square root of the power spectra).

36.2.1
Possible Reasons of the k^{-2} Frequency Scaling

The Wiener–Khinchine theorem states that the power spectrum is just the Fourier-transformed autocorrelation function [42]. In other words, the power

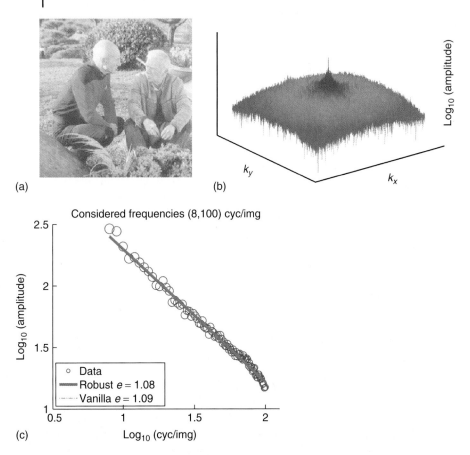

Figure 36.2 Amplitude spectrum of a natural image. (a) A natural or photographic image. (b) In a double-logarithmic representation, the power spectrum of many natural images falls approximately with the square of spatial frequency, which has been interpreted as a signature of scale invariance (Eq. (36.1)). The power spectrum is equal to the squared amplitude spectrum shown here. (c) Hence, the slope of a line fitted to the orientation-averaged amplitude spectrum falls with the inverse of spatial frequency k^e with $e \approx -1$. See Figure 36.6 for further details. (Please find a color version of this figure on the color plates.)

spectrum reflects second-order statistics, and contains information about the similarity of pairs of intensity values as a function of their mutual distance (e.g., [17], p197). On the average, the spectral power of natural images has a tendency to decrease as k^{-2} with spatial frequency k [27, 40, 41]. According to the Wiener–Khinchine theorem, this indicates strong spatial correlations in luminance values between pixel pairs [27, 41, 43–45] (see also Table 1 in Ref. [46], but see Ref. [47]). But how can the k^{-2} property be related to image features? Three principal hypotheses were suggested: *occlusion*, *scale invariance*, and *contrast edges*. Each of these is described briefly in turn.

36.2.1.1 Occlusion

Physical objects are composed of surfaces with overlaid texture and luminance gradients [48]. Surfaces have a characteristic appearance (glossy, matte, shiny), which is the result of the interaction of illumination, how the surface reflects light, and surface curvature. For reliable object recognition, it is definitely a good strategy to recover surface reflectance. This implies that the visual system has to get rid of "interfering" features such as texture and luminance gradients [49, 50]. If we consider intensity values from *within* a surface, then these usually have a smaller variation than *between* surfaces. Ideally, crossing surface boundaries will cause intensity discontinuities. We then arrive at one possible explanation for the k^{-2} relationship: objects in the foreground of a scene convey luminance correlations over larger distances than background objects. Surfaces from background objects are relatively smaller due to perspective and occlusion by foreground objects [51].

36.2.1.2 Scale Invariance

The k^{-2} decrease of the power spectrum implies that the total energy contained between two frequencies $k_{\min}$ and $k_{\max}$ is constant for any fixed number of octaves $\log_2(k_{\max}/k_{\min})$ ("equal energy in equal octaves" [27]):

$$\int_{k_0}^{nk_0} g(k) \cdot (2\pi k) dk \equiv \text{const} \tag{36.1}$$

implies that $g(k) \propto k^{-2}$, that is, the power spectrum is self-similar (assuming $k_{\min} = k_0$ and $k_{\max} = nk_0$). Self-similarity was taken as a signature for scale invariance in natural images – zooming into an image would not alter the distribution of sizes and luminance values [45, 51–53]. The effect of scale and occlusion is simulated in Figure 36.3.

36.2.1.3 Contrast Edges

The scale invariance hypothesis can be questioned, though, because power-law power spectra can give rise to *non*-power-law autocorrelation functions [54]. The latter observation prompted an alternative explanation for the k^{-2} property. It is based on the occurrence of contrast edges [40], where the overall spectral slope α is just the average of individual slopes α_i of image constituents considered in isolation [54]. The overall value of α then depends on the distribution of sizes of constituting image features "i" as follows.

Consider a spectrum that decreases with k^α, thus $\alpha \leq 0$. According to Balboa and Grzywacz [54], slope values cannot assume arbitrary small values, but are confined to the biggest objects in the image. On one hand, large objects considered in isolation give rise to the minimal slope $\alpha_i = -1.5$ (-3 for power spectra). On the other hand, small objects can contribute at most $\alpha_i = 0$. The slopes of objects with intermediate sizes lie somewhere between these extremes. The eventually observed spectral slope α is the result of averaging the individual α_i.

Figure 36.4 shows an illustration of this idea with an artificial image composed of just a few equally sized disks. The latter image should be compared

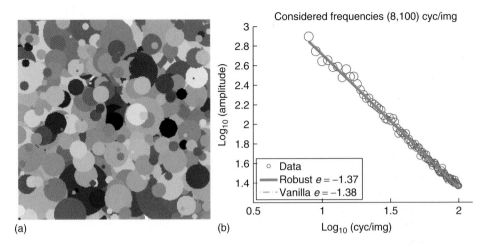

Figure 36.3 Occluding disks. The k^{-1} fall off of spectral amplitudes in the context of scale invariance and occlusion, respectively. (a) A "naturally looking" artificial image composed of mutually occluding disks of many sizes and intensity values. (b) The lines fitted to the orientation-averaged log–log amplitude spectrum. The fitting procedure is explained in Section 36.4.2.1. (Please find a color version of this figure on the color plates.)

with Figure 36.3, which is "naturally" looking, because of scale invariance and mutual occlusion. In this sense, the image of Figure 36.4 is "unnatural": it only contains isolated image structures at one scale, with neither occlusion nor perspective. Nevertheless, its frequency fall off is consistent with that of typical natural images.

36.2.2
The First Strategy: Spatial Decorrelation

Spatial decorrelation is based on insights from the Wiener–Khinchine theorem [42]. We have already seen that one possible explanation for the k^{-1} amplitude fall-off results from occlusion and perspective, leading to strong pairwise correlations in luminance. As a consequence, the visual stimulation from an object surface in the foreground is highly redundant: if one pixel is selected, then most of the other surface pixels will have similar luminance values. In the most fortunate case, we would even be able to predict all the other pixel values from just one [22]. At surface boundaries, typically abrupt changes in reflectance occur. In this sense, contours may be regarded as unpredictable image structures. Turning the last argument on its head means that spatial decorrelation can implement predictive coding.

The output of the retina is conveyed by activity patterns that travel along the axons of retinal ganglion cells. A number of these cells seem to favor image contrasts and perform boundary enhancement. The receptive fields of ganglion cells are known to have an antagonistic center-surround organization [55] (performing some sort of

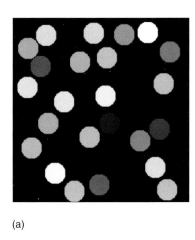 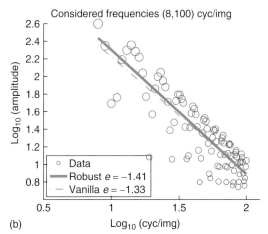

Figure 36.4 Single disks. The k^{-1} fall off of spectral amplitudes in the context of contrast edges. (a) An artificial image composed of nonoccluding, equally sized disks with various intensity values. The important point to realize here is that this image thus does not have the typical properties of natural images, that is, scale invariance, occlusion, perspective, and so on. (b) Despite being completely different, the lines fitted to the orientation-averaged amplitude spectrum have slope values that are consistent with natural image statistics (cf previous figure). (Please find a color version of this figure on the color plates.)

bandpass filtering). This receptive field organization prompted an interpretation in terms of predictive coding, where the surround computes a prediction of the center signal, for example, by computing the mean value [34]. Evidence for predictive coding in the retina has recently been found experimentally: it has been shown that the surround, in fact, reveals precise adaptation to local statistical properties [35]. The classical difference-of-Gaussian (*DoG*) model for the ganglion cell receptive field [56] is strictly valid only in the case of uniform surround stimulation.

Figure 36.5 aims to explain the connection between spatial decorrelation and center-surround receptive fields (DoG-like convolution kernel) in the spatial domain. The important point is that retinal ganglion cells do not fully whiten their input, as this would also cause the noise to be decorrelated, making it indistinguishable from the signal [17, 34].

Consider Figure 36.6, where just a random image with white noise is shown. "White noise" means that all spatial frequencies occur with equal probability. As there is no preference for any frequency, the corresponding amplitude spectrum is flat.

Figure 36.2 shows a natural image with a spectral slope that is considered as typical for natural images.

Now, we consider what would happen if we replaced the amplitude spectrum of the white noise image by one of a typical natural image and what would happen if we flattened (or decorrelated) the amplitude spectrum of the natural image. The

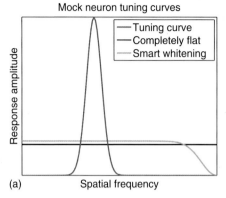
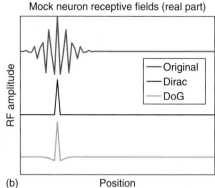

Figure 36.5 Tuning curves and whitening. Retinal ganglion cells (GCs) have a center-surround receptive field, and their experimentally measured tuning curves are suitable for compensation of the $1k^{-1}$ property of natural images. (a) In Fourier space, GC tuning curves do not increase with k up to the highest spatial frequencies, as this would cause the amplification of noise (curve "completely flat"). Rather, they exclude high spatial frequencies (curve "smart whitening"). The "Tuning curve" is the profile of a strict bandpass filter, which, in the spatial domain (b) resembles the Gabor-like receptive field of simple cells in the primary visual cortex (experimentally measured retinal filters are actually settled between a bandpass and a low-pass filter).

corresponding results are shown in Figure 36.7(a) and (b), respectively. The boundaries of the decorrelated natural image are enhanced. In other words, redundant information has been suppressed, and what remains are the unpredictable image structures.

Imposing a k^{-1} decrease on the amplitudes of the white noise image makes it appear as a photography of clouds. Thus, the perception of the noise image is no longer dominated by its high spatial frequencies. This demonstration may be considered as evidence that the visual system compensates in some way the k^{-1} amplitude fall off, either by decorrelation or response equalization.

36.2.3
The Second Strategy: Response Equalization ("Whitening")

The decorrelation hypothesis focuses on flattening the response spectrum of individual neurons when they are stimulated with natural images. This hypothesis, therefore, makes precise predictions about the tuning curves of neurons [17, 34]. With appropriate tuning curves and an appropriate sampling density, flattening of the response spectrum will lead to redundancy reduction in the responses of neighboring neurons [39].

Response equalization, on the other hand, makes no specific assumptions about the tuning curves of neurons. It is rather concerned with sensitivity adjustment of neurons as a function of their receptive field size. Its goal is to produce representations of natural scenes where – on an average – all neurons have approximately

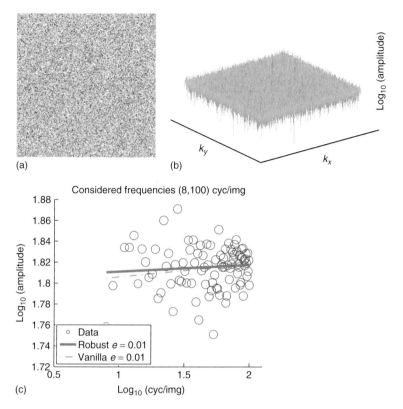

Figure 36.6 White noise: (a) in a white noise image, each 2D spatial frequency (k_x, k_y) occurs with the same probability. (b) 2D amplitude spectrum: this is equivalent to a flat amplitude spectrum. (c) 1D amplitude spectrum: the slope of a line fitted to orientation-averaged spectral amplitudes measures the flatness of the spectrum. With white noise, the *spectral slope* is approximately zero (legend symbol $e \equiv \alpha$ denotes slopes). Details on the fitting procedure can be found in Section 36.4.2.1. Notice that although white noise contains all frequencies equally, our perception of the image is nevertheless dominated by its high frequency content, indicating that the visual system attenuates all lower spatial frequencies. (Please find a color version of this figure on the color plates.)

the same response level [27, 57, 58]. Because natural scenes reveal a characteristic k^α amplitude spectrum (with spectral slope $\alpha \approx -1$) [40, 41], neurons tuned to high spatial frequencies should consequently increase their gain such that they produce the same response levels as the low-frequency neurons do. This is the *response equalization hypothesis*.

In what follows, we refer to response equalization as "whitening." Whitening may optimize the information throughput from one neuronal stage to another by adjusting the signal or response gain of one stage such that it matches the dynamic range of a successive stage [39]. The parameter values for this adjustment can be extracted from the statistical properties of the signal. This idea is illustrated below

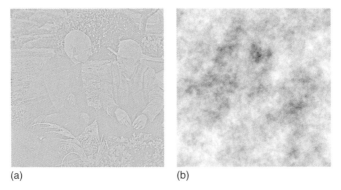

(a) (b)

Figure 36.7 Decorrelation. (a) Whitened amplitudes. The decorrelated version of the natural image shown in Figure 36.2a. Flattening of the amplitude spectrum was achieved by multiplying original amplitudes at frequency k by k^e ($e = +1.08$, see Figure 36.2c), while leaving phase relationships unaltered. As a result, all correlations between pixels are suppressed, and unpredictable image structures are enhanced. (b) Amplitudes $\propto k^{-1}$. This is the noise image of Figure 36.6a, but now with an amplitude spectrum similar to a natural image ($e = -1$), and with its original phase relationships. The modified noise image looks "natural" (reminiscent of clouds or marble), as we perceive also its features with low spatial frequencies.

where we derive psychophysical properties of face perception by using response equalization.

36.3
Face Image Statistics and Face Processing

The important points to remember from the earlier sections are as follows: (i) the first stages for processing visual information in the brain are optimized for matching statistical properties of natural images; (ii) face recognition in humans takes place in a preferred band of spatial frequencies ($\approx$ 8–16 cycles per face); and (iii) humans analyze an image by means of spatial frequency channels, which are the psychophysical correlate of neurons with a variety of receptive field sizes. With these three points in mind, we now describe a sequence of processing steps for deriving the preferred spatial frequencies for human face recognition directly from face images.

36.3.1
Face Image Dataset

Our face images were taken from the Face Recognition Grand Challenge (FRGC, http://www.frvt.org/FRGC, see supplementary figure 36.5 in Ref. [59]). Original images were converted to gray scale, and each face was rotated to align the eyes with a horizontal line (eye coordinates were available). The Matlab function $B =$

`imrotate(A,Θ, 'bicubic')` was used for rotation (*A* is the input image, *B* is the rotated output, and Θ is the rotation angle). In order to make computation feasible in a reasonable time with a typical desktop PC in 2009, face images were finally downsampled to 256 × 256 pixels via $B = $ `imresize(A,[256,256],'bilinear')`, and stored in TIFF format by `imwrite(A, filename,'tif')`.

From all gray-scale images, 868 images of female subjects and 868 images of male subjects were selected, which were free of rotation artifacts. Face images were approximately normalized with respect to interocular distance, with a mean value of 49.48 ± 0.941 pixels.

36.3.2
Dimension of Spatial Frequency

The spatial frequency content of objects depends on viewing distance. For studying face images, this would mean that possible results would depend on image size. In order to avoid that, we introduce an absolute (object centered) spatial frequency scale. Therefore, we need to establish a conversion from "cycles per image" (cyc/img) to "cycles per face" (cyc/face).

To this end, face dimensions were interactively determined with a graphical interface. A rectangular region was selected for each face. The ratio between width and height, respectively, of the rectangle, and image size (256 × 256 pixels), gives corresponding factors for converting cyc/img into cyc/face (Figure 36.8). If we think of conversion factors as defining the two main axes of an ellipse, we can also compute them at oblique orientations.

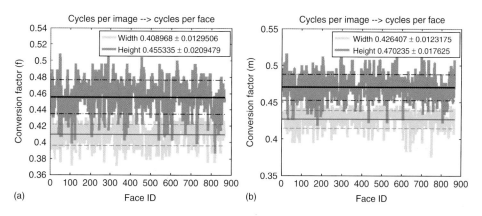

Figure 36.8 Spatial frequency conversion factors. The graphs show the factors for converting "cycles per image" (cyc/img) into "cycles per face" (cyc/face) for (a) female and (b) male face images. We have two factors for each gender, one for converting horizontal spatial frequencies, and another one for converting vertical spatial frequencies (figure legends: mean values ± s.d., corresponding to solid and broken lines, respectively). The abscissa indicates the number of each image (the "face ID").

36.4
Amplitude Spectra

For computing the amplitude spectra, we need to segregate internal from external face features. The face images show also the cropped shoulder region of each person, which gives rise to spectral artifacts; the Fourier transformation involves periodic boundary conditions, so cropping will generate spurious spatial frequencies. It turns out that we can kill three birds with one stone by windowing our face images with a *minimum four-term Blackman–Harris (BH) window* (supplementary figures 36.8 and 36.9 in Ref. [59], and equation 33 in Ref. [60]). The "third bird" has to do with spectral leakage, since Fourier windows are weighting functions that aim to reduce *spectral leakage* [60]. Images are represented in a finite spatial domain, and

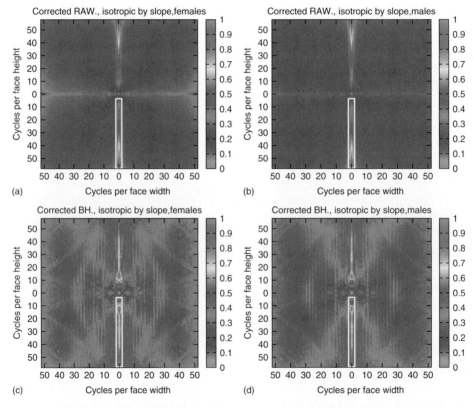

Figure 36.9 Slope-whitened mean amplitude spectra. (a,b) Slope-whitened, corrected mean raw spectra do not reveal amplitude enhancement as a consequence of whitening – (a = females, b = males). (c,d) However, when the mean-corrected BH spectra are whitened, clear maxima appear at horizontal feature orientations (marked by a white box), at $\approx 10-15$ cycles per face height. Thus, the maxima were induced by internal face features. (Please find a color version of this figure on the color plates.)

their size eventually defines the finite set of basis functions that is available for the discrete Fourier transformation. By definition, the discrete Fourier transformation assumes signals as being periodic, and therefore images have to be periodically extended ("tiling"). If the image content at opposite boundaries is discontinuous, then – through tiling – additional spatial frequencies are produced, which cannot be projected always on a single basis vector. Such nonprojectable frequencies result in nonzero contributions over the entire basis set ("leakage"). Here the term *truncation artifacts* refers to such spurious frequency content introduced by tiling. The cropped shoulder regions in our images would give rise to truncation artifacts.

36.4.1
Window Artifacts and Corrected Spectra

Notwithstanding all advantages, the BH window leaves a typical fingerprint (an artificial structure) in the amplitude spectra (supplementary figure 36.6a and b in Ref. [59]). Because of this complication, four types of amplitude spectra were eventually computed:

- amplitude spectra of the unmanipulated face images (i.e., with external features)
- *corrected* amplitude spectra of the none-manipulated images
- the amplitude spectra of the BH-windowed images
- *corrected* amplitude spectra of BH-windowed images

For correcting all types of artifacts, *inward diffusion* is applied to the amplitude spectra. Applying inward diffusion defines the corrected spectra. Details of this procedure can be found in Ref. [59]. "Windowing" refers to applying an adaptive *minimum 4-term BH window* to the face images ("adaptive" means that the window is centered at the nose).

36.4.2
Whitened Spectra of Face Images

In the following, we consider the *mean* amplitude spectra of the face images. Mean amplitude spectra are separately computed for each gender. The mean *raw spectrum* is obtained by computing the amplitude spectrum of each face image and just summing them up ("averaging"). Before averaging, we can correct the spectra by applying inward diffusion. The result of doing that is a mean *corrected raw spectrum*.

Two more types of mean amplitude spectra are computed by windowing. If these spectra are summed up right away, then we obtain the mean *BH spectrum*. If inward diffusion is applied before averaging, then we end up with the mean *corrected BH spectrum*. Notice that both of the windowed spectra are devoid of contributions from external face features.

Next, we *whiten* (response equalize) the spectra. To this end, three different procedures were devised. Three is better than one because then we can be pretty sure that our results do not depend in a significant way on the mathematical details of each respective procedure.

As defined earlier, $A \in \mathfrak{R}_+^{n \times m}$ is the amplitude spectrum of a face image I ($n = m = 256$). Spectral amplitudes $A(k_x, k_y)$ with equal spatial frequencies k lie on a circle with radius $k = \sqrt{k_x^2 + k_y^2}$ (spatial frequency coordinates $k_x, k_y \in [1, 127]$ cycles per image). Let n_k be the number of points on a circle containing amplitudes with frequency k. For our purposes, it is more comfortable to use polar coordinates $\mathcal{A}(k, \Theta)$. Then, a one-dimensional, isotropic amplitude spectrum $\mu(k)$ is defined as

$$\mu(k) = \sum_\Theta \mathcal{A}(k, \Theta)/n_k \tag{36.2}$$

Note that n_k monotonically increases as a function of k, as higher spatial frequencies in the spectrum correspond to bigger frequency radii. We use the terms horizontal and vertical spatial frequencies for designating the main orientation of image structures. For example, a horizontally oriented bar (with its wave vector pointing vertically) will give rise to "horizontal spatial frequencies". Note that horizontal frequencies vary along the vertical dimension (frequency ordinate) in the amplitude spectrum.

36.4.2.1 Slope Whitening

As spectra fall approximately with $k^{-|\alpha|}$, we can fit a line with slope $\alpha < 0$ to a double-logarithmic representation of $\mu(k)$, that is, $y = \alpha x + b$ with $y = \log(\mu)$ and $x = \log(k)$. The intercept b plays no essential role, and will not be further considered. The fitting algorithms are provided with Matlab's statistical toolbox[1]. The function "`robustfit`" implements linear regression with less sensitivity to outliers (label "robust" in figures). The figure label "vanilla" denotes fitting by an ordinary multiple linear regression procedure (Matlab function "`regress`"). Only spatial frequencies from $k_{\min} = 8$ to $k_{\max} = 100$ were used for line fitting, as higher frequencies are usually associated with more noise. Furthermore, too few points n_k are available for averaging at low spatial frequencies via Eq. (36.2).

The slope-whitened spectrum $\mathcal{W}$ is computed then as

$$\mathcal{W}(k, \Theta) = \mathcal{A}(k, \Theta) \cdot k^{|\alpha|} \tag{36.3}$$

Several slope-whitened spectra are shown in Figure 36.9. The whitened corrected raw spectra in (a) females and (b) males fail to reveal any structure that could be related to human face perception. However, the corrected windowed BH-spectra (c) females; (d) males have peaks at around 10 cycles per face height, which perfectly fit the above-mentioned psychophysical results.

36.4.2.2 Variance Whitening

Define the spectral variance $\sigma^2(k)$ of orientation-averaged amplitudes $\mu(k)$ as

$$\sigma^2(k) = \sum_\Theta (\mathcal{A}(k, \Theta) - \mu)^2/(n_k - 1) \tag{36.4}$$

---

1) *Matlab* version 7.1.0.183 R14 SP3, *Statistical Toolbox* version 5.1

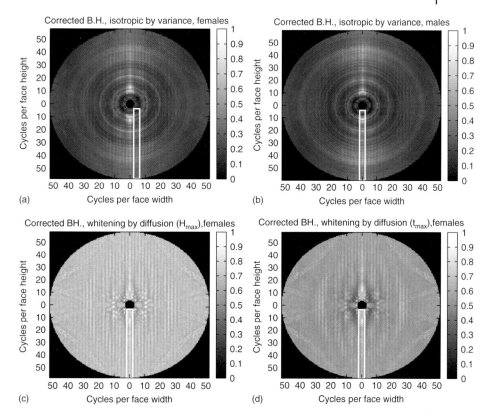

Figure 36.10 Variance-whitened and diffusion-whitened mean amplitude spectra. (a,b) Variance-whitened, corrected mean BH spectra (a = females, b = males). Considered frequency range from $k = 8$ to 127 cycles per image. Clear maxima are revealed at horizontal spatial frequencies around 10–15 cycles per face height. (c,d) Diffusion-whitened, corrected mean BH spectra. Both spectra show data for females. The considered frequency range was from $k = 8$ to 100 cycles per image. (c) shows the spectrum with the highest entropy ("H_{max}" criterion). Notice that there are several other "hot spots" in the spectrum corresponding largely to vertical spatial frequencies (i.e., ≈ 0 cycles per face height, ≈ 10 cycles per face *width*) apart from the one in the interesting frequency range surrounded by a white box. (d) shows the same spectrum at 200 iterations of the diffusion algorithm (Figure 36.12; "t_{max}" criterion). Although the spectrum's entropy now is slightly lower, the white box contains the maximum spectral amplitudes (in the 10–15 cycles per face height range), and relatively smaller "hot spot" amplitudes compared to (c).

Then, a variance-whitened spectrum is computed as

$$\mathcal{V} = \mathcal{A}/(\sigma^2(k) + \epsilon) \tag{36.5}$$

with a small positive constant $\epsilon \ll 1$. Examples of $\mathcal{V}$ are shown in Figure 36.10a (mean windowed BH-spectra for female face images) and b (male face images), respectively. Both spectra reveal maxima at the critical spatial frequencies for face identification at around 10 cycles per face height.

36.4.2.3 Whitening by Diffusion

Let $\mathcal{X}(k_x, k_y, t)$ be a sequence of amplitude spectra parameterized over time t, with the initial condition $\mathcal{X}(k_x, k_y, 0) \equiv \mathcal{A}(k_x, k_y)$. For $t > 0$, the $\mathcal{X}$ are defined according to the diffusion equation $\partial \mathcal{X}/\partial t = \nabla^2 \mathcal{X}$. The whitened spectrum then is $\mathcal{D} \equiv \mathcal{A}/(1 + \mathcal{X}(t))$. An optimally whitened spectrum is obtained at the instant $t \equiv t_H$ when the Shannon entropy of $\mathcal{D}$ is maximal (this defines the H_{max}-criterion). The algorithm shown in Figure 36.12 admits the specification of a frequency range $\mathcal{R}$, such that $(k_x, k_y) \in \mathcal{R}$.

Instead of "stopping" the algorithm when the spectrum's entropy reaches a peak, another possibility is to always apply the same number of diffusion steps, thus ignoring its momentary entropy. This defines the t_{max} criterion. This idea is based on the observation that H_{max} face spectra reveal local maxima at vertical spatial frequencies ("hot spots") besides the usual ones at horizontal spatial frequencies (see Figure 36.10c). Although these maxima are better compatible with psychophysical results of face perception (i.e., 10 cycles per face *width*), the maximum around 10 cycles per face height could be relatively enhanced by continuing with the diffusion process (the spectrum at $t_{max} = 200$ iterations is shown in Figure 36.10d). This curiosity is further detailed in Figure 36.11, where we see that entropy first shows a sharp increase and then slowly declines with time. In other words, once we passed the entropy peak, the spectrum may still be considered as being whitened, albeit not being the "whitest" one possible.

Figure 36.11 also shows how amplitudes at horizontal spatial frequencies evolve during the whitening process. The highest amplitude at 10 cycles per face height is

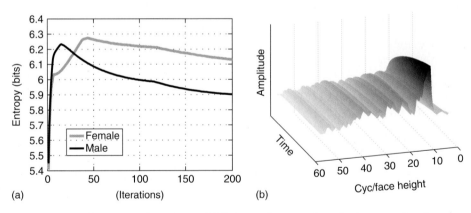

Figure 36.11 Time dependence of diffusion whitening. Both data correspond to mean BH-windowed spectra. (a) Time dependence of entropy for whitening by diffusion (males and females, see legend). After a strong initial increase in entropy, a peak is reached and entropy gradually decreases (entropy of male spectra peaks first). (b) Time evolution of a spectral amplitude trace (white box in Figure 36.10). Among all local maxima, the peak at around 10 cycles per face height is the last that reaches a plateau (i.e., where saturation occurs). The frontal side corresponds to 200 iterations of algorithm in Figure 36.12 (time increases from the back to the front).

```matlab
1  function [S_tmax S_hmax] = DiffusionWhitening (S,kVec)
2  %
3  % WHITENING BY DIFFUSION OF AMPLITUDE SPECTRUM 'S'
4  %
5  % 'kVec' is a row-vector with the considered frequency
6  % range, by default kVec = [8 100] was used
7  %
8  % S_tmax     Whitened spectrum at maximum iteration number
9  % S_hmax     Whitened spectrum at maximum entropy
10 %
11 % (C) Matthias S. Keil September 2009 (Helmetz)
12 %
13
14 % As the spectral DC is located at (129,129) for a 256 x 256
15 % spectrum we drop the first row and column in order to center
16 % the DC location symmetrically
17 %
18 S     =S(2:end,2:end);         % DC now @ (128,128)
19 rows = size (S,1);             % we assume rows = cols for 'S'
20
21 % Boolean matrix for selecting the frequency
22 % range specified by 'kVec'
23 rh    = (rows-1)/2;
24 [X,Y] = meshgrid(-rh:rh);
25 RAD   = sqrt(X.*X+Y.*Y);       % radial frequency map
26 KRANGE = (RAD≥kVec(1)) & (RAD ≤ kVec(end));
27
28 tmax = 200;                    % iteration limit
29 X    = S;                      % 'X' is subjected to heat diff.
30 H_max = entropy(S);            % current maximum entropy
31 S_hmax = S;                    % current most whitened spectrum
32
33 for t=1:tmax
34    X = X+0.5*del2(X);          % diffusion with Euler's method,
35                                % integration time constant 0.5
36    Y = S./(1+X);               % whitening of spectrum 'S' by
37                                % divisive normalization
38    H = entropy (Y(KRANGE));    % compute Shannon entropy across
39                                % the selected frequencies
40    if H >H_max                 % new highest entropy?
41      S_hmax = Y.*double (KRANGE); % yes, current spectrum is
42                                % the "whitest" one
43      H_max = H;                % store associated entropy
44    end
45 end
46 S_tmax = Y.*double (KRANGE);% whitenend spectrum at t=tmax
```

Figure 36.12 Matlab code for whitening by diffusion (Section 36.4.2.3). The function returns two spectra according to different criteria. S_{tmax} is the whitened spectrum at maximum iterations ($t_{max} = 200$), and S_{Hmax} as the flattest possible spectrum (i.e., the spectrum with the highest entropy for $t < t_{max}$). Examples for correspondingly whitened 2D spectra are shown in Figure 36.10c and d, respectively. The whitening-by-diffusion algorithm defined here represents an extension of the one used in Ref. [59].

the last one that "saturates." All smaller peaks already have reached their respective plateaus earlier, and some even begin to decrease again. Entropy is highest at some point when the global maximum still can grow relative to all other smaller peaks – but the best "signal-to-noise" ratio of the spectrum is obtained if we let the highest peak (our desired "signal") attain its maximum.

36.4.2.4 Summary and Conclusions

Our main result is that *windowed* and *whitened* amplitude spectra of face images revealed maxima in the spatial frequency range of 10–15 cycles per face height [59]. In the case of diffusion whitening (according to the H_{max} criterion), the spectra contain additional maxima at vertical spatial frequencies (10–15 cycles per face width). Furthermore, if face images were not windowed, then whitening failed in revealing maxima. Because windowing strips off the external face features (hair, throat, etc.), we can conclude that the maxima are induced by eyes, mouth, and the bottom termination of the nose (internal face features). Therefore, it seems that our brain sets its spatial frequency channels for face processing so as to match the resolution of internal face features [61].

36.5
Making Artificial Face Recognition "More Human"

Given that internal face features are "responsible" for the human spatial frequency preference for face identification (8–16 cycles per face) [3–10, 59, 61], we asked whether artificial face recognition could also benefit from this result [62].

Consider the hyperspace where each windowed face image is represented by a point (each pixel contributes a dimension to that space). Now we take various images from one and the same individual. If each of these images were exactly like the other (for different persons), then identification of individuals would be rather easy, since each point in our space would correspond to a different identity. However, certain factors (like viewing angle or illumination) map the images of one person not just on a single point, but rather generate a cloud of points. The problem is that the clouds for two persons may overlap in the hyperspace, giving rise to confusion of identity.

Any artificial face-recognition system must find hyperplanes that separate these point clouds (classes). Only then can we expect reliable face recognition.

Now we can ask how class separability depends on the images' spatial frequency content. As image size determines the highest spatial frequencies that can be contained, we resized the face images, and measured class separability of high-pass filtered images (equivalent to bandpass filtering the originally sized images) and unfiltered images (equivalent to low-pass filtering the original images). Notice that downsizing reduces the dimensionality of the hyperspace. The result suggests that class separability is optimal at intermediate spatial frequencies at about 16 cycles per face width, which agrees again very well with the psychophysical results mentioned above (but see Ref. [63]).

This means we have another good reason for choosing a certain frequency range for face recognition: making face recognition more reliable.

If we did it the other way around, and interpreted our results as clues about information processing in the brain, this would mean that the brain processes and memorizes faces as small thumbnail images (at least for recognition), with an approximate size of 32 × 32 pixels (assuming that the thumbnails contain exclusively internal face features). For artificial systems, on the other hand, this small size would accelerate the recognition process, as face identification usually relies on matching a target face against a database. Note, however, that faces can appear at any size in an image, and for face detection all scales have to be considered. To avoid ambiguity in terms of face versus nonface objects, external face features could be taken into account, that is, face shape, hair (although it is highly variable), and throat. At least with humans, the presence of a corresponding local context in images activated face-selective cells, albeit internal face features were actually unrecognizable [15].

Several psychophysical studies suggest that the brain encodes face identities as deviations from an average face (this may explain to a certain degree why – at first sight – all Chinese look the same to a "normal" Westerner), for example, Refs [64–67]. Perhaps the most similar artificial system to human average face encoding is the eigenface approach [68, 69]. It would nevertheless be interesting to implement the differential scheme of humans and evaluate its performance. After the averaging of 50–100 face images or so, the resulting average face seems to have converged when inspected by a human observer – averaging more images, or averaging different images seem to yield always the same average face.

The average face approach appears to be a quite powerful technique, which has been shown to boost recognition performance of already available artificial systems [70]. The authors used an online version of an artificial system with a database containing on the average 9 photographs from 3628 celebrities. This system was probed with the author's own collection of photographs of celebrities, where photographs that were already in the database of the online system were excluded. They achieved correct recognition in about 54% of their photographs. Now they did something very interesting. For each celebrity, an average image was created from the author's set of photographs. Then they probed the database with the averaged images. Surprisingly, the hit rate increased to 100%. Still a recognition rate of 80% was maintained if the average image was composed only of those photographs that could *not* be identified by the artificial system as single pictures.

36.6
Student Assignments

1) Write a Matlab function spatfreqnoise(I,k) that superimposes white noise with spatial frequency k on a gray-scale image I (assume a reasonable bandwidth of 1–2 octaves). Your results should look similar to the images in Figure 36.1. Now try an image that contains text. When trying several values of k for that

image, can you observe a similar masking effect with the letters as with face images?

2) Write a Matlab function $A = $ `ampspec1 d(I)` for computing the one-dimensional amplitude spectrum A of an image I according to formula 36.2. Now write another function $e = $ `line2ampspec(A)` that displays the one-dimensional spectrum A in a double-logarithmic representation and fits a line with slope e to it (cf Figure 36.2c). Try different "natural" images and compare their slopes. Scale invariance would imply $e = -1$. Are your slope values always settled around -1? Do you think that your values for e are still compatible with the notion of scale invariance, or would you instead prefer one of the alternative suggestions described in this chapter (occlusion or contrast edges, respectively)? Defend your proposal.

3) Show that Eq. (36.1) implies $g(k) \propto k^2$. This is the condition for having scale invariance.

4) Write a Matlab function `decorr(I,e)` for decorrelating an image I using the slope e. To do so, you may find it useful to decompose the Fourier spectrum of the image into its amplitude and phase. Decorrelate using "slope whitening" according to Eq. (36.3). Why can we assume that the e (computed with `line2ampspec(A)`) equals the exponent α?

5) What effect does decorrelation have on an image? Do you know a filter which, when convolved with an image, would give essentially the same result as decorrelation? How does the latter filter compare to the classical receptive fields of retinal ganglion cells? Would such a filter be a good idea if our image was noisy? How would you modify the filter kernel in the case of a noisy image? Do you think that the human retina faces similar "problems" and if so, under what conditions?

References

1. Sowden, P. and Schyns, P. (2006) Channel surfing in the visual brain. *Trends Cogn. Sci. (Regul. Ed.)*, **10**, 538–545.
2. Hubel, D. and Wiesel, T. (1978) Functional architecture of macaque monkey visual cortex. *Proc. R. Soc. London, B*, **198**, 1–59, ferrier Lecture.
3. Tieger, T. and Ganz, L. (1979) Recognition of faces in the presence of two-dimensional sinusoidal masks. *Percept. Psychophys.*, **26**, 163–167.
4. Fiorentini, A., Maffei, L., and Sandini, G. (1983) The role of high spatial frequencies in face perception. *Perception*, **12**, 195–201.
5. Hayes, A., Morrone, M., and Burr, D. (1986) Recognition of positive and negative band-pass filtered images. *Perception*, **15**, 595–602.
6. Costen, N., Parker, D., and Craw, I. (1996) Effects of high-pass and low-pass spatial filtering on face identification. *Percept. Psychophys.*, **58**, 602–612.
7. Näsänen, R. (1999) Spatial frequency bandwidth used in the recognition of facial images. *Vision Res.*, **39**, 3824–3833.
8. Gold, J., Bennett, P., and Sekuler, A. (1999) Identification of band-pass filtered letters and faces by human and ideal observers. *Vision Res.*, **39**, 3537–3560.
9. Ojanpää, H. and Näsänen, R. (2003) Utilisation of spatial frequency information in face search. *Vision Res.*, **43**, 2505–2515.

10. Tanskanen, T. et al. (2005) Face recognition and cortical responses show similar sensitivity to noise spatial frequency. *Cereb. Cortex*, **15**, 526–534.
11. Gaspar, C., Sekuler, A., and Bennett, P. (2008) Spatial frequency tuning of upright and inverse face identification. *Vision Res.*, **48**, 2817–2826.
12. Collin, C., Wang, K., and O'Byrne, B. (2006) Effects of image background on spatial-frequency threshold for face recognition. *Perception*, **35**, 1459–1472.
13. Ellis, H., Shepherd, J., and Davies, G. (1979) Identification of familiar and unfamiliar faces from internal and external face features: some implication for theories of face recognition. *Perception*, **8**, 431–439.
14. Young, A. et al. (1985) Matching familiar and unfamiliar faces on internal and external features. *Perception*, **14**, 737–746.
15. Cox, D., Meyers, E., and Sinha, P. (2004) Contextually evoked object-specific responses in human visual cortex. *Science*, **304**, 115–117.
16. Olshausen, B. and Field, D. (1996) Emergence of simple-cell receptive field properties by learning a sparse code for natural images. *Nature*, **381**, 607–609.
17. Atick, J. and Redlich, A. (1992) What does the retina know about natural scenes? *Neural Comput.*, **4**, 196–210.
18. Balboa, R. and Grzywacz, N. (2000) The minimum local-asperity hypothesis of early retinal lateral inhibition. *Neural Comput.*, **12**, 1485–1517.
19. Long, F., Yang, Z., and Purves, D. (2006) Spectral statistics in natural scenes predict hue, saturation, and brightness. *Proc. Natl. Acad. Sci. U.S.A.*, **103**, 6013–6018.
20. Clark, R. and Lott, R. (2009) Visual processing of the bee innately encodes higher-order image statistics when the information is consistent with natural ecology. *Vision Res.*, **49**, 1455–1464.
21. Karklin, Y. and Lewicki, M. (2009) Emergence of complex cell properties by learning to generalize in natural scenes. *Nature*, **457**, 83–86.
22. Attneave, F. (1954) Some informational aspects of visual perception. *Psychol. Rev.*, **61**, 183–193.
23. Sincich, L., Horton, J., and Sharpee, T. (2009) Preserving information in neural transmission. *J. Neurosci.*, **29**, 6207–6216.
24. Levy, W. and Baxter, R. (1996) Energy-efficient neural codes. *Neural Comput.*, **8**, 531–543.
25. Laughlin, S., de Ruyter van Steveninck, R., and Anderson, J. (1998) The metabolic cost of neural information. *Nat. Neurosci.*, **1**, 36–41.
26. Lenny, P. (2003) The cost of cortical computation. *Curr. Biol.*, **13**, 493–497.
27. Field, D. (1987) Relations between the statistics of natural images and the response properties of cortical cells. *J. Opt. Soc. Am. A*, **4**, 2379–2394.
28. Barlow, H. and Foldiak, P. (1989) Adaptation and decorrelation in the cortex, in *The Computing Neuron*, chapter 4 (eds R.R. Durbin, C. Miall, and G., Mitchinson), Addison-Wesley, New York, pp. 454–472.
29. Barlow, H. (1994) What is the computational goal of the neocortex? in *Large-Scale Neuronal Theories of the Brain* (eds C. Koch and J. Davis), MIT Press, Cambridge, MA, pp. 1–22.
30. Rolls, E. and Tovee, M. (1995) Sparseness of the neuronal representations of stimuli in the primate temporal visual cortex. *J. Neurophysiol.*, **73**, 713–726.
31. Olshausen, B. and Field, D. (2004) Sparse coding of sensory inputs. *Curr. Opin. Neurobiol.*, **14**, 481–487.
32. Lehky, S., Sejnowski, T., and Desimone, R. (2005) Selectivity and sparseness in the responses of striate complex cells. *Vision Res.*, **45**, 57–73.
33. Laughlin, S. and Sejnowski, T. (2003) Communication in neural networks. *Science*, **301**, 1870–1874.
34. Srinivasan, M., Laughlin, S., and Dubs, A. (1982) Predictive coding: a fresh view of inhibition in the retina. *Proc. R. Soc. London, B*, **216**, 427–459.
35. Hosoya, T., Baccus, S., and Meister, M. (2005) Dynamic predictive coding by the retina. *Nature*, **436**, 71–77.
36. Linsker, R. (1988) Self-organization in a perceptual network. *IEEE Trans. Comput.*, **21**, 105–117.
37. Linsker, R. (1992) Local synaptic learning rules are suffice to maximize mutual

information in a linear network. *Neural Comput.*, **4**, 691–702.

38. Wainwright, M. (1999) Visual adaptation as optimal information transmission. *Vision Res.*, **39**, 3960–3974.
39. Graham, D., Chandler, D., and Field, D. (2006) Can the theory of "whitening" explain the center-surround properties of retinal ganglion cell receptive fields? *Vision Res.*, **46**, 2901–2913.
40. Carlson, C. (1978) Thresholds for perceived image sharpness. *Photogr. Sci. Eng.*, **22**, 69–71.
41. Burton, G. and Moorhead, I. (1987) Color and spatial structure in natural scenes. *Appl. Opt.*, **26**, 157–170.
42. Wiener, N. (1964) *Extrapolation, Interpolation, and Smoothing of Stationary Time Series*, The MIT Press, Cambridge, MA.
43. Tolhurst, D., Tadmor, Y., and Chao, T. (1992) Amplitude spectra of natural images. *Ophthal. Physiol. Opt.*, **12**, 229–232.
44. van Hateren, H. (1992) Theoretical predictions of spatiotemporal receptive fields of fly LMCs, and experimental validation. *J. Comp. Physiol. A*, **171**, 157–170.
45. Ruderman, D. and Bialek, W. (1994) Statistics of natural images: scaling in the woods. *Phys. Rev. Lett.*, **73**, 814–817.
46. Bullock, V. (2000) Neural acclimation to 1/f spatial frequency spectra in natural images transduced by the human visual system. *Physica D*, **137**, 379–391.
47. Langer, M. (2000) Large-scale failures of f^α scaling in natural image spectra. *J. Opt. Soc. Am. A*, **17**, 28–33.
48. Keil, M. (2003) Neural architectures for unifying brightness perception and image processing. Ph.D. thesis. Universität Ulm, Faculty for Computer Science, Ulm, Germany, http://vts.uni-ulm.de/doc.asp?id=3042.
49. Keil, M. (2006) Smooth gradient representations as a unifying account of Chevreul's illusion, Mach bands, and a variant of the Ehrenstein disk. *Neural Comput.*, **18**, 871–903.
50. Keil, M., Cristóbal, G., and Neumann, H. (2006) Gradient representation and perception in the early visual system - a novel account to Mach band formation. *Vision Res.*, **46**, 2659–2674.
51. Ruderman, D. (1997) Origins of scaling in natural images. *Vision Res.*, **37**, 3385–3398.
52. Ruderman, D. (1994) The statistics of natural images. *Network: Comput. Neural Syst.*, **5**, 517–548.
53. Turiel, A., Parga, N., and Ruderman, D. (2000) Multiscaling and information content of natural color images. *Phys. Rev. E*, **62**, 1138–1148.
54. Balboa, R. and Grzywacz, N. (2001) Occlusions contribute to scaling in natural images. *Vision Res.*, **41**, 955–964.
55. Kuffler, S. (1953) Discharge patterns and functional organization of mammalian retina. *J. Neurophysiol.*, **16**, 37–68.
56. Rodieck, R.W. (1965) Quantitative analysis of cat retinal ganglion cell response to visual stimuli. *Vision Res.*, **5**, 583–601.
57. Field, D. and Brady, N. (1997) Wavelets, blur and the sources of variability in the amplitude spectra of natural scenes. *Vision Res.*, **37**, 3367–3383.
58. Brady, N. and Field, D. (2000) Local contrast in natural images: normalisation and coding efficiency. *Perception*, **29**, 1041–1055.
59. Keil, M. (2008) Does face image statistics predict a preferred spatial frequency for human face processing? *Proc. R. Soc. London, B*, **275**, 2095–2100.
60. Harris, F. (1978) On the use of windows for harmonic analysis with the discrete Fourier transform. *Proc. IEEE*, **66**, 51–84.
61. Keil, M. (2009) "I look in your eyes, honey": internal face features induce spatial frequency preference for human face processing. *PLoS Comput. Biol.*, **5**, e1000329.
62. Keil, M. *et al.* (2008) Preferred spatial frequencies for human face processing are associated with optimal class discrimination in the machine. *PLoS ONE*, **3**, e2590. DOI:10.1371/journal.pone.0003590.
63. Meytlis, M. and Sirovich, L. (2007) On the dimensionality of face space. *IEEE Trans. Pattern Anal. Mach. Intell.*, **29**, 1262–1267.
64. Leopold, D., Bondar, I., and Giese, M. (2006) Norm-based face encoding by

single neurons in the monkey inferotemporal cortex. *Nature*, **442**, 572–575.
65. Loeffler, G. *et al.* (2005) fMRI evidence for the neural representation of faces. *Nat. Neurosci.*, **8**, 1386–1391.
66. Valentine, T. (1991) A unified account of the effects of the distinctiveness, inversion, and race in face recognition. *Q. J. Exp. Psychol.*, **43**, 161–204.
67. Goldstein, A. and Chance, J. (1980) Memory for faces and schema theory. *J. Psychol.*, **105**, 47–59.
68. Kirby, M. and Sirovich, L. (1990) Application of the Karhunen-Loeve procedure for the characterization of human faces. *IEEE Trans. Pattern Anal. Mach. Intell.*, **12**, 103–108.
69. Turk, M. and Pentland, A. (1991) Eigenfaces for recognition. *J. Cogn. Neurosci.*, **3**, 71–86.
70. Jenkins, R. and Buron, A. (2008) 100% automatic face recognition. *Science*, **319**, 425.

37
Image Processing for Spacecraft Optical Navigation
Michael A. Paluszek and Pradeep Bhatta

37.1
Introduction

Spacecraft navigation has been done primarily by means of range and range rate measurements from ground-based antennae. This has proven to be very successful and reliable. The basis of these measurements is that the range

$$\rho = \sqrt{x^2 + y^2 + z^2} \tag{37.1}$$

and the range rate

$$\dot{\rho} = \frac{v^T r}{\rho} \tag{37.2}$$

contain information about all of the spacecraft states: the position in earth-centered-inertial (ECI) reference frame, $r = (x, y, z)$ and velocity $v = \dot{r}$. If the geometry changes sufficiently as measurements are taken, all six states can be recovered. Once the state is reasonably well known, a recursive estimator can track the state with single range and range rate measurements. However, since typically knowledge of the spacecraft state is only required for occasional course correcting maneuvers, a batch estimator on the ground is often used.

The major drawback with this approach is that it ties up ground antennae and facilities, thus causing it to be expensive. Furthermore, this method is not robust to loss of contact, space segment degradations, and other factors. Additionally, if frequent navigation solutions are required, it may not be possible to schedule the observations when needed. Some proposed missions, including solar sails, spacecraft with continuous propulsion, and formation flying missions require frequent updates. During the terminal approach in a rendezvous, it may be necessary to update the navigation solution at a very high rate. For this reason, optical navigation has been proposed for spacecraft navigation.

The process of optical navigation is shown in Figure 37.1. Each step in the process introduces errors that must be accounted for while computing the navigation solution. Not all navigation systems have all of these elements. For example, many

---

Optical and Digital Image Processing: Fundamentals and Applications, First Edition. Edited by Gabriel Cristóbal, Peter Schelkens, and Hugo Thienpont.
© 2011 Wiley-VCH Verlag GmbH & Co. KGaA. Published 2011 by Wiley-VCH Verlag GmbH & Co. KGaA.

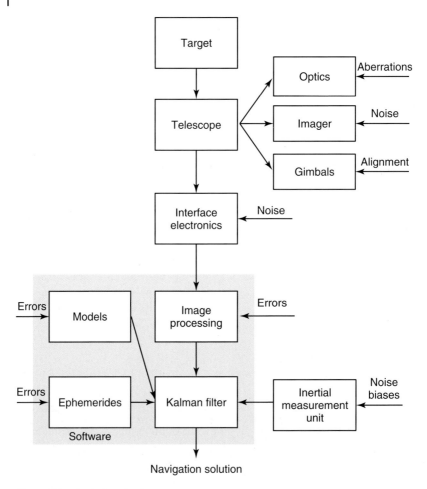

Figure 37.1 Optical navigation process.

systems use a telescope fixed to the spacecraft and do not have gimbals. An inertial measurement unit, which measures acceleration, is not used in all systems.

The position can be determined from the angle between a star and the vector to a landmark. This approach was used on Apollo [1] as a backup to the conventional range and range rate measurements, which used the telemetry, tracking, and control system. The range and range rate measurements were obtained from antennae on the ground.

The first major attempt to shift navigation from ground-based to onboard operations was made by NASA during the Deep Space 1 (DS-1) mission [2, 3] in 1999. During this mission, autonomous navigation was operational during the cruise phase, until the failure of the onboard star tracker. The principle adopted by DS-1 for the determination of autonomous orbit was optical triangulation.

Reduced state encounter navigation (RSEN), using optical images as the sole data type, was used on DS-1 to track comet Borrelly for 2 hours through closest approach. RSEN is initialized using the last ground or onboard estimate of spacecraft state relative to the target just before commencement of the final approach phase. RSEN was also used in the Stardust mission during comet Wild-2 flyby, and in the Deep Impact mission to navigate the impactor to hit an illuminated area on the comet Tempel 1 and by the flyby spacecraft to track the impact site.

The Japanese asteroid return mission Hayabusa [4, 5] employed wide-angle cameras for onboard navigation, in conjunction with light radio detecting and ranging (LIDAR) for the measurement of altitude with respect to the asteroid Itokawa. A target marker was dropped to act as a navigation aid by posing as an artificial landmark on the surface. A laser range finder was used to measure the altitude as the spacecraft got closer (less than 120 m) to the asteroid. Fan beam sensors detected potential obstacles that could hit the solar cell panels.

The European Space Agency's Small Missions for Advanced Research in Technology (SMART-1) mission employed the Advanced Moon Imaging Experiment (AMIE) camera for Onboard Autonomous Navigation (OBAN) [6, 7]. The moon and Earth were used as beacons. The beacons were too bright to enable observation of stars with the AMIE camera. A star tracker was used to estimate pointing direction. The OBAN framework is also being used for designing a vision-based autonomous navigation system for future manned Mars exploration missions [8].

The Near Earth Asteroid Rendezvous (NEAR) Shoemaker mission relied heavily on optical navigation [9]. The system located small craters on the surface of EROS and used them as references. This enabled rapid determination of the orbit near and around EROS. The more traditional approach of using the planet centroids was not feasible due to the irregular shape of EROS. The planet centroid technique was used on Mariner 9, Viking 1 and 2, Voyager 1 and 2, and Galileo. It was also used by NEAR on its approach to the asteroid Mathilde. More recently, this approach was used on the Mars Reconnaissance Orbiter (MRO).

37.2
Geometric Basis for Optical Navigation

Figure 37.2 shows the geometry of optical navigation in two dimensions. The angles θ_k are three possible types of angle measurements. θ_1 is the angle between two planets or two landmarks on the two planets. θ_2 is the angle between the two landmarks on a single planet or the chord width of the planet. θ_3 is the angle between a landmark vector and a star. The measurement equations are given below:

1) Angle between the two planets:

$$\cos\theta_1 = \frac{\rho_1^T \rho_2}{|\rho_1||\rho_2|} \qquad (37.3)$$

$$= \frac{(r+l_1)^T(r+l_2)}{|(r+l_1)||(r+l_2)|} \qquad (37.4)$$

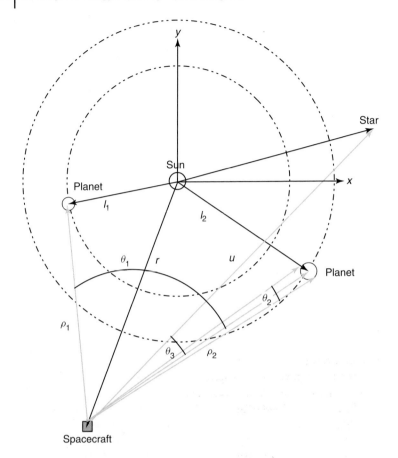

Figure 37.2 Measurement geometry.

where ρ_i is the vector from the spacecraft to either the planet i's landmark location or the planetary center, l_i is the vector from the Sun to either the planet i's landmark location or the planetary center, and r is the vector from the spacecraft to the Sun.

2) **Angle between landmarks on a planet or the angle subtended by the planetary chord:**

$$\sin\frac{\theta_2}{2} = \frac{a}{|l - r|} \qquad (37.5)$$

where a is either the planet radius or half the distance between the two landmarks, and l is the vector from the Sun to either the planet center or the center of the straight line joining the two landmarks.

3) **Angle between a landmark vector and a star:**

$$\cos\theta_3 = \frac{u^T(r - l)}{|r - l|} \qquad (37.6)$$

where u is the vector from the spacecraft to a reference star.

To compute the navigation solution, it is necessary to have at least one measurement of type θ_1 or θ_2, and two other measurements of any type. While these three measurements form the minimal required set, more measurements (whose statistics are known accurately) generally lead to better accuracy in the navigation solution.

37.2.1
Example: Analytical Solution of r

If we have one measurement of type θ_2 and three measurements of type θ_3, it is possible to compute the navigation position vector r analytically. Let θ_{21} be the angle subtended by the planetary chord, and θ_{3i} ($i = 1,2,3$) be angles between the vector from the center of the planet to the spacecraft and three stars. The range fix is found from

$$\rho = |r - l| = \frac{a}{\sin \frac{\theta_{21}}{2}} \tag{37.7}$$

Using ρ, θ_{3i} measurements, and the corresponding u_i vectors, r_x, r_y, and r_z from Eq. (37.6):

$$\begin{bmatrix} u_{1x} & u_{1y} & u_{1z} \\ u_{2x} & u_{2y} & u_{2z} \\ u_{3x} & u_{3y} & u_{3z} \end{bmatrix} \begin{bmatrix} r_x \\ r_y \\ r_z \end{bmatrix} = \begin{bmatrix} \rho \cos(\theta_{31}) + u_1^T l \\ \rho \cos(\theta_{32}) + u_2^T l \\ \rho \cos(\theta_{33}) + u_3^T l \end{bmatrix} \tag{37.8}$$

37.3
Optical Navigation Sensors and Models

37.3.1
Optics

The lowest order approximation to the optical system is a pinhole camera, as shown in Figure 37.3. The pinhole camera has a single ray per point on the target and that ray maps to a point on the focal plane.

A point $P(X, Y, Z)$ is mapped to the imaging plane by the relationships

$$u = \frac{fX}{Z}, \quad v = \frac{fY}{Z} \tag{37.9}$$

where u and v are coordinates in the focal plane, f is the focal length, and Z is the distance from the pinhole to the point along the axis normal to the focal plane. This assumes that the Z-axis of the coordinate frame X, Y, Z is aligned with the boresight of the camera.

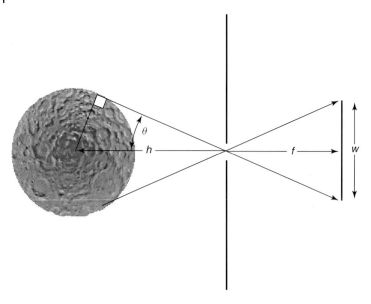

Figure 37.3 Pinhole camera.

In Figure 37.3, we see that a spherical object, whose center is at a distance h, is mapped to an angle θ seen by the imaging

$$\theta = \tan^{-1}\left(\frac{w}{2f}\right) \tag{37.10}$$

where f is the focal length. The above equation determines the relationship between the measured angles and the actual physical measurement. The shorter the focal length, the larger is the image and the corresponding angle.

37.3.2
Imaging System Resolution

The resolution of the imaging system will affect the accuracy of optical navigation. There are two independent factors. One is the aperture size and the other is the pixel size.

The angular resolution α_D of a telescope is limited by the aperture. The Rayleigh criterion for the resolution limit is given by

$$\alpha_D = 1.22\frac{\lambda}{D} \tag{37.11}$$

where D is the aperture diameter and λ is the wavelength of the light.

This limit is based on the first zero crossing of the Airy pattern for a point source (such as a star). It is possible to separate points that are closer than this value if the individual pixels are close enough together and the individual pixels have sufficient dynamic range. If multiple frames are combined, then the resolution can be improved using super-resolution techniques. The spatial resolution is

$$r_D = f\alpha_D \tag{37.12}$$

The image size w of an object of angular size θ is

$$w = f\theta \tag{37.13}$$

If the sensor has p pixels per unit length, then the angular resolution θ_p of each pixel is

$$\theta_p = f\theta/p \tag{37.14}$$

37.3.3
Basic Radiometry

Figure 37.4 illustrates the basic radiometric concepts. The simplest illumination model is the Lambertian surface reflection model:

$$L = \rho I^T n \tag{37.15}$$

where n is the normal to the surface, I is the vector to the illumination source, and $0 \le \rho \le 1$ is the surface albedo.

According to the Lambertian model, the Lambertian surface reflects light in a given direction d proportionally to the cosine of the angle between d and n. However, as the surface area seen from direction d is inversely proportionally to the angle between d and n, the angles cancel and d does not explicitly appear in the equation. In Figure 37.4, we have shown one source (the Sun); however, reflections

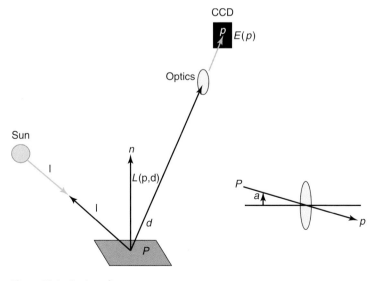

Figure 37.4 Basic radiometry.

from other surfaces will also be sources, thus complicating the illumination model. The energy received from the object is

$$E(p) = L(P) \frac{\pi}{4} \left(\frac{a}{f}\right)^2 \cos^4 \alpha \qquad (37.16)$$

where a is the diameter of the aperture and f is the focal length. α is the angle between the incoming light and the optical axis. f/a is the F-number of the optical system. For most optical systems, α is small, so the last factor is nearly 1. $E(p)$ is in units of watts per meter square, while $L(p)$ is in units of watts per meter square streadian.

Solar system bodies will be illuminated by the Sun. Table 37.1 lists the albedo and radiation reflected from each body at their mean distance from the Sun. The bond albedo is the ratio of the total radiation reflected to the incident radiation from the Sun. The geometric albedo is the fraction of the radiation relative to that from a flat Lambertian surface, which is an ideal reflector at all wavelengths.

The Sun's radiation (in W/m^2) is

$$p = \frac{1367}{r^2} \qquad (37.17)$$

where r is distance from the Sun in astronomical units.

37.3.4
Imaging

Four possible imaging technologies are used for imaging and star trackers:

1) CMOS (complementary metal oxide semiconductor)
2) APS (CMOS active pixel sensor)

Table 37.1 Reflected light from planets and moons.

Planet	Geometric albedo	Bond albedo	Albedo flux based on the bond albedos (W m^{-2})
Mercury	0.138	0.056	510.873
Venus	0.84	0.720	1881.160
Earth	0.367	0.390	533.130
Mars	0.15	0.160	94.213
Jupiter		0.700	35.343
Saturn		0.750	11.272
Uranus		0.900	3.340
Neptune		0.820	1.240
Pluto	0.44–0.61	0.145	0.127
Europa		0.670	33.828
Triton		0.760	1.149
Moon	0.113	0.120	164.040
Titan		0.210	3.156

3) CCDs (charge-coupled devices)
4) CIDs (charge injection devices).

CMOS are used in a number of commercial cameras and are also used in the advanced pixel sensor (APS). CCD imaging is used in spacecraft star trackers, spacecraft cameras, terrestrial astronomical cameras, and consumer products. CIDs are used in military star trackers.

In a CCD, each pixel captures photons that become hole–electron pairs. When the exposure time is reached, the electrons are transferred to an adjacent pixel and, eventually, to a readout register where they are converted to a digital signal. Some electrons are lost with each transfer. A CCD can be considered a hybrid device – the CCD sensor is analog and the readout is digital. A CCD requires complex timing and multiple clocks. CMOS and APS sensors image the same way, except that each pixel is converted to a digital signal locally. Unlike CCDs, which require multiple clocks, these require only a single clock and power supply. CCD sensors cannot produce color images except by having filtered pixels (which leads to large losses). CMOS can measure color by placing photodetectors at different depths in the material. An APS is a CMOS sensor with amplification at each pixel.

CCDs typically have relatively low noise levels [10]. Fixed pattern noise (FPN) is less than 1% of the saturation level and the photon response nonuniformity (PRNU) is less than 10% of the saturation level. The dark current is typically less than 10 pA cm^{-2}. The best CMOS has dark currents of 100 pA cm^{-2}. The fill factor for a CCD sensor is close to 100%. For an APS CMOS sensor it is around 55%. This means that a CCD will absorb 3 times as many photons as a CMOS device. For the same quantum efficiency (ratio of electrons to photons), the CCD chip will be 3 times as sensitive.

A CMOS device, since it is completely digital, can be fabricated on a single chip with additional data processing. CMOS is available as cameras on a chip with built in USB 2.0 and signal processing. CCDs require three different voltages, while CMOS operate at +5 V, the same as other electronics.

CIDs have some of the advantages of CMOS and CCD. Pixels can be addressed individually and the devices are inherently radiation hard. However, there are few producers of CID.

37.3.4.1 Noise and Performance Factors

Fill factor: Fill factor is the percentage of the sensor that is active. In a CCD, this is nearly 100%. In APS or CMOS sensors, this is considerably smaller owing to the additional support circuits. In addition, the sensing elements in an APS or CMOS chip occupy only a fraction of the pixel. The location of the sensing area must be factored into the centroiding algorithm; otherwise, a bias will result.

Dynamic range: This is the range of photon fluxes that can be accommodated without blooming.

Blooming: When a photelement is subject to strong illumination, it can overflow and spill charges into adjacent photoelements if their is a direct connection between photoelements as there is in a CCD. CMOS, APS, and CID sensors do not have such a path and are immune to blooming. Another effect is that the charge cannot be emptied in one transfer, leaving a "trail" of charges in that element. These leftover charges must be removed during successive readouts, but it means that each readout will be contaminated. CCDs can employ antiblooming devices to drain the overflow. Some of the excess charge will still remain. In addition, these devices take up space on the chip, reducing its overall light-gathering capability. Typically, such devices reduce the sensitivity by 10%.

Quantum efficiency: Quantum efficiency is the percentage of photons hitting a detector that produces a hole–electron pair output. CCDs have quantum efficiencies of over 90% at some wavelengths, making them well suited for low light (or deep sky) imaging. CMOS have a quantum efficiency of around 30%.

Dark current: Dark current is the constant output of a detector when it is not illuminated. The dark current will appear in every exposure. As it is constant, it can be measured and subtracted. This is best done on orbit. CCDs have very low dark current, as low as 1 electron per second at $-30\,°C$. CMOS cells are considerably higher. Part of this is due to the manufacturing of CMOS sensors in standard CMOS fabrication plants. CCD sensors that are used for imaging are fabricated on fabs specialized for imaging chips. As CMOS use increases, dark current will improve. With all sensors, dark current can be improved by cooling. For spacecraft applications and many terrestrial applications, thermoelectric coolers are used for imaging chip cooling. These are solid-state devices and can be integrated into the other sensor electronics.

The rate of dark current creation is [11]

$$S = AT^{3/2} e^{-\frac{V_g q}{2kT}} \tag{37.18}$$

where V_g is the gap voltage, T is the temperature in Kelvin, q is the electron's charge (1.6×10^{-19} C), and k is Boltzmann's constant (1.38×10^{-23} J K^{-1}); A is 3.36×10^6 when S is in volts. For a typical detector, V_g is 1.36 V. To convert this to a number of electrons,

$$e = \frac{St}{Cq} \tag{37.19}$$

where C is the imaging chip's capacitance since

$$S = C\frac{dV}{dt} \tag{37.20}$$

If the dark current is known, it can be subtracted. However, this requires measurement of the current in the dark for each pixel. This is known as a *thermal map*. This equation shows it is advantageous to cool the chip, which can be done with a thermoelectric cooler. There are limits to the extent to which the chip can be cooled since the spectral sensitivity changes. Charges of thermal origin are created in the imaging elements and transfer elements.

Fixed pattern noise: FPN is noise over the entire array that occurs in the same pattern on every exposure.

Readout noise: This is the noise added to a pixel output on reading it out.

Radiation hardness: Radiation hardness is the ability of the chip to withstand long-term and prompt radiation dosages. Prompt dosages can cause single event upsets, in which a switch is set incorrectly, for example, a readout switch. A greater problem is latchup, which requires power to be cycled to restore the functioning of the chip. Long-term damage can increase the noise and reduce the quantum efficiency.

Cosmic rays: A cosmic ray impact will cause the generation of charge in a pixel. Cosmic rays are indistinguishable from stars. While not an issue with planetary imaging, this can cause the sensor to misidentify stars. The simplest method is to take two images and compare them. The cosmic ray image will only appear in one of the frames. The star identification software will generally also ignore "false stars," particularly when the sensor is tracking known stars but the safest method is the two-frame method.

Thermal noise: Thermal noise is a function of the level of dark current. If N_t is the number of thermal charges created by the dark current, then

$$\sigma_o = \sqrt{N_t} \tag{37.21}$$

Transfer efficiency: Every time charge is transferred noise is created. The noise created is

$$\sigma_e = \sqrt{2 \epsilon n N_s} \tag{37.22}$$

where ϵ is the transfer efficiency, n is the number of transfers, and N_s is the charge transferred that includes thermal electrons. For an APS, a CID, or a CMOS sensor, $n = 1$. For a CCD, n is the number of cells between the cell of interest and the readout.

Reset noise: At the output of a CCD register, the floating readout diode is brought to a reference voltage before a packet is read out. This creates "reset noise," which is

$$\sigma_r = \frac{1}{q}\sqrt{kTC} \tag{37.23}$$

where C is the capacitance of the output diode. This noise can be eliminated by double correlated sampling.

Photon noise: Photon noise is a result of photons hitting the chip and is

$$\sigma_s = \sqrt{QN_p} \tag{37.24}$$

where Q is the quantum efficiency and N_p is the number of photons hitting the chip.

Quantization noise: Quantization noise is

$$\sigma_q = \frac{N}{2^n \sqrt{12}} \tag{37.25}$$

where N is the maximum number of electrons per pixel (full-well capacity) at the imaging chip's readout and n is the number of bits of the analog-to-digital converter.

Total noise: The total noise is found by summing the squares of the noise sources

$$\sigma_t = \sqrt{\sigma_o^2 + \sigma_e^2 + \sigma_r^2 + \sigma_s^2 + \sigma_q^2} \tag{37.26}$$

Linearity: The response curve of a photoelement is not entirely linear. At high levels, the elements saturates. At low levels, the response is not proportional to the input. In most photoelements, the nonlinearity at low signal levels is negligible. For example, the Cypress STAR 1000 sensor has a full-well capacity of 135 000 electrons but is linear to within 1% only with 95 000 electrons.

Amplifier noise: CMOS image sensors suffer from higher noise than CCDs because of the additional pixel and column amplifier transistor noise.

37.3.4.2 Data Reduction

The following steps are taken to reduce data for an imaging chip [12]. Not all are applicable to CMOS. Several of the steps either require calibration with uniform sources or no illumination.

1) Bias subtraction: This is the removal of a fixed voltage to all elements. This is done by the subtraction of a bias frame.
2) Dark current correction: This is the removal of the dark current. This is done by measuring the elements at the same temperature at which the exposure is taken and subtracting the dark frame. Hole accumulation diodes (HADs), patented by Sony, can reduce the dark current at the source.
3) Flat-field correction: Several images are taken under uniform illumination. These frames are averaged to produce the flat-field frame in which the frame contains the variations due to chip imperfections. The corrected frames (from the two steps above) are divided by this frame.
4) Cosmic ray correction: This is done by taking several images of the same scene and removing artifacts that do not appear in all of the images.

5) Saturation trail: If a pixel is saturated, the extra electrons will spill along the row. Any narrow feature along a row should be removed by postprocessing. This is a problem with CCD and does not appear in CMOS.
6) Diffraction features: Support structure, baffles, and other objects can cause diffraction features. These patterns can cause errors in centroiding. These errors can be minimized by employing a centroiding algorithm that uses the point-spread function for a point source, which will then ignore the spikes seen in the picture. This is also reduced if the image is defocused as only sharp objects have distinctive diffraction artifacts.
7) Bad column: In a CCD, if a column is bad, it will appear black and will need to be removed from the image. Any objects that intersect the bad column will be difficult to centroid. If a bad column is detected, then the telescope should be pointed so that the objects of interest (e.g., an asteroid) do not appear in that column. This does not happen in CMOS.
8) Trap: In a CCD, a damaged pixel may prevent reading of any pixels below that pixel. This will create artifacts in the image, which must be avoided. This does not happen in CMOS.
9) Dust: Dust on the lens will appear as constant dark areas. The flat-field correction will partially correct for these artifacts.
10) Photometric correction: Using objects of know brightness, this correction converts electrons produced by the detectors to flux.

37.3.4.3 Error Modeling

Using optical sensors for navigation requires knowledge of the error sources. They can be broadly classified as follows:

1) measurement source errors, that is, how well we know the physical quantities (e.g., planet radius);
2) alignment and placement errors of the instrument on the spacecraft;
3) alignment and placement of the telescope on the instrument;
4) optical errors such as aberrations;
5) imaging sensor errors; and
6) modeling errors in the software.

The accuracy of lunar landmarks from different missions is listed in Table 37.2 [13]. Accuracy and precision refer to measurement deviation from the actual value and its scatter, respectively. The error budget for the planets, Pluto, and the Sun are included in Table 37.3 [14–17].

37.4
Dynamical Models

The central body equations for orbital dynamics in Cartesian coordinates are

$$\ddot{r} + \mu \frac{r}{|r|^3} = a \qquad (37.27)$$

Table 37.2 Table of lunar landmark accuracies.

Mission	Horizontal accuracy	Vertical precision (m)	Vertical accuracy (m)
Apollo 15,16,17	30 km	4	400
Clementine	3 km	40	90
SLA-01	40 m	0.75	2.78
SLA-02	40 m	0.75	6.74
NLR	20 m	0.31	10
MOLA	100 m	0.38	1

Table 37.3 Table of errors.

Body	Mean radius (km)	Radius error (km)	Ephemeris error (km)
Mercury	2 439.7	± 1	1–2
Venus	6 051.8	± 1	1–2
Earth	6 371.00	± 0.01	1–2
Mars	3 389.50	± 0.1	1–2
Jupiter	69 911	± 6	100
Saturn	58 232	± 6	600–1 200
Uranus	25 362	± 7	1 250–2 500
Neptune	24 622	± 19	2 250–4 500
Pluto	1 151	± 6	2 800–5 600
Sun	695 690	± 140	–

where r is the position vector from the mass center of the central body. a is the acceleration vector of all other disturbances (such as thrusters, solar pressure, and gravitational accelerations due to other planets and planetary asymmetries), and μ is the gravitational parameter for the central body. The radius vector is

$$r = \begin{bmatrix} x \\ y \\ z \end{bmatrix} \tag{37.28}$$

and the radius magnitude is $|r| = \sqrt{x^2 + y^2 + z^2}$. The equations are nonlinear. However, when a is zero, r traces ellipses, parabolas, or hyperbolas. This equation is sufficient for the purposes of this chapter and we do not discuss perturbative accelerations.

37.5
Processing the Camera Data

The camera data are a pixel map containing an array of numbers from 0 to the maximum size of the word representing the pixel intensity. For example, with an 8-bit camera, this would be 255. The algorithms in this chapter require the identification of

- chord widths
- points on a planet as landmarks
- centroids.

Chord widths and centroids require that a fit be made to the shape of the planet that is independent of the lighting conditions of the planet. The amount of the planet seen will depend on the angle between the Sun, the planet target, and the spacecraft. The phases of the moon as seen from Earth are a result of this geometry. The general process is

- binary transform
- edge detection using a Canny filter
- fitting a circle or ellipsoid to a curve found through the edge detection.

Autotracking of landmarks is discussed further in Ref. [18]. This is complex combination of premission identification of landmarks and autonomous landmark identification and tracking.

37.6
Kalman Filtering

37.6.1
Introduction to Kalman Filtering

Given a physical system that can be modeled in the form

$$\dot{x} = f(x, u, t) \tag{37.29}$$
$$y = h(x, t)$$

the goal is to get the best possible estimate of x. If we assume that the noise sources are Gaussian and that the plant and measurement function can be linearized

$$\dot{x} = ax + bu \tag{37.30}$$
$$y = Hx$$

we get the standard Kalman filter algorithm

$$\hat{x}_k^- = \Phi \hat{x}_{k-1} + \Gamma u_k \tag{37.31}$$

$$P_k^- = \Phi P_{k-1} \Phi^T + Q \tag{37.32}$$

$$P_k^- = \frac{1}{2}(P_k^- + P_k^{-T}) \tag{37.33}$$

$$K = P_k^- H^T (H P_k^- H^T + R)^{-1} \tag{37.34}$$

$$\hat{x}_k = x_k^- + K(z_k - H\hat{x}_k^-) \tag{37.35}$$

$$P_k = (E - KH) P_k^- \tag{37.36}$$

$$P_k = \frac{1}{2}(P_k + P_k^T) \tag{37.37}$$

where Φ is the state transition matrix found by discretizing the dynamical equations, Γ is the input matrix also found as part of the discretizing process, z_k is the measurement vector at the kth time step, H is the measurement matrix, P_k is the state covariance matrix estimate at the kth time step, $\hat{x}_k$ is the state vector estimate at the kth time step, u_k is the vector of known inputs such as thruster firings estimate at the kth time step, Q is the plant noise covariance matrix, and E is the identity matrix. $\hat{x}_k^-$ and P_k^- are intermediate predictions of the state vector and state covariance matrix. Averaging the covariance matrix with its transpose (Eqs (37.33) and (37.37)) stabilizes the covariance propagation and eliminates the largest error term due to round off. Q encompasses both plant uncertainty and unmodeled disturbances. The Kalman filter algorithm is initialized using initial estimates of the state vector and the covariance matrix.

If the measurement matrix is nonlinear, the measurement equation can be replaced by the iterated extended Kalman filter (EKF). The recursive least squares method is implemented with an iterated EKF measurement update. The equation is

$$x_{k,i+1} = x_{k,0} + K_{k,i} \left[y - h(x_{k,i}) - H(x_{k,i})(x_{k,0} - x_{k,i}) \right] \tag{37.38}$$

$$K_{k,i} = P_{k,0} H^T(x_{k,i}) \left[H(x_{k,i}) P_{k,0} H^T(x_{k,i}) + R_k \right]^{-1}$$

$$P_{k,i+1} = \left[1 - K_{k,i} H(x_{k,i}) \right] P_{k,0}$$

where i denotes the iteration. Note that on the right-hand side P and R are always from step k and are not updated during the iteration.

37.6.2
The Unscented Kalman Filter

The unscented Kalman filter (UKF) is able to achieve greater estimation performance than the EKF by using the unscented transformation (UT). The UT allows the UKF to capture first- and second-order terms of the nonlinear system [19]. The state estimation algorithm employed is the one given by van der Merwe [20] and the parameter estimation algorithm that given by VanDyke et al. [19]. Unlike the EKF, the UKF does not require any derivatives or Jacobians of either the state equations or measurement equations. Instead of just propagating the state, the filter propagates the state and additional sigma points that are the states plus the square roots of rows or columns of the covariance matrix. Thus, the state and the state plus a standard deviation are propagated. This captures the uncertainty in the state. It is not necessary to numerically integrate the covariance matrix.

Two algorithms are presented in this section. The first is for the state estimation problem and the second is for the parameter estimation problem. The former assumes a nonlinear dynamical model. The latter assumes that the plant is represented by a constant, that is, the parameters do not change. This greatly simplifies the parameter estimation algorithms. Both algorithms can also be implemented as square root filters to reduce the computational load.

Both algorithms use the numerical simulation to propagate the state. The visibility of parameters will depend on the values of the system states and the dynamical coupling between states.

Weights: Both the state and parameter estimators use the same definitions for the weights:

$$W_0^m = \frac{\lambda}{L+\lambda} \tag{37.39}$$

$$W_0^c = \frac{\lambda}{L+\lambda} + 1 - \alpha^2 + \beta \tag{37.40}$$

$$W_i^m = W_i^c = \frac{1}{2(L+\lambda)} \tag{37.41}$$

where

$$\lambda = \alpha^2(L+\kappa) - L \tag{37.42}$$

$$\gamma = \sqrt{L+\lambda} \tag{37.43}$$

The constant α determines the spread of the sigma points around $\hat{x}$ and is usually set to between 10^{-4} and 1. β incorporates prior knowledge of the distribution of x and is 2 for a Gaussian distribution. κ is set to 0 for state estimation and $3 - L$ for parameter estimation. Care must be taken so that the weights are not negative.

Initialize the parameter filter with the expected value of the parameters: [20]

$$\hat{x}(t_0) = E\{\hat{x}_0\} \tag{37.44}$$

where E is expected value and the covariance for the parameters

$$P_{x_0} = E\{(x(t_0) - \hat{x}_0)(x(t_0) - \hat{x}_0)^T\} \tag{37.45}$$

The sigma points are then calculated. These are points found by adding the square root of the covariance matrix to the current estimate of the parameters:

$$x_\sigma = \begin{bmatrix} \hat{x} & \hat{x} + \gamma\sqrt{P} & \hat{x} - \gamma\sqrt{P} \end{bmatrix} \tag{37.46}$$

If there are L states, the P matrix is $L \times L$, so this array will be $L \times (2L+1)$.

The state equations are of the form

$$\chi_i = f(x_{\sigma,i}, u, w, t) \tag{37.47}$$

where χ is found by integrating the equations of motion.

The mean state is

$$\hat{x} = \sum_{i=0}^{2L} W_i^m \chi_i \qquad (37.48)$$

The state covariance is found from

$$\hat{P} = \sum_{i=0}^{2L} W_i^c (\chi_i - \hat{x})(\chi_i - \hat{x})^T \qquad (37.49)$$

Before incorporating the measurement, it is necessary to recompute the sigma points using the new covariance:

$$\chi_\sigma = \begin{bmatrix} \hat{x} & \hat{x} + \gamma\sqrt{P} & \hat{x} - \gamma\sqrt{P} \end{bmatrix} \qquad (37.50)$$

The expected measurements are

$$y_\sigma = h(x_\sigma) \qquad (37.51)$$

The mean measurement is

$$\hat{y} = \sum_{i=0}^{2L} W_i^m y_{\sigma,i} \qquad (37.52)$$

The covariances are

$$P_{yy} = \sum_{i=0}^{2L} W_i^c (y_{\sigma,i} - \hat{y})(y_{\sigma,i} - \hat{y})^T \qquad (37.53)$$

$$P_{xy} = \sum_{i=0}^{2L} W_i^c (\chi_i - \hat{x})(y_{\sigma,i} - \hat{y})^T \qquad (37.54)$$

The Kalman gain is

$$K = P_{xy} P_{yy}^{-1} \qquad (37.55)$$

The state update is

$$x = x + K(y - \hat{y}) \qquad (37.56)$$

where y is the actual measurement and the covariance update is

$$P = P - K P_{yy} K^T \qquad (37.57)$$

y must be the measurement that corresponds to the time of the updated state x.

37.7
Example Deep Space Mission

We demonstrate optical navigation using an example based on the trajectory of the NASA New Horizons spacecraft. Figure 37.2 shows the geometry of navigation. As discussed in Section 37.2, we need at least one measurement of type θ_1 or θ_2, and two other measurements of any type. Our system includes two telescopes both of which sight stars and planets. We measure the angle between the telescopes using the star measurements. We also get navigation updates from all three measurements:

1) planet to planet
2) planet to star
3) planet chord.

37.7.1
Sensor Model

Our sensor model is

$$y_k = \theta_k \tag{37.58}$$

where k is the telescope index. We use a central force model with the Sun as the only gravitational source. The simulation started on February 16, 2006 at midnight. The script has two options. In one option, only one planet is used as a measurement source. The chord width of the planet is measured along with the angle between a polar star and the planet vector. In this case, we rotate our telescope to look at different stars.

In the second option, two planets are measurement sources. In this simulation, they are Mars and Earth. The chord widths of both planets, angles between star vectors and the planet, and the angle between the planets are used as measurements. The stars used in the second case are within $8°$ of the planet vectors.

```
% Select the filter
%------------------

filter      = @UKF;   % Full covariance matrix filter
planets     = 1; % 1 or 2
starOffset  = 8*pi/180;
nSim        = 2000;
dT          = 2000; % sec
tEnd        = nSim*dT;
au          = 149597870; % km/au

if( planets == 2 )
    xP = zeros(17,nSim);
else
  xP  = zeros(14,nSim);
end
```

```
mu              = 132712438000;
jD0             = Date2JD( [2006 2 1 0 0 0] );

% Parameters for Earth and Mars
%-------------------------------
thetaE          = 2.30768364684019; % Initial
                    orbit angle of the earth (rad)
thetaM          = 1.56042659846335; % Initial orbit
                    angle of mars (rad)
omegaE          = sqrt(mu/au^3); % Earth orbit rate
omegaM          = sqrt(mu/(1.52*au)^3); % Mars orbit rate
t               = linspace(0,tEnd,nSim);
a               = omegaE*t + thetaE;
rE              = au*[cos(a);sin(a);zeros(1,nSim)]; % Earth
orbit
a               = omegaM*t + thetaM;
rM              = 1.52*au*[cos(a);sin(a);zeros(1,nSim)];
                    % Mars orbit

% Initial state [r;v]
%--------------------
x               = [rE(:,1) - [3e6;1e6;0];-33;-24;-10];

% Position and velocity uncertainty
%----------------------------------
r1Sigma         = 100; % km
v1Sigma         = 0.1; % km/s

% Measurement noise
% Errors: Earth radius 0.01 km, Mars radius 0.1 km,
ephem 2 km
% The elements are the noise for the following measurements
%    chord of planet 1
%    angle between star and planet 1
%    chord of planet 2
%    angle between star and planet 2
%    angle between planet 1 and planet 2
%----------------------------------------------------------
sigY            = 1e-1*[1e-6;1e-5;1e-6;1e-5;1e-5];

% State estimate at start
%------------------------
d.x             = x + [r1Sigma*randn(3,1);
                    v1Sigma*randn(3,1)];
```

```
% Covariance based on the uncertainty
%-----------------------------------
d.p            = diag([r1Sigma^2*ones(1,3) v1Sigma^2*ones
                 (1,3)]);
d.int          ='RK4';
d.rHSFun       ='RHSOrbitUKF';
d.measFun      ='OpticalNavigationMeasurement';
d.integrator   = @RK4;
d.alpha        = 0.8e-3; % UKF spread of sigma points
d.kappa        = 0; % UKF weighting factor
d.beta         = 2; % UKF incorporation of a priori
                 knowledge
d.dY           = (planets - 1)*3 + 2;
d.dT           = dT;
d.rHSFunData   = struct('mu',mu,'a',[0;0;0]);
d.rM           = diag(sigY(1:d.dY).^2); % Measurement noise
                 covariance
vecP           = [0 0 0 1e-6 1e-6 1e-6]';
d.rP           = diag(vecP.^2); % Plant noise covariance
d              = filter('initialize', d );
t              = 0;


% Measurement parameters that don't change
%----------------------------------------
clear g
g.a1           = 6378;
u1             = Unit([0 .1 -0.2;0 -0.3 0.1;1 1 1]);
g.u1           = u1(:,1);
g.l1           = rE(:,1);
j              = 1;


if( planets == 2 )
    g.a2 = 3397;
    g.u2 = [0;0;1];
    g.l2 = rM(:,1);
end


y = OpticalNavigationMeasurement( x(1:3), g );

for k = 1:nSim

  xP(:,k) = [d.x; x; y];
```

```
    x           = RK4( d.rHSFun, x, dT, t, d.rHSFunData ) +
                  vecP.*randn(6,1);
    t           = t + dT;

    g.l1        = rE(:,k);

    if( planets == 2 ) % Adding 2nd planet measurement
        g.l2 = rM(:,k);
        g.u1 = UnitVectorFromOffset( rE(:,k), x(1:3),
              starOffset );
        g.u2 = UnitVectorFromOffset( rM(:,k), x(1:3),
              starOffset );
    else
        g.u1 = u1(:,j);
        j    = j + 1;
        if( j > 3 )
            j = 1;
        end
    end

    d.measFunData = g;
    if( planets == 2 )
        y = OpticalNavigationMeasurement( x(1:3), g ) +
            sigY.*randn(5,1);
    else
        y = OpticalNavigationMeasurement( x(1:3), g ) +
            sigY(1:2).*randn(2,1);
    end

    % Kalman Filter
    %---------------
    d.t           = t;
    d             = filter( 'update', d, y );
end
```

37.7.2
Simulation Results

Figure 37.5 shows the heliocentric orbit. The Earth and Mars are nearly aligned, so the measurement geometry is not ideal. In addition, their positions do not change much during the simulation. Figure 37.6 shows the estimation results using one planet, the Earth, as the measurement source. Three near polar stars are sampled in sequence for the star/planet vector measurement. The measurement converges

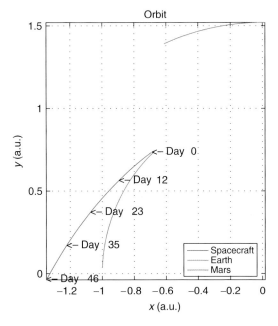

Figure 37.5 Planet orbit about the Sun. Only Mars and the Earth are used in the simulation. (Please find a color version of this figure on the color plates.)

quickly. Figure 37.7 shows when two planets are used as measurements. Although the measurements are all near the plane of the orbit, we see good results.

In practice, a combination of the two methods would be employed. In addition, one would employ a targeting strategy for picking targets that ensures good geometric information. The targeting method takes partial derivatives of the measurement equations and selects, through a search, a combination of measurements that are sensitive to x, y, and z.

37.8
Student Assignment

Design a heliocentric navigation system that uses the Sun and three stars. The system will measure the angles between each star and the line to the Sun plus the width of the Sun.

1) Write the measurement equations for the angles and the chord width of the Sun.
2) By taking partial derivatives with respect to the spacecraft vector, find the sensitivity of the x, y, and z measurements to the selected stars.

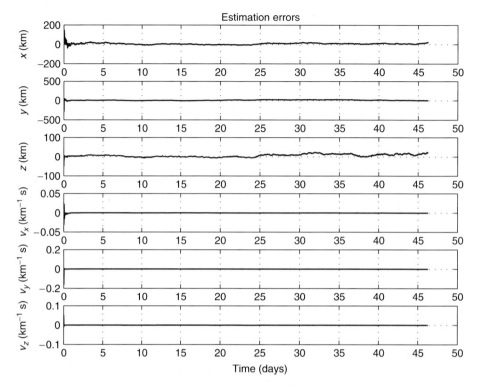

Figure 37.6 One planet measurement. This measures the chord of a single planet and the angle between that planet and a star visible in the same camera. A second measurement is just of the angle from stars to the planet.

3) Pick three stars (which need not be real) that give roughly equal sensitivity in all three axes.
4) Using a central force model for the Sun's gravity and the UKF, simulate the system in different orbits including a circular Sun orbit and an elliptical Sun orbit.
5) Using the simulation, study the sensitivity as a function of orbit.

37.9
Conclusion

Optical navigation is currently used for navigating near celestial objects such as planets, moons, and asteroids. Autonomous optical navigation using the UKF has the potential to supplement ground-based navigation, making possible more ambitious space missions.

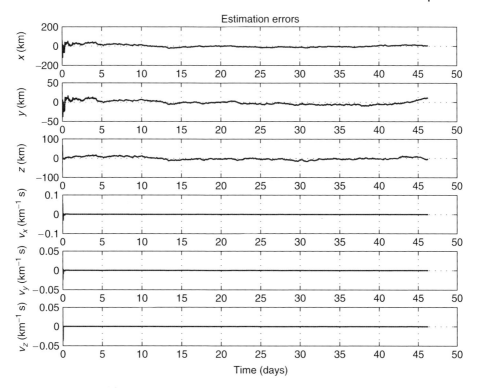

Figure 37.7 Two planets for measurements. In this case, two planets and angles to stars visible in the same camera frame are used.

References

1. Battin, R.A. (2004) *An Introduction to the Mathematics and Methods of Astrodynamics,* American Institute of Aeronautics and Astronautics, ISBN-13: 978-1563473425.
2. Bhaskaran, S., Reidel, J.E., and Synnott, S.P. (1996) Autonomous optical navigation for interplanetary missions. Proc. SPIE 2810, 32; doi: 10.1117/12.255151.
3. Reidel, J.E. et al. (1997) Navigation for the New Millennium: Autonomous Navigation for Deep Space 1. Proceedings of the 12th International Symposium on Space Flight Dynamics, Darmstadt, Germany, 303.
4. Kubota, T. et al (2003) An autonomous navigation and guidance system for muses-c asteroid landing. *Acta Astronaut.,* **52**, 125–131.
5. Kubota, T. et al. (2005) Robotics and Autonomous Technology for Asteroid Sample Return Mission. Proceedings of the 12th International Conference on Advanced Robotics.
6. Foing, B.H. et al. (2001) The ESA Smart-1 Mission to the Moon - Science and Technology Objectives. Proceedings of the National Space Society's 20th Annual International Space Development Conference, Houston, Texas, USA.
7. Polle, B. (2006) Low Thrust Missions Guidance and Navigation: The Smart-1 Oban Experiment Results and Perspectives. The 3rd International Workshop on Astrodynamics Tools and Techniques.
8. Polle, B. et al. (2003) Vision Based Navigation for Interplanetary Exploration

Opportunity for Aurora. Proceedings of the International Astronautical Congress.
9. Owen, W.M. Jr and Wang, T. (2001) Near optical navigation at Eros, *http://trsnew.jpl.nasa.gov/dspace/bitstream/2014/12992/1/01-1429.pdf*.
10. Hong, C.S. (2001) *On-chip spatial image processing with CMOS active pixel sensors*, PhD thesis, Univ. Waterloo, Canada.
11. Bull, C. (1991) *CCD Astronomy*, Willman-Bell, Inc, ISBN: 978-0943396286.
12. Hainaut, O. (1996) Basic image processing, *http://www.eso.org/~ohainaut/ccd/*.
13. Neumann, G.A. (2001) Some aspects of processing extraterrestrial lidar data: clementine, near, mola, International Archives of Photogrammetry and Remote Sensing, Vol. XXXIV-3, Annapolis, MD, pp. 73–80.
14. Standish, E. (2002) Comparison of old and new concepts: future of JPL ephemerides, Proceedings of the IERS Workshop on the Implementation of the New IAU Resolutions, Paris, France, pp. 61–63.
15. JPL Planets and pluto physical parameters (2010) *http://ssd.jpl.nasa.gov/?planet-phys-par*.
16. NASA Solar system exploration (2010) *http://solarsystem.nasa.gov/*.
17. Bhaskaran, S. (2006) Autonomous Navigation for Deep Space Missions. Aerospace Control and Guidance Systems Committee Meeting, Lake Tahoe, Nevada. 1–3 March (2006).
18. Riedel, J.E. et al. (2008) *Configuring the Deep Impact Autonav System for Lunar, Comet and Mars Landing*, AIAA-2008-6940; AIAA/AAS Astrodynamics Specialist Conference; Honolulu, HI.
19. VanDyke, M., Schwartz, J.L., and Hall, C.D. Unscented Kalman Filtering for Spacecraft Attitude State and Parameter Estimation, Proceedings of American Astronautical Society, Maui, Hawaii, USA, AAS-0115.
20. van der Merwe, R. and Wan, E., (2001) The square-root unscented Kalman filter for state and parameter-estimation, Proc. ICASSP, vol. 6, pp. 3461–3464.

38
ImageJ for Medical Microscopy Image Processing: An Introduction to Macro Development for Batch Processing
Tony Collins

38.1
Introduction

ImageJ is an open-source Java software application for scientific image processing and analysis. As it is useful for a broad range of imaging applications in the biomedical sciences, ImageJ is a key resource in laboratories across the world. Ninety percent of the image processing and analysis requirements in our core light microscopy facility at McMaster University are met using ImageJ. We also use ImageJ as a platform for algorithm development before translation to dedicated high-content screening packages. A summary of ImageJ's functionality is well documented in the ImageJ website and has been recently reviewed [1, 2].

This chapter introduces the basic structure of ImageJ and describes the development of an ImageJ macro for batch processing of light micrographs. New users who already possess some background in image processing and analysis can use this guide to walk through the creation of a macro in ImageJ. (It is recommended that this chapter is read at a computer with ImageJ installed and the images detailed in the chapter downloaded from the Wiley or McMaster Biophotonics Facility websites. Note that the images should be saved with their folder structure intact for this tutorial.)

38.2
Installation

ImageJ is "platform independent" and as such will run on all common operating systems (OSs) – this is because ImageJ is a Java application that requires only a Java runtime environment (JRE) specific to the OS. The correct JRE can be obtained from Sun Microsystems or by downloading the ImageJ version for a given OS from the ImageJ website (*http://rsb.info.nih.gov/ij/download.html*). These installation bundles contain the same ImageJ core files, with a range of JREs.

When installed, ImageJ puts the core program file, called *ij.jar*, into the program folder (e.g., typically *C:\ImageJ* in Windows) and creates a number of subfolders

---

Optical and Digital Image Processing: Fundamentals and Applications, First Edition. Edited by Gabriel Cristóbal, Peter Schelkens, and Hugo Thienpont.
© 2011 Wiley-VCH Verlag GmbH & Co. KGaA. Published 2011 by Wiley-VCH Verlag GmbH & Co. KGaA.

(\jre; \macros; \plugins; \luts) that contain the JRE, add-ons, and pseudocolor lookup-table files, respectively.

The core functionality of ImageJ is encoded in the IJ.jar file. Updating ImageJ simply requires running the menu command *Help>Update ImageJ*. This closes ImageJ and downloads the selected versions of IJ.jar. Updates are released every few weeks, so it is worth running the *Help>Update ImageJ* command on a regular basis. The latest IJ.jar file can be found in the ImageJ website (*http://rsb.info.nih.gov/ij/upgrade/*); details of the updates can be found on the news page.

38.2.1
Add-Ons

The core functions in ImageJ are extensively supplemented by a large number of freely available add-ons, called *Macros* and *Plugins*.

Macros are text files written in the ImageJ macro programming language. They can be as simple as a linear sequence of menu commands recorded by the software's *Macro Recorder* or as complex as almost fully fledged applications. Macros are saved as text files and can be run from any location with the menu command *Plugins>Macro>Run*.

A macro can also appear as a menu item in the ImageJ plugin menu if the given macro file name has an underscore in its name and it is located in ImageJ \plugins subfolder. A large number of macros are available from the ImageJ website (*http://rsbweb.nih.gov/ij/developer/index.html*).

Plugins are written in an extended version of the Java programming language and are generally more complex than macros, although the gap between the two is narrowing as the macro programming language evolves.

Plugins come as *.class files, which can be simply downloaded from where they are posted on the website. The *.class files need to be saved to the *ImageJ\Plugins* folder; when ImageJ is restarted, the plugin will appear as a new item in the Plugins menu. It will appear with the name of the *.class file; in other words, the plugin *My_plugin.class* saved in the *ImageJ\Plugins* subfolder will appear as a menu item *Plugins>My_plugin*.

Plugins are written as plain text files with the suffix *.java using text editors internal or external to ImageJ; they are compiled to become functional *.class files within ImageJ itself. The *.java files need to be located in the \plugins folder (or subfolder – but not a sub-subfolder) and compiled with the menu command *Plugins>Compile and Run*. Once compiled, the generated *.class file will be located in the *ImageJ* \plugins subfolder and will be added to the menu the next time ImageJ starts.

Many plugins are available from the ImageJ website: *http://rsbweb.nih.gov/ij/plugins/index.html*.

Complex plugins, or collections of plugins, that involve multiple *.class files can be collated into a *.jar file; these files names should also contain an underscore and be saved to the *ImageJ\Plugins* subfolder in order to be loaded by ImageJ on start-up. The *.jar files are zip-compressed files that have had their suffix changed from *.zip to *.jar. (i.e., they have been renamed from *my_jar.zip* to *my_jar.jar*). They may also contain a *plugin.config* file: this is a plain text file (with the suffix *.config*, not *.txt*), specifying which menu and by which name the plugin appears. For example, a *.jar file with a number of plugins including *my_plugin.class* could appear in the *Analyze>Tools* menu as a menu item (with a name such as "The plugin that I wrote"). This would be accomplished by adding the following line to a *plugins.config* file in the zipped *.jar file:

```
Analyze>Tools, "The plugin that I wrote", My_plugin
```

38.3
Plugin Collections

http://rsbweb.nih.gov/ij/plugins/index.html#collections Collecting plugins is a time-consuming process. Several plugin collections have been assembled to facilitate this task. We currently maintain a collection of plugins that are useful for our own light microscopy image processing and analysis (*http://www.macbiophotonics.ca/downloads.htm*). The MacBiophotonics (MBF) ImageJ collection of plugins and macros has been collated and organized by MBF; they have all been obtained free from the ImageJ website or elsewhere on the Internet, or we have written them ourselves. The credit for this work should go to the authors of the plugins. For researchers who will be primarily analyzing light microscopy images, downloading and installing the MBF_ImageJ collection is a good starting point for ImageJ.

38.4
Opening Images

ImageJ supports 8-, 16-, 24-, 32-, 48-bit TIFF images as well as most standard image formats such as JPG, GIF, BMP, PNG, and AVI. Image files from scientific images, however, contain not only image data, such as the pixel intensity values, but also metadata such as pixel size and other system settings. ImageJ can import images and metadata from image files created by most microscope systems, as well as DICOM and other medical imaging formats. For most systems, this means that original image data can be imported directly, without the extra step of exporting the raw data into an intermediary image format. ImageJ's support for microscopy image file formats is unparalleled, largely owing to the ongoing efforts of the LOCI group in Wisconsin (*http://www.loci.wisc.edu/*) as part of the Open Microscopy Environment project (*http://www.openmicroscopy.org*). The BioFormats series of plugins are collated as a *.jar file; this series is available from the LOCI (*http://www.loci.wisc.edu/ome/formats.html*) and ImageJ websites.

38.5
Developing a Macro

The remainder of this chapter introduces the macro programming language by considering a real-world application, in which a macro is developed to automate the analysis of microscope images data. This macro will introduce a number of approaches to macro writing to cover the basic skills for more elaborate programming.

The macro developed below analyzes images of adipocytes. The size of human adipocytes (fat cells) has profound implications in terms of function and pathology. Larger adipocytes release more breakdown products of fat (free fatty acids) and harmful proteins (adipokines) into the bloodstream, which adversely affects insulin resistance and causes inflammation in extra-adipose tissues such as muscle and liver, and arterial walls. This can lead to metabolic diseases such as cardiovascular disease and Type 2 diabetes. Here, we isolated human adipose tissue from individuals of varying ethnicities, ages (18–55), and body fatness (lean, overweight, and obese); fixed the cells; and stained them with hematoxylin/eosin. Approximately 50 000 adipocytes in total were imaged from 79 human subjects for automated cell sizing. Most previous studies investigating adipocyte size employed manual tracing of adipocytes. Here, we analyzed multiple size parameters of adipocytes using an ImageJ macro. From this, we found that the area, perimeter, maximum diameter (Feret's Diameter), width, and height of adipocytes increased markedly and significantly from lean to obese individuals, particularly those of South Asian descent.

38.5.1
Getting Started – Manually Planning the Analysis

Note that, at this point, it is helpful to access ImageJ as well as the images from the Wiley or MacBiophotonics website, and follow the steps outlined in the chapter.

The following list of steps describes the image processing and analysis routine that will ultimately be automated and used for batch processing in order to measure the average area of a large number of adipocytes. While there are many ways this can be done, our approach here illustrates some key aspects of the ImageJ macro language. We open the image file; set a "threshold" to identify background from foreground; separate the touching objects; set the measurements we want to make; and, finally, identify and analyze the cells.

> **Step 1: Open the file.** Open file 1.jpg, located in \images\Folder1, using the *File>Open* menu command or by dragging and dropping the file on to the ImageJ toolbar.
>
> **Step 2: Apply a smoothing filter.** For noisy images, it is worth applying a smoothing filter to the image to distinguish the objects of interest from the rest of the image (a process known as *segmentation*). A range of smoothing filters are available from the *Process>Filters* submenu; in this instance, we use the Gaussian filter. Run the menu command *Image>Filter>Gaussian Blur: Radius 2*.

Step 3: Set a threshold. To identify the foreground from the background, use the menu command *Image>Adjust>Threshold*. A threshold is automatically calculated based on the image histogram. One can manually adjust the threshold with the sliders in the threshold dialog.

Step 4: Apply the threshold. To convert the gray-scale image to a binary (black-or-white) image, apply the threshold *Process>Binary>Convert to mask*.

Step 5: Separate objects. Objects that are too close together can be further separated using the *Process>Binary>Watershed* menu command.

Step 6: Specify measurement. Specify which of the many potential measurements are needed, as well as defining several other settings, with the menu command *Analyze>Set Measurements*. For this example, select *Area*, *Mean Gray Level* (the intensity), *Shape Descriptors* (see the definitions on the ImageJ website *http://rsbweb.nih.gov/ij/docs/menus/analyze.html#set*), and *Display Label* to label each row in the subsequent results window with the file name.

Step 7: Determine the size range of particles. Before analzsing the particles, an important step is to first determine the range of sizes to include in the analysis, in order to exclude objects that are too big or too small to be genuine cells. To measure the size range of the particles to exclude, select a large particle that is *not* a cell and measure its area. To do this, either draw around using the circle-select tool on the tool bar or set a threshold (*Image>Adjust>Threshold*) and use the wand-tool on the toolbar to select an object. Once an object is selected, the size can be measured using the menu command *Analyze>Measure*. The size and other features of the object, including "roundness," that can also be used to include or exclude particles, are included in the results window.

Step 8: Identify and measure the objects. The particle analyzer dialog offers several options, most of which are self-explanatory. Here, we set the size range as 3000 to infinity. (The smallest size was manually determined in step 7.) At this stage, we want to "Show: Outlines" to be confident that the cells – and only cells – are identified as objects of interest. We also want to "display the results" and "exclude on edges" (but for now, ignore the other options). Click "OK" on the dialog to analyze the image.

Step 9: Check the results. Inspect the results to confirm that cells were correctly identified and analyzed.

38.5.2
Automating the Analysis: Recording Menu Commands

This section describes how to record the menu commands in a macro in order to automate the above steps.

Step 1. Close, but do not save, all the open images.

Step 2. Reopen image 1.jpg (*File>Open Recent>1.jpg*).

Step 3. To record the macro, open the *Macro recorder* using the menu command *Plugins>Macro>Record*.

Step 4. Select the menu commands to manually process the image a second time:

Process>Filter>Mean . . . : Radius 2>OK
Image>Adjust>Threshold
Process>Binary>Convert to mask
Process>Binary>Watershed
Analyze>Set Measurements>[Area/Mean Gray Level/Shape Descriptors]>OK
Analyze>Analyze Particles (it should have remembered your settings from the previous run)*>OK*

Step 5. Use the Recorder to create the macro below (then close the Recorder).

```
run("Mean...", "radius=2");
setAutoThreshold();
//run("Threshold...");
setThreshold(0, 127);
run("Convert to Mask");
run("Watershed");
run("Set Measurements...", "area mean shape display decimal=2");
run("Analyze Particles...", "size=3000-Infinity circularity=0.80-1.00
show=Outlines display exclude");
```

(see Appendix 38.A.)

The "//" characters prefixing the lines comment out the line, and as such the software ignores them. These are useful for annotating code: the macro recorder automatically inserts these when it translates a menu command to the equivalent macro language code. These can be deleted from the macro for clarity.

Step 6. Create a macro using the recorder's create button. Close the recorder (otherwise, it will record everything else one does).

Step 7. Save the macro as Macro1.txt using the Macro window's *File>Save* command.

Step 8. Close the image (without saving) and reopen image 1.jpg.

Step 9. Now run the macro from the macro text editor using the menu command *Macros>Run*.

38.5.3
Measuring Intensity

So far, we have quantified the area and shape of the objects of interest. One may have noticed, however, that the intensity of each object in the results table is 255. The reason for this is as follows: we converted the image to a black-and-white binary image in ImageJ and white pixels in a binary image in ImageJ are given a value of 255. We converted the image to a binary image so that we could use the *Watershed* function to separate touching objects. However, there is a potential problem here if the actual intensity itself of the original objects in the image is of interest. To address this issue, duplicate the image as an initial step: the duplicate can be used to identify and separate objects, and then use these objects used to measure the intensities in the original image. Open the "Set Measurements" dialog and specify that we want to "Redirect To" our original image.

Step 1. Close all images and reopen image 1.jpg.

Step 2. Open the macro recorder and use the following menu commands:

Image>Duplicate: "duplicate">OK
Process>Filter>Mean ... : Radius= 2>OK
Image>Adjust>Threshold
Process>Binary>Convert to mask
Process>Binary>Watershed
Analyze>Set Measurements>[Area/Mean Gray Level/Shape Descriptors, **Redirect To: 1.jpg**]>OK
Analyze>Analyze Particles>OK

Step 3. Create a macro; this should have the following text as the "set measurements" line :
```
run("Set Measurements..."," area mean shape display
redirect=1.jpg decimal=2");
```

Step 4. Save macro as Macro2.txt.

At this point, when we run the macro, our mean gray level (column heading "Mean") displays the mean intensity of the original image, not the binary object mask. However, at this point, the macro will fail when the analysis on Image 2.jpg is run, returning the following error message: *"1.jpg is not a valid choice for 'Redirect To:'"* If one does not close 1.jpg, the macro will use the objects from image 2.jpg to measure the intensities of image 1.jpg, which is not good.

The macro recorder converts menu commands to macro programming language code. To solve this current problem, we need to move beyond the "macro recorder" and learn some of ImageJ's macro programming language. We will keep coming back to the macro recorder to determine the syntax for menu commands which we can then add to our developing macro.

38.5.4
Basic Macro programming

Extensive processing and analysis protocols require the management of multiple images, so it is important to keep track of two or more image windows. This is possible by following either the window title of a given image or the hidden system identification number that ImageJ assigns images when multiple windows are created – the "image ID." Errors can be caused when window titles are used: it is possible, for example, to have multiple windows open with the same title. Image IDs are unique numbers assigned by ImageJ in the background. A search for "ID" on the online Macro Programming Language documentation (*http://rsbweb.nih.gov/ij/developer/macro/functions.html*) reveals a number of functions that relate to the image ID. One of these is *getImageID()* and another useful one is *selectImage*(id). There are also *getTitle*() and *selectWindow*() functions that use the window title as an input.

Generally, it is preferable to use image IDs rather than titles, given that there is less scope for selecting the wrong image. However, some functions such as the "Analyze Particle" command cannot receive an image ID as an input, so in this case, we need to use the image title in our macro and ensure that we do not have multiple images open with the same title. First, set a variable to equal the title of the original image. In the macro text editor, add to the very start of the macro:

```
origTitle = getTitle();
```

This sets the variable *origTitle* to equal the title of the image that was active when the macro was run. (Note that ImageJ is case sensitive and expects a semicolon at the end of a line, except for "if" and "for" statements; we discuss this later.) We can now add the *origTitle* variable, which is a "string" variable, to the "Set Measurements" command:

```
run("Set Measurements...", "area mean shape display
redirect=[" + origTitle +"] decimal=2");
```

Save as Macro2.txt. See Appendix 38.B.

Note that using the plus sign to concatenate strings, and the square parenthesis to ensure any spaces in the *origTitle* window name, does not cause errors. Open image 2.jpg and run the macro.

Now we have a macro that should work for all of our images placed, regardless of their name, and we can move on to batch processing. If the macro does not work, check that the text is in the correct case and make sure spaces, brackets, and quotes are correctly placed.

38.5.5
Batch Processing

38.5.5.1 Making a "Function"
To automate the analysis of a series of images in a folder, we need to initially identify the folder where the images are located; then, we obtain the list of the files that this folder contains. When this is in place, we can loop through this list to perform the analysis. Before we start with some "file handling" code, it is a good idea to split our analysis code off to a "function" that can later be referred to in

the main part of the macro. To this end, we add the following before our current code:

```
function analyseImage()
    {
```

At the very end of the code, the close curly bracket ("}") is added; the entire function will be enclosed within these brackets. Shuffle the function down the editor by adding some carriage-returns before the "function … " line and above it add "analyseImage()" to call the function.

```
analyseImage();
function analyseImage()
    {
    origTitle= getTitle();
    run("Duplicate...", "title=duplicate");
    run("Mean...", "radius=2");
    setAutoThreshold();
    setThreshold(0, 128);
    run("Convert to Mask");
    run("Watershed");
    run("Set Measurements...", "area mean shape display redirect=[" + origTitle + "] decimal=2");
    run("Analyze Particles...", "size=3000-Infinity circularity=0.80-1.00 show=Nothing display exclude");
    }
```

Save as Macro3.txt. See Appendix 38.C.

Later, when we run this macro with the 1.jpg image open, we should obtain the same result as that obtained before we turned the analysis into a "function." However, it also leaves us with three image windows open: this will potentially cause problems when we batch process multiple images due to both out-of-memory errors and conflicts between multiple windows called *duplicate*. To avoid this, add some code to close the images after they have been measured with the "Analyze Particles" command.

```
selectWindow(origTitle);
close();
selectImage("duplicate");
close();
```

Note that to refer to a specific window, we need to give the name in quotes; that is, the command selectImage("duplicate") selects the image *titled* "duplicate," whereas selectImage(duplicate) would look for a window equal to the *value* of the "duplicate variable." In this case, there is no such variable and an error will occur.

Rather than closing the "outlines" image from the particle analysis, we can simply prevent it from being shown to start with. Change the value of "show" in the Analyze Particles line to "show=Nothing":

```
run("Analyze Particles...", "size=3000-Infinity circularity=0.80-1.00
    show=Nothing display exclude");
```

We can determine the correct syntax/capitalization for this by recording a new macro, using the macro recorder, running the particle analysis, and selecting "Display=Nothing" in the dialog. We can then create a new macro and use this in our macro, using the same upper/lower case. (Copying and pasting can help prevent errors here.) Save as Macro3.txt.

38.5.5.2 Processing a List of Files with a "for" Loop

Retrieving the directory and file list allows us to loop through the analysis for each image. At the very start of each macro, we should add the following:

```
our_dir = getDirectory("Choose Source Directory");
```

At this point, a dialog opens requesting a choice of folder containing the images for analysis. Selecting the folder (and clicking "OK" for the dialog) sets the variable *our_dir* to the selected folder path. To obtain the list of files in from this folder, use the code:

```
our_list = getFileList(our_dir);
```

This sets the variable *our_list* to be a list (or, more correctly, an "array") of file names from our folder. We can loop through the file list by replacing the *analyseImage()* line with the following code:

```
for(i = 0; i<our_list.length; i++)
    {
    open(our_dir+ our_list[i]);
    analyseImage();
    }
```

Save as macro4.txt; see Appendix 38.D.

In this case, the *"for"* loop starts a new variable to initially equal zero ("i = 0") and loops through the sequence enclosed in the curl brackets. The value of "i" increases incrementally by a value of one each time through the loop (this is what "i++" means); this continues until "i" is no longer less than the length of *our_list* ("i<our_list.length").

We can test this with the sample images in folder 1. To run the macro, select the *Macro>Run* command from the macro text editor window and select the *Folder1* directory.

38.5.5.3 Filtering Files with an "if" Statement

In folder 2 we currently have a file that is not an image file, and this will cause our macro to fail. To add some error checking on this, we can use an "if" statement to check that the file is a JPG image. To do this, the operations to carry out if the criteria are met are enclosed in curly bracket.

```
if(endsWith(our_list[i],".JPG"))
    {
    open(our_dir+ our_list[i]);
    analyseImage();
    }
```

Save as macro4.txt.

If the file name (i.e., our_list[i]) is a TIFF file, it can be opened and analyzed; otherwise, nothing is done.

This can be tested with the sample images in folder 2. To run the macro, select the *Macro>Run* command from the macro text editor window and select the *Folder2* directory.

38.5.6
Adding a Dialog Box

At this point, we have set the analysis based on our preliminary manual walk-through for a randomly selected image. It is often useful to have the flexibility to set some analysis parameters for the batch processing. Instead of editing the macro on each occasion, it is possible to enter some user-defined values via a dialog box. For example, it may be useful to decide at different times whether to use a mean, median, or no filter at all. In addition, we may also want to have the option of displaying the outlines of the objects or set different size ranges for the particles.

For this, there are several short steps: first, generate and display a dialog prompting for these values; second, retrieve these values from the dialog as variables; and third, insert the variables into the code as with the "image title" variable above.

38.5.6.1 Creating a Dialog

The code for creating a dialog is quite simple. It should be inserted at the start of the macro, immediately following the "our_list=getFileList(our_dir)" line:

```
Dialog.create("Our Batch Analysis");
```

This creates an empty dialog that we need to populate with list choices, data-entry boxes, and options.

To enter a number value to the dialog, add a "number field" – a label that will appear in the dialog box – and its default value. As an example consider setting the values for the minimum and maximum particle size with the defaults as 1000 and infinity, respectively. Infinity is not a number and cannot be used, so instead set the default to an extremely high number.

```
Dialog.addNumber("Min Size", 1000);
Dialog.addNumber("Max Size", 9999999);
```

A check box can be added to offer a choice about displaying the outlines of the identified object, and its default choice, with

```
Dialog.addCheckbox("View Object Outlines", false);
```

To add a set of choices that enables a selection of smoothing filters is a little more complicated. First, create an array – a list – with the names of the choices; then, add this array to the dialog with a label and default value. The names in the list correspond exactly to the menu item names.

```
smoothArray = newArray("Mean...", "Median...", "None");
Dialog.addChoice("Smooth Filter", smoothArray, "Mean");
```

Finally, to open the dialog, add

```
Dialog.show();
```

Running the macro now will show the dialog, into which we can enter values, and click OK. Until we assign the dialog values to new variables in the macro, the values that are entered here will not be passed on to the macro code. Run the macro now and check for errors. (Common errors are due to incorrect capitalization, absent semicolons at the end of line, and typing errors). Save as macro5.txt.

38.5.6.2 Assigning Values to Variables from the Dialog

The following code should be added after the dialog is shown to assign values from the dialog to internal variables. We can assign values by stepping through the dialog values in the order that they were created.

```
our_min = Dialog.getNumber();
our_max = Dialog.getNumber();
our_outlines = Dialog.getCheckbox();
our_smooth = Dialog.getChoice();
```

Now we can insert these variables into our analysis function. To set the minimum and maximum particle values, change the Particle Analyzer line to the following:

```
run("Analyze Particles...", "size=" + our_min+ "-"
+our_max+" circularity=0.80-1.00 show=Nothing display
exclude");
```

Make sure to include the correct plus signs and quotation marks. One can test this by running the macro and setting the smallest size to 300 and the largest to 2000.

There are several ways to display the outlines of the objects using the macro. We can ask it to generate the Outlines image, as we have done earlier, and then close it (or link it to a true or false with the *our_outlines* variable). Alternatively, we can select whether or not the outline image is generated with an "if" statement. Note that two equal signs are needed to perform a comparison.

```
if(our_outlines==true)
        {
        run("Analyze Particles...", "size=" + our_min+"-"+ our_max+" circularity=0.80-1.00 show=Outlines display exclude");
        }
else
        {
        run("Analyze Particles...", "size=" + our_min+"-"+our_max+" circularity=0.80-1.00 show=Nothing display exclude");
        }
```

This leaves multiple images called *outlines of duplicate*. Here, add a line of code to rename these images after they have been generated by the "analyze particles" line in the macro.

```
rename(origTitle + " - outlines");
```

To change the smoothing filter, check whether the choice was "None." If the choice was not "None," then simply swap the our_smooth variable for the "Mean ... " value in the command. This is possible given that we have chosen the names of the variables to match the menu commands of the filter. (Note the " ... " after the name.)

```
if(our_smooth!="None")
        run(our_smooth, "radius=2");
```

It is important to be aware of the fact that the "!" before the "=" sign signifies that it is not "equal to"; if there is a single line following the "if" statement, then curly brackets are unnecessary. (However, it is a good idea to use them in the event that one later adds something to the "if" statement.)

Save as Macro5.txt; see Appendix 38.E.

We have discussed the basics of recording a macro, and enabling this macro to automatically analyze (or "batch process") the images in a user-selected folder. If the images are in subfolders, the folders and files can be looped through by adapting the *BatchProcessFolders.txt* available on the ImageJ website. (This has been included in the Appendix as Macro 6, as "Macro6 – processing subfolders.txt." To test this, run the Macro6 file and select the *images* folder that contains both *Folder1* and *Folder2*.)

38.6
Further Practical Exercises

Using *Macro5* as a starting point, try the following.

Easy: add options in the dialog box for setting the object circularity range.

Medium: adapt *Macro5* to analyze the images in *Folder3*. These fluorescent micrographs are Hoechst stained CHO cells. This requires us to change a few things in the macro: the threshold for object identification and, since these are TIFF and not JPEG images, the file suffix will be different.

Hard: find and examine the *CustomTabStatFromResults.txt* macro in the ImageJ website. Adapt this macro to generate a new summary results table for median intensity and area of all objects in each image in the folder (as opposed to the current table that shows values for each object). To do this, one will need to integrate some functions from the *CustomTabStatFromResults.txt* macro into *Macro5* as new functions and call them after analyzing each image. (Note that for those who need this functionality to calculate mean values, an easier way to achieve this is to use the *Summary* option in the Particle Analyzer.)

38.7
Important Websites

ImageJ – *rsbweb.nih.gov/ij*
LOCI Bioformats – *www.loci.wisc.edu*
ImageJ Document Wiki – *imagejdocu.tudor.lu*
Java Runtime libraries – *www.java.com*
ImageJ Collection for Microscopy – *www.macbiophotonics.ca*

Appendix 38.A: Analyzing a Single Image

```
run("Mean...", "radius=2");
setAutoThreshold();
//run("Threshold...");
setThreshold(0, 127);
run("Convert to Mask");
run("Watershed");
run("Set Measurements...", "area mean shape display decimal=2");
run("Analyze Particles...", "size=3000-Infinity circularity=0.80-1.00 show=Outlines display exclude");
```

Appendix 38.B: Including Intensity Measurements

```
origTitle = getTitle();
run("Duplicate...", "title=duplicate");
run("Mean...", "radius=2");
setAutoThreshold();
setThreshold(0, 128);
run("Convert to Mask");
run("Watershed");
run("Set Measurements...", "area mean shape display redirect=[" + origTitle + "] decimal=2");
run("Analyze Particles...", "size=3000-Infinity circularity=0.80-1.00 show=Outlines display exclude");
```

Appendix 38.C: Making a Function

```
analyseImage();
function analyseImage()
      {
      origTitle = getTitle();
      run("Duplicate...", "title=duplicate");
      run("Mean...", "radius=2");
      setAutoThreshold();
      setThreshold(0, 127);
      run("Convert to Mask");
      run("Watershed");
      run("Set Measurements...", "area mean shape display redirect=[" + origTitle + "] decimal=2");
      run("Analyze Particles...", "size=3000-Infinity circularity=0.80-1.00 show=Nothing display exclude");
      selectWindow(origTitle);
      close();
      selectImage("duplicate");
      close();
      }
```

Appendix 38.D: Batch Processing a Folder

```
our_dir = getDirectory("Choose Source Directory");
our_list = getFileList(our_dir);
for(i = 0; i<our_list.length; i++)
      {
      if(endsWith(our_list[i],".jpg"))
            {
            open(our_dir+ our_list[i]);
```

```
            analyseImage();
            }
      }
function analyseImage()
      {
      origTitle = getTitle();
      run("Duplicate...", "title=duplicate");
      run("Mean...", "radius=2");
      setAutoThreshold();
      setThreshold(0, 127);
      run("Convert to Mask");
      run("Watershed");
      run("Set Measurements...", "area mean shape display redirect=[" + origTitle + "] decimal=2");
      run("Analyze Particles...", "size=3000-Infinity circularity=0.80-1.00 show=Nothing display exclude");
      selectWindow(origTitle);
      close();
      selectImage("duplicate");
      close();
      }
```

Appendix 38.E: Adding a Dialog and Batch Processing a Folder

```
our_dir = getDirectory("Choose Source Directory");
our_list = getFileList(our_dir);
Dialog.create("Our Batch Analysis");
Dialog.addNumber("Min Size", 1000);
Dialog.addNumber("Max Size", 9999999);
Dialog.addCheckbox("View Object Outlines", false);
smoothArray = newArray("Mean...", "Median...", "None");
Dialog.addChoice("Smooth Filter", smoothArray, "Mean")
Dialog.show();
our_min = Dialog.getNumber();
our_max = Dialog.getNumber();
our_outlines = Dialog.getCheckbox();
our_smooth = Dialog.getChoice();
for(i = 0; i<our_list.length; i++)
      {
      if(endsWith(our_list[i],".jpg"))
            {
            open(our_dir+ our_list[i]);
            analyseImage();
            }
      }
```

```
function analyseImage()
    {
    origTitle = getTitle();
    run("Duplicate...", "title=duplicate");
    if(our_smooth!="None")
        run(our_smooth, "radius=2");
    setAutoThreshold();
    setThreshold(0, 127);
    run("Convert to Mask");
    run("Watershed");
    run("Set Measurements...", "area mean shape display redirect=[" + origTitle + "] decimal=2");
    if(our_outlines==true)
        {
        run("Analyze Particles...", "size=" + our_min + "-" + our_max+" circularity=0.80-1.00 show=Outlines display exclude");
        rename(origTitle + " - outlines");
        }
    else
        {
        run("Analyze Particles...", "size=" + our_min + "-"+our_max+" circularity=0.80-1.00 show=Nothing display exclude");
        }
    selectWindow(origTitle);
    close();
    selectImage("duplicate");
    close();
    }
```

Appendix 38.F: Batch Processing Subfolders

```
our_dir = getDirectory("Choose Source Directory");
our_list = getFileList(our_dir);
Dialog.create("Our analysis");
Dialog.addNumber("Min Size", 1000);
Dialog.addNumber("Max Size", 99999);
Dialog.addCheckbox("View Object Outlines", false);
smoothArray = newArray("Mean...", "Median...", "None");
Dialog.addChoice("Smooth Filter", smoothArray, "Mean")
Dialog.show();
our_min = Dialog.getNumber();
our_max = Dialog.getNumber();
```

```
our_outlines = Dialog.getCheckbox();
our_smooth = Dialog.getChoice();
setBatchMode(true);
processFolder(our_dir);
function analyseImage(path)
      {
      origTitle = getTitle();
      rename(path);
      origTitle = getTitle();
      run("Duplicate...", "title=duplicate");
      if(our_smooth!="None")
            run(our_smooth, "radius=2");
      setThreshold(0, 128);
      run("Convert to Mask");
      run("Watershed");
      run("Set Measurements...", "area mean shape dis-
play redirect=[" + origTitle + "] decimal=2");
      if(our_outlines==true)
            showWhat = "Outlines";
      else
            showWhat  = "Nothing";
      run("Analyze Particles...", "size=" + our_min + "-"
+ our_max + " circularity=0.7-1.00 show=" + showWhat + "
display exclude");
      if(our_outlines==true)
            rename(origTitle + " - outlines");
      selectWindow(origTitle);
      close();
      selectImage("duplicate");
      close();
      }
function processFolder(dir)
      {
      list = getFileList(dir);
            for (i=0; i<list.length; i++)
            {
            if (endsWith(list[i], "/"))
                        processFolder(""+dir+list[i]);
                  else  {
                        path = dir+list[i];
                        processFile(path);
                        }
            }
      }
function processFile(path)
```

```
    {
        if(endsWith(path,".jpg"))
        {
        open(path);
        analyseImage(path);
        }
    }
```

References

1. Abramoff, M., Magelhaes, P., and Ram, S. (2004) Image processing with imageJ. *Biophotonics Int.*, **11**, 36–42.

2. Collins, T.J. (2007) ImageJ for microscopy. *BioTechniques*, **43**(Suppl. 1), 25–30.

Index

a

aberrations
- chromatic 8
- correction in high-resolution optical microscopy 307–311
- – microscope configurations 307–308
- definition of 296–297
- defocus 8
- dynamics 318–319
- effects of 298–300
- – numerical aperture 300–301
- measurement 312–317
microscopy 296–301
- representation of 297–298
- sources of 300
- spherical 9
- third-order 8
absorption 44
active contours (ACs) models 288
active contours, applications of 645, 654, 657–658
active multiview 3D displays 378–379
active surfaces or balloons 654
adaptive Huffman coding 453
adaptive optics in microscopy 295–321
- control strategies for 317–320
- – aberration dynamics 318–319
- – field-dependent aberrations 319–320
- measurement 303–305
- – indirect wavefront measurement 304
- principles of 301–306
additive mapping 78
advanced moon imaging experiment (AMIE) 835
advanced video coding (AVC) standard 370
affine invariance 161
airy disk pattern 275
airy pattern 27

σ-algebra of events 49
aliasing problem 108
alternating minimization 535
ambiguity function 254
American Association of Physicists in Medicine (AAPM) 364
amplification 45
- pumping process 45
- spontaneous emission 45
- spontaneous photons 45
amplitude modulators 179, 192–193
amplitude spectra, face features 820–826
- corrected spectra 821
- whitened spectra, face images 821–822
- – diffusion, whitening by 824–826
- – slope whitening 822
- – variance whitening 822–824
- window artifacts 821
anaglyph method 371–372
analog and digital image processing, linking 397–417, *see also* continuous image models, building; discrete representation of images and transforms
analog-to-digital converter (ADC) 572
analytic image 104
angle multiplexing 230
angular multiplexing 595, 600–601
angular resolution 28–29
anti-alias filtering for multiview autostereoscopic displays 384–387
aperture synthesis 327–333
- earth rotation aperture synthesis, principles 327–330
- receiving system response 330–333
- single-aperture radio telescopes 326–327
approximated in frequency domain 386
approximated in spatial domain 384
approximation error 125

Optical and Digital Image Processing: Fundamentals and Applications, First Edition. Edited by Gabriel Cristóbal, Peter Schelkens, and Hugo Thienpont.
© 2011 Wiley-VCH Verlag GmbH & Co. KGaA. Published 2011 by Wiley-VCH Verlag GmbH & Co. KGaA.

Index

approximation scheme 145
architectural axioms 158
arcminute microkelvin imager (AMI) telescope 332
arithmetic coding 453–455
arithmetic operations 285
astigmatism 22
astronomical image formation 323–343
– CLEAN algorithm 338–339
– deconvolution techniques 337–339
– full-sky imaging 335–337
– image formation 333–342
– – derivation of intensity from visibility 333–337
astronomical image processing system (AIPS) 341
astronomical imaging 551, 563–564
atoms 508
autofocusing 285
automated target recognition (ATR) system 667–692, see also hybrid digital–optical correlator for ATR
automultiscopic displays 379
autoregressive moving average (ARMA) model 539
autostereoscopic displays (without glasses) 375–378
– microlenses 376
– pseudoscopic 376
– parallax barrier 376
– viewing zone 376
average feature-value invariance 161
axial super-resolution 607–608
axioms and properties 157–162
– architectural properties 160
– – convolution kernel 160
– – infinitesimal generator 160
– – linearity 160
– – semigroup property 160
– fundamental axioms 158–159
– – architectural axioms 158
– – causality 158
– – locality 158
– – Lyapunov functionals 159
– – maximum–minimum principle 159
– – nonenhancement of local extrema 159
– – preservation of positivity 159
– – regularity 158
– – stability axioms 158
– morphological properties 161–162
– – affine invariance 161
– – average feature-value invariance 161
– – feature scale invariance 161
– – isometry invariance 161
– – scale invariance 162

b

back focal plane 6
back focal point 6, 16
background 75
backlight unit 346
barrier layer 191
baseband 400
baseline 327
basis pursuit denoising problem (BPDP) 494
basket weaving 327
Bayesian approach 617–621, 661
Bayesian radio astronomer 339–341
Bayesian restoration 577–580
bays 76
Bennett–Brassard-1984 (BB84) QKD protocol 771–774
Bèzier splines 127
bidirectional prediction 448
biharmonic spline 128
binary symmetric channel 65
binary-phase only optical correlator (BPOC) 669
binocular stereopsis 369
bitwise holographic storage 229
blind deconvolution imaging 529–547, see also Image deconvolution
– multichannel deconvolution 539–542
– nonblind deconvolution 530
– space-variant convolution 531
– space-variant deconvolution 531
– space-variant extension 542–546
block-based motion estimation 446
block matching algorithm (BMA) 620
blue-green colors 41
blur kernel 530
border 76
Born approximation 701–704
– normalized 703–705
bragg planes 214
bragg-selective volume reflection hologram 215
Brenner's expression 106
Briggs pattern 363
bright-field microscopy 278–279
brightness histogram 77–78
B-splines 119
butterworth filters 92

c

cardinal function 121
CARS microscopy 717, 719

CARS microscopy image analysis, challenges 730–734
- CARS imaging chemometrics merging 734
- multiplex 734
- nonlinear microscopy particle identification 731
- nonresonant background numerical reduction 731–732
cascade recursive least squares (CRLS) 113
cathode ray tube (CRT) dispaly 347–349
cationic ring-opening polymerization (CROP) 245
cauchy distribution 54
central slice theorem 796
channel capacity 63–64
characteristic of random variable 54
charbonnier's diffusion filtering 168
charge-coupled devices (CCDs) 324, 571
chebyshev distance 74
chessboard 74
chiral nematic phase of LCs 180
chromatic aberration 8
chrominance indicator 426
1931 CIE Standard 38–39
CIE 1976 UCS Diagram 39–40
CIELAB diagram 42
circular aperture, diffraction at 27–28
city block 74
class of random events 49
classical restoration techniques 574–583
- constrained least squares (CLS) method 580
- maximum likelihood method 580
- method of generalized cross-validation 580
- Poisson noise model 580–583
- regularization parameter 578, 579
- SNR method 579
classical Richardson–Lucy (RL) deconvolution 573–574, 580
CLEAN algorithm 338–339
cliques 663
closed-loop video encoder 458
coded apertures 490–492
coherence 29–30
- length 29
- temporal or longitudinal coherence 29–30
- time 29
- transverse or spatial coherence 30
coherent light 275
coherent point spread function 18
collimated light 275
colocalization analysis 586

color 35–42, see also Theory of opponent colors
- color blindness 22
- color/spectral and multiple-component prediction and transforms 445
- images, representations for 155–176, see also under Scale-space representations
- spectrum of light 36
- tristimulus theory 36–40
columnar phase of LCs 180
complement event 50
complex amplitude 12
complex cells 421
compressed sensing (CS) 342, 507–526
- add or change priors 519–520
- compressive imagers 521–523
- compressive radio interferometry 523–524
- conventions 508
- greedy iterative methods 516, 518, 520–521
- information sources 524–525
- outside convexity 520
- random basis ensemble 517
- random convolution 517–518
- reconstructing from compressed information 512–515
- reproducible research 525–526
- science 2.0 effect 524–526
- sensing strategies market 515–515
- – random Fourier ensemble 516–517
- – random sub-Gaussian matrices 516
- single stage, sensing and compressing 510–512
- – new sensing model 511–512
- – Shannon–Nyquist sampling 510–511
- sparsity 508–510
compression 135
compression, digital image and video 441–461, see also entropy coding; quantization; video coding
- architecture 441–442
- color/spectral and multiple-component prediction and transforms 445
- data prediction and transformation 442–449
- data redundancy, removing 442–443
- JPEG 456
- JPEG 2000, 456–457
- spatial prediction 443–444
- spatial transforms 444–445
- temporal redundancy removal by motion estimation 445–449
- – bidirectional prediction 448
- – block-based motion estimation 446

compression, digital image and video (*contd.*)
– – improvements over basic approach 447–449
– – motion-compensated prediction 445–447
– – multiple reference frames 448
– – simple block-based motion model 446
– – subpixel accuracy 447
– – variable block sizes 448
compressive image acquisition 486, 488–494
– algorithms for restoring 494–499
– – for CS minimization problem solution 494–496
– – gradient projection for sparse reconstruction (GPSR) algorithm 495
– – iterative shrinkage/thresholding (IST) algorithms 496
– – least absolute shrinkage and selection operator (LASSO) 495
– – for nonnegativity constrained l_2–l_2 CS minimization 496–497
– – orthogonal matching pursuit (OMPs) 496
– – sparse reconstruction by separable approximation (SpaRSA) algorithm 495
– architectures for 488–494
– coded aperture imagers 490–492
– compressive-coded aperture (CCA) 492–494
– experimental results 499–501
– integration downsampling 492
– model-based sparsity 497–498
– noise 502
– physical constraints 486
– practical optical systems 486
– quantization 502
– Rice single-pixel camera 488
– spectral imagers 489
– stringent time constraints 486
– task-specific information compressive imagers 490
compressive radio interferometry 523–524
compressive sensing (CS) 485–488
– sparsity 487
compressive-coded aperture (CCA) 492–494
computer-generated holograms (CGHs) 202
computer graphics 72
computer vision 72
conditional entropy 58
conditional probabilities 51–52
– dependent events 51–52
– independent events 51–52
cones 22, 37, 421

confocal microscopy 283
constrained least squares (CLS) approach 580, 617–618
constructive interference 26
consumer drive implementation 233–234
contiguous pixels 75
continuous distribution 53
continuous faceplate devices 193
continuous Fourier transform (CFT) 621
continuous frequency response CFrR(p) 405
continuous image models, building 408–414
– discrete sinc interpolation 408–411
– image numerical differentiation and integration 411–414
– Newton–Cotes quadrature rules 412
continuous shearlet transform 151
continuous wavelet transform (CWT) 142–143
– dilation parameter 142
– translation parameter 142
continuum 27
contrast edges hypothesis 813–814
contrast masking 422
contrast parameter 169
contrast sensitivity functions (CSFs) 422
convergent wave, diffraction by 27–28
convex regions 75
convolution concept 78
convolution mask 79
convolution theorem 87
correlator 327
Crank–Nicholson method 164
critical voltage 183
cross entropy 581
cross talk functions 383
cross terms problem 108
cross Wigner distribution 103
cubic B-splines 119
curvature-based image quality assessment 426
curvature-based scale-space filters 172

d

1D directional PWVD 105–108
1D Fourier transform 80–85
2D Fourier transform 85–90
2D log-gabor filtering 110–112
3D displays 369–394, *see also* Planar multiview displays; Planar stereoscopic displays
– intermediate view interpolation 389–393
– – image-based rendering 389
– signal processing for 3D displays 381–393

– – 3D anaglyph visualization, enhancement   381–382
– – anti-alias filtering for multiview autostereoscopic displays   384–387
– – luminance/color balancing for stereo pairs   387–388
– – polarized 3D displays   382–384
– – shuttered 3D displays   382–384
2D-smoothed PWVD   108–110
– aliasing problem   108
– cross terms problem   108
– frequency-filtering   108
– oversampling   108
dark contrast microscopy   280–281
data distribution   573
data-fitting   531
data redundancy, removing   442–443
dataset   432
deblurring step   584–585
decoding stage   597
deconvolution from wavefront sensing (DWFS)   549–551
– image channel   550
– sensor channel   550
deconvolution   337–339, 529–547, *see also* blind deconvolution imaging; wavefront sensing
– maximum entropy deconvolution   339–341
deficit of convexity   75–76
deflation   113
defocus aberration   8
deformable mirrors   305
deformable models   645, 654–661
degree of coherence   602
Denisyuk holograms   209–210, 215
denoising   139
dependent events   51–52
destructive interference   26
detection mechanisms   423
DFT-based integration   412
dia-illumination microscopes   308
dichromated gelatin   240–241
dictionary   139
differential interference contrast (DIC) microscopy   273, 281
differentiation ramp filter   412
diffraction efficiency of hologram   213–214
diffraction   11–13, 25–28
– airy pattern   27
– at a circular aperture   27–28
– by convergent wave at   28
– continuum   27
– convergent   27
– Fraunhofer diffraction pattern   27
– Fresnel diffraction pattern   27
– integral   27
– of light   27
– at a one-dimensional slit   26–27
– pattern   27
– ringing   27
diffractive optical resolution   593–594
diffractive optical super-resolution   595–608
– limitations   595–597
– nonholographic approaches   594, 597–602
diffuse optical tomography (DOT)   695
diffusion equation   699
– infinite homogeneous media   699
diffusivity   167
digital cinema   365
digital grating   598
digital holography   603–607
– digital-to-analog conversion in   414–417
– – case study (reconstruction of Kinoform)   414–417
digital image processing in microscopy   284–291
– autofocusing   285
– – active methods   285
– – passive methods   285
– classification   289–291
– – nonparametric   290
– – nonsupervised   290
– – parametric   290
– – supervised   290
– illumination correction   285
– image enhancement   286–287
– – frequency domain methods   286
– – spatial domain methods   286
– mosaic generation   285
– preprocessing   284–286
– segmentation   287–289
digital image representation   73–78
– background   75
– bays   76
– border   76
– – inner border   76
– – outer border   76
– brightness histogram   77–78
– chessboard   74
– city block   74
– contiguous pixels   75
– convex regions   75
– deficit of convexity   75–76
– discretization   73
– distance between the samples   73
– edge   77
– holes   75
– image function   73

digital image representation (*contd.*)
– lakes   76
– metric properties   74–77
– 4-neighbors   74
– 6-neighbors   74
– 8-neighbors   74
– nonconvex regions   75
– objects   75
– picture element   74
– pixel   74
– pixel adjacency   74
– quantization   73
– segmentation   75
– simple contiguous   75
– spatial arrangement of samples   73
– – hexagonal grid   73
– – square grid   73
– topological properties   74–77
digital light projector (DLP) technology   185, 353–358
digital micromirror device (DMD)   192
digital postprocessing stage   597
digital video quality metric (DVQ)   424
dilation parameter   142
dirac delta functions   386
direct addressing   189
direct view displays   345–353
– types   346, *see also* emissive displays; reflective displays; transmissive displays
– working principle   345–346
direct wavefront sensing   312–314
directional transforms   148–152, *see also under* wavelets
– haar wavelet system   148
– separable transforms   148–149
dirichlet condition constrains   532
discrete channel   63
discrete convolution   79
discrete cosine transform (DCT)   402, 424, 464
discrete distribution   53
discrete Fourier transform (DFT)   85, 106, 402, 406, 621
– point spread function of   406–407
discrete representation of images and transforms   398–407
– 1D integral Fourier transform   403
– equivalent analog transforms, characterization   402–407
– mirror of digital computers, imaging transforms in   401–402
– signal discretization   399–401
discrete shearlet transform   152

discrete sinc interpolation   408–411
– as gold standard for image resampling   408–411
– signal recovery from sparse or nonuniformly sampled data   409–411
discrete-time Fourier transform (DTFT)   386
discrete wavelet transform (DWT)   143–144
– wavelet identity   143
– wavelet series   144
discretization   73
disjunctive events   50
disparity   390
displays   345–366, *see also* 3D displays; direct view displays; medical display systems
distortion measure based models on human visual sensitivity   425
distribution function   52–53
double-random phase encryption (DRPE)   740, 741–746
– linear method   743
– nonlinear methods   743
– optical setup for   743
– resistance against attacks   746
drizzle algorithm   325
dual-frequency effect   184
dual-tree complex wavelet transform (DT-CWT)   149–151
dynamic holography   220–224
– holographic cinematography   220–222
– holographic video   223–224
– real-time integral holography   222–223
dynamic range   365, 602

e

earth rotation aperture synthesis, principles   327–330
edge detection   93
edge   77
edge-affected variable conductance diffusion (EA-VCD)   168
edge-based segmentation   652–654
edge-stopping function   168
effective cutoff frequency   560–562
electrically addressed spatial light modulators (EA-SLMs)   189–190
electroholography   202
electromagnetic (EM) spectrum   1–3
– microwaves   2
– radio waves   2
– x-rays   2
– γ-rays   2
electromagnetic fields   10
electron microscopy   283–284
electronic paper (e-Paper)   352–353

electro-optic effect 242
elementary event 49
embedded block coding by optimized truncation (EBCOT) 457
embedded quantization 451–452
emissive displays 348–351
– CRT display 348–349
– light-emitting diodes (LEDs) display 347
– organic light-emitting diode (OLED) displays 348, 350–351
– plasma display 349–350
encoding–decoding process 601
encoding stage 597, 603
encryption algorithms optical techniques 740–755
– based on real functions 748–749
– color image encryption 750–751
– double-random phase encryption (DRPE) 741–746
– fractional Fourier transforms 753–755
– Fresnel domain 751–753
– holographic memory 749–750
– phase encoding 740–741
– random phase mask (RPM) 740–741
– wavelength multiplexing 750–751
energy density spectrum 99
entropy 54–62
– conditional entropy 58
– joint entropy 56–60
– properties 55–56
– relative entropy 62
– Rényi entropy 62
entropy coding 442, 452–455
– adaptive Huffman coding 453
– arithmetic coding 453–455
– Huffman coding 452–453
epidermal growth factor (EGF) 587
epi-illumination microscopes 308
equivalent analog transforms, characterization 402–407
– 2D integral Fourier transform 404
– point spread function (PSF) 406–407
error correction codes (ECCs) 236
étendue 355
euclidean distance 74
Euler–Lagrange equation 168, 532–533
excitance 31
expectation-maximization (EM) 649–650
expected value 54
eye fundus imaging 561, 564–567

f

face image statistics, *see* human face recognition and image statistics using Matlab
far field approximation 13
farsightedness 22
fast Fourier transform (FFT) Cooley–Tukey algorithm 517
fast Fourier transformation (FFT) 85
feature-based image quality models 425–427
– curvature-based image quality assessment 426
– distortion measure based 425
– perceptual video quality metric (PVQM) 426
– singular value decomposition (SVD) 425
– structural and information-theoretic models 427–430
– temporal trajectory aware quality measure 427
– temporal variations of spatial-distortion-based VQA 427
– video quality metric (VQM) 426–427
feature extraction methods 682–685
– horizontal and vertical binning 682–684
– principal component analysis (PCA) 684–685
feature scale invariance 161
feature space 156
feed-forward 686–687
Fergus's method 534–536
Fermat's principle 3
ferroelectric LCs 187–189
ferroelectric liquid crystal (FLC) 669
ferro-electric-liquid-crystal-onsilicon (FLCoS) 359
– architecture 360
Fick's law 703
field-dependent aberrations 319–320
filter cube 282
filtering in image domain 90–95
– butterworth filters 92
– edge detection 93
– Gaussian filters 92
– gradient operators 90
– median filtering 92
– noise filtering 90–91
– Prewitt operator 94
– robust filtering 92
– smoothing filters 91–92
– Sobel operator 95
filters 78
– linear 78
– nonlinear filters 78

Fisher discriminant analysis (FDA)   291
fluorescence microscopy   275, 282
fluorescence molecular tomography (FMT)   696–706
– Born approximation for excitation term   701–702
– – boundary conditions   702
– diffusion equation   699
– – for infinite homogeneous media   699
– excitation source term   700
– experimental setup   705–706
– fluorescence source term   700–702
– inverse problem   703
– normalized Born approximation   703
– optical parameters   698–699
– principle   696–697
– theoretical background   697–705
flux loss   332
fly's eye integrator   354
focal length   71
focal point   274
Fourier analysis, classical   140–141
Fourier coefficients   140
Fourier kernels   400
Fourier optics   14–21
– applications   14–21
– back focal point   16
– Fourier plane   17
– ideal thin lens as   14–17
– image plane   17
– imaging and optical image processing   17–19
– joint-transform correlator   19
– object plane   17
– optical correlator   19–21
– pupil function   17
– spectrum   17
Fourier plane   17
Fourier series   140
Fourier transform   80, 140
– properties of   82
fovea   37, 421
fractional Fourier transforms   753–755
frame rates   183–184
Fraunhofer approximation   13
Fraunhofer diffraction formula   13–14, 27
free-space spatial impulse response   13
frequency domain methods   80–90, 286, 621–622
– 1D Fourier transform   80–85
– 2D Fourier transform   85–90
– 2D signals   80
– discrete Fourier transform (DFT)   85
– fast Fourier transformation (FFT)   85

– Fourier transform   80
– spatial frequencies   85
frequency-filtering   108
Fresnel diffraction formula   12, 14, 27
Fresnel domain   751–753
front focal plane   6
full-reference (FR) algorithms   419
full-sky imaging   335–337
fully complex modulation methods   194–195
fundamental axioms   158–159

g

Gabor filters   431
Gabor function   110–112, 141
Gaussian channel   65–66
Gaussian filters   92
Gaussian noise model   575, 578–580
Gaussian scale mixture (GSM) model   429, 573
Gaussian scale space theory   165–167
Gaussian-modulated coherent state (GMCS) QKD protocol   776–777
general class distribution   97
generalized cross-validation method   580
generalized sampling theorem   121
generic image coding   455–456
geometrical optics   3–9, *see also* ray transfer matrix; two-lens imaging system
– aberrations   8–9
– Fermat's principle   3
– homogeneous medium   3
– photons   3
– ray optics   3
– ray transfer matrix   3–6
geometrical resolution   593–611
– fundamental limits   594–595
geometrical super-resolution   608–611
Geometry-driven scale space   171–173
– curvature-based scale-space filters   172
gerchberg–Saxton algorithm   265
global precedence   424
gold standard for image resampling   408–411
gradient-based method   288
gradient-based wavefront sensors   552–558
gradient operators   90, 93
gradient projection for sparse reconstruction (GPSR) algorithm   495
graphic user interface (GUI) of GOC System   674–675
grating light valve (GLV)   360–361
gray-scale images, representations for   155–176, *see also under* Scale-space representations

gray-scale optical correlator (GOC) 669–670, 674, 675–676
green fluorescent protein (GFP) 587
Green's function 163

h

Haar wavelets 146, 148
harmonic analysis 139
harmonic functions 81
Hartmann–Shack wavefront sensor 553
head-mounted displays (HMDs) 370
Heisenberg's uncertainty principle 141
Henyey–Greenstein function 699
Hermite splines 127
hexagonal grid 73
high-pass filter 145
high-resolution optical microscopy, aberration correction in 307–311
high-temperature polysilicon (HTPS) 353
Hilbert space 121, 139
Hilbert transform pair 149
Hogel 219
holes 75
holographic approaches 602–608
– holographic wavefront coding 602–603
holographic data storage technology 227–247
– bitwise 229
– consumer drive implementation 233–234
– data channel overview 236–237
– dichromated gelatin 240–241
– drive architecture 231–233
– light path
– – during read operation 232
– – during write operation 231
– materials for 237–242
– material for data storage 243–246
– media for 246
– microholographic storage 229
– pagewise 229
– photochromics 242
– photopolymers 239–240
– photorefractive 242
– photothermoplastics 241–242
– reading data 230
– silver halide photographic emulsion 238–239
– tolerances and basic servo 234–236
– – reference angle 235
– – tilt 235
– writing data 229
holographic memory 749–750
holographic stereograms 202
holographic visualization of 3D data, 201–225, see also Dynamic holography
– amplitude and phase, reproducing 203–207
– Bragg-selective volume reflection hologram 215
– color reproduction 210–215
– Denisyuk hologram 215
– holographic approximations 215–220
– holographic stereogram 217–220
– – integral photography 217
– – Lippmann's photography 217
– phase versus amplitude, diffraction efficiency 213–214
– rainbow hologram 216–217
– reflection 211–212
– surface relief holograms 214–215
– thin versus thick hologram 215
– transmission 212–213
– types of 207–215
– – Denisyuk holograms 209–210
– – transmission versus reflection 207–209
homogeneous mapping 78
homogeneous medium 3
homologous points 390
hough transform 288
hubble space telescope (HST) 324
Huber–Markov Gibbs prior model 620
Hue 41
Huffman coding 452–453
human face recognition and image statistics using Matlab 809–831
– image statistics 811–820, see also Amplitude spectra, face features
– – face image dataset 818–819
– – face processing 818
– – spatial frequency, dimension 819–820
– neural information-processing 811–818, see also individual entry
human visual system (HVS) 21–23, 420–422
– abnormal eyes 22
– – astigmatism 22
– – farsightedness 22
– – nearsightedness 22
– color blindness 22
– cones 22
– contrast sensitivity functions (CSFs) 422
– human-visual-system-based models 422–425
– – digital video quality metric (DVQ) 424
– – discrete cosine transform (DCT) domain 424
– – moving picture quality metric (MPQM) 424

human visual system (HVS) (contd.)
– – scalable wavelet-based distortion metric for VQA   425
– – Teo and Heeger model   423–424
– – visual difference predictor (VDP)   423
– – visual discrimination model (VDM)   423
– lateral geniculate nucleus (LGN)   421
– light adaptation   422
– masking   422
– – contrast   422
– – temporal   422
– – texture   422
– photoreceptors   421
– – cones   421
– – fovea   421
– – rods   421
– – saccades   421
– point-spread function (PSF)   420
– rods   22
Hurter–Driffield curve   635
Huygens' principle   25
hybrid digital–optical correlator for ATR   667–692
– ATR process, experimental demonstration of   687–690
– – sonar mine identification   689–690
– – vehicle identification   687–689
– gray-scale optical correlator system's space–bandwidth product matching   668–670
– input SLM selection   671–673
– miniaturized gray-scale optical correlator   673–676
– neural network for target verification   681–687
– OT-MACH filter, optimization of   677–681
– – optimization approach   677–681
hybrid optical–digital implementation   114–116
hybrid restoration methods   583

i
illumination correction   285
image analysis   643–664, *see also* intermediate-level vision
image and volumetric coding   455–457
– generic image coding   455–456
image deconvolution   530–534
image filtering paradigm   78–79
– additive mapping   78
– convolution concept   78
– – discrete convolution   79
– convolution mask   79
– homogeneous mapping   78

– linear combination   78
– linear filters   78
– nonlinear filters   78
image formation from optical telescopes   324–325
image interpretation   72
image plane   17
image processing, fundamentals   71–95, *see also* digital image representation; frequency domain
– image filtering paradigm   78–79
– physical process of image formation   72
image quality assessment/video quality assessment algorithms (IQA/VQA)   419
image registration   67
image restoration   571–589
– in biomedical processing   571–589
– classical restoration techniques   574–583
– colocalization analysis   586
– outline of algorithm   584–586
– – deblurring step   585
– – noise reduction   584–585
– photon or shot noise   571
– read-out or amplifier noise   572
– sensor noise   571
imageJ for medical microscopy image processing   859–877
– add-ons   860–861
– – macro recorder   860
– – macros   860
– – plug ins   860
– batch processing   866–869
– – filtering files   869
– – function   866–868
– – loop files, processing   868
– dialog box   869–872
– – adding   869
– – assigning values to variables   870–872
– – creating   869–870
– installation   859
– macro, developing   862–866
– – analysis planning   862–863
– – automating, recording menu commands   863–864
– – basic macro programming   865–866
– – intensity, measuring   864–865
– opening images   861
– plug in collections   861
impossible event   50
independent component analysis (ICA)   291
independent events   51–52
indirect wavefront sensing   314–317
information channel   62–66
– binary symmetric channel   65

– channel capacity   63–64
– discrete channel   63
– Gaussian channel   65–66
– symmetric channel   64–65
information fusion by segmentation   466–470
information theory, basics   49–68, *see also* probability
– Kullback–Leibler distance (KLD)   62
initial condition   12
inner border   76
integral photography   217–219
integration downsampling   492
interference   25–26
– constructive   26
– destructive   26
intermediate image plane   274
intermediate-level vision   643–664, *see also* segmentation
– active contours, applications of   657–658
– deformable models   654–661
– mathematical formulation   655–657
– mathematical morphology   651
– optimal edge detection   654
– scale-space approach   654
– seeded region growing   651
– supervised approaches   646–647
– unsupervised approaches   647–650
International Electrotechnical Commission (IEC)   455
International Standardization Organization (ISO)   455
interpolation splines   119
inverse filter   574–575
inverse Fourier transform (IFT)   468
inverse task   72
inversion   101
– inversion formula   140
iris recognition   67–68
irradiance   31
irregularly spaced unit   386
isometry invariance   161
iterative constraint Tikhonov–Miller (ICTM) algorithm   578–579
iterative shrinkage/thresholding (IST) algorithms   496

j

Jepson and Fleet algorithm   431
Jet Propulsion Laboratory (JPL)   669
joint entropy   56–60
joint photographic experts group (JPEG)   464
joint spatial/spatial-frequency representations   97–116, *see also* 2D log-Gabor filtering; pseudo-Wigner–Ville distribution (PWVD); Wigner distribution
– analytic image   104
– energy density spectrum   99
– fundamentals of   98–103
– hybrid optical–digital implementation   114–116
– notation   99–100
– spatial and spatial-frequency marginal conditions   99
– spectrogram   104
– texture segmentation   112–114
joint technical committee for information technology   455
joint-transform correlator   19
joint-transform power spectrum (JTPS)   19–20
jones matrix formulation for TN-LC   184
JPEG   473–474
just noticeable difference (JND) index   362–363

k

Karhunen–Loève transform (KLT)   113
kernel   103
– kernel function   111
– kernel regression (KR)   638
kinoform hologram-encoding method   414–417
K-means clustering   648–649
knots   125
Koenderink's formulation of causality   158
Köhler illumination   278
Kramers–Kronig relationship   228
Kullback–Leibler distance (KLD)   62, 535

l

L vanishing moments   145
labeling   663
laboratory for image and video engineering (LIVE)   433
lakes   76
Laplace distribution approximation   535
Laplacian based method   288
Laplacian of Gaussian (LoG)   653
lateral geniculate nucleus (LGN)   421
law of refraction   3–4
least absolute shrinkage and selection operator (LASSO)   495
leave-one-out principle   580
Lebesgue spaces   139
length, coherence   29
lenticular technology   377

lexicographical ordering 174
light 275
– adaptation 422
– coherent 275
– collimated 275
– interaction with matter 275
– light-emitting diodes (LEDs) display 347
– monochromatic 275
– polarized 275
light amplification by stimulated emission of radiation (LASER) 43
– basic elements of 46
– coherence 47
– direction 47
– intensity 47–48
– monochromatic 47
light oscillation by stimulated emission of radiation (LOSER) 45
light polarization 371, 373–375
– μPol™ (micropol) 374
– Planar's StereoMirror™ 374
– Zscreen® 374
light shuttering 375
linear (Pearson's) correlation coefficient (LCC) 432
linear discriminant analysis (LDA) 291
linear nematic (TN) SLMs 181
Lippmann's photography 217
liquid crystal displays (LCDs) 346
liquid crystal on silicon (LCoS) 353
liquid crystal TV (LCTV) 20
liquid crystals (LCs) 180–191
– addressing methods 189–191
– – direct addressing 189
– – electrical 189
– – multiplex addressing 189
– – optical 189
– – silicon backplane addressing 190
– chiral nematic phase 180
– columnar phase 180
– critical voltage 183
– ferroelectric 187–189
– frame rates 183–184
– microelectromechanical systems (MEMS) 185
– modulation methods 186–187
– nematic LC 180–183, see also Nematic liquid crystal SLMs
– optically addressed spatial light modulators (OA-SLMs) 190–191
– smectic LC 180
– thermotropic LC 180
Lloyd–Max quantizers 450–451
log-Gabor filters 110–112

Lorentzian distribution 54
lossless 135
lossy compression 135
low-density parity check (LDPC) 236
low-pass filter 145
Lukosz approach 603–604
lumen 34
luminance/color balancing for stereo pairs 387–388
luminous terms 33–34
– luminous flux 34
– visual or photometric quantity 34
– visual spectral flux 33
Lyapunov functionals 159, 170

m

Mach–Zehnder interferometer 466
marginal ordering 174
marginals 101
Markov chain Monte Carlo (MCMC) 341
Markov random field (MRF) approach 619, 661, 663–664
masking 422
– contrast 422
– temporal 422
– texture 422
matched probability model 469
mathematical morphology scale space 173–176
matrix formalism for phase-space rotations 255–257
maximizing the a posteriori probability (MAP) 646–647
maximum a posteriori (MAP) approach 535, 576
maximum differentiation competition (MAD) 433
maximum entropy deconvolution 339–341
maximum entropy method (MEM) 339–341, 578
maximum likelihood difference scaling (MLDS) 433
maximum likelihood estimation (ML) 577, 580
maximum principle 158
maximum–minimum principle 159
Maxwell's equations 1, 9–11
mean opinion score (MOS) 419
mean square error (MSE) 419
measurement of light 30–35
– radiometry versus photometry 30
median absolute deviation (MAD) 169
median filtering 90, 92

medical display systems 362–366
- applications 362–366
- - digital cinema 365
- - LCD displays 362
- - nighttime training 365
- calibration 362–364
- physical measurements 363
- quality assurance of 362–364
- visual inspection tests 363
metric algorithm 424
Mexican hat filter 653
microdisplays (MDs) 353
microelectromechanical systems (MEMS) 185, 193
microholographic storage 229
microlenses 376
microscopic imaging 273–292
- bright-field microscopy 278–279
- components of 276–277
- - adjustment knobs 276
- - camera 276
- - computer 277
- - condenser 277
- - eyepiece 276
- - illumination 277
- - objective 276
- - specimen holder or stage 277
- confocal microscopy 283
- continuous image 274
- dark contrast microscopy 280–281
- DIC microscopy 281
- digital image processing in 284–291
- digital image 274
- displayed image 274
- electron microscopy 283–284
- fluorescence microscopy 282
- image formation in 274–275
- image formation, basic concepts 274–276
- Köhler illumination 278
- optical image 274
- phase contrast microscopy 279–280
- resolution 275–276
- types of 277–284
- types of image 274
microwaves 2
millimeter wave (MMW) 2
miniaturized gray-scale optical correlator 673–676
- alignment mode 674
- 512 × 512 GOC system architecture 673–674
- graphic user interface (GUI) of GOC System 674–675
- run mode 674

minimum mean square error (MMSE) 577
Minkowski formulation 424
mirror devices 192–193
- amplitude modulators 192–193
- continuous faceplate devices 193
- digital micromirror device (DMD) 192
- MEMS deformable mirrors 193
- phase modulators 193
- segmentedmirrors 193
mirror of digital computers, imaging transforms in 401–402
mispick 114
model-based segmentation 661–664
model-based sparsity 497–498
modified peak signal-to-noise ratio (MPSNR) 628
modified uniformly redundant array (MURAs) 491
monochromatic light 275
monocular concept with multiplexing 233
mosaic generation 285
motion-based video integrity evaluation (MOVIE) 431–432
motion-compensated prediction 445–447
motion-modeling-based algorithms 430–435
- MOVIE 431–432
- performance evaluation 432–434
- reduced reference (RR) algorithms 432
- spatial quality assessment 431
- SW-SSIM 431
- validation 432–434
motion parallax 378
moving picture quality metric (MPQM) 424
multichannel deconvolution 539–542
multichannel-regularized super-resolution image reconstruction algorithm 624–628
multiconjugate adaptive optics (MCAO) 320
multiple quantum-well (MQW) SLMs 191–192
- barrier layer 191
- well layer 191
multiple reference frames 448
multiplex addressing 189
multiplexed holograms 603–604
multiplexing in other degrees of freedom 601–602
multiplexing techniques 230, 466–470
- angle multiplexing 230
- monocular concept with 233
- polytopic multiplexing 230
multiplicative bases 401
multiresolution analysis (MRA), wavelets 144–148
- approximation scheme 145

multiresolution analysis (MRA), wavelets (contd.)
– dilation equation or 2-scale relation   145
– Haar wavelet system   146
– high-pass filter   145
– L vanishing moments   145
– low-pass filter   145
– two-channel digital filter bank   145
multiscale analysis, splines for   129–131
– spline pyramids   129
– spline wavelets   129
multiscale structural similarity index (MS-SSIM)   428–429
multispectral fluorescence tomography   695–713
– application   695
– experimental setup   705
– Fluorescence molecular tomography (FMT)   696–706
– methodology   711–712
– multicolor phantom   709–711
– multitarget detection and separation   709–712
– spectral deconvolution   707–708
– spectral tomography   706–708
– theory   695
– *in vitro* fluophore unmixing   709
– *in vivo* fluophore unmixing   711
multiview autostereoscopic displays, anti-alias filtering for   384–387
multiview coding (MVC)   370
multiview displays   370, 378
mutual information   60–61
– application   67–68
– – image registration   67
– – iris recognition   67–68
mutually exclusive events   50

n

natural scene statistics (NSS)   422
Navier–Stokes equations   170
nearsightedness   22
4-neighbors   74
6-neighbors   74
8-neighbors   74
nematic liquid crystal SLMs   180–183
– dual-frequency effect   184
– linear   181
– temperature effects   184
– transient nematic effect   184
– twisted   181, 184–186, *see also* twisted nematic (TN) SLMs
neural information-processing   811–818
– contrast edges hypothesis   813–814

– k^{-2} frequency scaling   811–813
– occlusion hypothesis   813
– response equalization   811, 816–818
– scale invariance hypothesis   812–813
– spatial decorrelation   811, 814–816
neural network identification   686–687
neurons   686
Newton–Cotes quadrature rules   412
no-cloning theorem, quantum   770–771
noise   139
– noise equivalent resolution   594
– noise filtering   90
nonblind deconvolution   530
nonconvex regions   75
nonholographic approaches   597–602
– time multiplexing   597–600
nonnegative matrix factorization (NMF)   291
nonnegative sensing matrices, compensating for   498–499
nonnegativity constrained l_2–l_2 CS minimization, algorithms for   496–497
nonstationary signal   84
normal weighting   334
normalized Born approximation   703–705
normalized root-mean-square (NRMS)   465
numerical aperture   300–301
Nyquist sampling condition   608

o

object plane   17
objective quality assessment   419
objects   75
occlusion hypothesis   813
onboard autonomous navigation (OBAN)   835
One-Mile radio telescope   324
One-Mile telescope   324
one-panel reflective DLP with UHP lamp   358–359
one-panel reflective LCoS with high-brightness LEDs   359
one-time pad (OTP) encryption   769–770
'opponent colors' theory, *see* theory of opponent colors
optical and geometrical super-resolution   593–611
optical axis   3
optical correlator   19–21
optical data encryption   739–765
– security systems applications   755–765
– – biometrics verification   756–759
– – digital cryptography   755–756
– – multifactor identification   756–759
– – remote verification ID tags   759–765

– techniques   740–755, *see also* encryption algorithms optical techniques
optical homodyne detector   780–781
optical image compression methods   463–481
– background   464–466
– composite filters   464
– compressing interference patterns   465
– optical JPEG implementation results   472–473
– information fusion by segmentation   466–470
– normalized root-mean-square (NRMS)   465
– optical and digital JPEG comparison   473–474
– phase-shifting interferometry digital holography (PSIDH)   465
– postprocessing   464
– segmented filters   465
– by using JPEG and JPEG2000 standards   470–474
optical JPEG2000 implementation   474
optical path difference   279
optical region   2
optically addressed spatial light modulators (OA-SLMs)   190–191
optimal edge detection   654
order of approximation   125
organic light-emitting diode (OLED) displays   348, 350–351
– top-emitting   351
orthogonal matching pursuit (OMPs)   496
orthonormality   122
outer border   76
oversampling   108, 237

p

pagewise holographic storage   229
parallax barrier   376
paraxial optics   4
paraxial rays   3–4
parseval's theorem   83
partial differential equation (PDE)   157
partial volume interpolation   67
partially coherent light   252–253
passive multiview 3D displays   379–380
path compensation   329
peak to correlation energy (PCE)   480
peak-signal-to-noise ratio (PSNR)   420
perceptual video quality metric (PVQM)   426
periodicity   88
phantom edges   653
phase center   328
phase contrast microscopy   279–280

phase modulators   179
phase-shifting interferometry digital holography (PSIDH)   465
phase-space rotators   251–269
– applications   264–268
– matrix formalism for   255–257
– optical system design for   260–264
– signal representation in   252–255
– – partially coherent light   252–253
– – Wigner distribution   253–255
– for two-dimensional signals   257–260
phase-space tomography   789–807
– fundamentals   790–793
– Radon transform   794–796
– separable in cartesian coordinates   793–794
phasor   12
photobleaching   282
photochromics   242
photon or shot noise   571
photonics, fundamentals of   25–48, *see also* color
– absorption   44
– amplification   45
– basic laser Physics   43–46
– normal or spontaneous emission of light   43–44
– photometric terms   33–34
– – spectral terms   33
– photometric units   34–35
– stimulated emission of light   44
photon-number splitting (PNS) attack   778
photons   3
photopolymers   239–240
photoreceptors   421
photorefractive effect   242
photothermoplastics   241–242
physical process of image formation   72
Picket-fence effect, 380n4
pico projector   359
picture element   74
pixels   74, 645
– pixel adjacency   74
– pixel based segmentation   645–651
planar multiview displays   378–380
– active multiview 3D displays   378–379
– passive multiview 3D displays   379–380
planar stereoscopic displays   370–378
– autostereoscopic displays (without glasses)   375–378
– head-mounted displays (HMDs)   370
– stereoscopic displays with glasses   371–375
– – light polarization   371, 373–375

planar stereoscopic displays (*contd.*)
– – spectral filtering   371–373
– – temporal shuttering   371
Planck's equation   43–44
plasma display   349–350
point spread function (PSF)   325, 420, 530, 542, 558–559, 571, 628
– of discrete Fourier analysis   406–407
point-scanning microscopes   308–309
Poisson noise model   580–583
– Richardson–Lucy restoration   580–581
polarized 3D displays   382–384
polarized light   275
polyharmonic splines   120
polyharmonic splines   127–129
– biharmonic spline   128
– thin-plate spline   127
– triharmonic spline   128
polytopic multiplexing   230
positron emission tomography (PET)   3
preprocessing, image   284–286
Prewitt operator   94
primary beam   328
principal component analysis (PCA)   291, 684–685
probability   49–54, *see also* Conditional probabilities
– characteristic of random variable   54
– continuous distribution   53
– discrete distribution   53
– distribution function   52–53
– entropy   54–62
– expected value   54
– mutual information   54–62
– mutually exclusive events   50
– probability density function (PDF)   53
– probability mass function   53
– random variable   52
– several events   50–51
– sum of probabilities rule   50
projection displays   353–361
– architectures   356–361
– – one-panel reflective DLP with UHP lamp   358–359
– – one-panel reflective LCoS with high-brightness LEDs   359
– – pico projector   359
– – three-panel grating light valve projector   359–361
– – three-panel reflective   357–358
– – three-panel transmissive technology   356–357
– – wire-grid architecture   359
– – X-cube LCoS projection architecture   359

– basic concepts   353–356
– components   353–356
– digital light processing (DLP)   353
– Fly's eye integrator   354
– high-temperature polysilicon (HTPS)   353
– light source   354
– liquid crystal on silicon (LCoS)   353
– rod integrator   354
– UHP lamp technology   355
prolate spheroidal wavefunction (PSWF)   335
pseudoscopic display   376
pseudo-Wigner–Ville distribution (PWVD)   105–110
– 1D directional PWVD   105–108
– 1D-smoothed PWVD   105
– 2D-smoothed PWVD
– – definition   108–110
– – implementation   108–110
– Brenner's expression   106
– spatial window   105
– spatial-averaging window   105
pseudo-WVD   98
pumping process   45
pupil function   17

q

quality assessment   419–439, *see also* motion-modeling-based algorithms
– mean opinion score (MOS)   419
– objective   419
– subjective   419
– video VIF   430
quantization   73, 449–452
– embedded quantization   451–452
– Lloyd–Max quantizers   450–451
– principle   449–450
– scalar quantization   449
– vector quantization   449
quantum bit error rate (QBER)   773
quantum dots (QD)   587, 778
quantum encryption   769–788
– BB84 quantum key distribution protocol   771–774
– principle   770–777
– quantum channel   771
– quantum no-cloning theorem   770–771
– quantum nondemolition (QND) measurement   778
quantum key distribution (QKD) systems   770, 777–783
– continuous variable QKD protocol   776–777
– demonstrations   781–783
– – free space experiments   782–783

– – telecom fiber QKD experiments 781–782
– entanglement-based 774–776
– quantum random number generator (QRNC) 781
– quantum state detection 779–780
– – APD-SPD 779–780
– – optical homodyne detector 780–781
– – single-photon detector 779–780
– security of 783–785
– – quantum hacking and countermeasures 783–784
– – self-testing QKD 784–785
– sources 777–779
– – attenuated laser source 778
– – EPR photon pair 778–779
– – single-photon source 777–778

r

radiance 32
radiant energy 30
radiant exposure 32
radiant flux 31
radiant flux density 31
radiant intensity 31
radiative transfer equation (RTE) 697–699
radio waves 2
radiometry
– terms and units 30–32
– versus photometry 30
Radon transform 791, 794–796
– Radon–Wigner transform 792
rainbow hologram 216–217
random basis ensemble 517
random convolution (RC) 517–518
random Fourier ensemble 516–517
random phase mask (RPM) 740–741
random sub-Gaussian matrices 516
random variable 52
ray optics 3
ray transfer matrix 3–6
– back focal plane 6
– back focal point 6
– front focal plane 6
– optical axis 3
– paraxial rays 3
– thin-lens matrix 5–6
– translation matrix 5
rays 3
γ-rays 2
read-out or amplifier noise 572
real-time integral holography 222–223
real-time processing 638
recursive dyadic partitions (RDP) 497

reduced reference (RR) algorithms 432
reduced state encounter navigation (RSEN) 835
Reed–Solomon code 236
reference beam 204
reflection holograms 207–209
reflection 275
reflective displays 351–353
– electronic paper (e-Paper) 352–353
– reflective LCD 352
refraction 275
regions
– convex regions 75
– nonconvex regions 75
– region based segmentation 645–651
– region-based segmentation methods 288
– region growing 288
– regions of interest 72
regressive filter 560
regularly spaced interpolation 121
relative entropy 62
remote verification ID tags 759–765
– NIR receiver resolution 763–764
– triangular pattern 763
Rényi entropy 62
reserve blocks 237
resolution 28–29, 275–276
– angular 28–29
– spatial 29–30
response equalization 811
restoration filter, significance of 553, 559–560
restricted isometry property (RIP) 342, 487
retina 36
Rice single-pixel camera 488
Richardson–Lucy (RL) restoration 536, 580–581
– classical regularization of 581–583
– SPERRIL 583–589
– – origin and related methods 583–584
ringing 27
robust filtering 92
rod integrator 354
rods 22, 37, 421
ronchi grating 598
root-mean-squared error (RMSE) 432

s

saccades 421
sampling functions 399
sampling interval 400
saturation or chroma 41

scalable wavelet-based distortion metric for VQA   425
scalar quantization   449
scalar wave equation   11
scale (multiplicative) basis functions   399–401
scale invariance hypothesis   813
scale invariance   162
scale-space approach   654
scale-space representations for gray-scale and color images   155–176
– axioms and properties   157–162, *see also individual entry*
– definitions   156–157
– Gaussian scale space   165–167
– geometry-driven scale space   171–173
– mathematical morphology scale space   173–176
– – lexicographical ordering   174
– – marginal ordering   174
– multiscale representation   155
– original signal   157
– PDE equation-based formulation   162–165
– – classification   162–163
– – explicit time marching scheme   164
– – implicit time schemes   164
– – numerical approximation   164
– – solving   163–165
– – variational formulation   165
– – von Neumann boundary conditions   162
– – well-posedness   163
– representation   165–166
– scale-space theory   155
– variable conductance diffusion   167–171
scattered light   204
science 2.0 effect   524–526
second-generation hologram   204
security systems applications of encryption   755–765, *see also under* optical data encryption
segmentation   75, 287–289, 643
– edge-based   288
– – active contours (ACs) models   288
– – gradient based method   288
– – Hough transform   288
– – Laplacian based method   288
– – segmentation   652–654
– in intermediate-level vision context   644
– – edge based   645
– – model based   645
– – pixel and region based   645–651
– model-based segmentation   661–664
– – Bayesian decision theory   662
– – statistical labeling   662

– region-based   288
– – region growing   288
– – splitting   288
– – thresholding   288
– – watershed   288
segmented amplitude mask (SAM)   467
segmented composite phase-only filters (SCPOFs)   478–479
segmented mirrors   193
self-testing QKD   784–785
semantics   72
sensor noise   571
separable transforms   148–149
servo systems   233–236
– wobble servo   236
several events   50–51
Shack–Hartman wavefront sensor (SHWS)   304, 312
Shan's method   536
Shannon entropy   170
Shannon sampling theorem   121
Shannon–Nyquist sampling   510–511
shearlets   151–152
– continuous shearlet transform   151
– discrete shearlet transform   152
shift (convolutional) basis functions   399–400
short space Fourier transform   104
short time Fourier transformation (STFT)   84
shuttered 3D displays   382–384
sifted keys   772
signal discretization   399–401
– scale (multiplicative) basis functions   399
– shift (convolutional) basis functions   399
signal processing for 3D displays   381–393, *see also under* 3D displays
signal processing techniques   464
silicon backplane addressing   190
silver halide photographic emulsion   238–239
simple block-based motion model   446
simple cells   421
simple contiguous   75
single-aperture radio telescopes   326–327
– aperture distribution   326
– beam   326
single-channel (SC) deconvolution   534–539
single-photon source, QKD   777–778
'single-pixel' imaging system   489
single-scale structural similarity index (SS-SSIM)   428
singular value decomposition (SVD)   425
small missions for advanced research in technology (SMART-1)   835

smectic LC   180
smoothing filters   91–92
sobel operator   95
source plane   13
space–bandwidth product (SBP)   669–670
spacecraft optical navigation, image processing for   833–858
– geometric basis   835–837
– noise and performance factors   841–844
– – amplifier noise   844
– – blooming   842
– – camera data processing   847
– – cosmic rays   843
– – dark current   842
– – deep space mission   850–855
– – dynamic range   841
– – dynamical models   845–847
– – fill factor   841
– – fixed pattern noise (FPN)   843
– – Kalman filtering   847–854
– – linearity   844
– – photon noise   844
– – quantization noise   844
– – quantum efficiency   842
– – radiation hardness   843
– – readout noise   843
– – reset noise   843
– – thermal noise   843
– – total noise   844
– – transfer efficiency   843
– sensors and models   837–845
– – imaging system resolution   838
– – imaging   840–841
– – optics   837–838
space-variant convolution   531
space-variant deconvolution   531
space-variant extension   542–546
sparse reconstruction by separable approximation (SpaRSA) algorithm   495
sparse representation   135
spatial and spatial-frequency marginal conditions   99
spatial arrangement of samples   73
spatial-averaging window   105
spatial decorrelation   811
spatial domain   80, 139, 286
spatial frequency   85, 329
spatial light modulators (SLMs)   20, 179–198, 229, 260, 464, 667, *see also* Mirror devices
– amplitude modulators   179
– applications   196–197
– fully complex modulation methods   194–195
– multiple quantum-well (MQW) SLMs   191–192
– phase modulators   179
– types of   180–193, *see also* Liquid crystals (LCs)
spatial multiplexing   601
spatial prediction   443–444
spatial quality assessment   431
spatial resolution   29–30
spatial transforms   444–445
spatial window   105
Spearman's rank ordered correlation coefficient (SROCC)   432
speckle   355
– speckle holography   551
spectral filtering   371–373
spectral flux density   33
spectral imagers   489
spectral radiant flux   33
spectral sensitivity of the eye   33
spectrogram   104
spectrum   17, 36
speed-weighted structural similarity index (SW-SSIM)   431
spherical aberration   9
splines   119–133, *see also* Polyharmonic splines; Tensor product splines
– as interpolants and basis functions   120–129
– in biomedical image and volume registration   131–132
– B-splines   119
– cubic B-splines   119
– interpolation splines   119
– for multiscale analysis   129–131
– polyharmonic splines   120
– theoretical results about   120–131
splitting   288
spontaneous emission   45
spontaneous parametric down-conversion (SPDC)   778
spontaneous photons   45
spurious edges   653
square grid   73
square kilometer array (SKA) telescope   324
stability axioms   158
standard sensitivity curve   33
state-of-the-art SC algorithms   536
Stein unbiased risk estimate (SURE) threshold   573
stereoscope   370–378
stimulated emission of light   44
stimulus   38
Strehl ratio   302

subjective quality assessment 419
submillimeter wave (SMMW) 2
subpixel motion information 613–640
subpixel-accurate motion-compensated prediction 447–448
sum of probabilities rule 50
superposition 139
super-resolution image reconstruction 613–640
– applications of 631–640
– – for color imaging systems 632–634
– – in spatial resolution and dynamic range 634–636
– considering inaccurate subpixel motion estimation 623–631
– – misregistration error 623–624
– – multichannel-regularized algorithm 624–628
– development 631–640
– frequency domain interpretation 621–622
– fundamentals of 614–622
– observation model 616–617
– as an inverse problem 617–621
– for video systems 636–640
super-resolution 341
– for video systems 636–640
suprathreshold distortion 424
surface relief holograms 214–215
surface-stabilized ferroelectric liquid crystal (SS-FLC) 187
symmetric channel 64–65
synthesized beam 331
system étendue 355
system matrix 7–8

t

target verification, neural network for 681–687
task-specific information compressive imagers 490
temporal indicator 426
temporal masking 422
temporal or longitudinal coherence 29–30
temporal redundancy removal by motion estimation 445–449
temporal shuttering 371
temporal trajectory aware quality measure 427
temporal variations of spatial-distortion-based VQA 427
tensor product splines 120–127
– approximation error 125
– – order of approximation 125
– Bèzier splines 127

– cardinal function 121
– generalized sampling theorem 121
– Hermite splines 127
– Hilbert space 121
– interpolation context 120
– irregular interpolation problems 125
– regularly spaced interpolation 121
– Shannon sampling theorem 121
Teo and Heeger model 423–424
terahertz band (THz band) 2
terms, photonics 30–35
texture masking 422
texture segmentation 112–114
– cascade recursive least squares (CRLS) 113
– computational procedure consists 113
– Karhunen–Loève transform (KLT) 113
– neural networks-based methods 113
theory of opponent colors 40–42
– blue-green colors 41
– hue 41
– saturation or chroma 41
– visual observations 40–41
– 'yellow minus blue' 41
thermotropic LC 180
thick hologram 215
thin hologram 215
thin-lens matrix 5–6
thin-plate spline 127
third-order aberrations 8
three-panel grating light valve projector 359–361
three-panel reflective technology 357–358
three-panel transmissive technology 356–357
thresholding 288, 645, 652
Tikhonov model 535
Tikhonov energy potential model 168
Tikhonov–Miller (TM) restoration 578
– iterative constrained Tikhonov–Miller 579
time domain 139
time multiplexing 597–600
time, coherence 29
time–frequency window 141
time-scale analysis 141
Toeplitz-block-Toeplitz matrix 533
tomographic reconstruction 791, 794
total variation 531
transient nematic effect 184
transition operator 157
translation invariance 101–102
translation matrix 5
translation parameter 142
transmission holograms 207–209
transmission/absorption 275

transmissive displays   346–348
– backlight unit   346
– disadvantages   347
– liquid crystal displays (LCDs)   346–347
– placement   347
– thickness of   347
– twisted nematic (TN) LCDs   347
transverse or spatial coherence   30
triharmonic spline   128
tristimulus theory   36–40
– B-cones   37
– 1931 CIE Standard   38–39
– CIE 1976 UCS Diagram   39–40
– color-matching function   37–38
– cones   37
– fovea   37
– G-cones   37
– R-cones   37
– retina   36
– rods   37
– stimulus   38
twisted nematic (TN) LCDs   347
twisted nematic (TN) SLMs   181, 184–186
– Jones matrix formulation for   184
two-channel digital filter bank   145
two-chemistry approach   245
two-dimensional signals, basic phase-space rotators for   257–260
two-lens imaging system   6–8
two-photon   228
two-point correlation function   790
two-step optical imagery   228

u

uncertainty principle   84
uniform chromaticity space   40
unique calculus of rationality   339
units, photonics   30–35
universal quality index (UQI)   428

v

Van Cittert–Zernike theorem   255, 523
variable block sizes   448
variable conductance diffusion   167–171
– diffusivity   167
– edge-affected variable conductance diffusion (EA-VCD)   168
– Lyapunov functionals   170
– Navier–Stokes equations   170
– Shannon entropy   170
vector quantization   449
vector Wiener filter   560
vertical cavity surface emitting laser (VCSEL)   605

very long baseline interferometry (VLBI)   324
vibrational spectroscopic images, analysis   723–730
– background correction   724
– baseline correction   724
– bin   725
– classical image representations   725
– cluster analysis   728
– denoising   724
– exploratory image analysis   725
– functional group mapping   725
– image compression   725
– image segmentation   728
– linear discriminant analysis   729
– normalization   724
– preprocessing   723
– principal component analysis   726
– quality test   723
– supervised image segmentation   729
vibrational spectroscopy   717–739
– biomedical imaging based on   717–739
– – CARS microscopy image analysis, challenges   730–734
– – coherent anti-Stokes–Raman scattering microscopy   721
– – focal plane array (FPA)   719
– – infrared spectroscopy   718
– – Raman spectroscopy   720
– – vibrational imaging   718–722
– – vibrational spectroscopic images, analysis   723–730
– – vibrational spectroscopy   718–722
video coding   457–460
– closed-loop video encoder   458
– H.261   458–459
– H.264/AVC   459–460
video quality assessment (VQA) algorithms
– scalable wavelet-based distortion metric for   425
– structural similarity for   430
video quality metric (VQM)   426–427
video VIF   430
viewing zone   376
visual difference predictor (VDP)   423
visual discrimination model (VDM)   423
visual information fidelity (VIF)   429–430
visual noise   430
visual or photometric quantity   34
visual perception   419–439, *see also* Human visual system (HVS)
visual signal-to-noise ratio (VSNR)   424
visual spectral flux   33
Von Neumann boundary conditions   162

w

watershed   288
wave equation   9–11
wave optics   9, 11–13
wavefront sensing   312–317, 549–568
– application   563–567
– deconvolution from wavefront sensing (DWFS)   550–551
– direct   312–314
– gradient-based wavefront sensors   553–558
– implementation of deconvolution from   562–563
– indirect   314–317
– minimization criteria   560
– past and present   551–552
– resolution of restored image, effective cutoff frequency   560–562
– restoration filters   559–560
– restoration process   552–563
wavelength multiplexing   601
wavelength   36
wavelet basis functions   401
wavelet transform   141
wavelets   135–153
– classical Fourier analysis   140–141
– continuous wavelet transform   142–143
– decomposition   137
– directional transforms   148–152
– discrete wavelet transform   143–144
– dual-tree complex wavelet transform (DT-CWT)   149–151
– forces of nature   141–142
– inversion formula   140
– large-scale features   141
– multiresolution analysis   144–148
– scrutinizing an image   139–142
– shearlets   151–152
– small-scale features   141
– time–frequency analysis   141
– time-scale analysis   141
– wavelet identity   143
– wavelet series   144
– wavelet transform   141
– window function   140
wavenumber   11
well layer   191
well-posedness   163
widefield microscopes   309–311
Wiener filter   560, 575–577
Wiener–Hopf equations   575
Wiener–Khinchin relation   328, 811
Wigner distribution (WD)   100–103, 252–255, 789, 800–807
– basic properties   100
– cross Wigner distribution   103
– inversion   101
– marginals   101
– overlap of two images   102
– product of images   102
– real images   102
– reconstruction of   800–807
– translation invariance   101–102
– WD projections measurements   798–800
Wigner–Ville distribution (WVD)   97
– pseudo-WVD   98
window function   140
windows   206
wire-grid architecture   359
wobble servo   236

x

X-cube LCoS projection architecture   359
xenon lamps   354
X-rays   2

z

Zernike polynomials   297, 315, 555
zero-crossing detection   653
zero-mean Gaussian noise   616